Also Available from McGraw-Hill

Schaum's Outline Series in Electronics & Electrical Engineering

Most outlines include basic theory, definitions and hundreds of example problems solved in step-by-step detail, and supplementary problems with answers.

Related titles on the current list include:

Analog & Digital Communications
Basic Circuit Analysis
Basic Electrical Engineering
Basic Electricity
Basic Mathematics for Electricity & Electronics
Digital Principles
Electric Circuits
Electric Machines & Electromechanics
Electric Power Systems
Electromagnetics
Electronic Circuits
Electronic Communications
Electronic Devices & Circuits
Electronics Technology
Feedback & Control Systems
Introduction to Digital Systems
Microprocessor Fundamentals
Signals & Systems

Schaum's Solved Problems Books

Each title in this series is a complete and expert source of solved problems with solutions worked out in step-by-step detail.

Related titles on the current list include:

3000 Solved Problems in Calculus
3000 Solved Problems in Differential Equations
3000 Solved Problems in Electric Circuits
3000 Solved Problems in Electromagnetics
3000 Solved Problems in Linear Algebra
3000 Solved Problems in Numerical Analysis
3000 Solved Problems in Physics

Available at most college bookstores, or for a complete list of titles and prices, write to: Schaum Division
        McGraw-Hill, Inc.
        1221 Avenue of the Americas
        New York, NY 10020

LINEAR CONTROL SYSTEM ANALYSIS AND DESIGN
Conventional and Modern

Fourth Edition

John J. D'Azzo
Constantine H. Houpis

Air Force Institute of Technology
Graduate School of Engineering
Wright-Patterson Air Force Base

McGraw-Hill, Inc.

New York   St. Louis   San Francisco   Auckland   Bogotá
Caracas   Lisbon   London   Madrid   Mexico City   Milan
Montreal   New Delhi   San Juan   Singapore
Sydney   Tokyo   Toronto

This book was set in Times Roman by Publication Services, Inc.
The editors were George T. Hoffman and John M. Morriss;
the production supervisor was Annette Mayeski.
The cover was designed by Rafael Hernandez.
Project supervision was done by Publication Services, Inc.
R. R. Donnelley & Sons Company was printer and binder.

LINEAR CONTROL SYSTEM ANALYSIS AND DESIGN
Conventional and Modern

This book is printed on recycled, acid-free
paper containing 10% postconsumer waste.

1 2 3 4 5 6 7 8 9 0 DOC DOC 9 0 9 8 7 6 5

ISBN 0-07-016321-9

Library of Congress Cataloging-in-Publication Data

D'Azzo, John Joachim.
    Linear control system analysis and design: conventional and
modern / John J. D'Azzo, Constantine H. Houpis.—4th ed.
        p.   cm.—(McGraw-Hill series in electrical and computer
engineering. Control theory)
    Includes index.
    ISBN 0-07-016321-9
    1. Linear control systems.   2. Control theory.   I. Houpis,
Constantine H.   II. Title.   III. Series.
TJ213.D33   1995
629.8'32—dc20   94-25299

CONTENTS

Preface   xv

1 Introduction   1

  1.1   Introduction   1
  1.2   Introduction to Control Systems   1
  1.3   Definitions   6
  1.4   Historical Background   8
  1.5   Digital Control Development   12
  1.6   Mathematical Background   13
  1.7   General Nature of the Engineering Control Problem   15
  1.8   Computer Literacy   16
  1.9   Outline of Text   16

2 Writing System Equations   19

  2.1   Introduction   19
  2.2   Electric Circuits and Components   21
  2.3   Basic Linear Matrix Algebra   25
  2.4   State Concepts   28
  2.5   Transfer Function and Block Diagram   34
  2.6   Mechanical Translation Systems   35
  2.7   Analogous Circuits   41
  2.8   Mechanical Rotational Systems   41
  2.9   Thermal Systems   46
  2.10 Hydraulic Linear Actuator   48
  2.11 Liquid-Level System   53
  2.12 Rotating Power Amplifiers   54
  2.13 DC Servomotor   56
  2.14 AC Servomotor   57
  2.15 Lagrange's Equation   59
  2.16 Summary   63

3 Solution of Differential Equations   64

  3.1   Introduction   64
  3.2   Standard Inputs to Control Systems   65

3.3	Steady-State Response: Sinusoidal Input	66
3.4	Steady-State Response: Polynomial Input	67
3.5	Transient Response: Classical Method	70
3.6	Definition of Time Constant	73
3.7	Example: Second-Order System—Mechanical	74
3.8	Example: Second-Order System—Electrical	76
3.9	Second-Order Transients	77
3.10	Time-Response Specifications	80
3.11	CAD Accuracy Checks (CADAC)	82
3.12	State-Variable Equations	82
3.13	Characteristic Values	84
3.14	Evaluating the State Transition Matrix	85
3.15	Complete Solution of the State Equation	88
3.16	Summary	89

4   Laplace Transform   **91**

4.1	Introduction	91
4.2	Definition of the Laplace Transform	92
4.3	Derivation of Laplace Transforms of Simple Functions	92
4.4	Laplace Transform Theorems	94
4.5	CAD Accuracy Checks: CADAC	97
4.6	Application of the Laplace Transform to Differential Equations	97
4.7	Inverse Transformation	98
4.8	Heaviside Partial-Fraction Expansion Theorems	99
4.9	MATLAB Partial-Fraction Example	106
4.10	Partial-Fraction Shortcuts	107
4.11	Graphical Interpretation of Partial-Fraction Coefficients	109
4.12	Frequency Response from the Pole-Zero Diagram	113
4.13	Location of Poles and Stability	116
4.14	Laplace Transform of the Impulse Function	117
4.15	Second-Order System with Impulse Excitation	119
4.16	Additional Matrix Operations and Properties	120
4.17	Solution of State Equation	126
4.18	Evaluation of the Transfer-Function Matrix	128
4.19	Summary	129

5   System Representation   **131**

5.1	Introduction	131
5.2	Block Diagrams	131
5.3	Determination of the Overall Transfer Function	136
5.4	Standard Block Diagram Terminology	138
5.5	Position Control System	140
5.6	Simulation Diagrams	144
5.7	Signal Flow Graphs	149
5.8	State Transition Signal Flow Graph	154
5.9	Parallel State Diagrams from Transfer Functions	158
5.10	Diagonalizing the **A** Matrix	160
5.11	Use of State Transformation for the State Equation Solution	168

5.12	Transforming a Matrix with Complex Eigenvalues	169
5.13	Transforming an **A** Matrix into Companion Form	172
5.14	Summary	175

6 Control-System Characteristics — 176

6.1	Introduction	176
6.2	Routh's Stability Criterion	176
6.3	Mathematical and Physical Forms	182
6.4	Feedback System Types	183
6.5	Analysis of System Types	185
6.6	Example: Type 2 System	190
6.7	Steady-State Error Coefficients	192
6.8	CAD Accuracy Checks: CADAC	196
6.9	Use of Steady-State Error Coefficients	196
6.10	Nonunity-Feedback System	198
6.11	Summary	199

7 Root Locus — 200

7.1	Introduction	200
7.2	Plotting Roots of a Characteristic Equation	201
7.3	Qualitative Analysis of the Root Locus	204
7.4	Procedure Outline	207
7.5	Open-Loop Transfer Function	208
7.6	Poles of the Control Ratio $C(s)/R(s)$	209
7.7	Application of the Magnitude and Angle Conditions	211
7.8	Geometrical Properties (Construction Rules)	215
7.9	CAD Accuracy Checks (CADAC)	225
7.10	Examples	225
7.11	Example 1: MATLAB Root Locus	231
7.12	Performance Characteristics	234
7.13	Transport Lag	238
7.14	Synthesis	240
7.15	Summary of Root-Locus Construction Rules for Negative Feedback	241
7.16	Summary	242

8 Frequency Response — 244

8.1	Introduction	244
8.2	Correlation of the Sinusoidal and Time Responses	245
8.3	Frequency-Response Curves	246
8.4	Bode Plots (Logarithmic Plots)	247
8.5	General Frequency-Transfer-Function Relationships	249
8.6	Drawing the Bode Plots	250
8.7	Example of Drawing a Bode Plot	256
8.8	System Type and Gain as Related to Log Magnitude Curves	259
8.9	CAD Accuracy Check (CADAC)	262
8.10	Experimental Determination of Transfer Functions	262
8.11	Direct Polar Plots	263

8.12	Summary: Direct Polar Plots	269
8.13	Nyquist's Stability Criterion	270
8.14	Examples of Nyquist's Criterion Using Direct Polar Plot	278
8.15	Nyquist's Stability Criterion Applied to Systems Having Dead Time	281
8.16	Definitions of Phase Margin and Gain Margin and Their Relation to Stability	283
8.17	Stability Characteristics of the Log Magnitude and Phase Diagram	285
8.18	Stability from the Nichols Plot (Log Magnitude–Angle Diagram)	286
8.19	Summary	288

9 Closed-Loop Tracking Performance Based on the Frequency Response   290

9.1	Introduction	290
9.2	Direct Polar Plot	291
9.3	Determination of M_m and ω_m for a Simple Second-Order System	292
9.4	Correlation of Sinusoidal and Time Responses	295
9.5	Constant $M(\omega)$ and $\alpha(\omega)$ Contours of $C(j\omega)/R(j\omega)$ on the Complex Plane (Direct Plot)	296
9.6	Constant $1/M$ and α Contours (Unity Feedback) in the Inverse Polar Plane	303
9.7	Gain Adjustment for a Desired M_m of a Unity-Feedback System: Direct Polar Plot	304
9.8	Constant M and α Curves on the Log Magnitude–Angle Diagram (Nichols Chart)	307
9.9	Generation of MATLAB (1992 Student Version) Bode and Nyquist Plots	309
9.10	Adjustment of Gain by Use of the Log Magnitude–Angle Diagram	312
9.11	Correlation of Pole-Zero Diagram with Frequency and Time Responses	312
9.12	Summary	317

10 Root-Locus Compensation: Design   319

10.1	Introduction to Design	319
10.2	Transient Response: Dominant Complex Poles	321
10.3	Additional Significant Poles	326
10.4	Root-Locus Design Considerations	329
10.5	Reshaping the Root Locus	331
10.6	CAD Accuracy Checks (CADAC)	331
10.7	Ideal Integral Cascade Compensation (PI Controller)	332
10.8	Cascade Lag Compensation Design Using Passive Elements	333
10.9	Ideal Derivative Cascade Compensation (PD Controller)	339
10.10	Lead Compensation Design Using Passive Elements	340
10.11	General Lead-Compensator Design	345
10.12	Lag-Lead Cascade Compensation Design	346
10.13	Comparison of Cascade Compensators	349
10.14	PID Controller	352

10.15 Introduction to Feedback Compensation 353
10.16 Feedback Compensation: Design Procedures 355
10.17 Simplified Rate Feedback Compensation: A Design Approach 355
10.18 Design of Rate Feedback 358
10.19 Design: Feedback of Second Derivative of Output 362
10.20 Results of Feedback Compensation Design 364
10.21 Rate Feedback: Plants with Dominant Complex Poles 364
10.22 Summary 365

11 Frequency-Response Compensation Design 367
11.1 Introduction to Feedback Compensation Design 367
11.2 Selection of a Cascade Compensator 369
11.3 Cascade Lag Compensator 372
11.4 Design Example: Cascade Lag Compensation 375
11.5 Lead Compensator 379
11.6 Design Example: Cascade Lead Compensation 381
11.7 Lag-Lead Compensator 385
11.8 Design Example: Cascade Lag-Lead Compensation 387
11.9 Feedback Compensation Design Using Log Plots 390
11.10 Design Example: Feedback Compensation (Log Plots) 392
11.11 Application Guidelines: Basic Minor-Loop Feedback Compensators 397
11.12 Summary 398

12 Control-Ratio Modeling 401
12.1 Introduction 401
12.2 Modeling a Desired Tracking Control Ratio 402
12.3 Guillemin-Truxal Design Procedure 406
12.4 Introduction to Disturbance Rejection 408
12.5 A Second-Order Disturbance-Rejection Model 409
12.6 Disturbance-Rejection Design Principles for SISO Systems 411
12.7 Disturbance-Rejection Design Example 415
12.8 Disturbance-Rejection Models 418
12.9 Summary 422

**13 Design: Closed-Loop Pole-Zero Assignment
    (State-Variable Feedback)** 423
13.1 Introduction 423
13.2 Controllability and Observability 424
13.3 State Feedback for SISO Systems 431
13.4 State-Feedback Design for SISO Systems Using the Control
     Canonical (Phase-Variables) Form 433
13.5 State-Variable Feedback (Physical Variables) 436
13.6 General Properties of State Feedback (Using Phase Variables) 439
13.7 State-Variable Feedback: Steady-State Error Analysis 442
13.8 Use of Steady-State Error Coefficients 444
13.9 State-Variable Feedback: All-Pole Plant 448

13.10 Plants with Complex Poles 451
13.11 Compensator Containing a Zero 451
13.12 State-Variable Feedback: Pole-Zero Plant 453
13.13 Summary 460

14 Parameter Sensitivity and State Space Trajectories 462

14.1 Introduction 462
14.2 Sensitivity 462
14.3 Sensitivity Analysis 466
14.4 Parameter Sensitivity Examples 472
14.5 Inaccessible States 473
14.6 State-Space Trajectories 477
14.7 Linearization (Jacobian Matrix) 488
14.8 Summary 492

15 Digital Control Systems 493

15.1 Introduction 493
15.2 Sampling 494
15.3 Ideal Sampling 496
15.4 z-Transform Theorems 500
15.5 Synthesis in the z Domain (Direct Method) 500
15.6 The Inverse z Transform 503
15.7 Zero-Order Hold 504
15.8 Limitations 506
15.9 Tustin Transformation 507
15.10 Tustin Transformation Properties 509
15.11 Pseudo-Continuous-Time (PCT) Control System (DIG Method) 512
15.12 Analysis of a Basic (Uncompensated) System 514
15.13 Design of Digital Control Systems 519
15.14 Direct (DIR) Design Technique 520
15.15 Lead Controller (Compensator): DIR Design Method 521
15.16 Lag and Lag-Lead Controllers: DIR Design Method 523
15.17 Digitization (DIG) Design Technique 523
15.18 Summary 526

16 Entire Eigenstructure Assignment for Multivariable Systems 527

16.1 Introduction 527
16.2 Effect of Eigenstructure on Time Response 528
16.3 Entire Eigenstructure Assignment 530
16.4 Examples of Entire Eigenstructure Assignment for Regulators 531
16.5 MATLAB Eigenvectors 537
16.6 Uncontrollable Systems 539
16.7 Tracking Systems 541
16.8 Tracking-System Design Example 543
16.9 MATLAB Example of Tracker Design in Sec. 16.8 546
16.10 Summary 551

17   Design of Tracking Systems Using Output Feedback   554

   17.1   Introduction   554
   17.2   Output Feedback Tracking System   555
   17.3   Block Diagonalization   557
   17.4   Analysis of Closed-Loop System Performance   559
   17.5   Design Procedure for Regular Plants   562
   17.6   Regular System Design Example   563
   17.7   Irregular Plant Characteristics   566
   17.8   Irregular System Performance   568
   17.9   Design of the Measurement Matrix **M**   569
   17.10   Irregular System Design Example   571
   17.11   Tracker Simulation   574
   17.12   Summary   578

18   Quantitative Feedback Theory (QFT) Technique   580

   18.1   Introduction   580
   18.2   Frequency Responses with Parameter Variations   582
   18.3   Introduction to the QFT Method (Single-Loop System)   584
   18.4   Minimum-Phase System Performance Specifications   586
   18.5   Multiple-Input Multiple-Output (MIMO) Uncertain Plants   590
   18.6   Plant Templates of $P(s)$, $\mathcal{T}\mathbf{P}(j\omega_i)$ 592
   18.7 U-Contour   595
   18.8   Tracking Bounds Lm $\mathbf{B}_R(j\omega)$ on the NC   597
   18.9   Disturbance Bounds $\mathbf{B}_D(j\omega_i)$: Case 1 [$d_2(t) = D_o u_{-1}(t)$,
           $d_1(t) = 0$]   600
   18.10   Disturbance Bounds $\mathbf{B}_D(j\omega_i)$: Case 2 [$d_1(t) = D_o u_{-1}(t)$,
           $d_2(t) = 0$]   605
   18.11   The Composite Boundary $\mathbf{B}_o(j\omega_i)$ 607
   18.12   Shaping of $\mathbf{L}_o(j\omega)$ 608
   18.13   Guidelines for Shaping $\mathbf{L}_o(j\omega)$ 614
   18.14   Design of the Prefilter $F(s)$ 615
   18.15   Basic Design Procedure for a MISO System   617
   18.16   Design Example 1   618
   18.17   Design Example 2   629
   18.18   Template Generation for Unstable Plants   630
   18.19   Summary   632

   Appendixes   636

   A   Table of Laplace Transform Pairs   636
   B   Interactive Computer-Aided Design Programs for Digital
       and Continuous Control-System Analysis and Synthesis   640
       B.1   Introduction   640
       B.2   Overview of ICECAP-PC and TOTAL-PC   641
       B.3   Overview of MATLAB   645
       B.4   QFT CAD Packages   647

B.5   Computer-Aided Design Accuracy Checks (CADAC)   647
B.6   Other Computer-Aided Design Packages   649

Problems   651

Answers to Selected Problems   722

Index   739

PREFACE

The emphasis of this fourth edition is on strengthening its use as a textbook for undergraduates who desire to obtain a solid foundation in control theory. Both analog and sampled-data single-input single-output (SISO) feedback-control systems are covered in detail. This has been accomplished by deleting the optimal control chapters of the prior editions and by strengthening the digital control chapter. The text was further strengthened by improving the latter chapters, which present an introduction to advanced control system design techniques for both SISO and multiple-input and multiple-output (MIMO) control systems. These latter chapters are intended for advanced undergraduate and first-year graduate students majoring in control theory. A major revision has also been made in the problem section. To support this emphasis, revisions were incorporated throughout this new edition to enhance the clarity and to streamline for the student a comprehensive presentation of control theory and design. Extensive use is made of computer-aided-design (CAD) packages to assist and simplify the design process.

The authors have exerted meticulous care with explanations, diagrams, calculations, tables, and symbols. The student is made aware that rigor is necessary for design applications and advanced control work. Also stressed is the importance of clearly understanding the concepts that provide the rigorous foundations of modern control theory. The text provides a strong, comprehensive, and illuminating account of those elements of conventional control theory that have relevance in the analysis and design of control systems. The presentation of a variety of different techniques contributes to the development of the student's working understanding of what A. T. Fuller has called "the enigmatic control system." To provide a coherent development of the subject, an attempt is made to eschew formal proofs and lemmas, with an organization that draws the perceptive student steadily and surely into the demanding theory of multivariable control systems. Design examples are included throughout each chapter in order to enhance and solidify the student's understanding of the text material. It is the opinion of the authors that a student who has reached this point is fully equipped to undertake with confidence the challenges presented by more advanced control theories, as typified by Chapters 16 through 18. The importance and necessity of making extensive use of computers is emphasized by references to comprehensive computer-aided-design (CAD) programs. The book uses the computer programs MATLAB®, by The MathWorks, Inc.; ICECAP-PC; and TOTAL-PC (see App. B). These programs have viable control system computational tools.

After much consideration, it was decided to retain Chapter 2, which sets forth the appropriate differential equations to describe the performance of physical systems, networks, and devices. Some elementary matrix algebra, the block diagram, and the transfer function are also included. The essential concept of modern control theory, the state space, is also introduced. The approach used is the simultaneous derivation of the state-vector differential equation with the single-input single-output (SISO) differential equation for a chosen physical system. The relationship of the transfer function to the state equation of the system is deferred until Chapter 4. The derivation of a mathematical description of a physical system by using Lagrange equations is also given.

Chapter 3 serves as reference material for the reader, presenting the classical method of solving differential equations and the nature of the resulting response. Once the state-variable equation has been introduced, careful account is given of its solution. The central importance of the state transition matrix is brought out, and the state transition equation is derived. The idea of an eigenvalue is next explained, and this theory is used with the Cayley-Hamilton and Sylvester theorems to evaluate the state transition matrix.

The early part of Chapter 4 presents a comprehensive description of Laplace transform methods and pole-zero maps. Some further aspects of matrix algebra are introduced before the solution of the state equation by the use of Laplace transforms is dealt with. Finally, the evaluation of transfer matrices is clearly explained.

Chapter 5 begins with system representation by the conventional block-diagram approach. It is followed by a straightforward account of simulation diagrams and the determination of the state transition equation by the use of signal flow graphs. By deriving parallel state diagrams from system transfer functions, the advantages of having the state equation in uncoupled form are established. This is followed by the methods of diagonalizing the system matrix, a clear treatment of how to transform an **A** matrix that has complex eigenvalues into a suitable alternative block-diagonal form, and the transformation to companion form.

In Chapter 6 the basic feedback system characteristics are introduced. This includes the relationship between system type and the ability of the system to follow or track polynomial inputs.

In Chapter 7 the details of the root-locus method of analysis are thoroughly presented. Then the frequency-response method of analysis is given in Chapters 8 and 9, using both the log and the polar plots. These chapters include the following topics: Nyquist stability criterion; correlation between the s plane, frequency domain, and time domain; and gain setting to achieve a desired output response peak value. Chapters 10 and 11 describe the methods for improving system performance, along with examples of the techniques for applying cascade and feedback compensators. Both the root-locus and frequency-response methods of designing compensators are covered. Use of computer-aided-design (CAD) programs to assist the designer is illustrated throughout the text.

The concept of modeling a desired control ratio that has figures of merit that satisfy the system performance specifications is developed in Chapter 12. The system inputs generally fall into two categories: (1) a desired input, which the system output is to track (a tracking system), and (2) a disturbance input, for which the system output is to be minimal (a disturbance-rejection system). Desired control ratios for both types of systems are synthesized by the proper placement of poles and inclusion of zeros, if required. Chapter 12 also includes the Guillemin-Truxal design procedure for designing a tracking control system and a design procedure emphasizing disturbance rejection.

The technique of achieving desired system characteristics by using complete state-variable feedback is developed thoroughly and carefully in Chapter 13. The very important concepts of modern control theory—controllability and observability—are treated in a simple, straightforward, and correct manner. This provides a useful foundation for the work of Chapters 16 and 17.

Chapter 14 includes a presentation of the sensitivity concepts of Bode for the variation of system parameters. Also included is the method of using feedback transfer functions to form estimates of inaccessible states for use in state feedback. An account of trajectories in the state space and some associated phase-plane techniques are also given. A feature of this approach is that the arguments are extended to nonlinear systems. The treatment of such systems by linearization is presented.

Chapter 15 presents an introduction to the analysis and design of digital control systems. This permits an extension from linear control system design to the rapidly growing area of digital control. The effectiveness of digital compensation is clearly demonstrated. The concept of a pseudo-continuous-time (PCT) model of a digital system permits the use of continuous-time methods for the design of digital control systems.

Chapter 16 provides a thorough presentation of the principles and techniques of entire eigenstructure assignment for MIMO systems by means of state feedback. Eigenvalue assignment is extended to include the simultaneous assignment of the associated eigenvectors and provides the means for shaping the output response to meet design specifications.

There are many worthwhile control system design techniques available in the technical literature, which are based on both modern and conventional control theory. The applicability of each technique may be limited to certain classes of design problems. The control engineer must have a sufficiently broad perspective to be able to apply the right technique to the right design problem. For some techniques the designer is assisted by available computer-aided-design (CAD) packages. Design techniques such as LQR, LQG, or LTR are thoroughly covered in the literature. Two design techniques that have very desirable design characteristics for MIMO systems are covered in detail in this text. Chapter 17 presents an output feedback state-space technique based on singular perturbation theory. This leads to the design principles developed by Professor B. Porter, which incorporate proportional plus integral controllers. Application to a practical aircraft control system is demonstrated. Chapter 18 presents a control system design method based on quantitative feedback theory (QFT). This is a very powerful design method when plant parameters vary over a broad range of operating conditions. Chapter 18 presents an introduction to and lays the foundation for the quantitative feedback theory (QFT) developed by Professor I. M. Horowitz. This technique incorporates the concept of designing a robust control system that maintains the desired system performance over a prescribed region of structured plant parameter uncertainty. The authors feel that these methods have proven their applicability to the design of practical MIMO control systems. These chapters are intended to further strengthen the fundamentals presented earlier in the text and to "whet the appetite" of the budding control engineer. They are very effective in achieving robust output feedback high-gain systems.

This textbook provides students of control engineering with a clear, unambiguous, and relevant account of appropriate, contemporary, and state-of-the-art control theory. It is suitable as an introductory and bridging text for undergraduate and graduate students.

The text is arranged so that it can also be used for self-study by the engineer in practice. Included are many examples of feedback control systems in various areas of practice (electrical, aeronautical, mechanical, etc.). It is a strong basic feedback control text that can be used for study in any of the various branches of engineering. To make the text meaningful and valuable to all engineers, the authors have attempted to unify the treatment of physical control systems through use of mathematical and block-diagram models common to all. The text has been thoroughly class-tested, thus enhancing its value for classroom and self-study use. There are many computer-aided-design (CAD) packages available to assist a control engineer in the analysis, design, and simulation of control systems. Some of these are listed in App. B. The use of these packages, such as MATLAB, ICECAP-PC and TOTAL-PC, are stressed throughout the text. In adapting to the use of these packages, an engineer must become *computer literate*. Throughout the text the authors stress the use of *computer-aided-design accuracy checks* (CADAC), in order to assist the engineer in becoming computer literate. CADAC stresses the use of the appropriate fundamentals in assuring the viability of a computer solution.

The authors express their thanks to the students who have used this book and to the faculty who have reviewed it for their helpful comments and recommendations. Especial appreciation is expressed to Dr. R. E. Fontana, Professor Emeritus of Electrical Engineering, Air Force Institute of Technology, for the encouragement he has provided, and to Dr. T. J. Higgins, Professor Emeritus of Electrical Engineering, University of Wisconsin, for his thorough review of the earlier manuscripts.

Appreciation is expressed to Professor Donald McLean, the University of Southampton, England, formerly a visiting Professor at the Air Force Institute of Technology. Our association with him has been an enlightening and refreshing experience. Important advanced concepts are based on collaborative work with Professor Brian Porter, University of Salford, England, and Isaac M. Horowitz, Professor Emeritus, Weizmann Institute of Science, Rehovot, Israel and University of California, Davis. The personal relationship with them has been a source of inspiration and deep respect.

John J. D'Azzo
Constantine H. Houpis

LINEAR CONTROL SYSTEM
ANALYSIS AND DESIGN
Conventional and Modern

CHAPTER
1

INTRODUCTION

1.1   INTRODUCTION

Automatic control systems permeate life in all advanced societies today. Such systems act as catalysts in promoting progress and development. They are integral components of any industrial society and are necessary for the production of goods required by an increasing world population. Technological developments have made it possible to travel to the moon and to explore outer space. The successful operation of space vehicles, the space shuttle, space stations, and reconfigurable flight control systems depends on the proper functioning of the large number of control systems used in such ventures.

1.2   INTRODUCTION TO CONTROL SYSTEMS

The toaster in Fig. 1.1a can be set for the desired darkness of the toasted bread. The setting of the "darkness" knob, or timer, represents the input quantity, and the degree of darkness and crispness of the toast produced is the output quantity. If the degree of darkness is not satisfactory, because of the condition of the bread or some similar reason, this condition can in no way automatically alter the length of time that heat is applied. Since the output quantity has no influence on the input quantity, there is no *feedback* in this system. The heater portion of the toaster represents the dynamic part of the system, and the timer unit is the *reference selector*.

The DC shunt motor of Fig. 1.1b is another example. For a given value of field current, a required value of voltage is applied to the armature to produce the desired value of motor speed. In this case the motor is the dynamic part of the system, the applied armature voltage is the input quantity, and the speed of the shaft is the output quantity. A variation of the speed from the desired value, due to a change of

1

Desired toast
darkness setting

(*a*)

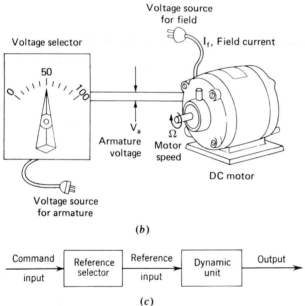

(*b*)

Command input → Reference selector → Reference input → Dynamic unit → Output

(*c*)

FIGURE 1.1
Open-loop control systems: (*a*) automatic toaster; (*b*) electric motor;
(*c*) functional block diagram.

mechanical load on the shaft, can in no way cause a change in the value of the applied armature voltage to maintain the desired speed. Therefore, the output quantity has no influence on the input quantity.

Systems in which the output quantity has no effect upon the input quantity are called *open-loop control systems*. The examples just cited are represented symbolically by a functional block diagram, as shown in Fig. 1.1c. The desired darkness of the toast or the desired speed of the motor is the command input; the selection of the value of time on the toaster timer or the value of voltage applied to the motor armature is represented by the reference-selector block; and the output of this block is identified as the reference input. The reference input is applied to the dynamic unit that performs the desired control function, and the output of this block is the desired output.

A person could be assigned the task of sensing the actual value of the output and comparing it with the command input. If the output does not have the desired value, the person can alter the reference-selector position to achieve this value. Introducing the person provides a means through which the output is *fed back* and is compared with the input. Any necessary change is then made in order to cause the output to equal the desired value. The *feedback* action, therefore, controls the input to the dynamic unit. Systems in which the output has a direct effect upon the input quantity are called *closed-loop control systems.*

To improve the performance of the closed-loop system so that the output quantity is as close as possible to the desired quantity, the person can be replaced by a mechanical, electrical, or other form of comparison unit. The functional block diagram of a *single-input single-output* (SISO) closed-loop control system is illustrated in Fig. 1.2. Comparison between the reference input and the feedback signal results in an *actuating signal* that is the difference between these two quantities. The actuating signal acts to maintain the output at the desired value. This system is called a *closed-loop control system.* The designation *closed-loop* implies the action resulting from the comparison between the output and input quantities in order to maintain the output at the desired value. Thus, the output is controlled in order to achieve the desired value.

Examples of closed-loop control systems are illustrated in Figs. 1.3 to 1.6. In a home heating system the desired room temperature (command input) is set on the thermostat in Fig.1.3 (reference selector). A bimetallic coil in the thermostat is affected by both the actual room temperature (output) and the reference-selector setting. If the room temperature is lower than the desired temperature, the coil strip alters its shape and causes a mercury switch to operate a relay, which turns on the furnace to produce heat in the room. When the room temperature reaches the desired temperature, the shape of the coil strip is again altered so that the mercury switch opens. This change deactivates the relay and in turn shuts off the furnace. In this example, the bimetallic coil performs the function of a comparator since the output

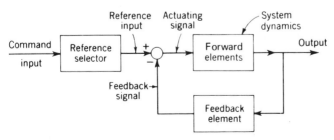

FIGURE 1.2
Functional block diagram of a closed-loop control system.

FIGURE 1.3
Home heating control system.

(room temperature) is fed back directly to the comparator. The switch, relay, and furnace are the dynamic elements of this closed-loop control system.

A closed-loop control system of great importance to all multistory buildings is the automatic elevator of Fig. 1.4. A person in the elevator presses the button corresponding to the desired floor, producing an actuating signal that indicates the desired floor and turns on the motor that raises or lowers the elevator. As the elevator approaches the desired floor, the actuating signal decreases in value and, with the proper switching sequences, the elevator stops at the desired floor and the actuating signal is reset to zero. The closed-loop control system for the express elevator in the Sears Tower building in Chicago is designed so that it ascends or descends the 103 floors in just under one minute with maximum passenger comfort.

The Air Force Materials Directorate, Wright-Patterson Air Force Base, Ohio, has developed robots under its Integrated Computer-Aided Manufacturing program

FIGURE 1.4
Automatic elevator.

FIGURE 1.5
Production-line robot.

Sun
sensor
Pyrheliometer
Declination
scale
Declination
drive motor
Hour-angle
drive motor
Boresight
Polar axis
Hour-angle
scale
Gear case
Latitude
adjustment
and scale
Level vial

FIGURE 1.6
Sun-following pyrheliometer.

for use on the production lines of aerospace manufacturing companies. The robot shown in Fig. 1.5 looks like a gigantic artificial limb. This robot performs a number of production functions, including drilling and routing operations on aircraft access panels that range in size from 12 in. to 1 yd square. It can also do additional jobs like painting, welding, manipulating parts, assembly, and handling. The development of small, reliable, versatile minicomputers has permitted the practical development of robots, which have an increasingly important role in manufacturing.[12,14,15] †

The interest in practical large-scale solar-energy systems has spurred the development of sun-following control systems (see Fig. 1.6).[11] The movement of the sun in relation to the earth is nonuniform and often obscured by clouds. This sun follower has been designed to point at the sun within 1 degree accuracy and to compensate for day-to-day changes in the sun's motion. This sun follower is essentially an open-loop drive actuated by a clock. There is a periodic sensing of the correct sun position, and corrections are made to reduce the sun-sensor error. The use of a microprocessor permits accurate sun following with trouble-free operation. This system can be used to control sun pointing of photovoltaic arrays, solar concentrators, or other devices that need to follow the sun in order to receive the maximum energy.

Design methods for analog SISO control systems represented by Fig 1.2 are covered in Chapters 6 to 14. Some systems require a precision in their performance that cannot be achieved by the structure of Fig. 1.2. Also, systems exist for which there are multiple inputs and/or multiple outputs, as discussed in Chapters 16 to 18. The design methods for such systems are often based on a representation of the system in terms of *state variables*. For example, position, velocity, and acceleration may represent the state variables of a position control system. The definition of state variables and their use in representing systems are contained in Chapters 2, 3, and 5. The use of state-variable methods for the design of control systems is presented in Chapters 13, 16, and 17. The design methods presented in Chapters 7 through 17 require knowledge of a fixed mathematical model of the system that is being controlled. The parameters of some systems change because of the range of conditions under which they operate. The quantitative feedback theory (QFT) is therefore presented in Chapter 18 for the design of control systems that have structured plant parameter uncertainty. When QFT is used, the parameter variations and performance specifications are included at the onset of the design process. The use of a digital computer to assist the engineer in the design process is emphasized throughout this book, and a sampling of available computer-aided-design (CAD) programs is given in Appendix B.

1.3 DEFINITIONS

From the preceding discussion the following definitions are evolved, based in part on the standards of the IEEE,[1] and are used in this text.

---

† References are indicated by superscript numbers and are found at the end of the chapter.

System. A combination of components that act together to perform a function not possible with any of the individual parts. The word *system* as used herein is interpreted to include physical, biological, organizational, and other entities, and combinations thereof, which can be represented through a common mathematical symbolism. The formal name *systems engineering* can also be assigned to this definition of the word *system.* Thus, the study of feedback control systems is essentially a study of an important aspect of systems engineering and its application.

Command input. The motivating input signal to the system, which is independent of the output of the system and exercises complete control over it (if the system is completely controllable).

Reference selector (reference input element). The unit that establishes the value of the reference input. The reference selector is calibrated in terms of the desired value of the system output.

Reference input. The reference signal produced by the reference selector, i.e., the command expressed in a form directly usable by the system. It is the actual signal input to the control system.

Disturbance input. A disturbance input signal to the system that has an unwanted effect on the system output.

Forward element (system dynamics). The unit that reacts to an actuating signal to produce a desired output. This unit does the work of controlling the output and thus may be a power amplifier.

Output (controlled variable). The quantity that must be maintained at a prescribed value, i.e., it must follow the command input without responding to disturbance inputs.

Open-loop control system. A system in which the output has no effect upon the input signal.

Feedback element. The unit that provides the means for feeding back the output quantity, or a function of the output, in order to compare it with the reference input.

Actuating signal. The signal that is the difference between the reference input and the feedback signal. It is the input to the control unit that causes the output to have the desired value.

Closed-loop control system. A system in which the output has an effect upon the input quantity in such a manner as to maintain the desired output value.

The fundamental difference between the open- and closed-loop systems is the *feedback action,* which may be continuous or discontinuous. Continuous control implies that the output is continuously fed back and compared with the reference input. In one form of discontinuous control the input and output quantities are periodically sampled and compared; i.e., the control action is discontinuous in time. This is commonly called a *digital, discrete-data,* or *sampled-data* feedback control system. A discrete-data control system may incorporate a digital computer, which improves the performance achievable by the system. In another form of discontinuous control system the actuating signal must reach a prescribed value before the system dynamics reacts to it; i.e., the control action is discontinuous in amplitude rather than in time.

This type of discontinuous control system is commonly called an *on-off* or *relay* feedback control system. Both forms may be present in a system. In this text continuous control systems are considered in detail since they lend themselves readily to a basic understanding of feedback control systems. Digital control systems are introduced in Chapter 15.

With the preceding introductory material, it is proper to state a definition[1] of a feedback control system: "A control system that operates to achieve prescribed relationships between selected system variables by comparing functions of these variables and using the comparison to effect control." The following definitions are also used.

Servomechanism (often abbreviated as *servo*). The term is often used to refer to a mechanical system in which the steady-state error is zero for a constant input signal. Sometimes, by generalization, it is used to refer to any feedback control system.

Regulator. This term is used to refer to systems in which there is a constant steady-state output for a constant signal. The name is derived from the early speed and voltage controls, called speed and voltage regulators.

1.4 HISTORICAL BACKGROUND[2]

The action of steering an automobile to maintain a prescribed direction of movement satisfies the definition of a feedback control system. In Fig. 1.7, the prescribed direction is the reference input. The eyes perform the function of comparing the actual direction of movement with the prescribed direction, the desired output. The eyes transmit a signal to the brain, which interprets this signal and transmits a signal to the arms to turn the steering wheel, adjusting the actual direction of movement to bring it in line with the desired direction. Thus, steering an automobile constitutes a feedback control system.

One of the earliest open-loop control systems was Hero's device for opening the doors of a temple. The command input to the system (see Fig. 1.8) was lighting a fire upon the altar. The expanding hot air under the fire drove the water from the container into the bucket. As the bucket became heavier, it descended and turned the door spindles by means of ropes, causing the counterweight to rise. The door could be closed by dousing the fire. As the air in the container cooled and the pressure was thereby reduced, the water from the bucket siphoned back into the storage container. Thus, the bucket became lighter and the counterweight, being heavier, moved down, thereby closing the door. This action occurs as long as the bucket is higher than the container. The device was probably actuated when the ruler and his entourage started to ascend the temple steps. The system for opening the door was not visible or known to the masses. Thus, it created an air of mystery and demonstrated the power of the Olympian gods.

James Watt's flyball governor for controlling speed, developed in 1788, can be considered the first widely used automatic feedback control system. Maxwell, in 1868, made an analytic study of the stability of the flyball governor. This was followed by a more detailed solution of the stability of a third-order flyball governor in 1876 by the Russian engineer Wischnegradsky.[3] Minorsky made one of the earlier

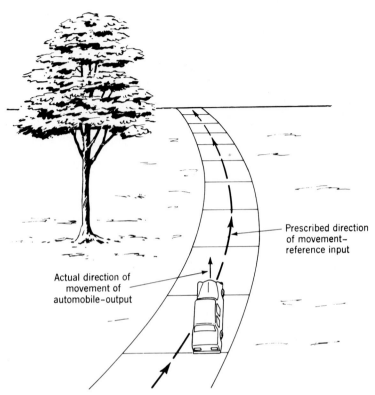

FIGURE 1.7
A pictorial demonstration of an automobile as a feedback control system.

deliberate applications of nonlinear elements in closed-loop systems in his study of automatic ship steering about 1922.[4]

A significant date in the history of automatic feedback control systems is 1934, when Hazen's paper "Theory of Servomechanisms" was published in the *Journal of the Franklin Institute,* marking the beginning of the very intense interest in this new field. It was in this paper that the word *servomechanism* originated, from the words *servant* (or slave) and *mechanism.* Black's important paper on feedback amplifiers appeared[5] in the same year. After World War II, control theory was studied intensively and applications have proliferated. Many books and thousands of articles and technical papers have been written, and the application of control systems in the industrial and military fields has been extensive. This rapid growth of feedback control systems was accelerated by the equally rapid development and widespread use of computers.

An early military application of a feedback control system is the antiaircraft radar-tracking control system shown in Fig. 1.9. The radar antenna detects the position and velocity of the target airplane, and the computer takes this information and determines the correct firing angle for the gun. This angle includes the necessary lead angle so that the shell reaches the projected position at the same time as the

FIGURE 1.8
Hero's device for opening temple doors.

airplane. The output signal of the computer, which is a function of the firing angle, is fed into an amplifier that provides power for the drive motor. The motor then aims the gun at the necessary firing angle. A feedback signal proportional to the gun position ensures correct alignment with the position determined by the computer. Since the gun must be positioned both horizontally and vertically, this system has two drive motors, which are parts of two coordinated feedback loops.

The advent of the nuclear reactor was a milestone in the advancement of science and technology. For proper operation of the reactor the power level must be maintained at a desired value or must vary in a prescribed manner. This must be accomplished automatically with minimum human supervision. Figure 1.10 is a simplified block diagram of a feedback control system for controlling the power output level of a reactor. If the power output level differs from the reference input value, the actuating signal produces a signal at the output of the control elements. This, in turn, moves the regulating rod in the proper direction to achieve the desired power level of the nuclear reactor. The position of the regulating rod determines the rate of nuclear fission and, therefore, the total power generated. This output nuclear power can be converted into steam power, for example, which is then used for generating electric energy.

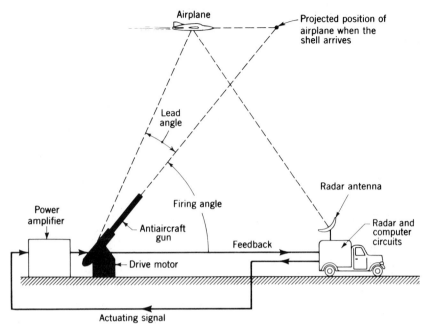

FIGURE 1.9
Antiaircraft radar-tracking control system.

The control theory developed through the late 1950s may be categorized as *conventional* control theory and is effectively applied to many control-design problems, especially to SISO systems. Since then, control theory has been developed for the design of more complicated systems and for *multiple-input multiple-output* (MIMO) systems. Space travel has become possible only because of the advent of modern control theory. Areas such as trajectory optimization and minimum-time and/or minimum-fuel problems, which are very important in space travel, can be readily handled by multivariable control theory. The introduction of microprocessors as control elements, i.e., performing control functions in contrast to being used solely as computational tools, has had an enormous impact on the design of feedback control systems that achieve desired control-system specifications.

The development of control concepts in the engineering field has been extended to the realm of human and biomedical engineering. For example, the basic concept of feedback control is used extensively in the field of business

FIGURE 1.10
A feedback system for controlling the power level of a nuclear reactor.

management. The field of medicine is also one to which the principles of control systems and systems engineering are being applied extensively. Thus, standards of optimum performance are established in all areas of endeavor. Then the actual performance is compared with the desired standard, and any difference between the two is used to bring them into closer agreement.

1.5   DIGITAL CONTROL DEVELOPMENT[13]

With the decreasing cost of digital hardware, economical digital control implementation is now used extensively. Applications include process control, automatic aircraft stabilization and control, guidance and control of aerospace vehicles, and numeric control of manufacturing machines. The development of digital control systems is illustrated by the following example of a digital flight control system.

Numerous changes have been made over time in aircraft flight control systems. Initially, they were purely mechanical, which was ideal for smaller, slow-speed, low-performance aircraft because they were easy to maintain. However, greater mechanical force is required to control modern high-performance airplanes. Thus, a hydraulic power boost system was added to the mechanical control. As aircraft became larger, faster, and heavier, they became harder to control because the pilot could not provide the necessary power to operate the control surfaces directly. Thus, the entire effort of moving the control surface had to be provided by the actuator. A stability augmentation system (SAS) was added to the hydraulically boosted mechanical system to make the aircraft flyable under all flight configurations. Motion sensors were used to detect aircraft perturbations and to provide electric signals to a SAS computer, which, in turn, calculated the proper amount of servo actuator force required. When a higher-authority SAS was required, as with advanced aircraft, both series- and parallel-pitch axis dampers were installed. This so-called command augmentation system (CAS) allowed greater flexibility in control because the parallel damper could provide full-authority travel without restricting the pilot's stick movements. Although planes were originally designed to be statically stable, longitudinally unstable aircraft are more agile. In these aircraft the flight control system must now provide the required stability.

The next step in the evolution of flight control systems was the use of a fly-by-wire (FBW) control system, shown in Fig. 1.11. In this design, all pilot commands are transmitted to the control-surface actuators through electric wires. Thus, all mechanical linkages from the pilot's control stick to the servo actuators are removed from the aircraft. The FBW system offered the advantages of reduced weight, improved survivability, and decreased maintenance. However, the pilot is required to believe in and accept the increased survivability that is provided by using redundancy throughout the entire flight control system. Individual component reliability is also increased by replacing the older analog circuitry with newer digital hardware. These updated systems are referred to as digital flight control systems (DFCS).

The use of airborne digital processors further reduces the cost, weight, and maintenance of modern aircraft. Other advantages associated with newer digital equipment include greater accuracy, increased modification flexibility through the use of software changes, improved in-flight reconfiguration techniques that

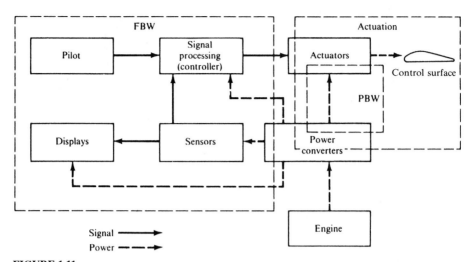

FIGURE 1.11
Fly-by-wire (FBW) and power-by-wire (PBW) control systems. (Control Systems Development Branch, Flight Dynamics Laboratory, Wright-Patterson Air Force Base, Ohio.)

compensate for physical damage and equipment failures, and more reliable preflight and postflight maintenance testing. An example of a modern high-performance aircraft is shown in Fig. 1.12. This aircraft has a three-axis fly-by-wire flight control system. It also includes a digital built-in task computer that runs through all the preflight tests to make sure that all equipment is functioning properly.

1.6   MATHEMATICAL BACKGROUND

The early studies of control systems were based on the solution of differential equations by classical means. Other than for simple systems, the analysis in this approach is tedious and does not readily indicate what changes should be made to improve system performance. Use of the Laplace transform simplifies this analysis somewhat. Nyquist's paper[6] published in 1932 dealt with the application of steady-state frequency-response techniques to feedback amplifier design. This work was extended by Black[5] and Bode.[7] Hall[8] and Harris[9] applied frequency-response analysis in the study of feedback control systems, which furthered the development of control theory as a whole.

Another advance occurred in 1948, when Evans[10] presented his root-locus theory. This theory affords a graphical display of the stability properties of a system and permits the graphical evaluation of both the time and the frequency response. Laplace transform theory and network theory are joined in the root-locus calculation. In the conventional control theory portion of this text the reader learns to appreciate the simplicity and value of the root-locus technique.

The Laplace transform and the principles of linear algebra are used in applying state variable methods to system analysis and design. The nth-order differential equation describing the system can be converted into a set of n first-order differential equations, expressed in terms of the state variables, that can be written

1 - Left Canard	5 - L. Elevator
2 - Right Canard	6 - R. Elevator
3 - Outboard Flaperon	7 - Inboard Flaperon
4 - L. Inboard Flaperon	8 - Outboard Flaperon

9 - Rudder

FIGURE 1.12
An aircraft designed for aero redundancy for reconfiguration to maintain desired flying qualities.

in matrix notation for simpler mathematical manipulation. Matrix equations lend themselves very well to computer solution. This characteristic has enabled modern control theory to solve many problems, such as nonlinear and optimization problems, that could not be solved by conventional control theory.

Mathematical models are used in the linear analysis presented in this text. Once a physical system has been described by a set of mathematical equations, they are manipulated to achieve an appropriate mathematical format. When this has been done, the subsequent method of analysis is independent of the nature of the physical system; i.e., it does not matter whether the system is electrical, mechanical, or something else. This technique helps the designer to spot similarities based upon previous experience.

The reader should recognize that no single design method is intended to be used to the exclusion of the others. Depending upon the known factors and the simplicity or complexity of a control-system problem, a designer may use one method exclusively or a combination of methods. With experience in the design of feedback control systems comes the ability to use the advantages of each method.

The modern control theory presented in this text includes the method for the assignment of the complete eigenstructure, which consists of both the closed-loop eigenvalue spectrum and the associated set of eigenvectors. This method is an extension of previous methods that assigned only the closed-loop eigenvalues. The eigenvectors can be selected from identifiable subspaces that are associated with the system plant and control matrices. This method provides great potential for shaping the system output response to meet desired performance standards. Additional state-of-the-art design techniques for MIMO control systems are presented that bring together many of the fundamentals presented earlier in the text. These chapters present the concepts of designing a robust control system in which the plant parameters may vary over specified ranges during the entire operating regime.

A control engineer must be proficient in the use of available comprehensive computer-aided design (CAD) programs such as MATLAB®, ICECAP-PC, and TOTAL-PC (see Appendix B),[12] which are control-system CAD programs. Many CAD packages for personal computers (PCs) and mainframe computers (see Appendix B) are available commercially. The use of a CAD package enhances the designer's control-system design proficiency, since it minimizes and expedites the tedious and repetitive calculations involved in the design of a satisfactory control system. To understand and use a computer-aided analysis and design package, one must first achieve a conceptual understanding of the theory and processes involved in the analysis and synthesis of control systems. Once the conceptual understanding is achieved by direct calculations, the reader is urged to use all available computer aids. For complicated design problems engineers must write their own digital computer program that is especially geared to help achieve a satisfactory system performance.

1.7 GENERAL NATURE OF THE ENGINEERING CONTROL PROBLEM

In general, a control problem can be divided into the following steps:

1. A set of performance specifications is established.
2. The performance specifications establish the control problem.
3. A set of differential equations that describe the physical system is formulated.
4. Using conventional control theory aided by available[12] or specially written computer programs, the following procedure is used:
    (a) The performance of the basic (original or uncompensated) system is determined by application of one of the available methods of analysis (or a combination of them).
    (b) If the performance of the original system does not meet the required specifications, cascade or feedback compensation must be added to improve the response.
    (c) For plants having structured parameter uncertainty, the QFT design technique may be used.
5. The designer may specify the desired closed-loop eigenvalue spectrum. Then the desired contribution of each mode to each state and output response is selected. The eigenvector spaces are identified, and the eigenvectors are assigned that best meet the selected modal composition of the states that are present in the outputs.

Design of the system to obtain the desired performance is the control problem. The necessary basic equipment is then assembled into a system to perform the desired control function. Although most systems are nonlinear, in many cases the nonlinearity is small enough to be neglected, or the limits of operation are small enough to allow a linear analysis to be used. In this textbook linear systems or those that can be approximated as linear systems are primarily considered. Because of the relative simplicity and straightforwardness of this approach, the reader can obtain a thorough understanding of linear systems. After mastering the terminology, definitions, and methods of analysis for linear control systems, the engineer will find it easier to undertake a study of nonlinear systems. A method of linearizing a nonlinear system is included in Chapter 14.

A basic system has the minimum amount of equipment necessary to accomplish the control function. After a control system has been synthesized to achieve the desired performance, based upon a linear analysis, final adjustments can be made in a simulation or on the actual system to take into account the nonlinearities that were neglected. A computer is generally used in the design, depending upon the complexity of the system.

1.8 COMPUTER LITERACY

In the mid-1960s, the first practical CAD package, *FREQR,* for frequency domain analysis, was developed at the Air Force Institute of Technology (AFIT). This was followed by the development of two other AFIT CAD packages: *PARTL* for partial-fraction expansion of transfer functions and for obtaining a time response, and *ROOTL* for obtaining root-locus data and plots. These CAD packages became the basis for the first comprehensive control system design CAD package, called *TOTAL,* the forerunner of *ICECAP,* which was developed in 1978 at AFIT. TOTAL became the catalyst, along with other control CAD packages developed by other individuals, for the development of the current highly developed, readily available commercial control system design CAD packages. One of these CAD packages is *MATLAB,* which has become a valuable tool for a control engineer and is illustrated in this text.

Becoming proficient (computer literate) in the use of these CAD packages (tools) is essential for a control systems engineer. It is also essential to develop procedures for checking the CAD results at each stage of the analysis and design. This is necessary in order to verify that these CAD tools have generated results that are consistent with theory. Whenever in doubt concerning the operation of a specific CAD tool, apply the CAD tool to a simple problem whose known analytical solution can be compared with the computer-generated output.

1.9 OUTLINE OF TEXT

The text is divided into three parts. The first part, consisting of Chapters 2 through 4, provides the mathematical foundation for modeling physical systems and obtaining time solutions using classical or Laplace transform methods. The second part consists of Chapters 5 through 15 and represents material that is usually covered in the first undergraduate course in control theory and control system design. The third part, Chapters 16 through 18, consists of advanced undergraduate or graduate topics. This partitioning provides the instructor with clear demarcation lines within the text material.

The first few chapters deal with the mathematics and physical system modeling that underlie the analysis of control systems. Once the technique of writing the system equations (and, in turn, their Laplace transforms) that describe the performance of a dynamic system has been mastered, the ideas of block and simulation diagrams and transfer functions are developed. When physical systems are described in terms of block diagrams and transfer functions, they exhibit basic servo characteristics. These characteristics are described and discussed. The concept of state is introduced, and system equations are developed in the standard matrix format. The necessary linear algebra required to manipulate the matrix equations is included. Then follows a presentation of the various methods of analysis that can be used in the study of feedback control systems. SISO systems are used initially to facilitate an understanding of the synthesis methods. These methods rely on root-locus and steady-state frequency-response analysis. Improvement of the basic system by using compensators, if it does not meet the desired specifications, is then presented. These compensators can be designed by analyzing an open-loop transfer function. Alternatively, a closed-loop transfer function can be synthesized that produces the desired overall system performance. Chapter 12 is devoted to modeling a desired closed-loop transfer function for tracking a desired input or for rejecting (not responding to) disturbance inputs.

The next portion of the text deals with SISO system design using state feedback. Topics such as controllability and observability, pole placement via state-variable feedback, parameter sensitivity, phase-plane analysis, and linearization of nonlinear systems are presented. The fundamentals of sampled-data control-system analysis and design are presented in Chapter 15. This permits the study of both continuous and discrete control systems and a comparison of their respective characteristics. Chapter 16 presents the importance of the eigenvector on system response and the method of entire eigenstructure assignment for MIMO systems. The subspaces in which the eigenvectors must be located are identified, and then selected eigenvectors are used to compute the required feedback matrix. Many worthwhile control-system design techniques are available in the technical literature, involving both modern and conventional control theory. Each technique has its advantages and disadvantages and may be applicable to only certain classes of design problems. Ease of design is directly dependent upon the available CAD packages. Chapters 17 and 18 present an introduction to two multivariable control-system design techniques that are applicable to the design of many practical control systems. These two design methods yield robust control systems that maintain the desired system performance over a prescribed region of structured plant parameter uncertainty. Appendix A gives a table of Laplace transform pairs. A description of CAD packages that are useful to a feedback control system engineer is presented in Appendix B.

In closing this introductory chapter, it is important to stress that feedback control engineers are essentially "system engineers," i.e., people whose primary concern is with the design and synthesis of an overall system. Control engineers rely on and work closely with other engineers in various recognized branches of engineering. This leads to the development of the transfer functions or system equations of various control systems. For example, the biomedical control engineer works closely with the medical profession in modeling biological functions, human-machine interface problems, etc.

The following design policy includes factors that are worthy of consideration in any control system design problem:

1. Use proven design methods.
2. Select the system design that has the minimum complexity.
3. Use the minimum specifications or requirements that yield a satisfactory system response. Compare the cost with the performance and select the fully justified system implementation.
4. Perform a complete and adequate simulation and testing of the system.

REFERENCES

1. *IEEE Standard Dictionary of Electrical and Electronics Terms,* Wiley-Interscience, New York, 1972.
2. Mayr, O.: *Origins of Feedback Control,* The MIT Press, Cambridge, Mass., 1971.
3. Trinks, W.: *Governors and the Governing of Prime Movers,* Van Nostrand, Princeton, N.J., 1919.
4. Minorsky, N.: "Directional Stability and Automatically Steered Bodies," *J. Am. Soc. Nav. Eng.,* vol. 34, p. 280, 1922.
5. Black, H. S.: "Stabilized Feedback Amplifiers," *Bell Syst. Tech. J.,* 1934.
6. Nyquist, H.: "Regeneration Theory," *Bell Syst. Tech. J.,* 1932.
7. Bode, H. W.: *Network Analysis and Feedback Amplifier Design,* Van Nostrand, Princeton, N.J., 1945.
8. Hall, A. C.: "Application of Circuit Theory to the Design of Servomechanisms," *J. Franklin Inst.,* 1946.
9. Harris, H.: "The Frequency Response of Automatic Control Systems," *Trans. AIEE,* vol. 65, pp. 539–545, 1946.
10. Evans, W. R.: "Graphical Analysis of Control Systems," *Trans. AIEE,* vol. 67, pt. II, pp. 547–551, 1948.
11. Much, C. H., P. L. Rossini, and D. G. Stuart: "A Microprocessor-Controlled Autonomous Sun-Following Mount for Solar Photovoltaic Energy Planning," *5th Annu. Conf. Ind. Control Appl. Microprocessors,* Philadelphia, March 1979.
12. Allan, R.: "Busy Robots Spur Productivity," *IEEE Spectrum,* vol. 16, pp. 31–36, 1979.
13. Houpis, C. H., and G. B. Lamont: *Digital Control Systems Theory, Hardware, Software,* 2d ed., McGraw-Hill, New York, 1992.
14. Dorf, R. C.: *Encyclopedia of Robotics,* Wiley-Interscience, New York, 1988.
15. Vidyasagar, M.: "System Theory and Robotics," *IEEE Control Systems,* pp. 16–17, April 1987.

CHAPTER
2

WRITING SYSTEM EQUATIONS

2.1 INTRODUCTION

An accurate mathematical model that describes a system completely must be determined in order to analyze a dynamic system. The derivation of this model is based upon the fact that the dynamic system can be completely described by known differential equations (Chapter 2) or by experimental test data (Sec. 8.9). The ability to analyze the system and determine its performance depends on how well the characteristics can be expressed mathematically. Techniques for solving linear differential equations with constant coefficients are presented in Chapters 3 and 4. However, the solution of a time-varying or nonlinear equation often requires a numerical, graphic, or computer procedure.[16] The systems considered in this chapter are described completely by a set of linear *constant-coefficient* differential equations. Such systems are said to be *linear time-invariant* (LTI),[1] i.e., the relationship between the system input and output is independent of time. Since the system does not change with time, the output is independent of the time at which the input is applied. Linear methods are used because there is extensive and elegant mathematics for solving linear equations. For many systems, there are regions of operation for which the linear representation works very well.

This chapter presents methods for writing the differential and state equations for a variety of electrical, mechanical, thermal, and hydraulic systems.[2] This step is the first that must be mastered by the control-systems engineer. The basic physical laws are given for each system, and the associated parameters are defined. Examples are included to show the application of the basic laws to physical equipment. The

19

result is a differential equation, or a set of differential equations, that describes the system. The equations derived are limited to linear systems or to systems that can be represented by linear equations over their useful operating range. The important concepts of system state and of state variables are also introduced. The system equations, expressed in terms of state variables, are called *state equations*. The analytical tools from linear algebra are introduced as they are needed in the development of the state equations and their solutions. The solutions of both the differential and the state equations are covered in Chapter 3.

The analysis of behavior of the system equations is enhanced by using the block diagram representation of the system. Complete drawings showing all the detailed parts are frequently too congested to show the specific functions that are performed by each system component. It is common to use a block diagram, in which each function is represented by a block, to simplify the picture of the complete system. Each block is labeled with the name of the component, and the blocks are appropriately interconnected by line segments. This type of diagram removes excess detail from the picture and shows the functional operation of the system. The use of a block diagram provides a simple means by which the functional relationship of the various components can be shown and reveals the operation of the system more readily than does observation of the physical system itself. The simple functional block diagram shows clearly that apparently different physical systems can be analyzed by the same techniques. Since a block diagram is involved not with the physical characteristics of the system but only with the functional relationship between various points in the system, it can reveal the similarity between apparently unrelated physical systems.

A block diagram[3] represents the flow of information and the functions performed by each component in the system. A further step taken to increase the information supplied by the block diagram is to label the input quantity into each block and the output quantity from each block. Arrows are used to show the direction of the flow of information. The block represents the function or dynamic characteristics of the component and is represented by a transfer function (Sec. 2.5). The complete block diagram shows how the functional components are connected and the mathematical equations that determine the response of each component. Examples of block diagrams are shown throughout this chapter.

In general, variables that are functions of time are represented by lowercase letters. These are sometimes indicated by the form $x(t)$, but more often this is written just as x. There are some exceptions, because of established convention, in the use of certain symbols.

To simplify the writing of differential equations, the D operator notation is used.[4] The symbols D and $1/D$ are defined by

$$Dy \equiv \frac{dy(t)}{dt} \qquad D^2 y \equiv \frac{d^2 y(t)}{dt^2} \tag{2.1}$$

$$D^{-1}y \equiv \frac{1}{D}y \equiv \int_0^t y(\tau)\, d\tau + \int_{-\infty}^0 y(\tau)\, d\tau = \int_0^t y(\tau)\, d\tau + Y_0 \tag{2.2}$$

where Y_0 represents the value of the integral at time $t = 0$, that is, the *initial value* of the integral.

2.2 ELECTRIC CIRCUITS AND COMPONENTS[5]

The equations for an electric circuit obey Kirchhoff's laws, which state the following:

1. The algebraic sum of the potential differences around a closed path equals zero. This can be restated as follows: in traversing any closed loop the sum of the voltage rises equals the sum of the voltage drops.
2. The algebraic sum of the currents entering (or leaving) a node is equal to zero. In other words, the sum of the currents entering the junction equals the sum of the currents leaving the junction.

The voltage sources are usually alternating-current (ac) or direct-current (dc) generators. The usual dc voltage source is a battery. The voltage drops appear across the three basic electrical elements: resistors, inductors, and capacitors. These elements have constant component values.

The voltage drop across a resistor is given by Ohm's law, which states that the voltage drop across a resistor is equal to the product of the current through the resistor and its resistance. Resistors absorb energy from the system. Symbolically, this voltage is written as

$$v_R = Ri \tag{2.3}$$

The voltage drop across an inductor is given by Faraday's law, which is written

$$v_L = L\frac{di}{dt} \equiv LDi \tag{2.4}$$

This equation states that the voltage drop across an inductor is equal to the product of the inductance and the time rate of increase of current. A positive-valued derivative Di implies an increasing current, and thus a positive voltage drop; a negative-valued derivative implies a decreasing current, and thus a negative voltage drop.

The positively directed voltage drop across a capacitor is defined as the ratio of the magnitude of the positive electric charge on its positive plate to the value of its capacitance. Its direction is from the positive plate to the negative plate. The charge on a capacitor plate is equal to the time integral of the current entering the plate from the initial instant to the arbitrary time t, plus the initial value of the charge. The capacitor voltage is written in the form

$$v_C = \frac{q}{C} = \frac{1}{C}\int_0^t i\,d\tau + \frac{Q_0}{C} \equiv \frac{i}{CD} \tag{2.5}$$

The mks units for these electrical quantities in the practical system are given in Table 2.1.

TABLE 2.1
Electrical symbols and units

Symbol	Quantity	Units
e or v	Voltage	Volts
i	Current	Amperes
L	Inductance	Henrys
C	Capacitance	Farads
R	Resistance	Ohms

Series Resistor-Inductor Circuit

The voltage source e in Fig. 2.1 is a function of time. When the switch is closed, setting the voltage rise equal to the sum of the voltage drops produces

$$v_R + v_L = e$$

$$Ri + L\frac{di}{dt} = Ri + LDi = e$$

(2.6)

When the applied voltage e is known, the equation can be solved for the current i as shown in Chapter 3. This equation can also be written in terms of the voltage v_L across the inductor in the following manner. The voltage across the inductor is $v_L = LDi$. Therefore, the current through the inductor is

$$i = \frac{1}{LD}v_L$$

Substituting this value into the original equation gives

$$\frac{R}{LD}v_L + v_L = e$$

(2.7)

The node method is also convenient for writing the system equations directly in terms of the voltages. The junctions of any two elements are called *nodes*. This circuit has three nodes, labeled a, b, and c (see Fig. 2.1). One node is used as a reference point; in this circuit node c is used as the reference. The voltages at all the other nodes are considered with respect to the reference node. Thus v_{ac} is the voltage drop from node a to node c, and v_{bc} is the voltage drop from node b to the reference node c. For simplicity, these voltages are written just as v_a and v_b.

The source voltage $v_a = e$ is known; therefore, there is only one unknown voltage, v_b, and only one node equation is necessary. Kirchhoff's second law, that

FIGURE 2.1
Series resistor-inductor circuit.

FIGURE 2.2
Series *RLC* circuit.

the algebraic sum of the currents entering (or leaving) a node must equal zero, is applied to node *b*. The current from node *b* to node *a*, through the resistor *R*, is $(v_b - v_a)/R$. The current from node *b* to node *c* through the inductor *L* is $(1/LD)v_b$. The sum of these currents must equal zero:

$$\frac{v_b - v_a}{R} + \frac{1}{LD}v_b = 0 \tag{2.8}$$

Rearranging terms gives

$$\left(\frac{1}{R} + \frac{1}{LD}\right)v_b - \frac{1}{R}v_a = 0 \tag{2.9}$$

Except for the use of different symbols for the voltages, this is the same as Eq. (2.7). Note that the node method requires writing only one equation.

Series Resistor-Inductor-Capacitor Circuit

For the series *RLC* circuit shown in Fig. 2.2, the applied voltage is equal to the sum of the voltage drops when the switch is closed:

$$v_L + v_R + v_C = e \qquad LDi + Ri + \frac{1}{CD}i = e \tag{2.10}$$

The circuit equation can be written in terms of the voltage drop across any circuit element. For example, in terms of the voltage across the resistor, $v_R = Ri$, Eqs. (2.10) become

$$v_R + \frac{L}{R}Dv_R + \frac{1}{RCD}v_R = e \tag{2.11}$$

Multiloop Electric Circuits

Multiloop electric circuits (see Fig. 2.3) can be solved by either loop or nodal equations. The following example illustrates both methods. The problem is to solve for the output voltage v_o.

LOOP METHOD. A loop current is drawn in each closed loop (usually in a clockwise direction); then Kirchhoff's voltage equation is written for each loop:

$$\left(R_1 + \frac{1}{CD}\right)i_1 - R_1i_2 - \frac{1}{CD}i_3 = e \tag{2.12}$$

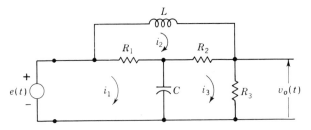

FIGURE 2.3
Multiloop network.

$$-R_1 i_1 + (R_1 + R_2 + LD)i_2 - R_2 i_3 = 0 \tag{2.13}$$

$$-\frac{1}{CD}i_1 - R_2 i_2 + \left(R_2 + R_3 + \frac{1}{CD}\right)i_3 = 0 \tag{2.14}$$

The output voltage is $\qquad v_o = R_3 i_3 \tag{2.15}$

These four equations must be solved simultaneously to obtain $v_o(t)$ in terms of the input voltage $e(t)$ and the circuit parameters.

NODE METHOD. The junctions, or nodes, are labeled by letters in Fig. 2.4. Kirchhoff's current equations are written for each node in terms of the node voltages, where node d is taken as reference. The voltage v_{bd} is the voltage of node b with reference to node d. For simplicity, the voltage v_{bd} is written just as v_b. Since there is one known node voltage $v_a = e$ and two unknown voltages v_b and v_o, only two equations are required:

For node b: $\qquad\qquad\qquad i_1 + i_2 + i_3 = 0 \tag{2.16}$

For node c: $\qquad\qquad\qquad -i_3 + i_4 + i_5 = 0 \tag{2.17}$

In terms of the node voltages, these equations are

$$\frac{v_b - v_a}{R_1} + CDv_b + \frac{v_b - v_o}{R_2} = 0 \tag{2.18}$$

$$\frac{v_o - v_b}{R_2} + \frac{v_o}{R_3} + \frac{1}{LD}(v_o - e) = 0 \tag{2.19}$$

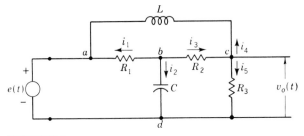

FIGURE 2.4
Multinode network.

Rearranging the terms in order to systematize the form of the equations gives

$$\left(\frac{1}{R_1} + CD + \frac{1}{R_2}\right)v_b - \frac{1}{R_2}v_o = \frac{1}{R_1}e \tag{2.20}$$

$$-\frac{1}{R_2}v_b + \left(\frac{1}{R_2} + \frac{1}{R_3} + \frac{1}{LD}\right)v_o = \frac{1}{LD}e \tag{2.21}$$

For this example, only two nodal equations are needed to solve for the potential at node c. An additional equation must be used if the current in R_3 is required. With the loop method, three equations must be solved simultaneously to obtain the current in any branch; an additional equation must be used if the voltage across R_3 is required. The method that requires the solution of the fewest equations should be used. This varies with the circuit.

The rules for writing the node equations are summarized as follows:

1. The number of equations required is equal to the number of unknown node voltages.
2. An equation is written for each node.
3. Each equation includes the following:
   (a) The node voltage multiplied by the sum of all the admittances that are connected to this node. This term is positive.
   (b) The node voltage at the other end of each branch multiplied by the admittance connected between the two nodes. This term is negative.

The reader should learn to apply these rules so that Eqs. (2.20) and (2.21) can be written directly from the circuit of Fig. 2.4.

2.3 BASIC LINEAR MATRIX ALGEBRA[6,17]

Before proceeding with a discussion of other physical systems, it is appropriate to introduce some basic matrix concepts needed in the development of the state method of describing and analyzing physical systems.

MATRIX. A *matrix* is a rectangular array of elements. The elements may be real or complex numbers or variables of time or frequency. A matrix with α rows and β columns is called an $\alpha \times \beta$ matrix or is said to be of *order* $\alpha \times \beta$. The matrix is also said to have size or dimension $\alpha \times \beta$. If $\alpha = \beta$, the matrix is called a *square matrix*. Boldface capital letters are used to denote rectangular matrices, and boldface small letters are used to denote column matrices. A general expression for an $\alpha \times \beta$ matrix is

$$\mathbf{M} = \begin{bmatrix} m_{11} & m_{12} & \cdots & m_{1\beta} \\ m_{21} & m_{22} & \cdots & m_{2\beta} \\ \vdots & & \ddots & \vdots \\ m_{\alpha 1} & m_{\alpha 2} & \cdots & m_{\alpha\beta} \end{bmatrix} = \begin{bmatrix} m_{ij} \end{bmatrix} \tag{2.22}$$

The elements are denoted by lowercase letters. A double-subscript notation is used to denote the location of the element in the matrix; thus, the element m_{ij} is located in the ith row and the jth column.

TRANSPOSE. The transpose of a matrix $\mathbf{M}$ is denoted by $\mathbf{M}^T$. The matrix $\mathbf{M}^T$ is obtained by interchanging the rows and columns of the matrix $\mathbf{M}$. In general, with $\mathbf{M} = [m_{ij}]$, the transpose matrix is $\mathbf{M}^T = [m_{ji}]$.

VECTOR. A vector is a matrix that has either one row or one column. An $\alpha \times 1$ matrix is called a *column vector,* written

$$\mathbf{x} = \begin{bmatrix} x_1 \\ x_2 \\ \vdots \\ x_\alpha \end{bmatrix} \tag{2.23}$$

and a $1 \times \beta$ matrix is called a *row vector,* written

$$\mathbf{x}^T = [\, x_1 \quad x_2 \quad \cdots \quad x_\beta \,] \tag{2.24}$$

Thus, the transpose of a column vector $\mathbf{x}$ yields the row vector $\mathbf{x}^T$.

ADDITION AND SUBTRACTION OF MATRICES. The sum or difference of two matrices $\mathbf{M}$ and $\mathbf{N}$, both of order $\alpha \times \beta$, is a matrix $\mathbf{W}$ of order $\alpha \times \beta$. The element w_{ij} of $\mathbf{W} = \mathbf{M} \pm \mathbf{N}$ is

$$w_{ij} = m_{ij} \pm n_{ij} \tag{2.25}$$

These operations are commutative and associative, i.e.,

Commutative: $\qquad\qquad \mathbf{M} \pm \mathbf{N} = \pm\mathbf{N} + \mathbf{M}$

Associative: $\qquad\qquad (\mathbf{M} + \mathbf{N}) + \mathbf{Q} = \mathbf{M} + (\mathbf{N} + \mathbf{Q})$

Example 1. Addition of matrices.

$$\begin{bmatrix} 3 & 1 & 2 \\ 1 & 0 & -4 \\ 0 & 5 & 7 \end{bmatrix} + \begin{bmatrix} 1 & -3 & 1 \\ 2 & 1 & 5 \\ -4 & -1 & 0 \end{bmatrix} = \begin{bmatrix} 4 & -2 & 3 \\ 3 & 1 & 1 \\ -4 & 4 & 7 \end{bmatrix}$$

MULTIPLICATION OF MATRICES. The multiplication of two matrices $\mathbf{MN}$ can be performed *if and only if* (iff) they *conform.* If the orders of $\mathbf{M}$ and $\mathbf{N}$ are $\alpha \times \beta$ and $\gamma \times \delta$, respectively, they are conformable iff $\beta = \gamma$; that is, the number of columns of $\mathbf{M}$ must be equal to the number of rows of $\mathbf{N}$. Under this condition $\mathbf{MN} = \mathbf{W}$, where the elements of $\mathbf{W}$ are defined by

$$w_{ij} = \sum_{k=1}^{\beta} m_{ik} n_{kj} \tag{2.26}$$

That is, each element of the ith row of $\mathbf{M}$ is multiplied by the corresponding element of the jth column of $\mathbf{N}$, and these products are summed to yield the ijth element of $\mathbf{W}$. The dimension or order of the resulting matrix $\mathbf{W}$ is $\alpha \times \delta$.

Matrix multiplication operations are summarized as follows:

1. An $\alpha \times \beta$ $\mathbf{M}$ matrix times a $\beta \times \gamma$ $\mathbf{N}$ matrix yields an $\alpha \times \gamma$ $\mathbf{W}$ matrix; that is, $\mathbf{MN} = \mathbf{W}$.
2. An $\alpha \times \beta$ $\mathbf{M}$ matrix times a $\beta \times \alpha$ $\mathbf{N}$ matrix yields an $\alpha \times \alpha$ $\mathbf{W}$ square matrix; that is, $\mathbf{MN} = \mathbf{W}$. Note that although $\mathbf{NM} = \mathbf{Y}$ is also a square matrix, it is of order $\beta \times \beta$.
3. When $\mathbf{M}$ is of order $\alpha \times \beta$ and $\mathbf{N}$ is of order $\beta \times \alpha$, each of the products $\mathbf{MN}$ and $\mathbf{NM}$ is conformable. However, in general,

$$\mathbf{MN} \neq \mathbf{NM}$$

Thus the product of $\mathbf{M}$ and $\mathbf{N}$ is said to be *noncommutable;* that is, the order of multiplication cannot be interchanged.
4. The product $\mathbf{MN}$ can be referred to as "$\mathbf{N}$ *premultiplied* by $\mathbf{M}$" or as "$\mathbf{M}$ *postmultiplied* by $\mathbf{N}$." In other words, premultiplication or postmultiplication is used to indicate whether one matrix is multiplied by another from the left or from the right.
5. A $1 \times \alpha$ row vector times an $\alpha \times \beta$ matrix yields a $1 \times \beta$ row vector; that is, $\mathbf{x}^T\mathbf{M} = \mathbf{y}^T$.
6. A $\beta \times \alpha$ matrix times an $\alpha \times 1$ column vector yields a $\beta \times 1$ column vector; that is, $\mathbf{Mx} = \mathbf{y}$.
7. A $1 \times \alpha$ row vector times an $\alpha \times 1$ column vector yields a 1×1 matrix, that is, $\mathbf{x}^T\mathbf{y} = \mathbf{w}$, which contains a single scalar element.
8. The k-fold multiplication of a square matrix $\mathbf{M}$ by itself is indicated by $\mathbf{M}^k$.

These operations are illustrated by the following examples.

Example 2. Matrix multiplication.

$$\begin{bmatrix} 2 & 1 & 3 \\ 1 & 4 & 1 \end{bmatrix}\begin{bmatrix} 2 & 1 & 0 \\ 0 & 3 & 1 \\ 1 & 2 & 1 \end{bmatrix} = \begin{bmatrix} 7 & 11 & 4 \\ 3 & 15 & 5 \end{bmatrix} \qquad \begin{bmatrix} 2 & 1 & 3 \\ 1 & 4 & 1 \end{bmatrix}\begin{bmatrix} 2 & 1 \\ 0 & 3 \\ 1 & 2 \end{bmatrix} = \begin{bmatrix} 7 & 11 \\ 3 & 15 \end{bmatrix}$$

$$\begin{bmatrix} 2 & 1 \\ 0 & 3 \\ 1 & 2 \end{bmatrix}\begin{bmatrix} 2 & 1 & 3 \\ 1 & 4 & 1 \end{bmatrix} = \begin{bmatrix} 5 & 6 & 7 \\ 3 & 12 & 3 \\ 4 & 9 & 5 \end{bmatrix} \qquad \begin{bmatrix} 2 & 1 \\ 1 & 1 \end{bmatrix}\begin{bmatrix} 1 & 0 \\ 2 & 3 \end{bmatrix} = \begin{bmatrix} 4 & 3 \\ 3 & 3 \end{bmatrix}$$

$$\begin{bmatrix} 2 & 1 \\ 0 & 3 \end{bmatrix}\begin{bmatrix} 1 \\ 3 \end{bmatrix} = \begin{bmatrix} 5 \\ 9 \end{bmatrix} \qquad \qquad \begin{bmatrix} 1 & 3 \end{bmatrix}\begin{bmatrix} 2 \\ 1 \end{bmatrix} = [5]$$

MULTIPLICATION OF A MATRIX BY A SCALAR. *The multiplication of matrix* $\mathbf{M}$ *by a scalar* k is effected by multiplying each element m_{ij} by k, that is,

$$k\mathbf{M} = \mathbf{M}k = [km_{ij}] \qquad\qquad (2.27)$$

Example 3.

$$2\begin{bmatrix} 1 & 0 \\ 5 & -7 \end{bmatrix} = \begin{bmatrix} 1 & 0 \\ 5 & -7 \end{bmatrix} 2 = \begin{bmatrix} 2 & 0 \\ 10 & -14 \end{bmatrix}$$

UNIT OR IDENTITY MATRIX.   A unit matrix, denoted $\mathbf{I}$, is a diagonal matrix in which each element on the principal diagonal is unity. Sometimes the notation $\mathbf{I}_n$ is used to indicate an identity matrix of order n. An example of a unit matrix is

$$\mathbf{I} = \begin{bmatrix} 1 & 0 & 0 & 0 \\ 0 & 1 & 0 & 0 \\ 0 & 0 & 1 & 0 \\ 0 & 0 & 0 & 1 \end{bmatrix}$$

The premultiplication or postmultiplication of a matrix $\mathbf{M}$ by the unit matrix $\mathbf{I}$ leaves the matrix unchanged.

$$\mathbf{MI} = \mathbf{IM} = \mathbf{M}$$

DIFFERENTIATION OF A MATRIX.   The differentiation of a matrix with respect to a scalar is effected by differentiating each element of the matrix with respect to the indicated variable:

$$\frac{d}{dt}[\mathbf{M}(t)] = \dot{\mathbf{M}}(t) = [\dot{m}_{ij}] = \begin{bmatrix} \dot{m}_{11} & \dot{m}_{12} & \cdots & \dot{m}_{1\beta} \\ \dot{m}_{21} & \dot{m}_{22} & \cdots & \dot{m}_{2\beta} \\ \vdots & \vdots & \ddots & \vdots \\ \dot{m}_{\alpha 1} & \dot{m}_{\alpha 2} & \cdots & \dot{m}_{\alpha\beta} \end{bmatrix}$$

The derivative of a product of matrices follows rules similar to those for the derivative of a scalar product, with preservation of order. Thus, typically,

$$\frac{d}{dt}[\mathbf{M}(t)\mathbf{N}(t)] = \mathbf{M}(t)\dot{\mathbf{N}}(t) + \dot{\mathbf{M}}(t)\mathbf{N}(t)$$

INTEGRATION OF A MATRIX.   The integration of a matrix is effected by integrating each element of the matrix with respect to the indicated variable:

$$\int \mathbf{M}(t) \, dt = \left[\int m_{ij} \, dt \right]$$

2.4   STATE CONCEPTS

Basic matrix properties are used to introduce the concept of state and the method of writing and solving the state equations. The state of a system (henceforth referred to only as state) is defined by Kalman[7] as follows:

STATE.   The *state* of a system is a mathematical structure containing a set of n variables $x_1(t), x_2(t), \ldots, x_i(t), \ldots, x_n(t)$, called the *state variables*, such that *the initial* *values* $x_i(t_0)$ of this set and the system inputs $u_j(t)$ are sufficient to describe uniquely the system's future response of $t \geq t_0$. A minimum set of state variables is required to represent the system accurately. The m inputs, $u_1(t), u_2(t), \ldots, u_j(t), \ldots, u_m(t)$, are deterministic; i.e., they have specific values for all values of time $t \geq t_0$.

Generally the initial starting time t_0 is taken to be zero. The state variables need not be physically observable and measurable quantities; they may be purely mathematical quantities. The following additional definitions apply:

STATE VECTOR. The set of state variables $x_i(t)$ represents the elements or components of the n-dimensional state vector $\mathbf{x}(t)$; that is,

$$\mathbf{x}(t) \equiv \begin{bmatrix} x_1(t) \\ x_2(t) \\ \vdots \\ x_n(t) \end{bmatrix} = \begin{bmatrix} x_1 \\ x_2 \\ \vdots \\ x_n \end{bmatrix} \equiv \mathbf{x} \qquad (2.28)$$

The order of the system characteristic equation is n, and the state equation representation of the system consists of n first-order differential equations. When all the inputs $u_j(t)$ to a given system are specified for $t > t_0$, the resulting state vector uniquely determines the system behavior for any $t > t_0$.

STATE SPACE. State space is defined as the n-dimensional space in which the components of the state vector represent its coordinate axes.

STATE TRAJECTORY. State trajectory is defined as the path produced in the state space by the state vector $\mathbf{x}(t)$ as it changes with the passage of time. State space and state trajectory in the two-dimensional case are referred to as the *phase plane* and *phase trajectory*, respectively.[16]

The first step in applying these definitions to a physical system is the selection of the system variables that are to represent the state of the system. Note that there is no unique way of making this selection. The three common representations for expressing the system state are the *physical, phase,* and *canonical state variables.* The physical state-variable representation is introduced in this chapter. The other two representations are introduced in later chapters.

The selection of the state variables for the physical-variable method is based upon the energy-storage elements of the system. Table 2.2 lists some common energy-storage elements that exist in physical systems and the corresponding energy equations. The physical variable in the energy equation for each energy-storage element *can* be selected as a state variable of the system. Only independent physical variables are chosen to be state variables. *Independent state variables* are those state variables that cannot be expressed in terms of the remaining assigned state variables. In some systems it may be necessary to identify more state variables than just the energy-storage variables. This situation is illustrated in some of the following examples, where velocity is a state variable. When position, the integral of this state variable, is of interest, it must also be assigned as a state variable.

Example 1. Series RL circuit (Fig. 2.1). Only one energy-storage element, the inductor, is present in this circuit; thus there is only one state variable. From Table 2.2, the state variable is $x_1 = i$. The equation desired is one that contains the first derivative of the state variable. Letting $u = e$, the loop equation, Eq. (2.6), can be rewritten as

$$Rx_1 + L\dot{x}_1 = u$$

$$\dot{x}_1 = -\frac{R}{L}x_1 + \frac{1}{L}u \qquad (2.29)$$

The letter u is the standard notation for the input forcing function and is called the *control variable*. Equation (2.29) is the standard form of the *state equation* of the system. There is only one state equation because this is a first-order system, $n = 1$.

Example 2. Series RLC circuit (Fig. 2.2). This circuit contains two energy-storage elements, the inductor and capacitor. From Table 2.2, the two assigned state variables are identified as $x_1 = v_c$ (the voltage across the capacitor) and $x_2 = i$ (the current in the inductor). Thus two state equations are required.

A loop or branch equation is written to obtain an equation containing the derivative of the current in the inductor. A node equation is written to obtain an equation containing the derivative of the capacitor voltage. The number of loop equations that must be written is equal to the number of state variables representing currents in inductors. The number of equations involving node voltages that must be written is equal to the number of state variables representing voltages across capacitors. These are usually, but not always, node equations. These loop and node equations are written in terms of the inductor *branch* currents and capacitor branch voltages. From these equations, it is necessary to determine which of the assigned physical variables are independent. When a variable of interest is not an energy physical variable, then this variable is identified as an *augmented state variable*.

TABLE 2.2
Energy-storage elements

Element	Energy	Physical variable
Capacitor C	$\dfrac{Cv^2}{2}$	Voltage v
Inductor L	$\dfrac{Li^2}{2}$	Current i
Mass M	$\dfrac{Mv^2}{2}$	Translational velocity v
Moment of inertia J	$\dfrac{J\omega^2}{2}$	Rotational velocity ω
Spring K	$\dfrac{Kx^2}{2}$	Displacement x
Fluid compressibility $\dfrac{V}{K_B}$	$\dfrac{VP_L^2}{2K_B}$	Pressure P_L
Fluid capacitor $C = \rho A$	$\dfrac{\rho Ah^2}{2}$	Height h
Thermal capacitor C	$\dfrac{C\theta^2}{2}$	Temperature θ

Figure 2.2 is redrawn in Fig. 2.5 with node b as the reference node. The node equation for node a and the loop equation are, respectively,

$$C\dot{x}_1 = x_2 \tag{2.30}$$

and

$$L\dot{x}_2 + Rx_2 + x_1 = u \tag{2.31}$$

Rearranging terms to the standard state equation format yields

$$\dot{x}_1 = \frac{1}{C}x_2 \tag{2.32}$$

$$\dot{x}_2 = -\frac{1}{L}x_1 - \frac{R}{L}x_2 + \frac{1}{L}u \tag{2.33}$$

Equations (2.32) and (2.33) represent the state equations of the system containing two independent state variables. Note that they are *first-order* linear differential equations and are $n = 2$ in number. They are the minimum number of state equations required to represent the system's future performance.

The following definition is based upon these two examples.

STATE EQUATION. The state equations of a system are a set of n first-order differential equations, where n is the number of independent states.

The state equations represented by Eqs. (2.32) and (2.33) are expressed in matrix notation as

$$\begin{bmatrix} \dot{x}_1 \\ \dot{x}_2 \end{bmatrix} = \begin{bmatrix} 0 & \dfrac{1}{C} \\ -\dfrac{1}{L} & -\dfrac{R}{L} \end{bmatrix} \begin{bmatrix} x_1 \\ x_2 \end{bmatrix} + \begin{bmatrix} 0 \\ \dfrac{1}{L} \end{bmatrix} [u] \tag{2.34}$$

It can be expressed in a more compact form as

$$\dot{\mathbf{x}} = \mathbf{A}\mathbf{x} + \mathbf{b}u \tag{2.35}$$

where $\quad \dot{\mathbf{x}} = \begin{bmatrix} \dot{x}_1 \\ \dot{x}_2 \end{bmatrix} \qquad\qquad$ an $n \times 1$ column vector

$$\mathbf{A} = \begin{bmatrix} a_{11} & a_{12} \\ a_{21} & a_{22} \end{bmatrix} = \begin{bmatrix} 0 & \dfrac{1}{C} \\ -\dfrac{1}{L} & -\dfrac{R}{L} \end{bmatrix} \qquad \text{an } n \times n \text{ plant coefficient matrix}$$

FIGURE 2.5
Series *RLC* circuit.

$$\mathbf{x} = \begin{bmatrix} x_1 \\ x_2 \end{bmatrix} \qquad \text{an } n \times 1 \text{ state vector}$$

$$\mathbf{b} = \begin{bmatrix} b_1 \\ b_2 \end{bmatrix} = \begin{bmatrix} 0 \\ \dfrac{1}{L} \end{bmatrix} \qquad \text{an } n \times 1 \text{ control matrix}$$

and, in this case, $\mathbf{u} = [u]$ is a one-dimensional control vector. In Eq. (2.35), matrices $\mathbf{A}$ and $\mathbf{x}$ are conformable.

If the output quantity $y(t)$ for the circuit of Fig. 2.2 is the voltage across the capacitor v_C, then

$$y(t) = v_C = x_1$$

Thus the matrix *system output* equation for this example is

$$\mathbf{y}(t) = \mathbf{c}^T \mathbf{x} + \mathbf{d}\mathbf{u} = \begin{bmatrix} 1 & 0 \end{bmatrix} \begin{bmatrix} x_1 \\ x_2 \end{bmatrix} + [0][u] \qquad (2.36)$$

where $\mathbf{c}^T = \begin{bmatrix} 1 & 0 \end{bmatrix}$ a 1×2 row vector

$\mathbf{y} = [y]$ a one-dimensional output vector

$\mathbf{d} = \mathbf{0}$

Equations (2.35) and (2.36) are for a single-input single-output (SISO) system. For a multiple-input multiple-output (MIMO) system, with m inputs and l outputs, these equations become

$$\dot{\mathbf{x}} = \mathbf{A}\mathbf{x} + \mathbf{B}\mathbf{u} \qquad (2.37)$$

$$\mathbf{y} = \mathbf{C}\mathbf{x} + \mathbf{D}\mathbf{u} \qquad (2.38)$$

where $\mathbf{B} = n \times m$ control matrix

$\mathbf{C} = l \times n$ output matrix

$\mathbf{D} = l \times m$ feedforward matrix

$\mathbf{u} = m$-dimensional control vector

$\mathbf{y} = l$-dimensional output vector

Example 3. Obtain the state equation for the circuit of Fig. 2.6, where i_2 is considered to be the output of this system. The assigned state variables are $x_1 = i_1$, $x_2 = i_2$, and $x_3 = v_C$. Thus two loop and one node equations are written

FIGURE 2.6
An electric circuit.

$$R_1 x_1 + L_1 \dot{x}_1 + x_3 = u \tag{2.39}$$

$$-x_3 + L_2 \dot{x}_2 + R_2 x_2 = 0 \tag{2.40}$$

$$-x_1 + x_2 + C \dot{x}_3 = 0 \tag{2.41}$$

The three state variables are independent, and the system state and output equations are

$$\dot{\mathbf{x}} = \begin{bmatrix} -\dfrac{R_1}{L_1} & 0 & -\dfrac{1}{L_1} \\[2mm] 0 & -\dfrac{R_2}{L_2} & \dfrac{1}{L_2} \\[2mm] \dfrac{1}{C} & -\dfrac{1}{C} & 0 \end{bmatrix} \mathbf{x} + \begin{bmatrix} \dfrac{1}{L_1} \\[2mm] 0 \\[2mm] 0 \end{bmatrix} \mathbf{u} \tag{2.42}$$

$$\mathbf{y} = \begin{bmatrix} 0 & 1 & 0 \end{bmatrix} \mathbf{x} \tag{2.43}$$

Example 4. Obtain the state equations for the circuit of Fig. 2.7. The output is the voltage v_1. The input or control variable is a current source $i(t)$. The assigned state variables are i_1, i_2, i_3, v_1, and v_2. Three loop equations and two node equations are written:

$$v_1 = L_1 \, Di_1 \tag{2.44}$$

$$v_2 = L_2 \, Di_2 + v_1 \tag{2.45}$$

$$v_2 = L_3 \, Di_3 \tag{2.46}$$

$$i_2 = C_1 \, Dv_1 + i_1 \tag{2.47}$$

$$i = i_3 + C_2 \, Dv_2 + i_2 \tag{2.48}$$

Substituting from Eqs. (2.44) and (2.46) into Eq. (2.45), or writing the loop equation through L_1, L_2, and L_3 and then integrating (multiplying by $1/D$), gives

$$L_3 i_3 = L_2 i_2 + L_1 i_1 + K \tag{2.49}$$

where K is a function of the initial conditions. This equation reveals that one inductor current is dependent upon the other two inductor currents. Thus, this circuit has only *four* independent physical state variables, two inductor currents and two capacitor voltages. The four independent state variables are designated as $x_1 = v_1$, $x_2 = v_2$, $x_3 = i_1$, and $x_4 = i_2$, and the control variable is $u = i$. Three state equations are obtainable from Eqs. (2.44), (2.45), and (2.47). The fourth equation is obtained by eliminating the dependent current i_3 from Eqs. (2.48) and (2.49). The result in matrix form is

FIGURE 2.7
Electric circuit.

$$\dot{\mathbf{x}} = \begin{bmatrix} 0 & 0 & -\dfrac{1}{C_1} & \dfrac{1}{C_1} \\ 0 & 0 & -\dfrac{L_1}{L_3 C_2} & -\dfrac{L_2 + L_3}{L_3 C_2} \\ \dfrac{1}{L_1} & 0 & 0 & 0 \\ -\dfrac{1}{L_2} & \dfrac{1}{L_2} & 0 & 0 \end{bmatrix} \mathbf{x} + \begin{bmatrix} 0 \\ \dfrac{1}{C_2} \\ 0 \\ 0 \end{bmatrix} = \mathbf{u} \tag{2.50}$$

$$\mathbf{y} = [\,1 \quad 0 \quad 0 \quad 0\,]\mathbf{x} \tag{2.51}$$

The dependence of i_3 on i_1 and i_2 as shown by Eq. (2.49) may not be readily observed. In that case the matrix state equation for this example would be written with five state variables.

The examples in this section are fairly straightforward. In general it is necessary to write more than just the number of state equations, because other system variables appear in them. These equations are solved simultaneously to eliminate all internal variables in the circuit except for the state variables. This procedure is necessary in some of the problems for this chapter. For more complex circuits it is possible to introduce more generalized and systematized linear graphs for obtaining the state equations. They are not included in this book.

2.5   TRANSFER FUNCTION AND BLOCK DIAGRAM

A quantity that plays a very important role in control theory is the system *transfer function*, defined as follows:

TRANSFER FUNCTION.   If the system differential equation is linear, the ratio of the output variable to the input variable, where the variables are expressed as functions of the D operator, is called the transfer function.

Consider the system output $v_C = y$ in the RLC circuit of Fig. 2.2. Substituting $i = C\,Dv_C$ into Eq. (2.10) yields

$$(LCD^2 + RCD + 1)v_C(t) = e(t) \tag{2.52}$$

The system transfer function is

$$G(D) = \frac{y(t)}{u(t)} = \frac{v_C(t)}{e(t)} = \frac{1}{LCD^2 + RCD + 1} \tag{2.53}$$

The notation $G(D)$ is used to denote a transfer function when it is expressed in terms of the D operator. It may also be written simply as G.

The *block diagram* representation of this system (Fig. 2.8) represents the mathematical operation $G(D)u(t) = y(t)$; that is, the transfer function times the input is equal to the output of the block. The resulting equation is the differential equation of the system.

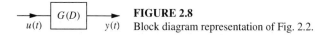

FIGURE 2.8
Block diagram representation of Fig. 2.2.

2.6 MECHANICAL TRANSLATION SYSTEMS[18,19]

Mechanical systems obey Newton's law that the sum of the forces equals zero; that is, the sum of the applied forces must be equal to the sum of the reactive forces. The three qualities characterizing elements in a mechanical translation[†] system are mass, elastance, and damping. The following analysis includes only linear functions. Static friction, Coulomb friction, and other nonlinear friction terms are not included. Basic elements entailing these qualities are represented as network elements,[8] and a mechanical network is drawn for each mechanical system to facilitate writing the differential equations.

The mass M is the inertial element. A force applied to a mass produces an acceleration of the mass. The reaction force f_M is equal to the product of mass and acceleration and is opposite in direction to the applied force. In terms of displacement x, velocity v, and acceleration a, the force equation is

$$f_M = Ma = M\,Dv = M\,D^2x \tag{2.54}$$

The network representation of mass is shown in Fig. 2.9a. One terminal, a, has the motion of the mass; and the other terminal, b, is considered to have the motion of the reference. The reaction force f_M is a function of time and acts "through" M.

The elastance, or stiffness, K provides a restoring force as represented by a spring. Thus, if stretched, the string tries to contract; if compressed, it tries to expand to its normal length. The reaction force f_K on each end of the spring is the same and is equal to the product of the stiffness K and the amount of deformation of the spring. The network representation of a spring is shown in Fig. 2.9b. The displacement of each end of the spring is measured from the original or equilibrium position. End c has a position x_c, and end d has a position x_d, measured from the respective equilibrium positions. The force equation, in accordance with Hooke's law, is

$$f_K = K(x_c - x_d) \tag{2.55}$$

(a)   (b)   (c)

FIGURE 2.9
Network elements of mechanical translation systems.

[†]*Translation* means motion in a straight line.

If the end d is stationary, then $x_d = 0$ and the preceding equation reduces to

$$f_K = K x_c \tag{2.56}$$

The plot f_K vs. x_c for a real spring is not usually a straight line, because the spring characteristic is nonlinear. However, over a limited region of operation, the linear approximation, i.e., a constant value for K, gives satisfactory results.

The damping, or viscous, friction B characterizes the element that absorbs energy. The damping force is proportional to the difference in velocity of two bodies. The assumption that the viscous friction is linear simplifies the solution of the dynamic equation. The network representation of damping action is a dashpot, as shown in Fig. 2.9c. Damping may either be intentional or occur unintentionally and is present because of physical construction. The reaction damping force f_B is approximated by the product of damping B and the relative velocity of the two ends of the dashpot. The direction of this force, given by Eq. (2.57), depends on the relative magnitudes and directions of the velocities Dx_e and Dx_f:

$$f_B = B(v_e - v_f) = B(Dx_e - Dx_f) \tag{2.57}$$

Damping may be added to a system by use of a dashpot. The basic operation of a dashpot, in which the housing is filled with an incompressible fluid, is shown in Fig. 2.10. If a force f is applied to the shaft, the piston presses against the fluid, increasing the pressure on side b and decreasing the pressure on side a. As a result, the fluid flows around the piston from side b to side a. If necessary, a small hole can be drilled through the piston to provide a positive path for the flow of fluid. The damping force required to move the piston inside the housing is given by Eq. (2.57), where the damping B depends on the dimensions and the fluid used.

Before the differential equations of a complete system can be written, the mechanical network must first be drawn. This is done by connecting the terminals of elements that have the same displacement. Then the force equation is written for each node or position by equating the sum of the forces at each position to zero. The equations are similar to the node equations in an electric circuit, with force analogous to current, velocity analogous to voltage, and the mechanical elements with their appropriate operators analogous to admittance. The reference position in all of the following examples should be taken from the static equilibrium positions. The force of gravity therefore does not appear in the system equations. The U.S. customary and metric systems of units are shown in Table 2.3.

FIGURE 2.10
Dashpot construction.

TABLE 2.3
Mechanical translation symbols and units

Symbol	Quantity	U.S. customary units	Metric units
f	Force	Pounds	Newtons
x	Distance	Feet	Meters
v	Velocity	Feet/second	Meters/second
a	Acceleration	Feet/second^2	Meters/second^2
M†	Mass	Slugs $= \dfrac{\text{pound-seconds}^2}{\text{foot}}$	Kilograms
K	Stiffness coefficient	Pounds/foot	Newtons/meter
B	Damping coefficient	Pounds/(foot/second)	Newtons/(meter/second)

†Mass M in the U.S. customary system above has the dimensions of slugs. Sometimes it is given in units of pounds. If so, then in order to use the consistent set of units above, the mass must be expressed in slugs by using the conversion factor 1 slug $= 32.2$ lb.

Simple Mechanical Translation System

The system shown in Fig. 2.11 is initially at rest. The end of the spring and the mass have positions denoted as the reference positions, and any displacements from these reference positions are labeled x_a and x_b, respectively. A force f applied at the end of the spring must be balanced by a compression of the spring. The same force is also transmitted through the spring and acts at point x_b.

To draw the mechanical network, the first step is to locate the points x_a and x_b and the reference. The network elements are then connected between these points. For example, one end of the spring has the position x_a, and the other end has the position x_b. Therefore, the spring is connected between these points. The complete mechanical network is drawn in Fig. 2.11b.

The displacements x_a and x_b are nodes of the circuit. At each node the sum of the forces must add to zero. Accordingly, the equations may be written for nodes a and b as

$$f = f_K = K(x_a - x_b) \tag{2.58}$$

$$f_K = f_M + f_B = MD^2 x_b + BD x_b \tag{2.59}$$

(a)　　　　　　　　　　　　　(b)

FIGURE 2.11
(a) Simple mass-spring-damper mechanical system; (b) corresponding mechanical network.

These two equations can be solved for the two displacements x_a and x_b and their respective velocities $v_a = Dx_a$ and $v_b = Dx_b$.

It is possible to obtain one equation relating x_a to f, x_b to x_a, or x_b to f by combining Eqs. (2.58) and (2.59):

$$K(MD^2 + BD)x_a = (MD^2 + BD + K)f \tag{2.60}$$

$$(MD^2 + BD + K)x_b = Kx_a \tag{2.61}$$

$$(MD^2 + BD)x_b = f \tag{2.62}$$

The solution of Eq. (2.61) shows the motion x_b resulting from a given motion x_a. Also, the solutions of Eqs. (2.60) and (2.62) show the motions x_a and x_b, respectively, resulting from a given force f. From each of these three equations the following transfer functions are obtained:

$$G_1 = \frac{x_a}{f} = \frac{MD^2 + BD + K}{K(MD^2 + BD)} \tag{2.63}$$

$$G_2 = \frac{x_b}{x_a} = \frac{K}{MD^2 + BD + K} \tag{2.64}$$

$$G = \frac{x_b}{f} = \frac{1}{MD^2 + BD} \tag{2.65}$$

Note that the last equation is equal to the product of the first two, i.e.,

$$G = G_1 G_2 = \frac{x_a}{f}\frac{x_b}{x_a} = \frac{x_b}{f} \tag{2.66}$$

The block diagram representing the mathematical operation of Eq. (2.66) is shown in Fig. 2.12. Figure 2.12a is a detailed representation that indicates all variables in the system. The two blocks G_1 and G_2 are said to be in *cascade*. Figure 2.12b, called the *overall block diagram* representation, shows only the input f and the output x_b, where x_b is considered the output variable of the system of Fig. 2.11.

The multiplication of transfer functions, as in Eq. (2.66), is valid as long as there is no coupling or loading between the two blocks in Fig. 2.12a. The signal x_a is unaffected by the presence of the block having the transfer function G_2; thus the multiplication is valid. When electric circuits are coupled, the transfer functions may not be independent unless they are isolated by an electronic amplifier with a very high input impedance.

(a) (b)

FIGURE 2.12
Block diagram representation of Fig. 2.11: (a) detailed; (b) overall.

Example 1. Determine the state equations for Eq. (2.59).

Equation (2.59) involves only one energy-storage element, the mass M, whose energy variable is v_b. The output quantity in this system is the position $y = x_b$. Since this quantity is not one of the physical or energy-related state variables, it is necessary to increase the number of state variables to 2, i.e., $x_b = x_1$ is an augmented state variable. Equation (2.59) is of second order, which confirms that two state variables are required. Note that the spring constant does not appear in this equation since the force f is transmitted through the spring to the mass. With $x_1 = x_b$, $x_2 = v_b = \dot{x}_1$, $u = f$, and $y = x_1$, the state and output equations are

$$\dot{\mathbf{x}} = \begin{bmatrix} 0 & 1 \\ 0 & -\dfrac{B}{M} \end{bmatrix} \mathbf{x} + \begin{bmatrix} 0 \\ \dfrac{1}{M} \end{bmatrix} u = \mathbf{Ax} + \mathbf{b}u \qquad y = \begin{bmatrix} 1 & 0 \end{bmatrix} \mathbf{x} = \mathbf{c}^T \mathbf{x} \qquad (2.67)$$

Example 2. Determine the state equations for Eq. (2.61).

Equation (2.61) involves two energy-storage elements, K and M, whose energy variables are x_b and v_b, respectively. Note that the spring constant K does appear in this equation since x_a is the input u. Therefore, a state variable must be associated with this energy element. Thus, the assigned state variables are x_b and $v_b = Dx_b$, which are independent state variables of this system. Let $x_1 = x_b = y$ and $x_2 = v_b = \dot{x}_1$. Converting Equation (2.61) into state-variable form yields the two equations

$$\dot{x}_1 = x_2 \qquad \dot{x}_2 = -\frac{K}{M}x_1 - \frac{B}{M}x_2 + \frac{K}{M}u \qquad (2.68)$$

They can be put into the matrix form

$$\dot{\mathbf{x}} = \mathbf{Ax} + \mathbf{b}u \qquad y = \mathbf{c}^T \mathbf{x} \qquad (2.69)$$

Multiple-Element Mechanical Translation System

A force $f(t)$ is applied to the mass M_1 of Fig. 2.13a. The sliding friction between the masses M_1 and M_2 and the surface is indicated by the viscous friction coefficients B_1 and B_2. The system equations can be written in terms of the two displacements x_a and x_b. The mechanical network is drawn by connecting the terminals of the elements that have the same displacement (see Fig. 2.13b). Since the sum of the forces at each node must add to zero, the equations are written according to the rules for node equations:

For node a: $\qquad (M_1 D^2 + B_1 D + B_3 D + K_1)x_a - (B_3 D)x_b = f \qquad (2.70)$

For node b: $\qquad -(B_3 D)x_a + (M_2 D^2 + B_2 D + B_3 D + K_2)x_b = 0 \qquad (2.71)$

A definite pattern to these equations can be detected. Observe that K_1, M_1, B_1, and B_3 are connected to node a and that Eq. (2.70), for node a, contains all four of these terms as coefficients of x_a. Notice that element B_3 is also connected to node b and that the term $-B_3$ appears as a coefficient of x_b. When this pattern is used, Eq. (2.71) can be written directly. Thus, since K_2, M_2, B_2, and B_3 are connected to node b, they appear as coefficients of x_b. Element B_3 is also connected to node a, and $-B_3$ appears as the coefficient of x_a. Each term in the equation must be a force.

The node equations for a mechanical system follow directly from the mechanical network of Fig. 2.13b. They are similar in form to the node equations for an electric circuit and follow the same rules.

The block diagram representing the system in Fig. 2.13 is also given by Fig. 2.12, where again the system output is x_b. The transfer functions $G_1 = x_a/f$ and $G_2 = x_b/x_a$ are obtained by solving the equations for this system.

Example 3. Obtain the state equations for the system of Fig. 2.13, where x_b is the system output. There are four energy-storage elements, thus the four assigned state variables are x_a, x_b, Dx_a, and Dx_b. Analyzing Eqs. (2.70), and (2.71) shows that this is a fourth-order system. Let

$$x_1 = x_b \quad \text{for spring } K_2 \qquad x_2 = \dot{x}_1 = v_b \quad \text{for mass } M_2$$

$$x_3 = x_a \quad \text{for spring } K_1 \qquad x_4 = \dot{x}_3 = v_a \quad \text{for mass } M_1$$

$$u = f \qquad y = x_b = x_1$$

The four state variables are independent, i.e., no one state variable can be expressed in terms of the remaining state variables. The system equations are

$$\dot{\mathbf{x}} = \mathbf{A}\mathbf{x} + \mathbf{b}u \qquad y = \mathbf{c}^T\mathbf{x} \tag{2.72}$$

$$\text{where} \quad \mathbf{A} = \begin{bmatrix} 0 & 1 & 0 & 0 \\ -\dfrac{K_2}{M_2} & -\dfrac{B_2+B_3}{M_2} & 0 & \dfrac{B_3}{M_2} \\ 0 & 0 & 0 & 1 \\ 0 & \dfrac{B_3}{M_1} & -\dfrac{K_1}{M_1} & -\dfrac{B_1+B_3}{M_1} \end{bmatrix}, \quad \mathbf{b} = \begin{bmatrix} 0 \\ 0 \\ 0 \\ \dfrac{1}{M_1} \end{bmatrix}$$

$$\mathbf{c}^T = \begin{bmatrix} 1 & 0 & 0 & 0 \end{bmatrix}$$

(a)

(b)

FIGURE 2.13
(a) Multiple-element mechanical system; (b) corresponding mechanical network.

2.7 ANALOGOUS CIRCUITS

Analogous circuits represent systems for which the differential equations have the same form. The corresponding variables and parameters in two circuits represented by equations of the same form are called *analogs*. An electric circuit can be drawn that looks like the mechanical circuit and is represented by node equations that have the same mathematical form as the mechanical equations. The analogs are listed in Table 2.4. In this table the force f and the current i are analogs and are classified as "through" variables. There is a physical similarity between the two, because a measuring instrument must be placed in series in both cases; i.e., an ammeter and a force indicator must be placed in series with the system. Also, the velocity "across" a mechanical element is analogous to voltage across an electrical element. Again, there is a physical similarity, because a measuring instrument must be placed across the system in both cases. A voltmeter must be placed across a circuit to measure voltage; it must have a point of reference. A velocity indicator must also have a point of reference. Nodes in the mechanical network are analogous to nodes in the electric network.

2.8 MECHANICAL ROTATIONAL SYSTEMS

The equations characterizing rotational systems are similar to those for translation systems. Writing torque equations parallels the writing of force equations, where the displacement, velocity, and acceleration terms are now angular quantities. The applied torque is equal to the sum of the reaction torques. The three elements in a rotational system are inertia, the spring, and the dashpot. The mechanical-network representation of these elements is shown in Fig. 2.14.

The torque applied to a body having a moment of inertia J produces an angular acceleration. The reaction torque T_J is opposite to the direction of the applied torque and is equal to the product of moment of inertia and acceleration. In terms of angular acceleration α, angular velocity ω, or angular displacement θ, the torque equation is

$$T_J = J\alpha = J D\omega = J D^2\theta \qquad (2.73)$$

When a torque is applied to a spring, the spring is twisted by an angle θ. The applied torque is transmitted through the spring and appears at the other end. The reaction spring torque T_K that is produced is equal to the product of the *stiffness*, or

TABLE 2.4
Electrical and mechanical analogs

Mechanical translation element		Electrical element	
Symbol	**Quantity**	**Symbol**	**Quantity**
f	Force	i	Current
$v = Dx$	Velocity	e or v	Voltage
M	Mass	C	Capacitance
K	Stiffness coefficient	$1/L$	Reciprocal inductance
B	Damping coefficient	$G = 1/R$	Conductance

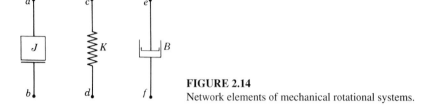

FIGURE 2.14
Network elements of mechanical rotational systems.

elastance, K of the spring and the angle of twist. By denoting the positions of the two ends of the spring, measured from the neutral position, as θ_c and θ_d, the reaction torque is

$$T_K = K(\theta_c - \theta_d) \tag{2.74}$$

Damping occurs whenever a body moves through a fluid, which may be a liquid or a gas such as air. To produce motion of the body, a torque must be applied to overcome the reaction damping torque. The damping is represented as a dashpot with a *viscous friction coefficient B*. The damping torque T_B is equal to the product of damping B and the relative angular velocity of the ends of the dashpot. The reaction torque of a damper is

$$T_B = B(\omega_e - \omega_f) = B(D\theta_e - D\theta_f) \tag{2.75}$$

Writing the differential equations is simplified by first drawing the mechanical network for the system. This is done by first designating nodes that correspond to each angular displacement. Then each element is connected between the nodes that correspond to the two motions at each end of that element. The inertia elements are connected from the reference node to the node representing its position. The spring and dashpot elements are connected to the two nodes that represent the position of each end of the element. Then the torque equation is written for each node by equating the sum of the torques at each node to zero. These equations are similar to those for mechanical translation and are analogous to those for electric circuits. The units for these elements are given in Table 2.5.

An electrical analog can be obtained for a mechanical rotational system in the manner described in Sec. 2.7. The changes to Table 2.4 are due only to the fact that rotational quantities are involved. Thus torque is the analog of current, and the moment of inertia is the analog of capacitance.

Simple Mechanical Rotational System

The system shown in Fig. 2.15 has a mass, with a moment of inertia J, immersed in a fluid. A torque T is applied to the mass. The wire produces a reactive torque proportional to its stiffness K and to the angle of twist. The fins moving through the fluid have a damping B, which requires a torque proportional to the rate at which

TABLE 2.5
Mechanical rotational symbols and units

Symbol	Quantity	U.S. customary units	Metric units
T	Torque	Pound-feet	Newton-meters
θ	Angle	Radians	Radians
ω	Angular velocity	Radians/second	Radians/second
α	Angular acceleration	Radians/second^2	Radians/second^2
$J^\dagger$	Moment of inertia	Slug-feet^2 (or pound-feet-seconds^2)	Kilogram-meters^2
K	Stiffness coefficient	Pound-feet/radian	Newton-meters/radian
B	Damping coefficient	Pound-feet/(radian/second)	Newton-meters/(radians/second)

†The moment of inertia J has the dimensions of mass-distance^2, which have the units slug feet^2 in the U.S. customary system. Sometimes the units are given as pound-feet^2. To use the consistent set of units above, the moment of inertia must be expressed in slug-feet^2 by using the conversion factor 1 slug = 32.2 lb.

they are moving. The mechanical network is drawn in Fig. 2.15b. There is one node having a displacement θ; therefore, only one equation is necessary:

$$J D^2\theta + B D\theta + K\theta = T(t) \tag{2.76}$$

Multiple-Element Mechanical Rotational System

The system represented by Fig. 2.16a has two disks that have damping between them and also between each of them and the frame. The mechanical network is drawn in Fig. 2.16b by first identifying the three angular displacements θ_1, θ_2, and θ_3. Then the elements are connected to the proper nodes. The equations are written directly in systematized form, since the torques at each node must add to zero, as follows:

$$\text{Node 1:} \quad K_1\theta_1 - K_1\theta_2 = T(t) \tag{2.77}$$

$$\text{Node 2:} \quad -K_1\theta_1 + [J_1D^2 + (B_1 + B_3)D + K_1]\theta_2 - (B_3D)\theta_3 = 0 \tag{2.78}$$

$$\text{Node 3:} \quad -(B_3D)\theta_2 + [J_2D^2 + (B_2 + B_3)D + K_2]\theta_3 = 0 \tag{2.79}$$

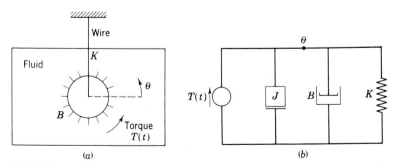

FIGURE 2.15
(a) Simple rotational system; (b) mechanical network.

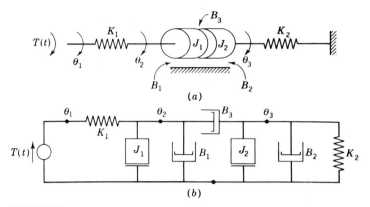

FIGURE 2.16
(a) Rotational system; (b) corresponding mechanical network.

These three equations can be solved simultaneously for θ_1, θ_2, and θ_3 as a function of the applied torque. If θ_3 is the system output, these equations can be solved for the following four transfer functions:

$$G_1 = \frac{\theta_1}{T}, \qquad G_2 = \frac{\theta_2}{\theta_1}, \qquad G_3 = \frac{\theta_3}{\theta_2}, \qquad G = G_1G_2G_3 = \frac{\theta_1\,\theta_2\,\theta_3}{T\,\theta_1\,\theta_2} = \frac{\theta_3}{T}$$

(2.80)

The detailed and overall block diagram representations of Fig. 2.16 are shown in Fig. 2.17. The overall transfer function G given by Eqs. (2.80) is the product of the transfer functions, which are said to be *in cascade*. This product of transfer functions applies in general to any number of elements in cascade when there is no loading between the blocks.

> **Example.** Determine the state variables for the system of Fig. 2.16, where θ_3 is the system output y and torque T is the system input u. The spring K_1 is effectively not in the system because the torque is transmitted through the spring. Since there are three energy-storage elements J_1, J_2, and K_2, the assigned state variables are $D\theta_2$, $D\theta_3$, and θ_3. These are independent state variables. Therefore, let $x_1 = \theta_3$, $x_2 = D\theta_3$, and $x_3 = D\theta_2$. If the system input is $u = \theta_1$, then the spring K_1 is included, resulting in four state variables (see Prob. 2.17).

Effective Moment of Inertia and Damping of a Gear Train

When the load is coupled to a drive motor through a gear train, the moment of inertia and damping relative to the motor are important. Since the shaft length between gears is very short, the stiffness may be considered infinite. A simple representation of a gear train is shown in Fig. 2.18a. The following definitions are used:

FIGURE 2.17
Detailed and overall block diagram representations of Fig. 2.16.

$$N = \text{number of teeth on each gear}$$

$$\omega = D\theta = \text{velocity of each gear}$$

$$n_a = \text{ratio of (speed of driving shaft)/(speed of driven shaft)}$$

$$\theta = \text{angular position}$$

Typically, gearing may not mesh perfectly, and the teeth of one gear may not fill the full space between the gear teeth of the matching gear. As a consequence, there can be times when the drive gear is not in contact with the other gear. This nonlinear characteristic is called *dead zone*. When the dead zone is very small, it is neglected for the linearized gearing model.

The mechanical network for the gear train is shown in Fig. 2.18b. At each gear pair, two torques are produced. For example, a restraining torque T_1 is the load on gear 1 produced by the rest of the gear train. A driving torque T_2 is also produced on gear 2. T_2 is the torque transmitted to gear 2 to drive the rest of the gear train. These torques are inversely proportional to the ratio of the speeds of the respective gears. The block labeled n_a between T_1 and T_2 is used to show the relationship between them; that is, $T_2 = n_a T_1$. A transformer may be considered as the electrical

(*a*)

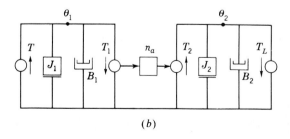

(*b*)

FIGURE 2.18
(*a*) Representation of a gear train; (*b*) corresponding mechanical network.

analog of the gear train, with angular velocity analogous to voltage and torque analogous to current. The equations describing the system are

$$J_1 D^2\theta_1 + B_1 D\theta_1 + T_1 = T \qquad J_2 D^2\theta_2 + B_2 D\theta_2 + T_L = T_2$$

$$n_a = \frac{\omega_1}{\omega_2} = \frac{\theta_1}{\theta_2} = \frac{N_2}{N_1} \qquad \theta_2 = \frac{\theta_1}{n_a} \qquad T_2 = n_a T_1 \tag{2.81}$$

The equations can be combined to produce

$$J_1 D^2\theta_1 + B_1 D\theta_1 + \frac{1}{n_a}(J_2 D^2\theta_2 + B_2 D\theta_2 + T_L) = T \tag{2.82}$$

This equation can be expressed in terms of the torque and the input position θ_1 only:

$$\left(J_1 + \frac{J_2}{n_a^2}\right) D^2\theta_1 + \left(B_1 + \frac{B_2}{n_a^2}\right) D\theta_1 + \frac{T_L}{n_a} = T \tag{2.83}$$

Equation (2.83) represents the system performance as a function of a single dependent variable. An equivalent system is one having an equivalent moment of inertia and damping equal to

$$J_{eq} = J_1 + \frac{J_2}{n_a^2} \tag{2.84}$$

$$B_{eq} = B_1 + \frac{B_2}{n_a^2} \tag{2.85}$$

If the solution for θ_2 is wanted, the equation can be altered by the substitution of $\theta_1 = n_a\theta_2$. This system can be generalized for any number of gear stages. When the gear-reduction ratio is large, the load moment of inertia may contribute a negligible value to the equivalent moment of inertia. When a motor rotor is geared to a long shaft that provides translational motion, then the elastance or stiffness K of the shaft must be taken into account in deriving the system's differential equations (see Prob. 2.25).

2.9   THERMAL SYSTEMS[9]

A limited number of thermal systems can be represented by linear differential equations. The basic requirement is that the temperature of a body be considered uniform. This approximation is valid when the body is small. Also, when the region consists of a body of air or liquid, the temperature can be considered uniform if there is perfect mixing of the fluid. The necessary condition of equilibrium requires that the heat added to the system equal the heat stored plus the heat carried away. This requirement can also be expressed in terms of rate of heat flow.

The symbols shown in Table 2.6 are used for thermal systems. A thermal-system network is drawn for each system in which the thermal capacitance and thermal resistance are represented by network elements. The differential equations can then be written from the thermal network.

TABLE 2.6
Thermal symbols and units

Symbol	Quantity	U.S. customary units	Metric units
q	Rate of heat flow	Btu/minute	Joules/second
M	Mass	Pounds	Kilograms
S	Specific heat	Btu/(pounds)(°F)	Joules/(kilogram)(°C)
C	Thermal capacitance $C = MS$	Btu/°F	Joules/°C
K	Thermal conductance	Btu/(minute)(°F)	Joules/(second)(°C)
R	Thermal resistance	Degrees/(Btu/minute)	Degrees/(joule/second)
θ	Temperature	°F	°C
h	Heat energy	Btu	Joules

The additional heat stored in a body whose temperature is raised from θ_1 to θ_2 is given by

$$h = \frac{q}{D} = C(\theta_2 - \theta_1) \tag{2.86}$$

In terms of rate of heat flow, this equation can be written as

$$q = CD(\theta_2 - \theta_1)$$

The thermal capacitance determines the amount of heat stored in a body. It is analogous to the electric capacitance of a capacitor in an electric circuit, which determines the amount of charge stored. The network representation of thermal capacitance is shown in Fig. 2.19a.

Rate of heat flow through a body in terms of the two boundary temperatures θ_3 and θ_4 is

$$q = \frac{\theta_3 - \theta_4}{R} \tag{2.87}$$

The thermal resistance determines the rate of heat flow through the body. This is analogous to the resistance of a resistor in an electric circuit, which determines the current flow. The network representation of thermal resistance is shown in Fig. 2.19b. In the thermal network the temperature is analogous to potential.

Simple Mercury Thermometer

Consider a thin glass-walled thermometer filled with mercury that has stabilized at a temperature θ_1. It is plunged into a bath of temperature θ_0 at $t = 0$. In its simplest

(a)   (b)

FIGURE 2.19
Network elements of thermal systems.

form, the thermometer can be considered to have a capacitance C that stores heat and a resistance R that limits the heat flow. The temperature of the mercury is θ_m. The flow of heat into the thermometer is

$$q = \frac{\theta_0 - \theta_m}{R}$$

The heat entering the thermometer is stored in the thermal capacitance and is given by

$$h = \frac{q}{D} = C(\theta_m - \theta_1)$$

These equations can be combined to form

$$\frac{\theta_0 - \theta_m}{RD} = C(\theta_m - \theta_1) \tag{2.88}$$

Differentiating Eq. (2.88) and rearranging terms gives

$$RC\,D\theta_m + \theta_m = \theta_0 \tag{2.89}$$

The thermal network is drawn in Fig. 2.20. The node equation for this circuit, with the temperature considered as a voltage, gives Eq. (2.89) directly. From this equation the transfer function $G = \theta_m/\theta_0$ may be obtained. Since there is one energy-storage element, the independent state variable is $x_1 = \theta_m$ and the input is $u = \theta_0$. Thus the state equation is

$$\dot{x}_1 = -\frac{1}{RC}x_1 + \frac{1}{RC}u \tag{2.90}$$

A more exact model of a mercury thermometer is presented in Reference 9. A simple transfer system for a water tank heater is also presented in this reference.

2.10   HYDRAULIC LINEAR ACTUATOR

The valve-controlled hydraulic actuator is used in many applications as a force amplifier. Very little force is required to position the valve, but a large output force is controlled. The hydraulic unit is relatively small, which makes its use very attractive. Figure 2.21 shows a simple hydraulic actuator in which motion of the valve regulates the flow of oil to either side of the main cylinder. An input motion x of a few thousandths of an inch results in a large change of oil flow. The resulting difference in pressure on the main piston causes motion of the output shaft. The oil flowing

FIGURE 2.20
Simple network representation of a thermometer.

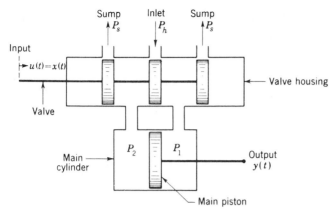

FIGURE 2.21
Hydraulic actuator.

in is supplied by a source that maintains a constant high pressure P_h, and the oil on the opposite side of the piston flows into the drain (sump) at low pressure P_s. The load-induced pressure P_L is the difference between the pressures on each side of the main piston:

$$P_L = P_1 - P_2 \tag{2.91}$$

The flow of fluid through an inlet orifice is given by[10]

$$q = ca\sqrt{2g\frac{\Delta p}{w}} \tag{2.92}$$

where
c = orifice coefficient

a = orifice area

Δp = pressure drop across orifice

w = specific weight of fluid

g = gravitational acceleration constant

q = rate of flow of fluid

Simplified Analysis

As a first-order approximation, it can be assumed that the orifice coefficient and the pressure drop across the orifice are constant and independent of valve position. Also, the orifice area a is proportional to the valve displacement x. Equation (2.92), which gives the rate of flow of hydraulic fluid through the valve, can be rewritten in linearized form as

$$q = C_x x \tag{2.93}$$

where x is the displacement of the valve. The displacement of the main piston is directly proportional to the volume of fluid that enters the main cylinder, or, equivalently, the rate of flow of fluid is proportional to the rate of displacement of the output. When the compressibility of the fluid and the leakage around the valve and main piston are neglected, the equation of motion of the main piston is

$$q = C_b \, Dy \tag{2.94}$$

Combining the two equations gives

$$Dy = \frac{C_x}{C_b} x = C_1 x \tag{2.95}$$

This analysis is essentially correct when the load reaction is small.

More Complete Analysis

When the load reaction is not negligible, a load pressure P_L must be produced across the piston. A more complete analysis takes into account the pressure drop across the orifice, the leakage of oil around the piston, and the compressibility of the oil. The pressure drop Δp across the orifice is a function of the source pressure P_h and the load pressure P_L. Since P_h is assumed constant, the flow equation for q is a function of valve displacement x and load pressure P_L:

$$q = f(x, P_L) \tag{2.96}$$

The differential rate of flow dq, expressed in terms of partial derivatives, is

$$dq = \frac{\partial q}{\partial x} \, dx + \frac{\partial q}{\partial P_L} \, dP_L \tag{2.97}$$

If q, x, and P_L are measured from zero values as reference points, and if the partial derivatives are constant at the values they have at zero, the integration of Eq. (2.97) gives

$$q = \left(\frac{\partial q}{\partial x}\right)_0 x + \left(\frac{\partial q}{\partial P_L}\right)_0 P_L \tag{2.98}$$

By defining $\qquad C_x \equiv \left(\dfrac{\partial p}{\partial x}\right)_0 \qquad$ and $\qquad C_p \equiv \left(\dfrac{-\partial q}{\partial P_L}\right)_0$

the flow equation for fluid entering the main cylinder can be written as

$$q = C_x x - C_p P_L \tag{2.99}$$

Both C_x and C_p have positive values. A comparison of Eq. (2.99) with Eq. (2.93) shows that the load pressure reduces the flow into the main cylinder. The flow of fluid into the cylinder must satisfy the continuity conditions of equilibrium. This flow is equal to the sum of the components:

$$q = q_0 + q_l + q_c \tag{2.100}$$

where $\quad q_0$ = incompressible component (causes motion of piston)
$\quad q_l$ = leakage component
$\quad q_c$ = compressible component

The component q_0, which produces a motion y of the main piston, is

$$q_0 = C_b \, Dy \tag{2.101}$$

The compressible component is derived in terms of the bulk modulus of elasticity of the fluid, which is defined as the ratio of incremental stress to incremental strain. Thus

$$K_B = \frac{\Delta P_L}{\Delta V/V}$$

Solving for ΔV and dividing both sides of the equation by Δt gives

$$\frac{\Delta V}{\Delta t} = \frac{V}{K_B} \frac{\Delta P_L}{\Delta t}$$

Taking the limit as Δt approaches zero and letting $q_c = dV/dt$ gives

$$q_c = \frac{V}{K_B} DP_L \tag{2.102}$$

where V is the effective volume of fluid under compression and K_B is the bulk modulus of the hydraulic oil. The volume V at the middle position of the piston stroke is often used in order to linearize the differential equation.

The leakage component is

$$q_l = LP_L \tag{2.103}$$

where L is the leakage coefficient of the whole system.

Combining these equations gives

$$q = C_x x - C_p P_L = C_b \, Dy + \frac{V}{K_B} DP_L + LP_L \tag{2.104}$$

and rearranging terms gives

$$C_b \, Dy + \frac{V}{K_B} DP_L + (L + C_p)P_L = C_x x \tag{2.105}$$

The force developed by the main piston is

$$F = n_F A P_L = CP_L \tag{2.106}$$

where n_F is the force conversion efficiency of the unit and A is the area of the main actuator piston.

An example of a specific type of load, consisting of a mass and a dashpot, is shown in Fig. 2.22. The equation for this system is obtained by equating the force produced by the piston, which is given by Eq. (2.106), to the reactive load forces:

$$F = M D^2 y + B \, Dy = CP_L \tag{2.107}$$

FIGURE 2.22
Load on a hydraulic piston.

Substituting the value of P_L from Eq. (2.107) into Eq. (2.105) gives the equation relating the input motion x to the response y:

$$\frac{MV}{CK_B} D^3 y + \left[\frac{BV}{CK_B} + \frac{M}{C}(L + C_p)\right] D^2 y + \left[C_b + \frac{B}{C}(L + C_p)\right] Dy = C_x x$$

$$(2.108)$$

The preceding analysis is based on perturbations about the reference set of values $x = 0$, $q = 0$, $P_L = 0$. For the entire range of motion x of the valve, the quantities $\partial q/\partial x$ and $-\partial q/\partial P_L$ can be determined experimentally. Although they are not constant at values equal to the values C_x and C_p at the zero reference point, average values can be assumed in order to simulate the system by linear equations. For conservative design the volume V is determined for the main piston at the midpoint.

To write the state equation for the hydraulic actuator and load of Figs. 2.21 and 2.22, the energy-related variables must be determined. The mass M has the output velocity Dy as an energy-storage variable. The compressible component q_c also represents an energy-storage element in a hydraulic system. The compression of a fluid produces stored energy, just as in the compression of a spring. The equation for hydraulic energy is

$$E(t) = \int_0^t P(\tau)q(\tau)\, d\tau \tag{2.109}$$

where $P(t)$ is the pressure and $q(t)$ is the rate of flow of fluid. The energy storage in a compressed fluid is obtained in terms of the bulk modulus of elasticity K_B. Combining Eq. (2.102) with Eq. (2.109) for a constant volume yields

$$E_c(P_L) = \int_0^{P_L} \frac{V}{K_B} P_L\, dP_L = \frac{V}{2K_B} P_L^2 \tag{2.110}$$

The stored energy in a compressed fluid is proportional to the pressure P_L squared; thus P_L may be used as a physical state variable.

Since the output quantity in this system is the position y, it is necessary to increase the state variables to three. Further evidence of the need for three state variables is the fact that Eq. (2.108) is a third-order equation. Therefore, in this example, let $x_1 = y$, $x_2 = Dy = \dot{x}_1$, $x_3 = P_L$, and $u = x$. Then, from Eqs. (2.105) and (2.107), the state and output equations are

$$\dot{\mathbf{x}} = \begin{bmatrix} 0 & 1 & 0 \\ 0 & -\dfrac{B}{M} & \dfrac{C}{M} \\ 0 & -\dfrac{C_b K_B}{V} & -\dfrac{K_B(L + C_p)}{V} \end{bmatrix} \mathbf{x} + \begin{bmatrix} 0 \\ 0 \\ \dfrac{K_B C_X}{V} \end{bmatrix} \mathbf{u} \tag{2.111}$$

$$\mathbf{y} = [y] = [1 \quad 0 \quad 0]\mathbf{x} = [x_1] \tag{2.112}$$

The effect of augmenting the state variables by adding the piston displacement $x_1 = y$ is to produce a singular system; that is, $|\mathbf{A}| = 0$. This property does not

(b)

FIGURE 2.23
Liquid-level system and its equivalent electrical analog.

appear if a spring is added to the load, as shown in Prob. 2.19. In that case $x_1 = y$ is an independent state variable.

A detailed derivation of a positive-displacement rotational hydraulic transmission is presented in Reference 16.

2.11 LIQUID-LEVEL SYSTEM[11]

Figure 2.23a represents a two-tank liquid-level control system. Definitions of the system parameters are

q_i, q_1, q_2 = rates of flow of fluid h_1, h_2 = heights of fluid level

R_1, R_2 = flow resistance A_1, A_2 = cross-sectional tank areas

The following basic linear relationships hold for this system:

$$q = \frac{h}{R} = \text{rate of flow through orifice} \tag{2.113}$$

$$q_n = \text{(tank input rate of flow)} - \text{(tank output rate of flow)}$$

$$= \text{net tank rate of flow} = A\,Dh \tag{2.114}$$

Applying Eq. (2.114) to tanks 1 and 2 yields, respectively,

$$q_{n_1} = A_1 Dh_1 = q_i - q_1 = q_i - \frac{h_1 - h_2}{R_1} \tag{2.115}$$

$$q_{n_2} = A_2 Dh_2 = q_1 - q_2 = \frac{h_1 - h_2}{R_1} - \frac{h_2}{R_2} \tag{2.116}$$

These equations can be solved simultaneously to obtain the transfer functions h_1/q_i and h_2/q_i.

The energy stored in each tank represents potential energy, which is equal to $\rho A h^2/2$, where ρ is the fluid density coefficient. Since there are two tanks, the system has two energy-storage elements, whose energy-storage variables are h_1 and h_2. Letting $x_1 = h_1$, $x_2 = h_2$, and $u = q_i$ in Eqs. (2.115) and (2.116) reveals that x_1 and x_2 are independent state variables. Thus, the state equation is

$$\dot{\mathbf{x}} = \begin{bmatrix} -\dfrac{1}{R_1 A_1} & \dfrac{1}{R_1 A_1} \\[2mm] \dfrac{1}{R_1 A_2} & -\dfrac{1}{R_1 A_2} - \dfrac{1}{R_2 A_2} \end{bmatrix} \mathbf{x} + \begin{bmatrix} \dfrac{1}{A_1} \\[2mm] 0 \end{bmatrix} \mathbf{u} \tag{2.117}$$

The levels of the two tanks are the outputs of the system. Letting $y_1 = x_1 = h_1$ and $y_2 = x_2 = h_2$ yields

$$\mathbf{y} = \begin{bmatrix} 1 & 0 \\ 0 & 1 \end{bmatrix} \mathbf{x} \tag{2.118}$$

The potential energy of a tank can be represented as a capacitor whose stored potential energy is $C v_c^2/2$; thus, the electrical analog of h is v_C. As a consequence, an analysis of Eqs. (2.115) and (2.116) yields the analogs between the hydraulic and electrical quantities listed in Table 2.7. The analogous electrical equations are

$$C_1 D v_1 = i_i - \frac{v_1 - v_2}{R_1} \tag{2.119}$$

$$C_2 D v_2 = \frac{v_1 - v_2}{R_1} - \frac{v_2}{R_2} \tag{2.120}$$

These two equations yield the analogous electric circuit of Fig. 2.23b.

2.12 ROTATING POWER AMPLIFIERS[12,13]

A dc generator can be used as a power amplifier, in which the power required to excite the field circuit is lower than the power output rating of the armature circuit. The voltage e_g induced in the armature circuit is directly proportional to the product

TABLE 2.7
Hydraulic and electrical analogs

Hydraulic element		Electrical element	
Symbol	Quantity	Symbol	Quantity
q_i	Input flow rate	i_i	Current source
h	Height	v_C	Capacitor voltage
A	Tank area	C	Capacitance
R	Flow resistance	R	Resistance

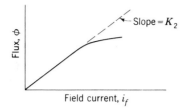

FIGURE 2.24
Magnetization curve.

of the magnetic flux ϕ set up by the field and the speed of rotation ω of the armature. This is expressed by

$$e_g = K_1 \phi \omega \qquad (2.121)$$

The flux is a function of the field current and the type of iron used in the field. A typical magnetization curve showing flux as a function of field current is given in Fig. 2.24. Up to saturation the relation is approximately linear, and the flux is directly proportional to field current:

$$\phi = K_2 i_f \qquad (2.122)$$

Combining these equations gives

$$e_g = K_1 K_2 \omega i_f \qquad (2.123)$$

When the generator is used as a power amplifier, the armature is driven at constant speed, and this equation becomes

$$e_g = K_g i_f \qquad (2.124)$$

A generator is represented schematically in Fig. 2.25, in which L_f and R_f and L_g and R_g are the inductance and resistance of the field and armature circuits, respectively. The equations for the generator are

$$e_f = (L_f D + R_f) i_f \qquad e_g = K_g i_f \qquad e_t = e_g - (L_g D + R_g) i_a \qquad (2.125)$$

The armature current depends on the load connected to the generator terminals. Combining the first two equations gives

$$(L_f D + R_f) e_g = K_g e_f \qquad (2.126)$$

Equation (2.126) relates the generated voltage e_g to the input field voltage e_f. The power output $p_o = e_t i_a$ is much larger than the power input $p_i = e_f i_f$. Thus, the ratio p_o/p_i represents the power gain.

FIGURE 2.25
Schematic diagram of a generator.

2.13 DC SERVOMOTOR

The developed torque of a dc motor is proportional to the magnitude of the flux due to the field current i_f and the armature current i_m. For any given motor the only two adjustable quantities are the flux and the armature current. The developed torque can therefore be expressed as

$$T(t) = K_3 \phi i_m \tag{2.127}$$

To avoid confusion with t for time, the capital letter T is used to indicate torque. It may denote either a constant or a function that varies with time. There are two modes of operation of a servomotor. For the armature control mode, the field current is held constant and an adjustable voltage is applied to the armature. In the field control mode, the armature current is held constant and an adjustable voltage is applied to the field. The armature control method of operation is considered in detail. The reader can refer to Reference 16 for the field control method of speed control.

A constant field current is obtained by separately exciting the field from a fixed dc source. The flux is produced by the field current and is, therefore, essentially constant. Thus the torque is proportional only to the armature current, and Eq. (2.127) becomes

$$T(t) = K_T i_m \tag{2.128}$$

When the motor armature is rotating, a voltage e_m is induced that is proportional to the product of flux and speed. Because the polarity of this voltage opposes the applied voltage e_a, it is commonly called the *back emf*. Since the flux is held constant, the induced voltage e_m is directly proportional to the speed ω_m:

$$e_m = K_1 \phi \omega_m = K_b \omega_m = K_b D\theta_m \tag{2.129}$$

The torque constant K_T and the generator constant K_b are the same for mks units. Control of the motor speed is obtained by adjusting the voltage applied to the armature. Its polarity determines the direction of the armature current and, therefore, the direction of the torque generated. This, in turn, determines the direction of rotation of the motor. A circuit diagram of the armature-controlled dc motor is shown in Fig. 2.26. The armature inductance and resistance are labeled L_m and R_m. The voltage equation of the armature circuit is

$$L_m D i_m + R_m i_m + e_m = e_a \tag{2.130}$$

The current in the armature produces the required torque according to Eq. (2.128). The required torque depends on the load connected to the motor shaft. If the load

FIGURE 2.26
Circuit diagram of a dc motor.

consists only of a moment of inertia and damper (friction), as shown in Fig. 2.27, the torque equation can be written:

$$J\,D\omega_m + B\omega_m = T(t) \tag{2.131}$$

The required armature current i_m can be obtained by equating the generated torque of Eq. (2.128) to the required load torque of Eq. (2.131). Inserting this current and the back emf from Eq. (2.129) into Eq. (2.130) produces the system equation in terms of the motor velocity ω_m:

$$\frac{L_m J}{K_T} D^2\omega_m + \frac{L_m B + R_m J}{K_T} D\omega_m + \left(\frac{R_m B}{K_T} + K_b\right)\omega_m = e_a \tag{2.132}$$

This equation can also be written in terms of motor position θ_m:

$$\frac{L_m J}{K_T} D^3\theta_m + \frac{L_m B + R_m J}{K_T} D^2\theta_m + \frac{R_m B + K_b K_T}{K_T} D\theta_m = e_a \tag{2.133}$$

There are two energy-storage elements, J and L_m, for the system represented by Figs. 2.26 and 2.27. Designating the independent state variables as $x_1 = \omega_m$ and $x_2 = i_m$ and the input as $u = e_a$ yields the state equation

$$\dot{\mathbf{x}} = \begin{bmatrix} -\dfrac{B}{J} & \dfrac{K_T}{J} \\ -\dfrac{K_b}{L_m} & -\dfrac{R_m}{L_m} \end{bmatrix} \mathbf{x} + \begin{bmatrix} 0 \\ \dfrac{1}{L_m} \end{bmatrix} \mathbf{u} \tag{2.134}$$

If motor position θ_m is the output, another differential equation is required; that is, $\dot{\theta}_m = \omega$. If the solution of θ_m is required, then the system is of third order. A third state variable, $x_3 = \theta_m$, must therefore be added.

The transfer functions ω_m/e_a and θ_m/e_a can be obtained from Eqs. (2.132) and (2.133). The armature inductance is small and can usually be neglected, $L_m \approx 0$. Equation (2.133) is thus reduced to a second-order equation. The corresponding transfer function has the form

$$G = \frac{\theta_m}{e_a} = \frac{K_M}{D(T_m D + 1)} \tag{2.135}$$

2.14   AC SERVOMOTOR[14]

An ac servomotor is basically a two-phase induction motor that has two stator field coils placed 90 electrical degrees apart, as shown in Fig. 2.28a. In a two-phase motor the ac voltages e and e_c are equal in magnitude and separated by a phase angle of 90°. A two-phase induction motor runs at a speed slightly below the synchronous speed

FIGURE 2.27
Inertia and friction as a motor load.

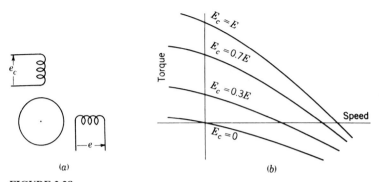

FIGURE 2.28
(*a*) Schematic diagram of a two-phase induction motor; (*b*) servomotor characteristics.

and is essentially a constant-speed motor. The synchronous speed n_s is determined by the number of poles P produced by the stator windings and the frequency f of the voltage applied to the stator windings: $n_s = 120 f/P$ revolutions per minute.

When the unit is used as a servomotor, the speed must be proportional to an input voltage. The two-phase motor can be used as a servomotor by applying an ac voltage e of fixed amplitude to one of the motor windings. When the other voltage e_c is varied, the torque and speed are a function of this voltage. Figure 2.28*b* shows a set of torque-speed curves for various control voltages.

Note that the curve for zero control-field voltage goes through the origin and that the slope is negative. This means that when the control-field voltage becomes zero, the motor develops a decelerating torque, causing it to stop. The curves show a large torque at zero speed. This is a requirement for a servomotor in order to provide rapid acceleration. It is accomplished in an induction motor by building the rotor with a high resistance.

The torque-speed curves are not straight lines. Therefore, a linear differential equation cannot be used to represent the exact motor characteristics. Sufficient accuracy may be obtained by approximating the characteristics by straight lines. The following analysis is based on this approximation.

The torque generated is a function of both the speed ω and the control-field voltage E_c. In terms of partial derivatives, the torque equation is approximated by effecting a double Taylor series expansion of $T(E_c, \omega)$ about the origin and keeping only the linear terms:

$$\left. \frac{\partial T}{\partial E_c} \right|_{\text{origin}} E_c + \left. \frac{\partial T}{\partial \omega} \right|_{\text{origin}} \omega = T(E_c, \omega) \tag{2.136}$$

If the torque-speed motor curves are approximated by parallel straight lines, the partial derivative coefficients of Eq. (2.136) are constants that can be evaluated from the graph. Let

$$\frac{\partial T}{\partial E_c} = K_c \quad \text{and} \quad \frac{\partial T}{\partial \omega} = K_\omega \tag{2.137}$$

For a load consisting of a moment of inertia and damping, the load torque required is

$$T_L = J\,D\omega + B\omega \tag{2.138}$$

Since the generated and load torques must be equal, Eqs. (2.136) and (2.138) are equated:

$$K_c E_c + K_\omega \omega = J D\omega + B\omega$$

Rearranging terms gives

$$J\,D\omega + (B - K_\omega)\omega = K_c E_c \tag{2.139}$$

In terms of position θ, this equation can be written as

$$J\,D^2\theta + (B - K_\omega)\,D\theta = K_c E_c \tag{2.140}$$

In order for the system to be stable (see Chapter 3) the coefficient $(B - K_\omega)$ must be positive. Observation of the motor characteristics shows that $K_\omega = \partial T/\partial\omega$ is negative; therefore, the stability requirement is satisfied.

Analyzing Eq. (2.139) reveals that this system has only one energy-storage element J. Thus the state equation, where $x_1 = \omega$ and $u = E_c$, is

$$\dot{x}_1 = -\frac{B - K_\omega}{J}x_1 + \frac{K_c}{J}u \tag{2.141}$$

2.15 LAGRANGE'S EQUATION

Previous sections show the application of Kirchhoff's laws for writing the differential equations of electric networks and the application of Newton's laws for writing the equations of motion of mechanical systems. These laws can be applied, depending on the complexity of the system, with relative ease. In many instances systems contain combinations of electric and mechanical components. The use of Lagrange's equation provides a systematic unified approach for handling a broad class of physical systems, no matter how complex their structure.[15]

Lagrange's equation is given by

$$\frac{d}{dt}\left(\frac{\partial T}{\partial \dot{q}_n}\right) - \frac{\partial T}{\partial q_n} + \frac{\partial D}{\partial \dot{q}_n} + \frac{\partial V}{\partial q_n} = Q_n \qquad n = 1, 2, 3, \ldots \tag{2.142}$$

where T = total kinetic energy of system
D = dissipation function of system
V = total potential energy of system
Q_n = generalized applied force at the coordinate n
q_n = generalized coordinate
$\dot{q}_n = dq_n/dt$ (generalized velocity)

and $n = 1, 2, 3, \ldots$ denotes the number of independent coordinates or degrees of freedom that exist in the system. The total kinetic energy T includes all energy terms, regardless of whether they are electrical or mechanical. The dissipation function D represents one-half the rate at which energy is dissipated as heat; dissipation is produced by friction in mechanical systems and by resistance in electric circuits. The total potential energy stored in the system is designated by V. The forcing functions applied to a system are designated by Q_n; they take the form of externally applied forces or torques in mechanical systems and appear as voltage or current sources in electric circuits. These quantities are illustrated for electrical and mechanical components in the following examples. Kinetic energy is associated with the generalized velocity and, for an inductor and mass, is given by $T_L = L\dot{q}^2/2 = Li^2/2$ and $T_M = M\dot{x}^2/2 = Mv^2/2$, respectively. Potential energy is associated with the generalized position and, for a capacitor and spring, is given by $V_C = q^2/2C$ and $V_K = Kx^2/2$, respectively. The dissipation function is always a function of the generalized velocity and, for an electrical resistor and mechanical friction, is given by $D_R = R\dot{q}^2/2 = Ri^2/2$ and $D_B = B\dot{x}^2/2 = Bv^2/2$, respectively. The generalized applied force for an electric circuit is the electromotive force of a generator, thus $Q_q = E$. For the earth's gravitational force applied to a mass, the generalized applied force is $Q_f = Mg$. Similar relationships can be developed for other physical systems. The application of Lagrange's equation is illustrated by the following example.

Example. Electromechanical system with capacitive coupling. Figure 2.29 shows a system in which mechanical motion is converted into electric energy. This represents the action that takes place in a capacitor microphone. Plate a of the capacitor is fastened rigidly to the frame. Sound waves impinge upon and exert a force on plate b of mass M, which is suspended from the frame by a spring K and which has damping B. The output voltage that appears across the resistor R is intended to reproduce electrically the sound-wave patterns which strike the plate b.

At equilibrium, with no external force extended on plate b, there is a charge q_0 on the capacitor. This produces a force of attraction between the plates so that the spring is stretched by an amount x_1 and the space between the plates is x_0. When sound waves exert a force on plate b there will be a resulting motion x that is measured from the equilibrium position. The distance between the plates will then be $x_0 - x$, and the charge on the plates will be $q_0 + q$.

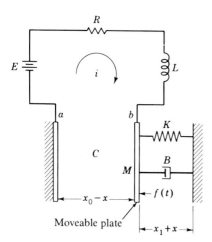

Moveable plate

FIGURE 2.29
Electromechanical system with capacitive coupling.

The capacitance is approximated by

$$C = \frac{\varepsilon A}{x_0 - x} \qquad \text{and} \qquad C_0 = \frac{\varepsilon A}{x_0}$$

where ε is the dielectric constant for air and A is the area of the plate.
The energy expressions for this system are

$$T = \frac{1}{2}L\dot{q}^2 + \frac{1}{2}M\dot{x}^2$$

$$D = \frac{1}{2}R\dot{q}^2 + \frac{1}{2}B\dot{x}^2$$

$$V = \frac{1}{2C}(q_0 + q)^2 + \frac{1}{2}K(x_1 + x)^2$$

$$= \frac{1}{2\varepsilon A}(x_0 - x)(q_0 + q)^2 + \frac{1}{2}K(x_1 + x)^2$$

The method is simple and direct. It is merely necessary to include all the energy terms, whether electrical or mechanical. The electromechanical coupling in this example appears in the potential energy. Here, the presence of charge on the plates of the capacitor exerts a force on the mechanical system. Also, motion of the mechanical system produces an equivalent emf in the electric circuit.

The two degrees of freedom are the displacement x of plate b and the charge q on the capacitor. Applying Lagrange's equation twice gives

$$M\ddot{x} + B\dot{x} - \frac{1}{2\varepsilon A}(q_0 + q)^2 + K(x_1 + x) = f(t) \tag{2.143}$$

$$L\ddot{q} + R\dot{q} + \frac{1}{\varepsilon A}(x_0 - x)(q_0 - q) = E \tag{2.144}$$

These equations are nonlinear. However, a good linear approximation can be obtained, because x and q are very small quantities and, therefore, the x^2, q^2, and xq terms can be neglected. This gives

$$(q_0 + q)^2 \approx q_0^2 + 2q_0 q$$

$$(x_0 - x)(q_0 - q) \approx x_0 q_0 - q_0 x + x_0 q$$

With these approximations the system equations become

$$M\ddot{x} + Kx_1 + Kx - \frac{q_0^2}{2\varepsilon A} - \frac{2q_0 q}{2\varepsilon A} + B\dot{x} = f(t) \tag{2.145}$$

$$L\ddot{q} + \frac{x_0 q_0}{\varepsilon A} - \frac{q_0 x}{\varepsilon A} + \frac{x_0 q}{\varepsilon A} + R\dot{q} = E \tag{2.146}$$

From Eq. (2.145), by setting $f(t) = 0$ and taking steady-state conditions, the result is

$$Kx_1 - \frac{q_0^2}{2\varepsilon A} = 0$$

This simply equates the force on the spring and the force due to the charges at the equilibrium condition. Similarly, in Eq. (2.146) at equilibrium

$$\frac{x_0 q_0}{\varepsilon A} = \frac{q_0}{C_0} = E$$

Therefore, the two system equations can be written in linearized form as

$$M\ddot{x} + B\dot{x} + Kx - \frac{q_0}{\varepsilon A}q = f(t) \tag{2.147}$$

$$L\ddot{q} + R\dot{q} + \frac{q}{C_0} - \frac{q_0}{\varepsilon A}x = 0 \tag{2.148}$$

These equations show that $q_0/\varepsilon A$ is the coupling factor between the electrical and mechanical portions of the system.

Another form of electromechanical coupling exists when current in a coil produces a force that is exerted on a mechanical system and, simultaneously, motion of a mass induces an emf in an electric circuit. The electromagnetic coupling is represented in Table 2.8. In that case the kinetic energy includes a term

$$T = l\beta N_c x i = U x i \tag{2.149}$$

The energy for this system, which is shown in Table 2.8, may also be considered potential energy. Since Eq. (2.142) contains the terms $-\partial T/\partial q_n$ and $\partial V/\partial q_n$, identification of the coupling energy as potential energy V would change the sign on the corresponding term in the differential equation.

The main advantage of Lagrange's equation is the use of a single systematic procedure, eliminating the need to consider separately Kirchhoff's laws for the electrical aspects and Newton's law for the mechanical aspects of the system in formulat-

TABLE 2.8
Identification of energy functions for electromechanical elements

Definition	Element and symbol	Kinetic energy T
x = relative motion i = current in coil $U = l\beta N_c$ (electromechanical coupling constant) β = flux density produced by permanent magnet l = length of coil N_c = number of turns in coil		Uxi

ing the statements of equilibrium. Once this procedure is mastered, the differential equations that describe the system are readily obtained.

2.16 SUMMARY

The examples in this chapter cover many of the basic elements of control systems. In order to write the differential and state equations, the basic laws governing performance are first stated for electrical, mechanical, thermal, and hydraulic systems. These basic laws are then applied to specific devices, and their differential equations of performance are obtained. The basic matrix, state, transfer function, and block diagram concepts are introduced. Lagrange's equation has been introduced to provide a systematized method for writing the differential equations of electrical, mechanical, and electromechanical systems. This chapter constitutes a reference for the reader who wants to review the fundamental concepts involved in writing differential equations of performance.

REFERENCES

1. Rugh, W. J.: *Linear System Theory,* Prentice Hall, Englewood Cliffs, N. J., 1993.
2. Blackburn, J. F., ed.: *Components Handbook,* McGraw-Hill, New York, 1948.
3. Stout, T. M.: "A Block Diagram Approach to Network Analysis," *Trans. AIEE,* vol. 71, pp. 255–260, 1952.
4. Kreyszig, E.: *Advanced Engineering Mathematics,* 7th ed., John Wiley, New York, 1993.
5. Nilson, J. W.: *Electric Circuits,* 2nd ed., Addison-Wesley, Reading, Mass., 1980.
6. Gantmacher, F. R.: *Applications of the Theory of Matrices,* Wiley-Interscience, New York, 1959.
7. Kalman, R. E.: "Mathematical Description of Linear Dynamical Systems," *J. Soc. Ind. Appl. Math., Ser. A, Control,* vol. 1, no. 2, pp. 152–192, 1963.
8. Gardner, M. F., and J. L. Barnes: *Transients in Linear Systems,* Wiley, New York, 1942, chap. 2.
9. Hornfeck, A. J.: "Response Characteristics of Thermometer Elements," *Trans. ASME,* vol. 71, pp. 121–132, 1949.
10. *Flow Meters: Their Theory and Application,* American Society of Mechanical Engineers, New York, 1937.
11. Takahashi, Y., M. J. Rabins, and D. M. Auslander: *Control and Dynamic Systems,* Addison-Wesley, Reading, Mass., 1970.
12. Saunders, R. M.: "The Dynamo Electric Amplifier: Class A Operation," *Trans. AIEE,* vol. 68, pp. 1368–1373, 1949.
13. Litman, B.: "An Analysis of Rotating Amplifiers," *Trans. AIEE,* vol. 68, pt. II, pp. 1111–1117, 1949.
14. Hopkin, A. M.: "Transient Response of Small Two-Phase Servomotors," *Trans. AIEE,* vol. 70, pp. 881–886, 1951.
15. Ogar, G. W., and J. J. D'Azzo: "A Unified Procedure for Deriving the Differential Equations of Electrical and Mechanical Systems," *IRE Trans. Educ.,* vol. E-5, no. 1, pp. 18–26, March 1962.
16. D'Azzo, J. J., and C. H. Houpis: *Linear Control System Analysis & Design: Conventional and Modern,* 3rd ed., McGraw-Hill, New York, 1988.
17. Strang, G.: *Linear Algebra and Its Applications,* 3rd ed., Harcourt, Brace, Jovanovich, San Diego, Calif., 1988.
18. Davis, S. A., and B. K. Ledgerwood: *Electromechanical Components for Servomechanisms,* McGraw-Hill, New York, 1961.
19. Cochin, I.: *Analysis and Design of Dynamic Systems,* Harper & Row, New York, 1980.

CHAPTER
3

SOLUTION OF DIFFERENTIAL EQUATIONS

3.1 INTRODUCTION

The general solution of a linear differential equation[1,2] is the sum of two components, the particular integral and the complementary function. Often the particular integral is the steady-state component of the solution of the differential equation; the complementary function, which is the solution of the corresponding homogeneous equation, is then the transient component of the solution. Often the steady-state component of the response has the same form as the driving function. In this book the particular integral is called the steady-state solution even when it is not periodic. The form of the transient component of the response depends only on the roots of the characteristic equation. For nonlinear differential equations the form of the response also depends on the initial or boundary conditions. The instantaneous value of the transient component depends on the boundary conditions, the roots of the characteristic equation, and the instantaneous value of the steady-state component.

This chapter covers methods of determining the steady-state and the transient components of the solution. These components are first determined separately and then added to form the complete solution. Analysis of the transient solution should develop in the student a feel for the solution to be expected.

The method of solution is next applied to the matrix state equation. A general format is obtained for the complementary solution in terms of the state transition matrix (STM). Then the complete solution is obtained as a function of the STM and the input forcing function.

3.2 STANDARD INPUTS TO CONTROL SYSTEMS

For some control systems the input has a specific form that may be represented either by an analytical expression or as a specific curve. An example of the latter is the pattern used in a machining operation, where the cutting tool is required to follow the path indicated by the pattern outline. For other control systems the input may be random in shape. In this case it cannot be expressed analytically and is not repetitive. An example is the camera platform used in a photographic airplane. The airplane flies at a fixed altitude and speed, and the camera takes a series of pictures of the terrain below it, which are then fitted together to form one large picture of the area. This task requires that the camera platform remain level regardless of the motion of the airplane. Since the attitude of the airplane varies with wind gusts and depends on the stability of the airplane itself, the input to the camera platform is obviously a random function.

It is important to have a basis of comparison for various systems. One way of doing this is by comparing the response with a standardized input. The input or inputs used as a basis of comparison must be determined from the required response of the system and the actual form of its input. The following standard inputs, with unit amplitude, are often used in checking the response of a system:

1. Sinusoidal function $\qquad\qquad r = \cos \omega t$

2. Power series function $\qquad\qquad r = a_0 + a_1 t + a_2 \dfrac{t^2}{2} + \cdots$

3. Step function $\qquad\qquad r = u_{-1}(t)$

4. Ramp (step velocity) function $\qquad r = u_{-2}(t) = t u_{-1}(t)$

5. Parabolic (step acceleration) function $\quad r = u_{-3}(t) = \dfrac{t^2}{2} u_{-1}(t)$

6. Impulse function $\qquad\qquad r = u_0(t)$

Functions 3 to 6 are called *singularity functions* (Fig. 3.1). The singularity functions can be obtained from one another by successive differentiation or integration. For example, the derivative of the parabolic function is the ramp function, the derivative

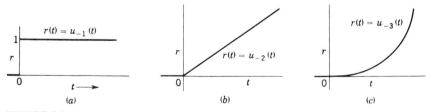

FIGURE 3.1
Singularity functions: (*a*) step function, $u_{-1}(t)$; (*b*) ramp function, $u_{-2}(t)$; (*c*) parabolic function, $u_{-3}(t)$.

of the ramp function is the step function, and the derivative of the step function is the impulse function.

For each of these inputs a complete solution of the differential equation is determined in this chapter. First, generalized methods are developed to determine the steady-state output $c(t)_{ss}$ for each type of input. These methods are applicable to linear differential equations of any order. Next, the method of evaluating the transient component of the response $c(t)_t$ is determined, and the form of the transient component of the response is shown to depend on the characteristic equation. Addition of the steady-state component and the transient component gives the complete solution, that is, $c(t) = c(t)_{ss} + c(t)_t$. The coefficients of the transient terms are determined by the instantaneous value of the steady-state component, the roots of the characteristic equation, and the initial conditions. Several examples are used to illustrate these principles. The solution of the differential equation with a pulse input is postponed until the next chapter.

3.3 STEADY-STATE RESPONSE: SINUSOIDAL INPUT

The input quantity r is assumed to be a sinusoidal function of the form

$$r(t) = R\cos(\omega t + \alpha) \tag{3.1}$$

The general integrodifferential equation to be solved is of the form

$$A_v D^v c + A_{v-1} D^{v-1} c + \cdots + A_0 D^0 c + A_{-1} D^{-1} c + \cdots + A_{-w} D^{-w} c = r \tag{3.2}$$

The steady-state solution can be obtained directly by use of Euler's identity,

$$e^{j\omega t} = \cos\omega t + j\sin\omega t$$

The input can then be written

$$r = R\cos(\omega t + \alpha) = \text{real part of } (Re^{j(\omega t + \alpha)})$$

$$= \text{Re}(Re^{j(\omega t + \alpha)}) = \text{Re}(Re^{j\alpha}e^{j\omega t}) = \text{Re}(\mathbf{R}e^{j\omega t}) \tag{3.3}$$

For simplicity, the phrase *real part of* or its symbolic equivalent Re is often omitted, but it must be remembered that the real part is intended. The quantity $\mathbf{R} = Re^{j\alpha}$ is the *phasor* representation of the input; i.e., it has both a magnitude R and an angle α. The magnitude R represents the maximum value of the input quantity $r(t)$. For simplicity the angle $\alpha = 0°$ usually is chosen for $\mathbf{R}$. The input r from Eq. (3.3) is inserted in Eq. (3.2). Then, for the expression to be an equality, the response c must be of the form

$$c(t)_{ss} = C\cos(\omega t + \phi) = \text{Re}(Ce^{j\phi}e^{j\omega t}) = \text{Re}(\mathbf{C}e^{j\omega t}) \tag{3.4}$$

where $\mathbf{C} = Ce^{j\phi}$ is a phasor quantity having the magnitude C and the angle ϕ. The nth derivative of c_{ss} with respect to time is

$$D^n c(t)_{ss} = \text{Re}[(j\omega)^n \mathbf{C}e^{j\omega t}] \tag{3.5}$$

Inserting c_{ss} and its derivatives from Eqs. (3.4) and (3.5) into Eq. (3.2) gives

$$\text{Re}[A_v(j\omega)^v \mathbf{C}e^{j\omega t} + A_{v-1}(j\omega)^{v-1}\mathbf{C}e^{j\omega t} + \cdots + A_{-w}(j\omega)^{-w}\mathbf{C}e^{j\omega t}] = \text{Re}(\mathbf{R}e^{j\omega t}) \tag{3.6}$$

Canceling $e^{j\omega t}$ from both sides of the equation and then solving for $\mathbf{C}$ gives

$$\mathbf{C} = \frac{\mathbf{R}}{A_v(j\omega)^v + A_{v-1}(j\omega)^{v-1} + \cdots + A_0 + A_{-1}(j\omega)^{-1} + \cdots + A_{-w}(j\omega)^{-w}} \tag{3.7}$$

where $\mathbf{C}$ is the phasor representation of the output; i.e., it has a magnitude C and an angle ϕ. Since the values of C and ϕ are functions of the frequency ω, the phasor output is written as $\mathbf{C}(j\omega)$ to show this relationship. Similarly, $\mathbf{R}(j\omega)$ denotes the fact that the input is sinusoidal and may be a function of frequency.

Comparing Eqs. (3.2) and (3.6), it can be seen that one equation can be determined easily from the other. Substituting $j\omega$ for D, $\mathbf{C}(j\omega)$ for c, and $\mathbf{R}(j\omega)$ for r in Eq. (3.2) results in Eq. (3.6). The reverse is also true and is independent of the order of the equation. It should be realized that this is simply a *rule of thumb* that yields the desired expression.

The time response can be obtained directly from the phasor response. The output is

$$c(t)_{ss} = \text{Re}(\mathbf{C}e^{j\omega t}) = |\mathbf{C}|\cos(\omega t + \phi) \tag{3.8}$$

As an example, consider the rotational hydraulic transmission described in Reference 9, Sec. 2.11, which has an input x and an output angular position θ_m. The equation that relates input to output in terms of system parameters is repeated here:

$$\frac{VJ}{K_BC}D^3\theta_m + \frac{LJ}{C}D^2\theta_m + d_m D\theta_m = d_p\omega_p x \tag{3.9}$$

When the input is the sinusoid, $x = X\sin\omega t$, the corresponding phasor equation can be obtained from Eq. (3.9) by replacing $x(t)$ by $\mathbf{X}(j\omega)$, θ_m by $\boldsymbol{\Theta}_m(j\omega)$, and D by $j\omega$. The ratio of phasor output to phasor input is termed the *frequency transfer function*, often designated by $\mathbf{G}(j\omega)$. This ratio, in terms of the steady-state sinusoidal phasors, is

$$\mathbf{G}(j\omega) = \frac{\boldsymbol{\Theta}_m(j\omega)}{\mathbf{X}(j\omega)} = \frac{d_p\omega_p/d_m}{j\omega[(VJ/K_BC\,d_m)(j\omega)^2 + (LJ/C\,d_m)(j\omega) + 1]} \tag{3.10}$$

3.4 STEADY-STATE RESPONSE: POLYNOMIAL INPUT

A development of the steady-state solution of a differential equation with a general polynomial input is presented first. Particular cases of the power series are then covered in detail.

The general differential equation is repeated here:

$$A_v D^v c + A_{v-1} D^{v-1} c + \cdots + A_0 c + A_{-1} D^{-1} c + \cdots + A_{-w} D^{-w} c = r \tag{3.11}$$

The polynomial input is of the form

$$r(t) = R_0 + R_1 t + \frac{R_2 t^2}{2!} + \cdots + \frac{R_k t^k}{k!} \tag{3.12}$$

where the highest-order term in the input is $R_k t^k / k!$. For $t < 0$ the value of $r(t)$ is zero. The object is to find the steady-state or particular solution of the dependent variable c. The method used is to assume a polynomial solution of the form

$$c(t)_{\text{ss}} = b_0 + b_1 t + \frac{b_2 t^2}{2!} + \cdots + \frac{b_q t^q}{q!} \tag{3.13}$$

where determination of the value of q is given as follows:

The assumed solution is then substituted into the differential equation. The coefficients $b_0, b_1, b_2, \ldots$ of the polynomial solution are evaluated by equating the coefficients of like powers of t on both sides of the equation. Inserting $r(t)$ from Eq. (3.12) into Eq. (3.11), the highest power of t on the right side of Eq. (3.11) is k; therefore t^k must also appear on the left side of this equation. The highest power of t on the left side of the equation is produced by the lowest-order derivative term $D^{-w} c$ and is equal to q plus the order of the lowest derivative. Therefore, the value of the highest-order exponent of t that appears in the assumed solution of Eq. (3.13) is

$$q = k - w \qquad q \geq 0 \tag{3.14}$$

where k is the highest exponent appearing in the input and w is the lowest-order exponent of the differential operator D appearing in the general differential equation (3.11). Equation (3.14) is valid only for positive values of q. For each of the following examples the response is a polynomial, because the input is of that form. However, the highest power in the response may not be the same as that of the input. When the lowest-order derivative is zero, that is, $w = 0$ in Eq. (3.11), then $q = k$. When integral terms are present, the highest order of t in $c(t)$ is smaller than k.

Step-Function Input

A convenient input $r(t)$ to a system is an abrupt change represented by a unit step function, as shown in Fig. 3.1a. This type of input cannot always be put into a system since it may take a definite length of time to make the change in input, but it represents a good mathematical input for checking system response.

The servomotor described in Sec. 2.13 is used as an example. The response of motor velocity ω_m in terms of the voltage e_a applied to the armature, as given by Eq. (2.132), is of the form

$$A_2 D^2 x + A_1 Dx + A_0 x = r \tag{3.15}$$

The unit step function $r = u_{-1}(t)$ is a polynomial in which the highest exponent of t is $k = 0$. When the input is a unit step function, the method of solution for a polynomial can therefore be used, that is, the steady-state response is also a polynomial. The lowest-order derivative in Eq. (3.15) is $w = 0$; therefore, $q = 0$ and the steady-state response has only one term of the form

$$x_{ss} = b_0 \qquad (3.16)$$

The derivatives are $Dx_{ss} = 0$ and $D^2 x_{ss} = 0$. Inserting these values into Eq. (3.15) yields

$$x(t)_{ss} = b_0 = \frac{1}{A_0} \qquad (3.17)$$

Ramp-Function Input (Step Function of Velocity)

The ramp function is a fixed rate of change of a variable as a function of time. This input is shown in Fig. 3.1b. The input and its derivative are expressed mathematically as

$$r(t) = u_{-2}(t) = tu_{-1}(t) \qquad Dr = u_{-1}(t)$$

A ramp input is a polynomial input where the highest power of t is $k = 1$. When the input $r(t)$ in Eq. (3.15) is the ramp function $u_{-2}(t)$, the highest power of t in the polynomial output is $q = k = 1$. The output is therefore

$$x(t)_{ss} = b_0 + b_1 t \qquad (3.18)$$

The derivatives are $Dx_{ss} = b_1$ and $D^2 x_{ss} = 0$. Inserting these values into Eq. (3.15), and then equating coefficients of t raised to the same power, yields

$$t^0 : \quad A_1 b_1 + A_0 b_0 = 0 \qquad b_0 = -\frac{A_1 b_1}{A_0} = -\frac{A_1}{A_0^2}$$

$$t^1 : \qquad A_0 b_1 = 1 \qquad b_1 = \frac{1}{A_0}$$

Thus, the steady-state solution is

$$x(t)_{ss} = -\frac{A_1}{A_0^2} + \frac{1}{A_0} t \qquad (3.19)$$

Parabolic-Function Input (Step Function of Acceleration)

A parabolic function has a constant second derivative, which means that the first derivative is a ramp. The input $r(t)$ is expressed mathematically as

$$r = u_{-3}(t) = \frac{t^2}{2} u_{-1}(t) \qquad Dr = u_{-2}(t) = tu_{-1}(t) \qquad D^2 r = u_{-1}(t)$$

A parabolic input is a polynomial input where the highest power of t is $k = 2$. A steady-state solution is found in the conventional manner for a polynomial input. Consider Eq. (3.15) with a parabolic input. The order of the lowest derivative in the system equation is $w = 0$. The value of q is therefore equal to 2, and the steady-state response is of the form

$$x(t)_{ss} = b_0 + b_1 t + \frac{b_2 t^2}{2} \tag{3.20}$$

The derivatives of x are $Dx_{ss} = b_1 + b_2 t$ and $D^2 x_{ss} = b_2$. Inserting these values into Eq. (3.15), followed by equating the coefficients of t raised to the same power, yields $b_2 = 1/A_0$, $b_1 = -A_1/A_0^2$, and $b_0 = A_1^2/A_0^3 - A_2/A_0^2$.

3.5   TRANSIENT RESPONSE: CLASSICAL METHOD

The classical method of solving for the complementary function or transient response of a differential equation requires, first, the writing of the homogeneous equation. The general differential equation has the form

$$b_v D^v c + b_{v-1} D^{v-1} c + \cdots + b_0 D^0 c + b_{-1} D^{-1} c + \cdots + b_{-w} D^{-w} c = r \tag{3.21}$$

where r is the forcing function and c is the response. The homogeneous equation is formed by letting the right side of the differential equation equal zero:

$$b_v D^v c_t + b_{v-1} D^{v-1} c_t + \cdots + b_0 c_t + b_{-1} D^{-1} c_t + \cdots + b_{-w} D^{-w} c_t = 0 \tag{3.22}$$

where c_t is the transient component of the general solution.

The general expression for the transient response, which is the solution of the homogeneous equation, is obtained by assuming a solution of the form

$$c_t = A_m e^{mt} \tag{3.23}$$

where m is a constant yet to be determined. Substituting this value of c_t into Eq. (3.22) and factoring $A_m e^{mt}$ from all terms gives

$$A_m e^{mt} (b_v m^v + b_{v-1} m^{v-1} + \cdots + b_0 + \cdots + b_{-w} m^{-w}) = 0 \tag{3.24}$$

Equation (3.24) must be satisfied for $A_m e^{mt}$ to be a solution. Since e^{mt} cannot be zero for all values of time t, it is necessary that

$$Q(m) = b_v m^v + b_{v-1} m^{v-1} + \cdots + b_0 + \cdots + b_{-w} m^{-w} = 0 \tag{3.25}$$

This equation is purely algebraic and is termed the *characteristic equation*. There are $v + w$ roots, or *eigenvalues*, of the characteristic equation; therefore, the complete transient solution contains the same number of terms of the form $A_m e^{mt}$ if all the roots are simple. Thus the transient component, when there are no multiple roots, is

$$c_t = A_1 e^{m_1 t} + A_2 e^{m_2 t} + \cdots + A_k e^{m_k t} + \cdots + A_{v+w} e^{m_{v+w} t} \tag{3.26}$$

where each $e^{m_k t}$ is described as a *mode* of the system. If there is a root m_q of multiplicity p, the transient includes corresponding terms of the form

$$A_{q1} e^{m_q t} + A_{q2} t e^{m_q t} + \cdots + A_{qp} t^{p-1} e^{m_q t} \tag{3.27}$$

Instead of using the detailed procedure just outlined, the characteristic equation of Eq. (3.25) is usually obtained directly from the homogeneous equation of Eq. (3.22) by substituting m for Dc_t, m^2 for $D^2 c_t$, etc.

Since the coefficients of the transient solution must be determined from the initial conditions, there must be $v + w$ known initial conditions. These conditions are values of the variable c and of its derivatives that are known at specific times. The $v + w$ initial conditions are used to set up $v + w$ simultaneous equations of c and its derivatives. The value of c includes both the steady-state and transient components. Since determination of the coefficients includes consideration of the steady-state component, the input affects the value of the coefficient of each exponential term.

Complex Roots

If all values of m_k are real, the transient terms can be evaluated as indicated above. Frequently, some values of m_k are complex. When this happens, they always occur in pairs that are complex conjugates and are of the form

$$m_k = \sigma + j\omega_d \qquad m_{k+1} = \sigma - j\omega_d \tag{3.28}$$

where σ is called the *damping coefficient* and ω_d is called the *damped natural frequency*. The transient terms corresponding to these values of m are

$$A_k e^{(\sigma + j\omega_d)t} + A_{k+1} e^{(\sigma - j\omega_d)t} \tag{3.29}$$

These terms are combined to a more useful form by factoring the term $e^{\sigma t}$:

$$e^{\sigma t}(A_k e^{j\omega_d t} + A_{k+1} e^{-j\omega_d t}) \tag{3.30}$$

By using the Euler identity $e^{\pm j\omega_d t} = \cos \omega_d t \pm j \sin \omega_d t$ and then combining terms, expression (3.30) can be put into the form

$$e^{\sigma t}(B_1 \cos \omega_d t + B_2 \sin \omega_d t) \tag{3.31}$$

This can be converted into the form

$$A e^{\sigma t} \sin(\omega_d t + \phi) \tag{3.32}$$

where $A = \sqrt{B_1^2 + B_2^2}$ and $\phi = \tan^{-1}(B_1/B_2)$. This form is very convenient for plotting the transient response. The student must learn to use this equation directly without deriving it each time. Often the constants in the transient term are evaluated more readily from the initial conditions by using the form of Eq. (3.31), and then the expression is converted to the more useful form of Eq. (3.32).

This transient term is called an *exponentially damped sinusoid;* it consists of a sine wave of frequency ω_d whose magnitude is $A e^{\sigma t}$; that is, it is decreasing exponentially with time if σ is negative. It has the form shown in Fig. 3.2, in which the curves $\pm A e^{\sigma t}$ constitute the *envelope.* The plot of the time solution always remains between the two branches of the envelope.

For the complex roots given in Eq. (3.28), where σ is negative, the transient decays with time and eventually dies out. A negative σ represents a stable system. When σ is positive, the transient increases with time and will destroy the equipment unless otherwise restrained. A positive σ represents the undesirable case of an unstable system. Control systems must be designed so that they are always stable.

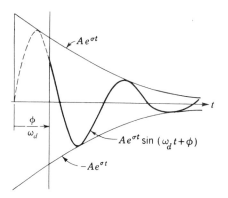

FIGURE 3.2
Sketch of an exponentially damped sinusoid.

Damping Ratio ζ and Undamped Natural Frequency ω_n

When the characteristic equation has a pair of complex-conjugate roots, it has a quadratic factor of the form $b_2 m^2 + b_1 m + b_0$. The roots of this factor are

$$m_{1,2} = -\frac{b_1}{2b_2} \pm j\sqrt{\frac{4b_2 b_0 - b_1^2}{4b_2^2}} = \sigma \pm j\omega_d \qquad (3.33)$$

The real part σ is recognized as the exponent of e, and ω_d is the frequency of the oscillatory portion of the component stemming from this pair of roots, as given by expression (3.32).

The quantity b_1 represents the effective damping constant of the system. If the numerator under the square root in Eq. (3.33) is zero, then b_1 has the value $2\sqrt{b_2 b_0}$ and the two roots $m_{1,2}$ are equal. This represents the critical value of the damping constant and is written $b_1' = 2\sqrt{b_2 b_0}$.

The *damping ratio* ζ is defined as the ratio of the actual damping constant to the critical value of the damping constant:

$$\zeta = \frac{\text{Actual damping constant}}{\text{Critical damping constant}} = \frac{b_1}{b_1'} = \frac{b_1}{2\sqrt{b_2 b_0}} \qquad (3.34)$$

When ζ is positive and less than unity, the roots are complex and the transient is a damped sinusoid of the form of expression (3.32). When ζ is less than unity, the response is said to be *underdamped*. When ζ is greater than unity, the roots are real and the response is *overdamped*; i.e., the transient solution consists of two exponential terms with real exponents. For $\zeta > 0$ the transients decay with time and the response $c(t)$ approaches the steady state value. For $\zeta < 0$ the transient increases with time and the response is unstable.

The undamped natural frequency ω_n is defined as the frequency of the sustained oscillation of the transient if the damping is zero:

$$\omega_n = \sqrt{\frac{b_0}{b_2}} \qquad (3.35)$$

The case of zero damping constant, $b_1 = 0$, means that the transient response does not die out; it is a sine wave of constant amplitude.

The quadratic factors are frequently written in terms of the damping ratio and the undamped natural frequency. Underdamped systems are generally analyzed in terms of these two parameters. After factoring b_0, the quadratic factor of the characteristic equation is

$$\frac{b_2}{b_0}m^2 + \frac{b_1}{b_0}m + 1 = \frac{1}{\omega_n^2}m^2 + \frac{2\zeta}{\omega_n}m + 1 \tag{3.36}$$

When it is multiplied through by ω_n^2, the quadratic appears in the form

$$m^2 + 2\zeta\omega_n m + \omega_n^2 \tag{3.37}$$

The two forms given by Eqs. (3.36) and (3.37) are called the *standard forms* of the quadratic factor, and the corresponding roots are

$$m_{1,2} = \sigma \pm j\omega_d = -\zeta\omega_n \pm j\omega_n \sqrt{1 - \zeta^2} \tag{3.38}$$

The transient response of Eq. (3.32) for the underdamped case, written in terms of ζ and ω_n, is

$$Ae^{\sigma t}\sin(\omega_d t + \phi) = Ae^{-\zeta\omega_n t}\sin(\omega_n \sqrt{1 - \zeta^2}\, t + \phi) \tag{3.39}$$

From this expression the effect on the transient of the terms ζ and ω_n can readily be seen. The larger the product $\zeta\omega_n$, the faster the transient will decay. These terms also affect the damped natural frequency of oscillation of the transient, $\omega_d = \omega_n\sqrt{1 - \zeta^2}$, which varies directly as the undamped natural frequency and decreases with an increase in the damping ratio.

3.6 DEFINITION OF TIME CONSTANT

The transient terms have the exponential form Ae^{mt}. When $m = -a$ is real and negative, the plot of Ae^{-at} has the form shown in Fig. 3.3. The value of time that makes the exponent of e equal to -1 is called the *time constant T*. Thus

$$-aT = -1 \quad \text{and} \quad T = \frac{1}{a} \tag{3.40}$$

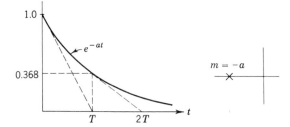

FIGURE 3.3
Plot of the exponential e^{-at} and the root location.

In a duration of time equal to one time constant the exponential e^{-at} decreases from the value 1 to the value 0.368. Geometrically, the tangent drawn to the curve Ae^{-at} at $t = 0$ intersects the time axis at the value of time equal to the time constant T.

When $m = \sigma \pm j\omega_d$ is a complex quantity, the transient has the form $Ae^{\sigma t} \sin(\omega_d t + \phi)$. A plot of this function is shown in Fig. 3.2. In the case of the damped sinusoid, the time constant is defined in terms of the parameter σ that characterizes the envelope $Ae^{-\sigma t}$. Thus the time constant T is equal to

$$T = \frac{1}{|\sigma|} \tag{3.41}$$

In terms of the damping ratio and the undamped natural frequency, the time constant is $T = 1/\zeta\omega_n$. Therefore, the larger the product $\zeta\omega_n$, the greater the instantaneous rate of decay of the transient.

3.7 EXAMPLE: SECOND-ORDER SYSTEM— MECHANICAL

The simple mechanical system of Sec. 2.6 is used as an example and is shown in Fig. 3.4. Equation (2.61) relates the displacement x_b to x_a:

$$M D^2 x_b + B D x_b + K x_b = K x_a \tag{3.42}$$

Consider the system to be originally at rest. Then the function x_a moves 1 unit at time $t = 0$; that is, the input is a unit step function $x_a(t) = u_{-1}(t)$. The problem is to find the motion $x_b(t)$.

The displacement of the mass is given by

$$x_b(t) = x_{b,ss} + x_{b,t}$$

where $x_{b,ss}$ is the steady-state solution and $x_{b,t}$ is the transient solution.

The steady-state solution is found first by using the method of Sec. 3.4. In this example it may be easier to consider the following:

1. Since the input x_a is a constant, the response x_b must reach a fixed steady-state position.

2. When x_b reaches a constant value, the velocity and acceleration become zero.

FIGURE 3.4
Simple mechanical system.

By putting $D^2 x_b = D x_b = 0$ into Eq. (3.42), the final or steady-state value of x_b is $x_{b,ss} = x_a$. After the steady-state solution has been found, the transient solution is determined. The characteristic equation is

$$M m^2 + B m + K = M \left(m^2 + \frac{B}{M} m + \frac{K}{M} \right) = 0$$

Putting this in terms of ζ and ω_n gives

$$m^2 + 2\zeta \omega_n m + \omega_n^2 = 0 \tag{3.43}$$

for which the roots are $m_{1,2} = -\zeta \omega_n \pm \omega_n \sqrt{\zeta^2 - 1}$. The transient solution depends on whether the damping ratio ζ is (1) greater than unity, (2) equal to unity, or (3) smaller than unity. For ζ greater than unity the roots are real and have the values $m_1 = -a$ and $m_2 = -b$, and the transient response is

$$x_{b,t} = A_1 e^{-at} + A_2 e^{-bt} \tag{3.44}$$

For ζ equal to unity, the roots are real and equal; that is, $m_1 = m_2 = -\zeta \omega_n$. Since there are multiple roots, the transient response is

$$x_{b,t} = A_1 e^{-\zeta \omega_n t} + A_2 t e^{-\zeta \omega_n t} \tag{3.45}$$

For ζ less than unity, the roots are complex,

$$m_{1,2} = -\zeta \omega_n \pm j \omega_n \sqrt{1 - \zeta^2}$$

and the transient solution, as outlined in Sec. 3.5, is

$$x_{b,t} = A e^{-\zeta \omega_n t} \sin(\omega_n \sqrt{1 - \zeta^2}\, t + \phi) \tag{3.46}$$

The complete solution is the sum of the steady-state and transient solutions. For the underdamped case, $\zeta < 1$, the complete solution of Eq. (3.42) with a unit step input is

$$x_b(t) = 1 + A e^{-\zeta \omega_n t} \sin(\omega_n \sqrt{1 - \zeta^2}\, t + \phi) \tag{3.47}$$

The two constants A and ϕ must next be determined from the initial conditions. In this example the system was initially at rest; therefore, $x_b(0) = 0$. The energy stored in a mass is $W = \frac{1}{2} M v^2$. From the principle of conservation of energy, the *velocity of a system with mass cannot change instantaneously;* thus $D x_b(0) = 0$. Two equations are necessary, one for $x_b(t)$ and one for $D x_b(t)$. Differentiating Eq. (3.47) yields

$$D x_b(t) = -\zeta \omega_n A e^{-\zeta \omega_n t} \sin(\omega_n \sqrt{1 - \zeta^2}\, t + \phi)$$

$$+ \omega_n \sqrt{1 - \zeta^2}\, A e^{-\zeta \omega_n t} \cos(\omega_n \sqrt{1 - \zeta^2}\, t + \phi) \tag{3.48}$$

Inserting into Eqs. (3.47) and (3.48) the initial conditions $x_b(0) = 0$, $D x_b(0) = 0$, and $t = 0$ yields the two simultaneous equations

$$0 = 1 + A \sin \phi \qquad 0 = -\zeta \omega_n A \sin \phi + \omega_n \sqrt{1 - \zeta^2}\, A \cos \phi$$

These equations are then solved for A and ϕ:

$$A = \frac{-1}{\sqrt{1-\zeta^2}} \qquad \phi = \tan^{-1}\frac{\sqrt{1-\zeta^2}}{\zeta} = \cos^{-1}\zeta$$

where the units of ϕ are radians. Thus the complete solution is

$$x_b(t) = 1 - \frac{e^{-\zeta\omega_n t}}{\sqrt{1-\zeta^2}}\sin(\omega_n\sqrt{1-\zeta^2}\,t + \cos^{-1}\zeta) \qquad (3.49)$$

When the complete solution has been obtained, it should be checked to see that it satisfies the known conditions. For example, putting $t = 0$ into Eq. (3.49) gives $x_b(0) = 0$; therefore, the solution checks. In a like manner, the constants can be evaluated for the other two cases of damping. Equation (3.49) shows that the steady-state value $x_{b,ss}$ is equal to x_a. Thus, the output *tracks* the input.

3.8 EXAMPLE: SECOND-ORDER SYSTEM— ELECTRICAL

The electric circuit of Fig. 3.5 is used to illustrate further the determination of initial conditions. The circuit, as shown, is in the steady state. At time $t = 0$ the switch is closed. The problem is to solve for the current $i_2(t)$ through the inductor for $t > 0$.

Two loop equations are written for this circuit:

$$10 = 20i_1 - 10i_2 \qquad (3.50)$$

$$0 = -10i_1 + \left(D + 25 + \frac{100}{D}\right)i_2 \qquad (3.51)$$

Eliminating i_1 from these equations yields

$$10 = \left(2D + 40 + \frac{200}{D}\right)i_2$$

Differentiating this equation yields

$$(2D^2 + 40D + 200)i_2 = 0 \qquad (3.52)$$

FIGURE 3.5
Electric circuit: $E = 10$ V, $R_1 = 10\ \Omega$, $R_2 = 15\ \Omega$, $R_3 = 10\ \Omega$, $L = 1$ H, $C = 0.01$ F.

The steady-state solution is found by using the method of Sec. 3.4. Since the input of Eq. (3.52) is a step function of value zero, the steady-state output is

$$i_{2,\text{ss}} = 0 \tag{3.53}$$

This can also be deduced from an inspection of the circuit. When a branch contains a capacitor, the steady-state current is always zero for a dc source.

Next the transient solution is determined. The characteristic equation is $m^2 + 20m + 100 = 0$, for which the roots are $m_{1,2} = -10$. Thus the circuit is critically damped, and the current through the inductor can be expressed by

$$i_2(t) = i_2(t)_t = A_1 e^{-10t} + A_2 t e^{-10t} \tag{3.54}$$

Two equations and two initial conditions are necessary to evaluate the two constants A_1 and A_2. Equation (3.54) and its derivative,

$$Di_2(t) = -10A_1 e^{-10t} + A_2(1 - 10t)e^{-10t} \tag{3.55}$$

are utilized in conjunction with the initial conditions $i_2(0^+)$ and $Di_2(0^+)$. In this example the currents just before the switch is closed are $i_1(0^-) = i_2(0^-) = 0$. The energy stored in the magnetic field of a single inductor is $W = \frac{1}{2}Li^2$. Since this energy cannot change instantly, the current through an inductor cannot change instantly either. Therefore, $i_2(0^+) = i_2(0^-) = 0$. $Di_2(t)$ is found from the original circuit equation, Eq. (3.51) which is rewritten as:

$$Di_2(t) = 10i_1(t) - 25i_2(t) - v_c(t) \tag{3.56}$$

To determine $Di_2(0^+)$, it is necessary first to determine $i_1(0^+)$, $i_2(0^+)$, and $v_c(0^+)$. Since $i_2(0^+) = 0$, Eq. (3.50) yields $i_1(0^+) = 0.5$ A. Since the energy $W = \frac{1}{2}Cv^2$ stored in a capacitor cannot change instantly, the voltage across the capacitor cannot change instantly either. The steady-state value of capacitor voltage for $t < 0$ is $v_c(0^-) = 10$ V; thus

$$v_c(0^-) = v_c(0^+) = 10 \text{ V} \tag{3.57}$$

Inserting these values into Eq. (3.56) yields $Di_2(0^+) = -5$. Substituting the initial conditions into Eqs. (3.54) and (3.55) results in $A_1 = 0$ and $A_2 = -5$. Therefore, the current through the inductor for $t \geq 0$ is

$$i_2(t) = -5te^{-10t} \tag{3.58}$$

3.9   SECOND-ORDER TRANSIENTS[2]

The response to a unit step-function input is usually used as a means of evaluating the response of a system. The example of Sec. 3.7 is used as an illustrative simple second-order system. The differential equation given by Eq. (3.42) can be expressed in the form

$$\frac{D^2 c}{\omega_n^2} + \frac{2\zeta}{\omega_n} Dc + c = r$$

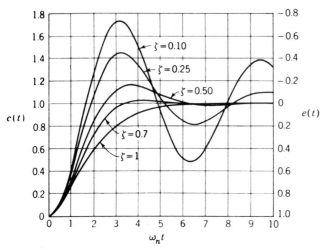

FIGURE 3.6
Simple second-order transients.

This is defined as a *simple* second-order equation because there are no derivatives of r on the right side of the equation. The underdamped response ($\zeta < 1$) to a unit step input, subject to zero initial conditions, is derived in Sec. 3.7 and is given by

$$c(t) = 1 - \frac{e^{-\zeta\omega_n t}}{\sqrt{1-\zeta^2}} \sin(\omega_n \sqrt{1-\zeta^2}\, t + \cos^{-1}\zeta) \tag{3.59}$$

A family of curves representing this equation is shown in Fig. 3.6, where the abscissa is the dimensionless variable $\omega_n t$. The curves are thus a function only of the damping ratio ζ.

These curves show that the amount of overshoot depends on the damping ratio ζ. For the overdamped and critically damped case, $\zeta \geq 1$, there is no overshoot. For the underdamped case, $\zeta < 1$, the system oscillates around the final value. The oscillations decrease with time, and the system response approaches the final value. The peak overshoot for the underdamped system is the first overshoot. The time at which the peak overshoot occurs, t_p, can be found by differentiating $c(t)$ from Eq. (3.59) with respect to time and setting this derivative equal to zero:

$$\frac{dc}{dt} = \frac{\zeta\omega_n e^{-\zeta\omega_n t}}{\sqrt{1-\zeta^2}} \sin(\omega_n \sqrt{1-\zeta^2}\, t + \cos^{-1}\zeta)$$

$$- \omega_n e^{-\zeta\omega_n t} \cos(\omega_n \sqrt{1-\zeta^2}\, t + \cos^{-1}\zeta) = 0$$

This derivative is zero at $\omega_n \sqrt{1-\zeta^2}\, t = 0, \pi, 2\pi, \ldots$. The peak overshoot occurs at the first value after zero, provided there are zero initial conditions; therefore,

$$t_p = \frac{\pi}{\omega_n \sqrt{1-\zeta^2}} \tag{3.60}$$

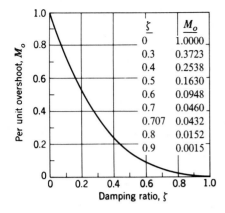

ζ	M_o
0	1.0000
0.3	0.3723
0.4	0.2538
0.5	0.1630
0.6	0.0948
0.7	0.0460
0.707	0.0432
0.8	0.0152
0.9	0.0015

FIGURE 3.7
Peak overshoot vs. damping ratio for a simple second-order equation $D^2 c/\omega_n^2 + (2\zeta/\omega_n) Dc + c = r(t)$.

Inserting this value of time in Eq. (3.59) gives the normalized peak overshoot with a step input as $M_p = c(t_p)/R_0$:

$$M_p = c_p = 1 + \exp\left(-\frac{\zeta \pi}{\sqrt{1 - \zeta^2}}\right) \tag{3.61}$$

The per unit overshoot M_o as a function of damping ratio is shown in Fig. 3.7, where

$$M_o = \frac{c_p - c_{ss}}{c_{ss}} \tag{3.62}$$

The variation of the frequency of oscillation of the transient with variation of damping ratio is also of interest. In order to represent this variation by one curve, the quantity ω_d/ω_n is plotted against ζ in Fig. 3.8. If the scales of ordinate and abscissa are equal, the curve is an arc of a circle. Note that this curve has been plotted for $\zeta \le 1$. Values of damped natural frequency for $\zeta > 1$ are mathematical only, not physical.

The error in the system is the input minus the output; thus the error equation is

$$e = r - c = \frac{e^{-\zeta \omega_n t}}{\sqrt{1 - \zeta^2}} \sin(\omega_n \sqrt{1 - \zeta^2}\, t + \cos^{-1} \zeta) \tag{3.63}$$

The variation of error with time is sometimes plotted. These curves can be obtained from Fig. 3.6 by realizing that the curves start at $e(0) = +1$ and have the final value $e(\infty) = 0$.

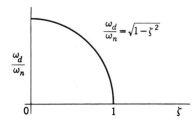

$$\frac{\omega_d}{\omega_n} = \sqrt{1 - \zeta^2}$$

FIGURE 3.8
Frequency of oscillation vs. damping ratio.

TABLE 3.1
Exponential values

t	$e^{-\zeta\omega_n t}$	Error, %
$1T$	0.368	36.8
$2T$	0.135	13.5
$3T$	0.050	5.0
$4T$	0.018	1.8
$5T$	0.007	0.7

Response Characteristics

The transient-response curves for the second-order system show a number of significant characteristics. The overdamped system is slow-acting and does not oscillate about the final position. For some applications the absence of oscillations may be necessary. For example, an elevator cannot be allowed to oscillate at each stop. But for systems where a fast response is necessary, the slow response of an overdamped system cannot be tolerated.

The underdamped system reaches the final value faster than the overdamped system, but the response oscillates about this final value. If this oscillation can be tolerated, the underdamped system is faster-acting. The amount of permissible overshoot determines the desirable value of the damping ratio. For example, a damping ratio $\zeta = 0.4$ has an overshoot of 25.4 percent, and a damping ratio $\zeta = 0.8$ has an overshoot of 1.52 percent.

The settling time is the time required for the oscillations to decrease to a specified absolute percentage of the final value and thereafter to remain less than this value. Errors of 2 or 5 percent are common values used to determine settling time. For second-order systems the value of the transient component at any time is equal to or less than the exponential $e^{-\zeta\omega_n t}$. The value of this term is given in Table 3.1 for several values of t expressed in a number of time constants T.

The settling time for a 2 percent error criterion is approximately 4 time constants; for a 5 percent error criterion, it is 3 time constants. The percent error criterion used must be determined from the response desired for the system. The time for the envelope of the transient to die out is

$$T_s = \frac{\text{Number of time constants}}{\zeta\omega_n} \tag{3.64}$$

Since ζ must be determined and adjusted for the permissible overshoot, the undamped natural frequency determines the settling time. When $c(t)_{ss} = 0$, then t_s may be based upon ± 2 percent of $c(t_p)$.

3.10   TIME-RESPONSE SPECIFICATIONS[3]

The desired performance characteristics of a tracking system of any order may be specified in terms of the transient response to a unit step-function input. The perfor-

FIGURE 3.9
Typical underdamped response to a step input.

mance of a system may be evaluated in terms of the following quantities, as shown in Fig. 3.9.

1. Peak overshoot c_p, which is the magnitude of the largest overshoot, often occurring at the first overshoot. This may also be expressed in percent of the final value.
2. Time to maximum overshoot t_p is the time to reach the maximum overshoot.
3. Time to first zero error t_0, which is the time required to reach the final value the first time. It is often referred to as *duplicating time*.
4. Settling time t_s, the time required for the output response first to reach and thereafter remain within a prescribed percentage of the final value. This percentage must be specified in the individual case. Common values used for settling time are 2 and 5 percent. As commonly used, the 2 or 5 percent is applied to the envelope that yields T_s. The actual t_s may be smaller than T_s.
5. Rise time t_r, defined as the time for the response, on its initial rise, to go from 0.1 to 0.9 times the steady-state value.
6. Frequency of oscillation ω_d of the transient depends on the physical characteristics of the system.

If the system has an initial constant output $c_1(t)_{ss}$ and then the input is changed to obtain a new constant value $c_2(t)_{ss}$, the transient response for an underdamped system is illustrated in Fig. 3.10. The per unit overshoot may then be redefined from the expression in Eq. (3.62) to the new value

$$M_o = \frac{\text{Maximum overshoot}}{\text{Signal transition}} = \frac{|\Delta c_p| - |\Delta c|}{|\Delta c|} \qquad (3.65)$$

FIGURE 3.10
Underdamped transient.

The time response differs for each set of initial conditions. Therefore, to compare the time responses of various systems it is necessary to start with standard initial conditions. The most practical standard is to start with the system at rest. Then the response characteristics, such as maximum overshoot and settling time, can be compared meaningfully.

For some systems these specifications are also applied for a ramp input. In such cases the plot of error with time is used with the definitions. For systems subject to shock inputs the response due to an impulse is used as a criterion of performance.

3.11 CAD ACCURACY CHECKS (CADAC)

A proficient control system engineer is an individual who has mastered control theory and its application in analysis and design. The competent engineer must become *computer literate* in the use of CAD packages like MATLAB[8] or ICECAP. The individual must fully comprehend the importance of the following factors:

1. It is essential to have a firm understanding of the CAD algorithm being used. This leads to a judgment about the reasonableness of the CAD program results.
2. CAD input data must be entered accurately. Remember: *garbage in—garbage out.*
3. No computer has infinite accuracy. Therefore, round-off must be considered.
4. CAD accuracy checks (CADAC) help to validate the accuracy of the CAD output data. *The CADAC are necessary, but not necessarily sufficient, conditions for ensuring the accuracy of the CAD output data.*

The figures of merit (FOM) equations for M_p, t_p, and T_s can serve as CADAC. They apply not only for simple second-order systems but also for systems in which the dominant roots (a pair of complex roots or a real root) of the characteristic equation dictate the time response characteristics. These CADAC are important for the design methods presented in later chapters.

3.12 STATE-VARIABLE EQUATIONS[4–7]

The differential equations and the corresponding state equations of various physical systems are derived in Chapter 2. The state variables selected in Chapter 2 are restricted to the energy-storage variables. In Chapter 5 different formulations of the state variables are presented. In large-scale systems, with many inputs and outputs, the state-variable approach can have distinct advantages over conventional methods, especially when digital computers are used to obtain the solutions. Although this text is restricted to linear time-invariant (LTI) systems, the state-variable approach is applicable to nonlinear and to time-varying systems. In these cases a computer is a practical method for obtaining the solution. A feature of the state-variable method is that it decomposes a complex system into a set of smaller systems that can be normalized to have a minimum interaction and that can be solved individually. Also, it provides a unified approach that is used extensively in modern control theory.

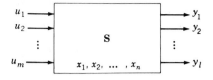

FIGURE 3.11
General system representation.

The block diagram of Fig. 3.11 represents a system **S** that has m inputs, l outputs, and n state variables. The coefficients in the equations representing an LTI system are constants. The matrix state and output equations are then

$$\dot{\mathbf{x}}(t) = \mathbf{A}\mathbf{x}(t) + \mathbf{B}\mathbf{u}(t) \qquad (3.66)$$

$$\mathbf{y}(t) = \mathbf{C}\mathbf{x}(t) + \mathbf{D}\mathbf{u}(t) \qquad (3.67)$$

The variables $\mathbf{x}(t)$, $\mathbf{u}(t)$, and $\mathbf{y}(t)$ are column vectors, and **A**, **B**, **C**, and **D** are matrices having constant elements. Equation (3.66) is solved first for the state vector $\mathbf{x}(t)$. This result is then used in Eq. (3.67) to determine the output $\mathbf{y}(t)$.

The homogeneous state equation, with the input $\mathbf{u}(t) = 0$, is

$$\dot{\mathbf{x}} = \mathbf{A}\mathbf{x} \qquad (3.68)$$

where **A** is a constant $n \times n$ matrix and **x** is an $n \times 1$ column vector.

For the scalar first-order equation $\dot{x} = ax$, the solution, in terms of the initial conditions at time $t = 0$, is

$$x(t) = e^{at}x(0) \qquad (3.69)$$

For any other initial condition, at time $t = t_0$, the solution is

$$x(t) = e^{a(t-t_0)}x(t_0) \qquad (3.70)$$

Comparing the scalar and the state equations shows the solution of Eq. (3.68) to be analogous to the solution given by Eq. (3.70); it is

$$\mathbf{x}(t) = \exp[\mathbf{A}(t - t_0)]\mathbf{x}(t_0) \qquad (3.71)$$

The exponential function of a scalar that appears in Eq. (3.69) can be expressed as the infinite series

$$e^{at} = \exp[at] = 1 + \frac{at}{1!} + \frac{(at)^2}{2!} + \frac{(at)^3}{3!} + \cdots + \frac{(at)^k}{k!} + \cdots \qquad (3.72)$$

The analogous exponential function of a square matrix **A** that appears in Eq. (3.71), with $t_0 = 0$, is

$$e^{\mathbf{A}t} = \exp[\mathbf{A}t] = \mathbf{I} + \frac{\mathbf{A}t}{1!} + \frac{(\mathbf{A}t)^2}{2!} + \frac{(\mathbf{A}t)^3}{3!} + \cdots + \frac{(\mathbf{A}t)^k}{k!} + \cdots \qquad (3.73)$$

Thus $\exp[\mathbf{A}t]$ is a square matrix of the same order as **A**. A more useful form for the infinite series of Eq. (3.73) is the closed form developed in Sec. 3.14. The closed form is called the *state transition matrix* (STM) or the *fundamental matrix* of the system and is denoted by

$$\mathbf{\Phi}(t) = e^{\mathbf{A}t} = \exp[\mathbf{A}t] \tag{3.74}$$

The term STM is descriptive of the unforced or natural response and is the expression preferred by engineers. The STM has the following properties:[6,10]

1. $\mathbf{\Phi}(t_2 - t_1)\mathbf{\Phi}(t_1 - t_0) = \mathbf{\Phi}(t_2 - t_0)$ for any t_0, t_1, t_2 (3.75)

2. $\mathbf{\Phi}(t)\mathbf{\Phi}(t)\cdots\mathbf{\Phi}(t) = \mathbf{\Phi}^q(t) = \mathbf{\Phi}(qt)$ q = positive integer   (3.76)

3. $\mathbf{\Phi}^{-1}(t) = \mathbf{\Phi}(-t)$ (3.77)

4. $\mathbf{\Phi}(0) = \mathbf{I}$ unity matrix   (3.78)

5. $\mathbf{\Phi}(t)$ is nonsingular for all finite values of t (3.79)

3.13   CHARACTERISTIC VALUES

Consider a system of equations represented by

$$\dot{\mathbf{x}} = \mathbf{A}\mathbf{x} \tag{3.80}$$

The signals $\dot{\mathbf{x}}$ and $\mathbf{x}$ are column vectors, and $\mathbf{A}$ is a square matrix of order n. One case for which a solution of this equation exists is if $\mathbf{x}$ and $\dot{\mathbf{x}}$ have the same direction in the state space but differ only in magnitude by a scalar proportionality factor λ. The solution must therefore have the form $\dot{\mathbf{x}} = \lambda\mathbf{x}$. Inserting this into Eq. (3.80) and rearranging terms yields

$$[\lambda\mathbf{I} - \mathbf{A}]\mathbf{x} = 0$$

This equation has a nontrivial solution only if $\mathbf{x}$ is not zero. It is therefore required that the matrix $[\lambda\mathbf{I} - \mathbf{A}]$ not have full rank; therefore, the determinant of the coefficients of $\mathbf{x}$ must be zero:

$$Q(\lambda) \equiv |\lambda\mathbf{I} - \mathbf{A}| = 0 \tag{3.81}$$

When $\mathbf{A}$ is of order n, the resulting polynomial equation is the *characteristic equation*

$$Q(\lambda) = \lambda^n + a_{n-1}\lambda^{n-1} + \cdots + a_1\lambda + a_0 = 0 \tag{3.82}$$

The roots λ_i of the characteristic equation are called the *characteristic values* or *eigenvalues* of $\mathbf{A}$. The roots may be distinct (simple) or repeated with a multiplicity p. Also, a root may be real or complex. Complex roots must appear in conjugate pairs, and the set of roots containing the complex-conjugate roots is said to be *self-conjugate*. The polynomial $Q(\lambda)$ may be written in factored form as

$$Q(\lambda) = (\lambda - \lambda_1)(\lambda - \lambda_2)\cdots(\lambda - \lambda_n) \tag{3.83}$$

The product of *eigenvalues* of a matrix $\mathbf{A}$ is equal to its determinant, that is, $\lambda_1 \lambda_2 \cdots \lambda_n = |\mathbf{A}|$. Also, the sum of the *eigenvalues* is equal to the sum of the elements on the main diagonal (the trace) of $\mathbf{A}$, that is,

$$\sum_{l}^{n} \lambda_i = \sum_{i=1}^{n} a_{ii} \equiv \text{trace } \mathbf{A}$$

3.14 EVALUATING THE STATE TRANSITION MATRIX

There are several methods for evaluating the STM $\mathbf{\Phi}(t) = \exp[\mathbf{A}t]$ in closed form for a given matrix $\mathbf{A}$. The method illustrated here is based on the Cayley-Hamilton theorem. The Laplace transform method is covered in the next chapter, and a state transformation method is covered in Chapter 5. Consider a general polynomial of the form

$$N(\lambda) = \lambda^m + C_{m-1}\lambda^{m-1} + \cdots + C_1\lambda + C_0 \qquad (3.84a)$$

When the polynomial $N(\lambda)$ is divided by the characteristic polynomial $Q(\lambda)$, the result is

$$\frac{N(\lambda)}{Q(\lambda)} = F(\lambda) + \frac{R(\lambda)}{Q(\lambda)}$$

or
$$N(\lambda) = F(\lambda)Q(\lambda) + R(\lambda) \qquad (3.84b)$$

The function $R(\lambda)$ is the remainder, and it is a polynomial whose maximum order is $n - 1$, or 1 less than the order of $Q(\lambda)$. For $\lambda = \lambda_i$ the value $Q(\lambda_i) = 0$; thus

$$N(\lambda_i) = R(\lambda_i) \qquad (3.85)$$

The matrix polynomial corresponding to Eq. (3.84a), using $\mathbf{A}$ as the variable, is

$$\mathbf{N(A)} = \mathbf{A}^m + C_{m-1}\mathbf{A}^{m-1} + \cdots + C_1\mathbf{A} + C_0\mathbf{I} \qquad (3.86)$$

Since the characteristic equation $Q(\lambda) = 0$ has n roots, there are n equations $Q(\lambda_1) = 0, Q(\lambda_2) = 0, \ldots, Q(\lambda_n) = 0$. The analogous matrix equation is

$$\mathbf{Q(A)} = \mathbf{A}^n + a_{n-1}\mathbf{A}^{n-1} + \cdots + a_1\mathbf{A} + a_0\mathbf{I} = \mathbf{0}$$

where $\mathbf{0}$ indicates a null matrix of the same order as $\mathbf{Q(A)}$. This equation implies the Cayley-Hamilton theorem, which is sometimes expressed as "every square matrix $\mathbf{A}$ satisfies its own characteristic equation." The matrix polynomial corresponding to Eqs. (3.84b) and (3.85) is therefore

$$\mathbf{N(A)} = \mathbf{F(A)Q(A)} + \mathbf{R(A)} = \mathbf{R(A)} \qquad (3.87)$$

Equations (3.85) and (3.87) are valid when $N(\lambda)$ is a polynomial of any order (or even an infinite series) as long as it is analytic. The exponential function $N(\lambda) = e^{\lambda t}$ is an analytic function that can be represented by an infinite series, as shown in Eq. (3.72). Since this function converges in the region of analyticity, it can be expressed in closed form by a polynomial in λ of degree $n - 1$. Thus, for each eigenvalue, from Eq. (3.85)

$$e^{\lambda_i t} = R(\lambda_i) = \alpha_0(t) + \alpha_1(t)\lambda_i + \cdots + \alpha_k(t)\lambda_i^k + \cdots + \alpha_{n-1}(t)\lambda_i^{n-1} \quad (3.88)$$

Inserting the n distinct roots λ_i into Eq. (3.88) yields n equations that can be solved simultaneously for the coefficients α_k. These coefficients may be inserted into the corresponding matrix equation:

$$e^{\mathbf{A}t} = N(\mathbf{A}) = R(\mathbf{A}) = \alpha_0(t)\mathbf{I} + \alpha_1(t)\mathbf{A} + \cdots + \alpha_k(t)\mathbf{A}^k + \cdots + \alpha_{n-1}(t)\mathbf{A}^{n-1}$$

$$(3.89)$$

For the STM this yields

$$\mathbf{\Phi}(t) = \exp[\mathbf{A}t] = \sum_{k=0}^{n-1} \alpha_k(t)\mathbf{A}^k \qquad (3.90)$$

For the case of multiple roots, refer to a text on linear algebra[4,7,10] for the evaluation of $\mathbf{\Phi}(t)$.

Example. Find $\exp[\mathbf{A}t]$ when

$$\mathbf{A} = \begin{bmatrix} 0 & 6 \\ -1 & -5 \end{bmatrix}$$

The characteristic equation is

$$Q(\lambda) = |\lambda\mathbf{I} - \mathbf{A}| = \begin{vmatrix} \lambda & -6 \\ 1 & \lambda + 5 \end{vmatrix} = \lambda^2 + 5\lambda + 6 = 0$$

The roots of this equation are $\lambda_1 = -2$ and $\lambda_2 = -3$. Since $\mathbf{A}$ is a second-order matrix, the remainder polynomial given by Eq. (3.88) is of first order:

$$N(\lambda_i) = R(\lambda_i) = \alpha_0 + \alpha_1\lambda_i \qquad (3.91)$$

Substituting $N(\lambda_i) = e^{\lambda_i t}$ and the two roots λ_1 and λ_2 into Eq. (3.91) yields

$$e^{-2t} = \alpha_0 - 2\alpha_1 \quad \text{and} \quad e^{-3t} = \alpha_0 - 3\alpha_1$$

Solving these equations for the coefficients α_0 and α_1 yields

$$\alpha_0 = 3e^{-2t} - 2e^{-3t} \quad \text{and} \quad \alpha_1 = e^{-2t} - e^{-3t}$$

The STM obtained by using Eq. (3.90) is

$$\mathbf{\Phi}(t) = \exp[\mathbf{A}t] = \alpha_0\mathbf{I} + \alpha_1\mathbf{A} = \begin{bmatrix} \alpha_0 & 0 \\ 0 & \alpha_0 \end{bmatrix} + \begin{bmatrix} 0 & 6\alpha_1 \\ -\alpha_1 & -5\alpha_1 \end{bmatrix}$$

$$= \begin{bmatrix} 3e^{-2t} - 2e^{-3t} & 6e^{-2t} - 6e^{-3t} \\ -e^{-2t} + e^{-3t} & -2e^{-2t} + 3e^{-3t} \end{bmatrix} \qquad (3.92)$$

Another procedure for evaluating the STM, $\mathbf{\Phi}(t)$, is by use of the Sylvester expansion. The format in this method makes it especially useful with a digital computer. The method is presented here without proof for the case when $\mathbf{A}$ is a square matrix with n distinct eigenvalues. The polynomial $\mathbf{N}(\mathbf{A})$ can be written

$$e^{\mathbf{A}t} = \mathbf{N}(\mathbf{A}) = \sum_{i=1}^{n} N(\lambda_i)\mathbf{Z}_i(\lambda) \tag{3.93}$$

where

$$\mathbf{Z}_i(\lambda) = \frac{\displaystyle\sum_{\substack{j=1 \\ j \neq i}}^{n} (\mathbf{A} - \lambda_j\mathbf{I})}{\displaystyle\sum_{\substack{j=1 \\ j \neq i}}^{n} (\lambda_i - \lambda_j)} \tag{3.94}$$

For the previous example there are two eigenvalues $\lambda_1 = -2$ and $\lambda_2 = -3$. The two matrices $\mathbf{Z}_1$ and $\mathbf{Z}_2$ are evaluated from Eq. (3.94):

$$\mathbf{Z}_1 = \frac{\mathbf{A} - \lambda_2\mathbf{I}}{\lambda_1 - \lambda_2} = \frac{\begin{bmatrix} 0+3 & 6 \\ -1 & -5+3 \end{bmatrix}}{-2 - (-3)} = \begin{bmatrix} 3 & 6 \\ -1 & -2 \end{bmatrix}$$

$$\mathbf{Z}_2 = \frac{\mathbf{A} - \lambda_1\mathbf{I}}{\lambda_2 - \lambda_1} = \frac{\begin{bmatrix} 0+2 & 6 \\ -1 & -5+2 \end{bmatrix}}{-3 - (-2)} = \begin{bmatrix} -2 & -6 \\ 1 & 3 \end{bmatrix}$$

Using these values in Eq. (3.93), where $N(\lambda_i) = e^{\lambda_i t}$, yields

$$\mathbf{\Phi}(t) = \exp[\mathbf{A}t] = e^{\lambda_1 t}\mathbf{Z}_1 + e^{\lambda_2 t}\mathbf{Z}_2$$

Equation (3.92) contains the value of $\mathbf{\Phi}(t)$.

The matrices $\mathbf{Z}_i(\lambda)$ are called the *constituent matrices* of $\mathbf{A}$ and have the following properties:

1. Orthogonality:

$$\mathbf{Z}_i\mathbf{Z}_j = 0 \qquad i \neq j \tag{3.95}$$

2. The sum of the complete set is equal to the unit matrix:

$$\sum_{i=1}^{n} \mathbf{Z}_i = \mathbf{I} \tag{3.96}$$

3. Idempotency:

$$\mathbf{Z}_i^r = \mathbf{Z}_i \tag{3.97}$$

where $r > 0$ is any positive integer.

Equation (3.93) can be conveniently applied to evaluate other functions of $\mathbf{A}$. For example, to obtain $\mathbf{A}^r$ use the form

$$\mathbf{A}^r = \lambda_1^r \mathbf{Z}_1 + \lambda_2^r \mathbf{Z}_2 + \cdots + \lambda_n^r \mathbf{Z}_n = \sum_{i=1}^{n} \lambda_i^r \mathbf{Z}_i \tag{3.98}$$

The operations involved in computing Eq. (3.98) consist primarily of addition, which is simpler than multiplying $\mathbf{A}$ by itself r times.

3.15   COMPLETE SOLUTION OF THE STATE EQUATION

When an input $\mathbf{u}(t)$ is present, the complete solution for $\mathbf{x}(t)$ is obtained from Eq. (3.66). The starting point for obtaining the solution is the following equality, obtained by applying the rule for the derivative of the product of two matrices (see Sec. 2.3):

$$\frac{d}{dt}[e^{-\mathbf{A}t}\mathbf{x}(t)] = e^{-\mathbf{A}t}[\dot{\mathbf{x}}(t) - \mathbf{A}\mathbf{x}(t)] \tag{3.99}$$

Utilizing Eq. (3.66), $\dot{\mathbf{x}} = \mathbf{A}\mathbf{x} + \mathbf{B}\mathbf{u}$, in the right side of Eq. (3.99) yields

$$\frac{d}{dt}[e^{-\mathbf{A}t}\mathbf{x}(t)] = e^{-\mathbf{A}t}\mathbf{B}\mathbf{u}(t)$$

Integrating this equation between 0 and t gives

$$e^{-\mathbf{A}t}\mathbf{x}(t) - \mathbf{x}(0) = \int_0^t e^{-\mathbf{A}\tau}\mathbf{B}\mathbf{u}(\tau)\,d\tau \tag{3.100}$$

Multiplying by $e^{\mathbf{A}t}$ on the left and rearranging terms produces

$$\mathbf{x}(t) = e^{\mathbf{A}t}\mathbf{x}(0) + \int_0^t e^{\mathbf{A}(t-\tau)}\mathbf{B}\mathbf{u}(\tau)\,d\tau = \mathbf{\Phi}(t)\mathbf{x}(0) + \int_0^t \mathbf{\Phi}(t-\tau)\mathbf{B}\mathbf{u}(\tau)\,d\tau \tag{3.101}$$

Using the STM and generalizing for initial conditions at time $t = t_0$ gives the solution to the state-variable equation with an input $\mathbf{u}(t)$ as

$$\mathbf{x}(t) = \mathbf{\Phi}(t - t_0)\mathbf{x}(t_0) + \int_{t_0}^t \mathbf{\Phi}(t - \tau)\mathbf{B}\mathbf{u}(\tau)\,d\tau \qquad t > t_0 \tag{3.102}$$

This equation is called the *state transition equation;* i.e., it describes the change of state relative to the initial conditions $\mathbf{x}(t_0)$ and the input $\mathbf{u}(t)$. It may be more convenient to change the variables in the integral of Eq. (3.101) by letting $\beta = t - \tau$, which gives

$$\mathbf{x}(t) = \mathbf{\Phi}(t)\mathbf{x}(0) + \int_0^t \mathbf{\Phi}(\beta)\mathbf{B}\mathbf{u}(t - \beta)\,d\beta \tag{3.103}$$

In modern terminology the state solution $\mathbf{x}(t)$ as given by Eqs. (3.101) to (3.103) is described by

$$\mathbf{x}(t) = \mathbf{x}_{zi}(t) + \mathbf{x}_{zs}(t) \tag{3.104}$$

where $\mathbf{x}_{zi}(t)$ is called the *zero-input* response, that is, $\mathbf{u}(t) = \mathbf{0}$, and $\mathbf{x}_{zs}(t)$ is called the *zero-state* response, that is, $\mathbf{x}(t_0) = \mathbf{0}$.

Example. Solve the following state equation for a unit step-function scalar input, $u(t) = u_{-1}(t)$:

$$\dot{\mathbf{x}} = \begin{bmatrix} 0 & 6 \\ -1 & -5 \end{bmatrix} \mathbf{x} + \begin{bmatrix} 0 \\ 1 \end{bmatrix} u(t)$$

The STM for this equation is given by Eq. (3.92). Thus, the total time solution is obtained by substituting this value of $\mathbf{\Phi}(t)$ into Eq. (3.103). Since $u(t - \beta)$ has the value of unity for $0 < \beta < t$, Eq. (3.103) yields

$$\mathbf{x}(t) = \mathbf{\Phi}(t)\mathbf{x}(0) + \int_0^t \begin{bmatrix} 6e^{-2\beta} - 6e^{-3\beta} \\ -2e^{-2\beta} + 3e^{-3\beta} \end{bmatrix} d\beta$$

$$= \begin{bmatrix} 3e^{-2t} - 2e^{-3t} & 6e^{-2t} - 6e^{-3t} \\ -e^{-2t} + e^{-3t} & -2e^{-2t} + 3e^{-3t} \end{bmatrix} \begin{bmatrix} x_1(0) \\ x_2(0) \end{bmatrix} + \begin{bmatrix} 1 - 3e^{-2t} + 2e^{-3t} \\ e^{-2t} - e^{-3t} \end{bmatrix} \tag{3.105}$$

The integral of a matrix is the integral of each element in the matrix with respect to the indicated scalar variable. This property is used in evaluating the particular integral in Eq. (3.105). Inserting the initial conditions produces the final solution for the state transition equation. The Laplace transform method presented in Chapter 4 is a more direct method for finding this solution. An additional method is shown in Chapter 5.

3.16 SUMMARY

This chapter establishes the manner of solving differential equations. The steady-state and transient solutions are determined separately and then added to form the total response. Then the constants of the transient component of the solution are evaluated to satisfy the initial conditions. The ability to anticipate the form of the response is very important. The solution of differential equations has been extended to include solution of the matrix state and output equations. This method is also applicable to multiple-input multiple-output, time-varying, and nonlinear systems. The matrix formulation of these systems lends itself to digital-computer solutions. Obtaining the complete solution by means of the Laplace transform is covered in the next chapter.

REFERENCES

1. Kreyszig, E.: *Advanced Engineering Mathematics*, 7th ed., John Wiley & Sons, New York, 1993.
2. Kinariwala, B. K., F. F. Kuo, and N. K. Tsao: *Linear Circuits and Computation*, John Wiley & Sons, New York, 1973.

3. D'Azzo, J. J., and C. H. Houpis: *Feedback Control System Analysis and Synthesis,* 2nd ed., McGraw-Hill, New York, 1966.
4. Rugh, W. J.: *Linear System Theory,* Prentice Hall, Englewood Cliffs, N.J., 1993.
5. Kailath, T.: *Linear Systems,* Prentice-Hall, Englewood Cliffs, N.J., 1980.
6. Ward, J. R., and R. D. Strum: *State Variable Analysis,* Prentice-Hall, Englewood Cliffs, N.J., 1970.
7. Wiberg, D. M.: *State Space and Linear Systems,* Schaum's Outline Series, McGraw-Hill, New York, 1971.
8. Saadat, H.: *Computational Aids in Control Systems Using MATLAB™*, McGraw-Hill, New York, 1993.
9. D'Azzo, J. J., and C. H. Houpis: *Linear Control System Analysis and Design*, 3rd Ed., McGraw-Hill, New York, 1988.
10. De Russo, P. M., R. J. Roy, and C. M. Close: *State Variables for Engineers*, Krieger, Malabar, Fla., 1990.

CHAPTER
4

LAPLACE
TRANSFORM

4.1 INTRODUCTION

The Laplace transform method is used extensively[1-3] to facilitate and systematize the solution of ordinary constant-coefficient differential equations. The advantages of this modern transform method for the analysis of linear-time-invariant (LTI) systems are the following:

1. It includes the boundary or initial conditions.
2. The work involved in the solution is simple algebra.
3. The work is systematized.
4. The use of a table of transforms reduces the labor required.
5. Discontinuous inputs can be treated.
6. The transient and steady-state components of the solution are obtained simultaneously.

The disadvantage of transform methods is that if they are used mechanically, without knowledge of the actual theory involved, they sometimes yield erroneous results. Also, a particular equation can sometimes be solved more simply and with less work by the classical method. Although an understanding of the Laplace transform method is essential, it must be emphasized that the solutions of differential equations are readily obtained by use of CAD packages such as MATLAB, ICECAP-PC,[8] and TOTAL-PC[8] (see Appendix B). Laplace transforms are also applied to the solution of system equations that are in matrix state-variable format. The method for using the state and output equations to obtain the system transfer function is presented.

4.2 DEFINITION OF THE LAPLACE TRANSFORM

The direct Laplace transformation of a function of time $f(t)$ is given by

$$\mathcal{L}[f(t)] = \int_0^\infty f(t)e^{-st}\,dt = F(s) \tag{4.1}$$

where $\mathcal{L}[f(t)]$ is a shorthand notation for the Laplace integral. Evaluation of the integral results in a function $F(s)$ that has s as the parameter. This parameter s is a complex quantity of the form $\sigma + j\omega$. Since the limits of integration are zero and infinity, it is immaterial what value $f(t)$ has for negative or zero time.

There are limitations on the functions $f(t)$ that are Laplace-transformable. Basically, the requirement is that the Laplace integral converge, which means that this integral has a definite functional value. To meet this requirement[3] the function $f(t)$ must be (1) piecewise continuous over every finite interval $0 \le t_1 \le t \le t_2$ and (2) of exponential order. A function is piecewise continuous in a finite interval if that interval can be divided into a finite number of subintervals, over each of which the function is continuous and at the ends of each of which $f(t)$ possesses finite right- and left-hand limits. A function $f(t)$ is of exponential order if there exists a constant a such that the product $e^{-at}|f(t)|$ is bounded for all values of t greater than some finite value T. This imposes the restriction that σ, the real part of s, must be greater than a lower bound σ_a for which the product $e^{-\sigma_a t}|f(t)|$ is of exponential order. A linear differential equation with constant coefficients and with a finite number of terms is Laplace-transformable if the driving function is Laplace-transformable.

All cases covered in this book are Laplace-transformable. The basic purpose in using the Laplace transform is to obtain a method of solving differential equations that involves only simple algebraic operations in conjunction with a table of transforms.

4.3 DERIVATION OF LAPLACE TRANSFORMS OF SIMPLE FUNCTIONS

A number of examples are presented to show the derivation of the Laplace transform of several time functions. A list of common transform pairs is given in Appendix A.

Step Function $u_{-1}(t)$

The Laplace transform of the unit step function $u_{-1}(t)$ (see Fig. 3.1a) is

$$\mathcal{L}[u_{-1}(t)] = \int_0^\infty u_{-1}(t)e^{-st}\,dt = U_{-1}(s) \tag{4.2}$$

Since $u_{-1}(t)$ has the value 1 over the limits of integration,

$$U_{-1}(s) = \int_0^\infty e^{-st}\,dt = -\left.\frac{e^{-st}}{s}\right|_0^\infty = \frac{1}{s} \qquad \text{if } \sigma > 0 \tag{4.3}$$

The step function is undefined at $t = 0$; but this is immaterial, for the integral is defined by a limit process

$$\int_0^\infty f(t)e^{-st}\, dt = \lim_{\substack{T\to\infty \\ \varepsilon\to 0}} \int_\varepsilon^T f(t)e^{-st}\, dt \tag{4.4}$$

and the explicit value at $t = 0$ does not affect the value of the integral. The value of the integral obtained by taking limits is implied in each case but is not written out explicitly.

Decaying Exponential $e^{-\alpha t}$

The exponent α is a positive real number.

$$\mathscr{L}[e^{-\alpha t}] = \int_0^\infty e^{-\alpha t} e^{-st}\, dt = \int_0^\infty e^{-(s+\alpha)t}\, dt$$

$$= -\frac{e^{-(s+\alpha)t}}{s+\alpha}\Big|_0^\infty = \frac{1}{s+\alpha} \qquad \sigma > -\alpha \tag{4.5}$$

Sinusoid $\cos \omega t$

Here ω is a positive real number.

$$\mathscr{L}[\cos \omega t] = \int_0^\infty \cos \omega t \, e^{-st}\, dt \tag{4.6}$$

Expressing $\cos \omega t$ in exponential form gives

$$\cos \omega t = \frac{e^{j\omega t} + e^{-j\omega t}}{2}$$

Then

$$\mathscr{L}[\cos \omega t] = \frac{1}{2}\left(\int_0^\infty e^{(j\omega - s)t}\, dt + \int_0^\infty e^{(-j\omega - s)t}\, dt \right)$$

$$= \frac{1}{2}\left[\frac{e^{(j\omega - s)t}}{j\omega - s} + \frac{e^{(-j\omega - s)t}}{-j\omega - s} \right]_0^\infty$$

$$= \frac{1}{2}\left(-\frac{1}{j\omega - s} - \frac{1}{-j\omega - s} \right) = \frac{s}{s^2 + \omega^2} \qquad \sigma > 0 \tag{4.7}$$

Ramp Function $u_{-2}(t) = tu_{-1}(t)$

$$\mathscr{L}[t] = \int_0^\infty te^{-st}\, dt \qquad \sigma > 0 \tag{4.8}$$

This expression is integrated by parts by using

$$\int_a^b u\,dv = uv\Big|_a^b - \int_a^b v\,du.$$

Let $u = t$ and $dv = e^{-st}\,dt$. Then $du = dt$ and $v = -e^{-st}/s$. Thus

$$\int_0^\infty te^{-st}\,dt = -\frac{te^{-st}}{s}\Big|_0^\infty - \int_0^\infty \left(-\frac{e^{-st}}{s}\right)dt$$

$$= 0 - \frac{e^{-st}}{s^2}\Big|_0^\infty = \frac{1}{s^2} \qquad \sigma > 0 \qquad\qquad (4.9)$$

4.4   LAPLACE TRANSFORM THEOREMS

Several theorems that are useful in applying the Laplace transform are presented in this section. In general, they are helpful in evaluating transforms.

Theorem 1. Linearity. If a is a constant or is independent of s and t, and if $f(t)$ is transformable, then

$$\mathscr{L}[af(t)] = a\mathscr{L}[f(t)] = aF(s) \qquad\qquad (4.10)$$

Theorem 2. Superposition. If $f_1(t)$ and $f_2(t)$ are both Laplace-transformable, the principle of superposition applies:

$$\mathscr{L}[f_1(t) \pm f_2(t)] = \mathscr{L}[f_1(t)] \pm \mathscr{L}[f_2(t)] = F_1(s) \pm F_2(s) \qquad\qquad (4.11)$$

Theorem 3. Translation in time. If the Laplace transform of $f(t)$ is $F(s)$ and a is a positive real number, the Laplace transform of the translated function $f(t - a)u_{-1}(t - a)$ is

$$\mathscr{L}[f(t - a)u_{-1}(t - a)] = e^{-as}F(s) \qquad\qquad (4.12)$$

Translation in the positive t direction in the real domain becomes multiplication by the exponential e^{-as} in the s domain.

Theorem 4. Complex differentiation. If the Laplace transform of $f(t)$ is $F(s)$, then

$$\mathscr{L}[tf(t)] = -\frac{d}{ds}F(s) \qquad\qquad (4.13)$$

Multiplication by time in the real domain entails differentiation with respect to s in the s domain.

Example 1. Using $\mathscr{L}[\cos\omega t]$ from Eq. (4.7),

$$\mathscr{L}[t\cos\omega t] = -\frac{d}{ds}\left[\frac{s}{s^2 + \omega^2}\right] = \frac{s^2 - \omega^2}{(s^2 + \omega^2)^2}$$

Example 2. Using $\mathscr{L}[e^{-\alpha t}]$ from Eq. (4.5),

$$\mathscr{L}[te^{-\alpha t}] = -\frac{d}{ds}\mathscr{L}[e^{-\alpha t}] = -\frac{d}{ds}\left(\frac{1}{s+\alpha}\right) = \frac{1}{(s+\alpha)^2}$$

Theorem 5. Translation in the s domain. If the Laplace transform of $f(t)$ is $F(s)$ and a is either real or complex, then

$$\mathscr{L}\left[e^{at}f(t)\right] = F(s-a) \tag{4.14}$$

Multiplication of e^{at} in the real domain becomes translation in the s domain.

Example 3. Starting with $\mathscr{L}[\sin \omega t] = \omega/(s^2 + \omega^2)$ and applying Theorem 5 gives

$$\mathscr{L}[e^{-\alpha t}\sin \omega t] = \frac{\omega}{(s+\alpha)^2 + \omega^2}$$

Theorem 6. Real differentiation. If the Laplace transform of $f(t)$ is $F(s)$, and if the first derivative of $f(t)$ with respect to time $Df(t)$ is transformable, then

$$\mathscr{L}[Df(t)] = sF(s) - f(0^+) \tag{4.15}$$

The term $f(0^+)$ is the value of the right-hand limit of the function $f(t)$ as the origin $t = 0$ is approached from the right side (thus through positive values of time). This includes functions, such as the step function, that may be undefined at $t = 0$. For simplicity, the plus sign following the zero is usually omitted, although its presence is implied.

The transform of the second derivative $D^2 f(t)$ is

$$\mathscr{L}[D^2 f(t)] = s^2 F(s) - sf(0) - Df(0) \tag{4.16}$$

where $Df(0)$ is the value of the limit of the derivative of $f(t)$ as the origin $t = 0$, is approached from the right side.

The transform of the nth derivative $D^n f(t)$ is

$$\mathscr{L}[D^n f(t)] = s^n F(s) - s^{n-1} f(0) - s^{n-2} Df(0) - \cdots - s\,D^{n-2} f(0) - D^{n-1} f(0) \tag{4.17}$$

Note that the transform includes the initial conditions, whereas in the classical method of solution the initial conditions are introduced separately to evaluate the coefficients of the solution of the differential equation. When all initial conditions are zero, the Laplace transform of the nth derivative of $f(t)$ is simply $s^n F(s)$.

Theorem 7. Real integration. If the Laplace transform of $f(t)$ is $F(s)$, its integral

$$D^{-1}f(t) = \int_0^t f(t)\, dt + D^{-1}f(0^+)$$

is transformable and the value of its transform is

$$\mathscr{L}[D^{-1}f(t)] = \frac{F(s)}{s} + \frac{D^{-1}f(0^+)}{s} \tag{4.18}$$

The term $D^{-1}f(0^+)$ is the constant of integration and is equal to the value of the integral as the origin is approached from the positive or right side. The plus sign is omitted in the remainder of this text.

The transform of the double integral $D^{-2}f(t)$ is

$$\mathcal{L}\left[D^{-2}f(t)\right] = \frac{F(s)}{s^2} + \frac{D^{-1}f(0)}{s^2} + \frac{D^{-2}f(0)}{s} \tag{4.19}$$

The transform of the nth-order integral $D^{-n}f(t)$ is

$$\mathcal{L}\left[D^{-n}f(t)\right] = \frac{F(s)}{s^n} + \frac{D^{-1}f(0)}{s^n} + \cdots + \frac{D^{-n}f(0)}{s} \tag{4.20}$$

Theorem 8. Final value. If $f(t)$ and $Df(t)$ are Laplace-transformable, if the Laplace transform of $f(t)$ is $F(s)$, and if the limit $f(t)$ as $t \to \infty$ exists, then

$$\lim_{s \to 0} sF(s) = \lim_{t \to \infty} f(t) \tag{4.21}$$

This theorem states that the behavior of $f(t)$ in the neighborhood of $t = \infty$ is related to the behavior of $sF(s)$ in the neighborhood of $s = 0$. If $sF(s)$ has poles [values of s for which $|sF(s)|$ becomes infinite] on the imaginary axis (excluding the origin) or in the right-half s plane, there is no finite final value of $f(t)$ and the theorem cannot be used. If $f(t)$ is sinusoidal, the theorem is invalid, since $\mathcal{L}[\sin \omega t]$ has poles at $s = \pm j\omega$ and $\lim_{t \to \infty} \sin \omega t$ does not exist. However, for poles of $sF(s)$ at the origin, $s = 0$, this theorem gives the final value of $f(\infty) = \infty$. This correctly describes the behavior of $f(t)$ as $t \to \infty$.

Theorem 9. Initial value. If the function $f(t)$ and its first derivative are Laplace-transformable, if the Laplace transform of $f(t)$ is $F(s)$, and if $\lim_{s \to \infty} sF(s)$ exists, then

$$\lim_{s \to \infty} sF(s) = \lim_{t \to 0} f(t) \tag{4.22}$$

This theorem states that the behavior of $f(t)$ in the neighborhood of $t = 0$ is related to the behavior of $sF(s)$ in the neighborhood of $|s| = \infty$. There are no limitations on the locations of the poles of $sF(s)$.

Theorem 10. Complex integration. If the Laplace transform of $f(t)$ is $F(s)$ and if $f(t)/t$ has a limit as $t \to 0^+$, then

$$\mathcal{L}\left[\frac{f(t)}{t}\right] = \int_s^\infty F(s)\,ds \tag{4.23}$$

This theorem states that division by the variable in the real domain entails integration with respect to s in the s domain.

4.5 CAD ACCURACY CHECKS: CADAC

The Laplace transform theorems 8 and 9 for the initial and final values are valuable CADAC. These theorems should be used, when appropriate, to assist in the validation of the computer solutions of control system problems. Additional CADAC are presented throughout this text.

4.6 APPLICATION OF THE LAPLACE TRANSFORM TO DIFFERENTIAL EQUATIONS

The Laplace transform is now applied to the solution of the differential equation for the simple mechanical system that is solved by the classical method in Sec. 3.7. The differential equation of the system is repeated here:

$$MD^2x_2 + BDx_2 + Kx_2 = Kx_1 \tag{4.24}$$

The position $x_1(t)$ undergoes a unit step displacement. This is the input and is called the *driving function*. The unknown quantity for which the equation is to be solved is the output displacement $x_2(t)$, called the *response function*. The Laplace transform of Eq. (4.24) is

$$\mathscr{L}[Kx_1] = \mathscr{L}[MD^2x_2 + BDx_2 + Kx_2] \tag{4.25}$$

The transform of each term is

$$\mathscr{L}[Kx_1] = KX_1(s)$$

$$\mathscr{L}[Kx_2] = KX_2(s)$$

$$\mathscr{L}[BDx_2] = B[sX_2(s) - x_2(0)]$$

$$\mathscr{L}[MD^2x_2] = M[s^2X_2(s) - sx_2(0) - Dx_2(0)]$$

Substituting these terms into Eq. (4.25) and collecting terms gives

$$KX_1(s) = (Ms^2 + Bs + K)X_2(s) - [Msx_2(0) + MDx_2(0) + Bx_2(0)] \tag{4.26}$$

Equation (4.26) is the transform equation and shows how the initial conditions— the initial position $x_2(0)$ and the initial velocity $Dx_2(0)$—are incorporated into the equation. The function $X_1(s)$ is called the *driving transform*; the function $X_2(s)$ is called the *response transform*. The coefficient of $X_2(s)$, which is $Ms^2 + Bs + K$, is called the *characteristic function*. The equation formed by setting the characteristic function equal to zero is called the *characteristic equation* of the system. Solving for $X_2(s)$ gives

$$X_2(s) = \frac{K}{Ms^2 + Bs + K}X_1(s) + \frac{Msx_2(0) + Bx_2(0) + MDx_2(0)}{Ms^2 + Bs + K} \tag{4.27}$$

The coefficient of $X_1(s)$ is defined as the *system transfer function*. The second term on the right side of the equation is called the *initial condition component*. Combining the terms of Eq. (4.27) yields

$$X_2(s) = \frac{KX_1(s) + Msx_2(0) + Bx_2(0) + MDx_2(0)}{Ms^2 + Bs + K} \tag{4.28}$$

Finding the function $x_2(t)$ whose transform is given by Eq. (4.28) is symbolized by the inverse transform operator $\mathcal{L}^{-1}$; thus

$$x_2(t) = \mathcal{L}^{-1}[X_2(s)]$$

$$= \mathcal{L}^{-1}\left[\frac{KX_1(s) + Msx_2(0) + Bx_2(0) + MDx_2(0)}{Ms^2 + Bs + K}\right] \quad (4.29)$$

The function $x_2(t)$ can be found in the table of Appendix A after inserting numerical values into Eq. (4.29). For this example the system is initially at rest. From the principle of conservation of energy the initial conditions are $x_2(0) = 0$ and $Dx_2(0) = 0$. Assume, as in Sec. 3.7, that the damping ratio ζ is less than unity. Since $x_1(t)$ is a step function, the time response function is

$$x_2(t) = \mathcal{L}^{-1}\left[\frac{K/M}{s(s^2 + Bs/M + K/M)}\right] = \mathcal{L}^{-1}\left[\frac{\omega_n^2}{s(s^2 + 2\zeta\omega_n s + \omega_n^2)}\right] \quad (4.30)$$

where $\omega_n = \sqrt{K/M}$ and $\zeta = B/2\sqrt{KM}$. Reference to transform pair 27a in Appendix A provides the solution directly as

$$x_2(t) = 1 - \frac{e^{-\zeta\omega_n t}}{\sqrt{1 - \zeta^2}} \sin\left(\omega_n \sqrt{1 - \zeta^2}\, t + \cos^{-1}\zeta\right)$$

4.7   INVERSE TRANSFORMATION

The application of Laplace transforms to a differential equation yields an algebraic equation. From the algebraic equation the transform of the response function is readily found. To complete the solution the inverse transform must be found. In some cases the inverse-transform operation

$$f(t) = \mathcal{L}^{-1}[F(s)] \quad (4.31)$$

can be performed by direct reference to transform tables or by use of a digital-computer program (see Appendix B). The linearity and translation theorems are useful in extending the tables. When the response transform cannot be found in the tables, the general procedure is to express $F(s)$ as the sum of partial fractions with constant coefficients. The partial fractions have a first-order or quadratic factor in the denominator and are readily found in the table of transforms. The complete inverse transform is the sum of the inverse transforms of each fraction. This procedure is illustrated next.

The response transform $F(s)$ can be expressed, in general, as the ratio of two polynomials $P(s)$ and $Q(s)$. Consider that these polynomials are of degree w and n, respectively, and are arranged in descending order of the powers of the variable s; thus,

$$F(s) = \frac{P(s)}{Q(s)} = \frac{a_w s^w + a_{w-1}s^{w-1} + \cdots + a_1 s + a_0}{s^n + b_{n-1}s^{n-1} + \cdots + b_1 s + b_0} \quad (4.32)$$

The a's and b's are real constants, and the coefficient of the highest power of s in the denominator has been made equal to unity. Only those $F(s)$ that are proper fractions are considered, i.e., those in which n is greater than w.[†] The first step is to factor $Q(s)$ into first-order and quadratic factors with real coefficients:

$$F(s) = \frac{P(s)}{Q(s)} = \frac{P(s)}{(s - s_1)(s - s_2) \cdots (s - s_k) \cdots (s - s_n)} \qquad (4.33)$$

The values $s_1, s_2, \ldots, s_n$ in the finite plane that make the denominator equal to zero are called the *zeros* of the denominator. These values of s, which may be either real or complex, also make $|F(s)|$ infinite, and so they are called *poles* of $F(s)$. Therefore, the values $s_1, s_2, \ldots, s_n$ are referred to as zeros of the denominator or poles of the complete function in the finite plane, i.e., there are n poles of $F(s)$. Methods of factoring polynomials exist in the literature. Digital-computer programs are available to perform this operation (see Appendix B).[8]

The transform $F(s)$ can be expressed as a series of fractions. If the poles are simple (nonrepeated), the number of fractions is equal to n, the number of poles of $F(s)$. In such case the function $F(s)$ can be expressed as

$$F(s) = \frac{P(s)}{Q(s)} = \frac{A_1}{s - s_1} + \frac{A_2}{s - s_2} + \cdots + \frac{A_k}{s - s_k} + \cdots + \frac{A_n}{s - s_n} \qquad (4.34)$$

The procedure is to evaluate the constants $A_1, A_2 \ldots, A_n$ corresponding to the poles $s_1, s_2, \ldots, s_n$. The coefficients $A_1, A_2, \ldots$ are termed the *residues*[‡] of $F(s)$ at the corresponding poles. Cases of repeated factors and complex factors are treated separately. Several ways of evaluating the constants are shown in the following section.

4.8 HEAVISIDE PARTIAL-FRACTION EXPANSION THEOREMS

The technique of partial-fraction expansion is set up to take care of all cases systematically. There are four classes of problems, depending on the denominator $Q(s)$. Each of these cases is illustrated separately.

Case 1: $F(s)$ has first-order real poles.

Case 2: $F(s)$ has repeated first-order real poles.

Case 3: $F(s)$ has a pair of complex-conjugate poles (a quadratic factor in the denominator).

Case 4: $F(s)$ has repeated pairs of complex-conjugate poles (a repeated quadratic factor in the denominator).

[†] If $n = w$, first divide $P(s)$ by $Q(s)$ to obtain $F(s) = a_w + P_1(s)/Q(s) = a_w + F_1(s)$. Then express $F_1(s)$ as the sum of partial fractions with constant coefficients.

[‡] More generally the residue is the coefficient of the $(s - s_i)^{-1}$ term in the Laurent expansion of $F(s)$ about $s = s_i$.

Case 1: First-Order Real Poles

The locations of three real poles of $F(s)$ in the s plane are shown in Fig. 4.1. The poles may be positive, zero, or negative, and they lie on the real axis in the s plane. In this example, s_1 is positive, s_0 is zero, and s_2 is negative. For the poles shown in Fig. 4.1 the transform $F(s)$ and its partial fractions are

$$F(s) = \frac{P(s)}{Q(s)} = \frac{P(s)}{s(s - s_1)(s - s_2)} = \frac{A_0}{s} + \frac{A_1}{s - s_1} + \frac{A_2}{s - s_2} \tag{4.35}$$

There are as many fractions as there are factors in the denominator of $F(s)$. Since $s_0 = 0$, the factor $s - s_0$ is written simply as s. The inverse transform of $F(s)$ is

$$f(t) = A_0 + A_1 e^{s_1 t} + A_2 e^{s_2 t} \tag{4.36}$$

The pole s_1 is positive; therefore, the term $A_1 e^{s_1 t}$ is an increasing exponential and the system is unstable. The pole s_2 is negative, and the term $A_2 e^{s_2 t}$ is a decaying exponential with a final value of zero. Therefore, for a system to be stable, all real poles that contribute to the complementary solution must be in the left half of the s plane.

To evaluate a typical coefficient A_k, multiply both sides of Eq. (4.34) by the factor $s - s_k$. The result is

$$(s - s_k)F(s) = (s - s_k)\frac{P(s)}{Q(s)}$$

$$= A_1 \frac{s - s_k}{s - s_1} + A_2 \frac{s - s_k}{s - s_2} + \cdots + A_k + \cdots + A_n \frac{s - s_k}{s - s_n} \tag{4.37}$$

The multiplying factor $s - s_k$ on the left side of the equation and the same factor of $Q(s)$ should be cancelled. By letting $s = s_k$, each term on the right side of the equation is zero except A_k. Thus, a general rule for evaluating the constants for single-order real poles is

$$A_k = \left[(s - s_k)\frac{P(s)}{Q(s)}\right]_{s = s_k} = \left[\frac{P(s)}{Q'(s)}\right]_{s = s_k} \tag{4.38}$$

where $Q'(s_k) = [dQ(s)/ds]_{s = s_k} = [Q(s)/(s - s_k)]_{s = s_k}$. The coefficients A_k are the *residues* of $F(s)$ at the corresponding poles. For the case of

$$F(s) = \frac{s + 2}{s(s + 1)(s + 3)} = \frac{A_0}{s} + \frac{A_1}{s + 1} + \frac{A_2}{s + 3}$$

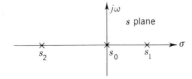

FIGURE 4.1
Location of real poles in the s plane.

the constants are

$$A_0 = [sF(s)]_{s=0} = \left[\frac{s+2}{(s+1)(s+3)}\right]_{s=0} = \frac{2}{3}$$

$$A_1 = [(s+1)F(s)]_{s=-1} = \left[\frac{s+2}{s(s+3)}\right]_{s=-1} = -\frac{1}{2}$$

$$A_2 = [(s+3)F(s)]_{s=-3} = \left[\frac{s+2}{s(s+1)}\right]_{s=-3} = -\frac{1}{6}$$

The solution as a function of time is

$$f(t) = \frac{2}{3} - \frac{1}{2}e^{-t} - \frac{1}{6}e^{-3t} \tag{4.39}$$

Case 2: Multiple-Order Real Poles

The position of real poles of $F(s)$, some of which are repeated, is shown in Fig. 4.2. The notation $]_r$ indicates a pole of order r. All real poles lie on the real axis of the s plane. For the poles shown in Fig. 4.2 the transform $F(s)$ and its partial fractions are

$$F(s) = \frac{P(s)}{Q(s)} = \frac{P(s)}{(s-s_1)^3(s-s_2)}$$

$$= \frac{A_{13}}{(s-s_1)^3} + \frac{A_{12}}{(s-s_1)^2} + \frac{A_{11}}{s-s_1} + \frac{A_2}{s-s_2} \tag{4.40}$$

The order of $Q(s)$ in this case is 4, and there are four fractions. Note that the multiple pole s_1, which is of order 3, has resulted in three fractions on the right side of Eq. (4.40). To designate the constants in the partial fractions, a single subscript is used for a first-order pole. For multiple-order poles a double-subscript notation is used. The first subscript designates the pole, and the second subscript designates the order of the pole in the partial fraction. The constants associated with first-order denominators in the partial-fraction expansion are called residues: therefore, only the constants A_{11} and A_2 are residues of Eq. (4.40).

The inverse transform of $F(s)$ (see transform pairs 7 and 8 in Appendix A) is

$$f(t) = A_{13}\frac{t^2}{2}e^{s_1t} + A_{12}te^{s_1t} + A_{11}e^{s_1t} + A_2e^{s_2t} \tag{4.41}$$

FIGURE 4.2
Location of real poles in the s plane.

For the general transform with repeated real roots,

$$F(s) = \frac{P(s)}{Q(s)} = \frac{P(s)}{(s - s_q)^r (s - s_1) \cdots}$$

$$= \frac{A_{qr}}{(s - s_q)^r} + \frac{A_{q(r-1)}}{(s - s_q)^{r-1}} + \cdots + \frac{A_{q(r-k)}}{(s - s_q)^{r-k}} + \cdots$$

$$+ \frac{A_{q1}}{s - s_q} + \frac{A_1}{s - s_1} + \cdots \tag{4.42}$$

The constant A_{qr} can be evaluated by simply multiplying both sides of Eq. (4.42) by $(s - s_q)^r$, giving

$$(s - s_q)^r F(s) = \frac{(s - s_q)^r P(s)}{Q(s)} = \frac{P(s)}{(s - s_1) \cdots}$$

$$= A_{qr} + A_{q(r-1)}(s - s_q) + \cdots$$

$$+ A_{q1}(s - s_q)^{r-1} + A_1 \frac{(s - s_q)^r}{s - s_1} + \cdots \tag{4.43}$$

Note that the factor $(s - s_q)^r$ must be divided out of the left side of the equation.

For $s = s_q$, all terms on the right side of the equation are zero except A_{qr}; therefore,

$$A_{qr} = \left[(s - s_q)^r \frac{P(s)}{Q(s)} \right]_{s = s_q} \tag{4.44}$$

Evaluation of $A_{q(r-1)}$ cannot be performed in a similar manner. Multiplying both sides of Eq. (4.42) by $(s - s_q)^{r-1}$ and letting $s = s_q$ would result in both sides being infinite, which leaves $A_{q(r-1)}$ indeterminate. If the term A_{qr} were eliminated from Eq. (4.43), $A_{q(r-1)}$ could be evaluated. This can be done by differentiating Eq. (4.43) with respect to s:

$$\frac{d}{ds} \left[(s - s_q)^r \frac{P(s)}{Q(s)} \right] = A_{q(r-1)} + 2A_{q(r-2)}(s - s_q) + \cdots \tag{4.45}$$

Letting $s = s_q$ gives

$$A_{q(r-1)} = \left\{ \frac{d}{ds} \left[(s - s_q)^r \frac{P(s)}{Q(s)} \right] \right\}_{s = s_q} \tag{4.46}$$

Repeating the differentiation gives the coefficient $A_{q(r-2)}$ as

$$A_{q(r-2)} = \left\{ \frac{1}{2} \frac{d^2}{ds^2} \left[(s - s_q)^r \frac{P(s)}{Q(s)} \right] \right\}_{s = s_q} \tag{4.47}$$

This process can be repeated until each constant is determined. A general formula for finding these coefficients associated with the repeated real pole of order r is

$$A_{q(r-k)} = \left\{ \frac{1}{k!} \frac{d^k}{ds^k} \left[(s - s_q)^r \frac{P(s)}{Q(s)} \right] \right\}_{s=s_q} \tag{4.48}$$

For the case of

$$F(s) = \frac{1}{(s+2)^3(s+3)} = \frac{A_{13}}{(s+2)^3} + \frac{A_{12}}{(s+2)^2} + \frac{A_{11}}{s+2} + \frac{A_2}{s+3} \tag{4.49}$$

the constants are

$$A_{13} = \left[(s+2)^3 F(s) \right]_{s=-2} = 1 \qquad A_{12} = \left\{ \frac{d}{ds} \left[(s+2)^3 F(s) \right] \right\}_{s=-2} = -1$$

$$A_{11} = \left\{ \frac{d^2}{2ds^2} \left[(s+2)^3 F(s) \right] \right\}_{s=-2} = 1 \qquad A_2 = \left[(s+3)F(s) \right]_{s=-3} = -1$$

and the solution as a function of time is

$$f(t) = \frac{t^2}{2} e^{-2t} - te^{-2t} + e^{-2t} - e^{-3t} \tag{4.50}$$

A recursive algorithm for computing the partial fraction expansion of rational functions having any number of multiple poles is readily programmed on a digital computer and is given in Ref. 11.

Case 3: Complex-Conjugate Poles

The position of complex poles of $F(s)$ in the s plane is shown in Fig. 4.3. Complex poles always are present in complex-conjugate pairs; their real part may be either positive or negative. For the poles shown in Fig. 4.3 the transform $F(s)$ and its partial fractions are

$$F(s) = \frac{P(s)}{Q(s)} = \frac{P(s)}{(s^2 + 2\zeta\omega_n s + \omega_n^2)(s - s_3)}$$

$$= \frac{A_1}{s - s_1} + \frac{A_2}{s - s_2} + \frac{A_3}{s - s_3}$$

$$= \frac{A_1}{s + \zeta\omega_n - j\omega_n\sqrt{1 - \zeta^2}} + \frac{A_2}{s + \zeta\omega_n + j\omega_n\sqrt{1 - \zeta^2}} + \frac{A_3}{s - s_3} \tag{4.51}$$

FIGURE 4.3
Location of complex-conjugate poles in the s plane.

The inverse transform of $F(s)$ is

$$f(t) = A_1 \exp\left[\left(-\zeta\omega_n + j\omega_n \sqrt{1 - \zeta^2}\right)t\right]$$

$$+ A_2 \exp\left[\left(-\zeta\omega_n - j\omega_n \sqrt{1 - \zeta^2}\right)t\right] + A_3 e^{s_3 t} \qquad (4.52)$$

Since the poles s_1 and s_2 are complex conjugates, and since $f(t)$ is a real quantity, the coefficients A_1 and A_2 must also be complex conjugates. Equation (4.52) can be written with the first two terms combined to a more useful damped sinusoidal form:

$$f(t) = 2|A_1|e^{-\zeta\omega_n t} \sin\left(\omega_n \sqrt{1 - \zeta^2}\, t + \phi\right) + A_3 e^{s_3 t}$$

$$= 2|A_1|e^{\sigma t} \sin(\omega_d t + \phi) + A_3 e^{s_3 t} \qquad (4.53)$$

where ϕ = angle of $A_1 + 90°$, $\omega_d = \omega_n \sqrt{1 - \zeta^2}$, and $\sigma = -\zeta\omega_n$.

The values of A_1 and A_3, as found in the manner shown previously, are

$$A_1 = [(s - s_1)F(s)]_{s=s_1} \qquad \text{and} \qquad A_3 = [(s - s_3)F(s)]_{s=s_3}$$

Since s_1 is complex, the constant A_1 is also complex. *Remember that A_1 is associated with the complex pole with the positive imaginary part.*

In Fig. 4.3 the complex poles have a negative real part, $\sigma = -\zeta\omega_n$, where the damping ratio ζ is positive. For this case the corresponding transient response is known as a damped sinusoid and is shown in Fig. 3.2. Its final value is zero. The angle η shown in Fig. 4.3 is measured from the negative real axis and is related to the damping ratio by

$$\cos \eta = \zeta$$

If the complex pole has a positive real part, the time response increases exponentially with time and the system is unstable. If the complex roots are in the right half of the s plane, the damping ratio ζ is negative. The angle η for this case is measured from the positive real axis and is given by

$$\cos \eta = |\zeta|$$

For the case of

$$F(s) = \frac{10}{(s^2 + 6s + 25)(s + 2)} = \frac{A_1}{s + 3 - j4} + \frac{A_2}{s + 3 + j4} + \frac{A_3}{s + 2} \qquad (4.54)$$

the constants are

$$A_1 = \left[(s + 3 - j4)\frac{10}{(s^2 + 6s + 25)(s + 2)}\right]_{s=-3+j4}$$

$$= \left[\frac{10}{(s + 3 + j4)(s + 2)}\right]_{s=-3+j4} = 0.303\underline{/-194°}$$

$$A_3 = \left[(s + 2)\frac{10}{(s^2 + 6s + 25)(s + 2)}\right]_{s=-2} = \left(\frac{10}{s^2 + 6s + 25}\right)_{s=-2} = 0.59$$

The solution is

$$f(t) = 0.606e^{-3t} \sin(4t - 104°) + 0.59e^{-2t} \tag{4.55}$$

The function $F(s)$ of Eq. (4.54) appears in Appendix A as transform pair 29. With the notation of Appendix A, the phase angle in the damped sinusoidal term is

$$\phi = \tan^{-1} \frac{b}{c-a} = \tan^{-1} \frac{4}{2-3} = \tan^{-1} \frac{4}{-1} = 104° \tag{4.56}$$

It is important to note that $\tan^{-1}[4/(-1)] \neq \tan^{-1}(-4/1)$. To get the correct value for the angle ϕ, it is useful to draw a sketch in order to avoid ambiguity and ensure that ϕ is evaluated correctly.

IMAGINARY POLES. The position of imaginary poles of $F(s)$ in the s plane is shown in Fig. 4.4. As the real part of the poles is zero, the poles lie on the imaginary axis. This situation is a special case of complex poles, i.e., the damping ratio $\zeta = 0$. For the poles shown in Fig. 4.4 the transform $F(s)$ and its partial fractions are

$$F(s) = \frac{P(s)}{Q(s)} = \frac{P(s)}{(s^2 + \omega_n^2)(s - s_3)} = \frac{A_1}{s - s_1} + \frac{A_2}{s - s_2} + \frac{A_3}{s - s_3} \tag{4.57}$$

The quadratic can be factored in terms of the poles s_1 and s_2; thus,

$$s^2 + \omega_n^2 = (s - j\omega_n)(s + j\omega_n) = (s - s_1)(s - s_2)$$

The inverse transform of Eq. (4.57) is

$$f(t) = A_1 e^{j\omega_n t} + A_2 e^{-j\omega_n t} + A_3 e^{s_3 t} \tag{4.58}$$

As $f(t)$ is a real quantity, the coefficients A_1 and A_2 are complex conjugates. The terms in Eq. (4.58) can be combined into the more useful form

$$f(t) = 2|A_1| \sin(\omega_n t + \phi) + A_3 e^{s_3 t} \tag{4.59}$$

Note that since there is no damping term multiplying the sinusoid, this term represents a steady-state value. The angle ϕ is given by

$$\phi = \text{angle of } A_1 + 90°$$

The values of A_1 and A_3, found in the conventional manner, are

$$A_1 = [(s - s_1)F(s)]_{s=s_1} \qquad A_3 = [(s - s_3)F(s)]_{s=s_3}$$

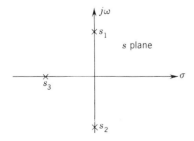

FIGURE 4.4
Poles of $F(s)$ containing imaginary conjugate poles in the s plane.

For the case where

$$F(s) = \frac{100}{(s^2 + 25)(s + 2)} = \frac{A_1}{s - j5} + \frac{A_2}{s + j5} + \frac{A_3}{s + 2} \tag{4.60}$$

the values of the coefficients are

$$A_1 = [(s - j5)F(s)]_{s = j5} = \left[\frac{100}{(s + j5)(s + 2)} \right]_{s = j5} = 1.86\underline{/-158.2°}$$

$$A_3 = [(s + 2)F(s)]_{s = -2} = \left(\frac{100}{s^2 + 25} \right)_{s = -2} = 3.45$$

The solution is

$$f(t) = 3.72 \sin (5t - 68.2°) + 3.45e^{-2t} \tag{4.61}$$

Case 4: Multiple-Order Complex Poles

Multiple-order complex-conjugate poles are rare. They can be treated in much the same fashion as repeated real poles.

4.9   MATLAB PARTIAL-FRACTION EXAMPLE

The MATLAB *RESIDUE* command is used to obtain the partial-fraction expansion of Eq. (4.54).

MATLAB Partial-Fraction m-file

```
>> Diary Part-frac
% Setup transfer function
% Numerator Constant of F(s).
k = 10;
% Poles of F(s).
p(1) = -3+j*4, p(2) = -3 -j*4, p(3) = -2;
num = k;
den = poly(p);
% Look at help.
help residue
% Use RESIDUE command to call for residues.
[r1, p1, k1] = residue(num, den);
% List residues in complex form.
r1
r1 =
      -2.9412e-001+ 7.3529e-002i
      -2.9412e-001 - 7.3529e-002i
       5.8824e-001
% List poles. This is a data entry check.
p1
```

```
p1 =
        -3.0000e+000 + 4.0000e+000i
        -3.0000e+000 - 4.0000e+000i
        -2.0000e+000
% Angle of residues in degrees.
angle(r1)*180/pi
ans =
        1.6596e+002
       -1.6596e+002
        0
% Magnitude of residues.
abs(r1)
ans =
        3.0317e-001
        3.0317e-001
        5.8824e-001
% CADAC (accuracy check)
[num1, den1] = residue(r1, p1, k1);
% Form numerator polynomial.
num1
num1 =
   0 -4.4409e-016 1.0000e+001
% Note false roots in numerator
% Form denominator polynomial.
den1
den1 =
   1.0000e+000 8.0000e+000 3.7000e+001 5.0000e+001
% Accuracy check on poles of F(s).
roots(den1)
ans =
        -3.0000e+000+ 4.0000e+000i
        -3.0000e+000 - 4.0000e+000i
        -2.0000e+000
% Time response plot is shown in Fig. 4.5.
% Plot duration of 3 seconds.
time = 0:.2:3;
impulse(num,den,t)
print
```

These MATLAB answers agree with those given previously for the solution of Eq. (4.54). The reader can use the residues to form the time response given in Eq. (4.55) by using Appendix A.

4.10 PARTIAL-FRACTION SHORTCUTS

Some shortcuts can be used to simplify the partial-fraction expansion procedures. They are most useful for transform functions that have multiple poles or complex poles. With multiple poles, the evaluation of the constants by the process of repeated differentiation given by Eq. (4.48) can be very tedious, particularly when a factor

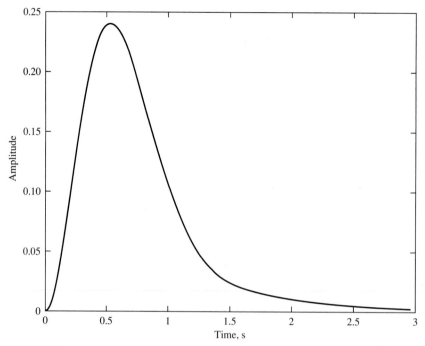

FIGURE 4.5
Response for Eq. (4.55) from MATLAB.

containing s appears in the numerator of $F(s)$. The modified procedure is to evaluate only those coefficients for real poles that can be obtained readily without differentiation. The corresponding partial fractions are then subtracted from the original function. The resultant function is easier to work with to get the additional partial fractions than the original function $F(s)$.

Example 1. For the function $F(s)$ of Eq. (4.49), which contains a pole $s_1 = -2$ of multiplicity 3, the coefficient $A_{13} = 1$ is readily found. Subtracting the associated fraction from $F(s)$ and putting the result over a common denominator yields

$$F_x(s) = \frac{1}{(s+2)^3(s+3)} - \frac{1}{(s+2)^3} = \frac{-1}{(s+2)^2(s+3)}$$

Note that the remainder $F_x(s)$ contains the pole $s_1 = -2$ with a multiplicity of only 2. Thus the coefficient A_{12} is now easily obtained from $F_x(s)$ without differentiation. The procedure is repeated to obtain all the coefficients associated with the multiple pole.

Example 2. Consider the function

$$F(s) = \frac{20}{(s^2 + 6s + 25)(s + 1)} = \frac{As + B}{s^2 + 6s + 25} + \frac{C}{s + 1} \qquad (4.62)$$

which has a pair of complex poles. Instead of partial fractions for each pole, leave the quadratic factor in one fraction. Note that the numerator of the fraction containing the quadratic is a polynomial in s and is one degree lower than the denominator. Residue C is $C = [(s + 1)F(s)]_{s=-1} = 1$. Subtracting this fraction from $F(s)$ gives

$$\frac{20}{(s^2 + 6s + 25)(s + 1)} - \frac{1}{s + 1} = \frac{-s^2 - 6s - 5}{(s^2 + 6s + 25)(s + 1)}$$

Since the pole at $s = -1$ has been removed, the numerator of the remainder must be divisible by $s + 1$. The simplified function must be equal to the remaining partial fraction; i.e.,

$$-\frac{s + 5}{s^2 + 6s + 25} = \frac{As + B}{s^2 + 6s + 25} = \frac{-(s + 5)}{(s + 3)^2 + 4^2}$$

The inverse transform for this fraction can be obtained directly from Appendix A, pair 26. The complete inverse transform of $F(s)$ is

$$f(t) = e^{-t} - 1.12e^{-3t} \sin(4t + 63.4°) \tag{4.63}$$

Leaving the complete quadratic in one fraction has resulted in real constants for the numerator. The chances for error are probably less than evaluating the residues, which are complex numbers, at the complex poles.

A rule presented by Hazony and Riley[4] is very useful for evaluating the coefficients of partial-fraction expansions. For a normalized[†] ratio of polynomials the rule is expressed as follows:

1. If the denominator is one degree higher than the numerator, the sum of the residues is 1.
2. If the denominator is two or more degrees higher than the numerator, the sum of the residues is 0.

Equation (4.32) with a_w factored from the numerator is a normalized ratio of polynomials. These rules are applied to the ratio $F(s)/a_w$. It should be noted that *residue* refers only to the coefficients of terms in a partial-fraction expansion with first-degree denominators. Coefficients of terms with higher-degree denominators are referred to only as coefficients. These rules can be used to simplify the work involved in evaluating the coefficients of partial-fraction expansions, particularly when the original function has a multiple-order pole. For example, only A_{11} and A_2 in Eq. (4.49) are residues, and therefore $A_{11} + A_2 = 0$. Since $A_2 = -1$, the value of $A_{11} = 1$ is obtained directly.

Digital-computer programs (see Appendix B) are readily available for evaluating the partial-fraction coefficients and obtaining a tabulation and plot of $f(t)$.[8]

4.11 GRAPHICAL INTERPRETATION OF PARTIAL-FRACTION COEFFICIENTS[5]

The preceding sections describe the analytical evaluation of the partial-fraction coefficients. These constants are directly related to the pole-zero pattern of the function

---

[†] A normalized polynomial is a monic polynomial, that is, the coefficient of the term containing the highest degree is equal to unity.

$F(s)$ and can be determined graphically, whether the poles and zeros are real or in complex-conjugate pairs. As long as $P(s)$ and $Q(s)$ are in factored form, the coefficients can be determined graphically by inspection. Rewriting Eq. (4.32) with the numerator and denominator in factored form and with $a_w = K$ gives

$$F(s) = \frac{P(s)}{Q(s)} = \frac{K(s - z_1)(s - z_2)\cdots(s - z_m)\cdots(s - z_w)}{(s - p_1)(s - p_2)\cdots(s - p_k)\cdots(s - p_n)}$$

$$= K\frac{\displaystyle\prod_{m=1}^{w}(s - z_m)}{\displaystyle\prod_{k=1}^{n}(s - p_k)} \tag{4.64}$$

The zeros of this function, $s = z_m$, are those values of s for which the function is zero; that is, $F(z_m) = 0$. Zeros are indicated by a small circle on the s plane. The poles of this function, $s = p_k$, are those values of s for which the function is infinite; that is, $|F(p_k)| = \infty$. Poles are indicated by a small $\times$ on the s plane. Poles are also known as singularities[†] of the function. When $F(s)$ has only simple poles (first-order poles), it can be expanded into partial fractions of the form

$$F(s) = \frac{A_1}{s - p_1} + \frac{A_2}{s - p_2} + \cdots + \frac{A_k}{s - p_k} + \cdots + \frac{A_n}{s - p_n} \tag{4.65}$$

The coefficients are obtained from Eq. (4.38) and are given by

$$A_k = [(s - p_k)F(s)]_{s = p_k} \tag{4.66}$$

The first coefficient is

$$A_1 = \frac{K(p_1 - z_1)(p_1 - z_2)\cdots(p_1 - z_w)}{(p_1 - p_2)(p_1 - p_3)\cdots(p_1 - p_n)} \tag{4.67}$$

Figure 4.6 shows the poles and zeros of a function $F(s)$. The quantity $s = p_1$ is drawn as an arrow from the origin of the s plane to the pole p_1. This arrow is called a *directed line segment*. It has a magnitude equal to $|p_1|$ and an angle of 180°. Similarly, the directed line segment $s = z_1$ is drawn as an arrow from the origin to the zero z_1 and has a corresponding magnitude and angle. A directed line segment is also drawn from the zero z_1 to the pole p_1. By the rules of vector addition, this directed line segment is equal to $p_1 - z_1$; in polar form it has a magnitude equal to its length and an angle ψ, as shown. By referring to Eq. (4.67), it is seen that $p_1 - z_1$ appears in the numerator of A_1. Whereas this quantity can be evaluated analytically,

---

[†] A singularity of a function $F(s)$ is a point where $F(s)$ does not have a derivative.

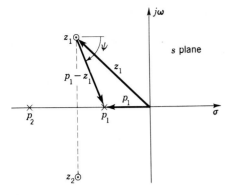

FIGURE 4.6
Directed line segments drawn on a pole-zero ·
diagram of a function $F(s)$.

it can also be measured from the pole-zero diagram. In a similar fashion, each factor in Eq. (4.67) can be obtained graphically; the directed line segments are drawn *from* the zeros and each of the other poles *to* the pole p_1. The angle from a zero is indicated by the symbol ψ, and the angle from a pole is indicated by the symbol θ.

The general rule for evaluating the coefficients in the partial-fraction expansion is quite simple. The value of A_k is the product of K and the directed distances from each zero to the pole p_k divided by the product of the directed distances from each of the other poles to the pole p_k. Each of these directed distances is characterized by a magnitude and an angle. This statement can be written in equation form as

$$A_k = K \frac{\text{product of directed distances from each zero to pole} p_k}{\text{product of directed distances from all other poles to pole} p_k}$$

$$= K \frac{\prod\limits_{m=1}^{w} |p_k - z_m| \bigg/ \sum\limits_{m=1}^{w} \psi(p_k - z_m)}{\prod\limits_{\substack{c=1 \\ c \neq k}}^{n} |p_k - p_c| \bigg/ \sum\limits_{c=1}^{n} \theta(p_k - p_c)} \qquad (4.68)$$

Equations (4.67) and (4.68) readily reveal that as the pole p_k comes closer to the zero z_m, the corresponding coefficient A_k becomes smaller (approaches a zero value). This characteristic is very important in analyzing system response characteristics. For real poles the values of A_k must be real but can be either positive or negative. A_k is positive if the total number of real poles and real zeros to the right of p_k is even, and it is negative if the total number of real poles and real zeros to the right of p_k is odd. For complex poles the values of A_k are complex. If $F(s)$ has no finite zeros, the numerator of Eq. (4.68) is equal to K. A repeated factor of the form $(s + a)^r$ in either the numerator or the denominator is included r times in Eq. (4.68), where the values of $|s + a|$ and $\angle s + a$ are obtained graphically. The application of the graphical technique is illustrated by the following example:

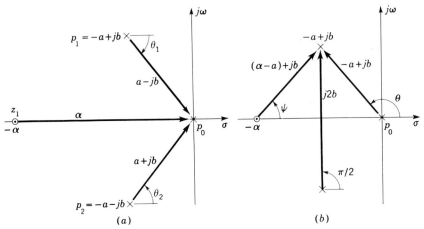

FIGURE 4.7
Directed line segments for evaluating (a) A_0 of Eq. (4.70) and (b) A_1 of Eq. (4.71).

Example. A function $F(s)$ with complex poles (step function response):

$$F(s) = \frac{K(s + \alpha)}{s\left[(s + a)^2 + b^2\right]} = \frac{K(s + \alpha)}{s(s + a - jb)(s + a + jb)}$$

$$= \frac{K(s - z_1)}{s(s - p_1)(s - p_2)} = \frac{A_0}{s} + \frac{A_1}{s + a - jb} + \frac{A_2}{s + a + jb} \qquad (4.69)$$

The coefficient A_0, for the pole $p_0 = 0$, is obtained by use of the directed line segments shown in Fig. 4.7a:

$$A_0 = \frac{K(\alpha)}{(a - jb)(a + jb)} = \frac{K\alpha}{\sqrt{a^2 + b^2}\, e^{j\theta_1} \sqrt{a^2 + b^2}\, e^{j\theta_2}} = \frac{K\alpha}{a^2 + b^2} \qquad (4.70)$$

Note that the angles θ_1 and θ_2 are equal in magnitude but opposite in sign. The coefficient for a real pole is therefore always a real number.

The coefficient A_1, for the pole $p_1 = -a + jb$, is obtained by use of the directed line segments shown in Fig. 4.7b:

$$A_1 = \frac{K[(\alpha - a) + jb]}{(-a + jb)(j2b)} = \frac{K\sqrt{(\alpha - a)^2 + b^2}\, e^{j\psi}}{\sqrt{a^2 + b^2}\, e^{j\theta} 2b e^{j\pi/2}}$$

$$= \frac{K}{2b} \sqrt{\frac{(\alpha - a)^2 + b^2}{a^2 + b^2}}\, e^{j(\psi - \theta - \pi/2)} \qquad (4.71)$$

where $\qquad \psi = \tan^{-1} \frac{b}{\alpha - a} \qquad$ and $\qquad \theta = \tan^{-1} \frac{b}{-a}$

Since the constants A_1 and A_2 are complex conjugates, it is not necessary to evaluate A_2. It should be noted that zeros in $F(s)$ affect both the magnitude and the angle of the coefficients in the time function.

The response as a function of time is therefore

$$f(t) = A_0 + 2|A_1|e^{-at} \sin\left(bt + \underline{/A_1} + 90°\right)$$

$$= \frac{K\alpha}{a^2 + b^2} + \frac{K}{b}\sqrt{\frac{(\alpha - a)^2 + b^2}{a^2 + b^2}}\, e^{-at} \sin(bt + \phi) \tag{4.72}$$

where $\quad \phi = \psi - \theta = \tan^{-1}\dfrac{b}{\alpha - a} - \tan^{-1}\dfrac{b}{-a} = $ angle of $A_1 + 90°$ $\quad$ (4.73)

This agrees with transform pair 28 in Appendix A. A_0 and A_1 can be evaluated by measuring the directed line segments directly on the pole-zero diagram. It is interesting to note that, for a given ζ, as the magnitude $\omega_n = \sqrt{a^2 + b^2}$ becomes larger, the coefficients A_0 and A_1 approach zero as revealed by an analysis of Eqs. (4.70) and (4.71), respectively. (The limiting value is $\lim_{\omega_n \to \infty}[A_1] = K/2b$.) This is an important characteristic when it is desired that $f(t)$ be very small. The application of this characteristic is used in disturbance rejection (Chapter 12). Also see Prob. 4.17 for further illustration.

4.12 FREQUENCY RESPONSE FROM THE POLE-ZERO DIAGRAM

The frequency response of a system is described as the steady-state response with a sine-wave forcing function for all values of frequency. This information is often presented graphically, using two curves. One curve shows the ratio of output amplitude to input amplitude M and the other curve shows the phase angle of the output α, where both are plotted as a function of frequency, often on a logarithmic scale.

Consider the input to a system as sinusoidal and given by

$$x_1(t) = X_1 \sin \omega t \tag{4.74}$$

and the output is

$$x_2(t) = X_2 \sin(\omega t + \alpha) \tag{4.75}$$

The magnitude of the frequency response for the function given in Eq. (4.64) is

$$M = \left|\frac{X_2(j\omega)}{X_1(j\omega)}\right| = \left|\frac{P(j\omega)}{Q(j\omega)}\right| = \left|\frac{K(j\omega - z_1)(j\omega - z_2)\cdots}{(j\omega - p_1)(j\omega - p_2)\cdots}\right| \tag{4.76}$$

and the angle is

$$\alpha = \underline{/P(j\omega)} - \underline{/Q(j\omega)}$$

$$= \underline{/K} + \underline{/j\omega - z_1} + \underline{/j\omega - z_2} + \cdots - \underline{/j\omega - p_1} - \underline{/j\omega - p_2} - \cdots \tag{4.77}$$

Figure 4.8 shows a pole-zero diagram and the directed line segments for evaluating M and α corresponding to a frequency ω_1. As ω increases, each of the directed

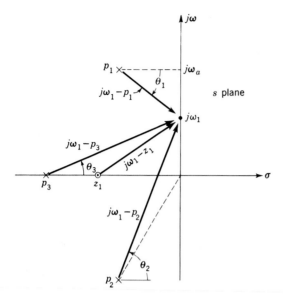

FIGURE 4.8
Pole-zero diagram of Eq. (4.76) showing the directed line segments for evaluating the frequency response $M\underline{/\alpha}$.

line segments changes in both magnitude and angle. Several characteristics can be noted by observing the change in the magnitude and angle of each directed line segment as the frequency ω increases from 0 to ∞. The magnitude and angle of the frequency response can be obtained graphically from the pole-zero diagram. The magnitude M and the angle α are a composite of all the effects of all the poles and zeros. In particular, for a system function that has all poles and zeros in the left half of the s plane, the following characteristics of the frequency response are noted:

1. At $\omega = 0$ the magnitude is a finite value, and the angle is $0°$.
2. If there are more poles than zeros, then as $\omega \to \infty$, the magnitude of M approaches zero and the angle is $-90°$ times the difference between the number of poles and zeros.
3. The magnitude can have a peak value M_m only if there are complex poles fairly close to the imaginary axis. It is shown later that the damping ratio ζ of the complex poles must be less than 0.707 to obtain a peak value for M. Of course, the presence of zeros could counteract the effect of the complex poles, and so it is possible that no peak would be present even if ζ of the poles were less than 0.707.

Typical frequency-response characteristics are illustrated for two common functions. As a first example,

$$\frac{X_2(s)}{X_1(s)} = \frac{P(s)}{Q(s)} = \frac{a}{s + a} \qquad (4.78)$$

The frequency response shown in Fig. 4.9b and c starts with a magnitude of unity and an angle of $0°$. As ω increases, the magnitude decreases and approaches zero while

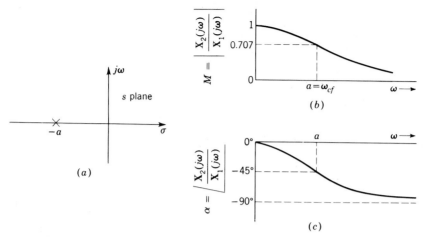

FIGURE 4.9
(a) Pole location of Eq. (4.78); (b) M vs. ω; (c) α vs. ω.

the angle approaches $-90°$. At $\omega = a$ the magnitude is 0.707 and the angle is $-45°$. The frequency $\omega = a$ is called the *corner frequency* ω_{cf} or the break frequency. Figure 4.9 shows the pole location, the frequency-response magnitude M, and the frequency-response angle α as a function of ω.

The second example is a system transform with a pair of conjugate poles:

$$\frac{X_2(s)}{X_1(s)} = \frac{\omega_n^2}{s^2 + 2\zeta\omega_n s + \omega_n^2} = \frac{\omega_n^2}{(s - p_1)(s - p_2)}$$

$$= \frac{\omega_n^2}{\left(s + \zeta\omega_n - j\omega_n\sqrt{1 - \zeta^2}\right)\left(s + \zeta\omega_n + j\omega_n\sqrt{1 - \zeta^2}\right)} \tag{4.79}$$

The location of the poles of Eq. (4.79) and the directed line segments for $s = j\omega$ are shown in Fig. 4.10a. At $\omega = 0$ the values are $M = 1$ and $\alpha = 0°$. As ω increases, the magnitude M first increases, because $j\omega - p_1$ is decreasing faster than $j\omega - p_2$ is increasing. By differentiating the magnitude of $|M(j\omega)|$, obtained from Eq. (4.79), with respect to ω, it can be shown (see Sec. 9.3) that the maximum value M_m and the frequency ω_m at which it occurs are given by

$$M_m = \frac{1}{2\zeta\sqrt{1 - \zeta^2}} \tag{4.80}$$

$$\omega_m = \omega_n\sqrt{1 - 2\zeta^2} \tag{4.81}$$

A circle drawn on the s plane in Fig. 4.10a, with the poles p_1 and p_2 as the end-points of the diameter, intersects the imaginary axis at the value ω_m given by Eq. (4.81). Both Eq. (4.81) and the geometrical construction of the circle show that the M curve has a maximum value, other than at $\omega = 0$, only if $\zeta < 0.707$. The angle α becomes more negative as ω increases. At $\omega = \omega_n$ the angle is equal to $-90°$. This

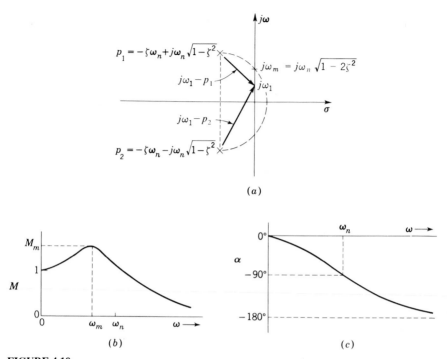

FIGURE 4.10
(a) Pole-zero diagram, $\zeta < 0.707$; (b) magnitude of frequency response; (c) angle of frequency response.

is the corner frequency for a quadratic factor with complex roots. As $\omega \to \infty$, the angle α approaches $-180°$. The magnitude and angle of the frequency response for $\zeta < 0.707$ are shown in Fig. 4.10. The M and α curves are continuous for all linear systems. The M vs. ω plot of Fig. 4.10b describes the frequency response characteristics for the undamped simple second-order system of Eq. (4.79). In later chapters this characteristic is used as the standard reference for analyzing and synthesizing higher-order systems.

4.13 LOCATION OF POLES AND STABILITY

The stability and the corresponding response of a system can be determined from the location of the poles of the response transform $F(s)$ in the s plane. The possible positions of the poles are shown in Fig. 4.11, and the responses are given in Table 4.1. These poles are the roots of the characteristic equation.

Poles of the response transform at the origin or on the imaginary axis that are not contributed by the forcing function result in a continuous output. These outputs are undesirable in a control system. Poles in the right-half s plane result in transient terms that increase with time. Such performance characterizes an unstable system; therefore, poles in the right-half s plane are not desirable, in general.

FIGURE 4.11
Location of poles in the s plane. (Numbers are used to identify the poles.)

4.14 LAPLACE TRANSFORM OF THE IMPULSE FUNCTION

Figure 4.12 shows a rectangular pulse of amplitude $1/a$ and of duration a. The analytical expression for this pulse is

$$f(t) = \frac{u_{-1}(t) - u_{-1}(t - a)}{a} \tag{4.82}$$

and its Laplace transform is

$$F(s) = \frac{1 - e^{-as}}{as} \tag{4.83}$$

If a is decreased, the amplitude increases and the duration of the pulse decreases but the area under the pulse, and thus its strength, remains unity. The limit of $f(t)$ as $a \to 0$ is termed a unit impulse and is designated by $\delta(t) = u_0(t)$. The Laplace transform of the unit impulse, as evaluated by use of L'Hospital's theorem, is $\Delta(s) = 1$.

When any function has a jump discontinuity, the derivative of the function has an impulse at the discontinuity. Some systems are subjected to shock inputs. For example, an airplane in flight may be jolted by a gust of wind of short duration. When a gun is fired, it has a reaction force of large magnitude and very short duration, and so does a steel ball bouncing off a steel plate. When the duration of a disturbance or input to a system is very short compared with the natural periods of the system, the

TABLE 4.1
Relation of response to location of poles

Position of pole	Form of response	Characteristics
1	Ae^{-at}	Damped exponential
2–2*	$Ae^{-bt} \sin(ct + \phi)$	Exponentially damped sinusoid
3	A	Constant
4–4*	$A \sin(dt + \phi)$	Constant-sinusoid
5	Ae^{et}	Increasing exponential (unstable)
6–6*	$Ae^{ft} \sin(gt + \phi)$	Exponentially increasing sinusoid (unstable)

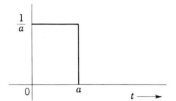

FIGURE 4.12
Rectangular pulse of unit area.

response can often be well approximated by considering the input to be an impulse of proper strength. An impulse of infinite magnitude and zero duration does not occur in nature; but if the pulse duration is much smaller than the time constants of the system, a representation of the input by an impulse is a good approximation. The shape of the impulse is unimportant as long as the strength of the equivalent impulse is equal to the strength of the actual pulse.

Since the Laplace transform of an impulse is defined, the approximate response of a system to a pulse is often found more easily by this method than by the classical method. Consider the RLC circuit shown in Fig. 4.13 when the input voltage is an impulse. The problem is to find the voltage e_c across the capacitor as a function of time. The differential equation relating e_c to the input voltage e and the impedances is written by the node method:

$$\left(CD + \frac{1}{R + LD}\right)e_c - \frac{1}{R + LD}e = 0 \tag{4.84}$$

Rationalizing the equation yields

$$(LCD^2 + RCD + 1)e_c = e \tag{4.85}$$

Taking the Laplace transform of this equation gives

$$(s^2LC + sRC + 1)E_c(s) - sLCe_c(0) - LCDe_c(0) - RCe_c(0) = E(s) \tag{4.86}$$

Consider an initial condition $e_c(0^-) = 0$. The voltage across the capacitor is related to the current in the series circuit by the expression $De_c = i/C$. Since the current cannot change instantaneously through an inductor, $De_c(0^-) = 0$. These initial conditions that exist at $t = 0^-$ are used in the Laplace-transformed equation when the forcing function is an impulse. The solution for the current shows a discontinuous step change at $t = 0$.

The input voltage e is an impulse; therefore, $E(s) = 1$. Solving for $E_c(s)$ gives

$$E_c(s) = \frac{1/LC}{s^2 + (R/L)s + 1/LC} = \frac{\omega_n^2}{s^2 + 2\zeta\omega_n s + \omega_n^2} \tag{4.87}$$

FIGURE 4.13
RLC series circuit.

Depending on the relative sizes of the parameters, the circuit may be overdamped or underdamped. In the overdamped case the voltage $e_c(t)$ rises to a peak value and then decays to zero. In the underdamped case the voltage oscillates with a frequency $\omega_d = \omega_n \sqrt{1 - \zeta^2}$ and eventually decays to zero. The form of the voltage $e_c(t)$ for several values of damping is shown in Sec. 4.15.

The stability of a system is revealed by the impulse response. With an impulse input the response transform contains only poles and zeros contributed by the system parameters. The impulse response is therefore the inverse transform of the system transfer function and is called the *weighting function:*

$$g(t) = \mathcal{L}^{-1}[G(s)] \tag{4.88}$$

The impulse response provides one method of evaluating the system function; although the method is not simple, it is possible to take the impulse response, which is a function of time, and convert it into the transfer function $G(s)$, which is a function of s.

The system response $g(t)$ to an impulse input $\delta(t)$ can be used to determine the response $c(t)$ to any input $r(t)$.[6] The response $c(t)$ is determined by the use of the convolution or superposition integral

$$c(t) = \int_0^t r(\tau)g(t - \tau)\,d\tau \tag{4.89}$$

This method is normally used for inputs that are not sinusoidal or a power series.

4.15   SECOND-ORDER SYSTEM WITH IMPULSE EXCITATION

The response of a second-order system to a step-function input is shown in Fig. 3.6. The impulse function is the derivative of the step function; therefore, the response to an impulse is the derivative of the response to a step function. Figure 4.14 shows the response for several values of damping ratio ζ for the second-order function of Eq. (4.87).

The first zero of the impulse response occurs at t_p, which is the time at which the maximum value of the step-function response occurs.

For the underdamped case the impulse response is

$$f(t) = \frac{\omega_n}{\sqrt{1 - \zeta^2}} e^{-\zeta \omega_n t} \sin \omega_n \sqrt{1 - \zeta^2}\, t \tag{4.90}$$

By setting the derivative of $f(t)$ with respect to time equal to zero, the maximum overshoot can be shown to occur at

$$t_m = \frac{\cos^{-1} \zeta}{\omega_n \sqrt{1 - \zeta^2}} \tag{4.91}$$

The maximum value of $f(t)$ is

$$f(t_m) = \omega_n \exp\left(-\frac{\zeta \cos^{-1} \zeta}{\sqrt{1 - \zeta^2}}\right) \tag{4.92}$$

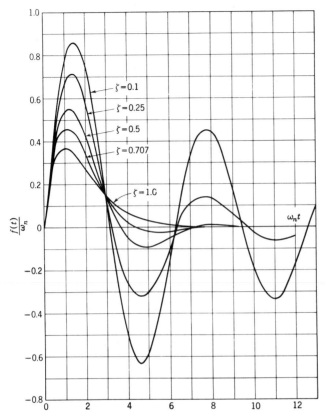

FIGURE 4.14
Response to an impulse for $F(s) = \omega_n^2/(s^2 + 2\zeta\omega_n s + \omega_n^2)$.

The impulse response $g(t)$ can be measured after applying an input pulse that has a short duration compared with the largest system time constant. Then it can be used in a deconvolution technique to determine the transfer function $G(s)$. It can also be used to determine the disturbance-rejection characteristics of control systems.

4.16 ADDITIONAL MATRIX OPERATIONS AND PROPERTIES

Additional characteristics of matrices required to solve matrix state equations by using the Laplace transform are presented in this section.

PRINCIPAL DIAGONAL.   The principal diagonal of a square matrix $\mathbf{M} = [m_{ij}]$ consists of the elements m_{ii}.

DIAGONAL MATRIX.   A diagonal matrix is a square matrix in which all the elements off the principal diagonal are zero. This fact can be expressed as $m_{ij} = 0$

for $i \neq j$ and the principal diagonal elements $m_{ii} = m_i$ are not all zero. When the diagonal elements are all equal, the matrix is called a *scalar* matrix.

TRACE. The *trace* of a square matrix $\mathbf{M}$ of order n is the sum of all the elements along the principal diagonal; that is,

$$\text{trace } \mathbf{M} = \sum_{i=1}^{n} m_{ii}$$

DETERMINANT. The determinant of a square matrix $\mathbf{M}$ of order n is the sum of all possible signed products of n elements containing one and only one element from every row and column in the matrix. The determinant of $\mathbf{M}$ may be denoted by $|\mathbf{M}|$, or det $\mathbf{M}$, or Δ_M. It is assumed that the reader knows how to evaluate determinants.

Example 1.

$$\mathbf{M} = \begin{bmatrix} 3 & 1 & 2 \\ 1 & 0 & -4 \\ 0 & 5 & 7 \end{bmatrix} \qquad |\mathbf{M}| = \begin{vmatrix} 3 & 1 & 2 \\ 1 & 0 & -4 \\ 0 & 5 & 7 \end{vmatrix} = 63$$

Characteristics of determinants are as follows:

1. The determinant of a unit matrix is $|\mathbf{I}| = 1$.

2. The determinant of a matrix is zero if (*a*) any row or column contains all zeros or (*b*) the elements of any two rows (or columns) have a common ratio.

SINGULAR MATRIX. A square matrix is said to be *singular* if the value of its determinant is zero. If the value is not zero, the matrix is *nonsingular.*

MINOR. The minor M_{ij} of a square matrix $\mathbf{M}$ of order n is the determinant formed after the ith row and jth column are deleted from $\mathbf{M}$.

PRINCIPAL MINOR. A principal minor is a minor M_{ii} whose diagonal elements are also the diagonal elements of the square matrix $\mathbf{M}$.

COFACTOR. A cofactor is a *signed* minor and is given by

$$C_{ij} = \Delta_{ij} = (-1)^{i+j} M_{ij} \qquad (4.93)$$

Example 2. For the matrix $\mathbf{M}$ of Example 1, the cofactor C_{21} is obtained by deleting the second row and first column, giving

$$C_{21} = (-1)^{2+1} \begin{vmatrix} 1 & 2 \\ 5 & 7 \end{vmatrix} = +3$$

ADJOINT MATRIX. The adjoint of square matrix $\mathbf{M}$, denoted as adj $\mathbf{M}$, is the transpose of the cofactor matrix. The cofactor matrix is formed by replacing each element of $\mathbf{M}$ by its cofactor.

$$\text{adj } \mathbf{M} = [C_{ji}] = [C_{ij}]^T = \begin{bmatrix} \text{array of} \\ \text{cofactors} \end{bmatrix}^T \tag{4.94}$$

Example 3. The adj $\mathbf{M}$ for the matrix $\mathbf{M}$ of Example 1 is

$$\text{adj } \mathbf{M} = \begin{bmatrix} 20 & -7 & 5 \\ 3 & 21 & -15 \\ -4 & 14 & -1 \end{bmatrix}^T = \begin{bmatrix} 20 & 3 & -4 \\ -7 & 21 & 14 \\ 5 & -15 & -1 \end{bmatrix}$$

INVERSE MATRIX. The product of a matrix and its adjoint is a scalar matrix, as illustrated by the following example.

Example 4. The product $\mathbf{M}$ adj $\mathbf{M}$, using values in Examples 1 and 3, is

$$\mathbf{M} \text{ adj } \mathbf{M} = \begin{bmatrix} 3 & 1 & 2 \\ 1 & 0 & -4 \\ 0 & 5 & 7 \end{bmatrix} \begin{bmatrix} 20 & 3 & -4 \\ -7 & 21 & 14 \\ 5 & -15 & -1 \end{bmatrix}$$

$$= \begin{bmatrix} 63 & 0 & 0 \\ 0 & 63 & 0 \\ 0 & 0 & 63 \end{bmatrix} = 63 \begin{bmatrix} 1 & 0 & 0 \\ 0 & 1 & 0 \\ 0 & 0 & 1 \end{bmatrix} = |\mathbf{M}| \, \mathbf{I} \tag{4.95}$$

The inverse of a square matrix $\mathbf{M}$ is denoted by $\mathbf{M}^{-1}$ and has the property

$$\mathbf{M}\mathbf{M}^{-1} = \mathbf{M}^{-1}\mathbf{M} = \mathbf{I} \tag{4.96}$$

The inverse matrix $\mathbf{M}^{-1}$ is defined from the result of Example 4 as

$$\mathbf{M}^{-1} = \frac{\text{adj } \mathbf{M}}{|\mathbf{M}|} \tag{4.97}$$

The inverse exists only if $|\mathbf{M}| \neq 0$, that is, the matrix $\mathbf{M}$ is nonsingular.

INVERSE OF A PRODUCT OF MATRICES. The inverse of a product of matrices is equal to the product of the inverses of the individual matrices in reverse order:

$$[\mathbf{A}\mathbf{B}]^{-1} = \mathbf{B}^{-1}\mathbf{A}^{-1}$$

PRODUCT OF DETERMINANTS. The determinant of the product of square matrices is equal to the product of the individual determinants:

$$|\mathbf{A}\mathbf{B}\mathbf{C}| = |\mathbf{A}| \cdot |\mathbf{B}| \cdot |\mathbf{C}| \tag{4.98}$$

RANK OF A MATRIX. The rank r of a matrix $\mathbf{M}$, not necessarily square, is the order of the largest square array contained in $\mathbf{M}$ that has a nonzero determinant.

Example 5.

$$\mathbf{M} = \begin{bmatrix} 1 & 2 & 3 \\ 2 & 3 & 4 \\ 3 & 5 & 7 \end{bmatrix}$$

The determinant $|\mathbf{M}| = 0$; that is, $\mathbf{M}$ is a singular matrix. The square array obtained by deleting the first row and second column has a nonzero determinant:

$$\begin{vmatrix} 2 & 4 \\ 3 & 7 \end{vmatrix} = 2$$

Therefore, $\mathbf{M}$ has a rank of 2.

DEGENERACY (OR NULLITY) OF A MATRIX. When a matrix $\mathbf{M}$ of order n has rank r, there are $q = n - r$ rows or columns that are linear combinations of the r rows or columns. The matrix $\mathbf{M}$ is then said to have *degeneracy q*. In Example 5 the matrix $\mathbf{M}$ is of order $n = 3$, the rank is $r = 2$, and the degeneracy is $q = 1$ (simple degeneracy). Note that the third row of $\mathbf{M}$ is the sum of the first two rows. Also, the third column is two times the second column minus the first column.

SYMMETRIC MATRIX. A square matrix containing only real elements is symmetric if it is equal to its transpose, that is, $\mathbf{A} = \mathbf{A}^T$. The relationship between the elements is

$$a_{ij} = a_{ji}$$

A symmetric matrix is symmetrical about its principal diagonal.

TRANSPOSE OF A PRODUCT OF MATRICES. The transpose of a product of matrices is the product of the transposed matrices in reverse order:

$$(\mathbf{AB})^T = \mathbf{B}^T \mathbf{A}^T \tag{4.99}$$

This relationship can be described by the statement that the product of transposed matrices is equal to the transpose of the product in reverse order of the original matrices.

HERMITE NORMAL FORM. A number of properties of a matrix can be determined by transforming the matrix to *Hermite normal form* (HNF). These properties include the determination of the rank of a matrix, the inverse of a matrix of full rank, the characteristic polynomial of a matrix, the eigenvectors associated with the eigenvalue, the controllability of a system, and the observability of a system. The transformation is accomplished on a matrix $\mathbf{M}$ by performing the following basic or *elementary* operations:

1. Interchanging the ith and jth rows
2. Multiplying the ith row by a nonzero scalar k
3. Adding to the elements of the ith row the corresponding elements of the jth row multiplied by a scalar k

The purpose of these operations is to produce a unit element as the first nonzero element in each row (from the left), with the remaining elements of the column all zero. These operations can be accomplished by premultiplying by *elementary matrices,* where the elementary matrices are formed by performing the desired basic operations on a unit matrix. For example, for a third-order matrix **M**, premultiplying by the following elementary matrices $\mathbf{E}_i$ produces the results indicated:

$$
\begin{bmatrix} 0 & 1 & 0 \\ 1 & 0 & 0 \\ 0 & 0 & 1 \end{bmatrix}
\begin{bmatrix} 1 & 0 & 0 \\ 0 & 1 & 0 \\ 0 & 0 & k \end{bmatrix}
\begin{bmatrix} 1 & 0 & 0 \\ 0 & 1 & k \\ 0 & 0 & 1 \end{bmatrix}
\tag{4.100}
$$

$$
\begin{array}{ccc}
\text{Interchanging} & \text{Multiplying} & \text{Adding } k \text{ times} \\
\text{rows 1 and 2} & \text{row 3 by } k & \text{row 3 to row 2}
\end{array}
$$

Example 6. The matrix of Example 5 is reduced to HNF by the following operations:

1. Subtract 2 times row 1 from row 2 and 3 times row 1 from row 3. Premultiplying by the corresponding elementary matrix $\mathbf{E}_1$ produces

$$
\mathbf{E}_1\mathbf{M} = \begin{bmatrix} 1 & 0 & 0 \\ -2 & 1 & 0 \\ -3 & 0 & 1 \end{bmatrix}\begin{bmatrix} 1 & 2 & 3 \\ 2 & 3 & 4 \\ 3 & 5 & 7 \end{bmatrix} = \begin{bmatrix} 1 & 2 & 3 \\ 0 & -1 & -2 \\ 0 & -1 & -2 \end{bmatrix} = \mathbf{M}_1
$$

2. Add 2 times row 2 to row 1, subtract row 2 from row 3, and multiply row 2 by -1. Premultiplying by the corresponding elementary matrix $\mathbf{E}_2$ produces

$$
\mathbf{E}_2\,\mathbf{M}_1 = \begin{bmatrix} 1 & 2 & 0 \\ 0 & -1 & 0 \\ 0 & -1 & 1 \end{bmatrix}\begin{bmatrix} 1 & 2 & 3 \\ 0 & -1 & -2 \\ 0 & -1 & -2 \end{bmatrix} = \begin{bmatrix} 1 & 0 & -1 \\ 0 & 1 & 2 \\ 0 & 0 & 0 \end{bmatrix} = \mathbf{M}_2
$$

The matrix $\mathbf{M}_2$ is in Hermite normal form and has the same rank as the matrix **M**. Since $\mathbf{M}_2$ has one zero row, it has a rank of 2, which agrees with the result of Example 5.

MATRIX INVERSION BY ROW OPERATIONS.   The inverse $\mathbf{M}^{-1}$ of a nonsingular matrix **M** can be obtained as follows:

1. Form the augmented matrix $[\,\mathbf{M} \mid \mathbf{I}\,]$.
2. Perform row operations on the augmented matrix so that the left portion **M** becomes the identity matrix **I**. When the left portion is so transformed, the right portion **I** is transformed into the matrix inverse $\mathbf{M}^{-1}$. Thus, the transformed augmented matrix becomes $[\,\mathbf{I} \mid \mathbf{M}^{-1}\,]$.

Example 7. For the matrix of Example 1, the augmented matrix is

$$
[\,\mathbf{M}\mid\mathbf{I}\,] = \begin{bmatrix} 3 & 1 & 2 & \vdots & 1 & 0 & 0 \\ 1 & 0 & -4 & \vdots & 0 & 1 & 0 \\ 0 & 5 & 7 & \vdots & 0 & 0 & 1 \end{bmatrix}
$$

A suitable set of basic row operations includes subtracting 3 times row 2 from row 1 and then interchanging rows 1 and 2, yielding

$$\left[\begin{array}{ccc|ccc} 1 & 0 & -4 & 0 & 1 & 0 \\ 0 & 1 & 14 & 1 & -3 & 0 \\ 0 & 5 & 7 & 0 & 0 & 1 \end{array}\right]$$

Subtracting 5 times row 2 from row 3 yields

$$\left[\begin{array}{ccc|ccc} 1 & 0 & -4 & 0 & 1 & 0 \\ 0 & 1 & 14 & 1 & -3 & 0 \\ 0 & 0 & -63 & -5 & 15 & 1 \end{array}\right]$$

Dividing row 3 by -63, adding 4 times the new row 3 to row 1, and subtracting 14 times the new row 3 from row 2 yields

$$\left[\begin{array}{ccc|ccc} 1 & 0 & 0 & 20/63 & 3/63 & -4/63 \\ 0 & 1 & 0 & -7/63 & 21/63 & 14/63 \\ 0 & 0 & 1 & 5/63 & -15/63 & -1/63 \end{array}\right] = [\,\mathbf{I} \,\vdots\, \mathbf{M}^{-1}\,]$$

The inverse matrix $\mathbf{M}^{-1}$ agrees with the value obtained from Eq. (4.97) and the results of Examples 1 and 3.

EVALUATION OF THE CHARACTERISTIC POLYNOMIAL. The characteristic polynomial for the matrix $\mathbf{A}$ appearing in the state equation

$$\dot{\mathbf{x}} = \mathbf{A}\mathbf{x} + \mathbf{b}u \tag{4.101}$$

is

$$Q(\lambda) = |\lambda\mathbf{I} - \mathbf{A}| = \lambda^n + a_{n-1}\lambda^{n-1} + \cdots + a_1\lambda + a_0 \tag{4.102}$$

When the $n \times (n + 1)$ matrix

$$\mathbf{M}_c = [\mathbf{b} \quad \mathbf{A}\mathbf{b} \quad \mathbf{A}^2\mathbf{b} \quad \cdots \quad \mathbf{A}^n\mathbf{b}] \tag{4.103}$$

is reduced to HNF, which must be full rank, the coefficients of the characteristic polynomial appear in the last column,[10] i.e.,

$$\left[\begin{array}{ccccccc} 1 & 0 & 0 & \cdots & 0 & 0 & -a_0 \\ 0 & 1 & 0 & \cdots & 0 & 0 & -a_1 \\ \vdots & \vdots & \vdots & \ddots & \vdots & \vdots & \vdots \\ 0 & 0 & 0 & \cdots & 0 & 1 & -a_{n-1} \end{array}\right] \tag{4.104}$$

This procedure may be simpler than forming the determinant $Q(\lambda) = |\lambda\mathbf{I} - \mathbf{A}|$.

Example 8.

$$\mathbf{A} = \left[\begin{array}{ccc} 1 & 6 & -3 \\ -1 & -1 & 1 \\ -2 & 2 & 0 \end{array}\right] \qquad \mathbf{b} = \left[\begin{array}{c} 1 \\ 1 \\ 1 \end{array}\right]$$

$$\mathbf{M}_c = \left[\begin{array}{cccc} 1 & 4 & -2 & 10 \\ 1 & -1 & -3 & -5 \\ 1 & 0 & -10 & -2 \end{array}\right] \tag{4.105}$$

Reducing $\mathbf{M}_c$ to HNF yields

$$\begin{bmatrix} 1 & 0 & 0 & -2 \\ 0 & 1 & 0 & 3 \\ 0 & 0 & 1 & 0 \end{bmatrix} \begin{matrix} \leftarrow -a_0 \\ \leftarrow -a_1 \\ \leftarrow -a_2 \end{matrix}$$

The characteristic polynomial is therefore

$$Q(\lambda) = \lambda^3 - 3\lambda + 2$$

LINEAR INDEPENDENCE.   A set of vectors $\mathbf{v}_1, \mathbf{v}_2, \ldots, \mathbf{v}_k, \ldots, \mathbf{v}_n$ is linearly independent provided there is no set of scalars $\alpha_i(i = 1, 2, \ldots, n)$, not all zero, that satisfies

$$\alpha_1 \mathbf{v}_1 + \cdots + \alpha_k \mathbf{v}_k + \cdots + \alpha_n \mathbf{v}_n = 0 \tag{4.106}$$

If this equation is satisfied with the α_i not all zero, then the vectors are *linearly dependent*. Linear independence can be checked by forming the matrix

$$\begin{bmatrix} \mathbf{v}_1^T \\ \mathbf{v}_2^T \\ \vdots \\ \mathbf{v}_n^T \end{bmatrix}$$

If this matrix has rank n, the vectors are linearly independent; otherwise they are linearly dependent. The rank of this matrix can be determined by reducing it to Hermite normal form.

4.17   SOLUTION OF STATE EQUATION[7,10]

The homogeneous state equation is

$$\dot{\mathbf{x}} = \mathbf{A}\mathbf{x} \tag{4.107}$$

The solution of this equation can be obtained by taking the Laplace transform

$$s\mathbf{X}(s) - \mathbf{x}(0) = \mathbf{A}\mathbf{X}(s)$$

Grouping of the terms containing $\mathbf{X}(s)$ yields

$$[s\mathbf{I} - \mathbf{A}]\mathbf{X}(s) = \mathbf{x}(0)$$

The unit matrix $\mathbf{I}$ is introduced so that all terms in the equations are proper matrices. Premultiplying both sides of the equation by $[s\mathbf{I} - \mathbf{A}]^{-1}$ yields

$$\mathbf{X}(s) = [s\mathbf{I} - \mathbf{A}]^{-1}\mathbf{x}(0) \tag{4.108}$$

The inverse Laplace transform of Eq. (4.108) gives the time response

$$\mathbf{x}(t) = \mathcal{L}^{-1}\left\{[s\mathbf{I} - \mathbf{A}]^{-1}\right\}\mathbf{x}(0) \tag{4.109}$$

Comparing this solution with Eq. (3.70) yields the following expression for the state transition matrix:

$$\Phi(t) = \mathcal{L}^{-1}[s\mathbf{I} - \mathbf{A}]^{-1} \tag{4.110}$$

The *resolvent matrix* is designated by $\Phi(s)$ and is defined by[†]

$$\Phi(s) = [s\mathbf{I} - \mathbf{A}]^{-1} \tag{4.111}$$

Example 1. Solve for $\Phi(t)$ (see the example of Sec. 3.13) for

$$\mathbf{A} = \begin{bmatrix} 0 & 6 \\ -1 & -5 \end{bmatrix}$$

Using Eq. (4.111) gives

$$\Phi(s) = \left[\begin{bmatrix} s & 0 \\ 0 & s \end{bmatrix} - \begin{bmatrix} 0 & 6 \\ -1 & -5 \end{bmatrix}\right]^{-1} = \begin{bmatrix} s & -6 \\ 1 & s+5 \end{bmatrix}^{-1}$$

Using Eq. (4.97) gives

$$\Phi(s) = \frac{1}{s^2 + 5s + 6}\begin{bmatrix} s+5 & 6 \\ -1 & s \end{bmatrix} = \begin{bmatrix} \dfrac{s+5}{(s+2)(s+3)} & \dfrac{6}{(s+2)(s+3)} \\ \dfrac{-1}{(s+2)(s+3)} & \dfrac{s}{(s+2)(s+3)} \end{bmatrix} \tag{4.112}$$

Using transform pairs 12 to 14 in Appendix A yields the inverse transform

$$\Phi(t) = \begin{bmatrix} 3e^{-2t} - 2e^{-3t} & 6e^{-2t} - 6e^{-3t} \\ -e^{-2t} + e^{-3t} & -2e^{-2t} + 3e^{-3t} \end{bmatrix}$$

This is the same as the value obtained in Eq. (3.92) and is an alternate method of obtaining the same result.

The state equation with an input present is

$$\dot{\mathbf{x}} = \mathbf{A}\mathbf{x} + \mathbf{B}\mathbf{u}$$

Solving by means of the Laplace transform yields

$$\mathbf{X}(s) = [s\mathbf{I} - \mathbf{A}]^{-1}\mathbf{x}(0) + [s\mathbf{I} - \mathbf{A}]^{-1}\mathbf{B}\mathbf{U}(s) = \Phi(s)\mathbf{x}(0) + \Phi(s)\mathbf{B}\mathbf{U}(s) \tag{4.113}$$

Example 2. Solve for the state equation in the example of Sec. 3.14 for a step-function input. The function $\Phi(s)$ is given by Eq. (4.112) in Example 1 and $\mathbf{U}(s) = 1/s$. Inserting these values into Eq. (4.113) gives

$$\mathbf{X}(s) = \Phi(s)\mathbf{x}(0) + \begin{bmatrix} \dfrac{6}{s(s+2)(s+3)} \\ \dfrac{1}{(s+2)(s+3)} \end{bmatrix}$$

[†]When $\mathbf{A}$ is a companion matrix, a direct algorithm for evaluating $[s\mathbf{I} - \mathbf{A}]^{-1}$ and $\Phi(t)$ is presented in Ref. 9.

Applying the transform pairs 10 and 12 of Appendix A yields the same result for $\mathbf{x}(t)$ as in Eq. (3.105). The Laplace transform method is a direct method of solving the state equation.

The possibility exists in the elements of $\mathbf{\Phi}(s)$ that a numerator factor of the form $s + a$ will cancel a similar term in the denominator. In that case the transient mode e^{-at} does not appear in the corresponding element of $\mathbf{\Phi}(t)$. This has a very important theoretical significance since it affects the *controllability and observability* of a control system. These properties are described in Sec. 13.7. Also, as described in Chapter 7, the ability to design a tracking control system requires that the system be observable and controllable.

4.18   EVALUATION OF THE TRANSFER-FUNCTION MATRIX

The transfer-function matrix of a system with multiple inputs and outputs is evaluated from the state and output equations

$$\dot{\mathbf{x}} = \mathbf{Ax} + \mathbf{Bu} \qquad \mathbf{y} = \mathbf{Cx} + \mathbf{Du}$$

Taking the Laplace transform of these equations and solving for the output $\mathbf{Y}(s)$ in terms of the input $\mathbf{U}(s)$ yields

$$\mathbf{Y}(s) = \mathbf{C\Phi}(s)\mathbf{x}(0) + \mathbf{C\Phi}(s)\mathbf{BU}(s) + \mathbf{DU}(s)$$

Since the transfer-function relationship is $\mathbf{Y}(s) = \mathbf{G}(s)\mathbf{U}(s)$ and is defined for zero initial conditions, the system transfer-function matrix is given by analogy as

$$\mathbf{G}(s) = \mathbf{C\Phi}(s)\mathbf{B} + \mathbf{D} \tag{4.114}$$

For multiple-input multiple-output systems $\mathbf{G}(s)$ is a matrix in which the elements are the transfer functions between each output and each input of the system.

Example. Determine the transfer functions and draw a block diagram for the two-input two-output system represented by

$$\dot{\mathbf{x}} = \begin{bmatrix} 0 & 1 \\ -2 & -3 \end{bmatrix}\mathbf{x} + \begin{bmatrix} 1 & 1 \\ 0 & -2 \end{bmatrix}\mathbf{u} \qquad \mathbf{y} = \begin{bmatrix} 0 & -2 \\ 1 & 0 \end{bmatrix}\mathbf{x}$$

The characteristic polynomial is

$$|s\mathbf{I} - \mathbf{A}| = (s + 1)(s + 2)$$

From Eq. (4.111), the resolvent matrix is

$$\mathbf{\Phi}(s) = \frac{1}{(s + 1)(s + 2)}\begin{bmatrix} s + 3 & 1 \\ -2 & s \end{bmatrix}$$

Then, from Eq. (4.114)

$$\mathbf{G}(s) = \mathbf{C\Phi}(s)\mathbf{B} = \begin{bmatrix} \dfrac{4}{(s + 1)(s + 2)} & \dfrac{4}{s + 2} \\ \dfrac{s + 3}{(s + 1)(s + 2)} & \dfrac{1}{s + 2} \end{bmatrix} = \begin{bmatrix} G_{11}(s) & G_{12}(s) \\ G_{21}(s) & G_{22}(s) \end{bmatrix} \tag{4.115}$$

The block diagram for the system is shown in Fig. 4.15.

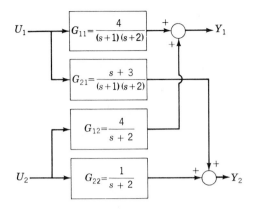

FIGURE 4.15
Block diagram for Eq. (4.115).

4.19   SUMMARY

This chapter discusses the important characteristics and the use of the Laplace transform, which is employed extensively with differential equations because it systematizes their solution. Also it is used extensively in feedback-system synthesis. The pole-zero pattern has been introduced to represent a system function. The pole-zero pattern is significant because it determines the amplitudes of all the time-response terms. The frequency response has also been shown to be a function of the pole-zero pattern. The solution of linear time-invariant (LTI) state equations can be obtained by means of the Laplace transform. The procedure is readily adapted for the use of a digital computer,[8] which is advantageous for multiple-input multiple-output (MIMO) systems.

Later chapters cover feedback-system analysis and synthesis by three methods. The first of these is the root-locus method, which locates the poles and zeros of the system in the s plane. Knowing the poles and zeros permits an exact determination of the time response. The second method is based on the frequency response. Since the frequency response is a function of the pole-zero pattern, the two methods are complementary and give equivalent information in different forms. The root-locus and frequency-response methods are primarily applicable to SISO systems and rely on use of transfer functions to represent the system. They can be adapted for the synthesis of MIMO systems. The state-feedback method is readily used for both SISO and MIMO systems. The necessary linear algebra operations for state-feedback synthesis are presented in this chapter in preparation for their use in later chapters.

System stability requires that all roots of the characteristic equation be located in the left half of the s plane. They can be identified on the pole-zero diagram and are the poles of the overall transfer function. The transfer function can be obtained from the overall differential equation relating an input to an output. The transfer function can also be obtained from the state-equation formulation, as shown in this chapter.

REFERENCES

1. Churchill, R. V.: *Operational Mathematics,* 3d ed., McGraw-Hill, New York, 1972.
2. Thomson, W. T.: *Laplace Transformation,* 2nd ed., Prentice-Hall, Englewood Cliffs, N.J., 1966.

3. Kreyszig, E.: *Advanced Engineering Mathematics,* 7th ed., John Wiley & Sons, New York, 1993.
4. Hazony, D., and I. Riley: "Simplified Technique for Evaluating Residues in the Presence of High Order Poles," paper presented at the Western Electric Show and Convention, August 1959.
5. Aseltine, J. A.: *Transform Method in Linear System Analysis,* McGraw-Hill, New York, 1958.
6. Nilson, J. W.: *Electric Circuits,* 2nd ed., Addison-Wesley, Reading, Mass., 1980.
7. Ward, J. R., and R. D. Strum: *State Variable Analysis,* Prentice-Hall, Englewood Cliffs, N.J., 1970.
8. *ICECAP-PC* by Dr. G. B. Lamont and *TOTAL-PC* by Dr. R. Ewing, Graduate School of Engineering. Air Force Institute of Technology, AFIT/ENG, Bldg. 642, 2950 P St., Wright-Patterson Air Force Base, Ohio 45433-7765.
9. Taylor, F. J.: "A Novel Inversion of $(sI - A)$," *Int. J. Systems Science,* vol. 5, no. 2, pp. 153–160, Feb. 1974.
10. Strang, G.: *Linear Algebra and Its Applications,* 3rd ed., Harcourt, Brace, Jovanovich, New York, 1988.
11. Bongiorno, J. J., Jr.: "A Recursive Algorithm for Computing the Partial Fraction Expansion of Rational Functions Having Multiple Poles," *IEEE Trans. Autom. Control,* vol. AC-29, pp. 650–652, 1984.

CHAPTER
5

SYSTEM REPRESENTATION

5.1 INTRODUCTION

This chapter introduces the basic principles of system representation. From the concepts introduced in earlier chapters, a number of systems are represented in block-diagram form. Feedback is included in these systems in order to achieve the desired performance. Also, the standard symbols and definitions are presented. These are extensively used in the technical literature on control systems and form a common basis for understanding and clarity. While block diagrams simplify the representation of functional relationships within a system, the use of signal flow graphs (SFG) provides further simplification for larger systems that contain intercoupled relationships. Simulation diagrams are presented as a means of representing a system for which the overall differential equation is known. Then the inclusion of initial conditions in the SFG leads to the state-diagram representation. To provide flexibility in the method used by the designer for system analysis and design, the system representation may be in either the transfer-function or the state-equation form. The procedures for converting from one representation to the other are presented in Chapter 4 and in this chapter. When the state-variable format is used, the mathematical equations may be written so that the states are uncoupled. This simplifies the equations, leading to an **A** matrix that is diagonal. The techniques for transforming the state vector in order to achieve different representations are presented.

5.2 BLOCK DIAGRAMS

The representation of physical components by blocks is shown in Chapter 2. For each block the transfer function provides the dynamic mathematical relationship between the input and output quantities. Also, Chapter 1 and Fig. 1.2 describe the concept of feedback, which is used to achieve a better response of a control system to a

131

command input. This section presents several examples of control systems and their representation by block diagrams. The blocks represent the functions performed rather than the components of the system.

Example 1. A temperature-control system.   The first example is an industrial process, temperature control system, shown in Fig. 5.1a. The voltage r, obtained from a potentiometer, is calibrated in terms of the desired temperature θ_{comm} to be produced within the tank. This voltage represents the input quantity to the feedback control system. The actual temperature θ, the output quantity, is measured by means of a thermocouple immersed in the tank. The voltage e_{th} produced in the thermocouple is proportional to θ. The voltage e_{th} is amplified to produce a voltage b, which is the feedback quantity. The voltage $e = r - b$ is the actuating signal and is amplified to produce the solenoid voltage e_1. The current i_s in the solenoid, which results from applying e_1, produces a proportional force f_s that acts on the solenoid armature and valve to control the valve position x. The valve position in turn controls the flow q of hot steam into the heating coil in the tank. The resulting temperature θ of the tank is directly proportional to the steam flow with a time delay that depends on the specific heat of the fluid and the mixing rate. The block diagram representation for this system (Fig. 5.1b) shows the functions of each unit and the signal flow through the system.

To show the operation of the system, consider that an increase in the tank temperature is required. The voltage r is increased to a value that represents the value of the desired temperature. This change in r causes an increase in the actuating signal e. This increase in e, through its effect on the solenoid and valve, causes an increase in the amount of hot steam flowing through the heating coil in the tank. The temperature θ therefore increases in proportion to the steam flow. When the output temperature rises to a value essentially equal to the desired temperature, the feedback voltage b is equal

FIGURE 5.1
An industrial process temperature control system.

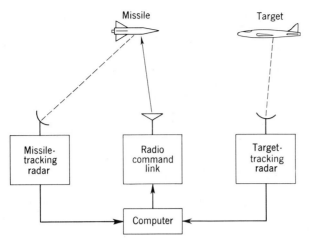

FIGURE 5.2
Command guidance interceptor system.

to the reference input r, $b \approx r$. The flow of steam through the heating coil is stabilized at a steady-state value that maintains θ at the desired value.

Example 2. Command guidance interceptor system. A more complex system is the command guidance system shown in Fig. 5.2, which directs the flight of a missile in space in order to intercept a moving target. The target may be an enemy bomber whose aim is to drop bombs at some position. The defense uses the missile with the objective of intercepting and destroying the bomber before it launches its bombs. The target-tracking radar is used first for detecting and then for tracking the target. It supplies information on target range and angle and their rates of change (time derivatives). This information is continuously fed into the computer, which calculates a predicted course for the target. The missile-tracking radar supplies similar information that is used by the computer to determine its flight path. The computer compares the two flight paths and determines the necessary change in missile flight path to produce a collision course. The necessary flight path changes are supplied to the radio command link, which transmits this information to the missile. This electrical information containing corrections in flight path is used by a control system in the missile. The missile control system converts the error signals into mechanical displacements of the missile airframe control surfaces by means of actuators. The missile responds to the positions of the aerodynamic control surfaces to follow the prescribed flight path, which is intended to produce a collision with the target. Monitoring of the target is continuous so that changes in the missile course can be corrected up to the point of impact. The block diagram of Fig. 5.3 depicts the functions of this command guidance system. Many individual functions are performed within each block. Some of the components of the missile control system are shown within the block representing the missile.

Example 3. Aircraft control system.[1] The feedback control system used to keep an airplane on a predetermined course or heading is necessary for the navigation of commercial airliners. Despite poor weather conditions and lack of visibility the airplane

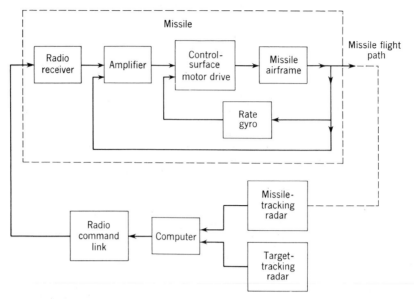

FIGURE 5.3
Block diagram of a generalized command guidance interceptor system.

must maintain a specified heading and altitude in order to reach its destination safely. In addition, in spite of rough air, the trip must be made as smooth and comfortable as possible for the passengers and crew. The problem is considerably complicated by the fact that the airplane has six degrees of freedom. This fact makes control more difficult than the control of a ship, whose motion is limited to the surface of the water. A *flight controller* is the feedback control system used to control the aircraft motion.

Two typical signals to the system are the correct flight path, which is set by the pilot, and the level position (attitude) of the airplane. The ultimately controlled variable is the actual course and position of the airplane. The output of the control system, the controlled variable, is the aircraft heading. In conventional aircraft three primary control surfaces are used to control the physical three-dimensional attitude of the airplane, the elevators, rudder, and ailerons. The axes used for an airplane and the motions produced by the control surfaces are shown in Fig. 5.4.

The directional gyroscope is used as the error-measuring device. Two gyros must be used to provide control of both heading and attitude (level position) of the airplane. The error that appears in the gyro as an angular displacement between the rotor and case is translated into a voltage by various methods, including the use of transducers such as potentiometers, synchros, transformers, or microsyns.

Additional stabilization for the aircraft can be provided in the control system by rate feedback. In other words, in addition to the primary feedback, which is the position of the airplane, another signal proportional to the angular rate of rotation of the airplane around the vertical axis is fed back in order to achieve a stable response. A "rate" gyro is used to supply this signal. This additional stabilization may be absolutely necessary for some of the newer high-speed aircraft.

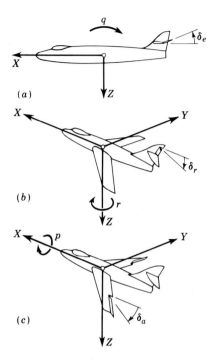

FIGURE 5.4

Airplane control surfaces: (*a*) elevator deflection produces pitching velocity q; (*b*) rudder deflection produces yawing velocity r; (*c*) aileron deflection produces rolling velocity p.

A block diagram of the aircraft control system (Fig. 5.5) illustrates control of the airplane heading by controlling the rudder position. In this system the heading that is controlled is the direction the airplane would travel in still air. The pilot corrects this heading, depending on the crosswinds, so that the actual course of the airplane coincides with the desired path. Another control included in the complete airplane control system controls the ailerons and elevators to keep the airplane in level flight.

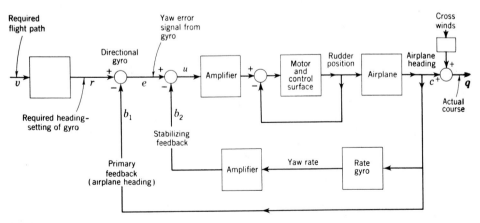

FIGURE 5.5
Airplane directional-control system.

5.3 DETERMINATION OF THE OVERALL TRANSFER FUNCTION

The block diagram of a control system with negative feedback can often be simplified to the form shown in Fig. 5.6, where the standard symbols and definitions used in feedback systems are indicated. In this feedback control system the output is the controlled variable C. This output is measured by a feedback element H to produce the primary feedback signal B, which is then compared with the reference input R. The difference E, between the reference input R and the feedback signal B, is the input to the controlled system G and is referred to as the *actuating signal*. For unity-feedback systems where $H = 1$, the actuating signal is equal to the error signal, which is the difference $R - C$. The transfer functions of the forward and feedback components of the system are G and H, respectively.

In using the block diagram to represent a linear feedback control system where the transfer functions of the components are known, the letter symbol is capitalized, indicating that it is a transformed quantity; i.e., it is a function of the operator D, the complex parameter s, or the frequency parameter $j\omega$. This representation applies for the transfer function, where G is used to represent $G(D), G(s)$, or $\mathbf{G}(j\omega)$. It also applies to all variable quantities, such as C, which represents $C(D), C(s)$, or $\mathbf{C}(j\omega)$. Lowercase symbols are used, as in Fig. 5.5, to represent any function in the time domain. For example, the symbol c represents $c(t)$.

The important characteristic of such a system is the *overall transfer function*, which is the ratio of the transform of the controlled variable C to the transform of the reference input R. This ratio may be expressed in operational, Laplace transform, or frequency (phasor) form. The overall transfer function is also referred to as the *control ratio*. The terms are used interchangeably in this text.

The equations describing this system in terms of the transform variable are

$$C(s) = G(s)E(s) \tag{5.1}$$

$$B(s) = H(s)C(s) \tag{5.2}$$

$$E(s) = R(s) - B(s) \tag{5.3}$$

FIGURE 5.6
Block diagram of a feedback system.

Combining these equations produces the *control ratio,* or *overall transfer function,*

$$\frac{C(s)}{R(s)} = \frac{G(s)}{1 + G(s)H(s)} \tag{5.4}$$

The characteristic equation of the closed-loop system is obtained from the denominator of the control ratio:

$$1 + G(s)H(s) = 0 \tag{5.5}$$

The stability and response of the closed-loop system, as determined by analysis of the characteristic equation, are discussed more fully in later chapters.

For simplified systems where the feedback is unity, that is, $H(s) = 1$, the actuating signal, given by Eq. (5.3), is now the error present in the system, i.e., the *reference input minus the controlled variable, expressed by*

$$E(s) = R(s) - C(s) \tag{5.6}$$

The control ratio with unity feedback is

$$\frac{C(s)}{R(s)} = \frac{G(s)}{1 + G(s)} \tag{5.7}$$

The *open-loop transfer function* is defined as the ratio of the output of the feedback path $B(s)$ to the actuating signal $E(s)$ for any given feedback loop. In terms of Fig. 5.6, the open-loop transfer function is

$$\frac{B(s)}{E(s)} = G(s)H(s) \tag{5.8}$$

The *forward transfer function* is defined as the ratio of the controlled variable $C(s)$ to the actuating signal $E(s)$. For the system shown in Fig. 5.5 the forward transfer function is

$$\frac{C(s)}{E(s)} = G(s) \tag{5.9}$$

In the case of unity feedback, where $H(s) = 1$, the open-loop and the forward transfer functions are the same. The forward transfer function $G(s)$ may be made up not only of elements in cascade but may also contain internal, or *minor,* feedback loops. The algebra of combining these internal feedback loops is similar to that used previously. An example of a controlled system with an internal feedback loop is shown in Fig. 5.5.

It is often useful to express the actuating signal E in terms of the input R. Solving from Eqs. (5.1) to (5.3) gives

$$\frac{E(s)}{R(s)} = \frac{1}{1 + G(s)H(s)} \tag{5.10}$$

The concept of system error y_e is defined as the ideal or desired system value minus the actual system output. The ideal value establishes the desired performance of the system. For unity-feedback systems the actuating signal is an actual measure of the error and is directly proportional to the system error (see Fig. 5.7).

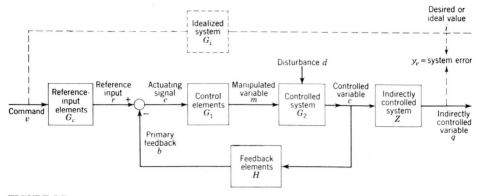

FIGURE 5.7
Block diagram of feedback control system containing all basic elements.

5.4   STANDARD BLOCK DIAGRAM TERMINOLOGY[2]

Figure 5.7 shows a block diagram representation of a feedback control system containing the basic elements. Figure 5.8 shows the block diagram and symbols of a more complicated system with multiple paths and inputs. Numerical subscripts are used to distinguish between blocks of similar functions in the circuit. For example, in Fig. 5.7 the control elements are designated by G_1 and the controlled system by G_2. In Fig. 5.8 the control elements are divided into two blocks, G_1 and G_2, to aid in

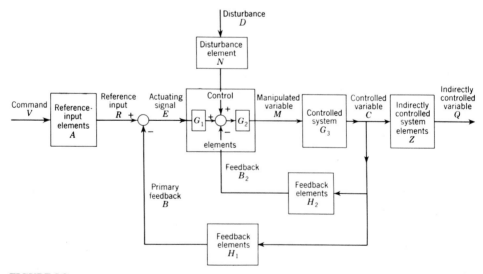

FIGURE 5.8
Block diagram of representative feedback control system showing multiple feedback loops. (All uppercase letters denote transformation.)

representing relations between parts of the system. Also, in this figure the primary feedback is represented by H_1. Additional minor feedback loops might be designated by H_2, H_3, and so forth.

The idealized system represented by the block enclosed with dashed lines in Fig. 5.7 can be understood to show the relation between the basic input to the system and the performance of the system in terms of the desired output. This would be the system agreed upon to establish the ideal value for the output of the system. In systems where the command is actually the desired value or ideal value, the idealized system would be represented by unity. The arrows and block associated with the idealized system, the ideal value, and the system error are shown in dashed lines on the block diagram because they do not exist physically in any feedback control system. For any specific problem, it represents the system (conceived in the mind of the designer) that gives the best approach, when considered as a perfect system, to the desired output or ideal value.

Definitions: Variables in the System

The *command v* is the input that is established by some means external to, and independent of, the feedback control system.

The *reference input r* is derived from the command and is the actual signal input to the system.

The *controlled variable c* is the quantity that is directly measured and controlled. It is the output of the controlled system.

The *primary feedback b* is a signal that is a function of the controlled variable and that is compared with the reference input to obtain the actuating signal.

The *actuating signal e* is obtained from a comparison measuring device and is the reference input minus the primary feedback. This signal, usually at a low energy level, is the input to the control elements that produce the manipulated variable.

The *manipulated variable m* is the quantity obtained from the control elements that is applied to the controlled system. The manipulated variable is generally at a higher energy level than the actuating signal and may also be modified in form.

The *indirectly controlled variable q* is the output quantity that is related through the indirectly controlled system to the controlled variable. It is outside the closed loop and is not directly measured for control.

The *ultimately controlled variable* is a general term that refers to the indirectly controlled variable. In the absence of the indirectly controlled variable, it refers to the controlled variable.

The *ideal value i* is the value of the ultimately controlled variable that would result from an idealized system operating with the same command as the actual system.

The *system error y_e* is the ideal value minus the value of the ultimately controlled variable.

The *disturbance d* is the unwanted signal that tends to affect the controlled variable. The disturbance may be introduced into the system at many places.

Definitions: System Components

The *reference input elements* G_v produce a signal r proportional to the command.

The *control elements* G_c produce the manipulated variable m from the actuating signal.

The *controlled system* G is the device that is to be controlled. This is frequently a high-power element.

The *feedback element* H produces the primary feedback b from the controlled variable. This is generally a proportionality device but may also modify the characteristics of the controlled variable.

The *indirectly controlled system* Z relates the indirectly controlled variable q to the controlled quantity c. This component is outside the feedback loop.

The *idealized system* G_i is one whose performance is agreed upon to define the relationship between the ideal value and the command. This is often called the *model* or *desired* system.

The *disturbance element* N denotes the functional relationship between the variable representing the disturbance and its effect on the control system.

5.5 POSITION CONTROL SYSTEM

Figure 5.9 shows a simplified block diagram of an angular position control system. The reference selector and the sensor, which produce the reference input $R = \theta_R$ and the controlled output position $C = \theta_o$, respectively, consist of rotational potentiometers. The combination of these units represents a rotational comparison unit that generates the actuating signal E for the position control system, as shown in Fig. 5.10a, where K_θ, in volts per radian, is the potentiometer sensitivity constant. The symbolic comparator for this system is shown in Fig. 5.10b.

FIGURE 5.9
Position-control system.

FIGURE 5.10
Rotational position comparison and its block diagram representation.

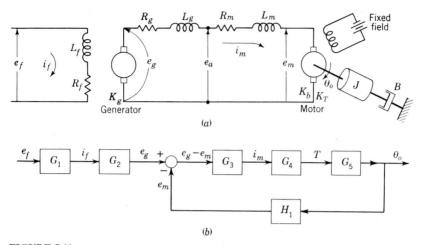

FIGURE 5.11
(a) Motor-generator control, (b) block diagram.

The transfer function of the motor-generator control is obtained by writing the equations for the schematic diagram shown in Fig. 5.11a. This figure shows a dc motor that has a constant field excitation and drives an inertia and friction load. The armature voltage for the motor is furnished by the generator, which is driven at constant speed by a prime mover. The generator voltage e_g is determined by the voltage e_f applied to the generator field. The generator is acting as a power amplifier for the signal voltage e_f. The equations for this system are as follows:

$$e_f = L_f D i_f + R_f i_f \tag{5.11}$$

$$e_g = K_g i_f \tag{5.12}$$

$$e_g - e_m = (L_g + L_m)D i_m + (R_g + R_m)i_m \tag{5.13}$$

$$e_m = K_b D\theta_o \tag{5.14}$$

$$T = K_T i_m = JD^2\theta_o + BD\theta_o \tag{5.15}$$

The block diagram drawn in Fig. 5.11b is based on these equations. Starting with the input quantity e_f, Eq. (5.11) shows that a current i_f is produced. Therefore, a block is drawn with e_f as the input and i_f as the output. Equation (5.12) shows that a voltage e_g is generated as a function of the current i_f. Therefore, a second block is drawn with the current i_f as the input and e_g as the output. Equation (5.13) relates the current i_m in the motor to the difference of two voltages, $e_g - e_m$. To obtain this difference a summation point is introduced. The quantity e_g from the previous block enters this summation point. To obtain the quantity $e_g - e_m$ there must be added the quantity e_m, entering the summation point with a minus sign. Up to this point the manner in which e_m is obtained has not yet been determined. The output of this summation point is used as the input to the next block, from which the current i_m is the output. Similarly, the block with current i_m as the input and the generated torque T as the output and the block with the torque input and the resultant motor position as the output are drawn. There must be no loose ends in the complete diagram; i.e., every dependent variable must be connected through a block or blocks into the system. Therefore, e_m is obtained from Eq. (5.14), and a block representing this relationship is drawn with θ_o as the input and e_m as the output. By using this procedure, the block diagram is completed and the functional relationships in the system are described. Note that the generator and the motor are no longer separately distinguishable. Also, the input and the output of each block are not in the same units, as both electrical and mechanical quantities are included.

The transfer functions of each block, as determined in terms of the pertinent Laplace transforms, are as follows:

$$G_1(s) = \frac{I_f(s)}{E_f(s)} = \frac{1/R_f}{1 + (L_f/R_f)s} = \frac{1/R_f}{1 + T_f s} \tag{5.16}$$

$$G_2(s) = \frac{E_g(s)}{I_f(s)} = K_g \tag{5.17}$$

$$G_3(s) = \frac{I_m(s)}{E_g(s) - E_m(s)} = \frac{1/(R_g + R_m)}{1 + \left[(L_g + L_m)/(R_g + R_m)\right]s}$$

$$= \frac{1/R_{gm}}{1 + (L_{gm}/R_{gm})s} = \frac{1/R_{gm}}{1 + T_{gm}s} \tag{5.18}$$

$$G_4(s) = \frac{T(s)}{I_m(s)} = K_T \tag{5.19}$$

$$G_5(s) = \frac{\Theta_o(s)}{T(s)} = \frac{1/B}{s[1 + (J/B)s]} = \frac{1/B}{s(1 + T_n s)} \tag{5.20}$$

$$H_1(s) = \frac{E_m(s)}{\Theta_o(s)} = K_b s \tag{5.21}$$

The block diagram can be simplified, as shown in Fig. 5.12, by combining the blocks in cascade. The block diagram is further simplified, as shown in Fig. 5.13, by evaluating an equivalent block from $E_g(s)$ to $\Theta_o(s)$, using the principle of Eq. (5.4). The final simplification results in Fig. 5.14. The overall transfer function $G_x(s)$ is

FIGURE 5.12
Simplified block diagram.

The block diagram shows:

$$E_f(s) \rightarrow \boxed{\dfrac{K_g/R_f}{1+T_f s}} \xrightarrow{E_g(s)} \boxed{\dfrac{K_T/R_{gm}B}{s[(1+T_{gm}s)(1+T_n s)+K_T K_b/R_{gm}B]}} \rightarrow \theta_o(s)$$

FIGURE 5.13
Reduced block diagram.

$$G_x(s) = \frac{\Theta_o(s)}{E_f(s)} = \frac{K_g K_T / R_f B R_{gm}}{s(1 + T_f s)\left[(1 + K_T K_b/B R_{gm}) + (T_{gm} + T_n)s + T_{gm}T_n s^2\right]}$$

$$(5.22)$$

This expression is the exact transfer function for the entire motor and generator combination. Certain approximations, if valid, may be made to simplify this expression. The first approximation is that the inductance of the generator and motor armatures is very small ($T_{gm} \approx 0$). With this approximation, the transfer function reduces to

$$\frac{\Theta_o(s)}{E_f(s)} = G_x(s) \approx \frac{K_x}{s(1 + T_f s)(1 + T_m s)} \tag{5.23}$$

where

$$K_x = \frac{K_g K_T}{R_f(B R_{gm} + K_T K_b)} \tag{5.24}$$

$$T_f = \frac{L_f}{R_f} \tag{5.25}$$

$$T_m = \frac{J R_{gm}}{B R_{gm} + K_T K_b} \tag{5.26}$$

If the frictional effect of the load is very small, the approximation can be made that $B \approx 0$. With this additional approximation, the transfer function has the same form as Eq. (5.23), but the constants are now

$$K_x = \frac{K_g}{R_f K_b} \tag{5.27}$$

$$T_f = \frac{L_f}{R_f} \tag{5.28}$$

$$T_m = \frac{J R_{gm}}{K_T K_b} \tag{5.29}$$

For simple components, as in this case, an overall transfer function can often be derived more easily by combining the original system equations. This can be done without the intermediate steps shown in this example. However, the purpose of this

$$E_f(s) \rightarrow \boxed{G_x(s)} \xrightarrow{\theta_o(s)}$$

FIGURE 5.14
Simplified block diagram for Fig. 5.11.

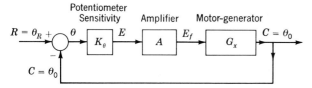

FIGURE 5.15
Equivalent representation of Fig. 5.9.

example is to show the representation of dynamic components by individual blocks and the combination of blocks.

A new block diagram representing the position-control system of Fig. 5.9 is redrawn in Fig. 5.15. Since the transfer function of the amplifier of Fig. 5.9 is $E_f(s)/E(s) = A$, the forward transfer function of the position-control system of Fig. 5.15 is

$$G(s) = \frac{\Theta_o(s)}{\Theta(s)} = \frac{AK_xK_\theta}{s(1 + T_fs)(1 + T_ms)} \tag{5.30}$$

The overall transfer function (control ratio) for this system, with $H(s) = 1$, is

$$\frac{C(s)}{R(s)} = \frac{\Theta_o(s)}{\Theta_R(s)} = \frac{G(s)}{1 + G(s)H(s)} = \frac{AK_xK_\theta}{s(1 + T_fs)(1 + T_ms) + AK_xK_\theta} \tag{5.31}$$

5.6 SIMULATION DIAGRAMS[10,11]

The simulation used to represent the dynamic equations of a system may show the actual physical variables that appear in the system, or it may show variables that are used purely for mathematical convenience. In either case the overall response of the system is the same. The simulation diagram is similar to the diagram used to represent the system on an analog computer. The basic elements used are ideal integrators, ideal amplifiers, and ideal summers, shown in Fig. 5.16. Additional elements such as multipliers and dividers may be used for nonlinear systems.

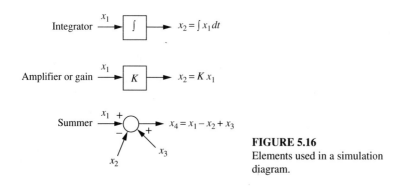

FIGURE 5.16
Elements used in a simulation diagram.

One of the methods used to obtain a simulation diagram includes the following steps:

1. Start with the differential equation.
2. On the left side of the equation put the highest-order derivative of the *dependent* variable. A first-order or higher-order derivative of the input may appear in the equation. In this case the highest-order derivative of the input is also placed on the left side of the equation. All other terms are put on the right side.
3. Start the diagram by assuming that the signal, represented by the terms on the left side of the equation, is available. Then integrate it as many times as needed to obtain all the lower-order derivatives. It may be necessary to add a summer in the simulation diagram to obtain the dependent variable explicitly.
4. Complete the diagram by feeding back the appropriate outputs of the integrators to a summer to generate the original signal of step 2. Include the input function if it is required.

Example. Draw the simulation diagram for the series RLC circuit of Fig. 2.2 in which the output is the voltage across the capacitor.

Step 1. When $y = v_c$ and $u = e$ are used, Eq. (2.10) becomes

$$LC\ddot{y} + RC\dot{y} + y = u \qquad (5.32)$$

Step 2. Rearrange terms to the form

$$\ddot{y} = bu - a\dot{y} - by \qquad (5.33)$$

where $a = R/L$ and $b = 1/LC$.

Step 3. The signal $\ddot{y}$ is integrated twice, as shown in Fig. 5.17a.

Step 4. The block or simulation diagram is completed as shown in Fig. 5.17b in order to satisfy Eq. (5.33).

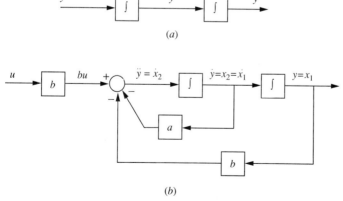

FIGURE 5.17
Simulation diagram for Eq. (5.32).

The state variables are often selected as the outputs of the integrators in the simulation diagram. In this case they are $y = x_1$ and $\dot{y} = x_2 = \dot{x}_1$. These values and $\ddot{y} = \dot{x}_2$ are shown in Fig. 5.17b. The state and output equations are therefore

$$\begin{bmatrix} \dot{x}_1 \\ \dot{x}_2 \end{bmatrix} = \begin{bmatrix} 0 & 1 \\ -\dfrac{1}{LC} & -\dfrac{R}{L} \end{bmatrix} \begin{bmatrix} x_1 \\ x_2 \end{bmatrix} + \begin{bmatrix} 0 \\ \dfrac{1}{LC} \end{bmatrix} u = \mathbf{Ax} + \mathbf{B}u \tag{5.34}$$

$$\mathbf{y} = \begin{bmatrix} 1 & 0 \end{bmatrix} \begin{bmatrix} x_1 \\ x_2 \end{bmatrix} + \mathbf{0}u = \mathbf{Cx} + \mathbf{D}u \tag{5.35}$$

It is common in a physical system for the $\mathbf{D}$ matrix to be zero, as in this example. These equations are different from Eqs. (2.30) and (2.32), yet they represent the same system. This difference illustrates that state variables are not unique. When the state variables are the dependent variable and the derivatives of the dependent variable, as in this example, they are called *phase variables*. The phase variables are applicable to differential equations of any order and can be used without drawing the simulation diagram, as shown below.

CASE 1. The general differential equation that contains no derivatives of the input is

$$D^n y + a_{n-1} D^{n-1} y + \cdots + a_1 Dy + a_0 y = u \tag{5.36}$$

The state variables are selected as the *phase variables*, which are defined by $x_1 = y$, $x_2 = Dx_1 = \dot{x}_1 = Dy$, $x_3 = \dot{x}_2 = D^2 y, \ldots, x_n = \dot{x}_{n-1} = D^{n-1} y$. Thus, $D^n y = \dot{x}_n$. When these state variables are used, Eq. (5.36) becomes

$$\dot{x}_n + a_{n-1} x_n + a_{n-2} x_{n-1} + \cdots + a_1 x_2 + a_0 x_1 = u \tag{5.37}$$

The resulting state and output equations using phase variables are

$$\begin{bmatrix} \dot{x}_1 \\ \dot{x}_2 \\ \vdots \\ \dot{x}_{n-1} \\ \dot{x}_n \end{bmatrix} = \begin{bmatrix} 0 & 1 & & & & \\ & 0 & 1 & & \mathbf{0} & \\ & & 0 & 1 & & \\ & & & 0 & \ddots & \\ & & & & \ddots & \\ & \mathbf{0} & & & 0 & 1 \\ -a_0 & -a_1 & -a_2 & -a_3 & \cdots & -a_{n-2} & -a_{n-1} \end{bmatrix} \mathbf{x} + \begin{bmatrix} 0 \\ 0 \\ \vdots \\ 0 \\ 1 \end{bmatrix} \mathbf{u}$$

$$\dot{\mathbf{x}} = \mathbf{A}_c \mathbf{x} + \mathbf{b}_c \mathbf{u} \tag{5.38}$$

$$\mathbf{y} = [x_1] = \begin{bmatrix} 1 & 0 & 0 & \cdots \end{bmatrix} \mathbf{x} = \mathbf{c}_c^T \mathbf{x} \tag{5.39}$$

These equations can be written directly from the original differential equation (5.36). The plant matrix $\mathbf{A}_c$ contains the number 1 in the *superdiagonal* and the negative of the coefficients of the original differential equation in the nth row. In this simple form the matrix $\mathbf{A}_c$ is called the *companion matrix*. Also, the $\mathbf{B}$ matrix takes on the form shown in Eq. (5.38) and is indicated by $\mathbf{b}_c$.

CASE 2. When derivatives of $u(t)$ appear in the differential equation, the phase variables may be used as the state variables and the state equation given in Eq. (5.38) still applies; however, the output equation is no longer given by Eq. (5.39). This is shown by considering the differential equation

$$(D^n + a_{n-1}D^{n-1} + \cdots + a_1 D + a_0)y$$
$$= (c_w D^w + c_{w-1}D^{w-1} + \cdots + c_1 D + c_0)u \qquad w \le n \quad (5.40)$$

The output y is specified as

$$y = (c_w D^w + c_{w-1}D^{w-1} + \cdots + c_1 D + c_0)x_1 \qquad (5.41)$$

In terms of the state variables this output equation becomes

$$y = c_w \dot{x}_w + c_{w-1}x_w + \cdots + c_1 x_2 + c_0 x_1 \qquad (5.42)$$

Substituting Eq. (5.41) into Eq. (5.40) and canceling common terms on both sides of the equation yields

$$(D^n + a_{n-1}D^{n-1} + \cdots + a_1 D + a_0)x_1 = u \qquad (5.43)$$

This equation has the same form as Eq. (5.36). Thus, in terms of phase variables, Eq. (5.43) can be represented in the form $\dot{\mathbf{x}} = \mathbf{A}_c \mathbf{x} + \mathbf{b}_c \mathbf{u}$, where the matrices $\mathbf{A}_c$ and $\mathbf{b}_c$ are the same as in Eq. (5.38). The cases for two values of w are:

(1) For $w = n$, the output equation (5.42) becomes

$$y = c_n \dot{x}_n + c_{n-1}x_n + \cdots + c_1 x_2 + c_0 x_1$$

Using Eq. (5.37) to eliminate $\dot{x}_n$ yields

$$y = \left[(c_0 - a_0 c_n) \quad (c_1 - a_1 c_n) \quad \cdots \quad (c_{n-1} - a_{n-1}c_n) \right] \mathbf{x} + [c_n]\mathbf{u}$$

$$= \mathbf{c}_c^T \mathbf{x} + \mathbf{d}_c \mathbf{u} \qquad (5.44)$$

(2) For $w < n$, $\dot{x}_w = x_{w+1}$, and $c_{w+i} = 0$ for $i = 1, 2 \ldots, n - w$. In this case the output $\mathbf{y}$ in Eq. (5.44) reduces to

$$y = \left[c_0 \quad c_1 \quad \cdots \quad c_w \quad 0 \quad \cdots \quad 0 \right] \mathbf{x} \qquad (5.45)$$

Figure 5.18 shows the simulation diagram that represents the system of Eq. (5.40) in terms of the state equation, Eq. (5.38), and the output equation, Eq. (5.45). It has the desirable feature that differentiation of the input signal $u(t)$ is not required to satisfy the differential equation. The simulation diagram has the advantage of requiring only two summers, regardless of the order of the system, one at the input and one at the output. This representation of a differential equation is considered again in Sec. 5.8. Differentiators are avoided in analog- and digital-computer simulations, because they accentuate any noise present in the signals.

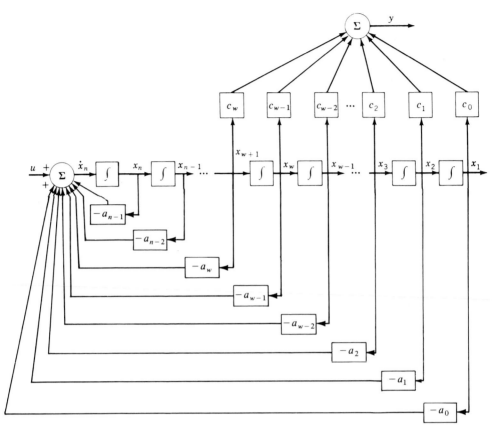

FIGURE 5.18
Simulation diagram representing the system of Eq. (5.40) in terms of Eqs. (5.38) and (5.45).

Different sets of state variables may be selected to represent a system. The selection of the state variables determines the $\mathbf{A}, \mathbf{B}, \mathbf{C}$, and $\mathbf{D}$ matrices of Eqs. (2.37) and (2.38). A general matrix block diagram representing the state and the output equations is shown in Fig. 5.19. The matrix $\mathbf{D}$ is called the *feedforward matrix*. The analysis and design of multiple-input multiple-output (MIMO) systems are covered in detail in Chapters 16 and 17.

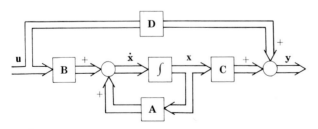

FIGURE 5.19
General matrix block diagram representing the state and output equations.

5.7 SIGNAL FLOW GRAPHS[3,4]

The block diagram is a useful tool for simplifying the representation of a system. The block diagrams of Figs. 5.9 and 5.15 have only one feedback loop and may be categorized as simple block diagrams. Figure 5.5 has three feedback loops; thus it is not a simple system. When intercoupling exists between feedback loops, and when a system has more than one input and one output, the control system and block diagram are more complex. Having the block diagram simplifies the analysis of a complex system. Such an analysis can be even further simplified by using a *signal flow graph* (SFG), which looks like a simplified block diagram.

An SFG is a diagram that represents a set of simultaneous equations. It consists of a *graph* in which *nodes* are connected by directed *branches*. The nodes represent each of the system variables. A branch connected between two nodes acts as a one-way signal multiplier: the direction of signal flow is indicated by an arrow placed on the branch, and the multiplication factor (transmittance or transfer function) is indicated by a letter placed near the arrow. Thus, in Fig. 5.20, the branch transmits the signal x_1 from left to right and multiplies it by the quantity a in the process. The quantity a is the transmittance, or transfer function. It may also be indicated by $a = t_{12}$, where the subscripts show that the signal flow is from node 1 to node 2.

Flow-Graph Definitions

A *node* performs two functions:

1. *Addition* of the signals on all incoming branches
2. *Transmission* of the total node signal (the sum of all incoming signals) to all outgoing branches

These functions are illustrated in the graph of Fig. 5.21, which represents the equations

$$w = au + bv \qquad x = cw \qquad y = dw \qquad (5.46)$$

Three types of nodes are of particular interest:

Source nodes (independent nodes). These represent independent variables and have only outgoing branches. In Fig. 5.21, nodes u and v are source nodes.

FIGURE 5.20
Signal flow graph for $x_2 = ax_1$.

FIGURE 5.21
Signal flow graph for Eqs. (5.46).

FIGURE 5.22

Mixed and sink nodes for a variable.

Sink nodes (dependent nodes). These represent dependent variables and have only incoming branches. In Fig. 5.21, nodes x and y are sink nodes.

Mixed nodes (general nodes). These have both incoming and outgoing branches. In Fig. 5.21, node w is a mixed node. A mixed node may be treated as a sink node by adding an outgoing branch of unity transmittance, as shown in Fig. 5.22, for the equation $x = au + bv$ and $w = cx = cau + cbv$.

A *path* is any connected sequence of branches whose arrows are in the same direction.

A *forward path* between two nodes is one that follows the arrows of successive branches and in which a node appears only once. In Fig. 5.21 the path uwx is a forward path between the nodes u and x.

Flow-Graph Algebra

The following rules are useful for simplifying a signal flow graph:

Series paths (cascade nodes). Series paths can be combined into a single path by multiplying the transmittances as shown in Fig. 5.23a.

FIGURE 5.23

Flow-graph simplifications.

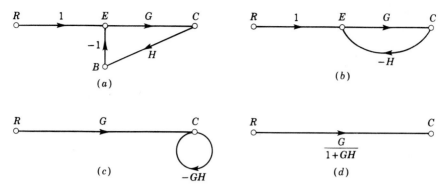

FIGURE 5.24

Successive reduction of the flow graph for the feedback system of Fig. 5.6.

Path gain. The product of the transmittances in a series path.

Parallel paths. Parallel paths can be combined by adding the transmittances as shown in Fig. 5.23b.

Node absorption. A node representing a variable other than a source or sink can be eliminated as shown in Fig. 5.23c.

Feedback loop. A closed path that starts at a node and ends at the same node.

Loop gain. The product of the transmittances of a feedback loop.

The equations for the feedback system of Fig. 5.6 are as follows:

$$C = GE \tag{5.47}$$

$$B = HC \tag{5.48}$$

$$E = R - B \tag{5.49}$$

Note that an equation is written for each dependent variable. The corresponding SFG is shown in Fig. 5.24a. The nodes B and E can be eliminated in turn to produce Fig. 5.24b and Fig. 5.24c, respectively. Figure 5.24c has a *self-loop* of value $-GH$. The final simplification is to eliminate the self-loop to produce the overall transmittance from the input R to the output C. This is obtained by summing signals at node C in Fig. 5.24c, yielding $C = GR - GHC$. Solving for C produces Fig. 5.24d.

General Flow-Graph Analysis

If all the source nodes are brought to the left and all the sink nodes are brought to the right, the SFG for an arbitrarily complex system can be represented by Fig. 5.25a. The effect of the *internal* nodes can be factored out by ordinary algebraic processes to yield the equivalent graph represented by Fig. 5.25b. This simplified graph is represented by

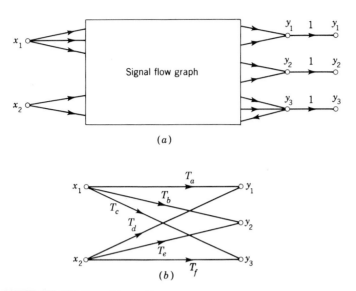

FIGURE 5.25
Equivalent signal flow graphs.

$$y_1 = T_a x_1 + T_d x_2 \tag{5.50}$$

$$y_2 = T_b x_1 + T_e x_2 \tag{5.51}$$

$$y_3 = T_c x_1 + T_f x_2 \tag{5.52}$$

The T's, called overall graph transmittances, are the overall transmittances from a specified source node to a specified dependent node. For linear systems the principle of superposition can be used to "solve" the graph. That is, the sources can be considered one at a time. Then the output signal is equal to the sum of the contributions produced by each input. The overall transmittances can be found by the ordinary processes of linear algebra, i.e., by the solution of the set of simultaneous equations representing the system. However, the same results can be obtained directly from the SFG. The fact that they can produce answers to large sets of linear equations *by inspection* gives the SFGs their power and usefulness.

The Mason Gain Rule

The overall transmittance can be obtained from the Mason gain formula[3]. The formula and definitions are followed by an example to show its application. The overall transmittance T is given by

$$T = \frac{\sum T_n \Delta_n}{\Delta} \tag{5.53}$$

where T_n is the transmittance of each forward path between a source and a sink node and Δ is the graph determinant, found from

$$\Delta = 1 - \sum L_1 + \sum L_2 - \sum L_3 + \cdots \qquad (5.54)$$

In this equation L_1 is the transmittance of each closed path, and $\sum L_1$ is the sum of the transmittances of all closed paths in the graph. L_2 is the product of the transmittances of two *nontouching* loops; loops are nontouching if they do not have any common nodes. $\sum L_2$ is the sum of the products of transmittances in all possible combinations of nontouching loops taken two at a time. L_3 is the product of the transmittances of three nontouching loops. $\sum L_3$ is the sum of the products of transmittances in all possible combinations of nontouching loops taken three at a time.

In Eq. (5.53), Δ_n is the cofactor of T_n. It is the determinant of the remaining subgraph when the forward path that produces T_n is removed. Thus, Δ_n does not include any loops that touch the forward path in question. Δ_n is equal to unity when the forward path touches all the loops in the graph or when the graph contains no loops. Δ_n has the same form as Eq. (5.54).

Example. Figure 5.26a shows a block diagram. Since $E_1 = M_1 - B_1 - B_3 = G_1 E - H_1 M_2 - H_3 C$, it is not necessary to show M_1, $-B_1$, and B_3 explicitly. The SFG is shown in Fig. 5.26b. Since this is a fairly complex system, the resulting equation is expected to be complex. However, the application of Mason's rule produces the resulting overall transmittance in a systematic manner. This system has four loops, whose transmittances are $-G_2H_1$, $-G_5H_2$, $-G_1G_2G_3G_5$, and $-G_2G_3G_5H_3$. Therefore

$$\sum L_1 = -G_2H_1 - G_5H_2 - G_1G_2G_3G_5 - G_2G_3G_5H_3 \qquad (5.55)$$

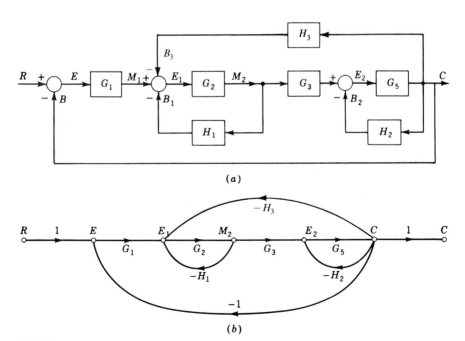

(a)

(b)

FIGURE 5.26
Block diagram and its signal flow graph.

Only two loops are nontouching; therefore,

$$\sum L_2 = (-G_2 H_1)(-G_5 H_2) \qquad (5.56)$$

Although there are four loops, there is no set of three loops that are nontouching; therefore,

$$\sum L_3 = 0 \qquad (5.57)$$

The system determinant can therefore be obtained from Eq. (5.54).

There is only one forward path between R and C. The corresponding forward transmittance is $G_1 G_2 G_3 G_5$. If this path, with its corresponding nodes, is removed from the graph, the remaining subgraph has no loops. The cofactor Δ_n is therefore equal to unity. The complete overall transmittance from R to C, obtained from Eq. (5.53), is

$$T = \frac{G_1 G_2 G_3 G_5}{1 + G_2 H_1 + G_5 H_2 + G_1 G_2 G_3 G_5 + G_2 G_3 G_5 H_3 + G_2 G_5 H_1 H_2} \qquad (5.58)$$

The complete expression for T is obtained by inspection from the SFG. The SFG is simpler than solving the five simultaneous equations that represent this system.

5.8   STATE TRANSITION SIGNAL FLOW GRAPH[5]

The state transition SFG or, more simply, the *state diagram,* is a simulation diagram for a system of equations and includes the initial conditions of the states. Since the state diagram in the Laplace domain satisfies the rules of Mason's SFG, it can be used to obtain the transfer function of the system and the state transition equation. As described in Sec. 5.6, the basic elements used in a simulation diagram are a gain, a summer, and an integrator. The signal-flow representation in the Laplace domain for an integrator is obtained as follows:

$$\dot{x}_1(t) = x_2(t)$$

$$X_1(s) = \frac{X_2(s)}{s} + \frac{x_1(t_0)}{s} \qquad (5.59)$$

Equation (5.59) may be represented either by Fig. 5.27a or Fig. 5.27b.

(a)   (b)

FIGURE 5.27
Representations of an integrator in the Laplace domain in a signal flow graph.

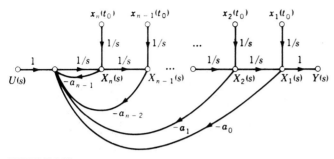

FIGURE 5.28
State diagram for Eq. (5.60).

A differential equation that contains no derivatives of the input, as given by Eq. (5.36), is repeated here:

$$D^n y + a_{n-1} D^{n-1} y + \cdots + a_1 D y + a_0 y = u \qquad (5.60)$$

In terms of the phase variables, with $x_1 = y$ and $\dot{x}_i = x_{i+1}$, this equation becomes

$$\dot{x}_n + a_{n-1} x_n + \cdots + a_1 x_2 + a_0 x_1 = u \qquad (5.60a)$$

The state diagram is drawn in Fig. 5.28 with phase variables. It is obtained by first drawing the number of integrator branches $1/s$ equal to the order of the differential equation. The outputs of the integrators are designated as the state variables, and the initial conditions of the n states are included as inputs in accordance with Fig. 5.27a. Taking the Laplace transform of Eq. (5.60a) yields, after dividing by s,

$$X_n(s) = \frac{1}{s}\left[-a_0 X_1(s) - a_1 X_2(s) - \cdots - a_{n-1} X_n(s) + U(s)\right] + \frac{1}{s} x_n(t_0)$$

The node $X_n(s)$ in Fig. 5.28 satisfies this equation. The signal at the unlabeled node between $U(s)$ and $X_n(s)$ is equal to $sX_n(s) - x_n(t_0)$.

The overall transfer function $Y(s)/U(s)$ is defined with all initial values of the states equal to zero. Using this condition and applying the Mason gain formula given by Eq. (5.53) yields

$$G(s) = \frac{Y(s)}{U(s)} = \frac{s^{-n}}{\Delta(s)} = \frac{s^{-n}}{1 + a_{n-1}s^{-1} + \cdots + a_1 s^{-(n-1)} + a_0 s^{-n}}$$

$$= \frac{1}{s^n + a_{n-1}s^{n-1} + \cdots + a_1 s + a_0} \qquad (5.61)$$

The state transition equation of the system is obtained from the state diagram by applying the Mason gain formula and considering each initial condition as a source. This is illustrated by the following example.

Example 1. Equation (5.32) can be expressed as

$$\ddot{y} + \frac{R}{L}\dot{y} + \frac{1}{LC}y = \frac{1}{LC}u$$

(*a*) Draw the state diagram. (*b*) Determine the state transition equation.

Solution

(*a*) The state diagram, Fig. 5.29, includes two integrators, because this is a second-order equation. The state variables are selected as the phase variables that are the outputs of the integrators, that is, $x_1 = y$ and $x_2 = \dot{x}_1$.

(*b*) The state transition equations are obtained by applying the Mason gain formula with the three inputs u, $x_1(t_0)$, and $x_2(t_0)$:

$$X_1(s) = \frac{s^{-1}(1 + s^{-1}R/L)}{\Delta(s)}x_1(t_0) + \frac{s^{-2}}{\Delta(s)}x_2(t_0) + \frac{s^{-2}/LC}{\Delta(s)}U(s)$$

$$X_2(s) = \frac{-s^{-2}/LC}{\Delta(s)}x_1(t_0) + \frac{s^{-1}}{\Delta(s)}x_2(t_0) + \frac{s^{-1}/LC}{\Delta(s)}U(s)$$

$$\Delta(s) = 1 + \frac{s^{-1}R}{L} + \frac{s^{-2}}{LC}$$

After simplification, these equations become

$$\mathbf{X}(s) = \begin{bmatrix} X_1(s) \\ X_2(s) \end{bmatrix}$$

$$= \frac{1}{s^2 + (R/L)s + 1/LC}\left\{ \begin{bmatrix} s + \dfrac{R}{L} & 1 \\ -\dfrac{1}{LC} & s \end{bmatrix}\begin{bmatrix} x_1(t_0) \\ x_2(t_0) \end{bmatrix} + \begin{bmatrix} \dfrac{1}{LC} \\ \dfrac{s}{LC} \end{bmatrix}U(s) \right\}$$

This equation is of the form given by Eq. (4.113), i.e.,

$$\mathbf{X}(s) = \mathbf{\Phi}(s)\mathbf{x}(t_0) + \mathbf{\Phi}(s)\mathbf{B}U(s)$$

Thus the resolvent matrix $\mathbf{\Phi}(s)$ is readily identified and, by use of the inverse Laplace transform, yields the state transition matrix $\mathbf{\Phi}(t)$. The elements of the resolvent matrix can be obtained directly from the SFG with $U(s) = 0$ and only one $x_j(t_0)$ considered at a time, i.e, all other initial conditions set to zero. Each element of $\mathbf{\Phi}(s)$ is

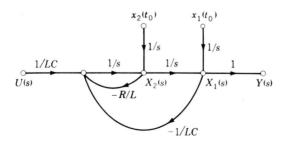

FIGURE 5.29
State diagram for Example 1.

$$\phi_{ij}(s) = \frac{X_i(s)}{x_j(t_0)}$$

For example,

$$\phi_{11}(s) = \frac{X_1(s)}{x_1(t_0)} = \frac{s^{-1}(1 + s^{-1}R/L)}{\Delta(s)} = \frac{s + R/L}{s^2\Delta(s)}$$

The complete state transition equation $\mathbf{x}(t)$ is obtainable through use of the state diagram. Therefore, $\mathbf{\Phi}(s)$ is obtained without performing the inverse operation $[s\mathbf{I} - \mathbf{A}]^{-1}$. Since $\mathbf{x}(t)$ represents phase variables, the system output is $y(t) = x_1(t)$.

Example 2. A differential equation containing derivatives of the input, given by Eq. (5.40), is repeated here.

$$(D^n + a_{n-1}D^{n-1} + \cdots + a_1 D + a_0)y$$

$$= (c_w D^w + c_{w-1}D^{w-1} + \cdots + c_1 D + c_0)u \qquad w \le n \quad (5.62)$$

Phase variables can be specified as the state variables, provided the output is identified by Eq. (5.41) as

$$y = (c_w D^w + c_{w-1}D^{w-1} + \cdots + c_1 D + c_0)x_1 \qquad (5.63)$$

Equation (5.62) then reduces to

$$(D^n + a_{n-1}D^{n-1} + \cdots + a_1 D + a_0)x_1 = u \qquad (5.64)$$

The resulting state diagram for Eqs. (5.63) and (5.64) is shown in Fig. 5.30 for $w = n$. The state equations obtained from the state diagram are given by Eq. (5.38). The output

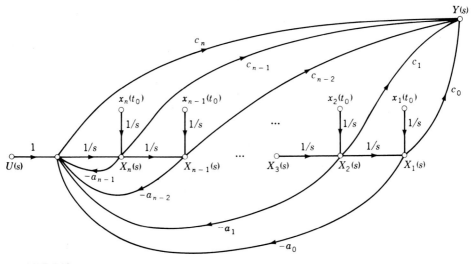

FIGURE 5.30
State diagram for Eq. (5.62) using phase variables, for $w = n$.

equation in matrix form is readily obtained from Fig. 5.30 in terms of the transformed variables $X(s)$ and the input $U(s)$ as

$$\mathbf{Y}(s) = \Big[(c_0 - a_0 c_n) \quad (c_1 - a_1 c_n) \quad \cdots$$

$$(c_{n-2} - a_{n-2} c_n) \quad (c_{n-1} - a_{n-1} c_n)\Big]\mathbf{X}(s) + c_n \mathbf{U}(s)$$

$$= \mathbf{c}_c^T \mathbf{X}(s) + \mathbf{d}_c \mathbf{U}(s) \tag{5.65}$$

Equation (5.65) is expressed in terms of the state $\mathbf{X}(s)$ and the input $\mathbf{U}(s)$. To obtain Eq. (5.65) from Fig. 5.30, delete all the branches having a transmittance $1/s$ and all the initial conditions in Fig. 5.30. The Mason gain formula is then used to obtain the output in terms of the state variables and the input. Note: there are two forward paths from each state variable to the output when $w = n$. If $w < n$, Eq. (5.65) becomes

$$\mathbf{Y}(s) = \Big[c_0 \quad c_1 \quad \cdots \quad c_w \quad 0 \quad \cdots \quad 0\Big]\mathbf{X}(s) \tag{5.66}$$

5.9 PARALLEL STATE DIAGRAMS FROM TRANSFER FUNCTIONS

A single-input single-output (SISO) system represented by the differential equation (5.62) may be represented by an overall transfer function of the form

$$\frac{Y(s)}{U(s)} = G(s) = \frac{c_w s^w + c_{w-1} s^{w-1} + \cdots + c_1 s + c_0}{s^n + a_{n-1} s^{n-1} + \cdots + a_1 s + a_0} \qquad w \le n \tag{5.67}$$

An alternate method of determining a simulation diagram and a set of state variables is to factor the denominator and to express $G(s)$ in partial fractions. When there are no repeated roots and $w = n$, the form is

$$G(s) = \frac{c_n s^n + c_{n-1} s^{n-1} + \cdots + c_1 s + c_0}{(s - \lambda_1)(s - \lambda_2) \cdots (s - \lambda_n)}$$

$$= c_n + \frac{f_1}{s - \lambda_1} + \frac{f_2}{s - \lambda_2} + \cdots + \frac{f_n}{s - \lambda_n} = c_n + \sum_{i=1}^{n} G_i(s)$$

where each

$$G_i(s) = \frac{f_i Z_i(s)}{U(s)} = \frac{f_i}{s - \lambda_i}$$

The symbols z_i and their Laplace transforms $Z_i(s)$ are used for the state variables in order to distinguish the form of the associated diagonal matrix. The output $Y(s)$ produced by the input $U(s)$ is therefore

$$Y(s) = c_n U(s) + \frac{f_1 U(s)}{s - \lambda_1} + \frac{f_2 U(s)}{s - \lambda_2} + \cdots$$

$$= c_n U(s) + f_1 Z_1(s) + f_2 Z_2(s) + \cdots \tag{5.68}$$

The state variables $Z_i(s)$ are selected to satisfy this equation. Each fraction represents a first-order differential equation of the form

$$\dot{z}_i - \lambda_i z_i = u \tag{5.69}$$

This expression can be simulated by an integrator with a feedback path of gain equal to λ_i, followed by a gain f_i. Therefore, the complete simulation diagram is drawn in Fig. 5.31. The reader may add the initial conditions of the states, $z_i(t_0)$, to Fig. 5.31. The $z_i(t_0)$ are the inputs to additional branches having transmittances of value $1/s$ and terminating at the nodes $Z_i(s)$. The term $c_n U(s)$ in Eq. (5.68) is satisfied in Fig. 5.31 by the feedforward path of gain c_n. This term appears only when the numerator and denominator of $G(s)$ have the same degree. The output of each integrator is defined as a state variable. The state equations are therefore of the form

$$\dot{\mathbf{z}} = \begin{bmatrix} \lambda_1 & & & \mathbf{0} \\ & \lambda_2 & & \\ & & \ddots & \\ \mathbf{0} & & & \lambda_n \end{bmatrix} \mathbf{z} + \begin{bmatrix} 1 \\ 1 \\ \vdots \\ 1 \end{bmatrix} \mathbf{u} = \mathbf{\Lambda z} + \mathbf{b}_n \mathbf{u} \tag{5.70}$$

$$\mathbf{y} = \begin{bmatrix} f_1 & f_2 & \cdots & f_n \end{bmatrix} \mathbf{z} + c_n \mathbf{u} = \mathbf{c}_n^T \mathbf{z} + \mathbf{d}_n \mathbf{u} \tag{5.71}$$

An important feature of these equations is that the **A** matrix appears in diagonal or *normal* or *canonical form*. This diagonal form is indicated by $\mathbf{\Lambda}$ (or $\mathbf{A}^*$), and the corresponding state variables are often called *canonical variables*. The elements of the **B** vector (indicated by $\mathbf{b}_n$) are all unity, and the elements of the **C** row matrix (indicated by $[\, f_1, \quad f_2, \quad \ldots, \quad f_n \,] = \mathbf{c}_n^T$) are the coefficients of the partial fractions in Eq. (5.68). The state diagram of Fig. 5.31 and the state equation of Eq. (5.70) represent the Jordan form for the system with distinct eigenvalues. The state and output equations for a multiple-input multiple-output (MIMO) system can be expressed as

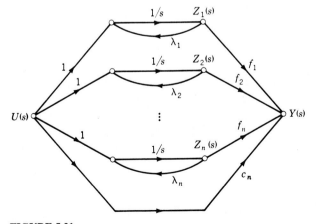

FIGURE 5.31
Simulation of Eq. (5.67) by parallel decomposition for $w = n$ (Jordan diagram for distinct roots).

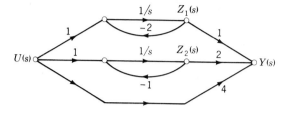

FIGURE 5.32
Simulation diagram for example of Sec. 5.9.

$$\dot{\mathbf{z}} = \mathbf{\Lambda}\mathbf{z} + \mathbf{B}_n\mathbf{u} \tag{5.72}$$

$$\mathbf{y} = \mathbf{C}_n\mathbf{z} + \mathbf{D}_n\mathbf{u} \tag{5.73}$$

The diagonal matrix $\mathbf{A} = \mathbf{\Lambda}$ means that each state equation is *uncoupled;* i.e., each state z_i can be solved independently of the other states. This fact simplifies the procedure for finding the state transition matrix $\mathbf{\Phi}(t)$. This form is also useful for studying the observability and controllability of a system, as discussed in Chapter 13. When the state equations are expressed in normal form, the state variables are often denoted by z_i. Transfer functions with repeated roots are not encountered very frequently and are not considered in this text.[10]

> **Example.** For the given transfer function $G(s)$, draw the parallel state diagram and determine the state equation. A partial-fraction expansion is performed on $G(s)$, and the state diagram is drawn in Fig. 5.32. Then the state and output equations are written.
>
> $$G(s) = \frac{4s^2 + 15s + 13}{s^2 + 3s + 2} = 4 + \frac{1}{s+2} + \frac{2}{s+1}$$
>
> $$\dot{\mathbf{z}}(t) = \begin{bmatrix} -2 & 0 \\ 0 & -1 \end{bmatrix}\mathbf{z}(t) + \begin{bmatrix} 1 \\ 1 \end{bmatrix}\mathbf{u}(t)$$
>
> $$\mathbf{y}(t) = \begin{bmatrix} 1 & 2 \end{bmatrix}\mathbf{z}(t) + 4\mathbf{u}(t)$$

5.10 DIAGONALIZING THE A MATRIX[6,7]

In Sec. 5.9 the method of partial-fraction expansion is shown to lead to the desirable normal form of the state equation, in which the $\mathbf{A}$ matrix is diagonal. The partial-fraction method is not convenient for multiple-input multiple-output systems or when the system equations are already given in state form. Therefore, this section presents a more general method for converting the state equation by means of a linear *similarity transformation.* Since the state variables are not unique, the intention is to transform the state vector $\mathbf{x}$ to a new state vector $\mathbf{z}$ by means of a constant, square, nonsingular transformation matrix $\mathbf{T}$ so that

$$\mathbf{x} = \mathbf{T}\mathbf{z} \tag{5.74}$$

Since $\mathbf{T}$ is a constant matrix, the differentiation of this equation yields

$$\dot{\mathbf{x}} = \mathbf{T}\dot{\mathbf{z}} \tag{5.75}$$

Substituting these values into the state equation $\dot{\mathbf{x}} = \mathbf{Ax} + \mathbf{Bu}$ produces

$$\mathbf{T\dot{z}} = \mathbf{ATz} + \mathbf{Bu} \tag{5.76}$$

Premultiplying by $\mathbf{T}^{-1}$ gives

$$\dot{\mathbf{z}} = \mathbf{T}^{-1}\mathbf{ATz} + \mathbf{T}^{-1}\mathbf{Bu} \tag{5.77}$$

The corresponding output equation is

$$\mathbf{y} = \mathbf{CTz} + \mathbf{Du} \tag{5.78}$$

It is easily shown that the eigenvalues are the same for both the original and the transformed equations. From Eq. (5.77) the new characteristic equation is

$$|\lambda\mathbf{I} - \mathbf{T}^{-1}\mathbf{AT}| = 0 \tag{5.79}$$

In this equation the unit matrix can be replaced by $\mathbf{I} = \mathbf{T}^{-1}\mathbf{T}$. Then, after prefactoring $\mathbf{T}^{-1}$ and postfactoring $\mathbf{T}$, the characteristic equation becomes

$$\left|\mathbf{T}^{-1}(\lambda\mathbf{I} - \mathbf{A})\mathbf{T}\right| = 0 \tag{5.80}$$

Using the property [see Eq. (4.98)] that the determinant of the product of matrices is equal to the product of the determinants of the individual matrices, Eq. (5.80) is equal to

$$|\mathbf{T}^{-1}| \cdot |\lambda\mathbf{I} - \mathbf{A}| \cdot |\mathbf{T}| = |\mathbf{T}^{-1}| \cdot |\mathbf{T}| \cdot |\lambda\mathbf{I} - \mathbf{A}| = 0$$

Since $|\mathbf{T}^{-1}| \cdot |\mathbf{T}| = |\mathbf{I}| = 1$, the characteristic equation of the transformed equation as given by Eq. (5.79) is equal to the characteristic equation of the original system, i.e.,

$$|\lambda\mathbf{I} - \mathbf{A}| = 0 \tag{5.81}$$

Therefore, it is concluded that the eigenvalues are invariant in a linear transformation given by Eq. (5.74).

The matrix $\mathbf{T}$ is called the *modal* matrix when it is selected so that $\mathbf{T}^{-1}\mathbf{AT}$ is diagonal, i.e.,

$$\mathbf{T}^{-1}\mathbf{AT} = \mathbf{\Lambda} = \begin{bmatrix} \lambda_1 & & & 0 \\ & \lambda_2 & & \\ & & \ddots & \\ 0 & & & \lambda_n \end{bmatrix} \tag{5.82}$$

This supposes that the eigenvalues are distinct. With this transformation the system equations are

$$\dot{\mathbf{z}} = \mathbf{\Lambda z} + \mathbf{B}'\mathbf{u} \tag{5.83}$$

$$\mathbf{y} = \mathbf{C}'\mathbf{z} + \mathbf{D}'\mathbf{u} \tag{5.84}$$

where $\quad \mathbf{\Lambda} = \mathbf{T}^{-1}\mathbf{AT} \quad\quad \mathbf{B}' = \mathbf{T}^{-1}\mathbf{B} \quad\quad \mathbf{C}' = \mathbf{CT} \quad\quad \mathbf{D}' = \mathbf{D}$

When $\mathbf{T}$ is selected so that $\mathbf{B}' = \mathbf{T}^{-1}\mathbf{B}$ contains only unit elements, then it is denoted as $\mathbf{B}_n$ as in Eq. (5.70) and in method 3 (see following). When $\mathbf{x} = \mathbf{Tz}$ is used to transform the equations into any other state-variable representation $\mathbf{T}^{-1}\mathbf{AT} = \mathbf{A}'$, then $\mathbf{T}$ is just called a **transformation matrix.** The case where $\mathbf{A}'$ is the companion matrix $\mathbf{A}_c$ is covered in Sec. 5.13.

Four methods are presented for obtaining the modal matrix $\mathbf{T}$ for the case where the matrix $\mathbf{A}$ has distinct eigenvalues. When $\mathbf{A}$ has multiple eigenvalues, this matrix can be transformed into diagonal form only if the number of independent eigenvectors is equal to the multiplicity of the eigenvalues. In such cases, the columns of $\mathbf{T}$ may be determined by methods of 2, 3, and 4.

Method 1

When there are distinct eigenvalues $\lambda_1, \lambda_2, \ldots, \lambda_n$ for the matrix $\mathbf{A}$ and it is in companion form ($\mathbf{A} = \mathbf{A}_c$), the Vandermonde matrix, which is easily obtained, is the modal matrix. The Vandermonde matrix is defined by

$$\mathbf{T} = \begin{bmatrix} 1 & 1 & \cdots & 1 \\ \lambda_1 & \lambda_2 & \cdots & \lambda_n \\ \lambda_1^2 & \lambda_2^2 & \cdots & \lambda_n^2 \\ \vdots & \vdots & \ddots & \vdots \\ \lambda_1^{n-1} & \lambda_2^{n-1} & \cdots & \lambda_n^{n-1} \end{bmatrix} \tag{5.85}$$

Example 1. Transform the state variables in the following equation in order to uncouple the states:

$$\dot{\mathbf{x}} = \begin{bmatrix} 0 & 1 & 0 \\ 0 & 0 & 1 \\ -24 & -26 & -9 \end{bmatrix}\mathbf{x} + \begin{bmatrix} 1 \\ 0 \\ 2 \end{bmatrix}\mathbf{u} \tag{5.86}$$

$$\mathbf{y} = \begin{bmatrix} 3 & 3 & 1 \end{bmatrix}\mathbf{x} \tag{5.87}$$

The characteristic equation $|\lambda\mathbf{I} - \mathbf{A}_c| = 0$ yields the roots $\lambda_1 = -2, \lambda_2 = -3$, and $\lambda_3 = -4$. Since these eigenvalues are distinct and the $\mathbf{A}$ matrix is in companion form, the Vandermonde matrix can be used as the modal matrix. Using Eq. (5.85) yields

$$\mathbf{T} = \begin{bmatrix} 1 & 1 & 1 \\ 2 & -3 & -4 \\ 4 & 9 & 16 \end{bmatrix}$$

A matrix inversion routine (using a programmable calculator or computer) yields

$$\mathbf{T}^{-1} = \frac{1}{2} \begin{bmatrix} 12 & 7 & 1 \\ -16 & -12 & -2 \\ 6 & 5 & 1 \end{bmatrix}$$

To show that this procedure does uncouple the states, the matrix system equations, in terms of the new state vector **z**, are obtained from Eqs. (5.77) and (5.78) as

$$\dot{\mathbf{z}} = \begin{bmatrix} -2 & 0 & 0 \\ 0 & -3 & 0 \\ 0 & 0 & -4 \end{bmatrix} \mathbf{z} + \begin{bmatrix} 7 \\ -10 \\ 4 \end{bmatrix} \mathbf{u} \tag{5.88}$$

$$\mathbf{y} = \begin{bmatrix} 1 & 3 & 7 \end{bmatrix} \mathbf{z} \tag{5.89}$$

This confirms that

$$\mathbf{T}^{-1}\mathbf{A}\mathbf{T} = \mathbf{\Lambda} \tag{5.90}$$

Method 2

The second method does not require the **A** matrix to be in companion form. Pre-multiplying Eq. (5.82) by **T**, where the elements of the modal matrix are defined by $\mathbf{T} = [v_{ij}]$, yields

$$\mathbf{A}\mathbf{T} = \mathbf{T}\mathbf{\Lambda} \tag{5.91}$$

The right side of this equation can be expressed in the form

$$\mathbf{T}\mathbf{\Lambda} = \begin{bmatrix} v_{11} & v_{12} & \cdots & v_{1n} \\ v_{21} & v_{22} & \cdots & v_{2n} \\ \vdots & \vdots & \ddots & \vdots \\ v_{n1} & v_{n2} & \cdots & v_{nn} \end{bmatrix} \begin{bmatrix} \lambda_1 & & & 0 \\ & \lambda_2 & & \\ & & \ddots & \\ 0 & & & \lambda_n \end{bmatrix}$$

$$= \begin{bmatrix} \lambda_1 v_{11} & \lambda_2 v_{12} & \cdots & \lambda_n v_{1n} \\ \lambda_1 v_{21} & \lambda_2 v_{22} & \cdots & \lambda_n v_{2n} \\ \vdots & \vdots & \ddots & \vdots \\ \lambda_1 v_{n1} & \lambda_2 v_{n2} & \cdots & \lambda_n v_{nn} \end{bmatrix} = \begin{bmatrix} \lambda_1 \mathbf{v}_1 & \lambda_2 \mathbf{v}_2 & \cdots & \lambda_n \mathbf{v}_n \end{bmatrix}$$

The column vectors $\mathbf{v}_i$ of the modal matrix

$$\mathbf{T} = \begin{bmatrix} \mathbf{v}_1 & \mathbf{v}_2 & \cdots & \mathbf{v}_n \end{bmatrix} \tag{5.92}$$

are called the eigenvectors. In terms of the $\mathbf{v}_i$, Eq. (5.91) is written as

$$\begin{bmatrix} \mathbf{A}\mathbf{v}_1 & \mathbf{A}\mathbf{v}_2 & \cdots & \mathbf{A}\mathbf{v}_n \end{bmatrix} = \begin{bmatrix} \lambda_1 \mathbf{v}_1 & \lambda_2 \mathbf{v}_2 & \cdots & \lambda_n \mathbf{v}_n \end{bmatrix} \tag{5.93}$$

Equating columns on the left and right sides of Eq. (5.93) yields

$$\mathbf{A}\mathbf{v}_i = \lambda_i \mathbf{v}_i \tag{5.94}$$

This equation can be put in the form

$$[\lambda_i \mathbf{I} - \mathbf{A}] \mathbf{v}_i = 0 \tag{5.95}$$

Equation (5.95) identifies that the eigenvectors [†] $\mathbf{v}_i$ are in the *null space* of the matrix $[\lambda_i\mathbf{I} - \mathbf{A}]$. Using the property $\mathbf{M}$ adj $\mathbf{M} = |\mathbf{M}|\mathbf{I}$ (see Sec. 4.16) and letting $\mathbf{M} = \lambda_i\mathbf{I} - \mathbf{A}$ yields $[\lambda_i\mathbf{I} - \mathbf{A}]$ adj $[\lambda_i\mathbf{I} - \mathbf{A}] = |\lambda_i\mathbf{I} - \mathbf{A}|\mathbf{I}$. Since λ_i is an eigenvalue, the characteristic equation is $|\lambda_i\mathbf{I} - \mathbf{A}| = 0$. Therefore, this equation becomes

$$[\lambda_i\mathbf{I} - \mathbf{A}]\,\text{adj}\,[\lambda_i\mathbf{I} - \mathbf{A}] = 0 \tag{5.96}$$

A comparison of Eqs. (5.95) and (5.96) shows that the eigenvector

$$\mathbf{v}_i \text{ is proportional to any nonzero column of adj}\,[\lambda_i\mathbf{I} - \mathbf{A}] \tag{5.97}$$

Since Eq. (5.77) contains $\mathbf{T}^{-1}$, a necessary condition is that $\mathbf{T}$ must be nonsingular. This is satisfied if the $\mathbf{v}_i$ are linearly independent. When the eigenvalues are distinct, Eq. (5.95) is a homogeneous equation whose rank is $n - 1$, and each value of λ_i yields only one eigenvector $\mathbf{v}_i$. When the eigenvalue λ_i has multiplicity r, the *rank deficiency* α of Eq. (5.95) is $1 \le \alpha \le r$. Rank deficiency denotes that the rank of Eq. (5.95) is $n - \alpha$. Then the linearly independent eigenvectors $\mathbf{v}_i$ that satisfy Eq. (5.95) are α in number. If $\alpha < r$, then the matrix $\mathbf{A}$ cannot be diagonalized.

Example 2. Use adj $[\lambda_i\mathbf{I} - \mathbf{A}]$ to obtain the modal matrix for the equations

$$\dot{\mathbf{x}} = \begin{bmatrix} -9 & 1 & 0 \\ -26 & 0 & 1 \\ -24 & 0 & 0 \end{bmatrix}\mathbf{x} + \begin{bmatrix} 2 \\ 5 \\ 0 \end{bmatrix}u \tag{5.98}$$

$$y = \begin{bmatrix} 1 & 2 & -1 \end{bmatrix}\mathbf{x} \tag{5.99}$$

The characteristic equation $|\lambda\mathbf{I} - \mathbf{A}| = 0$ yields the roots $\lambda_1 = -2, \lambda_2 = -3$, and $\lambda_3 = -4$. These eigenvalues are distinct, but $\mathbf{A}$ is not in companion form. Therefore, the Vandermonde matrix is not the modal matrix. Using the adjoint method gives

$$\text{adj}\,[\lambda\mathbf{I} - \mathbf{A}] = \text{adj}\begin{bmatrix} \lambda+9 & -1 & 0 \\ 26 & \lambda & -1 \\ 24 & 0 & \lambda \end{bmatrix}$$

$$= \begin{bmatrix} \lambda^2 & \lambda & 1 \\ -26\lambda - 24 & \lambda^2 + 9\lambda & \lambda + 9 \\ -24\lambda & -24 & \lambda^2 + 9\lambda + 26 \end{bmatrix}$$

For $\lambda_1 = -2$: $\text{adj}\,[-2\mathbf{I} - \mathbf{A}] = \begin{bmatrix} 4 & -2 & 1 \\ 28 & -14 & 7 \\ 48 & -24 & 12 \end{bmatrix}$ $\mathbf{v}_1 = \begin{bmatrix} 1 \\ 7 \\ 12 \end{bmatrix}$

For $\lambda_2 = -3$: $\text{adj}\,[-3\mathbf{I} - \mathbf{A}] = \begin{bmatrix} 9 & -3 & 1 \\ 54 & -18 & 6 \\ 72 & -24 & 8 \end{bmatrix}$ $\mathbf{v}_2 = \begin{bmatrix} 1 \\ 6 \\ 8 \end{bmatrix}$

For $\lambda_3 = -4$: $\text{adj}\,[-4\mathbf{I} - \mathbf{A}] = \begin{bmatrix} 16 & -4 & 1 \\ 80 & -20 & 5 \\ 96 & -24 & 6 \end{bmatrix}$ $\mathbf{v}_3 = \begin{bmatrix} 1 \\ 5 \\ 6 \end{bmatrix}$

[†] These are sometimes called the right eigenvectors because $[\lambda_i\mathbf{I} - \mathbf{A}]$ is multiplied on the right by $\mathbf{v}_i$ in Eq. (5.95).

For each λ_i the columns of adj $[\lambda_i \mathbf{I} - \mathbf{A}]$ are linearly related; i.e., they are proportional. The $\mathbf{v}_i$ may be multiplied by a constant and are selected to contain the smallest integers; often the leading term is reduced to 1. In practice, it is necessary to calculate only one column of the adjoint matrix. The modal matrix is

$$\mathbf{T} = [\mathbf{v}_1 \quad \mathbf{v}_2 \quad \mathbf{v}_3] = \begin{bmatrix} 1 & 1 & 1 \\ 7 & 6 & 5 \\ 12 & 8 & 6 \end{bmatrix}$$

The inverse is

$$\mathbf{T}^{-1} = -\frac{1}{2} \begin{bmatrix} -4 & 2 & -1 \\ 18 & -6 & 2 \\ -16 & 4 & -1 \end{bmatrix}$$

The matrix equations, in terms of the new state vector $\mathbf{z}$, are

$$\dot{\mathbf{z}} = \begin{bmatrix} -2 & 0 & 0 \\ 0 & -3 & 0 \\ 0 & 0 & -4 \end{bmatrix} \mathbf{z} + \begin{bmatrix} -1 \\ -3 \\ 6 \end{bmatrix} u = \mathbf{\Lambda} \mathbf{z} + \mathbf{B}' u \tag{5.100}$$

$$y = [3 \quad 5 \quad 5] \mathbf{z} = \mathbf{C}' \mathbf{z} \tag{5.101}$$

Once each eigenvalue is assigned the designation λ_1, λ_2, and λ_3, they appear in this order along the diagonal of $\mathbf{\Lambda}$. Note that the systems of Examples 1 and 2 have the same eigenvalues, but different modal matrices are required to uncouple the states.

Method 3

An alternate method for evaluating the n elements of each $\mathbf{v}_i$ is to form a set of n equations from the matrix equation, Eq. (5.94). This is illustrated by the following example.

Example 3. Rework Example 2 using Eq. (5.94). The matrices formed by using the **A** matrix of Eq. (5.98) are

$$\begin{bmatrix} -9 & 1 & 0 \\ -26 & 0 & 1 \\ -24 & 0 & 0 \end{bmatrix} \begin{bmatrix} v_{1i} \\ v_{2i} \\ v_{3i} \end{bmatrix} = \lambda_i \begin{bmatrix} v_{1i} \\ v_{2i} \\ v_{3i} \end{bmatrix} \tag{5.102}$$

Performing the multiplication yields

$$\begin{bmatrix} -9v_{1i} + v_{2i} \\ -26v_{1i} + v_{3i} \\ -24v_{1i} \end{bmatrix} = \begin{bmatrix} \lambda_i v_{1i} \\ \lambda_i v_{2i} \\ \lambda_i v_{3i} \end{bmatrix} \tag{5.103}$$

Each value of λ_i is inserted in this matrix, and the corresponding elements are equated to form three equations. For $\lambda_1 = -2$, the equations are

$$-9v_{11} + v_{21} = -2v_{11}$$

$$-26v_{11} + v_{31} = -2v_{21} \tag{5.104}$$

$$-24v_{11} = -2v_{31}$$

Only two of these equations are independent. This fact can be demonstrated by inserting v_{21} from the first equation and v_{31} from the third equation into the second equation. The result is the identity $v_{11} = v_{11}$. This merely confirms that $|\lambda_i \mathbf{I} - \mathbf{A}|$ is of rank $n - 1 = 2$. Since there are three elements (v_{1i}, v_{2i}, and v_{3i}) in the eigenvector $\mathbf{v}_i$ and only two independent equations are obtained from Eq. (5.102), the procedure is arbitrarily to set one of the elements equal to unity. Thus, after letting $v_{11} = 1$, these equations yield

$$\mathbf{v}_1 = \begin{bmatrix} v_{11} \\ v_{21} \\ v_{31} \end{bmatrix} = \begin{bmatrix} v_{11} \\ 7v_{11} \\ 12v_{11} \end{bmatrix} = \begin{bmatrix} 1 \\ 7 \\ 12 \end{bmatrix}$$

This result is the same as obtained in Example 2. The procedure is repeated to evaluate $\mathbf{v}_2$ and $\mathbf{v}_3$.

The need to arbitrarily select one of the elements in each column vector v_i can be eliminated by specifying that the desired matrix $\mathbf{B}' = \mathbf{B}_n$. In Sec. 5.9 the partial-fraction expansion of the transfer function leads to a $\mathbf{b}_n$ matrix containing all 1s, that is,

$$\mathbf{b}_n = \begin{bmatrix} 1 & 1 & \cdots & 1 \end{bmatrix}^T \tag{5.105}$$

To achieve this form for $\mathbf{b}_n$ the restriction is imposed that $\mathbf{b} = \mathbf{T}\mathbf{b}_n$. This produces n equations that are added to the n^2 equations formed by equating the elements of $\mathbf{A}\mathbf{T} = \mathbf{T}\mathbf{\Lambda}$. Only n^2 of these equations are independent, and they can be solved simultaneously for the n^2 elements of $\mathbf{T}$. The restriction that $\mathbf{b} = \mathbf{T}\mathbf{b}_n$ can be satisfied only if the system is controllable (see Sec. 13.2).

Method 4

The eigenvectors[†] $\mathbf{v}_i$ that satisfy the equation

$$[\lambda_i \mathbf{I} - \mathbf{A}] \mathbf{v}_i = 0 \tag{5.106}$$

are said to lie in the *null* space of the matrix $[\lambda_i \mathbf{I} - \mathbf{A}]$. These eigenvectors can be computed by the following procedure:[9]

1. Use elementary row operations (Sec. 4.16) to transform $[\lambda_i \mathbf{I} - \mathbf{A}]$ into HNF.
2. Rearrange the rows of this matrix if necessary so that the leading unit elements appear on the principal diagonal.
3. Identify the column of this matrix that contains a zero on the principal diagonal. After replacing the zero on the principal diagonal by -1, this column vector is identified as the basis vector for the eigenvector space.

[†]The eigenvectors $\mathbf{v}_i$ are not unique, because they can be formed by any combination of vectors in the null space and can be multiplied by any nonzero constant.

Example 4. Rework Example 2 using method 4.

$$[\lambda_i \mathbf{I} - \mathbf{A}] = \begin{bmatrix} \lambda_i + 9 & -1 & 0 \\ 26 & \lambda_i & -1 \\ 24 & 0 & \lambda_i \end{bmatrix} \tag{5.107}$$

For $\lambda_i = -2$ this matrix and its HNF normal form are

$$\begin{bmatrix} 7 & -1 & 0 \\ 26 & -2 & -1 \\ 24 & 0 & -2 \end{bmatrix} \sim \begin{bmatrix} 1 & 0 & -\dfrac{1}{12} \\ 0 & 1 & -\dfrac{7}{12} \\ 0 & 0 & 0 \end{bmatrix}$$

where $\sim$ indicates the result of elementary operations. The eigenvector $\mathbf{v}_1$ lies in the space defined by the third column after inserting -1 on the principal diagonal. This is written as

$$\mathbf{v}_1 \in \text{span} \left\{ \begin{bmatrix} -\dfrac{1}{12} \\ -\dfrac{7}{12} \\ -1 \end{bmatrix} \right\} \tag{5.108}$$

Similarly, for $\lambda_2 = -3$

$$\begin{bmatrix} 6 & -1 & 0 \\ 26 & -3 & -1 \\ 24 & 0 & -3 \end{bmatrix} \sim \begin{bmatrix} 1 & 0 & -\dfrac{1}{8} \\ 0 & 1 & \dfrac{6}{8} \\ 0 & 0 & 0 \end{bmatrix}$$

The eigenvector $\mathbf{v}_2$ lies in the space defined by the third column after inserting -1 on the principal diagonal, i.e.,

$$\mathbf{v}_2 \in \text{span} \left\{ \begin{bmatrix} -\dfrac{1}{8} \\ -\dfrac{6}{8} \\ -1 \end{bmatrix} \right\} \tag{5.109}$$

Also, for $\lambda_3 = -4$

$$\begin{bmatrix} 5 & -1 & 0 \\ 26 & -4 & -1 \\ 24 & 0 & -4 \end{bmatrix} \sim \begin{bmatrix} 1 & 0 & -\dfrac{1}{6} \\ 0 & 1 & -\dfrac{5}{6} \\ 0 & 0 & 0 \end{bmatrix}$$

The eigenvector $\mathbf{v}_3$ lies in the space defined by the third column after inserting -1 on the principal diagonal, i.e.,

$$\mathbf{v}_3 \in \text{span} \left\{ \begin{bmatrix} -\dfrac{1}{6} \\[6pt] -\dfrac{5}{6} \\[6pt] -1 \end{bmatrix} \right\} \tag{5.110}$$

The eigenvectors are selected from the one-dimensional spaces indicated in Eqs. (5.108) to (5.110). These vectors are linearly independent. For convenience, the basic vectors can be multiplied by a negative constant such that the elements are positive integers. Therefore, one selection for the modal matrix is

$$\mathbf{T} = [\mathbf{v}_1 \quad \mathbf{v}_2 \quad \mathbf{v}_3] = \begin{bmatrix} 1 & 1 & 1 \\ 7 & 6 & 5 \\ 12 & 8 & 6 \end{bmatrix} \tag{5.111}$$

which coincides with the results of method 2.

When the matrix $\mathbf{A}$ has *multiple eigenvalues,* the number of independent eigenvectors associated with the eigenvalue may be equal to the multiplicity of the eigenvalue. This property can be determined by using method 4. The number of columns of the modified HNF of $[\lambda_i \mathbf{I} - \mathbf{A}]$ containing zeros on the principal diagonal may be equal to r, the multiplicity of λ_i. When this occurs, the independent eigenvectors are used to form the modal matrix $\mathbf{T}$, with the result that $\mathbf{T}^{-1}\mathbf{A}\mathbf{T} = \mathbf{\Lambda}$ is a diagonal matrix (see Prob. 5.20).

5.11   USE OF STATE TRANSFORMATION FOR THE STATE EQUATION SOLUTION

The usefulness of the transformation of a matrix to diagonal form is now illustrated in solving the state equation. In Sec. 3.15 the state equation and its solution are given, respectively, by

$$\dot{\mathbf{x}}(t) = \mathbf{A}\mathbf{x}(t) + \mathbf{B}\mathbf{u}(t) \tag{5.112}$$

and

$$\mathbf{x}(t) = e^{\mathbf{A}t}\mathbf{x}(0) + \int_0^t e^{\mathbf{A}(t-\tau)}\mathbf{B}\mathbf{u}(\tau)\, d\tau \tag{5.113}$$

The evaluation of the state transition matrix $e^{\mathbf{A}t}$ is computed in Sec. 3.14 by using the Cayley-Hamilton theorem. This procedure is simplified by first expressing the matrix $\mathbf{A}$ in terms of the diagonalized matrix $\mathbf{\Lambda}$ and the modal matrix $\mathbf{T}$. Thus, from the development in Sec. 5.10

$$\mathbf{A} = \mathbf{T}\mathbf{\Lambda}\mathbf{T}^{-1} \tag{5.114}$$

where $\mathbf{\Lambda}$ is the diagonal matrix given in Eq. (5.82) and $\mathbf{T}$ is the modal matrix given in Eq. (5.92). Using the series representation of $e^{\mathbf{A}t}$ given in Eq. (3.72) and then using Eq. (5.114) yields

$$e^{\mathbf{A}t} = \mathbf{I} + \mathbf{A}t + \frac{\mathbf{A}^2 t^2}{2!} + \frac{\mathbf{A}^3 t^3}{3!} + \cdots$$

$$= \mathbf{I} + (\mathbf{T\Lambda T}^{-1})t + \frac{(\mathbf{T\Lambda T}^{-1})^2 t^2}{2!} + \frac{(\mathbf{T\Lambda T}^{-1})^3 t^3}{3!} + \cdots$$

$$= \mathbf{T}\left(\mathbf{I} + \mathbf{\Lambda}t + \frac{\mathbf{\Lambda}^2 t^2}{2!} + \frac{\mathbf{\Lambda}^3 t^3}{3!} + \cdots\right)\mathbf{T}^{-1} = \mathbf{T}e^{\mathbf{\Lambda}t}\mathbf{T}^{-1} \qquad (5.115)$$

Since $\mathbf{\Lambda}$ is a diagonal matrix, the matrix $e^{\mathbf{\Lambda}t}$ is also diagonal and has the form

$$e^{\mathbf{\Lambda}t} = \begin{bmatrix} e^{\lambda_1 t} & & & \mathbf{0} \\ & e^{\lambda_2 t} & & \\ & & \ddots & \\ \mathbf{0} & & & e^{\lambda_n t} \end{bmatrix} \qquad (5.116)$$

The work of finding the modal matrix $\mathbf{T}$ is therefore balanced by the ease of obtaining $e^{\mathbf{A}t}$. This is illustrated in the following example.

Example. Evaluate $e^{\mathbf{A}t}$ for the example of Sec. 3.13. The plant matrix $\mathbf{A}$ is

$$\mathbf{A} = \begin{bmatrix} 0 & 6 \\ -1 & -5 \end{bmatrix}$$

and the eigenvalues are $\lambda_1 = -2$ and $\lambda_2 = -3$. Since $\mathbf{A}$ is not in companion form, the modal matrix is evaluated using method 2, 3, or 4 of Sec. 5.10:

$$\mathbf{T} = [\mathbf{v}_1 \quad \mathbf{v}_2] = \begin{bmatrix} 3 & 2 \\ -1 & -1 \end{bmatrix}$$

Now, using Eq. (5.115) yields

$$e^{\mathbf{A}t} = \mathbf{T}e^{\mathbf{\Lambda}t}\mathbf{T}^{-1} = \mathbf{T}\begin{bmatrix} e^{\lambda_1 t} & 0 \\ 0 & e^{\lambda_2 t} \end{bmatrix}\mathbf{T}^{-1}$$

$$= \begin{bmatrix} 3 & 2 \\ -1 & -1 \end{bmatrix}\begin{bmatrix} e^{-2t} & 0 \\ 0 & e^{-3t} \end{bmatrix}\begin{bmatrix} 1 & 2 \\ -1 & -3 \end{bmatrix}$$

$$= \begin{bmatrix} 3e^{-2t} - 2e^{-3t} & 6e^{-2t} - 6e^{-3t} \\ -e^{-2t} + e^{-3t} & -2e^{-2t} + 3e^{-3t} \end{bmatrix}$$

This result is the same as that given in Eq. (3.92).

5.12 TRANSFORMING A MATRIX WITH COMPLEX EIGENVALUES

The transformation of variables described in Sec. 5.10 simplifies the system representation mathematically by removing any coupling between the system modes. Thus, diagonalizing the $\mathbf{A}$ matrix produces the decoupled form of system representation shown in Fig. 5.31. Each state equation corresponding to the normal form of the state equation, Eq. (5.83), has the form

$$\dot{z}_i(t) = \lambda_i z_i(t) + b_i u \tag{5.117}$$

Since only one state variable z_i appears in this equation, it can be solved independently of all the other states.

When the eigenvalues are complex, the matrix $\mathbf{T}$ that produces the diagonal matrix $\mathbf{\Lambda}$ contains complex numbers. This property is illustrated for the differential equation

$$\ddot{x} - 2\sigma \dot{x} + \left(\sigma^2 + \omega_d^2\right)x = u(t) \tag{5.118}$$

where $\omega_n^2 = \sigma^2 + \omega_d^2$ and the eigenvalues are $\lambda_{1,2} = \sigma \pm j\omega_d$. Using the phase variables $x_1 = x$ and $x_2 = \dot{x}_1$, the state and output equations are

$$\dot{\mathbf{x}} = \begin{bmatrix} 0 & 1 \\ -\omega_n^2 & 2\sigma \end{bmatrix} \mathbf{x} + \begin{bmatrix} 0 \\ 1 \end{bmatrix} \mathbf{u} \tag{5.119}$$

$$\mathbf{y} = [1 \quad 0]\mathbf{x} \tag{5.120}$$

The modes can be uncoupled, but doing so is undesirable, as shown next. Since the eigenvalues are distinct and $\mathbf{A}$ is a companion matrix, the modal matrix $\mathbf{T}$ is the Vandermonde matrix

$$\mathbf{T} = \begin{bmatrix} 1 & 1 \\ \sigma + j\omega_d & \sigma - j\omega_d \end{bmatrix} = [\mathbf{v}_1 \quad \mathbf{v}_2] \tag{5.121}$$

When the transformation $\mathbf{x} = \mathbf{Tz}$ is used,[8] the normal form of the state and output equations obtained from Eqs. (5.83) and (5.84) are

$$\dot{\mathbf{z}} = \begin{bmatrix} \sigma + j\omega_d & 0 \\ 0 & \sigma - j\omega_d \end{bmatrix} \mathbf{z} + \begin{bmatrix} \dfrac{1}{j2\omega_d} \\ -\dfrac{1}{j2\omega_d} \end{bmatrix} \mathbf{u} \tag{5.122}$$

$$\mathbf{y} = [1 \quad 1]\mathbf{z} \tag{5.123}$$

The simulation diagram based on these equations is shown in Fig. 5.33. The parameters in this diagram are complex quantities, which increases the difficulty when obtaining the mathematical solution. Thus, it may be desirable to perform another transformation in order to obtain a simulation diagram that contains only real quan-

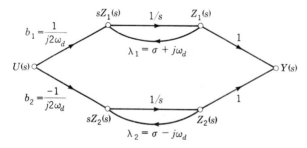

FIGURE 5.33
Simulation diagram for Eqs. (5.122) and (5.123).

tities. The pair of complex-conjugate eigenvalues λ_1 and λ_2 jointly contribute to one transient mode that has the form of a damped sinusoid. The additional transformation is $\mathbf{z} = \mathbf{Q}\mathbf{w}$, where

$$\mathbf{Q} = \begin{bmatrix} \dfrac{1}{2} & -\dfrac{j}{2} \\[2mm] \dfrac{1}{2} & \dfrac{j}{2} \end{bmatrix} \tag{5.124}$$

The new state and output equations are

$$\dot{\mathbf{w}} = \mathbf{Q}^{-1}\mathbf{\Lambda}\mathbf{Q}\mathbf{w} + \mathbf{Q}^{-1}\mathbf{T}^{-1}\mathbf{B}\mathbf{u} = \mathbf{\Lambda}_m\mathbf{w} + \mathbf{B}_m\mathbf{u} \tag{5.125}$$

$$\mathbf{y} = \mathbf{C}\mathbf{T}\mathbf{Q}\mathbf{w} + \mathbf{D}\mathbf{u} = \mathbf{C}_m\mathbf{w} + \mathbf{D}_m\mathbf{u} \tag{5.126}$$

For this example these equations[†] are in the modified canonical form

$$\dot{\mathbf{w}} = \begin{bmatrix} \sigma & \omega_d \\ -\omega_d & \sigma \end{bmatrix}\mathbf{w} + \begin{bmatrix} 0 \\ 1/\omega_d \end{bmatrix}\mathbf{u} \tag{5.127}$$

$$\mathbf{y} = \begin{bmatrix} 1 & 0 \end{bmatrix}\mathbf{w} \tag{5.128}$$

The simulation diagram for these equations is shown in Fig. 5.34. Note that only real quantities appear. Also, the two modes have not been isolated, which is an advantage for complex-conjugate roots.

The effect of the two transformations $\mathbf{x} = \mathbf{T}\mathbf{z}$ and $\mathbf{z} = \mathbf{Q}\mathbf{w}$ can be considered as one transformation given by

$$\mathbf{x} = \mathbf{T}\mathbf{Q}\mathbf{w} = \begin{bmatrix} 1 & 0 \\ \sigma & \omega_d \end{bmatrix}\mathbf{w} = \mathbf{T}_m\mathbf{w} \tag{5.129}$$

Comparison of the modified modal matrix $\mathbf{T}_m$ in Eq. (5.129) with the original matrix in Eq. (5.121) shows that

$$\mathbf{T}_m = \begin{bmatrix} \mathrm{Re}\,(\mathbf{v}_1) & \mathrm{Im}\,(\mathbf{v}_1) \end{bmatrix} \tag{5.130}$$

Therefore, $\mathbf{T}_m$ can be obtained directly from $\mathbf{T}$ without using the $\mathbf{Q}$ transformation.

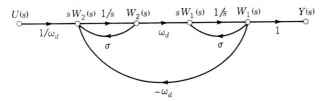

FIGURE 5.34
Simulation diagram for Eqs. (5.127) and (5.128).

[†] Although the trajectories in the state space are not drawn here, the result of this transformation produces trajectories that are symmetrical with respect to the w axes; see Sec. 15.2.

In the general case containing both complex and real eigenvalues the modified block diagonal matrix Λ_m and the resulting matrices $\mathbf{B}_m$ and $\mathbf{C}_m$ have the form

$$\Lambda_m = \begin{bmatrix} \sigma_1 & \omega_1 & & & & & \\ -\omega_1 & \sigma_1 & & & & & \mathbf{0} \\ & & \sigma_2 & \omega_2 & & & \\ & & -\omega_2 & \sigma_2 & & & \\ & & & & \lambda_5 & & \\ & & & & & \ddots & \\ \mathbf{0} & & & & & & \lambda_n \end{bmatrix} \qquad \mathbf{B}_m = \mathbf{T}_m^{-1}\mathbf{B} \qquad (5.131)$$

$$\mathbf{C}_m = \mathbf{C}\mathbf{T}_m \qquad (5.132)$$

With this modified matrix Λ_m the two oscillating modes produced by the two sets of complex eigenvalues $\sigma_1 \pm j\omega_1$ and $\sigma_2 \pm j\omega_2$ are uncoupled. The column vectors contained in $\mathbf{T}$ for the conjugate eigenvalues are also conjugates. Thus

$$\mathbf{T} = \begin{bmatrix} \mathbf{v}_1 & \mathbf{v}_1^* & \mathbf{v}_3 & \mathbf{v}_3^* & \mathbf{v}_5 & \cdots & \mathbf{v}_n \end{bmatrix} \qquad (5.133)$$

The modified modal matrix $\mathbf{T}_m$ can then be obtained directly by using

$$\mathbf{T}_m = \begin{bmatrix} \text{Re}(\mathbf{v}_1) & \text{Im}(\mathbf{v}_1) & \text{Re}(\mathbf{v}_3) & \text{Im}(\mathbf{v}_3) & \mathbf{v}_5 & \cdots & \mathbf{v}_n \end{bmatrix} \qquad (5.134)$$

5.13 TRANSFORMING AN A MATRIX INTO COMPANION FORM

The synthesis of feedback control systems and the analysis of their response characteristics is often considerably facilitated when the plant and control matrices in the matrix state equation have particular forms. The transformation from a physical or a general state variable $\mathbf{x}_p$ to the phase variable $\mathbf{x}$ is presented in this section.[10] In this case the state equation in terms of the general state variable is

$$\dot{\mathbf{x}}_p = \mathbf{A}_p\mathbf{x}_p + \mathbf{b}_p\mathbf{u} \qquad (5.135)$$

and the transformed state equation in terms of the phase variables is

$$\dot{\mathbf{x}} = \mathbf{A}_c\mathbf{x} + \mathbf{b}_c\mathbf{u} \qquad (5.136)$$

where the companion matrix $\mathbf{A}_c$ and the vector $\mathbf{b}_c$ have the respective forms

$$\mathbf{A}_c = \begin{bmatrix} 0 & 1 & & & & & \\ & 0 & 1 & & & \mathbf{0} & \\ & & 0 & \ddots & & & \\ & & & \ddots & \ddots & & \\ \mathbf{0} & & & \cdots & 0 & 1 \\ -a_0 & -a_1 & -a_2 & \cdots & -a_{n-2} & -a_{n-1} \end{bmatrix} \qquad \mathbf{b}_c = \begin{bmatrix} 0 \\ 0 \\ \vdots \\ 0 \\ 1 \end{bmatrix} \qquad (5.137)$$

The state equation given by Eq. (5.136) is often described in the literature as being in the *control canonical form*. The general state and phase variables are related by

$$\mathbf{x}_p = \mathbf{T}\mathbf{x} \tag{5.138}$$

so that
$$\mathbf{A}_c = \mathbf{T}^{-1}\mathbf{A}_p\mathbf{T} \qquad \mathbf{b}_c = \mathbf{T}^{-1}\mathbf{b}_p \tag{5.139}$$

Since the characteristic polynomial is invariant under a similarity transformation, it can be obtained from $\mathbf{Q}(\lambda) = |\lambda\mathbf{I} - \mathbf{A}_p|$ or by the method of Sec. 4.16 and has the form

$$\mathbf{Q}(\lambda) = \lambda^n + a_{n-1}\lambda^{n-1} + \cdots + a_1\lambda + a_0 \tag{5.140}$$

Thus, by using the coefficients of Eq. (5.140), the companion matrix can readily be formed, placing the negatives of the coefficients along the bottom row of $\mathbf{A}_c$.

The controllability matrix [see Eq. (13.68)] for the phase-variable state equation is

$$\mathbf{M}_{cc} = \begin{bmatrix} \mathbf{b}_c & \mathbf{A}_c\mathbf{B}_c & \cdots & \mathbf{A}_c^{n-1}\mathbf{b}_c \end{bmatrix}$$

$$= \begin{bmatrix} \mathbf{0} & & & & & 1 \\ & & & 0 & & \cdot^{\cdot^{\cdot}} \\ & & 0 & 1 & \cdots \\ & 0 & 1 & -a_{n-1} & \cdots \\ 0 & 1 & -a_{n-1} & a_{n-1}^2 - a_{n-2} & \cdots \\ 1 & -a_{n-1} & a_{n-1}^2 - a_{n-2} & -a_{n-1}^3 + 2a_{n-1}a_{n-2} - a_{n-3} & \cdots \end{bmatrix} \tag{5.141}$$

The matrix $\mathbf{M}_{cc}{}^{\dagger}$ in Eq. (5.141) can be formed since the coefficients of the characteristic polynomial $\mathbf{Q}(\lambda)$ are known. The matrix $\mathbf{M}_{cc}$ expressed in terms of the transformed matrices of Eq. (5.139) is

$$\mathbf{M}_{cc} = \begin{bmatrix} \mathbf{T}^{-1}\mathbf{b}_p & \mathbf{T}^{-1}\mathbf{A}_p\mathbf{b}_p & \mathbf{T}^{-1}\mathbf{A}_p^2\mathbf{b}_p & \cdots & \mathbf{T}^{-1}\mathbf{A}_p^{n-1}\mathbf{b}_p \end{bmatrix}$$

$$= \mathbf{T}^{-1}\begin{bmatrix} \mathbf{b}_p & \mathbf{A}_p\mathbf{b}_p & \mathbf{A}_p^2\mathbf{b}_p & \cdots & \mathbf{A}_p^{n-1}\mathbf{b}_p \end{bmatrix} = \mathbf{T}^{-1}\mathbf{M}_{cp} \tag{5.142}$$

The matrix $\mathbf{M}_{c_p}$ is obtained by using $\mathbf{b}_p$ and $\mathbf{A}_p$ of the general state-variable equation (5.135). Equation (5.142) yields

$$\mathbf{T} = \mathbf{M}_{cp}\mathbf{M}_{cc}^{-1} \tag{5.143}$$

An alternative procedure is to determine $\mathbf{T}^{-1}$ as follows:

---

†The matrix $\mathbf{M}_{cc}$ has *striped* form, that is, the elements constituting the secondary diagonal (from the lower left to the upper right) are all equal. The same is true of all elements in the lines parallel to it.

1. Form the augmented matrix $\begin{bmatrix} \mathbf{M}_{cp} & \mathbf{I} \end{bmatrix}$.
2. Perform elementary row operations (see Sec. 4.16) until this matrix has the form $\begin{bmatrix} \mathbf{M}_{cc} & \mathbf{T}^{-1} \end{bmatrix}$.
3. The right half of this matrix identifies the required transformation matrix.

Example. A general state-variable equation contains the matrices

$$\mathbf{A}_p = \begin{bmatrix} 1 & 6 & -3 \\ -1 & -1 & 1 \\ -2 & 2 & 0 \end{bmatrix} \qquad \mathbf{b}_p = \begin{bmatrix} 1 \\ 1 \\ 1 \end{bmatrix} \tag{5.144}$$

for which the characteristic polynomial is evaluated in Sec. 4.16 as

$$\mathbf{Q}(\lambda) = \lambda^3 - 3\lambda + 2 = (\lambda - 1)^2(\lambda + 2) \tag{5.145}$$

The matrix $\mathbf{M}_{cc}$ obtained in accordance with Eq. (5.141) is

$$\mathbf{M}_{cc} = \begin{bmatrix} 0 & 0 & 1 \\ 0 & 1 & 0 \\ 1 & 0 & 3 \end{bmatrix} \tag{5.146}$$

The augmented matrix obtained in accordance with step 1 of the procedure is

$$\begin{bmatrix} \mathbf{M}_{cp} & \mathbf{I} \end{bmatrix} = \begin{bmatrix} 1 & 4 & -2 & 1 & 0 & 0 \\ 1 & -1 & -3 & 0 & 1 & 0 \\ 1 & 0 & -10 & 0 & 0 & 1 \end{bmatrix} \tag{5.147}$$

Performing row operations until this matrix has the form $[\mathbf{M}_{cc} \quad \mathbf{T}^{-1}]$ yields

$$\begin{bmatrix} 0 & 0 & 1 & \vdots & 1/36 & 4/36 & -5/36 \\ 0 & 1 & 0 & \vdots & 7/36 & -8/36 & 1/36 \\ 1 & 0 & 3 & \vdots & 13/36 & 52/36 & -29/36 \end{bmatrix} = \begin{bmatrix} \mathbf{M}_{cc} & \mathbf{T}^{-1} \end{bmatrix} \tag{5.148}$$

The matrix $\mathbf{T}^{-1}$ from Eq. (5.148) yields the transformation matrix

$$\mathbf{T} = \begin{bmatrix} -5 & 4 & 1 \\ -6 & -1 & 1 \\ -13 & 0 & 1 \end{bmatrix} \tag{5.149}$$

which produces the desired forms of the matrices

$$\mathbf{A}_c = \mathbf{T}^{-1}\mathbf{A}_p\mathbf{T} = \begin{bmatrix} 0 & 1 & 0 \\ 0 & 0 & 1 \\ -2 & 3 & 0 \end{bmatrix} \tag{5.150}$$

and

$$\mathbf{b}_c = \mathbf{T}^{-1}\mathbf{b}_p = \begin{bmatrix} 0 \\ 0 \\ 1 \end{bmatrix} \tag{5.151}$$

The transformation from the general state variable $\mathbf{x}_p$ to the phase variable $\mathbf{x}$ can also be accomplished by first obtaining the transfer function $G(s) = \mathbf{c}_p^T[s\mathbf{I} - \mathbf{A}_p]^{-1}\mathbf{b}_p$. Then the state and output equations can be obtained directly by using Eqs. (5.38) and (5.44), respectively. The advantage of the method presented in this section is that it avoids having to obtain the inverse of the *polynomial matrix* $[s\mathbf{I} - \mathbf{A}_p]^{-1}$.

5.14 SUMMARY

The foundation has been completed in Chapters 2 to 5 for analyzing the performance of feedback control systems. The analysis of such systems and the adjustments of parameters to achieve the desired performance are covered in the following chapters. In order to provide clarity in understanding the principles and to avoid ambiguity, the systems considered in Chapters 6 to 15 are restricted primarily to single-input, single-output (SISO) systems. The procedures can be extended to multiple-input multiple-output (MIMO) systems, and some synthesis techniques for multivariable systems are presented in Chapters 16 through 18. The block diagram representation containing a single feedback loop is the basic unit used to develop the various techniques of analysis.

Since state variables are not unique, there are several ways of selecting them. One method is the use of physical variables, as in Chapter 2. The selection of the state variables as phase or canonical variables is demonstrated in this chapter through the medium of the simulation diagram. This diagram is similar to an analog-computer diagram and uses integrators, amplifiers, and summers. The outputs of the integrators in the simulation diagram are defined as the state variables.

Preparation for overall system analysis has been accomplished by writing the conventional differential and state equations in Chapter 2, obtaining the solution of these equations in Chapter 3, introducing the Laplace transform in Chapter 4, and incorporating this into the representation of a complete system in Chapter 5. Matrix methods are also presented in these chapters in preparation for system synthesis via the use of modern control theory.

REFERENCES

1. Etkin, B.: *Dynamics of Atmospheric Flight,* John Wiley & Sons, New York, 1972.
2. *IEEE Standard Dictionary of Electrical and Electronics Terms,* Wiley-Interscience, New York, 1972.
3. Mason, S. J.: "Feedback Theory: Further Properties of Signal Flow Graphs," *Proc. IRE,* vol. 44, no. 7, pp. 920–926, July 1956.
4. Mason, S. J., and H. J. Zimmerman: *Electronic Circuits, Signals, and Systems,* The M.I.T. Press, Cambridge, Mass., 1960.
5. Kuo, B. C.: *Linear Networks and Systems,* McGraw-Hill, New York, 1967.
6. Trimmer, J. D.: *Response of Physical Systems,* John Wiley & Sons, New York, 1950, p. 17.
7. Langholz, G., and S. Frankenthal: "Reduction to Normal Form of a State Equation in the Presence of Input Derivatives," *Intern. J. Systems Science,* vol. 5, no. 7, pp. 705–706, July 1974.
8. Crossley, T. R., and B. Porter: "Inversion of Complex Matrices," *Electronics Letters,* vol. 6, no. 4, pp. 90–91, February 1970.
9. Reid, J. G.: *Linear System Fundamentals: Continuous and Discrete, Classic and Modern,* McGraw-Hill, New York, 1983.
10. DeRusso, P. M., R. J. Roy, and C. M. Close: *State Variables for Engineers,* Krieger, Malabar, Fla., 1990.
11. Wiberg, D. M.: *State Space and Linear Systems,* Schaum's Outline Series, McGraw-Hill, New York, 1971.

CHAPTER
6

CONTROL-SYSTEM CHARACTERISTICS

6.1 INTRODUCTION

Open- and closed-loop transfer functions have certain basic characteristics that permit transient and steady-state analyses of the feedback-controlled system. Five factors of prime importance in feedback-control systems are *stability, the existence and magnitude of the steady-state error, controllability, observability, and parameter sensitivity.* The stability characteristic of a linear time-invariant system is determined from the system's characteristic equation. Routh's stability criterion provides a means for determining stability without evaluating the roots of this equation. The steady-state characteristics are obtainable from the open-loop transfer function for unity-feedback systems (or equivalent unity-feedback systems), yielding figures of merit and a ready means for classifying systems. The first two properties are developed in this chapter. The remaining properties are presented in Chapters 12 through 14. Computer-aided-design (CAD) programs like MATLAB[6], ICECAP-PC, and TOTAL-PC (see Appendix B) are available to assist in the determination of these system characteristics.

6.2 ROUTH'S STABILITY CRITERION[1,3,8,9]

The *response transform* $X_2(s)$ has the general form given by Eq. (4.32), which is repeated here in slightly modified form. $X_1(s)$ is the driving transform.

$$X_2(s) = \frac{P(s)}{Q(s)} X_1(s) = \frac{P(s)X_1(s)}{b_n s^n + b_{n-1} s^{n-1} + b_{n-2} s^{n-2} + \cdots + b_1 s + b_0} \quad (6.1)$$

Sections 4.6, 4.7, and 4.10 describe the methods used to evaluate the inverse transform $\mathcal{L}^{-1}[F(s)] = f(t)$. However, before the inverse transformation can be performed, the polynomial $Q(s)$ must be factored. CAD programs are readily avail-able for obtaining the roots of a polynomial.[5,6] Section 4.13 shows that stability of the response $x_2(t)$ requires that all zeros of $Q(s)$ have negative real parts. Since it is usually not necessary to find the exact solution when the response is unstable, a sim-ple procedure to determine the existence of zeros with positive real parts is needed. If such zeros of $Q(s)$ with positive real parts are found, the system is unstable and must be modified. Routh's criterion is a simple method of determining the number of zeros with positive real parts without actually solving for the zeros of $Q(s)$. Note that zeros of $Q(s)$ are poles of $X_2(s)$. The characteristic equation is

$$Q(s) = b_n s^n + b_{n-1} s^{n-1} + b_{n-2} s^{n-2} + \cdots + b_1 s + b_0 = 0 \qquad (6.2)$$

If the b_0 term is zero, divide by s to obtain the equation in the form of Eq. (6.2). The b's are real coefficients, and all powers of s from s^n to s^0 must be present in the characteristic equation. A necessary but not sufficient condition for stable roots is that all the coefficients in Eq. (6.2) be positive. If any coefficients other than b_0 are zero, or if all the coefficients do not have the same sign, then there are pure imaginary roots or roots with positive real parts and the system is unstable. In that case it is unnecessary to continue if only stability or instability is to be determined. When all the coefficients are present and positive, the system may or may not be stable because there still may be roots on the imaginary axis or in the right-half s plane. Routh's criterion is mainly used to determine stability. In special situations it may be necessary to determine the actual number of roots in the right-half s plane. For these situations the procedure described in this section can be used.

The coefficients of the characteristic equation are arranged in the pattern shown in the first two rows of the following Routhian array. These coefficients are then used to evaluate the rest of the constants to complete the array.

s^n	b_n	b_{n-2}	b_{n-4}	b_{n-6} $\cdots$
s^{n-1}	b_{n-1}	b_{n-3}	b_{n-5}	b_{n-7} $\cdots$
s^{n-2}	c_1	c_2	c_3	$\cdots$
s^{n-3}	d_1	d_2	$\cdots$	
$\cdots$	$\cdots$	$\cdots$		
s^1	j_1			
s^0	k_1			

The constants c_1, c_2, c_3, etc., in the third row are evaluated as follows:

$$c_1 = \frac{b_{n-1} b_{n-2} - b_n b_{n-3}}{b_{n-1}} \qquad (6.3)$$

$$c_2 = \frac{b_{n-1} b_{n-4} - b_n b_{n-5}}{b_{n-1}} \qquad (6.4)$$

$$c_3 = \frac{b_{n-1} b_{n-6} - b_n b_{n-7}}{b_{n-1}} \qquad (6.5)$$

This pattern is continued until the rest of the c's are all equal to zero. Then the d row is formed by using the s^{n-1} and s^{n-2} row. The constants are

$$d_1 = \frac{c_1 b_{n-3} - b_{n-1} c_2}{c_1} \tag{6.6}$$

$$d_2 = \frac{c_1 b_{n-5} - b_{n-1} c_3}{c_1} \tag{6.7}$$

$$d_3 = \frac{c_1 b_{n-7} - b_{n-1} c_4}{c_1} \tag{6.8}$$

This process is continued until no more d terms are present. The rest of the rows are formed in this way down to the s^0 row. The complete array is triangular, ending with the s^0 row. Notice that the s^1 and s^0 rows contain only one term each. Once the array has been found, *Routh's criterion states that the number of roots of the characteristic equation with positive real parts is equal to the number of changes of sign of the coefficients in the first column.* Therefore, the system is stable if all terms in the first column have the same sign.[†]

The following example illustrates this criterion:

$$Q(s) = s^5 + s^4 + 10s^3 + 72s^2 + 152s + 240 \tag{6.9}$$

The Routhian array is formed by using the procedure described above:

s^5	1	10	152
s^4	1	72	240
s^3	-62	-88	
s^2	70.6	240	
s^1	122.6		
s^0	240		

In the first column there are two changes of sign, from 1 to -62 and from -62 to 70.6; therefore, $Q(s)$ has two roots in the right-half s plane (RHP). Note that this criterion gives the number of roots with positive real parts but does not tell the values of the roots. If Eq. (6.9) is factored, the roots are $s_1 = -3, s_{2,3} = -1 \pm j\sqrt{3}$, and $s_{4,5} = +2 \pm j4$. This calculation confirms that there are two roots with positive real parts. The Routh criterion does not distinguish between real and complex roots.

Theorem 1.   Division of a row.   The coefficients of any row may be multiplied or divided by a positive number without changing the signs of the first column. The labor of evaluating the coefficients in Routh's array can be reduced by multiplying or dividing any row by a constant. This may result, for example, in reducing the size of the coefficients and therefore simplifying the evaluation of the remaining coefficients.

The following example illustrates this theorem:

$$Q(s) = s^6 + 3s^5 + 2s^4 + 9s^3 + 5s^2 + 12s + 20 \tag{6.10}$$

[†]Reference 7 shows that a system is unstable if there is a negative element in any position in any row. Thus, it is not necessary to complete the Routhian array if any negative number is encountered.

The Routhian array is

$$
\begin{array}{c|cccc}
s^6 & 1 & 2 & 5 & 20 \\
s^5 & \cancel{3} & \cancel{9} & \cancel{12} & \\
 & 1 & 3 & 4 & \text{(after dividing by 3)} \\
s^4 & -1 & 1 & 20 & \\
s^3 & \cancel{4} & \cancel{24} & & \\
 & 1 & 6 & & \text{(after dividing by 4)} \\
s^2 & 7 & 20 & & \\
s^1 & 22 & & & \text{(after multiplying by 7)} \\
s^0 & 20 & & &
\end{array}
$$

Notice that the size of the numbers has been reduced by dividing the s^5 row by 3 and the s^3 row by 4. The result is unchanged; i.e., there are two changes of signs in the first column and, therefore, there are two roots with positive real parts.

> **Theorem 2.   A zero coefficient in the first column.**   When the first term in a row is zero but not all the other terms are zero, the following methods can be used:
>
> 1. Substitute $s = 1/x$ in the original equation; then solve for the roots of x with positive real parts. The number of roots x with positive real parts will be the same as the number of s roots with positive real parts.
>
> 2. Multiply the original polynomial by the factor $(s+1)$, which introduces an additional negative root. Then form the Routhian array for the new polynomial.

Both of these methods are illustrated in the following example:

$$Q(s) = s^4 + s^3 + 2s^2 + 2s + 5 \tag{6.11}$$

The Routhian array is

$$
\begin{array}{c|ccc}
s^4 & 1 & 2 & 5 \\
s^3 & 1 & 2 & \\
s^2 & 0 & 5 &
\end{array}
$$

The zero in the first column prevents completion of the array. The following methods overcome this problem.

METHOD 1.   Letting $s = 1/x$ and rearranging the polynomial gives

$$Q(x) = 5x^4 + 2x^3 + 2x^2 + x + 1 \tag{6.12}$$

The new Routhian array is

$$
\begin{array}{c|ccc}
x^4 & 5 & 2 & 1 \\
x^3 & 2 & 1 & \\
x^2 & -1 & 2 & \\
x^1 & 5 & & \\
x^0 & 2 & &
\end{array}
$$

There are two changes of sign; therefore, there are two roots of x in the RHP. The number of roots s with positive real parts is also two. This method does not work when the coefficients of $Q(s)$ and of $Q(x)$ are identical.

METHOD 2.

$$Q_1(s) = Q(s)(s + 1) = s^5 + 2s^4 + 3s^3 + 4s^2 + 7s + 5$$

$$
\begin{array}{c|ccc}
s^5 & 1 & 3 & 7 \\
s^4 & 2 & 4 & 5 \\
s^3 & 2 & 9 & \\
s^2 & -10 & 10 & \\
s^1 & 11 & & \\
s^0 & 10 & &
\end{array}
$$

The same result is obtained by both methods. There are two changes of sign in the first column, so there are two zeros of $Q(s)$ with positive real parts. An additional method is described in Ref. 5.

Theorem 3.   A zero row.   When all the coefficients of one row are zero, the procedure is as follows:

1. The auxiliary equation can be formed from the preceding row, as shown below.
2. The Routhian array can be completed by replacing the all-zero row with the coefficients obtained by differentiating the auxiliary equation.
3. The roots of the auxiliary equation are also roots of the original equation. These roots occur in pairs and are the negative of each other. Therefore, these roots may be imaginary (complex conjugates) or real (one positive and one negative), may lie in quadruplets (two pairs of complex-conjugate roots), etc.

Consider the system that has the characteristic equation

$$Q(s) = s^4 + 2s^3 + 11s^2 + 18s + 18 = 0 \qquad (6.13)$$

The Routhian array is

$$
\begin{array}{c|ccc}
s^4 & 1 & 11 & 18 \\
s^3 & \cancel{2} & \cancel{18} & \\
& 1 & 9 & \text{(after dividing by 2)} \\
s^2 & \cancel{2} & \cancel{18} & \\
& 1 & 9 & \text{(after dividing by 2)} \\
s^1 & 0 & &
\end{array}
$$

The presence of a zero row (the s^1 row) indicates that there are roots that are the negatives of each other. The next step is to form the auxiliary equation from the preceding row, which is the s^2 row. The highest power of s is s^2, and only even powers of s appear. Therefore, the auxiliary equation is

$$s^2 + 9 = 0 \qquad (6.14)$$

The roots of this equation are

$$s = \pm j3$$

These are also roots of the original equation. The presence of imaginary roots indicates that the output includes a sinusoidally oscillating component.

To complete the Routhian array, the auxiliary equation is differentiated and is

$$2s + 0 = 0 \tag{6.15}$$

The coefficients of this equation are inserted in the s^1 row, and the array is then completed:

$$
\begin{array}{c|c}
s^1 & 2 \\
s^0 & 9
\end{array}
$$

Since there are no changes of sign in the first column, there are no roots with positive real parts.

In feedback systems, covered in detail in the following chapters, the ratio of the output to the input does not have an explicitly factored denominator. An example of such a function is

$$\frac{X_2(s)}{X_1(s)} = \frac{P(s)}{Q(s)} = \frac{K(s + 2)}{s(s + 5)(s^2 + 2s + 5) + K(s + 2)} \tag{6.16}$$

The value of K is an adjustable parameter in the system and may be positive or negative. The value of K determines the location of the poles and therefore the stability of the system. It is important to know the range of values of K for which the system is stable. This information must be obtained from the characteristic equation, which is

$$Q(s) = s^4 + 7s^3 + 15s^2 + (25 + K)s + 2K = 0 \tag{6.17}$$

The coefficients must all be positive in order for the zeros of $Q(s)$ to lie in the left half of the s plane (LHP), but this condition is not sufficient for stability. The Routhian array permits evaluation of precise boundaries for K:

$$
\begin{array}{c|ccc}
s^4 & 1 & 15 & 2K \\
s^3 & 7 & 25 + K & \\
s^2 & 80 - K & 14K & \\
s^1 & \dfrac{(80 - K)(25 + K) - 98K}{80 - K} & & \\
s^0 & 14K & &
\end{array}
$$

The term $80 - K$ for the s^2 row imposes the restriction $K < 80$, and the s^0 row requires $K > 0$. The numerator of the first term in the s^1 row is equal to $-K^2 - 43K + 2000$, and this function must be positive for a stable system. By use of the quadratic formula the zeros of this function are $K = 28.1$ and $K = -71.1$, and the numerator of the s^1 row is positive between these values. The combined restrictions on K for stability of the system are therefore $0 < K < 28.1$. For the value $K = 28.1$ the characteristic equation has imaginary roots that can be evaluated by applying

Theorem 3 to form the auxiliary equation. Also, for $K = 0$, it can be seen from Eq. (6.16) that there is no output. The methods for selecting the "best" value of K in the range between 0 and 28.1 are contained in later chapters. It is important to note that the Routh criterion provides useful but restricted information. Another method of determining stability is Hurwitz's criterion,[1,4] which establishes the necessary conditions in terms of the system determinants.

6.3 MATHEMATICAL AND PHYSICAL FORMS

In various systems the controlled variable, labeled C, shown in Fig. 6.1 may have the physical form of position, speed, temperature, rate of change of temperature, voltage, rate of flow, pressure, etc. Once the blocks in the diagram are related to transfer functions, it is immaterial to the analysis of the system what the physical form of the controlled variable may be. Generally, the important quantities are the controlled quantity c, its rate of change Dc, and its second derivative D^2c, that is, the first several derivatives of c, including the zeroth derivative. For any specific control system each of these "mathematical" functions has a definite "physical" meaning. For example, if the controlled variable c is position, then Dc is velocity and D^2c is acceleration. As a second example, if the controlled variable c is velocity, then Dc is acceleration and D^2c is the rate of change of acceleration.

Often the input signal to a system has an irregular form, such as that shown in Fig. 6.2, that cannot be expressed by any simple equation. This prevents a straightforward analysis of system response. It is noted, though, that the signal form shown in Fig. 6.2 may be considered to be composed of three basic forms of known types of input signals, i.e., a step in the region cde, a ramp in the region $0b$, and a parabola in the region ef. Thus, if the given linear system is analyzed separately for each of these types of input signals, there is then established a fair measure of performance with the irregular input. For example, consider the control system for a missile

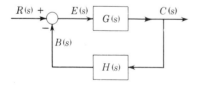

FIGURE 6.1
Simple feedback system.

FIGURE 6.2
Input signal to a system.

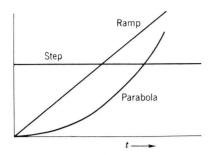

FIGURE 6.3
Graphical forms of step, ramp, and parabolic input functions.

that receives its command input from an interceptor guidance system (see Fig. 5.3). This input changes slowly compared with the control-system dynamics. Thus, step and ramp inputs adequately approximate most command inputs. Therefore, the three standard inputs not only approximate most command inputs but also provide a means for comparing the performance of different systems.

Consider that the system shown in Fig. 6.1 is a position-control system. Feedback control systems are often analyzed on the basis of a unit step input signal (see Fig. 6.3). This system can first be analyzed on the basis that the unit step input signal $r(t)$ represents position. Since this gives only a limited idea of how the system responds to the actual input signal, the system can then be analyzed on the basis that the unit step signal represents a constant velocity $Dr(t) = u_{-1}(t)$. This in reality gives an input position signal of the form of a ramp (Fig. 6.3) and thus a closer idea of how the system responds to the actual input signal. In the same manner the unit step input signal can represent a constant acceleration, $D^2r(t) = u_{-1}(t)$, to obtain the system's performance to a parabolic position input signal. The curves shown in Fig. 6.3 then represent acceleration, velocity, and position.

6.4 FEEDBACK SYSTEM TYPES

The simple closed-loop feedback system, with unity feedback, shown in Fig. 6.4, may be called a *tracker* since the output $c(t)$ is expected to track or follow the input $r(t)$. The open-loop transfer function for this system is $G(s) = C(s)/E(s)$, which is determined by the components of the actual control system. Generally $G(s)$ has one of the following mathematical forms:

FIGURE 6.4
Unity-feedback control system.

$$G(s) = \frac{K_0(1 + T_1s)(1 + T_2s)\cdots}{(1 + T_as)(1 + T_bs)\cdots} \tag{6.18}$$

$$G(s) = \frac{K_1(1 + T_1s)(1 + T_2s)\cdots}{s(1 + T_as)(1 + T_bs)\cdots} \tag{6.19}$$

$$G(s) = \frac{K_2(1 + T_1s)(1 + T_2s)\cdots}{s^2(1 + T_as)(1 + T_bs)\cdots} \tag{6.20}$$

Note that the constant term in each factor is equal to unity. The preceding equations are expressed in a more generalized manner by defining the standard form of the transfer function as

$$G(s) = \frac{K_m(1 + b_1s + b_2s^2 + \cdots + b_ws^w)}{s^m(1 + a_1s + a_2s^2 + \cdots + a_us^u)} = K_mG'(s) \tag{6.21}$$

where
$$a_1, a_2, \ldots = \text{constant coefficients}$$
$$b_1, b_2, \ldots = \text{constant coefficients}$$
$$K_m = \text{gain constant of the transfer function } G(s)$$
$$m = 0, 1, 2, \ldots \text{ denotes the transfer function type}$$
$$G'(s) = \text{forward transfer function with unity gain}$$

The degree of the denominator is $n = m + u$. For a unity-feedback system, E and C have the same units. Therefore, K_0 is nondimensional, K_1 has the units of seconds^{-1}, and K_2 has the units of seconds^{-2}.

In order to analyze each control system, a "type" designation is introduced. The designation is based upon the value of the exponent m of s in Eq. (6.21). Thus, when $m = 0$, the system represented by this equation is called a Type 0 system; when $m = 1$, it is called a Type 1 system; when $m = 2$, it is called a Type 2 system; etc. [See Prob. 6.20 for a Type -1 ($m = -1$) system.] Once a physical system has been expressed mathematically, the analysis is independent of the nature of the physical system. It is immaterial whether the system is electrical, mechanical, hydraulic, thermal, or a combination of these. The most common feedback control systems have Type 0, 1, or 2 open-loop transfer functions, as shown in Eqs. (6.18) to (6.20). It is important to analyze each type thoroughly and to relate it as closely as possible to its transient and steady-state solution.

The various types exhibit the following steady-state properties:

Type 0: A constant actuating signal results in a constant value for the controlled variable.

Type 1: A constant actuating signal results in a constant rate of change (constant velocity) of the controlled variable.

Type 2: A constant actuating signal results in a constant second derivative (constant acceleration) of the controlled variable.

Type 3: A constant actuating signal results in a constant rate of change of acceleration of the controlled variable.

These classifications lend themselves to definition in terms of the differential equations of the system and to identification in terms of the forward transfer function. For all classifications the degree of the denominator of $G(s)H(s)$ usually is equal to or greater than the degree of the numerator because of the physical nature of feedback control systems. That is, in every physical system there are energy-storage and dissipative elements such that there can be no instantaneous transfer of energy from the input to the output. However, exceptions do occur.

6.5 ANALYSIS OF SYSTEM TYPES

The properties presented in the preceding section are now examined in detail for each type of $G(s)$ appearing in stable unity-feedback tracking systems. First, remember the following theorems:

Final-value theorem:

$$\lim_{t \to \infty} f(t) = \lim_{s \to 0} sF(s) \tag{6.22}$$

Differentiation theorem:

$$\mathscr{L}[D^m c(t)] = s^m C(s) \qquad \text{when all initial conditions are zero} \tag{6.23}$$

Also, when the input is a polynomial, the steady-state output of a stable closed-loop system that has *unity feedback* has the same form as the input. Therefore, if the input is a ramp function, the steady-state output must also include a ramp function plus a constant, and so forth.

From the preceding section, it is seen that the forward transfer function defines the system type and is generally in the factored form

$$G(s) = \frac{C(s)}{E(s)} = \frac{K_m(1 + T_1 s)(1 + T_2 s) \cdots}{s^m (1 + T_a s)(1 + T_b s)(1 + T_c s) \cdots} \tag{6.24}$$

Solving this equation for $E(s)$ yields

$$E(s) = \frac{(1 + T_a s)(1 + T_b s)(1 + T_c s) \cdots}{K_m(1 + T_1 s)(1 + T_2 s) \cdots} s^m C(s) \tag{6.25}$$

Thus, applying the final-value theorem gives

$$e(t)_{ss} = \lim_{s \to 0} [sE(s)] = \lim_{s \to 0} \left[\frac{s(1 + T_a s)(1 + T_b s)(1 + T_c s) \cdots}{K_m(1 + T_1 s)(1 + T_2 s) \cdots} s^m C(s) \right]$$

$$= \lim_{s \to 0} \frac{s[s^m C(s)]}{K_m} \tag{6.26}$$

But applying the final-value theorem to Eq. (6.23) gives

$$\lim_{s \to 0} s[s^m C(s)] = D^m c(t)_{ss} \tag{6.27}$$

Therefore, Eq. (6.26) can be written as

$$e(t)_{ss} = \frac{D^m c(t)_{ss}}{K_m} \tag{6.28}$$

or
$$K_m e(t)_{ss} = D^m c(t)_{ss} \tag{6.29}$$

This equation relates a derivative of the output to the error; it is most useful for the case where $D^m c(t)_{ss} =$ constant. Then $e(t)_{ss}$ must also equal a constant, that is, $e(t)_{ss} = E_0$, and Eq. (6.29) may be expressed as

$$K_m E_0 = D^m c(t)_{ss} = \text{constant} = C_m \tag{6.30}$$

Note that $C(s)$ has the form

$$
\begin{aligned}
C(s) &= \frac{G(s)}{1 + G(s)} R(s) \\
&= \frac{K_m[(1 + T_1 s)(1 + T_2 s) \cdots]}{s^m(1 + T_a s)(1 + T_b s) \cdots + K_m(1 + T_1 s)(1 + T_2 s) \cdots} R(s) \quad (6.31)
\end{aligned}
$$

The expression for $E(s)$ in terms of the input $R(s)$ is obtained as follows:

$$E(s) = \frac{C(s)}{G(s)} = \frac{1}{G(s)} \frac{G(s)R(s)}{1 + G(s)} = \frac{R(s)}{1 + G(s)} \tag{6.32}$$

With $G(s)$ given by Eq. (6.24), the expression for $E(s)$ is

$$E(s) = \frac{s^m(1 + T_a s)(1 + T_b s) \cdots R(s)}{s^m(1 + T_a s)(1 + T_b s) \cdots + K_m(1 + T_1 s)(1 + T_2 s) \cdots} \tag{6.33}$$

Applying the final-value theorem to Eq. (6.33) yields

$$e(t)_{ss} = \lim_{s \to 0} s \left[\frac{s^m(1 + T_a s)(1 + T_b s) \cdots R(s)}{s^m(1 + T_a s)(1 + T_b s) \cdots + K_m(1 + T_1 s)(1 + T_2 s) \cdots} \right] \tag{6.34}$$

Equation (6.34) relates the steady-state error to the input; it is now analyzed for various system types and for step, ramp, and parabolic inputs.

Case 1: $m = 0$ (Type 0 System)

STEP INPUT $r(t) = R_0 u_{-1}(t), R(s) = R_0/s$. From Eq. (6.34):

$$e(t)_{ss} = \frac{R_0}{1 + K_0} = \text{constant} = E_0 \neq 0 \tag{6.35}$$

Applying the final-value theorem to Eq. (6.31) yields:

$$c(t)_{ss} = \frac{K_0}{1 + K_0} R_0 \tag{6.36}$$

From Eq. (6.35) it is seen that a Type 0 system with a constant input produces a constant value of the output and a constant actuating signal. [The same results can be obtained by applying Eq. (6.30).] Thus in a Type 0 system a fixed error E_0 is

FIGURE 6.5
Steady-state response of a Type 0 system with a step input.

required to produce a desired constant output C_0; that is, $K_0 E_0 = c(t)_{ss} = $ constant $= C_0$. For steady-state conditions:

$$e(t)_{ss} = r(t)_{ss} - c(t)_{ss} = R_0 - C_0 = E_0 \qquad (6.37)$$

Differentiating the preceding equation yields $Dr(t)_{ss} = Dc(t)_{ss} = 0$. Figure 6.5 illustrates these results.

RAMP INPUT $r(t) = R_1 t u_{-1}(t), R(s) = R_1/s^2$. From Eq. (6.34), $e(t)_{ss} = \infty$. Also, by the use of the Heaviside partial-fraction expansion, the particular solution of $e(t)$ obtained from Eq. (6.33) contains the term $[R_1/(1 + K_0)] t$. Therefore, the conclusion is that a Type 0 system with a ramp-function input produces a ramp output with a smaller slope; thus, there is an error that increases with time and approaches a value of infinity, meaning that a Type 0 system cannot follow a ramp input.

Similarly, a Type 0 system cannot follow the parabolic input

$$r(t) = \frac{R_2 t^2}{2} u_{-1}(t) \qquad (6.38)$$

since $e(t)_{ss} = r(t)_{ss} - c(t)_{ss}$ approaches a value of infinity.

Case 2: $m = 1$ **(Type 1 System)**

STEP INPUT $R(s) = R_0/s.$ From Eq. (6.34), $e(t)_{ss} = 0$. Therefore, a Type 1 system with a constant input produces a steady-state constant output of value identical with the input. That is, for a Type 1 system there is zero steady-state error between the output and input for a step input, i.e.,

$$e(t)_{ss} = r(t)_{ss} - c(t)_{ss} = 0 \qquad (6.39)$$

The preceding analysis is in agreement with Eq. (6.30). That is, for a step input, the steady-state output must also be a constant $c(t) = C_0 u_{-1}(t)$ so that

$$Dc(t)_{ss} = 0 = K_1 E_0 \qquad (6.40)$$

or

$$E_0 = 0 \qquad (6.41)$$

RAMP INPUT $R(s) = R_1/s^2.$ From Eq. (6.34)

$$e(t)_{ss} = \frac{R_1}{K_1} = \text{constant} = E_0 \neq 0 \qquad (6.42)$$

From Eq. (6.42) it can be seen that a Type 1 system with a ramp input produces a constant actuating signal. That is, in a Type 1 system a fixed error E_0 is required to produce a ramp output. This result can also be obtained from Eq. (6.30), that is, $K_1 E_0 = Dc(t)_{ss} = \text{constant} = C_1$. For steady-state conditions:

$$e(t)_{ss} = r(t)_{ss} - c(t)_{ss} = E_0 \tag{6.43}$$

For the ramp input

$$r(t) = R_1 t u_{-1}(t) \tag{6.44}$$

the particular solution for the output has the form of a power series

$$c(t)_{ss} = C_0 + C_1 t \tag{6.45}$$

Substituting Eqs. (6.44) and (6.45) into Eq. (6.43) yields:

$$E_0 = R_1 t - C_0 - C_1 t \tag{6.46}$$

This result can occur only with

$$R_1 = C_1 \tag{6.47}$$

This result signifies that the slope of the ramp input and the ramp output are equal. Of course, this condition is necessary if the difference between input and output is a constant. Figure 6.6 illustrates these results. The delay of the steady-state output ramp shown in Fig. 6.6 is equal to

$$\text{Delay} = \frac{e(t)_{ss}}{Dr(t)} = \frac{1}{K_1} \tag{6.48}$$

PARABOLIC INPUT $r(t) = (R_2 t^2/2) u_{-1}(t), R(s) = R_2/S^3$. The particular solution of $e(t)$ obtained from Eq. (6.33) contains the term

$$e(t) = \frac{R_2 t}{K_1} \tag{6.49}$$

Therefore, a Type 1 system with a parabolic input produces a parabolic output, but with an error that increases with time and approaches a value of infinity. This limit means that a Type 1 system cannot follow a parabolic input.

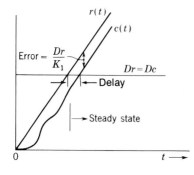

FIGURE 6.6
Steady-state response of a Type 1 system with a ramp input.

Case 3: $m = 2$ (Type 2 System)

STEP INPUT $R(s) = R_0/s$. Performing a corresponding analysis in the same manner as for Cases 1 and 2 reveals that a Type 2 system can follow a step input with zero steady-state error, i.e., $c(t)_{ss} = r(t) = R_0 u_{-1}(t)$ and $e(t)_{ss} = 0$.

RAMP INPUT $R(s) = R_1/s^2$. Performing a corresponding analysis in the same manner as for Cases 1 and 2 reveals that a Type 2 system can follow a ramp input with zero steady-state error, i.e., $c(t)_{ss} = r(t) = R_1 t$ and $e(t)_{ss} = 0$.

PARABOLIC INPUT $R(s) = R_2/s^3$. From Eq. (6.34)

$$e(t)_{ss} = \frac{R_2}{K_2} = \text{constant} = E_0 \neq 0 \tag{6.50}$$

From Eq. (6.50) it can be seen that a Type 2 system with a parabolic input produces a parabolic output with a constant actuating signal. That is, in a Type 2 system a fixed error E_0 is required to produce a parabolic output. This result is further confirmed by applying Eq. (6.30), i.e.,

$$K_2 E_0 = D^2 c(t)_{ss} = \text{constant} = C_2 \tag{6.51}$$

Thus, for steady-state conditions

$$e(t)_{ss} = r(t)_{ss} - c(t)_{ss} = E_0 \tag{6.52}$$

For a parabolic input given by

$$r(t) = \frac{R_2 t^2}{2} u_{-1}(t) \tag{6.53}$$

the particular solution of the output must be given by the power series

$$c(t)_{ss} = \frac{C_2}{2} t^2 + C_1 t + C_0 \tag{6.54}$$

Substituting Eqs. (6.53) and (6.54) into Eq. (6.52) yields

$$E_0 = \frac{R_2 t^2}{2} - \frac{C_2 t^2}{2} - C_1 t - C_0 \tag{6.55}$$

Differentiating Eq. (6.55) twice results in $C_1 = 0$ and $R_2 = C_2$. Therefore, $E_0 = -C_0$ and the input and output curves have the same shape but are displaced by a constant error. Figure 6.7 illustrates the results obtained above.

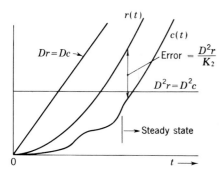

FIGURE 6.7
Steady-state response of a Type 2 system with a parabolic input.

TABLE 6.1
Steady-state response characteristics for stable unity-feedback systems

System type m	$r(t)_{SS}$	$c(t)_{SS}$	$e(t)_{SS}$	$e(\infty)$	Derivatives
0	$R_0 u_{-1}(t)$	$\frac{K_0}{1+K_0}R_0$	$\frac{R_0}{1+K_0}$	$\frac{R_0}{1+K_0}$	$Dr = Dc = 0$
	$R_1 t u_{-1}(t)$	$\frac{K_0 R_1}{1+K_0}t + C_0$	$\frac{R_1}{1+K_0}t - C_0$	∞	$Dr \neq Dc$
	$\frac{R_2 t^2}{2}u_{-1}(t)$	$\frac{K_0 R_2}{2(1+K_0)}t^2 + C_1 t + C_0$	$\frac{R_2}{2(1+K_0)}t^2 - C_1 t - C_0$	∞	$Dr \neq Dc$
1	$R_0 u_{-1}(t)$	R_0	0	0	$Dr = Dc = 0$
	$R_1 t u_{-1}(t)$	$R_1 t - \frac{R_1}{K_1}$	$\frac{R_1}{K_1}$	$\frac{R_1}{K_1}$	$Dr = Dc = R_1$
	$\frac{R_2 t^2}{2}u_{-1}(t)$	$\frac{R_2 t^2}{2} - \frac{R_2}{K_1}t + C_0$	$\frac{R_2}{K_1}t - C_0$	∞	$Dr \neq Dc$
2	$R_0 u_{-1}(t)$	R_0	0	0	$Dr = Dc = 0$
	$R_1 t u_{-1}(t)$	$R_1 t$	0	0	$Dr = Dc = R_1$
	$\frac{R_2 t^2}{2}u_{-1}(t)$	$\frac{R_2 t^2}{2} - \frac{R_2}{K_2}$	$\frac{R_2}{K_2}$	$\frac{R_2}{K_2}$	$D^2 r = D^2 c = R_2$ $Dr = Dc = R_2 t$

The results determined in this section verify the properties stated in the previous section for the system types. The steady-state response characteristics for Types 0, 1, and 2 unity-feedback systems are given in Table 6.1 and apply only to stable systems.

6.6 EXAMPLE: TYPE 2 SYSTEM

A Type 2 system is one in which the second derivative of the output is maintained constant by a constant actuating signal. In Fig. 6.8 is shown a positioning system that is a Type 2 system, where

K_b = motor back-emf constant, V/(rad/s)

K_x = potentiometer constant, V/rad

K_T = motor torque constant, lb · ft/A

A = integrator amplifier gain, s^{-1}

K_g = generator constant, V/A

$T_m = JR_{gm}/(K_b K_T + R_{gm}B)$ = motor mechanical constant, s

T_f = generator field constant, s

$K_M = K_T/(BR_{gm} + K_T K_b)$ = overall motor constant, rad/V · s

FIGURE 6.8
Position control of a space-vehicle camera (Type 2 system).

Assume that the inductance L_{g_m} of the motor-generator control unit, described in Secs. 2.13 and 2.14, is negligible. The transfer function of this unit, with the inductance neglected, is given by Eq. (2.135). Figure 6.9 is the block diagram representation of the Type 2 position-control system shown in Fig. 6.8. Thus the open-loop transfer function is

$$G(s) = \frac{\theta_o(s)}{E(s)} = \frac{K_2}{s^2(1 + T_f s)(1 + T_m s)} \tag{6.56}$$

where $K_2 = K_x A K_g K_M / R_f$ has the units of seconds^{-2}.

This Type 2 control system is unstable. As shown in Chapter 10, an appropriate cascade compensator can be added to the forward path to produce a stable system. The new transfer function, with $T_1 > T_2$, has the form

$$G_0(s) = G_c(s)G(s) = \frac{K_2'(1 + T_1 s)}{s^2(1 + T_f s)(1 + T_m s)(1 + T_2 s)} \tag{6.57}$$

For a constant input $r(t) = R_0 u_{-1}(t)$ the steady-state value of $c(t)_{ss}$ is

$$c(t)_{ss} = \lim_{s \to 0}\left[s \frac{C(s)}{R(s)} \frac{R_0}{s}\right] = \lim_{s \to 0}\left[\frac{G_0(s)}{1 + G_0(s)} R_0\right] = R_0 \tag{6.58}$$

Thus
$$E_0 = r(t)_{ss} - c(t)_{ss} = R_0 - R_0 = 0 \tag{6.59}$$

As expected, a Type 2 system follows a step-function input with no error.

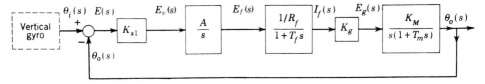

FIGURE 6.9
Block diagram representation of the system of Fig. 6.8.

6.7 STEADY-STATE ERROR COEFFICIENTS[2]

The definition of system types is the first step toward establishing a set of standard characteristics that permit the engineer to obtain as much information as possible about a given system with a minimum amount of calculation. Also, these standard characteristics must point the direction in which a given system must be modified to meet a given set of performance specifications. An important characteristic is the ability of a system to achieve the desired steady-state output with a minimum error. Thus, in this section there are defined *system error coefficients* that are a measure of a stable unity-feedback control system's steady-state accuracy for a given desired output that is relatively constant or slowly varying.

In Eq. (6.30) it is shown that a constant actuating signal exists when the derivative of the output is constant. This derivative is proportional to the actuating signal E_0 and to a constant K_m, which is the gain of the forward transfer function. The conventional names of these constants for the Type 0, 1, and 2 systems are *position, velocity,* and *acceleration error coefficients,* respectively. Since the conventional names were originally selected for application to position-control systems (servomechanisms), these names referred to the actual physical form of $c(t)$ or $r(t)$, which represent position, as well as to the mathematical form of $c(t)$, that is, c, Dc, and D^2c. These names are ambiguous when the analysis is extended to cover control of temperature, velocity, etc. In order to define general terms that are universally applicable, the authors have selected the terminology *step, ramp,* and *parabolic steady-state error coefficients.* Table 6.2 lists the conventional and the author's designation of the error coefficients.

The *error coefficients are independent of the system type. They apply to any system type and are defined for specific forms of the input, i.e., for a step, ramp, or parabolic input. These error coefficients are applicable only for stable unity-feedback systems.* The results are summarized in Table 6.3.

Steady-State Step Error Coefficient

The step error coefficient is defined as

$$\text{Step error coefficient} = \frac{\text{Steady-state value of output } c(t)_{ss}}{\text{Steady-state actuating signal } e(t)_{ss}} = K_p \quad (6.60)$$

TABLE 6.2
Correspondence between the conventional and the author's designation of steady-state error coefficients

Symbol	Conventional designation of error coefficients	Authors' designation of error coefficients
K_p	Position	Step
K_v	Velocity	Ramp
K_a	Acceleration	Parabolic

TABLE 6.3
Definitions of steady-state error coefficients for stable unity-feedback systems

Error coefficient	Definition of error coefficient	Value of error coefficient	Form of input signal $r(t)$
Step (K_P)	$\dfrac{c(t)_{ss}}{e(t)_{ss}}$	$\lim\limits_{s \to 0} G(s)$	$R_0 u_{-1}(t)$
Ramp (K_V)	$\dfrac{(Dc)_{ss}}{e(t)_{ss}}$	$\lim\limits_{s \to 0} sG(s)$	$R_1 t u_{-1}(t)$
Parabolic (K_a)	$\dfrac{(D^2 c)_{ss}}{e(t)_{ss}}$	$\lim\limits_{s \to 0} s^2 G(s)$	$\dfrac{R_1 t^2}{2} u_{-1}(t)$

and applies only for a step input, $r(t) = R_0 u_{-1}(t)$. The steady-state value of the output is obtained by applying the final-value theorem to Eq. (6.31):

$$c(t)_{ss} = \lim_{s \to 0} sC(s) = \lim_{s \to 0} \left[\frac{sG(s)}{1 + G(s)} \frac{R_0}{s} \right] = \lim_{s \to 0} \left[\frac{G(s)}{1 + G(s)} R_0 \right] \quad (6.61)$$

Similarly, from Eq. (6.32), for a unity-feedback system

$$e(t)_{ss} = \lim_{s \to 0} \left[s \frac{1}{1 + G(s)} \frac{R_0}{s} \right] = \lim_{s \to 0} \left[\frac{1}{1 + G(s)} R_0 \right] \quad (6.62)$$

Substituting Eqs. (6.61) and (6.62) into Eq. (6.60) yields

$$\text{Step error coefficient} = \frac{\lim\limits_{s \to 0} \left[\dfrac{G(s)}{1 + G(s)} R_0 \right]}{\lim\limits_{s \to 0} \left[\dfrac{1}{1 + G(s)} R_0 \right]} \quad (6.63)$$

Since both the numerator and the denominator of Eq. (6.63) in the limit can never be zero or infinity simultaneously, where $K_m \neq 0$, the indeterminate forms 0/0 and ∞/∞ never occur. Thus, this equation reduces to

$$\text{Step error coefficient} = \lim_{s \to 0} G(s) = K_p \quad (6.64)$$

Therefore, applying Eq. (6.64) to each type system yields

$$K_p = \begin{cases} \lim\limits_{s \to 0} \dfrac{K_0(1 + T_1 s)(1 + T_2 s) \cdots}{(1 + T_a s)(1 + T_b s)(1 + T_c s) \cdots} = K_0 & \text{Type 0 system} \quad (6.65) \\[4mm] \infty & \text{Type 1 system} \quad (6.66) \\[2mm] \infty & \text{Type 2 system} \quad (6.67) \end{cases}$$

Steady-State Ramp Error Coefficient

The ramp error coefficient is defined as

$$\text{Ramp error coefficient} = \frac{\text{Steady-state value of derivative of output } Dc(t)_{ss}}{\text{Steady-state actuating signal } e(t)_{ss}}$$

$$= K_v \tag{6.68}$$

and applies only for a ramp input, $r(t) = R_1 t u_{-1}(t)$. The first derivative of the output is given by

$$\mathcal{L}[Dc] = sC(s) = \frac{sG(s)}{1 + G(s)}R(s) \tag{6.69}$$

The steady-state value of the derivative of the output is obtained by using the final-value theorem:

$$Dc(t)_{ss} = \lim_{s \to 0} s\,[sC(s)] = \lim_{s \to 0}\left[\frac{s^2 G(s)}{1 + G(s)}\frac{R_1}{s^2}\right] = \lim_{s \to 0}\left[\frac{G(s)}{1 + G(s)}R_1\right] \tag{6.70}$$

Similarly, from Eq. (6.32), for a unity-feedback system

$$e(t)_{ss} = \lim_{s \to 0}\left[s\frac{1}{1 + G(s)}\frac{R_1}{s^2}\right] = \lim_{s \to 0}\left[\frac{1}{1 + G(s)}\frac{R_1}{s}\right] \tag{6.71}$$

Substituting Eqs. (6.70) and (6.71) into Eq. (6.68) yields

$$\text{Ramp error coefficient} = \frac{\displaystyle\lim_{s \to 0}\left[\frac{G(s)}{1 + G(s)}R_1\right]}{\displaystyle\lim_{s \to 0}\left[\frac{1}{1 + G(s)}\frac{R_1}{s}\right]} \tag{6.72}$$

Since this equation never has the indeterminate form 0/0 or ∞/∞, it can be simplified to

$$\text{Ramp error coefficient} = \lim_{s \to 0} sG(s) = K_v \tag{6.73}$$

Therefore, applying Eq. (6.73) to each type system yields

$$K_v = \begin{cases} \displaystyle\lim_{s \to 0}\frac{sK_0(1 + T_1 s)(1 + T_2 s)\cdots}{(1 + T_a s)(1 + T_b s)(1 + T_c s)\cdots} = 0 & \text{Type 0 system} \quad (6.74) \\[4mm] K_1 & \text{Type 1 system} \quad (6.75) \\[2mm] \infty & \text{Type 2 system} \quad (6.76) \end{cases}$$

Steady-State Parabolic Error Coefficient

The parabolic error coefficient is defined as

Parabolic error coefficient =

$$\frac{\text{Steady-state value of second derivative of output } D^2c(t)_{ss}}{\text{Steady-state actuating signal } e(t)_{ss}} = K_a \qquad (6.77)$$

and applies only for a parabolic input, $r(t) = (R_2t^2/2)u_{-1}(t)$. The second derivative of the output is given by

$$\mathcal{L}\left[D^2c\right] = s^2C(s) = \frac{s^2G(s)}{1 + G(s)}R(s) \qquad (6.78)$$

The steady-state value of the second derivative of the output is obtained by using the final-value theorem:

$$D^2c(t)_{ss} = \lim_{s \to 0} s\left[s^2C(s)\right] = \lim_{s \to 0}\left[\frac{s^3G(s)}{1 + G(s)}\frac{R_2}{s^3}\right] = \lim_{s \to 0}\left[\frac{G(s)}{1 + G(s)}R_2\right] \ (6.79)$$

Similarly, from Eq. (6.32), for a unity-feedback system

$$e(t)_{ss} = \lim_{s \to 0}\left[s\frac{1}{1 + G(s)}\frac{R_2}{s^3}\right] = \lim_{s \to 0}\left[\frac{1}{1 + G(s)}\frac{R_2}{s^2}\right] \qquad (6.80)$$

Substituting Eqs. (6.79) and (6.80) into Eq. (6.77) yields

$$\text{Parabolic error coefficient} = \frac{\lim\limits_{s \to 0}\left[\dfrac{G(s)}{1 + G(s)}R_2\right]}{\lim\limits_{s \to 0}\left[\dfrac{1}{1 + G(s)}\dfrac{R_2}{s^2}\right]} \qquad (6.81)$$

Since this equation never has the indeterminate form 0/0 or ∞/∞, it can be simplified to

$$\text{Parabolic error coefficient} = \lim_{s \to 0} s^2G(s) = K_a \qquad (6.82)$$

Therefore, applying Eq. (6.82) to each type system yields

$$K_a = \begin{cases} \lim\limits_{s \to 0} \dfrac{s^2K_0(1 + T_1s)(1 + T_2s)\cdots}{(1 + T_as)(1 + T_bs)(1 + T_cs)\cdots} = 0 & \text{Type 0 system} \quad (6.83) \\[2mm] 0 & \text{Type 1 system} \quad (6.84) \\[2mm] K_2 & \text{Type 2 system} \quad (6.85) \end{cases}$$

Based upon the steady-state error coefficient definitions and referring to the corresponding Figs. 6.5 through 6.7, the values of K_0, K_1, and K_2 can also be determined from the computer simulation of the control system. For example, from Fig. 6.6 and using Eq. (6.48), the gain is $K_1 = 1/\text{delay}$.

6.8 CAD ACCURACY CHECKS: CADAC

When appropriate, the steady-state error coefficient definitions can serve as CADAC. For example, before entering the $G(s)$ function into a CAD package, determine the values of K_m. This step is followed by obtaining the values of K_m from the expression of $G(s)$ on the computer screen or printed out by the CAD package. If the two sets of values agree, then the next required CAD manipulation is performed.

6.9 USE OF STEADY-STATE ERROR COEFFICIENTS

The use of steady-state error coefficients is discussed for stable unity-feedback systems, as illustrated in Fig. 6.10. Note that the steady-state error coefficients *can only be used* to find errors for unity-feedback systems that are *also* stable.

Type 1 System

For a Type 1 system with a step or ramp input considered at steady state, the value of $(Dc)_{ss}$ is

$$Dc(t)_{ss} = K_1 E_0 \tag{6.86}$$

Thus, the larger, the value of K_1, the smaller the size of the actuating signal e necessary to maintain a constant rate of change of the output. From the standpoint of trying to maintain $c(t) = r(t)$ at all times, a larger K_1 results in a more sensitive system. In other words, a larger K_1 results in a greater speed of response of the system to a given actuating signal $e(t)$. Therefore, the gain K_1 is another standard characteristic of a system's performance. The maximum value of K_1 is limited by stability considerations and is discussed in later chapters.

For the Type 1 system with a step input, the step error coefficient is equal to infinity, and the steady-state error is therefore zero. The steady-state output $c(t)_{ss}$ for a Type 1 system is equal to the input when $r(t) = $ constant.

Consider now a ramp input $r(t) = R_1 t u_{-1}(t)$. The steady-state value of $Dc(t)_{ss}$ is found by using the final-value theorem:

$$Dc(t)_{ss} = \lim_{s \to 0} s\,[sC(s)] = \lim_{s \to 0}\left[s\frac{sG(s)}{1 + G(s)}R(s)\right] \tag{6.87}$$

where $R(s) = R_1/s^2$ and

$$G(s) = \frac{K_1(1 + T_1 s)(1 + T_2 s)\cdots(1 + T_w s)}{s(1 + T_a s)(1 + T_b s)\cdots(1 + T_u s)}$$

FIGURE 6.10
Unity-feedback system.

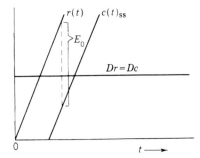

FIGURE 6.11
Steady-state response of a Type 1 system for $Dr =$ constant.

Inserting these values into Eq. (6.87) gives

$$
Dc(t)_{ss} = \lim_{s \to 0} \left[s \frac{\dfrac{sK_1(1 + T_1 s)(1 + T_2 s) \cdots (1 + T_w s)}{s(1 + T_a s)(1 + T_b s) \cdots (1 + T_u s)}}{+ K_1(1 + T_1 s)(1 + T_2 s) \cdots (1 + T_w s)} \frac{R_1}{s^2} \right] = R_1
$$

Therefore, $$(Dc)_{ss} = (Dr)_{ss} \tag{6.88}$$

The magnitude of the steady-state error is found by using the ramp error coefficient. From Eq. (6.73),

$$
K_v = \lim_{s \to 0} sG(s) = K_1
$$

From the definition of ramp error coefficient, the steady-state error is

$$
e(t)_{ss} = \frac{Dc(t)_{ss}}{K_1} \tag{6.89}
$$

Since $Dc(t)_{ss} = Dr(t)$,

$$
e(t)_{ss} = \frac{Dr}{K_1} = \frac{R_1}{K_1} = E_0 \tag{6.90}
$$

Therefore, a Type 1 system follows a ramp input with a constant error E_0. Figure 6.11 illustrates these conditions graphically.

Table of Steady-State Error Coefficients

Table 6.4 gives the values of the error coefficients for the Type 0, 1, and 2 systems. These values are determined from Table 6.3. The reader should be able to make ready use of Table 6.4 for evaluating the appropriate error coefficient. The error coefficient is used with the definitions given in Table 6.3 to evaluate the magnitude of the steady-state error.

TABLE 6.4
Steady-state error coefficients for stable systems

System type	Step error coefficient K_p	Ramp error coefficient K_v	Parabolic error coefficient K_a
0	K_0	0	0
1	∞	K_1	0
2	∞	∞	K_2

Polynomial Input: t^{m+1}

Note that a Type m system can follow an input of the form t^{m-1} with zero steady-state error. It can follow an input t^m, but there is a constant steady-state error. It cannot follow an input t^{m+1} because the steady-state error approaches infinity. However, if the input is present only for a finite length of time, the error is also finite. Then the error can be evaluated by taking the inverse Laplace transform of Eq. (6.33) and inserting the value of time. The maximum permissible error limits the time $(0 < t < t_1)$ that an input t^{m+1} can be applied to a control system.

6.10   NONUNITY-FEEDBACK SYSTEM

The nonunity-feedback system of Fig. 6.1 may be mathematically converted to an equivalent unity-feedback system from which its "effective" system type and steady-state error coefficient can be determined. The control ratio for Fig. 6.1 is

$$\frac{C(s)}{R(s)} = \frac{G(s)}{1 + G(s)H(s)} = \frac{N(s)}{D(s)} \tag{6.91}$$

and for the *equivalent* unity-feedback control system of the form of Fig. 6.4, the control ratio is

$$\frac{C(s)}{R(s)} = \frac{G_{eq}(s)}{1 + G_{eq}(s)} \tag{6.92}$$

Since the transfer functions $G(s)$ and $H(s)$ are known, equating Eqs. (6.91) and (6.92) yields

$$G_{eq}(s) = \frac{N(s)}{D(s) - N(s)} \tag{6.93}$$

When the nonunity-feedback system is stable, its steady-state performance characteristics can be determined based on Eq. (6.93).

Equation (6.91) can also be expressed in the form

$$\frac{C(s)}{R(s)} = \left[\frac{G(s)H(s)}{1 + G(s)H(s)} \right] \frac{1}{H(s)} \tag{6.94}$$

Thus, the nonunity feedback system of Fig. 6.1 can be represented by the block diagram of Fig. 6.12 in accordance with Eq. (6.94). This diagram shows a unity

FIGURE 6.12
Equivalent nonunity feedback representation of Fig. 6.1.

feedback system in cascade with $1/H(s)$. When H is a constant, Fig. 6.12 is an especially useful representation that permits the design to be performed using unity feedback methods.

6.10 SUMMARY

In this chapter the physical forms of the reference input and the controlled variable are related to the mathematical representations of these quantities. Since, in general, the forward transfer functions of most unity feedback control systems fall into three categories, they can be identified as Type 0, 1, and 2 systems. The associated definitions of the steady-state error coefficients readily provide information that is indicative of a stable system's steady-state performance. Thus, a start has been made in developing a set of standard characteristics. A nonunity-feedback system can also be represented by an equivalent unity-feedback system having the forward transfer function $G_{eq}(s)$ (see Sec. 6.9). Then the steady-state performance can be evaluated on the basis of the system type and the error coefficients determined from $G_{eq}(s)$.

REFERENCES

1. Guillemin, E. A.: *The Mathematics of Circuit Analysis,* Wiley, New York, 1949.
2. Chestnut, H., and R. W. Mayer: *Servomechanisms and Regulating System Design,* 2d ed., vol. 1, Wiley, New York, 1959.
3. Singh, V.: "Comments on the Routh-Hurwitz Criterion," *IEEE Trans. Autom. Control,* vol. AC-26, p. 612, 1981.
4. Porter, Brian: *Stability Criteria for Linear Dynamical Systems,* Academic, New York, 1968.
5. Gantmacher, F. R.: *Applications of the Theory of Matrices,* Wiley Interscience, New York, 1959.
6. Etter, D. M.:*Engineering Problem Solving with MATLAB®,* Prentice-Hall, Englewood Cliffs, N.J., 1993.
7. Krishnamurthi, V.: "Implications of Routh Stability Criteria," *IEEE Trans. Autom. Control,* vol. AC-25, pp. 554–555, 1980.
8. Khatwani, K. J.: "On Routh-Hurwitz Criterion," *IEEE Trans. Autom. Control,* vol. AC-26, pp. 583–584, 1981.
9. Pillai, S. K.: "On the ε-Method of the Routh-Hurwitz Criterion," *IEEE Trans. Autom. Control,* vol. AC-26, p. 584, 1981.

CHAPTER
7

ROOT LOCUS

7.1 INTRODUCTION

A designer uses the time response of a system to determine if his or her design of a control system meets specifications. An accurate solution of the system's performance can be obtained by deriving the differential equations for the control system and solving them. However, if the response does not meet specifications, it is not easy to determine from this solution just what physical parameters in the system should be changed to improve the response. A designer wishes to be able to predict a system's performance by an analysis that does not require the actual solution of the differential equations. Also, the designer would like this analysis to indicate readily the manner or method by which this system must be adjusted or compensated to produce the desired performance characteristics.

The first thing that a designer wants to know about a given system is whether or not it is stable. This can be determined by examining the roots obtained from the characteristic equation $1 + G(s)H(s) = 0$. By applying Routh's criterion to this equation, it is possible in short order to determine stability without solving for the roots. Yet this does not satisfy the designer because it does not indicate the degree of stability of the system, i.e., the amount of overshoot and the settling time of the controlled variable. Not only must the system be stable, but also the overshoot must be maintained within prescribed limits and transients must die out in a sufficiently short time. The graphical methods described in this text not only indicate whether a system is stable or unstable but, for a stable system, also show the degree of stability. A number of commercially available computer-aided-design (CAD) programs can be used to obtain the solution.[3] In addition to the time response, a designer can choose to analyze and interpret the steady-state sinusoidal response of the transfer function of the system to obtain an idea of the system's response. This method is based upon the interpretation of a Nyquist plot, discussed in Chapters 8 and 9. The frequency-response approach yields enough information to indicate whether the system needs to be adjusted or compensated and how the system should be compensated.

This chapter deals with *the root-locus method,*[1,2] which incorporates the more desirable features of both the classical and the frequency-response methods. *The root locus is a plot of the roots of the characteristic equation of the closed-loop system as a function of the gain of the open-loop transfer function.* With relatively small effort this graphical approach yields a clear indication of the effect of gain adjustment. The underlying principle is that the poles of $C(s)/R(s)$ (transient-response modes) are related to the zeros and poles of the open-loop transfer function $G(s)H(s)$ and also to the gain. This relationship is shown by Eq. (5.4), which is repeated here:

$$\frac{C(s)}{R(s)} = \frac{G(s)}{1 + G(s)H(s)} \tag{7.1}$$

An important advantage of the root-locus method is that the roots of the characteristic equation of the system can be obtained directly. This yields a complete and accurate solution of the transient and steady-state response of the controlled variable. Also, an approximate solution can be obtained with a reduction of the work required. A person can readily obtain proficiency with the root-locus method. With the help of a CAD program, such as MATLAB, ICECAP-PC or TOTAL-PC (see Appendix B), it is possible to design a system with relative ease.

7.2 PLOTTING ROOTS OF A CHARACTERISTIC EQUATION

To give a better insight into root-locus plots, consider the position-control system shown in Fig. 7.1. The motor produces a torque $T(s)$ that is proportional to the actuating signal $E(s)$. The load consists of the combined motor and load inertia J and viscous friction B. The forward transfer function is

$$G(s) = \frac{\theta_o(s)}{E(s)} = \frac{A/J}{s(s + B/J)} = \frac{K}{s(s + a)} \tag{7.2}$$

where $K = A/J$ and $a = B/J$. Assume that $a = 2$. Thus

$$G(s) = \frac{C(s)}{E(s)} = \frac{K}{s(s + 2)} \tag{7.3}$$

FIGURE 7.1
A position-control system.

When the transfer function is expressed with the coefficients of the highest powers of s in both the numerator and the denominator polynomials equal to unity, the value of K is defined as the *static loop sensitivity*. The control ratio (closed-loop transfer function), using Eq. (7.1), is

$$\frac{C(s)}{R(s)} = \frac{K}{s(s+2)+K} = \frac{K}{s^2 + 2s + K} = \frac{\omega_n^2}{s^2 + 2\zeta\omega_n s + \omega_n^2} \qquad (7.4)$$

where $\omega_n = \sqrt{K}$, $\zeta = 1/\sqrt{K}$, and K is considered to be adjustable from zero to an infinite value. The design problem is to determine the roots of the characteristic equation for all values of K and to plot these roots in the s plane. The roots of the characteristic equation, $s^2 + 2s + K = 0$, are given by

$$s_{1,2} = -1 \pm \sqrt{1 - K} \qquad (7.5)$$

For $K = 0$, the roots are $s_1 = 0$ and $s_2 = -2$, which also are the poles of the open-loop transfer function given by Eq. (7.3). When $K = 1$, then $s_{1,2} = -1$. Thus, when $0 < K < 1$, the roots $s_{1,2}$ are real and lie on the negative real axis of the s plane between -2 and -1 and 0 to -1, respectively. For the case where $K > 1$, the roots are complex and are given by

$$s_{1,2} = \sigma \pm j\omega_d = -\zeta\omega_n \pm j\omega_n\sqrt{1-\zeta^2} = -1 \pm j\sqrt{K-1} \qquad (7.6)$$

Note that the real part of all the roots is constant for all values of $K > 1$.

The roots of the characteristic equation are determined for a number of values of K (see Table 7.1) and are plotted in Fig. 7.2. Curves are drawn through these plotted points. On these curves, containing two branches, lie all possible roots of the characteristic equation for all values of K from zero to infinity. Note that each branch is calibrated with K as a parameter and the values of K at points on the locus are underlined; the arrows show the direction of increasing values of K. *These curves are defined as the root-locus plot of Eq. (7.4).* Once this plot is obtained, the roots that best fit the system performance specifications can be selected. Corresponding to the selected roots there is a required value of K that can be determined from the plot. When the roots have been selected, the time response can be obtained. Since this process of finding the root locus by calculating the roots for various values of

TABLE 7.1
Location of roots for the characteristic equation $s^2 + 2s + K = 0$

K	s_1	s_2
0	$-0 + j0$	$-2.0 - j0$
0.5	$-0.293 + j0$	$-1.707 - j0$
0.75	$-0.5 + j0$	$-1.5 - j0$
1.0	$-1.0 + j0$	$-1.0 - j0$
2.0	$-1.0 + j1.0$	$-1.0 - j1.0$
3.0	$-1.0 + j1.414$	$-1.0 - j1.414$
50.0	$-1.0 + j7.0$	$-1.0 - j7.0$

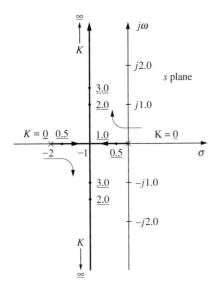

FIGURE 7.2
Plot of all roots of the characteristic equation $s^2 + 2s + K = 0$ for $0 \le K < \infty$. Values of K are underlined.

K becomes tedious for characteristic equations of order higher than second, a simpler method of obtaining the root locus would be preferable. The graphical methods for determining the root-locus plot are the subject of this chapter. CAD methods[3] are used to facilitate obtaining the root locus.

The value of K is normally considered to be positive. However, it is possible for K to be negative. For the example in this section, if the value of K is negative, Eq. (7.5) gives only real roots. Thus, the entire locus lies on the real axis, that is, $0 \le s_1 < +\infty$ and $-2 \ge s_2 > -\infty$ for $0 \ge K > -\infty$. For any negative value of K there is always a root in the right half of the s plane, and the system is unstable. Once the root locus has been obtained for a control system, it is possible to determine the variation in system performance with respect to a variation in sensitivity K. For the example of Fig. 7.1, the control ratio is written in terms of its roots, for $K > 1$, as

$$\frac{C(s)}{R(s)} = \frac{K}{(s - \sigma - j\omega_d)(s - \sigma + j\omega_d)} \tag{7.7}$$

Note, as defined in Fig. 4.3, that a root with a damping ratio ζ lies on a line making the angle $\eta = \cos^{-1} \zeta$ with the negative real axis. The damping ratio of several roots is indicated in Fig. 7.3. For an increase in the gain K of the system, analysis of the root locus reveals the following characteristics:

1. A decrease in the damping ratio ζ. This increases the overshoot of the time response.

2. An increase in the undamped natural frequency ω_n. The value of ω_n is the distance from the origin to the complex root.

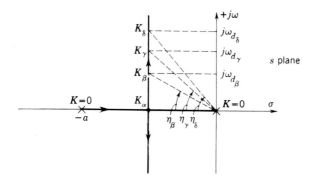

FIGURE 7.3
Root-locus plot of the position-control system of Fig. 7.1. $\eta = \cos^{-1}\zeta$.

3. An increase in the damped natural frequency ω_d. The value of ω_d is the imaginary component of the complex root and is equal to the frequency of the transient response.

4. No effect on the rate of decay σ; that is, it remains constant for all values of gain equal to or greater than K_α. For higher-order systems this is not the case.

5. The root locus is a vertical line for $K \geq K_\alpha$, and $\sigma = -\zeta\omega_n$ is constant. This means that no matter how much the gain is increased in a linear *simple* second-order system, the system can never become unstable. The time response of this system with a step-function input, for $\zeta < 1$, is of the form

$$c(t) = A_0 + A_1 \exp(-\zeta\omega_n t)\sin(\omega_d t + \phi)$$

The root locus of each control system can be analyzed in a similar manner to determine the variation in its time response that results from a variation in its loop sensitivity K.

7.3 QUALITATIVE ANALYSIS OF THE ROOT LOCUS

A zero is added to the simple second-order system of the preceding section so that the open-loop transfer function is

$$G(s) = \frac{K(s + 1/T_2)}{s(s + 1/T_1)} \tag{7.8}$$

The root locus of the control system having this transfer function is shown in Fig. 7.4b. When this root locus is compared with that of the original system, shown in Fig. 7.4a, it is seen that the branches have been "pulled to the left," or farther from the imaginary axis. For values of static loop sensitivity greater than K_α, the roots are farther to the left than for the original system. Therefore, the transients will decay faster, yielding a *more stable* system.

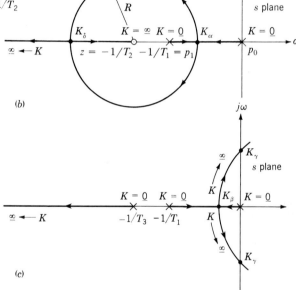

FIGURE 7.4

Various root-locus configurations for $H(s) = 1$: (a) Root locus of basic transfer function:

$$G(s) = \frac{K}{s(s + 1/T_1)}$$

(b) Root locus with additional zero:

$$G(s) = \frac{K(s + 1/T_2)}{s(s + 1/T_1)}$$

(c) Root locus with additional pole:

$$G(s) = \frac{K}{s(s + 1/T_1)(s + 1/T_3)}$$

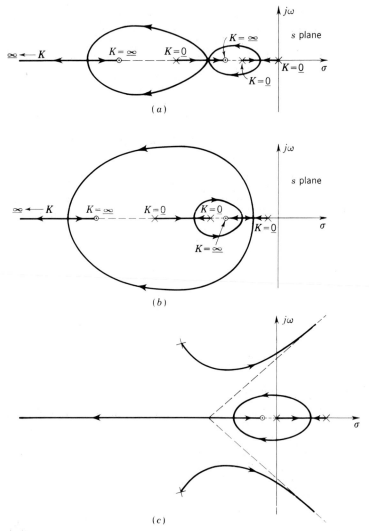

FIGURE 7.5
Various root-locus configurations:

$$(a)\ G(s)H(s) = \frac{K(s + 1/T_2)(s + 1/T_4)}{s(s + 1/T_1)(s + 1/T_3)}$$

$$(b)\ G(s)H(s) = \frac{K(s + 1/T_2)(s + 1/T_4)}{(s + 1/T_1)(s + 1/T_3)(s + 1/T_5)}$$

$$(c)\ G(s)H(s) = \frac{K(s + 1/T_2)}{s(s - 1/T)(s^2 + 2\zeta\omega_n s + \omega_n^2)}$$

These figures are not drawn to scale. Several other root-locus shapes are possible for a given pole-zero arrangement, depending on the specific values of the poles and zeros. Variations of the possible root-locus plots for a given pole-zero arrangement are shown in Reference 9.

If a pole, instead of a zero, is added to the original system the resulting transfer function is

$$G(s) = \frac{K}{s(s + 1/T_1)(s + 1/T_3)} \tag{7.9}$$

Figure 7.4c shows the corresponding root locus of the closed-loop control system. Note that the addition of a pole has "pulled the locus to the right" so that two branches cross the imaginary axis. For values of static loop sensitivity greater than K_β, the roots are closer to the imaginary axis than for the original system. Therefore, the transients will decay more slowly, yielding a *less stable* system. Also, for values of $K > K_\gamma$, two of the three roots lie in the right half of the s plane, resulting in an unstable system. The addition of the pole has resulted in a less stable system, compared with the original second-order system. Thus, the following general conclusions can be drawn:

1. The addition of a zero to a system has the general effect of pulling the root locus to the left, tending to make it a more stable and faster-responding system (shorter settling time T_s).

2. The addition of a pole to a system has the effect of pulling the root locus to the right, tending to make it a less stable and slower-responding system.

Figure 7.5 illustrates the root-locus configurations for negative-feedback control systems for more complex transfer functions. Note that the third system contains a pole in the right half of the s plane (RHP). It represents the performance of an airplane with an autopilot in the longitudinal mode.

The root-locus method is a graphical technique for readily determining the location of all possible roots of a characteristic equation as the gain is varied from zero to infinity. Also, it can readily be determined how the locus should be altered in order to improve the system's performance, based upon the knowledge of the effect of the addition of poles or zeros.

7.4 PROCEDURE OUTLINE

The procedure to be followed in applying the root-locus method is outlined. This procedure is modified when a CAD program[3] is used to obtain the root locus. Such a program can provide the desired data for the root locus in plotted or tabular form. The procedure outlined next and discussed in the following sections is intended to establish firmly for the reader the fundamentals of the root-locus method and to enhance the interpretation of the data obtained from the CAD program.

Step 1. Derive the open-loop transfer function $G(s)H(s)$ of the system.

Step 2. Factor the numerator and denominator of the transfer function into linear factors of the form $s + a$, where a may be real or complex.

Step 3. Plot the zeros and poles of the open-loop transfer function in the $s = \sigma + j\omega$ plane.

Step 4. The plotted zeros and poles of the open-loop function determine the roots of the characteristic equation of the closed-loop system $[1 + G(s)H(s) = 0]$. Use the geometrical shortcuts summarized in Sec. 7.8 or a CAD program[3] to determine the locus that describes the roots of the closed-loop characteristic equation.

Step 5. Calibrate the locus in terms of the loop sensitivity K. If the gain of the open-loop system is predetermined, the locations of the exact roots of $1 + G(s)H(s)$ are immediately known. If the location of the roots is specified, the required value of K can be determined.

Step 6. Once the roots have been found in step 5, the system's time response can be calculated by taking the inverse Laplace transform of $C(s)$, either manually or by use of a CAD program.

Step 7. If the response does not meet the desired specifications, determine the shape that the root locus must have to meet these specifications.

Step 8. If gain adjustment alone does not yield the desired performance, synthesize the cascade compensator that must be inserted into the system. This process, called *compensation,* is described in later chapters.

7.5   OPEN-LOOP TRANSFER FUNCTION

In securing the open-loop transfer function, keep the terms in the factored form of $(s + a)$ or $(s^2 + 2\zeta\omega_n s + \omega_n^2)$. For unity feedback the open-loop function is equal to the forward transfer function $G(s)$. For nonunity feedback it also includes the transfer function of the feedback path. This open-loop transfer function is of the form

$$G(s)H(s) = \frac{K(s + a_1)\cdots(s + a_h)\cdots(s + a_w)}{s^m(s + b_1)(s + b_2)\cdots(s + b_c)\cdots(s + b_u)} \qquad (7.10)$$

where a_h and b_c may be real or complex numbers and may lie in either the left-half plane (LHP) or right-half plane (RHP). The value of K may be either positive or negative. For example, consider

$$G(s)H(s) = \frac{K(s + a_1)}{s(s + b_1)(s + b_2)} \qquad (7.11)$$

When the transfer function is in this form (with all the coefficients of s equal to unity), the K is defined as the *static loop sensitivity.* For this example, a zero of the open-loop transfer function exists at $s = -a_1$ and the poles are at $s = 0, s = -b_1$, and $s = -b_2$. Now let the zeros and poles of $G(s)H(s)$ in Eq. (7.10) be denoted by the letters z and p, respectively, i.e.,

$$z_1 = -a_1 \qquad z_2 = -a_2 \qquad \cdots \qquad z_w = -a_w$$

$$p_1 = -b_1 \qquad p_2 = -b_2 \qquad \cdots \qquad p_u = -b_u$$

Equation (7.10) can then be rewritten as

$$G(s)H(s) = \frac{K(s - z_1)\cdots(s - z_w)}{s^m(s - p_1)\cdots(s - p_u)} = \frac{K \prod\limits_{h=1}^{w} (s - z_h)}{s^m \prod\limits_{c=1}^{u} (s - p_c)} \tag{7.12}$$

where $\prod$ indicates a product of terms. The degree of the numerator is w and that of the denominator is $m + u = n$.

7.6 POLES OF THE CONTROL RATIO $C(s)/R(s)$

The underlying principle of the root-locus method is that the poles of the control ratio $C(s)/R(s)$ are related to the zeros and poles of the open-loop transfer function $G(s)H(s)$ and to the loop sensitivity K. This relationship can be shown as follows. Let

$$G(s) = \frac{N_1(s)}{D_1(s)} \tag{7.13}$$

and

$$H(s) = \frac{N_2(s)}{D_2(s)} \tag{7.14}$$

Then

$$G(s)H(s) = \frac{N_1 N_2}{D_1 D_2} \tag{7.15}$$

Thus,

$$\frac{C(s)}{R(s)} = M(s) = \frac{A(s)}{B(s)} = \frac{G(s)}{1 + G(s)H(s)} = \frac{N_1/D_1}{1 + N_1 N_2/D_1 D_2} \tag{7.16}$$

where

$$B(s) \equiv 1 + G(s)H(s) = 1 + \frac{N_1 N_2}{D_1 D_2} = \frac{D_1 D_2 + N_1 N_2}{D_1 D_2} \tag{7.17}$$

Rationalizing Eq. (7.16) gives

$$\frac{C(s)}{R(s)} = M(s) = \frac{P(s)}{Q(s)} = \frac{N_1 D_2}{D_1 D_2 + N_1 N_2} \tag{7.18}$$

From Eqs. (7.17) and (7.18) it is seen that the zeros of $B(s)$ are the poles of $M(s)$ and determine the form of the system's transient response. In terms of $G(s)H(s)$, given by Eq. (7.12), the degree of $B(s)$ is equal to $m + u$; therefore, $B(s)$ has $n = m + u$ finite zeros. As shown in Chapter 4 on Laplace transforms, the factors of $Q(s)$ produce transient components of $c(t)$ that fall into the categories shown in Table 7.2. The numerator $P(s)$ of Eq. (7.18) merely modifies the constant multipliers of these transient components. The roots of $B(s) = 0$, which is the characteristic equation of the system, must satisfy the equation

TABLE 7.2
Time response terms in $c(t)$

Denominator factor of $C(s)$	Corresponding inverse	Form
s	$u_{-1}(t)$	Step function
$s + \dfrac{1}{T}$	$e^{-t/T}$	Decaying exponential
$s^2 + 2\zeta\omega_m s + \omega_n^2$	$\exp(-\zeta\omega_n t)\sin(\omega_n \sqrt{1 - \zeta^2}\, t + \phi)$ where $\zeta < 1$	Damped sinusoid

$$B(s) \equiv 1 + G(s)H(s) = 0 \qquad (7.19)$$

These roots must also satisfy the equation:

$$G(s)H(s) = \frac{K(s - z_1)\cdots(s - z_w)}{s^m(s - p_1)\cdots(s - p_u)} = -1 \qquad (7.20)$$

Thus, as the loop sensitivity K assumes values from zero to infinity, the transfer function $G(s)H(s)$ must always be equal to -1. The corresponding values of s that satisfy Eq. (7.20), for any value of K, are the poles of $M(s)$. The plots of these values of s are defined as the root locus of $M(s)$.

Conditions that determine the root locus for *positive values* of loop sensitivity are now determined. The general form of $G(s)H(s)$ for *any* value of s is

$$G(s)H(s) = Fe^{-j\beta} \qquad (7.21)$$

The right side of Eq. (7.20), -1, can be written as

$$-1 = e^{j(1+2h)\pi} \qquad h = 0, \pm 1, \pm 2, \ldots$$

Equation (7.20) is satisfied *only* for those values of s for which

$$Fe^{-j\beta} = e^{j(1+2h)\pi}$$

where

$$F = \left| G(s)H(s) \right| = 1$$

and

$$-\beta = (1 + 2h)\pi \qquad (7.22)$$

From the preceding it can be concluded that the magnitude of $G(s)H(s)$, a function of the complex variable s, must always be unity and its phase angle must be an odd multiple of π if the particular value of s is a zero of $B(s) = 1 + G(s)H(s)$. Consequently, the following two conditions are formalized for the root locus for all positive values of K from zero to infinity:

For K > 0

Magnitude condition $\qquad\qquad \left| G(s)H(s) \right| = 1 \qquad (7.23)$

Angle condition $\quad \underline{/G(s)H(s)} = (1 + 2h)180° \qquad$ for $h = 0, \pm 1, \pm 2, \ldots \quad (7.24)$

In a similar manner, the conditions for *negative* values of loop sensitivity $(-\infty < K < 0)$ can be determined. [This corresponds to positive feedback, $e(t) = r(t) + b(t)$, and positive values of K.] The root locus must satisfy the following conditions.

For K < 0

Magnitude condition $\left|G(s)H(s)\right| = 1$ (7.25)

Angle condition $\underline{/G(s)H(s)} = h360°$ for $h = 0, \pm 1, \pm 2, \ldots$ (7.26)

Thus the root-locus method provides a plot of the variation of each of the poles of $C(s)/R(s)$ in the complex s plane as the loop sensitivity is varied from $K = 0$ to $K = \pm\infty$.

7.7 APPLICATION OF THE MAGNITUDE AND ANGLE CONDITIONS

Once the open-loop transfer function $G(s)H(s)$ has been determined and put into the proper form, the poles and zeros of this function are plotted in the $s = \sigma + j\omega$ plane. As an example, consider

$$G(s)H(s) = \frac{K(s + 1/T_1)^2}{s(s + 1/T_2)(s^2 + 2\zeta\omega_n s + \omega_n^2)} = \frac{K(s - z_1)^2}{s(s - p_1)(s - p_2)(s - p_3)} \quad (7.27)$$

For the quadratic factor $s^2 + 2\zeta\omega_n s + \omega_n^2$ with the damping ratio $\zeta < 1$, the complex-conjugate poles of Eq. (7.27) are

$$p_{2,3} = -\zeta\omega_n \pm j\omega_n \sqrt{1 - \zeta^2} = \sigma \pm j\omega_d$$

The plot of the poles and zeros of Eq. (7.27) is shown in Fig. 7.6. Remember that complex poles or zeros always occur in conjugate pairs, that σ is the damping constant, and that ω_d is the damped natural frequency of oscillation. A multiple pole or zero is indicated on the pole-zero diagram by $\times]_q$ or $\odot]_q$, where $q = 1, 2, 3, \ldots$ is the multiplicity of the pole or zero. The open-loop poles and zeros are plotted, and they are used in the graphical construction of the locus of the poles (the roots of the characteristic equation) of the closed-loop control ratio.

For any particular value (real or complex) of s the terms $s, s - p_1, s - p_2, s - z_1, \ldots$ are complex numbers designating *directed line segments*. For example, if $s = -4 + j4$ and $p_1 = -1$, then $s - p_1 = -3 + j4$ or $\left|s - p_1\right| = 5$, and

$$\phi_1 = \underline{/s - p_1} = 126.8° \quad \text{(see Fig. 7.6)} \quad (7.28)$$

Recall that it is shown in Sec. 7.6 that the roots of the characteristic equation $1 + G(s)H(s) = 0$ are all the values of s that satisfy the two conditions:

$$\left|G(s)H(s)\right| = 1 \quad (7.29)$$

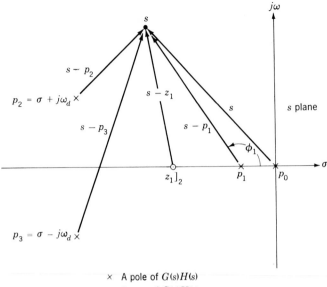

× A pole of $G(s)H(s)$
○ A zero of $G(s)H(s)$

FIGURE 7.6
Pole-zero diagram for Eq. (7.27).

$$\underline{/G(s)H(s)} = \begin{cases} (1 + 2h)180° & \text{for } K > 0 \\ h360° & \text{for } K < 0 \end{cases} \qquad h = 0, \pm 1, \pm 2, \ldots \qquad (7.30)$$

These are labeled as the magnitude and angle conditions, respectively. Therefore, applying these two conditions to the general Eq. (7.12) results in

$$\frac{|K| \cdot |s - z_1| \cdots |s - z_w|}{|s^m| \cdot |s - p_1| \cdot |s - p_2| \cdots |s - p_u|} = 1 \qquad (7.31)$$

and

$$-\beta = \underline{/s - z_1} + \cdots + \underline{/s - z_w} - m\angle s - \underline{/s - p_1} - \cdots - \underline{/s - p_u}$$

$$(7.32)$$

$$= \begin{cases} (1 + 2h)180° & \text{for } K > 0 \\ h360° & \text{for } K < 0 \end{cases}$$

Solving Eq. (7.31) for $|K|$ gives

$$|K| = \frac{|s^m| \cdot |s - p_1| \cdot |s - p_2| \cdots |s - p_u|}{|s - z_1| \cdots |s - z_w|} = \text{loop sensitivity} \qquad (7.33)$$

Multiplying Eq. (7.32) by -1 gives

$$\beta = \sum (\text{angles of denominator terms}) - \sum (\text{angles of numerator terms})$$

$$= \begin{cases} (1 + 2h)180° & \text{for } K > 0 \\ h360° & \text{for } K < 0 \end{cases} \tag{7.34}$$

All angles are considered positive, measured in the counterclockwise (CCW) sense. This form is convenient since $G(s)H(s)$ usually has more poles than zeros. Equations (7.33) and (7.34) are in the form used in the graphical construction of the root locus. In other words, there are particular values of s for which $G(s)H(s)$ satisfies the angle condition. For these values of s, Eq. (7.33) is used to determine the corresponding magnitude K. Those values of s that satisfy both the angle and the magnitude conditions are the roots of the characteristic equation and are $n = m + u$ in number. Thus, corresponding to step 4 in Sec. 7.4, the locus of all possible roots is obtained by applying the angle condition. This root locus can be calibrated in terms of the loop sensitivity K by using the magnitude condition.

Example. Determine the locus of all the closed-loop poles of $C(s)/R(s)$, for

$$G(s)H(s) = \frac{K_0(1 + 0.25s)}{(1 + s)(1 + 0.5s)(1 + 0.2s)} \tag{7.35}$$

$$G(s)H(s) = \frac{K(s + 4)}{(s + 1)(s + 2)(s + 5)} \qquad K = 2.5K_0 \tag{7.36}$$

Step 1. The poles and zeros are plotted in Fig. 7.7.

Step 2. In Fig. 7.7 the ϕ's are denominator angles and ψ's are numerator angles. Also, the l's are the lengths of the directed segments stemming from the denominator factors, and (l)'s are the lengths of the directed segments stemming from the numerator factors. After plotting the poles and zeros of the open-loop transfer functions, arbitrarily choose a search point. To this point, draw directed line segments from all the open-loop

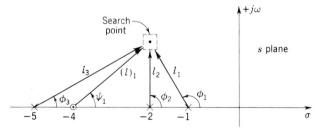

FIGURE 7.7
Construction of the root locus.

poles and zeros and label as indicated. For this search point to be a point on the locus, the following angle condition must be true:

$$\beta = \phi_1 + \phi_2 + \phi_3 - \psi_1 = \begin{cases} (1 + 2h)180° & \text{for } K > 0 \\ h360° & \text{for } K < 0 \end{cases} \qquad (7.37)$$

If this equation is not satisfied, select another search point until it is satisfied. Locate a sufficient number of points in the s plane that satisfy the angle condition. In the next section additional information is given that lessens and systematizes the work involved in this trial-and-error approach.

Step 3.   Once the complete locus has been determined, the locus can be calibrated in terms of the loop sensitivity for any root s_1 as follows:

$$|K| = \frac{l_1 l_2 l_3}{(l)_1} \qquad (7.38)$$

where $l_1 = |s_1 + 1|; l_2 = |s_1 + 2|; l_3 = |s_1 + 5|;$ and $(l)_1 = |s_1 + 4|$.

FIGURE 7.8
The complete root locus of Eq. (7.36); (a) for $K > 0$; (b) for $K < 0$.

In other words, the values of l_1, l_2, l_3, and $(l)_1$ can be measured for a given point s_1 that satisfies the angle condition; thus, the value of $|K|$ for this point can be calculated. The appropriate sign must be given to the magnitude of K, compatible with the particular angle condition that is utilized to obtain the root locus. Note that since complex roots must occur in conjugate pairs, the locus is symmetrical about the real axis. Thus, the bottom half of the locus can be drawn once the locus above the real axis has been determined. The root locus for this system is shown in Fig. 7.8. Note that for negative values of K, the three branches of the root locus lie in the left half plane for $K > -2.5$.

$W(s)$ *Plane*

From Eqs. (7.36) and (7.20)

$$W(s) = u_x + jv_y = \frac{(s+1)(s+2)(s+5)}{s+4} = -K \qquad (7.39)$$

The line $u_x = -K$ in the $W(s)$ plane maps into the curves indicated in Fig. 7.8. That is, for each value of u_x in the $W(s)$ plane there is a particular value or a set of values of s in the s plane.

7.8   GEOMETRICAL PROPERTIES (CONSTRUCTION RULES)

To facilitate the application of the root-locus method, the following rules are established for $K > 0$. These rules are based upon the interpretation of the angle condition and an analysis of the characteristic equation. These rules can be extended for the case where $K < 0$. The rules for both $K > 0$ and $K < 0$ are listed in Sec. 7.15 for easy reference. The rules presented aid in obtaining the root locus by expediting the plotting of the locus. The root locus can also be obtained by using a CAD program.[3] These rules provide *checkpoints* to ensure that the computer solution is correct. They also permit rapid sketching of the root locus, which provides a qualitative idea of achievable closed-loop system performance.

Rule 1: Number of Branches of the Locus

The characteristic equation $B(s) = 1 + G(s)H(s) = 0$ is of degree $n = m + u$; therefore, there are n roots. As the open-loop sensitivity K is varied from zero to infinity, each root traces a continuous curve. Since there are n roots, there are the same number of curves or branches in the complete root locus. Since the degree of the polynomial $B(s)$ is determined by the poles of the open-loop transfer function, *the number of branches of the root locus is equal to the number of poles of the open-loop transfer function.*

Rule 2: Real-Axis Locus

In Fig. 7.9 are shown a number of open-loop poles and zeros. If the angle condition is applied to any search point such as s_1 on the real axis, the angular contribution of

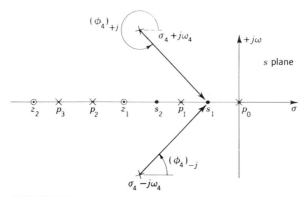

FIGURE 7.9
Determination of the real-axis locus.

all the poles and zeros on the real axis to the left of this point is zero. The angular contribution of the complex-conjugate poles to this point is $360°$. (This is also true for complex-conjugate zeros.) Finally, the poles and zeros on the real axis to the right of this point each contribute $180°$ (with the appropriate sign included). From Eq. (7.34) the angle of $G(s)H(s)$ to the point s_1 is given by

$$\phi_0 + \phi_1 + \phi_2 + \phi_3 + [(\phi_4)_{+j} + (\phi_4)_{-j}] - (\psi_1 + \psi_2) = (1 + 2h)180° \quad (7.40)$$

or $\qquad 180° + 0° + 0° + 0° + 360° - 0° - 0° = (1 + 2h)180° \qquad (7.41)$

Therefore, s_1 is a point on a branch of the locus. Similarly, it can be shown that the point s_2 is not a point on the locus. The poles and zeros to the left of a point s on the real axis and the $360°$ contributed by the complex-conjugate poles or zeros do not affect the odd-multiple-of-$180°$ requirement. Thus, *if the total number of real poles and zeros to the right of a search point s on the real axis is odd, this point lies on the locus.* In Fig. 7.9 the root locus exists on the real axis from p_0 to p_1, z_1 to p_2, and p_3 to z_2.

All points on the real axis between z_1 and p_2 in Fig. 7.9 satisfy the angle condition and are therefore points on the root locus. However, there is no guarantee that this section of the real axis is part of just one branch. Figure 7.5b and Prob. 7.4 illustrate the situation where part of the real axis between a pole and a zero is divided into three sections that are parts of three different branches.

Rule 3: Locus End Points

The magnitude of the loop sensitivity that satisfies the magnitude condition is given by Eq. (7.33) and has the general form

$$|W(s)| = K = \frac{\displaystyle\prod_{c=1}^{n} |s - p_c|}{\displaystyle\prod_{h=1}^{w} |s - z_h|} \qquad (7.42)$$

Since the numerator and denominator factors of Eq. (7.42) locate the poles and zeros, respectively, of the open-loop transfer function, the following conclusions can be drawn:

1. When $s = p_c$ (the open-loop poles), the loop sensitivity K is zero.
2. When $s = z_h$ (the open-loop zeros), the loop sensitivity K is infinite. When the numerator of Eq. (7.42) is of higher order than the denominator, then $s = \infty$ also makes K infinite, thus being equivalent in effect to a zero.

Thus, the locus starting points ($K = 0$) are at the open-loop poles and the locus ending points ($K = \infty$) are at the open-loop zeros (the point at infinity being considered as an equivalent zero of multiplicity equal to the quantity $n - w$).

Rule 4: Asymptotes of Locus as s Approaches Infinity

Plotting of the locus is greatly facilitated by evaluating the asymptotes approached by the various branches as s takes on large values. Taking the limit of $G(s)H(s)$ as s approaches infinity, based on Eqs. (7.12) and (7.20), yields

$$\lim_{s \to \infty} G(s)H(s) = \lim_{s \to \infty} \left[K \frac{\prod_{h=1}^{w}(s - z_h)}{\prod_{c=1}^{n}(s - p_c)} \right] = \lim_{s \to \infty} \frac{K}{s^{n-w}} = -1 \qquad (7.43)$$

Remember that K in Eq. (7.43) is still a variable in the manner prescribed previously, thus allowing the magnitude condition to be met. Therefore, as $s \to \infty$,

$$-K = s^{n-w} \qquad (7.44)$$

$$|-K| = |s^{n-w}| \qquad \text{magnitude condition} \qquad (7.45)$$

$$\underline{/-K} = \underline{/s^{n-w}} = (1 + 2h)180° \qquad \text{angle condition} \qquad (7.46)$$

Rewriting Eq. (7.46) gives $(n - w)\angle s = (1 + 2h)180°$ or

$$\gamma = \frac{(1 + 2h)180°}{n - w} \qquad \text{as } s \to \infty \qquad (7.47)$$

There are $n - w$ asymptotes of the root locus, and their angles are given by

$$\gamma = \frac{(1 + 2h)180°}{[\text{Number of poles of } G(s)H(s)] - [\text{Number of zeros of } G(s)H(s)]} \qquad (7.48)$$

Equation (7.48) reveals that, no matter what magnitude s may have, after a sufficiently large value has been reached, the argument (angle) of s on the root locus remains constant. For a search point that has a sufficiently large magnitude, the open-loop poles and zeros appear to it as if they had collapsed into a single point. Therefore, the branches are asymptotic to straight lines whose slopes and directions are given

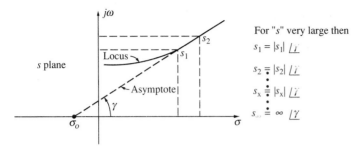

FIGURE 7.10
Asymptotic condition for large values of s.

by Eq. (7.48) (see Fig. 7.10). These asymptotes usually do not go through the origin. The correct real-axis intercept of the asymptotes is obtained from Rule 5.

Rule 5: Real-Axis Intercept of the Asymptotes

The real-axis crossing σ_o of the asymptotes can be obtained by applying the theory of equations. The result is

$$\sigma_o = \frac{\sum_{c=1}^{n} \mathrm{Re}(p_c) - \sum_{h=1}^{w} \mathrm{Re}(z_h)}{n - w} \tag{7.49}$$

The asymptotes are not dividing lines, and a locus may cross its asymptote. It may be valuable to know from which side the root locus approaches its asymptote. Lorens and Titsworth[4] present a method for obtaining this information. The locus lies exactly along the asymptote if the pole-zero pattern is symmetric about the asymptote line extended through the point σ_o (see Prob. 7.3h).

Rule 6: Breakaway Point on the Real Axis[10]

The branches of the root locus start at the open-loop poles where $K = 0$ and end at the finite open-loop zeros or at $s = \infty$. When the root locus has branches on the real axis between two poles, there must be a point at which the two branches *break away* from the real axis and enter the complex region of the s plane in order to approach zeros or the point at infinity. (Examples are shown in Fig. 7.11a-3: between p_0 and p_1, and in Fig. 7.11b-2: between p_2 and p_3.) For two finite zeros (see Fig. 7.11b-1) or one finite zero and one at infinity (see Fig. 7.11a-1) the branches are coming from the complex region and enter the real axis.

In Fig. 7.11a-3 between two poles there is a point s_a for which the loop sensitivity K_z is greater than for points on either side of s_a on the real axis. In other words, since K starts with a value of zero at the poles and increases in value as the locus moves away from the poles, there is a point somewhere in between where the K's for the two branches simultaneously reach a maximum value. This point is called the *breakaway point*. Plots of K vs. σ utilizing Eq. (7.33) are shown in Fig. 7.11 for the portions of the root locus that exist on the real axis for $K > 0$. The point s_b for which

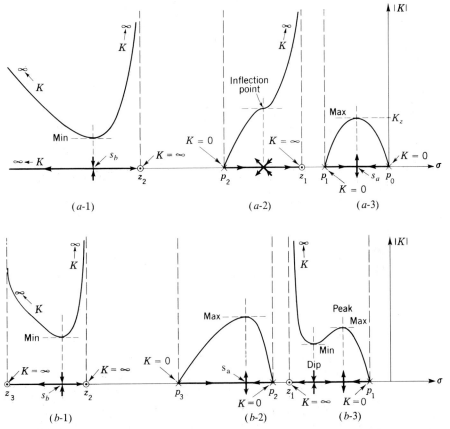

FIGURE 7.11
Plots of K and the corresponding real-axis locus for (a) Fig. 7.5a and (b) Fig. 7.5b.

the value of K is a minimum between two zeros is called the *break-in point*. The breakaway and break-in points can easily be calculated for an open-loop pole-zero combination for which the derivative of $W(s) = -K$ is of the second order. As an example, if

$$G(s)H(s) = \frac{K}{s(s + 1)(s + 2)} \tag{7.50}$$

then

$$W(s) = s(s + 1)(s + 2) = -K \tag{7.51}$$

Multiplying the factors together gives

$$W(s) = s^3 + 3s^2 + 2s = -K \tag{7.52}$$

When $s^3 + 3s^2 + 2s$ is a minimum, $-K$ is a minimum and K is a maximum. Thus, by taking the derivative of this function and setting it equal to zero, the points can be determined:

$$\frac{dW(s)}{ds} = 3s^2 + 6s + 2 = 0 \qquad (7.53)$$

or $\qquad s_{a,\,b} = -1 \pm 0.5743 = -0.4257,\, -1.5743$

Since the breakaway points s_a for $K > 0$ must lie between $s = 0$ and $s = -1$ in order to satisfy the angle condition, the value is

$$s_a = -0.4257 \qquad (7.54)$$

The other point, $s_b = -1.5743$, is the break-in point on the root locus for $K < 0$. Substituting $s_a = -0.4257$ into Eq. (7.52) gives the value of K at the breakaway point for $K > 0$ as

$$K = -[(-0.426)^3 + (3)(-0.426)^2 + (2)(-0.426)] = 0.385 \qquad (7.55)$$

When the derivative of $W(s)$ is of higher order than 2, a digital-computer program can be used to calculate the roots of the numerator polynomial of $dW(s)/ds$; these roots locate the breakaway and break-in points. Note that it is possible to have both a breakaway and a break-in point between a pole and zero (finite or infinite) on the real axis, as shown in Figs. 7.5, 7.11a-2, and 7.11b-3. The plot of K vs. σ for a locus between a pole and zero falls into one of the following categories:

1. The plot clearly indicates a peak and a dip, as illustrated between p_1 and z_1 in Fig. 7.11b-3. The peak represents a "maximum" value of K that identifies a breakaway point. The dip represents a "minimum" value of K that identifies a break-in point.

2. The plot contains an inflection point. This occurs when the breakaway and break-in points coincide, as is the case between p_2 and z_1 in Fig. 7.11a-2.

3. The plot does not indicate a dip-and-peak combination or an inflection point. For this situation there are no break-in or breakaway points.

Rule 7: Complex Pole (or Zero): Angle of Departure

The next geometrical shortcut is the rapid determination of the direction in which the locus leaves a complex pole or enters a complex zero. Although in Fig. 7.12 a complex pole is considered, the results also hold for a complex zero.

In Fig. 7.12a, an area about p_2 is chosen so that l_2 is very much smaller than l_0, l_1, l_3, and $(l)_1$. For illustrative purposes, this area has been enlarged many times in Fig. 7.12b. Under these conditions the angular contributions from all the other poles and zeros, except p_2, to a search point anywhere in this area are approximately constant. They can be considered to have values determined as if the search point were right at p_2. Applying the angle condition to this small area yields

$$\phi_0 + \phi_1 + \phi_2 + \phi_3 - \psi_1 = (1 + 2h)180° \qquad (7.56)$$

or the departure angle is

$$\phi_{2_D} = (1 + 2h)180° - (\phi_0 + \phi_1 + 90° - \psi_1)$$

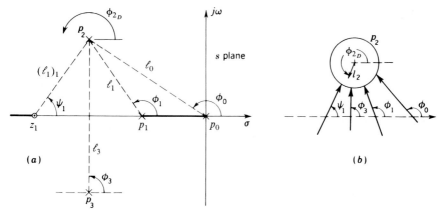

FIGURE 7.12
Angle condition in the vicinity of a complex pole.

In a similar manner the approach angle to a complex zero can be determined. For an open-loop transfer function having the pole-zero arrangement shown in Fig. 7.13, the approach angle ψ_1 to the zero z_1 is given by

$$\psi_{1_A} = (\phi_0 + \phi_1 + \phi_2 - 90°) - (1 + 2h)180° \qquad (7.57)$$

In other words, the direction of the locus as it leaves a pole or approaches a zero can be determined by adding up, according to the angle condition, all the angles of all vectors from all the other poles and zeros to the pole or zero in question. Subtracting this sum from $(1 + 2h)180°$ gives the required direction.

Rule 8: Imaginary-Axis Crossing Point

In cases where the locus crosses the imaginary axis into the right-half s plane, the crossover point can usually be determined by Routh's method or by similar means. For example, if the closed-loop characteristic equation $D_1D_2 + N_1N_2 = 0$ is of the form

$$s^3 + bs^2 + cs + Kd = 0$$

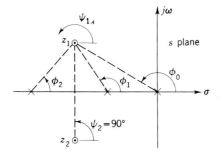

FIGURE 7.13
Angle condition in the vicinity of a complex zero.

the Routhian array is

$$
\begin{array}{c|cc}
s^3 & 1 & c \\
s^2 & b & Kd \\
s^1 & (bc - Kd)/b & \\
s^0 & Kd &
\end{array}
$$

An undamped oscillation may exist if the s^{-1} row in the array equals zero. For this condition the *auxiliary equation* obtained from the s^2 row is

$$bs^2 + Kd = 0 \tag{7.58}$$

and its roots are

$$s_{1,2} = \pm j \sqrt{\frac{Kd}{b}} = \pm j\omega_n \tag{7.59}$$

The loop sensitivity term K is determined by setting the s^1 row to zero:

$$K = \frac{bc}{d} \tag{7.60}$$

For $K > 0$, Eq. (7.59) gives the natural frequency of the undamped oscillation. This corresponds to the point on the imaginary axis where the locus crosses over into the right-half s plane. The imaginary axis divides the s plane into stable and unstable regions. Also, the value of K from Eq. (7.60) determines the value of the loop sensitivity at the crossover point. For values of $K < 0$ the term in the s^0 row is negative, thus characterizing an unstable response. The limiting values for a stable response are therefore

$$0 < K < \frac{bc}{d} \tag{7.61}$$

In like manner, the crossover point can be determined for higher-order characteristic equations. For these higher-order systems care must be exercised in *analyzing all terms in the first column that contain the term K* in order to obtain the correct range of values of gain for stability.

Rule 9: Intersection or Nonintersection of Root-Locus Branches[5]

The theory of complex variables yields the following properties:

1. A value of s that satisfies the angle condition of Eq. (7.34) is a point on the root locus. If $dW(s)/ds \neq 0$ at this point, there is one and only one branch of the root locus through the point.

2. If the first $y - 1$ derivatives of $W(s)$ vanish at a given point on the root locus, there are y branches approaching and y branches leaving this point; thus, there are root-locus intersections at this point. The angle between two adjacent *approaching* branches is given by

$$\lambda_y = \pm \frac{360°}{y} \tag{7.62}$$

s plane

FIGURE 7.14
Root locus for

$$G(s)H(s) = \frac{K}{(s + 2)(s + 4)(s^2 + 6s + 10)}$$

Also, the angle between a branch *leaving* and an adjacent branch that is *approaching* the same point is given by

$$\theta_y = \pm \frac{180°}{y} \tag{7.63}$$

Figure 7.14 illustrates these angles, with $\theta_y = 45°$ and $\lambda_y = 90°$.

Rule 10: Conservation of the Sum of the System Roots[6]

The technique described by this rule aids in the determination of the general shape of the root locus. Consider the general open-loop transfer function in the form

$$G(s)H(s) = \frac{K \prod\limits_{h=1}^{w} (s - z_h)}{s^m \prod\limits_{c=1}^{u} (s - p_c)} \tag{7.64}$$

Recalling that for physical systems $w \leq n = u + m$, the denominator of $C(s)/R(s)$ can be written as

$$B(s) = 1 + G(s)H(s) = \frac{\prod\limits_{j=1}^{n} (s - r_j)}{s^m \prod\limits_{c=1}^{u} (s - p_c)} \tag{7.65}$$

where r_j are the roots described by the root locus.

By substituting from Eq. (7.64) into Eq. (7.65) and equating numerators on each side of the resulting equation, the result is

$$s^m \prod\limits_{c=1}^{u} (s - p_c) + K \prod\limits_{h=1}^{w} (s - z_h) = \prod\limits_{j=1}^{n} (s - r_j) \tag{7.66}$$

Expanding both sides of this equation gives

$$\left(s^n - \sum_{c=1}^{u} p_c s^{n-1} + \cdots \right) + K \left(s^w - \sum_{h=1}^{w} z_h s^{w-1} + \cdots \right)$$

$$= s^n - \sum_{j=1}^{n} r_j s^{n-1} + \cdots$$

(7.67)

For those open-loop transfer functions in which $w \le n - 2$, the following is obtained by equating the coefficients of s^{n-1} of Eq. (7.67):

$$\sum_{c=1}^{u} p_c = \sum_{j=1}^{n} r_j$$

Since m open-loop poles have values of zero, this equation can also be written as

$$\sum_{j=1}^{n} p_j = \sum_{j=1}^{n} r_j$$

(7.68)

where p_j now represents all the open-loop poles, including those at the origin, and r_j are the roots of the characteristic equation. This equation reveals that as the system gain is varied from zero to infinity, the sum of the system roots is constant. In other words, the sum of the system roots is conserved and is independent of K. When a system has several root-locus branches that go to infinity (as $K \to \infty$), the directions of the branches are such that the sum of the roots is constant. A branch going to the right therefore requires that there will be a branch going to the left. The root locus of Fig. 7.2 satisfies the conservancy law for the root locus. The sum of the roots is a constant for all values of K.

For a unity-feedback system, Rao[7] has shown that the closed-loop pole and zero locations of

$$\frac{C(s)}{R(s)} = \frac{K(s - z_1) \cdots (s - z_w)}{(s - p_1) \cdots (s - p_n)}$$

(7.69)

satisfy the relation

$$\sum_{i=1}^{w} (-z_i)^{-q} = \sum_{j=1}^{n} (-p_j)^{-q} \qquad q = 1, 2, \ldots, m - 1$$

(7.70)

for $m > 1$, that is, for a Type 2 (or higher) system. This serves as a check on the accuracy of the root determination.

Rule 11: Determination of Roots on the Root Locus

After the root locus has been plotted, the specifications for system performance are used to determine the dominant roots. The *dominant branch* is the branch that is the closest to the imaginary axis. This root has the largest influence on the time response. When this branch yields complex dominant roots, the time response is oscillatory and the figures of merit (see Sec. 3.10) are the peak overshoot M_p, the time t_p at which the peak overshoot occurs, and the settling time t_s. An additional figure of merit, the gain K_m, significantly affects the steady-state error (see Chapter 6). All of these quantities can be used to select the *dominant roots*. Thus, the designer may use the damping ratio ζ, the undamped natural frequency ω_n, the damped natural frequency ω_d, the damping coefficient σ, or the gain K_m to select the dominant roots. When the dominant roots are selected, the required loop sensitivity can be determined by applying the magnitude condition, as shown in Eq. (7.33). The remaining roots on each of the other branches can be determined by any of the following methods:

Method 1. Determine the point on each branch of the locus that satisfies the same value of loop sensitivity as for the dominant roots.

Method 2. If all except one real or a complex pair of roots are known, either of the following procedures can be used.

Procedure 1. Divide the characteristic polynomial by the factors representing the known roots. The remainder gives the remaining roots.

Procedure 2. Equation (7.68), known as *Grant's rule,* can be used to find some of the roots.[6] A necessary condition is that the denominator of $G(s)H(s)$ be at least of degree 2 higher than the numerator. If all the roots except one real root are known, application of Eq. (7.68) yields directly the value of the real root. However, for complex roots of the form $r = \sigma \pm j\omega_d$ it yields only the value of the real component σ.

CAD programs[3] are available that yield the loop sensitivity and the roots of the characteristic equation of the system. When the damping ratio is specified for the dominant roots, these roots determine the value of K and all the remaining roots.

7.9   CAD ACCURACY CHECKS (CADAC)

The first seven geometrical construction rules, which can be readily evaluated, should be used as *CADAC* to assist in the validation of the root locus data being obtained from a CAD package.

7.10   EXAMPLES

Example 1. Find $C(s)/R(s)$ with $\zeta = 0.5$ for the dominant roots (roots closest to the imaginary axis) for the feedback control system represented by

$$G(s) = \frac{K_1}{s(s^2/2600 + s/26 + 1)} \quad \text{and} \quad H(s) = \frac{1}{0.04s + 1}$$

Rearranging gives

$$G(s) = \frac{2600K_1}{s(s^2 + 100s + 2600)} = \frac{N_1}{D_1} \quad \text{and} \quad H(s) = \frac{25}{s + 25} = \frac{N_2}{D_2}$$

Thus,

$$G(s)H(s) = \frac{65,000K_1}{s(s + 25)(s^2 + 100s + 2600)} = \frac{K}{s^4 + 125s^3 + 5100s^2 + 65,000s}$$

where $K = 65,000K_1$.

1. The poles of $G(s)H(s)$ are plotted on the s plane in Fig. 7.15; the values of these poles are $s = 0, -25, -50 + j10, -50 - j10$. The system is completely unstable for $K < 0$. Therefore, this example is solved only for the condition $K > 0$.
2. There are four branches of the root locus.
3. The locus exists on the real axis between 0 and -25.
4. The angles of the asymptotes are

$$\gamma = \frac{(1 + 2h)180°}{4} = \pm 45°, \pm 135°$$

5. The real-axis intercept of the asymptotes is

$$\sigma_o = \frac{0 - 25 - 50 - 50}{4} = -31.25$$

6. The breakaway point s_a (see Fig. 7.15) on the real axis between 0 and -25 is found by solving $dW(s)/ds = 0$ [see Eq. (7.39)]:

$$-K = s^4 + 125s^3 + 5100s^2 + 65,000s$$

$$\frac{d(-K)}{ds} = 4s^3 + 375s^2 + 10,200s + 65,000 = 0$$

$$s_a = -9.15$$

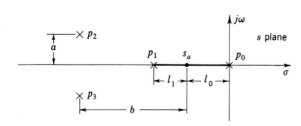

FIGURE 7.15
Location of the breakaway point.

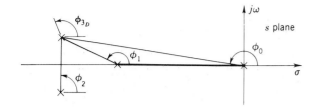

FIGURE 7.16
Determination of the departure angle.

7. The angle of departure ϕ_{3_D} (see Fig. 7.16) from the pole $-50 + j10$ is obtained from

$$\phi_0 + \phi_1 + \phi_2 + \phi_{3_D} = (1 + 2h)180°$$

$$168.7° + 158.2° + 90° + \phi_{3_D} = (1 + 2h)180°$$

$$\phi_{3_D} = 123.1°$$

Similarly, the angle of departure from the pole $-50 - j10$ is $-123.1°$.

8. The imaginary-axis intercepts are obtained from

$$\frac{C(s)}{R(s)} = \frac{2600K_1(s + 25)}{s^4 + 125s^3 + 5100s^2 + 65{,}000s + 65{,}000K_1} \qquad (7.71)$$

The Routhian array for the denominator of $C(s)/R(s)$, which is the characteristic polynomial, is (see Sec. 6.2, Theorem 1)

$$
\begin{array}{c|lcc}
s^4 & 1 & 5100 & 65{,}000\,K_1 \\
s^3 & 1 & 520 \text{ (after division by 125)} & \\
s^2 & 1 & 14.2\,K_1 \text{ (after division by 4580)} & \\
s^1 & 520 - 14.2\,K_1 & & \\
s^0 & 14.2\,K_1 & &
\end{array}
$$

Pure imaginary roots exist when the s^1 row is zero. This occurs when $K_1 = 520/14.2 = 36.6$. The auxiliary equation is formed from the s^2 row:

$$s^2 + 14.2K_1 = 0$$

and the imaginary roots are

$$s = \pm j\sqrt{14.2K_1} = \pm j\sqrt{520} = \pm j22.8$$

9. Additional points on the root locus are found by locating points that satisfy the angle condition

$$\underline{/s} + \underline{/s + 25} + \underline{/s + 50 - j10} + \underline{/s + 50 + j10} = (1 + 2m)180°$$

The root locus is shown in Fig. 7.17.

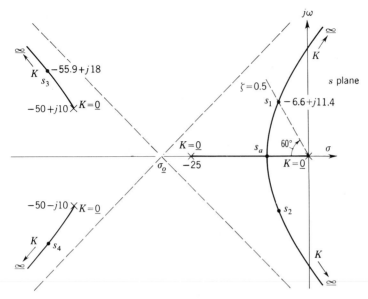

FIGURE 7.17
Root locus for

$$G(s)H(s) = \frac{65,000K_1}{s(s + 25)(s^2 + 100s + 2600)}$$

10. The radial line for $\zeta = 0.5$ is drawn on the graph of Fig. 7.17 at the angle (see Fig. 4.3 for definition of η)

$$\eta = \cos^{-1} 0.5 = 60°$$

The dominant roots obtained from the graph are

$$s_{1,2} = -6.6 \pm j11.4$$

11. The gain is obtained from the expression

$$K = 65,000K_1 = |s| \cdot |s + 25| \cdot |s + 50 - j10| \cdot |s + 50 + j10|$$

by using a calculator. For $s_1 = -6.6 + j11.4$,

$$K = 65,000K_1 = 598,800$$
$$K_1 = 9.25$$

12. The other roots are evaluated to satisfy the magnitude condition $K = 598,800$. The remaining roots of the characteristic equation are

$$s_{3,4} = -55.9 \pm j18.0$$

The real part of the additional roots can also be determined by using Grant's rule. From Eq. (7.68),

$$0 - 25 + (-50 + j10) + (-50 - j10)$$

$$= (-6.6 + j11.4) + (-6.6 - j11.4) + (\sigma + j\omega_d) + (\sigma - j\omega_d)$$

This gives $$\sigma = -55.9$$

By using this value, the roots can be determined from the root locus as $-55.9 \pm j18.0$.

13. The control ratio, using values of the roots obtained in steps 10 and 12, is

$$\frac{C(s)}{R(s)} = \frac{N_1 D_2}{\text{factors determined from root locus}}$$

$$= \frac{24{,}040(s + 25)}{(s + 6.6 + j11.4)(s + 6.6 - j11.4)(s + 55.9 + j18)(s + 55.9 - j18)}$$

$$= \frac{24{,}040(s + 25)}{(s^2 + 13.2s + 173.5)(s^2 + 111.8s + 3450)}$$

14. The response $c(t)$ for a unit step input is found from[3]

$$C(s) = \frac{24{,}040(s + 25)}{s(s^2 + 13.2s + 173.5)(s^2 + 111.8s + 3450)}$$

$$= \frac{A_0}{s} + \frac{A_1}{s + 6.6 - j11.4} + \frac{A_2}{s + 6.6 + j11.4}$$

$$+ \frac{A_3}{s + 55.9 - j18} + \frac{A_4}{s + 55.9 + j18}$$

The constants are

$$A_0 = 1.0 \qquad A_1 = 0.604\underline{/-201.7°} \qquad A_3 = 0.14\underline{/-63.9°}$$

Note that A_0 must be exactly 1, since $G(s)$ is Type 1 and the gain of $H(s)$ is unity. It can also be obtained from Eq. (7.71) for $C(s)/R(s)$ in step 8. Inserting $R(s) = 1/s$ and finding the final value gives $c(t)_{ss} = 1.0$. The response $c(t)$ is

$$c(t) = 1 + 1.21e^{-6.6t} \sin(11.4t - 111.7°) + 0.28e^{-55.9t} \sin(18t + 26.1°) \quad (7.72)$$

A plot of $c(t)$ is shown in Figs. 7.22 and 7.26.

15. The solution of the root locus design problem by use of MATLAB is presented in Sec. 7.11.

Example 2. Many aircraft flight control systems use an adjustable gain in the feedback path. Also, the forward transfer function may be nonminimum phase (nmp; see Sec. 8.9) and have a negative gain. This is illustrated by the control system shown in Fig. 7.18, where

$$G(s) = \frac{-2(s + 6)(s - 6)}{s(s + 3)(s + 4 - j4)(s + 4 + j4)} \qquad H(s) = K_h > 0$$

FIGURE 7.18
Nonunity-feedback control system.

Thus the open-loop transfer function is

$$G(s)H(s) = \frac{K(s + 6)(s - 6)}{s(s + 3)(s + 4 - j4)(s + 4 + j4)} \qquad (7.73)$$

where $K = -2K_h$. The root-locus characteristics are summarized as follows:

1. $G(s)H(s)$ has zeros located at $s = -6$ and $+6$. The zero at $s = 6$ means that this transfer function is nonminimum phase. The poles are located at $s = 0, -3, -4 + j4$ and $-4 - j4$.
2. There are four branches of the root locus.
3. For positive values of K, a branch of the root locus is located on the positive real axis from $s = 0$ to $s = 6$. Thus, there is a positive real root and the system is unstable. For $K < 0$ (the values of K_h are positive), the root locus must satisfy the 360° angle condition given in Eq. (7.26). The root locus is shown in Fig. 7.19. This plot is readily obtained using a CAD program.
4. The closed-loop transfer function is

$$\frac{C(s)}{R(s)} = \frac{-2(s + 6)(s - 6)}{s^4 + 11s^3 + (56 + K)s^2 + 96s - 36K} \qquad (7.74)$$

The Routhian array for the denominator polynomial shows that the imaginary-axis crossing occurs at $s = \pm j2.954$ for $K = -9.224$. Thus, the closed-loop system is stable for $K > -9.224$.

5. Using $\zeta = 0.707$ for selecting the dominant poles yields $s_{1,2} = -1.1806 \pm j1.1810$ and $s_{3,4} = -4.3194 \pm j3.4347$, with $K_h = 1.18(K = -2.36)$. The closed-loop transfer function is

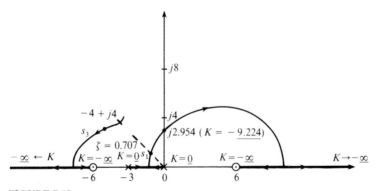

FIGURE 7.19
Root locus for Eq. (7.73) for negative values of K.

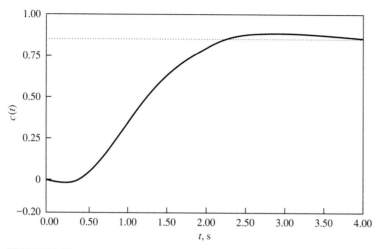

FIGURE 7.20
Step response for the closed-loop system given in Eq. (7.75).

$$\frac{C(s)}{R(s)} = \frac{-2(s + 6)(s - 6)}{(s + 1.1806 \pm j1.1810)(s + 4.3194 \pm j3.4347)} \qquad (7.75)$$

6. The output response for a unit-step input is

$$c(t) = 0.84782 - 1.6978e^{-1.1806t}\sin(1.181t + 29.33°)$$
$$-0.20346e^{-4.3194t}\sin(3.4347t + 175.47°) \qquad (7.76)$$

The step response for this system is plotted in Fig. 7.20. Note that the slight initial undershoot is due to the zero in the RHP. The figures of merit are $M_p = 0.8873$, $t_p = 2.89\text{s}$, $t_s = 3.85\text{s}$, and $c(\infty) = 0.8478$. Note that the forward transfer function $G(s)$ is Type 1, but the steady-state value of the output is not equal to the input because of the nonunity feedback. Also note that the numerator of $C(s)/R(s)$ is the same as the numerator of $G(s)$; thus, it is not affected by the value selected for K.

7.11 EXAMPLE 1: MATLAB ROOT LOCUS[11]

The `zetaex1.m` file

```
% Example of use of rloczeta
% Written by James M. Brown II, Graduate School of
% Engineering, Air Force Institute of Technology; 17 December
% 1993
format short e
% Form G, H, and GH numerators and denominators
gNum  =  2600;
gDen  =  [1  100  2600  0];
hNum  =  25;
hDen  =  [1 25];
```

```
ghNum   =   conv(gNum, hNum);
ghDen   =   conv(gDen, hDen);
% Calculate and display the gain required and all closed-loop
% poles
zeta  =  0.5;
tol   =  1e-12;
% The macro rloczeta was designed to compute the value of K for
% a given zeta.
[k,r]  = rloczeta(ghNum, ghDen, zeta, tol)
k =  9.2435e+00
r =  -5.5902e+01 - 1.8041e + 01i
      -5.5902e+01 + 1.8041e + 01i
      -6.5978e+00 - 1.1428e + 01i
      -6.5978e+00 + 1.1428e + 01i
% Calculate and display the damping ratio for each closed-loop
% pole
zta = -real(r)./abs(r)
zta =  9.5167e-01
       9.5167e-01
       5.0000e-01
       5.0000e-01
% Make a plot, Fig. 7.21, including the closed-loop roots
rlocus(ghNum,ghDen)
axis('square')
print -dps zetaex1a
% Axes can be changed to enlarge plot
axis([-30 2 -16 16])
axis('square')
hold on
% Roots can be located on plot
rlocus(ghNum,ghDen,k)
sgrid
hold off
print -dps zetaex1b
% Form the closed-loop transfer function
[cltfNum, cltfDen]  = feedback(k*gNum, gDen, hNum, hDen, -1);
% Calculate and plot the step response (Fig. 7.22)
% The macro fom was designed to compute the figures of merit.
fom(cltfNum, cltfDen)
Figures of merit for a unit step input
      rise time       =   0.12
      peak value      =   1.186
      peak time       =   0.26
      settling time   =   0.61
      final value     =   1
ans =  1.1860e+00   7.5783e-01   1.0000e+00
       6.1000e-01   1.2000e-01   2.6000e-01
print -dps zetaex1c
>> diary off
```

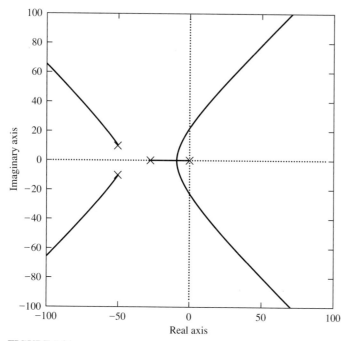

FIGURE 7.21
MATLAB root-locus plot, corresponding to Fig. 7.17.

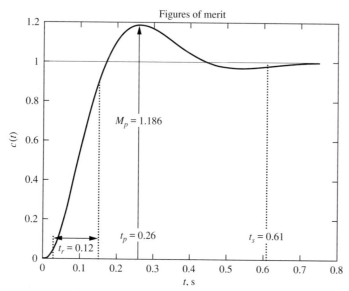

FIGURE 7.22
MATLAB time response, corresponding to Eq. (7.72) and Fig. 7.26.

7.12   PERFORMANCE CHARACTERISTICS

As pointed out early in this chapter, the root-locus method incorporates the more desirable features of both the classical method and the steady-state sinusoidal phasor analysis. In the examples of Sec. 7.10 a direct relationship is noted between the root locus and the time solution. This section is devoted to strengthening this relationship to enable the designer to synthesize and/or compensate a system.

General Introduction

In review, consider a simple second-order system whose control ratio is

$$\frac{C(s)}{R(s)} = \frac{K}{s^2 + 2\zeta\omega_n s + \omega_n^2} \tag{7.77}$$

and whose transient component of the response to a step input is

$$c(t)_t = C_1 e^{s_1 t} + C_2 e^{s_2 t} \tag{7.78}$$

where for $\zeta < 1$

$$s_{1,2} = -\zeta\omega_n \pm j\omega_n \sqrt{1 - \zeta^2} = \sigma \pm j\omega_d \tag{7.79}$$

Thus

$$c(t)_t = A e^{\sigma t} \sin(\omega_d t + \phi) \tag{7.80}$$

Consider now a plot of the roots in the s plane and their correlation with the transient solution in the time domain for a step input. Six cases are illustrated in Fig. 7.23 for different values of damping, showing both the locations of the roots and the corresponding transient plots. In the case of $\zeta = 0$, the roots lie on the $\pm j\omega$ axis and the response has sustained oscillations (Fig. 7.23d). Those portions of the root locus that yield roots in the RHP result in unstable operation (Fig. 7.23e and f). Thus, the desirable roots are on that portion of the root locus in the LHP. (As shown in Chap. 9, the $\pm j\omega$ axis of the s plane corresponds to the $-1 + j0$ point of the Nyquist plot.)

Table 4.1 summarizes the information available from Fig. 7.23; i.e., it shows the correlation between the location of the closed-loop poles and the corresponding transient component of the response. Thus, the value of the root-locus method is that it is possible to determine all the forms of the transient component of the response that a control system may have.

Plot of Characteristic Roots for $0 < \zeta < 1$

The important desired roots lie in the region in which $0 < \zeta < 1$ (generally between 0.4 and 0.8). In Fig. 7.23a, the radius r from the origin to the root s_1 is

$$r = \sqrt{\omega_{d_1}^2 + \sigma_{1,2}^2} = \sqrt{\omega_n^2(1 - \zeta^2) + \omega_n^2\zeta^2} = \omega_n \tag{7.81}$$

and

$$\cos\eta = \left|\frac{-\sigma_{1,2}}{r}\right| = \frac{\zeta\omega_n}{\omega_n} = \zeta \tag{7.82}$$

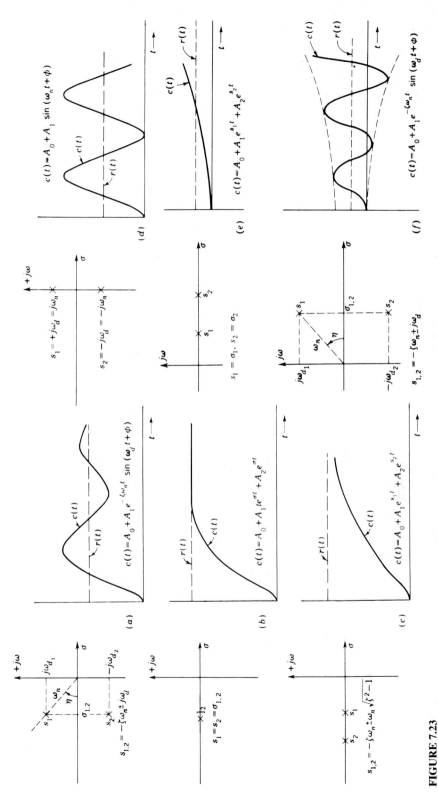

FIGURE 7.23

Plot of roots of a second-order characteristic equation in the s plane and their correlation to the transient solution in the time domain for a step input; (a) underdamped, stable, $0 < \zeta < 1$; (b) critically damped, stable, $\zeta = 1$; (c) real roots, overdamped, stable, $\zeta > 1$; (d) undamped, sustained oscillations, $\zeta = 0$; (e) real roots, unstable, $\zeta < -1$; (f) underdamped, unstable, $0 > \zeta > -1$.

235

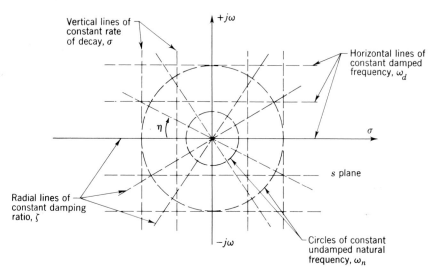

FIGURE 7.24
Constant-parameter curves on the s plane.

or

$$\eta = \cos^{-1} \zeta \qquad (7.83)$$

Based on the preceding equations, the constant-parameter loci are drawn in Fig. 7.24. From Figs. 7.23 and 7.24 and Eqs. (7.81) to (7.83) the following characteristics are summarized for the s plane:

1. Horizontal lines represent lines of constant damped natural frequency ω_d. The closer these lines are to the real axis, the lower the value of ω_d. For roots lying on the real axis ($\omega_d = 0$) there is no oscillation in the transient.
2. Vertical lines represent lines of constant damping or constant rate of decay of the transient. The closer these lines (or the characteristic roots) are to the imaginary axis, the longer it takes for the transient response to die out.
3. Circles about the origin are circles of constant undamped natural frequency ω_n. Since $\sigma^2 + \omega_d^2 = \omega_n^2$, the locus of the roots of constant ω_n is a circle in the s plane; that is, $|s_1| = |s_2| = \omega_n$. The smaller the circles, the lower the ω_n.
4. Radial lines passing through the origin with the angle η are lines of constant damping ratio ζ. The angle η is measured clockwise from the negative real axis for positive ζ, as shown in Figs. 7.23a and 7.24.

Variation of Roots with ζ

Note in Fig. 7.25 the following characteristics:

For $\zeta > 1$, the roots $s_{1,2} = -\zeta \omega_n \pm \omega_n \sqrt{\zeta^2 - 1}$ are real.
For $\zeta = 1$, the roots $s_{1,2} = -\zeta \omega_n$ are real and equal.

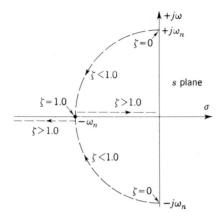

FIGURE 7.25

Variation of roots of a simple second-order system for a constant ω_n.

For $\zeta < 1$, the roots $s_{1,2} = -\zeta\omega_n \pm j\omega_n\sqrt{1-\zeta^2}$ are complex conjugates.
For $\zeta = 0$, the roots $s_{1,2} = \pm j\omega_n$ are imaginary.

Higher-Order Systems

Consider the control ratio of a system of order n given by

$$\frac{C(s)}{R(s)} = \frac{P(s)}{s^n + a_{n-1}s^{n-1} + \cdots + a_0} \tag{7.84}$$

In a system having one or more sets of complex-conjugate roots in the characteristic equation, each quadratic factor is of the form

$$s^2 + 2\zeta\omega_n s + \omega_n^2 \tag{7.85}$$

and the roots are

$$s_{1,2} = \sigma \pm j\omega_d \tag{7.86}$$

The relationships developed for ω_n and ζ earlier in this section apply equally well for each complex-conjugate pair of roots of an nth-order system. The distinction is that the *dominant* ζ and ω_n apply for that pair of complex-conjugate roots that lies closest to the imaginary axis. These values of ζ and ω_n are dominant because the corresponding transient term has the longest settling time and the largest magnitude. Thus, the dominant values of ζ and ω_n are selected from the root locus for the desired response. Remember that in selecting the values of the dominant ζ and ω_n, the other roots are automatically set. Depending on the location of the other roots, they modify the solution obtained from the dominant roots.

In Example 1 of Sec. 7.10 the response $c(t)$ with a step input given by Eq. (7.72) is

$$c(t) = 1 + 1.21e^{-6.6t}\sin(11.4t - 111.7°) + 0.28e^{-55.9t}\sin(18t + 26.1°) \tag{7.87}$$

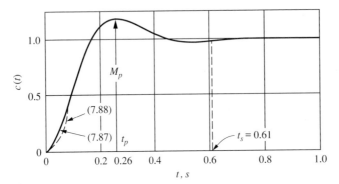

FIGURE 7.26
Plot of $c(t)$ vs. t for Eqs. (7.87) and (7.88).

Note that the transient term due to the roots of $s = -55.9 \pm j18$ dies out in approximately one-tenth the time of the transient term due to the dominant roots $s = -6.6 \pm j11.4$. Therefore, the solution can be approximated by the simplified equation

$$c(t) \approx 1 + 1.21 e^{-6.6t} \sin(11.4t - 111.7°) \tag{7.88}$$

This expression is valid except for a short initial period of time while the other transient term dies out. Equation (7.88) suffices if an approximate solution is desired, i.e., for determining M_p, t_p, and T_s, since the neglected term does not appreciably affect these three quantities. In general, the designer can judge from the location of the roots which ones may be neglected. Plots of $c(t)$ corresponding to Eqs. (7.87) and (7.88) are shown in Fig. 7.26. Since there is little difference between the two curves, the system can be approximated as a second-order system. The significant characteristics of the time response are $M_p = 1.19$, $t_p = 0.26$s, $t_s = 0.61$s, and $\omega_d \approx 11.4$ rad/s.

The transient response of any complex system, with the effect of all the roots taken into account, can often be considered to be the result of an *equivalent* second-order system. On this basis, it is possible to define effective (or equivalent) values of ζ and ω_n. Realize that in order to alter either effective quantity, the location of one or more of the roots must be altered.

7.13 TRANSPORT LAG[8]

Some elements in control systems are characterized by dead time or transport lag. This appears as a dead interval for which the output is delayed in response to an input. Figure 7.27a shows the block diagram of a transport-lag element. In Fig. 7.27b a typical resultant output $e_o(t) = e_i(t-\tau)u_{-1}(t-\tau)$ is shown, which has the same form as the input but is delayed by a time interval τ. Dead time is a nonlinear characteristic that can be represented precisely in terms of the Laplace transform (see Sec. 4.4) as a transcendental function. Thus,

$$E_o(s) = e^{-\tau s} E_i(s) \tag{7.89}$$

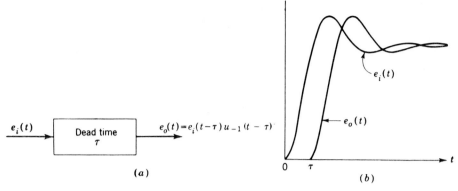

FIGURE 7.27
(a) Transport lag τ; (b) time characteristic.

The transfer function of the transport-lag element, where $s = \sigma + j\omega$, is

$$G_\tau(s) = \frac{E_o(s)}{E_i(s)} = e^{-\tau s} = e^{-\sigma\tau}\underline{/-\omega\tau} \qquad (7.90)$$

It has a negative angle that is directly proportional to the frequency.

The root locus for a system containing a transport lag is illustrated by the following example:[8]

$$G(s)H(s) = \frac{Ke^{-\tau s}}{s(s - p_1)} = \frac{Ke^{-\sigma\tau}e^{-j\omega\tau}}{s(s - p_1)} \qquad (7.91)$$

For $K > 0$ the angle condition is

$$\underline{/s} + \underline{/s - p_1} + \underline{/\omega\tau} = (1 + 2h)180° \qquad (7.92)$$

The root locus for the system with transport lag is drawn in Fig. 7.28. The magnitude condition used to calibrate the root locus is

$$K = |s| \cdot |s - p_1| e^{\sigma\tau} \qquad (7.93)$$

The same system without transport lag has only two branches (see Fig. 7.2) and is stable for all $K > 0$. The system with transport lag has an infinite number of branches and the asymptotes are all parallel to the real axis. The imaginary axis crossings of the asymptotes are

$$\omega = \frac{(1 + 2h)\pi}{\tau} \qquad (7.94)$$

where $h = 0, \pm1, \pm2, \ldots$. The values of K at the points where the branches cross the imaginary axis are

$$K = \omega\sqrt{p_1^2 + \omega^2} \qquad (7.95)$$

The two branches closest to the origin thus have the largest influence on system stability and are therefore the principal branches. The maximum value of loop sensitivity K given by Eq. (7.95) is therefore determined by the frequency ω_1 (shown in Fig. 7.28) that satisfies the angle condition $\omega_1\tau = \tan^{-1}|p_1/\omega_1|$.

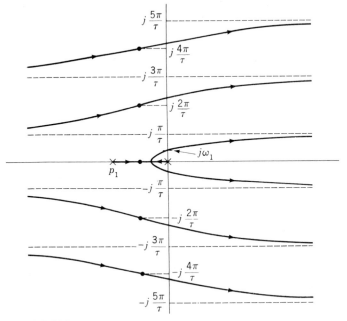

FIGURE 7.28
Root locus for

$$G(s)H(s) = \frac{Ke^{-\tau s}}{s(s - p_1)}$$

7.14 SYNTHESIS

The root-locus method lends itself very readily to the synthesis problem because of the direct relationship between the frequency and the time domains to the s domain. The desired response can be achieved by keeping in mind the following five points:

1. First plot the root locus; then the locations of the dominant closed-loop poles must be specified, based on the desired transient response.
2. Use these guidelines for selecting the roots:
   (a) Since the desired time response is often specified to have a peak value M_p between 1.0 and 1.4, the locus has at least one set of complex-conjugate poles. Often there is a pair of complex poles near the imaginary axis that dominates the time response of the system; if so, these poles are referred to as the *dominant-pole pair.*
   (b) To set the system gain, any of the following characteristics must be specified for the dominant-pole pair: the damping ratio ζ, the settling time T_s (for 2 percent: $T_s = 4/\zeta\omega_n$), the undamped natural frequency ω_n, or the damped natural frequency ω_d. As previously determined (see Fig. 7.24), each of these factors corresponds to either a line or a circle in the s domain.

(*c*) When the line or circle corresponding to the given information has been drawn, as stated in point 2*b*, its intersection with the locus determines the dominant-pole pair and fixes the value of the system gain. By applying the magnitude condition to this point, the value of the gain required can be determined. Setting the value of the gain to this value results in the transient-response terms, corresponding to the dominant-pole pair, having the desired characterizing form.

3. When the gain corresponding to the dominant roots has been determined, the remaining roots can be found by applying the magnitude condition to each branch of the locus or by using a CAD program. Once the value of the gain on the dominant branch is fixed, thus fixing the dominant poles, the locations of all remaining roots on the other branches are also fixed.

4. If the root locus does not yield the desired response, the locus must be altered by compensation to achieve the desired results. The subject of compensation in the *s* domain is discussed in later chapters.

7.15 SUMMARY OF ROOT-LOCUS CONSTRUCTION RULES FOR NEGATIVE FEEDBACK

RULE 1. The number of branches of the root locus is equal to the number of poles of the open-loop transfer function.

RULE 2. For *positive* values of K, the root locus exists on those portions of the real axis for which the total number of real poles and zeros to the right is an odd number. For *negative* values of K, the root locus exists on those portions of the real axis for which the total number of real poles and zeros to the right is an even number (including zero).

RULE 3. The root locus starts ($K = 0$) at the open-loop poles and terminates ($K = \pm\infty$) at the open-loop zeros or at infinity.

RULE 4. The angles of the asymptotes for those branches of the root locus that end at infinity are determined by

$$\gamma = \frac{(1 + 2h)180° \text{ for } K > 0 \quad \text{or} \quad h360° \text{ for } K < 0}{[\text{Number of poles of} G(s)H(s)] - [\text{Number of zeros of } G(s)H(s)]}$$

RULE 5. The real-axis intercept of the asymptotes is

$$\sigma_o = \frac{\sum_{c=1}^{n} \text{Re}(p_c) - \sum_{h=1}^{w} \text{Re}(z_h)}{n - w}$$

RULE 6. The breakaway point for the locus between two poles on the real axis (or the break-in point for the locus between two zeros on the real axis) can be determined by taking the derivative of the loop sensitivity K with respect to s. Equate this derivative to zero and find the roots of the resulting equation. The root that occurs between the poles (or the zeros) is the breakaway (or break-in) point.

RULE 7.   For $K > 0$ the angle of departure from a complex pole is equal to $180°$ minus the sum of the angles from the other poles plus the sum of the angles from all the zeros. Any of these angles may be positive or negative. For $K < 0$ the departure angle is $180°$ from that obtained for $K > 0$.

For $K > 0$ the angle of approach to a complex zero is equal to the sum of the angles from all the poles minus the sum of the angles from the other zeros minus $180°$. For $K < 0$ the approach angle is $180°$ from that obtained from $K > 0$.

RULE 8.   The imaginary-axis crossing of the root locus can be determined by setting up the Routhian array from the closed-loop characteristic polynomial. Equate the s^1 row to zero and form the auxiliary equation from the s^2 row. The roots of the auxiliary equation are the imaginary-axis crossover points.

RULE 9.   The selection of the dominant roots of the characteristic equation is based on the specifications that give the required system performance; i.e., it is possible to evaluate σ, ω_d, and ζ from Eqs. (3.60), (3.61), and (3.64), which in turn determine the location of the desired dominant roots. The loop sensitivity for these roots is determined by applying the magnitude condition. The remaining roots are then determined to satisfy the same magnitude condition.

RULE 10.   For those open-loop transfer functions for which $w \leq n-2$, Grant's rule states that the sum of the closed-loop roots is equal to the sum of the open-loop poles. Thus, once the dominant roots have been located, Grant's rule

$$\sum_{j=1}^{n} p_j = \sum_{j=1}^{n} r_j$$

can be used to find one real or two complex roots. Factoring known roots from the characteristic equation can simplify the work of finding the remaining roots.

A root-locus CAD program[3] will produce an accurate calibrated root locus. This considerably simplifies the work required for the system design. By specifying ζ for the dominant roots or the gain K_m, a computer program[3] can determine all the roots of the characteristic equation.

7.16   SUMMARY

In this chapter the root-locus method is developed to solve graphically for the roots of the characteristic equation. This method can be extended to solving for the roots of any polynomial. Any polynomial can be rearranged and put into the mathematical form of a ratio of factored polynomials that is equal to plus or minus unity, i.e.,

$$\frac{N(s)}{D(s)} = \pm 1$$

Once this form has been obtained, the procedures given in this chapter can be utilized to locate the roots of the polynomial.

The root locus permits an analysis of the performance of feedback systems and provides a basis for selecting the gain in order to best achieve the performance specifications. Since the closed-loop poles are obtained explicitly, the form of the time response is directly available. CAD programs are available for obtaining $c(t)$ vs. t. Widely used CAD packages for control systems simplify considerably the analysis and design process. They provide information including root-locus plots, roots of interest for a given gain, closed-loop transfer functions, and the time response with a specified input. If the performance specifications cannot be met, the root locus can be analyzed to determine the appropriate compensation to yield the desired results. This is covered in Chapter 10.

REFERENCES

1. Truxal, J. G.: *Automatic Feedback Control System Synthesis,* McGraw-Hill, New York, 1955.
2. Evans, W. R.: *Control-System Dynamics,* McGraw-Hill, New York, 1954.
3. A comprehensive list of CAD programs is given in Appendix B.
4. Lorens, C. S., and R. C. Titsworth: "Properties of Root Locus Asymptotes," letter in *IRE Trans. Autom. Control,* vol. AC-5, pp. 71–72, January 1960.
5. Wilts, C. H.: *Principles of Feedback Control,* Addison-Wesley, Reading, Mass., 1960.
6. Grant, A. J.: "The Conservation of the Sum of the System Roots as Applied to the Root Locus Method," unpublished paper, North American Aviation, Inc., Apr. 10, 1953.
7. Rao, S. N.: "A Relation between Closed Loop Pole and Zeros Locations," *Int. J. Control,* vol. 24, p. 147, 1976.
8. Chang, C. S.: "Analytical Method for Obtaining the Root Locus with Positive and Negative Gain," *IEEE Trans. Autom. Control,* vol. AC-10, pp. 92–94, 1965.
9. Yeh, V. C. M.: "The Study of Transients in Linear Feedback Systems by Conformal Mapping and Root-Locus Method," *Trans. ASME,* vol. 76, pp. 349–361, 1954.
10. Yeung, K. S.: "A Remark on the Use of Remec's Method of Finding Breakaway Points," *IEEE Trans. Autom. Control,* vol. AC-26, pp. 940–941, 1981.
11. Saadat, H.: *Computational Aids in Control Systems Using MATLAB®,* McGraw-Hill, New York, 1993.

CHAPTER
8

FREQUENCY RESPONSE

8.1 INTRODUCTION

In conventional control-system analysis there are two basic methods for predicting and adjusting a system's performance without resorting to the solution of the system's differential equation. One of these is the root-locus method, which is presented in Chapter 7. The frequency-response method, to which the Nyquist stability criterion may be applied, is developed in this chapter.[1,2] For the comprehensive study of a system by conventional methods it is necessary to use both methods of analysis. The principal advantage of the root-locus method is that the actual time response is easily obtained by means of the inverse Laplace transform because the precise root locations are known. However, it is sometimes necessary to express performance requirements in terms of the frequency response. Also, the noise, which is always present in any system, can result in poor overall performance. The frequency response of a system permits evaluation of the effect of noise. The design of a passband for the system response may exclude the noise and therefore improve the system performance as long as the dynamic performance specifications are met. The frequency response is also useful in situations for which the transfer functions of some or all of the components in a system are unknown. The frequency response can be determined experimentally for these situations, and an approximate expression for the transfer function can be obtained from the graphical plot of the experimental data. The frequency-response method is also a very powerful method for analyzing and designing a robust MIMO system with uncertain plant parameters (see Chapter 18). Thus, no particular method can be judged superior to the rest. Each has its particular use and advantage in a particular situation. As the reader becomes fully acquainted with all methods, the potentialities of each method become more apparent and the more appropriate method can be selected.

Design of control systems by state variable techniques is often based upon achieving an optimum performance according to a specified performance index PI; for example, minimizing the integral of squared error (ISE): PI $= \int_0^\infty e(t)^2 \, dt$, where $e \equiv r - c$. Both frequency-response and root-locus methods are valuable complementary tools for many of the techniques of modern control theory.

In this chapter two graphical representations of transfer functions are presented, the logarithmic and the polar plots. These plots are used to develop Nyquist's stability criterion and closed-loop design procedures. Also, the plots are readily obtained by the use of CAD packages such as MATLAB or ICECAP.[4]

8.2 CORRELATION OF THE SINUSOIDAL AND TIME RESPONSES[2]

As pointed out earlier, solving for $c(t)$ by the classical method is laborious and impractical for synthesis purposes, especially when the input is not a simple analytical function. The use of Laplace transform theory lessens the work involved and permits the engineer to synthesize and improve a system. The root-locus method illustrates this fact. The advantages of the graphical representations in the frequency domain of the transfer functions are developed in the following pages.

Once the frequency response of a system has been determined, the time response can be determined by inverting the corresponding Fourier transform. The behavior in the frequency domain for a given driving function $r(t)$ can be determined by the Fourier transform as

$$\mathbf{R}(j\omega) = \int_{-\infty}^{\infty} r(t)e^{-j\omega t} \, dt \tag{8.1}$$

For a given control system the frequency response of the controlled variable is

$$\mathbf{C}(j\omega) = \frac{\mathbf{G}(j\omega)}{1 + \mathbf{G}(j\omega)\mathbf{H}(j\omega)}\mathbf{R}(j\omega) \tag{8.2}$$

By use of the inverse Fourier transform, which is much used in practice, the controlled variable as a function of time is

$$c(t) = \frac{1}{2\pi} \int_{-\infty}^{\infty} \mathbf{C}(j\omega)e^{j\omega t} \, d\omega \tag{8.3}$$

If the design engineer cannot evaluate Eq. (8.3) by reference to a table of definite integrals, this equation can be evaluated by numerical or graphical integration. This is necessary if $\mathbf{C}(j\omega)$ is available only as a curve and cannot be simply expressed in analytical form, as is often the case. The procedure is described in several books.[10] In addition, methods have been developed based on the Fourier transform and a step input signal, relating $\mathbf{C}(j\omega)$ qualitatively to the time solution without actually taking the inverse Fourier transform. These methods permit the engineer to make an approximate determination of the system response through the interpretation of graphical plots in the frequency domain. This makes the design and improvement of feedback systems possible with a minimum effort and is emphasized in this chapter.

Section 4.12 shows that the frequency response is a function of the pole-zero pattern in the s plane. It is therefore related to the time response of the system. Two features of the frequency response are the maximum value M_m and the resonant frequency ω_m. Section 9.3 describes the qualitative relationship between the time response and the values M_m and ω_m. Since the location of the closed-loop poles can be determined from the root locus, there is a direct relationship between the root-locus and frequency-response methods.

8.3 FREQUENCY-RESPONSE CURVES

The frequency domain plots that have found great use in graphical analysis in the design of feedback control systems belong to two categories. The first category is the plot of the magnitude of the output-input ratio vs. frequency in rectangular coordinates, as illustrated in Sec. 4.12. In logarithmic coordinates these are known as *Bode plots*. Associated with this plot is a second plot of the corresponding phase angle vs. frequency. In the second category the output-input ratio may be plotted in polar coordinates with frequency as a parameter. There are two types of polar plots, direct and inverse. Polar plots are generally used only for the open-loop response and are commonly referred to as *Nyquist plots*.[3] The plots can be obtained experimentally or by a CAD package.[4] These programs assist in obtaining plots and the designs developed in this text. When a CAD program is not available, the Bode plots are easily obtained by a graphical procedure. The other plots can then be obtained from the Bode plots.

For a given sinusoidal input signal, the input and steady-state output are of the following forms:

$$r(t) = R \sin \omega t \tag{8.4}$$

$$c(t) = C \sin(\omega t + \alpha) \tag{8.5}$$

The closed-loop frequency response is given by

$$\frac{C(j\omega)}{R(j\omega)} = \frac{G(j\omega)}{1 + G(j\omega)H(j\omega)} = M(\omega)\underline{/\alpha(\omega)} \tag{8.6}$$

For each value of frequency, Eq. (8.6) yields a phasor quantity whose magnitude is M and whose phase angle α is the angle between $C(j\omega)$ and $R(j\omega)$.

An ideal system may be defined as one where $\alpha = 0°$ and $R(j\omega) = C(j\omega)$ for $0 < \omega < \infty$. (See curves 1 in Fig. 8.1.) However, this definition implies an instantaneous transfer of energy from the input to the output. Such a transfer cannot be achieved in practice since any physical system has some energy dissipation and some energy-storage elements. Curves 2 and 3 in Fig. 8.1 represent the frequency responses of practical control systems. The passband, or bandwidth, of the frequency response is defined as the range of frequencies from 0 to the frequency ω_b, where $M = 0.707$ of the value at $\omega = 0$. However, the frequency ω_m is more easily obtained than ω_b. The values M_m and ω_m are often used as figures of merit.

In any system the input signal may contain spurious noise signals in addition to the true signal input, or there may be sources of noise within the closed-loop system.

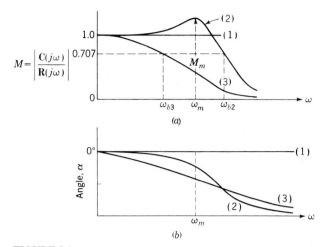

FIGURE 8.1
Frequency-response characteristics of $\mathbf{C}(j\omega)/\mathbf{R}(j\omega)$ in rectangular coordinates.

This noise is generally in a band of frequencies above the dominant frequency band of the true signal. Thus, to reproduce the true signal and attenuate the noise, feedback control systems are designed to have a definite passband. In certain cases the noise frequency may exist in the same frequency band as the true signal. When this occurs, the problem of estimating the desired signal is more complicated. Therefore, even if the ideal system were possible, it would not be desirable.

8.4 BODE PLOTS (LOGARITHMIC PLOTS)

The plotting of the frequency transfer function can be systematized and simplified by using logarithmic plots. The use of semilog paper eliminates the need to take logarithms of very many numbers and also expands the low-frequency range, which is of primary importance. The advantages of logarithmic plots are that (1) the mathematical operations of multiplication and division are transformed to addition and subtraction and (2) the work of obtaining the transfer function is largely graphical instead of analytical. The basic factors of the transfer function fall into three categories, and these can easily be plotted by means of straight-line asymptotic approximations.

In preliminary design studies the straight-line approximations are used to obtain approximate performance characteristics very quickly or to check values obtained from the computer. As the design becomes more firmly specified, the straight-line curves can be corrected for greater accuracy. From these logarithmic plots enough data in the frequency range of concern can readily be obtained to determine the corresponding polar plots.

Some basic definitions of logarithmic terms follow.

Logarithm. The logarithm of a complex number is itself a complex number. The abbreviation "log" is used to indicate the logarithm to the base 10:

$$\log \left|\mathbf{G}(j\omega)\right| e^{j\phi(\omega)} = \log \left|\mathbf{G}(j\omega)\right| + \log e^{j\phi(\omega)}$$

$$= \log \left|\mathbf{G}(j\omega)\right| + j0.434\phi(\omega) \qquad (8.7)$$

The real part is equal to the logarithm of the magnitude, $\log \left|\mathbf{G}(j\omega)\right|$, and the imaginary part is proportional to the angle, $0.434\phi(\omega)$. In the rest of this book the factor 0.434 is omitted and only the angle $\phi(\omega)$ is used.

Decibel. In feedback-system work the unit commonly used for the logarithm of the magnitude is the *decibel* (dB). When logarithms of transfer functions are used, the input and output variables are not necessarily in the same units; e.g., the output may be speed in radians per second, and the input may be voltage in volts.

Log magnitude. The logarithm of the magnitude of a transfer function $\mathbf{G}(j\omega)$ expressed in decibels is

$$20 \log \left|\mathbf{G}(j\omega)\right| \qquad \text{dB}$$

This quantity is called the *log magnitude,* abbreviated Lm. Thus

$$\text{Lm } \mathbf{G}(j\omega) = 20 \log \left|\mathbf{G}(j\omega)\right| \qquad \text{dB} \qquad (8.8)$$

Since the transfer function is a function of frequency, the Lm is also a function of frequency.

Octave and decade. Two units used to express frequency bands or frequency ratios are the octave and the decade. An octave is a frequency band from f_1 to f_2, where $f_2/f_1 = 2$. Thus, the frequency band from 1 to 2 Hz is 1 octave in width, and the frequency band from 17.4 to 34.8 Hz is also 1 octave in width. Note that 1 octave is not a fixed frequency bandwidth but depends on the frequency range being considered. The number of octaves in the frequency range from f_1 to f_2 is

$$\frac{\log(f_2/f_1)}{\log 2} = 3.32 \log \frac{f_2}{f_1} \qquad \text{octaves} \qquad (8.9)$$

There is an increase of 1 decade from f_1 to f_2 when $f_2/f_1 = 10$. The frequency band from 1 to 10 Hz or from 2.5 to 25 Hz is 1 decade in width. The number of decades from f_1 to f_2 is given by

$$\log \frac{f_2}{f_1} \qquad \text{decades} \qquad (8.10)$$

The dB values of some common numbers are given in Table 8.1. Note that the reciprocals of numbers differ only in sign. Thus, the dB value of 2 is $+6$ dB and the dB value of $\frac{1}{2}$ is -6 dB. Two properties are illustrated in Table 8.1.

Property 1. As a number doubles, the decibel value increases by 6 dB. The number 2.70 is twice as big as 1.35, and its decibel value is 6 dB more. The number 200 is twice as big as 100, and its decibel value is 6 dB greater.

Property 2. As a number increases by a factor of 10, the decibel value increases by 20 dB. The number 100 is ten times as large as the number 10, and its decibel value

TABLE 8.1
Decibel values of some common numbers

Number	Decibels	Number	Decibels
0.01	−40	2.0	6
0.1	−20	10.0	20
0.5	−6	100.0	40
1.0	0	200.0	46

is 20 dB more. The number 200 is one hundred times larger than the number 2, and its decibel value is 40 dB greater.

8.5   GENERAL FREQUENCY-TRANSFER-FUNCTION RELATIONSHIPS

The frequency transfer function can be written in generalized form as the ratio of polynomials:

$$\mathbf{G}(j\omega) = \frac{K_m(1 + j\omega T_1)(1 + j\omega T_2)^r \cdots}{(j\omega)^m(1 + j\omega T_a)[1 + (2\zeta/\omega_n)j\omega + (1/\omega_n^2)(j\omega)^2] \cdots}$$

$$= K_m\mathbf{G}'(j\omega) \tag{8.11}$$

where K_m is the gain constant. The logarithm of the transfer function is a complex quantity; the real portion is proportional to the log of the magnitude, and the complex portion is proportional to the angle. Two separate equations are written, one for the log magnitude and one for the angle:

$$\text{Lm } \mathbf{G}(j\omega) = \text{Lm } K_m + \text{Lm}(1 + j\omega T_1) + r\,\text{Lm}\,(1 + j\omega T_2) + \cdots - m\,\text{Lm}\,j\omega$$

$$- \text{Lm}(1 + j\omega T_a) - \text{Lm}\left[1 + \frac{2\zeta}{\omega_n}j\omega + \frac{1}{\omega_n^2}(j\omega)^2\right] - \cdots \tag{8.12}$$

$$\underline{/\mathbf{G}(j\omega)} = \underline{/K_m} + \underline{/1 + j\omega T_1} + r\underline{/1 + j\omega T_2} + \cdots - m\underline{/j\omega}$$

$$- \underline{/1 + j\omega T_a} - \underline{/1 + \frac{2\zeta}{\omega_n}j\omega + \frac{1}{\omega_n^2}(j\omega)^2} - \cdots \tag{8.13}$$

The angle equation may be rewritten as

$$\underline{/\mathbf{G}(j\omega)} = \underline{/K_m} + \tan^{-1}\omega T_1 + r\tan^{-1}\omega T_2 + \cdots - m90°$$

$$- \tan^{-1}\omega T_a - \tan^{-1}\frac{2\zeta\omega/\omega_n}{1 - \omega^2/\omega_n^2} - \cdots \tag{8.14}$$

The gain K_m is a real number but may be positive or negative; therefore, its angle is correspondingly 0° or 180°. Unless otherwise indicated, a positive value of gain is assumed in this book. Both the log magnitude and the angle given by these equations are functions of frequency. When the log magnitude and the angle are plotted as functions of the log of frequency, the resulting pair of curves are referred to as the Bode plots or the *log magnitude diagram and phase diagram*. Equations (8.12) and (8.13) show that the resultant curves are obtained by the addition and subtraction of the corresponding individual terms in the transfer function equation. The two curves can be combined into a single curve of log magnitude vs. angle, with frequency ω as a parameter. This curve, called the Nichols chart or the *log magnitude–angle diagram*, or the corresponding Nyquist polar plot, is used for the quantitative design of the feedback system to meet specifications of required performance.

8.6   DRAWING THE BODE PLOTS

The properties of frequency-response plots are presented in this section, but the data for these plots usually are obtained from a CAD program. The generalized form of the transfer function as given by Eq. (8.11) shows that the numerator and denominator have four basic types of factors:

$$K_m \tag{8.15}$$

$$(j\omega)^{\pm m} \tag{8.16}$$

$$(1 + j\omega T)^{\pm r} \tag{8.17}$$

$$\left[1 + \frac{2\zeta}{\omega_n}j\omega + \frac{1}{\omega_n^2}(j\omega)^2\right]^{\pm p} \tag{8.18}$$

Each of these terms except K_m may appear raised to an integral power other than 1. The curves of Lm and angle vs. the log frequency can easily be drawn for each factor. Then these curves for each factor can be added together graphically to get the curves for the complete transfer function. The procedure can be further simplified by using asymptotic approximations to these curves, as shown in the following pages.

Constants

Since the constant K_m is frequency-invariant, the plot of

$$\text{Lm } K_m = 20 \log K_m \qquad \text{dB}$$

is a horizontal straight line. The constant raises or lowers the Lm curve of the complete transfer function by a fixed amount. The angle, of course, is zero as long as K_m is positive.

$j\omega$ Factors

The factor $j\omega$ appearing in the denominator has a log magnitude

$$\text{Lm}(j\omega)^{-1} = 20 \log \left|(j\omega)^{-1}\right| = -20 \log \omega \tag{8.19}$$

When plotted against $\log \omega$, this curve is a straight line with a negative slope of 6 dB/octave or 20 dB/decade. Values of this function can be obtained from Table 8.1 for several values of ω. The angle is constant and equal to $-90°$.

When the factor $j\omega$ appears in the numerator, the Lm is

$$\text{Lm}(j\omega) = 20 \log |j\omega| = 20 \log \omega \qquad (8.20)$$

This curve is a straight line with a positive slope of 6 dB/octave or 20 dB/decade. The angle is constant and equal to $+90°$. Notice that the only difference between the curves for $j\omega$ and for $1/j\omega$ is a change in the sign of the slope of the Lm and a change in the sign of the angle. Both curves go through the point 0 dB at $\omega = 1$.

For the factor $(j\omega)^{\pm m}$ the Lm curve has a slope of $\pm 6m$ dB/octave or $\pm 20m$ dB/decade, and the angle is constant and equal to $\pm m90°$.

$1 + j\omega T$ **Factors**

The factor $1 + j\omega T$ appearing in the denominator has a log magnitude

$$\text{Lm}(1 + j\omega T)^{-1} = 20 \log |1 + j\omega T|^{-1} = -20 \log \sqrt{1 + \omega^2 T^2} \qquad (8.21)$$

For very small values of ω, that is, $\omega T \ll 1$,

$$\text{Lm}(1 + j\omega T)^{-1} \approx \log 1 = 0 \text{ dB} \qquad (8.22)$$

Thus, the plot of the Lm at small frequencies is the 0-dB line. For very large values of ω, that is, $\omega T \gg 1$,

$$\text{Lm}(1 + j\omega T)^{-1} \approx 20 \log |j\omega T|^{-1} = -20 \log \omega T \qquad (8.23)$$

The value of Eq. (8.23) at $\omega = 1/T$ is 0. For values of $\omega > 1/T$ this function is a straight line with a negative slope of 6 dB/octave. Therefore, the asymptotes of the plot of $\text{Lm}(1 + j\omega T)^{-1}$ are two straight lines, one of zero slope below $\omega = 1/T$ and one of –6 dB/octave slope above $\omega = 1/T$. These asymptotes are drawn in Fig. 8.2.

The frequency at which the asymptotes to the log magnitude curve intersect is defined as the corner frequency ω_{cf}. The value $\omega_{cf} = 1/T$ is the corner frequency for the function $(1 + j\omega T)^{\pm r} = (1 + j\omega/\omega_{cf})^{\pm r}$.

The exact values of $\text{Lm}(1 + j\omega T)^{-1}$ are given in Table 8.2 for several frequencies in the range a decade above and below the corner frequency. The exact curve is also drawn in Fig. 8.2. The error, in dB, between the exact curve and the asymptotes is approximately as follows:

1. At the corner frequency: 3 dB
2. One octave above and below the corner frequency: 1 dB
3. Two octaves from the corner frequency: 0.26 dB

Frequently the preliminary design studies are made by using the asymptotes only. The correction to the straight-line approximation to yield the true Lm curve is shown in Fig. 8.3. For more exact studies the corrections are put in at the corner

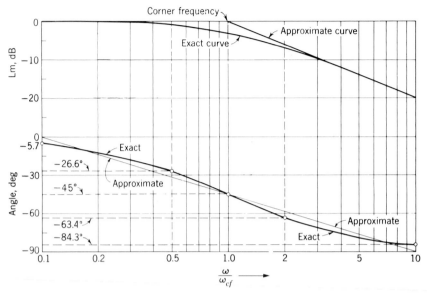

FIGURE 8.2
Log magnitude and phase diagram for

$$(1 + j\omega T)^{-1} = [1 + j(\omega/\omega_{cf})]^{-1}$$

frequency and at 1 octave above and below the corner frequency, and the new curve is drawn with a French curve. For a more accurate curve a CAD program should be used.

The phase angle curve for this function is also plotted in Fig. 8.2. At zero frequency the angle is $0°$; at the corner frequency $\omega = \omega_{cf}$ the angle is $-45°$; and at infinite frequency the angle is $-90°$. The angle curve is symmetrical about the corner-frequency value when plotted against $\log(\omega/\omega_{cf})$ or $\log \omega$. Since the abscissa

TABLE 8.2
Values of $\mathbf{Lm}(1 + j\omega T)^{-1}$ for several frequencies

ω/ω_{cf}	Exact value, dB	Value of the asymptote, dB	Error, dB
0.1	−0.04	0	−0.04
0.25	−0.26	0	−0.26
0.5	−0.97	0	−0.97
0.76	−2.00	0	−2.00
1	−3.01	0	−3.01
1.31	−4.35	−2.35	−2.00
2	−6.99	−6.02	−0.97
4	−12.30	−12.04	−0.26
10	−20.04	−20.0	−0.04

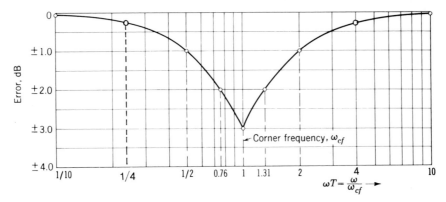

FIGURE 8.3
Log magnitude correction for $(1 + j\omega T)^{\pm 1}$.

of the curves in Fig. 8.2 is ω/ω_{cf}, the shapes of the angle and Lm curves are independent of the time constant T. Thus, when the curves are plotted with the abscissa in terms of ω, changing T just "slides" the Lm and the angle curves left or right so that the -3 dB and the $-45°$ points occur at the frequency $\omega = \omega_{cf}$.

For preliminary design studies, straight-line approximations of the phase curve can be used. The approximation is a straight line drawn through the following three points:

ω/ω_{cf}	0.1	1.0	10
Angle	0°	−45°	−90°

The maximum error resulting from this approximation is about $\pm 6°$. For greater accuracy, a smooth curve is drawn through the points given in Table 8.3.

The factor $1 + j\omega T$ appearing in the numerator has the log magnitude

$$\text{Lm}(1 + j\omega T) = 20 \log \sqrt{1 + \omega^2 T^2}$$

TABLE 8.3
Angles of $(1 + j\omega/\omega_{cf})^{-1}$
for key frequency points

ω/ω_{cf}	Angle, deg
0.1	−5.7
0.5	−26.6
1.0	−45.0
2.0	−63.4
10.0	−84.3

This is the same function as its inverse $\text{Lm}(1 + j\omega T)^{-1}$ except that it is positive. The corner frequency is the same, and the angle varies from 0 to 90° as the frequency increases from zero to infinity. The Lm and angle curves for the function $1 + j\omega T$ are symmetrical about the abscissa to the curves for $(1 + j\omega T)^{-1}$.

Quadratic Factors

Quadratic factors in the denominator of the transfer function have the form

$$\left[1 + \frac{2\zeta}{\omega_n} j\omega + \frac{1}{\omega_n^2}(j\omega)^2 \right]^{-1} \tag{8.24}$$

For $\zeta > 1$ the quadratic can be factored into two first-order factors with real zeros that can be plotted in the manner shown previously. But for $\zeta < 1$ Eq. (8.24) contains conjugate-complex factors, and the entire quadratic is plotted without factoring:

$$\text{Lm}\left[1 + \frac{2\zeta}{\omega_n} j\omega + \frac{1}{\omega_n^2}(j\omega)^2 \right]^{-1} = -20\log\left[\left(1 - \frac{\omega^2}{\omega_n^2} \right)^2 + \left(\frac{2\zeta\omega}{\omega_n} \right)^2 \right]^{1/2} \tag{8.25}$$

$$\text{Angle}\left[1 + \frac{2\zeta}{\omega_n} j\omega + \frac{1}{\omega_n^2}(j\omega)^2 \right]^{-1} = -\tan^{-1}\frac{2\zeta\omega/\omega_n}{1 - \omega^2/\omega_n^2} \tag{8.26}$$

From Eq. (8.25) it is seen that for very small values of ω the low-frequency asymptote is represented by Lm = 0 dB. For very high values of frequency, the Lm is approximately

$$-20\log\frac{\omega^2}{\omega_n^2} = -40\log\frac{\omega}{\omega_n}$$

and the high-frequency asymptote has a slope of –40 dB/decade. The asymptotes cross at the corner frequency $\omega_{cf} = \omega_n$.

From Eq. (8.25) it is seen that a resonant condition exists in the vicinity of $\omega = \omega_n$, where the peak value of the Lm is greater than 0 dB. Therefore, there may be a substantial deviation of the Lm curve from the straight-line asymptotes, depending on the value of ζ. A family of curves for several values of $\zeta < 1$ is plotted in Fig. 8.4. For the appropriate ζ the curve can be drawn by selecting sufficient points from Fig. 8.4 or computed from Eq. (8.25).

The phase-angle curve for this function also varies with ζ. At zero frequency the angle is 0°, at the corner frequency the angle is $-90°$, and at infinite frequency the angle is $-180°$. A family of curves for various values of $\zeta < 1$ is plotted in Fig. 8.4. Enough values to draw the appropriate curve can be taken from Fig. 8.4 or computed from Eq. (8.26). When the quadratic factor appears in the numerator, the magnitudes of the Lm and phase angle are the same as those in Fig. 8.4 except that they are changed in sign.

The $\text{Lm}[1 + j2\zeta\omega/\omega_n + (j\omega/\omega_n)^2]^{-1}$ with $\zeta < 0.707$ has a peak value. The magnitude of this peak value and the frequency at which it occurs are important terms. These values, given earlier in Sec. 4.12 and derived in Sec. 9.3, are repeated here:

FIGURE 8.4
Log magnitude and phase diagram for $[1 + j2\zeta\omega/\omega_n + (j\omega/\omega_n)^2]^{-1}$.

$$M_m = \frac{1}{2\zeta\sqrt{1 - \zeta^2}} \tag{8.27}$$

$$\omega_m = \omega_n\sqrt{1 - 2\zeta^2} \tag{8.28}$$

Note that the peak value M_m depends only on the damping ratio ζ. Since Eq. (8.28) is meaningful only for real values of ω_m, the curve of M vs. ω has a peak value greater than unity only for $\zeta < 0.707$. The frequency at which the peak value occurs depends on both the damping ratio ζ and the undamped natural frequency

ω_n. This information is used when adjusting a control system for good response characteristics. These characteristics are discussed in Chap. 9.

The discussion thus far has dealt with poles and zeros that are located in the left-half (LH) s plane. The Lm curves for poles and zeros lying in the right-half (RH) s plane are the same as those for poles and zeros located in the LH s plane. However, the angle curves are different. For example, the angle for the factor $(1 - j\omega T)$ varies from 0 to $-90°$ as ω varies from zero to infinity. Also, if ζ is negative, the quadratic factor of expression (8.24) contains RH s plane poles or zeros. Its angle varies from $-360°$ at $\omega = 0$ to $-180°$ at $\omega = \infty$. This information can be obtained from the pole-zero diagram discussed in Sec. 4.12, with all angles measured in a *counterclockwise* (ccw) direction. Some CAD packages do not consistently use a ccw measurement direction and thus yield inaccurate angle values (see Sec. 18.18).

8.7   EXAMPLE OF DRAWING A BODE PLOT

Figure 8.5 is the block diagram of a feedback control system with unity feedback. In this section the Lm and phase diagram is drawn in a systematic manner for the open-loop transfer function of this system. The Lm curve is drawn both for the straight-line approximation and for the exact curve. Table 8.4 lists the pertinent characteristics for each factor. The Lm asymptotes and angle curves for each factor are shown in Figs. 8.6 and 8.7, respectively. They are added algebraically to obtain the composite curve. The composite Lm curve using straight-line approximations is drawn directly, as outlined below.

Step 1. At frequencies less than ω_1, the first corner frequency, only the factors Lm 4 and Lm$(j\omega)^{-1}$ are effective. All the other factors have zero value. At ω_1, Lm 4 $= 12$ dB and Lm$(j\omega_1)^{-1} = 6$ dB; thus, at this frequency the composite curve has the value of 18 dB. Below ω_1 the composite curve has a slope of -20 dB/decade because of the Lm$(j\omega)^{-1}$ term.

Step 2. Above ω_1, the factor Lm$(1 + j2\omega)^{-1}$ has a slope of -20 dB/decade and must be added to the terms in step 1. When the slopes are added, the composite curve has a total slope of -40 dB/decade in the frequency band from ω_1 to ω_2. Since this bandwidth is 2 octaves, the value of the composite curve at ω_2 is -6 dB.

Step 3. Above ω_2, the numerator factor Lm$(1 + j0.5\omega)$ is effective. This factor has a slope of $+20$ dB/decade above ω_2 and must be added to obtain the composite curve.

FIGURE 8.5
Block diagram of a control system with unity feedback.

TABLE 8.4
Characteristics of log magnitude and angle diagrams for various factors

Factor	Corner frequency, ω_{cf}	Log magnitude	Angle characteristics
4	None	Constant magnitude of $+12$ dB	Constant $0°$
$(j\omega)^{-1}$	None	Constant slope of -20 dB/decade (0 dB at $\omega = 1$)	Constant $-90°$
$(1 + j2\omega)^{-1}$	$\omega_1 = 0.5$	0 slope below corner frequency; -20 dB/decade slope above corner frequency. (18 dB at ω_1)	Varies from 0 to $-90°$
$1 + j0.5\omega$	$\omega_2 = 2.0$	0 slope below corner frequency; $+20$ dB/decade slope above corner frequency. (-6 dB at ω_2)	Varies from 0 to $+90°$
$[1 + j0.05\omega + (j0.125\omega)^2]^{-1}$ $\zeta = 0.2$ $\omega_n = 8$	$\omega_3 = 8.0$	0 slope below corner frequency; -40 dB/decade slope above corner frequency. (-18 dB at ω_3)	Varies from 0 to $-180°$

The composite curve now has a total slope of -20 dB/decade in the frequency band from ω_2 to ω_3. The bandwidth from ω_2 to ω_3 is 2 octaves; therefore, the value of the composite curve at ω_3 is -18 dB.

Step 4. Above ω_3 the last term $\text{Lm}[1 + j0.05\omega + (j0.125\omega)^2]^{-1}$ must be added. This factor has a slope of -40 dB/decade; therefore, the total slope of the composite curve above ω_3 is -60 dB/decade.

Step 5. Once the asymptotic plot of $\text{Lm } \mathbf{G}(j\omega)$ has been drawn, the corrections can be added if desired. The corrections at each corner frequency and at an octave above and below the corner frequency are usually sufficient. For first-order terms the corrections are ± 3 dB at the corner frequencies and ± 1 dB at an octave above and below the corner frequency. The values for quadratic terms can be obtained from Fig. 8.4 since they are a function of ζ. If the correction curve for the quadratic factor is not available, the correction at the frequencies $\omega = \omega_n$ and $\omega = 0.707\omega_n$ can easily be calculated from Eq. (8.25). The correction at ω_m can also be obtained using Eqs. (8.27) and (8.28).

The corrected Lm curves for each factor and for the composite curve are shown in Fig. 8.6.

FIGURE 8.6
Log magnitude curve for

$$G(j\omega) = \frac{4(1 + j0.5\omega)}{j\omega(1 + j2\omega)[1 + j0.05\omega + (j0.125\omega)^2]}$$

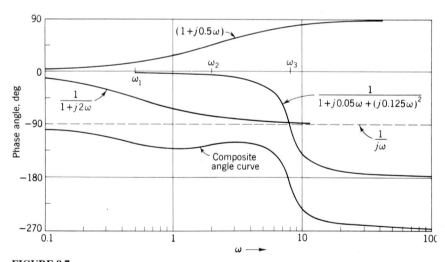

FIGURE 8.7
Phase-angle curve for

$$G(j\omega) = \frac{4(1 + j0.5\omega)}{j\omega(1 + j2\omega)[1 + j0.05\omega + (j0.125\omega)^2]}$$

Determination of the phase-angle curve of $G(j\omega)$ can be simplified by using the following procedure.

1. For the $(j\omega)^{-m}$ term, draw a line at the angle of $(-m)90°$.
2. For each $(1 + j\omega T)^{\pm 1}$ term, locate the angles from Table 8.4 at the corner frequency, an octave above and an octave below the corner frequency, and a decade above and below the corner frequency. Then draw a curve through these points for each $(1 + j\omega T)^{\pm 1}$ term.
3. For each $1 + j2\zeta\omega/\omega_n + (j\omega/\omega_n)^2$ term:
   (a) Locate the $\pm 90°$ point at the corner frequency $\omega = \omega_n$.
   (b) From Fig. 8.4, for the respective ζ, obtain a few points to draw the phase plot for each term with the aid of a French curve. The angle at $\omega = 0.707\omega_n$ may be sufficient and can be evaluated easily from Eq. (8.26) as $\tan^{-1} 2.828\zeta$.
4. Once the phase plot of each term of $G(j\omega)$ has been drawn, the composite phase plot of $G(j\omega)$ is determined by adding the individual phase curves.
   (a) Use the line representing the angle equal to

$$\underline{/\!\!\lim_{\omega\to 0} G(j\omega)} = (-m)90°$$

   as the base line, where m is the system type. Add or subtract the angles of each factor from this reference line.
   (b) At a particular frequency on the graph, measure the angle for each single-order and quadratic factor. Add and/or subtract them from the base line until all terms have been accounted for. The number of frequency points used is determined by the desired accuracy of the phase plots.
   (c) At $\omega = \infty$ the phase is 90° times the difference of the orders of the numerator and denominator $[-(n - w)90°]$.

8.8 SYSTEM TYPE AND GAIN AS RELATED TO LOG MAGNITUDE CURVES

The steady-state error of a closed-loop system depends on the system type and the gain. The system error coefficients are determined by these two characteristics, as noted in Chap. 6. For any given Lm curve the system type and gain can be determined. Also, with the transfer function given so that the system type and gain are known, they can expedite drawing the Lm curve. This is described for Type 0, 1, and 2 systems.

Type 0 System

A Type 0 system has a transfer function of the form

$$G(j\omega) = \frac{K_0}{1 + j\omega T_a}$$

At low frequencies, $\omega < 1/T_a$, Lm $G(j\omega) = 20\log K_0$, which is a constant. The slope of the Lm curve is zero below the corner frequency $\omega_1 = 1/T_a$ and -20 dB/decade above the corner frequency. The Lm curve is shown in Fig. 8.8.

FIGURE 8.8
Log magnitude plot for $\mathbf{G}(j\omega) = K_0/(1 + j\omega T_a)$.

For a Type 0 system the characteristics are as follows:

1. The slope at low frequencies is zero.
2. The magnitude at low frequencies is $20 \log K_0$.
3. The gain K_0 is the steady-state step error coefficient.

Type 1 System

A Type 1 system has a transfer function of the form

$$\mathbf{G}(j\omega) = \frac{K_1}{j\omega(1 + j\omega T_a)}$$

At low frequencies, $\omega < 1/T_a$, $\mathrm{Lm}\,\mathbf{G}(j\omega) \approx \mathrm{Lm}\,(K_1/j\omega) = \mathrm{Lm}\,K_1 - \mathrm{Lm}\,j\omega$, which has a slope of -20 dB/decade. At $\omega = K_1$, $\mathrm{Lm}(K_1/j\omega) = 0$. If the corner frequency $\omega_1 = 1/T_a$ is greater than K_1, the low-frequency portion of the curve of slope -20 dB/decade crosses the 0-dB axis at a value of $\omega_x = K_1$, as shown in Fig. 8.9a. If the corner frequency is less than K_1, the low-frequency portion of the curve of slope -20 dB/decade may be extended until it does cross the 0-dB axis. The value of the frequency at which the extension crosses the 0-dB axis is $\omega_x = K_1$. In other words, the plot $\mathrm{Lm}(K_1/j\omega)$ crosses the 0-dB value at $\omega_x = K_1$, as illustrated in Fig. 8.9b.

At $\omega = 1$, $\mathrm{Lm}\,j\omega = 0$; therefore, $\mathrm{Lm}(K_1/j\omega)_{\omega=1} = 20 \log K_1$. For $T_a < 1$ this value is a point on the slope of -20 dB/decade. For $T_a > 1$ this value is a point on the extension of the initial slope, as shown in Fig. 8.9b. The frequency ω_x is smaller or larger than unity according as K_1 is smaller or larger than unity.

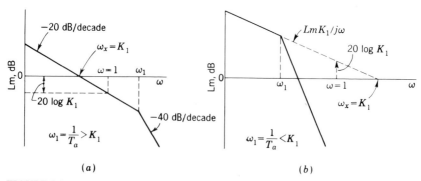

FIGURE 8.9
Log magnitude plot for $\mathbf{G}(j\omega) = K_1/j\omega(1 + j\omega T_a)$.

For a Type 1 system the characteristics are as follows:

1. The slope at low frequencies is –20 dB/decade.

2. The intercept of the low-frequency slope of –20 dB/decade (or its extension) with the 0-dB axis occurs at the frequency ω_x, where $\omega_x = K_1$.

3. The value of the low-frequency slope of –20 dB/decade (or its extension) at the frequency $\omega = 1$ is equal to $20 \log K_1$.

4. The gain K_1 is the steady-state ramp error coefficient.

Type 2 System

A Type 2 system has a transfer function of the form

$$G(j\omega) = \frac{K_2}{(j\omega)^2(1 + j\omega T_a)}$$

At low frequencies, $\omega < 1/T_a$, $\operatorname{Lm} G(j\omega) = \operatorname{Lm}[K_2/(j\omega)^2] = \operatorname{Lm} K_2 - \operatorname{Lm}(j\omega)^2$, for which the slope is -40 dB/decade. At $\omega^2 = K_2$, $\operatorname{Lm}[K_2/(j\omega)^2] = 0$; therefore, the intercept of the initial slope of -40 dB/decade (or its extension, if necessary) with the 0-dB axis occurs at the frequency ω_y, where $\omega_y^2 = K_2$.

At $\omega = 1$, $\operatorname{Lm}(j\omega)^2 = 0$; therefore, $\operatorname{Lm} [K_2/(j\omega)^2]_{\omega=1} = 20 \log K_2$. This point occurs on the initial slope or on its extension, according to whether $\omega_1 = 1/T_a$ is larger or smaller than $\sqrt{K_2}$. If $K_2 > 1$, the quantity $20 \log K_2$ is positive, and if $K_2 < 1$, the quantity $20 \log K_2$ is negative.

The Lm curve for a Type 2 transfer function is shown in Fig. 8.10. The determination of gain K_2 from the graph is shown.

For a Type 2 system the characteristics are as follows:

1. The slope at low frequencies is -40 dB/decade.

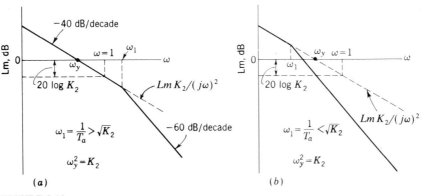

FIGURE 8.10
Log magnitude plot for $G(j\omega) = K_2/(j\omega)^2(1 + j\omega T_a)$.

2. The intercept of the low-frequency slope of -40 dB/decade (or its extension, if necessary) with the 0-dB axis occurs at a frequency ω_y, where $\omega_y^2 = K_2$.

3. The value on the low-frequency slope of -40 dB/decade (or its extension) at the frequency $\omega = 1$ is equal to $20 \log K_2$.

4. The gain K_2 is the steady-state parabolic error coefficient.

8.9 CAD ACCURACY CHECK (CADAC)

CADAC checks that apply for CAD package frequency domain analysis are the following:

1. Based on the given $\mathbf{G}(j\omega)$, determine the angle of this function, assuming a minimum-phase stable plant, at $\omega = 0$ and at $\omega = \infty$ (see Sec. 8.7). If the plant has any RH plane poles and/or zeros, it is necessary to ascertain whether the CAD package being used correctly measures all angles in the ccw (or cw) direction as ω goes from 0 to ∞ (see Sec. 18.18).

2. Based on the system type of $G(s)$, the Lm $\mathbf{G}(j\omega)$ can be readily determined at $\omega = 0$ for a Type 0 plant. The value of ω_x for a Type 1 plant or ω_y for a Type 2 plant should be verified.

8.10 EXPERIMENTAL DETERMINATION OF TRANSFER FUNCTIONS[5,9]

The Lm and phase-angle diagram is of great value when the mathematical expression for the transfer function of a given system is not known. The magnitude and angle of the ratio of the output to the input can be obtained experimentally for a steady-state sinusoidal input signal at a number of frequencies. These data are used to obtain the exact Lm and angle diagram. Asymptotes are drawn on the exact Lm curve, using the fact that their slopes must be multiples of ± 20 dB/decade. From these asymptotes the system type and the approximate time constants are determined. In this manner, the transfer function of the system can be synthesized.

Care must be exercised in determining whether any zeros of the transfer function are in the RH s plane. A system that has no open-loop zeros in the RH s plane is defined as a *minimum-phase* system,[6,12] and all factors have the form $1 + Ts$ and/or $1 + As + Bs^2$. A system that has open-loop zeros in the RH s plane is a *nonminimum-phase* system.[13] The stability is determined by the location of the poles and does not affect the designation of minimum or nonminimum phase.

The angular variation for poles or zeros in the RH s plane is different from those in the LH plane (see Sec. 18.18). For this situation, one or more terms in the transfer function have the form $1 - Ts$ and/or $1 \pm As \pm Bs^2$. As an example, consider the functions $1 + j\omega T$ and $1 - j\omega T$. The Lm plots of these functions are identical, but the angle diagram for the former goes from 0 to 90° whereas for the latter it goes from 0 to $-90°$. Therefore, care must be exercised in interpreting the angle plot to determine whether any factors of the transfer function lie in the RH s plane. Many practical systems are in the minimum-phase category.

8.11 DIRECT POLAR PLOTS

The earlier sections of this chapter present a simple graphical method for plotting the characteristic curves of the transfer functions. These curves are the Lm and the angle of $G(j\omega)$ vs. ω, plotted on semilog graph paper. The reason for presenting this method first is that these curves can be constructed easily and rapidly. However, frequently the polar plot of the transfer function is desired. The magnitude and angle of $G(j\omega)$, for sufficient frequency points, are readily obtainable from the Lm $G(j\omega)$ and $/G(j\omega)$ vs. $\log \omega$ curves. This approach often takes less time than calculating $|G(j\omega)|$ and $\phi(\omega)$ analytically for each frequency point desired, unless a CAD program is available.[4]

The data for drawing the polar plot of the frequency response can also be obtained from the pole-zero diagram, as described in Sec. 4.12. It is possible to visualize the complete shape of the frequency-response curve from the pole-zero diagram because the angular contribution of each pole and zero is readily apparent. The polar plot of $G(j\omega)$ is called the *direct polar plot*. The polar plot of $[G(j\omega)]^{-1}$ is called the *inverse polar plot*,[12] which is discussed briefly in Chap. 9.

Complex *RC* Network (Lag-Lead Compensator)

The circuit of Fig. 8.11 is used in later chapters as a compensator:

$$G(s) = \frac{E_o(s)}{E_{in}(s)} = \frac{1 + (T_1 + T_2)s + T_1 T_2 s^2}{1 + (T_1 + T_2 + T_{12})s + T_1 T_2 s^2} \qquad (8.29)$$

where the time constants are

$$T_1 = R_1 C_1 \qquad T_2 = R_2 C_2 \qquad T_{12} = R_1 C_2$$

As a function of frequency, the transfer function is

$$G(j\omega) = \frac{E_o(j\omega)}{E_{in}(j\omega)} = \frac{(1 - \omega^2 T_1 T_2) + j\omega(T_1 + T_2)}{(1 - \omega^2 T_1 T_2) + j\omega(T_1 + T_2 + T_{12})} \qquad (8.30)$$

By the proper choice of the time constants, the circuit acts as a lag network in the lower-frequency range of 0 to ω_x and as a lead network in the higher-frequency range of ω_x to ∞. That is, the steady-state sinusoidal output E_o lags or leads the sinusoidal input E_{in}, according to whether ω is smaller or larger than ω_x. The polar plot of this transfer function is a circle with its center on the real axis and lying in the first and fourth quadrants. Figure 8.12 illustrates the polar plot in nondimensionalized form for a typical circuit. When T_2 and T_{12} in Eq. (8.30) are expressed in terms of

FIGURE 8.11
An RC circuit used as a lag-lead compensator.

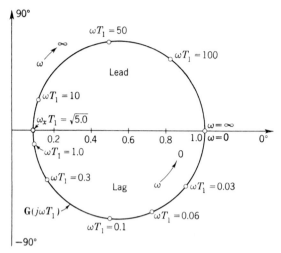

FIGURE 8.12
Polar plot of

$$G(j\omega T_1) = \frac{(1 + j\omega T_1)(1 + j0.2\omega T_1)}{(1 + j11.1\omega T_1)(1 + j0.0179\omega T_1)}$$

$$T_2 = 0.2T_1 \qquad T_{12} = 10.0T_1$$

T_1, the expression of the form given in the title of Fig. 8.12 results. Its properties are the following:

1. $\displaystyle\lim_{\omega \to 0} G(j\omega T_1) \to 1\underline{/0°}$

2. $\displaystyle\lim_{\omega \to \infty} G(j\omega T_1) \to 1\underline{/0°}$

3. At the frequency $\omega = \omega_x$, for which $\omega_x^2 T_1 T_2 = 1$, Eq. (8.30) becomes

$$G(j\omega_x T_1) = \frac{T_1 + T_2}{T_1 + T_2 + T_{12}} = |G(j\omega_x T_1)|\underline{/0°} \qquad (8.31)$$

Note that Eq. (8.31) represents the minimum value of the transfer function in the whole frequency spectrum. From Fig. 8.12 it is seen that for frequencies below ω_x the transfer function has a negative or lag angle. For frequencies above ω_x it has a positive or lead angle. The applications and advantages of this circuit are discussed in later chapters on compensation.

Type 0 Feedback Control System

The field-controlled servomotor[11] illustrates a typical Type 0 device. It has the transfer function

$$G(j\omega) = \frac{C(j\omega)}{E(j\omega)} = \frac{K_0}{(1 + j\omega T_f)(1 + j\omega T_m)} \qquad (8.32)$$

Note from Eq. (8.32) that

$$\mathbf{G}(j\omega) \rightarrow \begin{cases} K_0 \underline{/0^\circ} & \omega \rightarrow 0^+ \\ 0 \underline{/-180^\circ} & \omega \rightarrow \infty \end{cases}$$

Also, for each term in the denominator the angular contribution to $\mathbf{G}(j\omega)$, as ω goes from 0 to ∞, goes from 0 to -90°. Thus, the polar plot of this transfer function must start at $\mathbf{G}(j\omega) = K_0 \underline{/0^\circ}$ for $\omega = 0$ and proceeds first through the fourth and then through the third quadrants to $\lim_{\omega \rightarrow \infty} \mathbf{G}(j\omega) = 0 \underline{/-180^\circ}$ as the frequency approaches infinity. In other words, the angular variation of $\mathbf{G}(j\omega)$ is continuously decreasing, going in a clockwise (cw) direction from 0° to -180°. The exact shape of this plot is determined by the particular values of the time constants T_f and T_m.

If Eq. (8.32) had another term of the form $1 + j\omega T$ in the denominator, the transfer function would be

$$\mathbf{G}(j\omega) = \frac{K_0}{(1 + j\omega T_f)(1 + j\omega T_m)(1 + j\omega T)} \tag{8.33}$$

The point $\mathbf{G}(j\omega)|_{\omega = \infty}$ rotates cw by an additional 90°. In other words, when $\omega \rightarrow \infty$, $\mathbf{G}(j\omega) \rightarrow 0 \underline{/-270^\circ}$. In this case the curve crosses the real axis at a frequency ω_x for which the imaginary part of the transfer function is zero.

When a term of the form $1 + j\omega T$ appears in the numerator, the transfer function experiences an angular variation of 0 to 90° (a ccw rotation) as the frequency is varied from 0 to ∞. Thus, with increasing frequency, the angle of $\mathbf{G}(j\omega)$ may not change continuously in one direction. Also, the resultant polar plot may not be as smooth as the one shown in Fig. 8.13. For example, the transfer function

$$\mathbf{G}(j\omega) = \frac{K_0(1 + j\omega T_1)^2}{(1 + j\omega T_2)(1 + j\omega T_3)(1 + j\omega T_4)^2} \tag{8.34}$$

whose polar plot, shown in Fig. 8.14, has a "dent" when the time constants T_2 and T_3 are greater than T_1, and T_1 is greater than T_4. From the angular contribution of each factor and from the preceding analysis, it can be surmised that the polar plot has the general shape shown in the figure. In the event that T_1 is smaller than all the other time constants, the polar plot is similar in shape to the one shown in Fig. 8.13a.

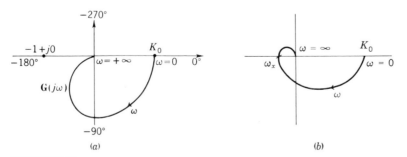

FIGURE 8.13
Polar plot for typical Type 0 transfer functions: (a) Eq. (8.32); (b) Eq. (8.33).

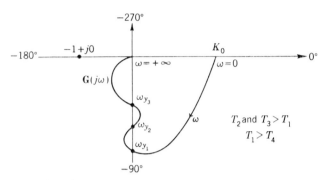

FIGURE 8.14
Polar plot for a more complex Type 0 transfer function [Eq. (8.34)].

In the same manner, a quadratic in either the numerator or the denominator of a transfer function results in an angular contribution of 0 to $\pm 180°$, respectively, and the polar plot of $G(j\omega)$ is affected accordingly. It can be seen from the examples that the polar plot of a Type 0 system always starts at a value K_0 (step error coefficient) on the positive real axis for $\omega = 0$ and ends at zero magnitude (for $n > w$) and tangent to one of the major axes at $\omega = \infty$. The final angle is $-90°$ times the order of the denominator minus the order of the numerator of $G(j\omega)$. The arrows on the plots of $G(j\omega)$ in Figs. 8.13 and 8.14 indicate the direction of increasing frequency.

Type 1 Feedback Control System

A typical Type 1 system is

$$G(j\omega) = \frac{C(j\omega)}{E(j\omega)} = \frac{K_1}{j\omega(1 + j\omega T_m)(1 + j\omega T_c)(1 + j\omega T_q)} \qquad (8.35)$$

Note from Eq. (8.35) that

$$G(j\omega) \to \begin{cases} \infty\underline{/-90°} & \omega \to 0^+ \qquad (8.36) \\ 0\underline{/-360°} & \omega \to \infty \qquad (8.37) \end{cases}$$

Note that the $j\omega$ term in the denominator contributes the angle $-90°$ to the total angle of $G(j\omega)$ for all frequencies. Thus, the basic difference between Eqs. (8.33) and (8.35) is the presence of the term $j\omega$ in the denominator of the latter equation. Since all the $1 + j\omega T$ terms of Eq. (8.35) appear in the denominator, its polar plot, as shown in Fig. 8.15, has no dents. From the remarks of this and previous sections, it can be seen that the angular variation of $G(j\omega)$ decreases continuously (ccw) in the same direction from -90 to $-360°$ as ω increases from 0 to ∞. The presence of any frequency-dependent factor in the numerator has the same general effect on the polar plot as that described previously for the Type 0 system.

From Eq. (8.36) it is seen that the magnitude of the function $G(j\omega)$ approaches infinity as the value of ω approaches zero. There is a line parallel to the $-90°$ axis but

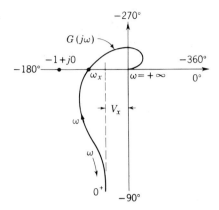

FIGURE 8.15
Polar plot for a typical Type 1 transfer function [Eq. (8.35)].

displaced to the left of the origin, to which $\mathbf{G}(j\omega)$ tends asymptotically as $\omega \rightarrow 0$. The true asymptote is determined by finding the value of the real part of $\mathbf{G}(j\omega)$ as ω approaches zero. Thus

$$V_x = \lim_{\omega \to 0} \text{Re}[\mathbf{G}(j\omega)] \tag{8.38}$$

or, for this particular transfer function,

$$V_x = -K_1(T_q + T_c + T_m) \tag{8.39}$$

Equation (8.39) shows that the magnitude of $\mathbf{G}(j\omega)$ approaches infinity asymptotically to a vertical line whose real-axis intercept equals V_x, as illustrated in Fig. 8.15. Note that the value of V_x is a direct function of the ramp error coefficient K_1.

The frequency of the crossing point on the negative real axis of the $\mathbf{G}(j\omega)$ function is that value of frequency ω_x for which the imaginary part of $\mathbf{G}(j\omega)$ is equal to zero. Thus,

$$\text{Im}[\mathbf{G}(j\omega_x)] = 0 \tag{8.40}$$

or, for this particular transfer function,

$$\omega_x = (T_c T_q + T_q T_m + T_m T_c)^{-1/2} \tag{8.41}$$

The real-axis crossing point is very important because it determines closed-loop stability, as described in later sections dealing with system stability.

Type 2 Feedback Control System

The transfer function of the Type 2 system illustrated in Sec. 6.6 and represented by Eq. (6.56) is

$$\mathbf{G}(j\omega) = \frac{\mathbf{C}(j\omega)}{\mathbf{E}(j\omega)} = \frac{K_2}{(j\omega)^2(1 + j\omega T_f)(1 + j\omega T_m)} \tag{8.42}$$

Its properties are
$$\mathbf{G}(j\omega) \rightarrow \begin{cases} \infty \underline{/-180°} & \omega \to 0^+ \tag{8.43} \\ 0 \underline{/-360°} & \omega \to +\infty \tag{8.44} \end{cases}$$

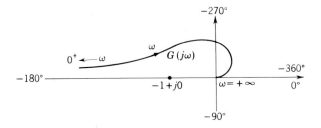

FIGURE 8.16
Polar plot of Eq. (8.42), a typical Type 2 transfer function resulting in an unstable feedback control system.

The presence of the $(j\omega)^2$ term in the denominator contributes $-180°$ to the total angle of $\mathbf{G}(j\omega)$ for all frequencies. For the transfer function of Eq. (8.42) the polar plot (Fig. 8.16) is a smooth curve whose angle $\phi(\omega)$ decreases continuously from -180 to $-360°$.

The introduction of an additional pole and a zero can alter the shape of the polar plot. Consider the transfer function

$$\mathbf{G}_o(j\omega) = \frac{K_2'(1 + j\omega T_1)}{(j\omega)^2(1 + j\omega T_f)(1 + j\omega T_m)(1 + j\omega T_2)} \qquad (8.45)$$

where $T_1 > T_2$. The polar plot can be obtained from the pole-zero diagram shown in Fig. 8.17. At $s = j\omega = j0^+$ the angle of each factor is zero except for the double pole at the origin. The angle at $\omega = 0^+$ is therefore $-180°$, as given by Eq. (8.43), which is still applicable. As ω increases from zero, the angle of $j\omega + 1/T_1$ increases faster than the angles of the other poles. In fact, at low frequencies the angle due to the zero is larger than the sum of the angles due to the poles located to the left of the zero. This fact is shown qualitatively at the frequency ω_1 in Fig. 8.17. Therefore, the angle of $\mathbf{G}(j\omega)$ at low frequencies is greater than $-180°$. As the frequency increases to a value ω_x, the sum of the component angles of $\mathbf{G}(j\omega)$ is $-180°$ and the polar plot crosses the real axis, as shown in Fig. 8.18. As ω increases further, the angle of $j\omega + 1/T_1$ shows only a small increase, but the angles from the poles increase rapidly. In the limit, as $\omega \to \infty$, the angles of $j\omega + 1/T_1$ and $j\omega + 1/T_2$ are equal and opposite in sign, so the angle of $\mathbf{G}(j\omega)$ approaches $-360°$, as given by Eq. (8.44).

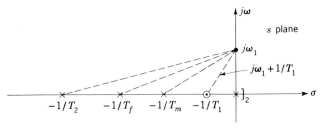

FIGURE 8.17
Pole-zero diagram for

$$G_o(s) = \frac{K_2'T_1}{T_fT_mT_2} \frac{s + 1/T_1}{s^2(s + 1/T_f)(s + 1/T_m)(s + 1/T_2)}$$

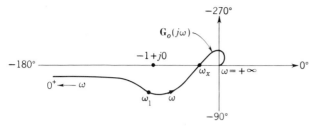

FIGURE 8.18
Polar plot for a typical Type 2 transfer function resulting in a stable
system [Eq. (8.45)].

Fig. 8.18 shows the complete polar plot of $G(j\omega)$. A comparison of Figs. 8.16
and 8.18 shows that both curves approach $-180°$ at $\omega = 0^+$, which is typical of Type
2 systems. As $\omega \to \infty$, the angle approaches $-360°$ since both $G(j\omega)$ and $G_o(j\omega)$
have the same degree, $n - w = 4$. When the Nyquist stability criterion described in
Secs. 8.13 and 8.14 is used, the feedback system containing $G(j\omega)$ can be shown to
be unstable, whereas the closed-loop system containing $G_o(j\omega)$ can be stable.

It can be shown that as $\omega \to 0^+$, the polar plot of a Type 2 system is below
the real axis if $\sum(T_{\text{numerator}}) - \sum(T_{\text{denominator}})$ is a positive value and is above the
real axis if it is a negative value. Thus, for this example the necessary condition is
$T_1 > (T_f + T_m + T_2)$.

8.12 SUMMARY: DIRECT POLAR PLOTS

To obtain the direct polar plot of a system's forward transfer function, the following
criteria are used to determine the key parts of the curve.

Step 1. The forward transfer function has the general form

$$G(j\omega) = \frac{K_m(1 + j\omega T_a)(1 + j\omega T_b)\cdots(1 + j\omega T_w)}{(j\omega)^m(1 + j\omega T_1)(1 + j\omega T_2)\cdots(1 + j\omega T_u)} \tag{8.46}$$

For this transfer function the system type is equal to the value of m and determines the
portion of the polar plot representing the $\lim_{\omega \to 0} G(j\omega)$. The low-frequency polar-
plot characteristics (as $\omega \to 0$) of the different system types are summarized in Fig.
8.19. The angle at $\omega = 0$ is $m(-90°)$. The arrow on the polar plots indicates the
direction of increasing frequency.

Step 2. The high-frequency end of the polar plot can be determined as follows:

$$\lim_{\omega \to +\infty} G(j\omega) = 0 \underline{/(w - m - u)90°} \tag{8.47}$$

Note that since the degree of the denominator of Eq. (8.46) is always greater than
the degree of the numerator, the high-frequency point ($\omega = \infty$) is approached (i.e.,
the angular condition) in the cw sense. The plot ends at the origin tangent to the axis
determined by Eq. (8.47). Tangency may occur on either side of the axis.

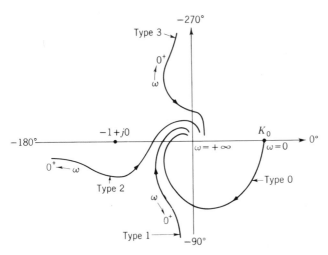

FIGURE 8.19
A summary of direct polar plots of different types of systems.

Step 3. The asymptote that the low-frequency end approaches, for a Type 1 system, is determined by taking the limit as $\omega \to 0$ of the real part of the transfer function.

Step 4. The frequencies at the points of intersection of the polar plot with the negative real axis and the imaginary axis are determined, respectively, by setting

$$\text{Im}[\mathbf{G}(j\omega)] = 0 \qquad (8.48)$$

$$\text{Re}[\mathbf{G}(j\omega)] = 0 \qquad (8.49)$$

Step 5. If there are no frequency-dependent terms in the numerator of the transfer function, the curve is a smooth one in which the angle of $\mathbf{G}(j\omega)$ continuously decreases as ω goes from 0 to ∞. With time constants in the numerator, and depending upon their values, the angle may not continuously vary in the same direction, thus creating "dents" in the polar plot.

Step 6. As is seen later in this chapter, it is important to know the exact shape of the polar plot of $\mathbf{G}(j\omega)$ in the vicinity of the $-1 + j0$ point and the crossing point on the negative real axis.

8.13   NYQUIST'S STABILITY CRITERION

A system designer must be sure that the closed-loop system he or she designs is stable. The Nyquist stability criterion[2,3,7] provides a simple graphical procedure for determining closed-loop stability from the frequency-response curves of the open-loop transfer function $\mathbf{G}(j\omega)\mathbf{H}(j\omega)$. The application of this method in terms of the polar plot is covered in this section; application in terms of the log magnitude–angle (Nichols) diagram is covered in Sec. 8.16.

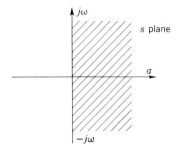

FIGURE 8.20
Prohibited region in the s plane.

For a stable system the roots of the characteristic equation

$$B(s) = 1 + G(s)H(s) = 0 \tag{8.50}$$

cannot be permitted to lie in the RH s plane or on the $j\omega$ axis, as shown in Fig. 8.20. In terms of $G = N_1/D_1$ and $H = N_2/D_2$, Eq. (8.50) becomes

$$B(s) = 1 + \frac{N_1 N_2}{D_1 D_2} = \frac{D_1 D_2 + N_1 N_2}{D_1 D_2} \tag{8.51}$$

Note that the numerator and denominator of $B(s)$ have the same degree, and the *poles of the open-loop transfer function $G(s)H(s)$ are the poles* of $B(s)$. The closed-loop transfer function of the system is

$$\frac{C(s)}{R(s)} = \frac{G(s)}{1 + G(s)H(s)} = \frac{N_1 D_2}{D_1 D_2 + N_1 N_2} \tag{8.52}$$

The denominator of $C(s)/R(s)$ is the same as the numerator of $B(s)$. The condition for stability may therefore be restated as follows: For a stable system none of the zeros of $B(s)$ can lie in the RH s plane or on the imaginary axis. Nyquist's stability criterion relates the number of zeros and poles of $B(s)$ that lie in the RH s plane to the polar plot of $G(s)H(s)$.

Limitations

In this analysis the assumption is that all the control systems are inherently linear or that their limits of operation are confined to give a linear operation. This yields a set of linear differential equations that describe the dynamic performance of the systems. Because of the physical nature of feedback control systems, the degree of the denominator $D_1 D_2$ is equal to or greater than the degree of the numerator $N_1 N_2$ of the open-loop transfer function $G(s)H(s)$. Mathematically, this statement means that $\lim_{s\to\infty} G(s)H(s) \to 0$ or a constant. These two factors satisfy the necessary limitations to the generalized Nyquist stability criterion.

Mathematical Basis for Nyquist's Stability Criterion

A rigorous mathematical derivation of Nyquist's stability criterion involves complex-variable theory and is available in the literature.[11] Fortunately, the result of the derivation is simple and is readily applied. A qualitative approach is now presented for the special case that $B(s)$ is a rational fraction. The characteristic function $B(s)$ given by Eq. (8.51) is rationalized, factored, and then written in the form

$$B(s) = \frac{(s - Z_1)(s - Z_2) \cdots (s - Z_n)}{(s - p_1)(s - p_2) \cdots (s - p_n)} \tag{8.53}$$

where $Z_1, Z_2, \ldots, Z_n$ are the zeros and $p_1, p_2, \ldots, p_n$ are the poles. [Remember that z_i is used to denote a zero of $G(s)H(s)$.] The poles p_i are the same as the poles of the open-loop transfer function $G(s)H(s)$ and include the s term for which $p = 0$, if it is present.

In Fig. 8.21 some of the poles and zeros of a generalized function $B(s)$ are arbitrarily drawn on the s plane. Also, an arbitrary *closed contour* Q' is drawn that encloses the zero Z_1. To the point O' on Q', whose coordinates are $s = \sigma + j\omega$, are drawn directed line segments from all the poles and zeros. The lengths of these directed line segments are given by $s - Z_1, s - Z_2, s - p_1, s - p_2$, etc. Not all the directed segments from the poles and zeros are indicated in the figure, as they are not necessary to proceed with this development. *As the point O' is rotated cw once around the closed contour Q', the line of length $(s - Z_1)$ rotates through a net cw angle of 360°. All the other directed* segments have rotated through a *net angle of 0°.* Thus, by referring to Eq. (8.53) it is seen that the cw rotation of 360° for the length $(s - Z_1)$ *must simultaneously be realized* by the function $B(s)$ for the enclosure of the zero Z_1 by the path Q'.

Consider now a larger closed contour Q'' that includes the zeros Z_1, Z_2, Z_3, and the pole p_5. *As a point O'' is rotated cw once around the closed curve Q'', each of the directed line segments from the enclosed pole and zeros rotates through a net cw angle of* 360°. Since the angular rotation of the pole is experienced by the characteristic function $B(s)$ in its denominator, the net angular rotation realized by Eq. (8.53) must be equal to the net angular rotations due to the pole p_5 minus the

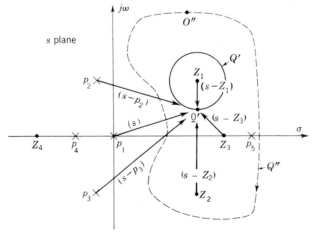

FIGURE 8.21
A plot of some poles and zeros of Eq. (8.53).

net angular rotations due to the zeros Z_1, Z_2, and Z_3. In other words, the net angular rotation experienced by $1 + G(s)H(s)$ is $360° - (3)(360°) = -720°$. Therefore, for this case, the net number of rotations N experienced by $B(s) = 1 + G(s)H(s)$ for the cw movement of point O'' *once* about the closed contour Q'' is equal to -2; that is,

$$N = \text{(number of poles enclosed)} - \text{(number of zeros enclosed)} = 1 - 3 = -2$$

where the minus sign denotes cw rotation. Note that if the contour Q'' includes only the pole p_5, $B(s)$ experiences one ccw rotation as the point O'' is moved cw around the contour. Also, for *any closed path* that may be chosen, all the poles and zeros that lie outside the closed path each contribute a net angular rotation of $0°$ to $B(s)$ as a point is moved once around this contour.

Generalizing Nyquist's Stability Criterion

Consider now a closed contour Q such that the whole RH s plane is encircled (see Fig. 8.22), thus enclosing all zeros and poles of $B(s)$ that have positive real parts. The theory of complex variables used in the rigorous derivation requires that the contour Q must not pass *through any poles or zeros of $B(s)$*. When the results of the preceding discussion are applied to the contour Q, the following properties are noted:

1. The total number of cw rotations of $B(s)$ due to its zeros is equal to the total number of zeros Z_R in the RH s plane.
2. The total number of ccw rotations of $B(s)$ due to its poles is equal to the total number of poles P_R in the RH s plane.
3. The *net* number of rotations N of $B(s) = 1 + G(s)H(s)$ about the origin is equal to its total number of poles P_R minus its total number of zeros Z_R in the RH s plane. N may be positive (ccw), negative (cw), or zero.

The essence of these three conclusions can be represented by the equation

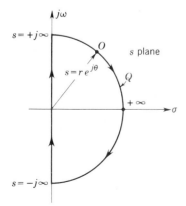

FIGURE 8.22
The contour that encloses the entire right-half s plane.

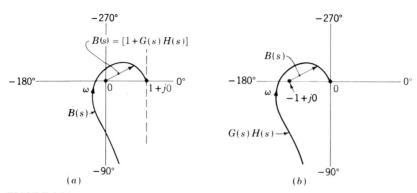

FIGURE 8.23
A change of reference for $B(s)$.

$$N = \frac{\text{Change in phase of } [1 + G(s)H(s)]}{2\pi} = P_R - Z_R \qquad (8.54)$$

where ccw rotation is defined as being positive and cw rotation is negative. In order for the characteristic function $B(s)$ to realize a net rotation N, the directed line segment representing $B(s)$ (see Fig. 8.23a) must rotate about the origin $360N$ degrees, or N complete revolutions. Solving for Z_R in Eq. (8.54) yields

$$Z_R = P_R - N \qquad (8.55)$$

Since $B(s)$ can have no zeros Z_R in the RH s plane for a stable system, it is therefore concluded that, *for a stable system, the net number of rotations of $B(s)$ about the origin must be ccw and equal to the number of poles P_R that lie in the RH s plane.* In other words, if $B(s)$ experiences a net cw rotation, this indicates that $Z_R > P_R$, where $P_R \geq 0$, and thus the closed-loop system is unstable. If there are zero net rotations, then $Z_R = P_R$ and the system may or may not be stable, according to whether $P_R = 0$ or $P_R > 0$.

Obtaining a Plot of $B(s)$

Figure 8.23a and b shows a plot of $B(s)$ and a plot of $G(s)H(s)$, respectively. By moving the origin of Fig. 8.23b to the $-1 + j0$ point, the curve is now equal to $1 + G(s)H(s)$, which is $B(s)$. Since $G(s)H(s)$ is known, this function is plotted, and then the origin is moved to the -1 point to obtain $B(s)$.

In general, the open-loop transfer functions of many physical systems do not have any poles P_R in the RH s plane. In this case, $Z_R = N$. *Thus, for a stable system the net number of rotations about the $-1 + j0$ point must be zero when there are no poles of $G(s)H(s)$ in the RH s plane.*

In the event that the function $G(s)H(s)$ has some poles in the RH s plane and the denominator is not in factored form, the number P_R can be determined by applying Routh's criterion to $D_1 D_2$. The Routhian array gives the number of roots in the RH s plane by the number of sign changes in the first column.

Analysis of Path Q

In applying Nyquist's criterion, the whole RH s plane must be encircled to ensure the inclusion of all poles or zeros of $B(s)$ in this portion of the plane. In Fig. 8.22 the entire RH s plane is enclosed by the closed path Q, which is composed of the following two segments:

1. One segment is the imaginary axis from $-j\infty$ to $+j\infty$.
2. The other segment is a semicircle of infinite radius that encircles the entire RH s plane.

   The portion of the path along the imaginary axis is represented mathematically by $s = j\omega$. Thus, replacing s by $j\omega$ in Eq. (8.53), followed by letting ω take on all values from $-\infty$ to $+\infty$, gives the portion of the $B(s)$ plot corresponding to that portion of the closed contour Q that lies on the imaginary axis.

   One of the requirements of the Nyquist criterion is that $\lim_{s\to\infty} G(s)H(s) \to 0$ or a constant. Thus $\lim_{s\to\infty} B(s) = \lim_{s\to\infty}[1+G(s)H(s)] \to 1$ or 1 plus the constant. As a consequence, as the point O moves along the segment of the closed contour represented by the semicircle of infinite radius, the corresponding portion of the $B(s)$ plot is a fixed point. As a result, the movement of point O along only the imaginary axis from $-j\infty$ to $+j\infty$ results in the same net rotation of $B(s)$ as if the whole contour Q were considered. *In other words, all the rotation of $B(s)$ occurs while the point O goes from $-j\infty$ to $+j\infty$ along the imaginary axis.* More generally, this statement applies only to those transfer functions $G(s)H(s)$ that conform to the limitations stated earlier in this section.[†]

Effect of Poles at the Origin on the Rotation of $B(s)$

Some transfer functions $G(s)H(s)$ have an s^m in the denominator. Since no poles or zeros can lie on the contour Q, the contour shown in Fig. 8.22 must be modified. For these cases the manner in which the $\omega = 0^-$ and $\omega = 0^+$ portions of the plot are joined is now investigated. This can best be done by taking an example. Consider the transfer function with positive values of T_1 and T_2:

$$G(s)H(s) = \frac{K_1}{s(1 + T_1 s)(1 + T_2 s)} \tag{8.56}$$

The direct polar plot of $\mathbf{G}(j\omega)\mathbf{H}(j\omega)$ of this function is obtained by substituting $s = j\omega$ into Eq. (8.56). Figure 8.24 represents the plot of $G(s)H(s)$ for values of s along the imaginary axis; that is, $j0^+ < j\omega < j\infty$ and $-j\infty < j\omega < j0^-$. The plot is drawn for both positive and negative frequency values. *The polar plot drawn for negative frequencies is the conjugate of the plot drawn for positive frequencies.*

[†] A transfer function that does not conform to these limitations and to which the immediately preceding italicized statement does not apply is $G(s)H(s) = s$.

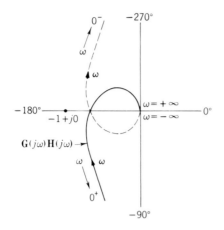

FIGURE 8.24
A plot of the transfer function $\mathbf{G}(j\omega)\mathbf{H}(j\omega)$ for Eq. (8.56).

That is, the curve for negative frequencies is symmetrical to the curve for positive frequencies, with the real axis as the axis of symmetry.

The closed contour Q of Fig. 8.22, in the vicinity of $s = 0$, is modified as shown in Fig. 8.25a to avoid passing through the origin. In other words, the point O is moved along the negative imaginary axis from $s = -j\infty$ to a point where $s = -j\varepsilon = 0^-\underline{/-\pi/2}$ becomes very small. Then the point O moves along a semi-circular path of radius $s = \varepsilon e^{j\theta}$ in the RH s plane with a very small radius ε until it reaches the positive imaginary axis at $s = +j\varepsilon = j0^+ = 0^+\underline{/\pi/2}$. From here the point O proceeds along the positive imaginary axis to $s = +j\infty$. Then, letting the radius approach zero, $\varepsilon \rightarrow 0$, for the semicircle around the origin ensures the inclusion of all poles and zeros in the RH s plane. To complete the plot of $B(s)$ in Fig. 8.24, it is necessary to investigate the effect of moving point O on this semicircle around the origin.

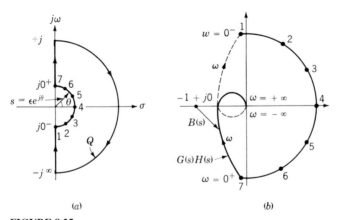

(a)   (b)

FIGURE 8.25
(a) The contour Q that encircles the right-half s plane, (b) complete plot for Eq. (8.56).

For the semicircular portion of the path Q represented by $s = \varepsilon e^{j\theta}$, where $\varepsilon \to 0$ and $-\pi/2 \leq \theta \leq \pi/2$, Eq. (8.56) becomes

$$G(s)H(s) = \frac{K_1}{s} = \frac{K_1}{\varepsilon e^{j\theta}} = \frac{K_1}{\varepsilon}e^{-j\theta} = \frac{K_1}{\varepsilon}e^{j\psi} \qquad (8.57)$$

where $K_1/\varepsilon \to \infty$ as $\varepsilon \to 0$, and $\psi = -\theta$ goes from $\pi/2$ to $-\pi/2$ as the directed segment s goes ccw from $\varepsilon\underline{/-\pi/2}$ to $\varepsilon\underline{/+\pi/2}$. Thus, in Fig. 8.24, the endpoints from $\omega \to 0^-$ and $\omega \to 0^+$ are joined by a semicircle of infinite radius in the first and fourth quadrants. Figure 8.25 illustrates this procedure and shows the completed contour of $G(s)H(s)$ as the point O moves along the modified contour Q in the s plane from point 1 to point 7. When the origin is moved to the $-1 + j0$ point, the curve becomes $B(s)$, as shown in Fig. 8.25b.

The plot of $B(s)$ in Fig. 8.25b does not encircle the $-1 + j0$ point; therefore, the encirclement N is zero. From Eq. (8.56) there are no poles within Q; that is, $P_R = 0$. Thus, when Eq. (8.55) is applied, $Z_R = 0$ and the closed-loop system is stable.

Transfer functions that have the term s^m in the denominator have the general form, as $\varepsilon \to 0$,

$$G(s)H(s) = \frac{K_m}{s^m} = \frac{K_m}{(\varepsilon^m)e^{jm\theta}} = \frac{K_m}{\varepsilon^m}e^{-jm\theta} = \frac{K_m}{\varepsilon^m}e^{jm\psi} \qquad (8.58)$$

where $m = 1, 2, 3, 4, \ldots$. With the reasoning used in the preceding example, it is seen from Eq. (8.58) that, as s moves from 0^- to 0^+, the plot of $G(s)H(s)$ traces m cw semicircles of infinite radius about the origin. If $m = 2$, then, as θ goes from $-\pi/2$ to $\pi/2$ in the s plane with radius ε, $G(s)H(s)$ experiences a net rotation of $(2)(180°) = 360°$.

Since the polar plots are symmetrical about the real axis, it is only necessary to determine the shape of the plot of $G(s)H(s)$ for a range of values of $0 < \omega < +\infty$. The net rotation of the plot for the range of $-\infty < \omega < +\infty$ is twice that of the plot for the range of $0 < \omega < +\infty$.

When $\mathbf{G}(j\omega)\mathbf{H}(j\omega)$ Passes through the Point $-1 + j0$

When the curve of $\mathbf{G}(j\omega)\mathbf{H}(j\omega)$ passes through the $-1 + j0$ point, the number of encirclements N is indeterminate. This situation corresponds to the condition where $B(s)$ has zeros on the imaginary axis. A necessary condition for applying the Nyquist criterion is that the path encircling the specified area must not pass through any poles or zeros of $B(s)$. When this condition is violated, the value for N becomes indeterminate and the Nyquist stability criterion cannot be applied. Simple imaginary zeros of $B(s)$ mean that the closed-loop system will have a continuous steady-state sinusoidal component in its output that is independent of the form of the input. Unless otherwise stated, this condition is considered unstable.

8.14 EXAMPLES OF NYQUIST'S CRITERION USING DIRECT POLAR PLOT

Several polar plots are illustrated in this section. These plots can be obtained with the aid of pole-zero diagrams, from calculations for a few key frequencies, or from a CAD program.[4] The angular variation of each term of a $G(s)H(s)$ function, as the contour Q is traversed, is readily determined from its pole-zero diagram. Both stable and unstable open-loop transfer functions $G(s)H(s)$ are illustrated in the following examples for which all time constants T_i are positive.

Example 1. Type 0. The direct polar plot is shown in Fig. 8.26 for the following Type 0 open-loop transfer function:

$$G(s)H(s) = \frac{K_0}{(1 + T_1 s)(1 + T_2 s)} \tag{8.59}$$

From this plot it is seen that $N = 0$, and Eq. (8.59) yields the value $P_R = 0$. Therefore, the value of Z_R is $Z_R = P_R - N = 0$ and the closed-loop system is stable. Thus, the Nyquist stability criterion indicates that, no matter how much the gain K_0 is increased, this system is always stable.

Example 2. Type 1. The example in Fig. 8.25 for the transfer function of the form

$$G(s)H(s) = \frac{K_1}{s(1 + T_1 s)(1 + T_2 s)} \tag{8.60}$$

is shown in the preceding section to be stable. However, if the gain K_1 is increased, the system is made unstable, as seen from Fig. 8.27. Note that the rotation of $B(s)$, in the direction of increasing frequency, produces a net angular rotation of $-720°$, or two complete cw rotations ($N = -2$), so that $Z_R = P_R - N = 2$.

The number of rotations N can be determined by drawing a radial line from the $-1 + j0$ point. Then, at each intersection of this radial line with the curve of $G(s)H(s)$, the direction of increasing frequency is noted. If the frequency along the curve of $G(s)H(s)$ is increasing as the curve crosses the radial line in a ccw direction, this crossing is positive. Similarly, if the frequency along the curve of $G(s)H(s)$ is increasing as the curve crosses the radial line in a cw direction, this crossing is negative. The sum of the crossings, including the sign, is equal to N. In Fig. 8.27 both crossings of the radial line are cw and negative. Thus, the sum of the crossings is $N = -2$.

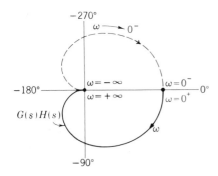

FIGURE 8.26
Plot of a typical Type 0 transfer function $G(s)H(s)$ for $-\infty < \omega < \infty$.

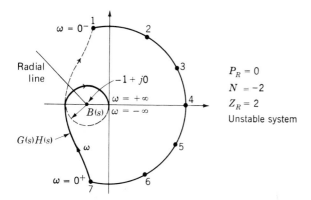

$P_R = 0$

$N = -2$

$Z_R = 2$

Unstable system

FIGURE 8.27
Polar plot of Fig. 8.25b with
increased gain.

Example 3. Type 2. Consider the transfer function

$$G(s)H(s) = \frac{K_2(1 + T_4 s)}{s^2(1 + T_1 s)(1 + T_2 s)(1 + T_3 s)} \qquad (8.61)$$

where $T_4 > T_1 + T_2 + T_3$. Figure 8.28 shows the mapping of $G(s)H(s)$ for the contour Q of the s plane. The word *mapping,* as used here, means that for a given point in the s plane there corresponds a given value of $G(s)H(s)$ or $B(s)$.

As pointed out in the preceding section, the presence of the s^2 term in the denominator of Eq. (8.61) results in a net rotation of 360° in the vicinity of $\omega = 0$, as shown in Fig. 8.28. For the complete contour of Fig. 8.25a the net rotation is zero; thus, since $P_R = 0$, the system is stable. The value of N can be determined by drawing the line radiating from the $-1 + j0$ point (see Fig. 8.28) and noting one cw and one ccw

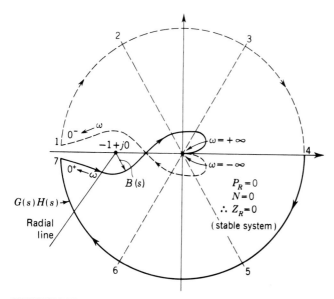

FIGURE 8.28
The complete polar plot of Eq. (8.61).

crossing; thus, $N = 0$. Like the previous example, this system can be made unstable by increasing the gain sufficiently for the $G(s)H(s)$ plot to cross the negative real axis to the left of the $-1 + j0$ point.

Example 4. Conditionally stable system. In this instance,

$$G(s)H(s) = \frac{K_0(1 + T_1 s)^2}{(1 + T_2 s)(1 + T_3 s)(1 + T_4 s)(1 + T_5 s)^2} \tag{8.62}$$

where $T_5 < T_1 < T_2, T_3$, and T_4. The complete polar plot of Eq. (8.62) is illustrated in Fig. 8.29 for a particular value of gain. For the radial line drawn from the $-1 + j0$ point the number of rotations is readily determined as $N = 0$.

It is seen in this example that the system can be made unstable not only by increasing the gain but also by decreasing the gain. If the gain is increased sufficiently for the $-1 + j0$ point to lie between the points c and d of the polar plot, the net cw rotation is equal to 2. Therefore, $Z_R = 2$ and the system is unstable. On the other hand, if the gain is decreased so that the $-1 + j0$ point lies between the points a and b of the polar plot, the net cw rotation is again 2 and the system is unstable. The gain can be further decreased so that the $-1 + j0$ point lies to the left of point a of the polar plot, resulting in a stable system. *This system is therefore conditionally stable.* A conditionally stable system is stable for a given range of values of gain but becomes unstable if the gain is either reduced or increased sufficiently. Such a system places a greater restriction on the stability and drift of amplifier gain and system characteristics. In addition, an effective gain reduction occurs in amplifiers that reach saturation with large input signals. This, in turn, may result in an unstable operation for the conditionally stable system.

Example 5. Open-loop unstable. For this case

$$G(s)H(s) = \frac{K_1(T_2 s + 1)}{s(T_1 s - 1)} \tag{8.63}$$

The complete polar plot for this equation is illustrated in Fig. 8.30 for a particular value of gain K_1. There is one cw intersection with the radial line drawn from the $-1 + j0$ point, and thus $N = -1$. Therefore, since $P_R = 1, Z_R = 2$, and the system is unstable. In this example the system is unstable for low values of gain ($0 < K_1 < K_{1x}$) for which the real-axis crossing b is to the right of the $-1 + j0$ point. For the range

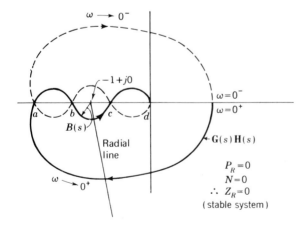

FIGURE 8.29
The complete polar plot of Eq. (8.62).

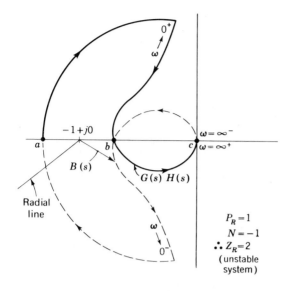

FIGURE 8.30

The complete polar plot of the open-loop unstable transfer function

$$G(s)H(s) = \frac{K_1(T_2s + 1)}{s(T_1s - 1)}$$

$P_R = 1$
$N = -1$
$\therefore Z_R = 2$
(unstable system)

$0 < K_1 < K_{1x}$ the $-1 + j0$ point is located, as shown in Fig. 8.30, between the points a and b. However, for the range $K_{1x} < K_1 < \infty$ the $-1 + j0$ point lies between the points b and c, which yields $N = +1$, thus resulting in $Z_R = 0$ and a stable closed-loop system.

8.15 NYQUIST'S STABILITY CRITERION APPLIED TO SYSTEMS HAVING DEAD TIME

Transport lag (dead time), as described in Sec. 7.13, is represented by the transfer function $G_\tau(s) = e^{-\tau s}$. The frequency transfer function is

$$\mathbf{G}_\tau(j\omega) = e^{-j\omega\tau} = 1\underline{/-\omega\tau} \tag{8.64}$$

It has a magnitude of unity and a negative angle whose magnitude increases directly in proportion to frequency. The polar plot of Eq. (8.64) is a unit circle that is traced indefinitely, as shown in Fig. 8.31. The Lm and phase-angle diagram shows a constant value of 0 dB and a phase angle that decreases with frequency.

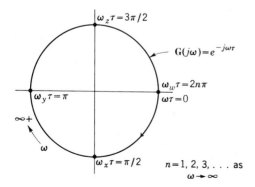

$n = 1, 2, 3, \ldots$ as
$\omega \rightarrow \infty$

FIGURE 8.31
Polar plot characteristic for transport lag.

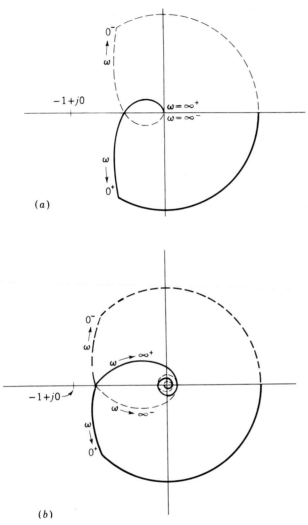

FIGURE 8.32
Complete polar plots of (*a*) Eq. (8.65) and (*b*) Eq. (8.66).

An example can best illustrate the application of the Nyquist criterion to a control system having transport lag. Figure 8.32*a* illustrates the complete polar plot of a stable system having a specified value of gain and the transfer function

$$G_x(s)H(s) = \frac{K_1}{s(1 + T_1 s)(1 + T_2 s)} \tag{8.65}$$

If dead time is added to this system, its transfer function becomes

$$G(s)H(s) = \frac{K_1 e^{-\tau s}}{s(1 + T_1 s)(1 + T_2 s)} \tag{8.66}$$

and the resulting complete polar plot is shown in Fig. 8.32*b*. When the contour Q is traversed and the polar plot characteristic of dead time, shown in Fig. 8.31, is included, the effects on the complete polar plot are as follows:

1. In traversing the imaginary axis of the contour Q between $0^+ < \omega < +\infty$, the polar plot of $\mathbf{G}(j\omega)\mathbf{H}(j\omega)$ in the third quadrant is shifted clockwise, closer to the $-1 + j0$ point. Thus, if the dead time is increased sufficiently, the $-1 + j0$ point will be enclosed by the polar plot and the system becomes unstable.
2. As $\omega \to +\infty$, the magnitude of the angle contributed by the transport lag increases indefinitely. This yields a spiraling curve as $|\mathbf{G}(j\omega)\mathbf{H}(j\omega)| \to 0$.

A transport lag therefore tends to make a system less stable.

8.16 DEFINITIONS OF PHASE MARGIN AND GAIN MARGIN AND THEIR RELATION TO STABILITY[8]

The stability and approximate degree of stability can be determined from the Lm and phase diagram. The stability characteristic is specified in terms of the following quantities:

Gain crossover. This is the point on the plot of the transfer function at which the magnitude of $\mathbf{G}(j\omega)$ is unity [Lm $\mathbf{G}(j\omega) = 0$ dB]. The frequency at gain crossover is called the *phase-margin frequency* ω_ϕ.

Phase margin angle. This is 180° plus the negative trigonometrically considered angle of the transfer function at the gain-crossover point. It is designated as the angle γ, which can be expressed as $\gamma = 180° + \phi$, where $\underline{/\mathbf{G}(j\omega_\phi)} = \phi$ is negative.

Phase crossover. This is the point on the plot of the transfer function at which the phase angle is $-180°$. The frequency at which phase crossover occurs is called the *gain-margin frequency* ω_c.

Gain margin. The gain margin is the factor a by which the gain must be changed in order to produce instability. Expressed in terms of the transfer function at the frequency ω_c, it is

$$\left|\mathbf{G}(j\omega_c)\right| a = 1 \tag{8.67}$$

On the polar plot of $\mathbf{G}(j\omega)$ the value at ω_c is

$$\left|\mathbf{G}(j\omega_c)\right| = \frac{1}{a} \tag{8.68}$$

In terms of the Lm, in decibels, this is

$$\mathrm{Lm}\, a = -\mathrm{Lm}\, \mathbf{G}(j\omega_c) \tag{8.69}$$

which identifies the gain margin on the Lm diagram.

These quantities are illustrated in Fig. 8.33 on both the log and the polar curves. Note the algebraic sign associated with these two quantities as marked on the curves. Figure 8.33*a* and *b* represents a stable system, and Fig. 8.33*c* and *d* represents an unstable system.

FIGURE 8.33
Log magnitude and phase diagram and polar plots of $G(j\omega)$, showing gain margin and phase margin angle: (*a*) and (*b*) stable; (*c*) and (*d*) unstable.

The phase margin angle is the amount of phase shift at the frequency ω_ϕ that would just produce instability. This angle would make the polar plot go through the -1 point. The phase margin angle for minimum-phase (m.p.) systems must be positive for a stable system, whereas a negative phase margin angle means that the system is unstable.

It can be shown that the phase margin angle is related to the effective damping ratio ζ of the system. This topic is discussed qualitatively in the next chapter. Satisfactory response is usually obtained with a phase margin of 45° to 60°. As an individual gains experience and develops his or her own particular technique, a desirable value of γ to be used for a particular system becomes more evident. This guideline for system performance applies only to those systems where behavior is that of an equivalent second-order system. The gain margin must be positive when expressed in decibels (greater than unity as a numeric) for a stable system. A negative gain margin means that the system is unstable. The damping ratio ζ of the system is also related to the gain margin. However, the phase margin gives a better estimate of damping ratio, and therefore of the transient overshoot of the system, than the gain margin.

The phase-margin frequency ω_ϕ, phase margin angle γ, crossover frequency ω_c, and the gain margin Lm a are also readily identified on the Nichols plot as shown in Fig. 8.34 and described in Sec. 8.18. Further information about the speed of response of the system can be obtained from the maximum value of the control ratio and the frequency at which this maximum occurs. The Lm–angle diagram or a CAD program can be used to obtain this data (see Chap. 9).

The relationship of stability and gain margin is modified for a conditionally stable system. This can be seen by reference to the polar plot of Fig. 8.29. The system for the particular gain represented in Fig. 8.29 is stable. Instability can occur with an increase or a decrease in gain. Therefore, both "upper" and "lower" gain margins must be identified, corresponding to the upper crossover frequency ω_{cu} and the lower crossover frequency ω_{cl}. In Fig. 8.29 the upper gain margin is larger than unity and the lower gain margin is less than unity.

8.17 STABILITY CHARACTERISTICS OF THE LOG MAGNITUDE AND PHASE DIAGRAM

The earlier sections of this chapter discuss how the asymptotes of the Lm vs. ω curve are related to each factor of the transfer function. For example, factors of the form $(1 + j\omega)^{-1}$ have a negative slope equal to -20 dB/decade at high frequencies. Also, the angle for this factor varies from $0°$ at low frequencies to $-90°$ at high frequencies.

The total phase angle of the transfer function at any frequency is closely related to the slope of the Lm curve at that frequency. A slope of -20 dB/decade is related to an angle of $-90°$; a slope of -40 dB/decade is related to an angle of $-180°$; a slope of -60 dB/decade is related to an angle of $-270°$; etc. Changes of slope at higher and lower corner frequencies, around the particular frequency being considered, contribute to the total angle at that frequency. The farther away the changes of slope are from the particular frequency, the less they contribute to the total angle at that frequency.

By observing the asymptotes of the Lm curve, it is possible to estimate the approximate value of the angle. With reference to Fig. 8.6 for the example of Sec. 8.7, the angle at $\omega = 4$ is now investigated. The slope of the curve at $\omega = 4$ is -20 dB/decade; therefore, the angle is near $-90°$. The slope changes at the corner frequencies $\omega = 2$ and $\omega = 8$, and so the slopes beyond these frequencies contribute to the total angle at $\omega = 4$. The actual angle, as read from Fig. 8.7, is $-122°$. The farther away the corner frequencies occur, the closer the angle is to $-90°$.

As has been seen, the stability of an m.p. system requires that the phase margin angle be positive. For this to be true, the angle at the gain crossover [Lm $\mathbf{G}(j\omega) = 0$ dB] must be greater than $-180°$. This requirement places a limit on the slope of the Lm curve at the gain crossover. *The slope at the gain crossover should be more positive than -40 dB/decade if the adjacent corner frequencies are not close.* A slope of -20 dB/decade is preferable. This value is derived from the consideration of a theorem by Bode. However, it can be seen qualitatively from the association of the slope of the Lm curve with the value of the phase angle. This guide should be used to assist in system design.

The Lm and phase diagram reveals some pertinent information, just as the polar plots do. For example, the gain can be adjusted (which raises or lowers the Lm curve) to produce a phase margin angle in the desirable range of 45 to 60°. The shape of the low-frequency portion of the curve determines system type and therefore the degree of steady-state accuracy. The system type and the gain determine the error coefficients and therefore the steady-state error. The phase-margin frequency ω_ϕ gives a qualitative indication of the speed of response of a system. However, this is only a qualitative relationship. A more detailed analysis of this relationship is made in the next chapter.

8.18   STABILITY FROM THE NICHOLS PLOT (LOG MAGNITUDE–ANGLE DIAGRAM)

The Lm–angle diagram is drawn by picking for each frequency the values of Lm and angle from the Lm and phase diagram. The resultant curve has frequency as a parameter. The curve for the example of Sec. 8.7, sketched in Fig. 8.34, shows a positive gain margin and phase margin angle; therefore, this represents a stable system. Changing the gain raises or lowers the curve without changing the angle characteristics. Increasing the gain raises the curve, thereby decreasing the gain margin and phase margin angle, with the result that the stability is decreased. Increasing the gain

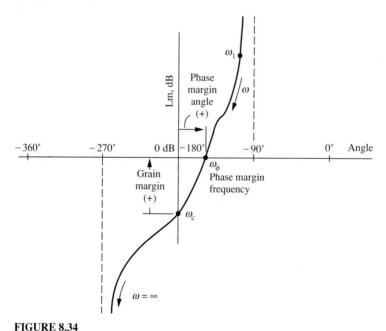

FIGURE 8.34
Example of a log magnitude–angle diagram for a minimum-phase open-loop transfer function.

so that the curve has a positive Lm at $-180°$ results in negative gain and phase margins; therefore, an unstable system results. Decreasing the gain lowers the curve and increases stability. However, a large gain is desired to reduce steady-state errors, as shown in Chap. 6.

The Lm–angle diagram for $G(s)H(s)$ can be drawn for all values of s on the contour Q of Fig. 8.25a. The resultant curve for m.p. systems is a closed contour. Nyquist's criterion can be applied to this contour by determining the number of points (having the values 0 dB and odd multiples of $180°$) enclosed by the curve of $G(s)H(s)$. This number is the value of N that is used in the equation $Z_R = N - P_R$ to determine the value of Z_R. As an example, consider a control system whose transfer function is given by

$$G(s) = \frac{K_1}{s(1 + Ts)} \tag{8.70}$$

Its Lm–angle diagram, for the contour Q, is shown in Fig. 8.35. From this figure it is seen that the value of N is zero and the system is stable. The Lm–angle contour for a nonminimum-phase system does not close; thus, it is more difficult to determine the value of N. For these cases the polar plot is easier to use to determine stability.

In the case of m.p. systems, it is not necessary to obtain the complete Lm–angle contour to determine stability. Only that portion of the contour is drawn representing $G(j\omega)$ for the range of values $0^+ < \omega < \infty$. The stability is then determined from the position of the curve of $G(j\omega)$ relative to the $(0\text{ dB}, -180°)$ point. In other words, the curve is traced in the direction of increasing frequency, i.e., "walking" along the

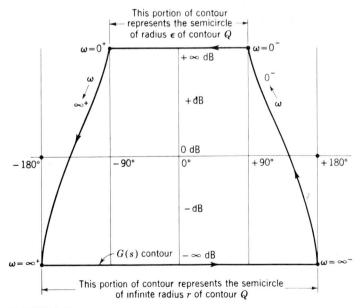

FIGURE 8.35
The log magnitude–angle contour for the minimum-phase system of Eq. (8.70).

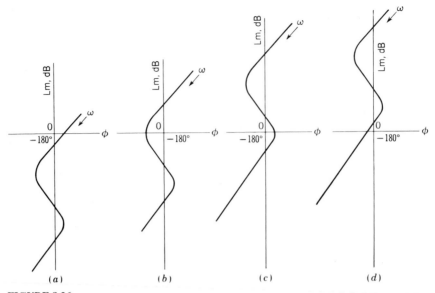

FIGURE 8.36
Log magnitude–angle diagram for a conditionally stable system: $(a),(c)$ stable; $(b),(d)$ unstable.

curve in the direction of increasing frequency. The system is stable if the (0 dB, $-180°$) point is to the right of the curve. This simplified rule of thumb is based on Nyquist's stability criterion for an m.p. system.

A conditionally stable system (as defined in Sec. 8.14, Example 4) is one in which the curve crosses the $-180°$ axis at more than one point. Figure 8.36 shows the transfer-function plot for such a system with two stable and two unstable regions. The gain determines whether the system is stable or unstable. The stable system represented by Fig. 8.36c has two values of gain margin, since instability can occur with either an increase or a decrease in gain. Corresponding to the upper and lower crossover frequencies ω_{cu} and ω_{cl}, the gain margins Lm a_u and Lm a_l are positive and negative, respectively. Additional stability information can be obtained from the Lm–angle diagram (see Chapter 9).

8.18   SUMMARY

In this chapter different types of frequency-response plots are introduced. From the Lm and phase-angle diagrams the polar plot can be obtained rapidly and with ease. All of these plots indicate the type of system that is represented and the necessary gain adjustments that must be made to improve the response. How these adjustments are made is discussed in the following chapter. The methods presented for obtaining the Lm frequency-response plots stress graphical techniques. For greater accuracy a CAD program can be used to calculate these data (see App. B).

The methods described in this chapter for obtaining frequency-response plots are based upon the condition that the transfer function of a given system is known.

These plots can also be obtained from experimental data, which do not require the analytical expression for the transfer function of a system to be known. These experimental data can be used to synthesize an analytical expression for the transfer function using the Lm plot (see Sec. 8.10).

This chapter shows that the polar plot of the transfer function $G(s)H(s)$, in conjunction with Nyquist's stability criterion, gives a rapid means of determining whether a system is stable or unstable. The same information is obtained from the Lm and phase-angle diagrams and the Lm–angle diagram (Nichols plot); the phase margin angle and gain margin are also used as a means of indicating the degree of stability. In the next chapter it is shown that other key information related to the time domain is readily obtained from any of these plots. Thus, the designer can determine whether the given system is satisfactory.

Another useful application of the Nyquist criterion is the analysis of systems having the characteristic of dead time. The Nyquist criterion has one disadvantage in that open-loop poles or zeros on the $j\omega$ axis require special treatment. This treatment is not expounded since, in general, most problems are free of poles or zeros on the $j\omega$ axis.

REFERENCES

1. Maccoll, L. A.: *Fundamental Theory of Servomechanisms,* Van Nostrand, Princeton, N.J., 1945.
2. James, H. M., N. B. Nichols, and R. S. Phillips: *Theory of Servomechanisms,* McGraw-Hill, New York, 1947.
3. Nyquist, H.: "Regeneration Theory," *Bell Syst. Tech. J.,* vol. 11, pp. 126–147, 1932.
4. A comprehensive list of CAD programs is given in App. B.
5. Bruns, R. A., and R. M. Saunders: *Analysis of Feedback Control Systems,* McGraw-Hill, New York, 1955, chap. 14.
6. Balabanian, N., and W. R. LePage: "What Is a Minimum-Phase Network?" *Trans. AIEE,* vol. 74, pt. II, pp. 785–788, January 1956.
7. Bode, H. W.: *Network Analysis and Feedback Amplifier Design,* Van Nostrand, Princeton, N.J., 1945, chap. 8.
8. Chestnut, H., and R. W. Mayer: *Servomechanisms and Regulating System Design,* 2nd ed., vol. 1, Wiley, New York, 1959.
9. Sanathanan, C. K., and H. Tsukui: "Synthesis of Transfer Function from Frequency Response Data," *Intern. J. Systems Science,* vol. 5, no. 1, pp. 41–54, January 1974.
10. Brown, G. S., and D. P. Campbell: *Principles of Servomechanisms,* Wiley, New York, 1948.
11. D'Azzo, J. J., and C. H. Houpis: *Feedback Control System Analysis and Synthesis,* 2nd ed., McGraw-Hill, New York, 1966.
12. D'Azzo, J. J., and C. H. Houpis: *Linear Control System Analysis and Design: Conventional and Modern,* 3rd ed., McGraw- Hill, New York, 1988.
13. Freudenberg, J. S., and D. P. Looze: "Right Half-Plane Poles and Zeros and Design Trade-Offs in Feedback Systems," *IEEE Trans. Automatic Control,* AC-30, pp. 555–565, 1985.

CHAPTER
9

CLOSED-LOOP TRACKING PERFORMANCE BASED ON THE FREQUENCY RESPONSE

9.1 INTRODUCTION

Chapter 8 is devoted to plotting the open-loop transfer function $\mathbf{G}(j\omega)\mathbf{H}(j\omega)$. The two types of plots that are emphasized are the direct polar plot (Nyquist plot) and the log magnitude–angle plot (Nichols plot). Also included in Chapter 8 is the determination of closed-loop stability in terms of the open-loop transfer function by use of the Nyquist stability criterion, which is illustrated in terms of both types of plot. The result of the stability study places bounds on the permitted range of values of gain.

This chapter develops a correlation between the frequency and the time responses of a system, leading to a method of gain setting in order to achieve a specified closed-loop frequency response.[1] The closed-loop frequency response is obtained as a function of the open-loop frequency response. Through the graphical representations in this chapter, the reader should obtain an appreciation of the responses that can be achieved by gain adjustment. CAD programs[2] like MATLAB can be used interactively to perform a number of designs rapidly. In addition, although the design is performed in the frequency domain, the closed-loop responses in the time domain are also obtained. Then a "best" design is selected by considering both the frequency and the time responses. CAD programs considerably reduce the computational time of the designer in achieving a system that meets the design specifications.

FIGURE 9.1
Block diagram of a simple control system.

9.2 DIRECT POLAR PLOT

Consider a simple control system with unity feedback, as shown in Fig. 9.1. The performance of this system with a sinusoidal input $\mathbf{R}(j\omega)$ is represented by the following equations:

$$\mathbf{E}(j\omega) = \mathbf{R}(j\omega) - \mathbf{C}(j\omega) \tag{9.1}$$

$$\frac{\mathbf{C}(j\omega)}{\mathbf{E}(j\omega)} = \mathbf{G}(j\omega) = |\mathbf{G}(j\omega)| e^{j\phi} \tag{9.2}$$

$$\frac{\mathbf{C}(j\omega)}{\mathbf{R}(j\omega)} = \frac{\mathbf{G}(j\omega)}{1 + \mathbf{G}(j\omega)} = \mathbf{M}(j\omega) = M(\omega)e^{j\alpha(\omega)} \tag{9.3}$$

$$\frac{\mathbf{E}(j\omega)}{\mathbf{R}(j\omega)} = \frac{1}{1 + \mathbf{G}(j\omega)} \tag{9.4}$$

Figure 9.2 is the polar plot of $\mathbf{G}(j\omega)$ for this control system. The directed line segment drawn from the $-1 + j0$ point to any point on the $\mathbf{G}(j\omega)\mathbf{H}(j\omega)$ curve represents the quantity $\mathbf{B}(j\omega) = 1 + \mathbf{G}(j\omega)\mathbf{H}(j\omega)$. For a unity-feedback system, $\mathbf{H}(j\omega) = 1$; therefore,

$$\mathbf{B}(j\omega) = |\mathbf{B}(j\omega)| e^{j\lambda(\omega)} = 1 + \mathbf{G}(j\omega) \tag{9.5}$$

The frequency control ratio $\mathbf{C}(j\omega)/\mathbf{R}(j\omega)$ is therefore equal to the ratio

$$\frac{\mathbf{C}(j\omega)}{\mathbf{R}(j\omega)} = \frac{\mathbf{A}(j\omega)}{\mathbf{B}(j\omega)} = \frac{|\mathbf{A}(j\omega)| e^{j\phi(\omega)}}{|\mathbf{B}(j\omega)| e^{j\lambda(\omega)}} = \frac{\mathbf{G}(j\omega)}{1 + \mathbf{G}(j\omega)} \tag{9.6}$$

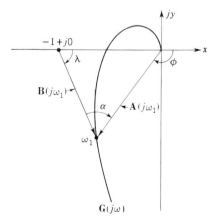

FIGURE 9.2
Polar plot of $\mathbf{G}(j\omega)$ for the control system of Fig. 9.1.

$$\frac{C(j\omega)}{R(j\omega)} = \frac{|A(j\omega)|}{|B(j\omega)|} e^{j(\phi-\lambda)} = M(\omega)e^{j\alpha(\omega)} \tag{9.7}$$

Note that the angle $\alpha(\omega)$ can be determined directly from the construction shown in Fig. 9.2. Since the magnitude of the angle $\phi(\omega)$ is greater than the magnitude of the angle $\lambda(\omega)$, the value of the angle $\alpha(\omega)$ is negative. Remember that ccw rotation is taken as positive.

Combining Eqs. (9.4) and (9.5) gives

$$\frac{E(j\omega)}{R(j\omega)} = \frac{1}{1 + G(j\omega)} = \frac{1}{|B(j\omega)| e^{j\lambda}} \tag{9.8}$$

From Eq. (9.8) it is seen that the greater the distance from the $-1 + j0$ point to a point on the $G(j\omega)$ locus, for a given frequency, the smaller the steady-state sinusoidal error for a stated sinusoidal input. Thus, the usefulness and importance of the polar plot of $G(j\omega)$ have been enhanced.

9.3   DETERMINATION OF M_m AND ω_m FOR A SIMPLE SECOND-ORDER SYSTEM

The frequency at which the maximum value of $|C(j\omega)/R(j\omega)|$ occurs is referred to as the *resonant frequency* ω_m. The maximum value is labeled M_m. These two quantities are figures of merit of a system. As shown in later chapters, compensation to improve system performance is based upon a knowledge of ω_m and M_m.

For a *simple second-order system* a direct and simple relationship can be obtained for M_m and ω_m in terms of the system parameters. Consider the position-control system of Fig. 9.3, which contains the servomotor described by Eq. (2.135). The forward and closed-loop transfer functions are, respectively,

$$\frac{C(s)}{E(s)} = \frac{K_1}{s(T_m s + 1)} \tag{9.9}$$

$$\frac{C(s)}{R(s)} = \frac{1}{(T_m/K_1)s^2 + (1/K_1)s + 1} = \frac{1}{s^2/\omega_n^2 + (2\zeta/\omega_n)s + 1} \tag{9.10}$$

where $K_1 = AK_M$. The damping ratio and the undamped natural frequency for this system, as determined from Eq. (9.10), are $\zeta = 1/(2\sqrt{K_1 T_m})$ and $\omega_n = \sqrt{K_1/T_m}$. The control ratio as a function of frequency is therefore

$$\frac{C(j\omega)}{R(j\omega)} = \frac{1}{(1 - \omega^2/\omega_n^2) + j2\zeta(\omega/\omega_n)} = M(\omega)e^{j\alpha(\omega)}. \tag{9.11}$$

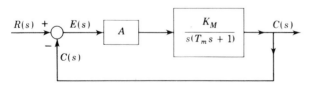

FIGURE 9.3
A position-control system.

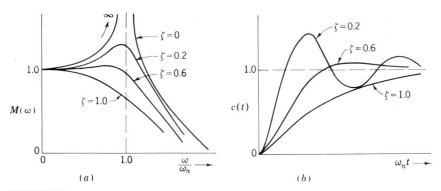

FIGURE 9.4
(a) Plots of M vs. ω/ω_n for a simple second-order system; (b) corresponding time plots for a step input.

Plots of $M(\omega)$ and $\alpha(\omega)$ vs. ω, for a particular value of ω_n, can be obtained for different values of ζ (see Fig. 8.4). In order to obtain a family of curves that are applicable to all simple second-order systems with different values of ω_n, they are plotted vs. ω/ω_n in Fig. 9.4a.

The inverse Laplace transform of $C(s)$, for a unit-step input, gives the time response, as derived in Sec. 3.9 for $\zeta < 1$, as

$$c(t) = 1 - \frac{1}{\sqrt{1 - \zeta^2}} e^{-\zeta \omega_n t} \sin(\omega_n \sqrt{1 - \zeta^2} t + \cos^{-1} \zeta) \qquad (9.12)$$

The plot of $c(t)$ for several values of damping ratio ζ is drawn in Fig. 9.4b.

Next, consider the magnitude M^2, as derived from Eq. (9.11):

$$M^2 = \frac{1}{(1 - \omega^2/\omega_n^2)^2 + 4\zeta^2(\omega^2/\omega_n^2)} \qquad (9.13)$$

To find the maximum value of M and the frequency at which it occurs, Eq. (9.13) is differentiated with respect to frequency and set equal to zero:

$$\frac{dM^2}{d\omega} = -\frac{-4(1 - \omega^2/\omega_n^2)(\omega/\omega_n^2) + 8\zeta^2(\omega/\omega_n^2)}{[(1 - \omega^2/\omega_n^2)^2 + 4\zeta^2(\omega^2/\omega_n^2)]^2} = 0 \qquad (9.14)$$

The frequency ω_m at which the value M exhibits a peak (see Fig. 9.5), as found from Eq. (9.14), is

$$\omega_m = \omega_n \sqrt{1 - 2\zeta^2} \qquad (9.15)$$

FIGURE 9.5
A closed-loop frequency-response curve indicating M_m and ω_m.

This value of frequency is substituted into Eq. (9.13) to yield

$$M_m = \frac{1}{2\zeta\sqrt{1-\zeta^2}} \tag{9.16}$$

From these equations it is seen that the curve of M vs. ω has a peak value, other than at $\omega = 0$, only for $\zeta < 0.707$. Figure 9.6 is a plot of M_m vs. ζ for a simple second-order system. For values of $\zeta < 0.4$, M_m increases very rapidly in magnitude; the transient oscillatory response is therefore excessively large and might damage the physical equipment.

In Secs. 3.5 and 7.12 the damped natural frequency ω_d for the transient of the simple second-order system is shown to be

$$\omega_d = \omega_n\sqrt{1-\zeta^2} \tag{9.17}$$

Also, in Sec. 3.9 the expression for the peak value M_p for a unit-step input to this simple system is determined. It is repeated here:

$$M_p = \frac{c_p}{r} = 1 + \exp\left(-\frac{\zeta\pi}{\sqrt{1-\zeta^2}}\right) \tag{9.18}$$

Therefore, for this simple second-order system the following conclusions are obtained in correlating the frequency and time responses:

1. Inspection of Eq. (9.15) reveals that ω_m is a function of both ω_n and ζ. Thus, for a given ζ, the larger the value of ω_m, the larger ω_n, and the faster the transient time of response for this system.
2. Inspection of Eqs. (9.16) and (9.18) shows that both M_m and M_p are functions of ζ. The smaller ζ becomes, the larger in value M_m and M_p become. Thus, it can be concluded that the larger the value of M_m, the larger the value of M_p. For values of $\zeta < 0.4$ the correspondence between M_m and M_p is only qualitative for this simple case. In other words, for $\zeta = 0$ the time domain yields $M_p = 2$, whereas the frequency domain yields $M_m = \infty$. When $\zeta > 0.4$, there is a close correspondence between M_m and M_p. As an example, for ζ equal to 0.6, $M_m = 1.04$ and $M_p = 1.09$.

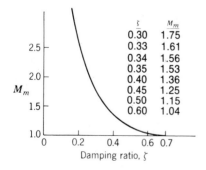

ζ	M_m
0.30	1.75
0.33	1.61
0.34	1.56
0.35	1.53
0.40	1.36
0.45	1.25
0.50	1.15
0.60	1.04

FIGURE 9.6
A plot of M_m vs. ζ for a simple second-order system.

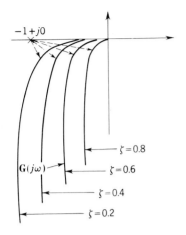

FIGURE 9.7
Polar plots of $\mathbf{G}(j\omega)$ for different damping ratios for the system shown in Fig. 9.3.

3. In Fig. 9.7 are shown polar plots for different damping ratios for the simple second-order system. Note that the shorter the distance between the $-1 + j0$ point and a particular $\mathbf{G}(j\omega)$ plot, as indicated by the dashed lines, the smaller the damping ratio. Thus, M_m is larger and consequently M_p is also larger.

From these characteristics a designer can obtain a good approximation of the time response of a simple second-order system by knowing only the M_m and ω_m of its frequency response.

The procedure of setting the derivative, with respect to ω, of $\mathbf{C}(j\omega)/\mathbf{R}(j\omega)$ to zero works very well with a simple system. But the differentiation of $\mathbf{C}(j\omega)/\mathbf{R}(j\omega)$ and solution for ω_m and M_m become tedious for more complex systems. Therefore, a graphic procedure is generally used, as shown in the following sections.[3]

9.4 CORRELATION OF SINUSOIDAL AND TIME RESPONSES[4]

Although the correlation in the preceding section is for a simple second-order system, it has been found by experience that M_m is also a function of the *effective* ζ and ω_n for higher-order systems. The effective ζ and ω_n of a higher-order system (see Sec. 7.12) is dependent upon the combined effect of the ζ and ω_n of each second-order term and the values of the real roots of the characteristic equation of $C(s)/R(s)$. Thus, in order to alter the M_m, the location of some of the roots must be changed. Which ones should be altered depends on which are dominant in the time domain.

From the analysis for a simple second-order system, whenever the frequency response has the shape shown in Fig. 9.5, the following correlation exists between the frequency and time responses for systems of any order:

1. The larger ω_m is made, the faster the time of response for the system.

2. The value of M_m gives a qualitative measure of M_p within the acceptable range of the *effective* damping ratio $0.4 < \zeta < 0.707$. In terms of M_m, the acceptable range is $1 < M_m < 1.4$.

3. The closer the $\mathbf{G}(j\omega)$ curve comes to the $-1 + j0$ point, the larger the value of M_m.

The characteristics developed in Chapter 6 can be added to these three items, i.e., the larger K_p, K_v, or K_a is made, the greater the steady-state accuracy for a step, a ramp, and a parabolic input, respectively. In terms of the polar plot, the farther the point $\mathbf{G}(j\omega)|_{\omega=0} = K_0$ for a Type 0 system is from the origin, the more accurate is the steady-state time response for a step input. For a Type 1 system, the farther the low-frequency asymptote (as $\omega \to 0$) is from the imaginary axis, the more accurate is the steady-state time response for a ramp input.

All the factors just mentioned are merely *guideposts* in the *frequency domain* to assist the designer in obtaining an *approximate* idea of the time response of a system. They serve as "stop-and-go signals" to indicate if the design is headed in the right direction in achieving the desired time response. If the desired performance specifications are not satisfactorily met, compensation techniques (see later chapters) must be used. After compensation the exact time response can be obtained by taking the inverse Laplace transform of $C(s)$. The approximate approach saves much valuable time. Exceptions to the preceding analysis do occur for higher-order systems. The digital computer is a valuable tool for obtaining the exact closed-loop frequency and time responses. Thus a direct correlation between them can easily be made.

9.5   CONSTANT $M(\omega)$ AND $\alpha(\omega)$ CONTOURS OF $C(j\omega)/R(j\omega)$ ON THE COMPLEX PLANE (DIRECT PLOT)

The open-loop transfer function and its polar plot for a given feedback control system have provided the following information so far:

1. The stability or instability of the system
2. If the system is stable, the degree of its stability
3. The system type
4. The degree of steady-state accuracy
5. A graphical method of determining $C(j\omega)/R(j\omega)$

These items provide a qualitative idea of the system's time response. The polar plot is useful since it permits the determination of M_m and ω_m. Also, it is used to determine the gain required to produce a specified value of M_m.

The contours of constant values of M drawn in the complex plane yield a rapid means of determining the values of M_m and ω_m and the value of gain required to achieve a desired value of M_m. In conjunction with the contours of constant values of $\alpha(\omega)$, also drawn in the complex plane, the plot of $C(j\omega)/R(j\omega)$ can be obtained rapidly. The M and α contours are developed only for unity-feedback systems.

Equation of a Circle

The equation of a circle with its center at the point (a, b) and having radius r is

$$(x - a)^2 + (y - b)^2 = r^2 \tag{9.19}$$

The circle is displaced to the left of the origin for a negative value of a, and for negative values of b the circle is displaced below the origin. This equation of a circle in the xy plane is used later in this section to express contours of constant M and α.

$M(\omega)$ Contours

Figure 9.2 is the polar plot of a forward transfer function $\mathbf{G}(j\omega)$ of a *unity-feedback system*. From Eq. (9.3), the magnitude M of the control ratio is

$$M(\omega) = \frac{|\mathbf{A}(j\omega)|}{|\mathbf{B}(j\omega)|} = \frac{|\mathbf{G}(j\omega)|}{|1 + \mathbf{G}(j\omega)|} \qquad (9.20)$$

The question at hand is: How many points are there in the complex plane for which the ratios of the magnitudes of the phasors $\mathbf{A}(j\omega)$ and $\mathbf{B}(j\omega)$ have the same value of $M(\omega)$? For example, by referring to Fig. 9.2, it can be seen that for the frequency $\omega = \omega_1$, M has a value of M_a. It is desirable to determine all the other points in the complex plane for which

$$\frac{|\mathbf{A}(j\omega)|}{|\mathbf{B}(j\omega)|} = M_a$$

To derive the constant M locus, express the transfer function in rectangular coordinates. That is,

$$\mathbf{G}(j\omega) = x + jy \qquad (9.21)$$

Substituting this equation into Eq. (9.20) gives

$$M = \frac{|x + jy|}{|1 + x + jy|} = \left[\frac{x^2 + y^2}{(1 + x)^2 + y^2} \right]^{1/2} \qquad \text{or} \qquad M^2 = \frac{x^2 + y^2}{(1 + x)^2 + y^2}$$

Rearranging the terms of this equation yields

$$\left(x + \frac{M^2}{M^2 - 1} \right)^2 + y^2 = \frac{M^2}{(M^2 - 1)^2} \qquad (9.22)$$

By comparison with Eq. (9.19) it is seen that Eq. (9.22) is the equation of a circle with its center on the real axis with M as a parameter. The center is located at

$$x_0 = -\frac{M^2}{M^2 - 1} \qquad (9.23)$$

$$y_0 = 0 \qquad (9.24)$$

and the radius is

$$r_0 = \left| \frac{M}{M^2 - 1} \right| \qquad (9.25)$$

Inserting a given value of $M = M_a$ into Eq. (9.22) results in a circle in the complex plane having a radius r_0 and its center at (x_0, y_0). This circle is called a constant M contour for $M = M_a$.

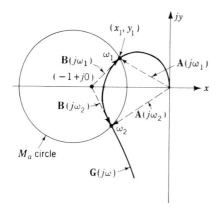

FIGURE 9.8
A plot of the transfer function $\mathbf{G}(j\omega)$ and the M_a circle.

The ratio of magnitudes of the phasors $\mathbf{A}(j\omega)$ and $\mathbf{B}(j\omega)$ drawn to any point on the M_a circle has the same value. As an example, refer to Fig. 9.8, in which the M_a circle and the $\mathbf{G}(j\omega)$ function are plotted. Since the circle intersects the $\mathbf{G}(j\omega)$ plot at the two frequencies ω_1 and ω_2, there are two frequencies for which

$$\frac{|\mathbf{A}(j\omega_1)|}{|\mathbf{B}(j\omega_1)|} = \frac{|\mathbf{A}(j\omega_2)|}{|\mathbf{B}(j\omega_2)|} = M_a$$

In other words, a given point (x_1, y_1) is simultaneously a point on a particular transfer function $\mathbf{G}(j\omega)$ and a point on the M circle passing through it.

The plot of Fig. 9.8 is redrawn in Fig. 9.9 with two more M circles added. In this figure the circle $M = M_b$ is just tangent to the $\mathbf{G}(j\omega)$ plot. There is only one point (x_3, y_3) for $\mathbf{G}(j\omega_3)$ in the complex plane for which the ratio $|\mathbf{A}(j\omega)/\mathbf{B}(j\omega)|$ is equal to M_b. Also, the M_c circle does not intersect and is not tangent to the $\mathbf{G}(j\omega)$ plot, indicating that there are no points in the plane that can simultaneously satisfy Eqs. (9.20) and (9.22) for $M = M_c$.

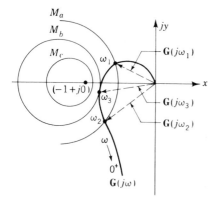

FIGURE 9.9
M contours and a $\mathbf{G}(j\omega)$ plot.

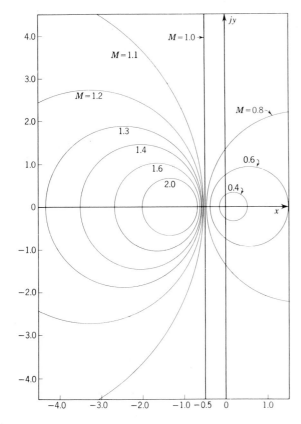

FIGURE 9.10
Constant M contours.

Figure 9.10 shows a family of circles in the complex plane for different values of M. In this figure, notice that the larger the value M, the smaller is its corresponding M circle. Thus, for the example shown in Fig. 9.9 the ratio $\mathbf{C}(j\omega)/\mathbf{R}(j\omega)$, for a unity-feedback control system, has a maximum value of M equal to $M_m = M_b$. A further inspection of Fig. 9.10 and Eq. (9.23) reveals the following:

1. For $M \to \infty$, which represents a condition of oscillation ($\zeta \to 0$), the center of the M circle $x_0 \to -1 + j0$ and the radius $r_0 \to 0$. This agrees with the statement made previously that as the $\mathbf{G}(j\omega)$ plot comes closer to the $-1 + j0$ point, the system's effective ζ becomes smaller and the degree of its stability decreases.

2. For $M(\omega) = 1$, which represents the condition where $\mathbf{C}(j\omega) = \mathbf{R}(j\omega)$, $r_0 \to \infty$ and the M contour becomes a straight line perpendicular to the real axis at $x = -\frac{1}{2}$.

3. For $M \to 0$, the center of the M circle $x_0 \to 0$ and the radius $r_0 \to 0$.

4. For $M > 1$ the centers of the circles lie to the left of $x = -1 + j0$, and for $M < 1$ the centers of the circles lie to the right of $x = 0$. All centers are on the real axis.

$\alpha(\omega)$ Contours

The $\alpha(\omega)$ contours, representing constant values of phase angle $\alpha(\omega)$ for $C(j\omega)/R(j\omega)$, can also be determined in the same manner as the M contours. Substituting Eq. (9.21) into Eq. (9.3) yields

$$M(\omega)e^{j\alpha(\omega)} = \frac{x + jy}{(1 + x) + jy} = \frac{A(j\omega)}{B(j\omega)} \tag{9.26}$$

The question at hand now is: How many points are there in the complex plane for which the ratio of the phasors $A(j\omega)$ and $B(j\omega)$ yields the same angle α? To answer this question, express the angle α obtained from Eq. (9.26) as follows:

$$\alpha = \tan^{-1}\frac{y}{x} - \tan^{-1}\frac{y}{1 + x}$$

$$= \tan^{-1}\frac{y/x - y/(1 + x)}{1 + (y/x)[y/(1 + x)]} = \tan^{-1}\frac{y}{x^2 + x + y^2}$$

$$\tan\alpha = \frac{y}{x^2 + x + y^2} = N \tag{9.27}$$

For a constant value of the angle α, $\tan\alpha = N$ is also constant. Rearranging Eq. (9.27) results in

$$\left(x + \frac{1}{2}\right)^2 + \left(y - \frac{1}{2N}\right)^2 = \frac{1}{4}\frac{N^2 + 1}{N^2} \tag{9.28}$$

By comparing with Eq. (9.19) it is seen that Eq. (9.28) is the equation of a circle with N as a parameter. It has its center at

$$x_q = -\frac{1}{2} \quad (9.29) \qquad y_q = \frac{1}{2N} \quad (9.30)$$

and has a radius of

$$r_q = \frac{1}{2}\left(\frac{N^2 + 1}{N^2}\right)^{1/2} \tag{9.31}$$

Inserting a given value of $N_q = \tan\alpha_q$ into Eq. (9.28) results in a circle (see Fig. 9.11) of radius r_q with its center at (x_q, y_q).

The tangent of angles in the first and third quadrant is positive. Therefore, the y_q coordinate given by Eq. (9.30) is the same for an angle in the first quadrant and for the negative of its supplement, which is in the third quadrant. As a result, the constant α contour is only an arc of the circle. As an example, the $\alpha = -310°$ and $-130°$ arcs are part of the same circle, as shown in Fig. 9.11a. Similarly, angle α in the second and fourth quadrants have the same value y_q if they are negativ· supplements of each other. The constant α contours for these angles are shown in Fig. 9.11b.

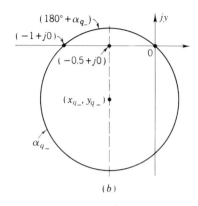

FIGURE 9.11
Arcs of constant α.

The ratio of the complex quantities $\mathbf{A}(j\omega)$ and $\mathbf{B}(j\omega)$, for all points on the α_q arc, yields the same phase angle α_q, that is,

$$\underline{/\mathbf{A}(j\omega)} - \underline{/\mathbf{B}(j\omega)} = \alpha_q \tag{9.32}$$

A good procedure, in order to avoid ambiguity in determining the angle α is to start at zero frequency, where $\alpha = 0$, and to proceed to higher frequencies, knowing that the angle is a continuous function.

Different values of α result in a family of circles in the complex plane with centers on the line represented by $\left(-\frac{1}{2}, y\right)$, as illustrated in Fig. 9.12.

Tangents to the *M* Circles

The line drawn through the origin of the complex plane tangent to a given M circle plays an important part in setting the gain of $\mathbf{G}(j\omega)$. Referring to Fig. 9.13 and recognizing that the line $bc = r_0$ is the radius and $ob = x_0$ is the distance to the center of the particular M circle, it becomes evident that

$$\sin\psi = \frac{bc}{ob} = \frac{M/(M^2 - 1)}{M^2/(M^2 - 1)} = \frac{1}{M} \tag{9.33}$$

This relationship is utilized in the section on gain adjustment. Also, the point a in Fig. 9.13 is the $-1 + j0$ point. This statement is proved from

$$(oc)^2 = (ob)^2 - (bc)^2 \qquad ac = oc\sin\psi \qquad \text{and} \qquad (oa)^2 = (oc)^2 - (ac)^2$$

Combining these equations yields

$$(oa)^2 = \frac{M^2 - 1}{M^2}\left[(ob)^2 - (bc)^2\right] \tag{9.34}$$

The values of the distances to the center ob and of the radius bc of the M circle [see Eqs. (9.23) and (9.25)] are then substituted into Eq. (9.34), which results in

$$oa = 1 \tag{9.35}$$

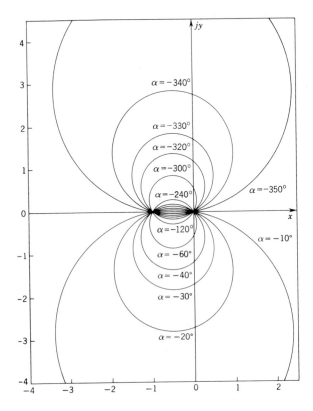

FIGURE 9.12
Constant α contours.

Some typical values necessary for constructing M and α contours are as follows:

$$M_o = 1.1, \ x_o = -5.76, \ r_o = 5.24, \ \psi = 65.4°$$

$$M_o = 1.3, \ x_o = -2.45, \ r_o = 1.88, \ \psi = 50.3°$$

$$\alpha = -60°, \quad N = -1.73, \quad r_q = 0.577, \quad y_q = -0.289$$

$$\alpha = 30°, \quad N = 0.577, \quad r_q = 1.000, \quad y_q = 0.866$$

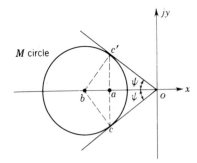

FIGURE 9.13
Determination of $\sin \psi$.

9.6 CONSTANT $1/M$ AND α CONTOURS (UNITY FEEDBACK) IN THE INVERSE POLAR PLANE

The closed-loop unity-feedback system can be represented by either of the following equations:

$$\frac{\mathbf{C}(j\omega)}{\mathbf{R}(j\omega)} = \frac{\mathbf{G}(j\omega)}{1 + \mathbf{G}(j\omega)} = Me^{j\alpha} \tag{9.36}$$

$$\frac{\mathbf{R}(j\omega)}{\mathbf{C}(j\omega)} = \frac{1}{\mathbf{G}(j\omega)} + 1 = \frac{1}{M}e^{-j\alpha} \tag{9.37}$$

Note that Eq. (9.37) is composed of two complex quantities that can be readily plotted in the complex plane, as shown in Fig. 9.14. Thus, $\mathbf{R}(j\omega)/\mathbf{C}(j\omega)$ can be obtained graphically by plotting the complex quantities $1/\mathbf{G}(j\omega)$ and 1.

Constant $1/M$ and α contours for the inverse plots are developed for unity-feedback systems. Figure 9.14 illustrates the quantities in Eq. (9.37) for a unity-feedback system, with $\mathbf{H}(j\omega) = 1$. Note that, because of the geometry of construction, the directed line segment drawn from the $-1 + j0$ point is also the quantity $1 + 1/\mathbf{G}(j\omega)$ or $(1/M)e^{-j\alpha}$. Thus, the inverse plot yields very readily the form and location of the contours of constant $1/M$ and α. In other words, in the inverse plane it can be seen that:

1. Contours of constant values of M are circles whose centers are at the $-1 + j0$ point, and the radii are equal to $1/M$.
2. Contours of constant values of $-\alpha$ are radial lines that pass through the $-1 + j0$ point.

Figure 9.15 shows the contour for a particular magnitude of $\mathbf{R}(j\omega)/\mathbf{C}(j\omega)$ and a tangent drawn to the M circle from the origin. In Fig. 9.15

$$\sin\psi = \left|\frac{ab}{ob}\right| = \frac{1/M}{1} = \frac{1}{M}$$

which is the same as for the contours in the direct plane. The directed line segment from $-1 + j0$ to the point (x_1, y_1) has a value of $1/M_q$ and an angle of $-\alpha_q$. If the polar

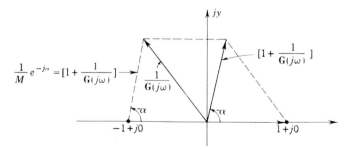

FIGURE 9.14
Representation of the quantities in Eq. (9.37) with $\mathbf{H}(j\omega) = 1$.

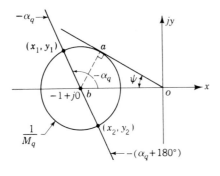

FIGURE 9.15
Contour for a particular magnitude
$|\mathbf{R}(j\omega)/\mathbf{C}(j\omega)| = 1/M_q$.

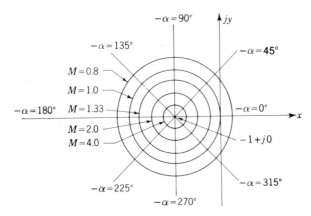

FIGURE 9.16

Typical M and α contours on the inverse polar plane for unity feedback.

plot of $1/\mathbf{G}(j\omega)$ passes through the (x_1, y_1) point, then $\mathbf{R}(j\omega_1)/\mathbf{C}(j\omega_1)$ has these values. The directed line segment from $-1 + j0$ to the point (x_2, y_2) has a magnitude of $1/M_q$ and an angle of $-\alpha_q - 180°$. If the polar plot of $1/\mathbf{G}(j\omega)$ passes through the (x_2, y_2) point, then $|\mathbf{R}(j\omega_2)/\mathbf{C}(j\omega_2)|$ has these values. Figure 9.16 illustrates families of typical M and α contours.

9.7 GAIN ADJUSTMENT FOR A DESIRED M_m OF A UNITY-FEEDBACK SYSTEM: DIRECT POLAR PLOT

Gain adjustment is the first step in adjusting the system for the desired performance. The procedure for adjusting the gain is outlined in this section. Figure 9.17a shows $\mathbf{G}_x(j\omega)$ with its respective M_m circle in the complex plane. Since

$$\mathbf{G}_x(j\omega) = x + jy = K_x\mathbf{G}'_x(j\omega) = K_x(x' + jy') \tag{9.38}$$

it follows that
$$x' + jy' = \frac{x}{K_x} + j\frac{y}{K_x}$$

where $\mathbf{G}'_x(j\omega) = \mathbf{G}_x(j\omega)/K_x$ is defined as the frequency-sensitive portion of $\mathbf{G}_x(j\omega)$ with unity gain. Note that changing the gain merely changes the amplitude and not the angle of the locus of points of $\mathbf{G}_x(j\omega)$. Thus, if in Fig. 9.17a a

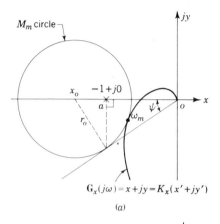

$$G_x(j\omega) = x + jy = K_x(x' + jy')$$

(a)

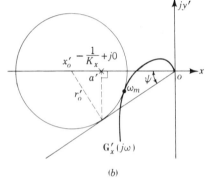

$G_x'(j\omega)$

(b)

FIGURE 9.17
(a) Plot of $\mathbf{G}_x(j\omega)$ with the respective M_m circle. (b) Circle drawn tangent both to the plot of $\mathbf{G}_x'(j\omega)$ and to the line representing the angle $\psi = \sin^{-1}(1/M_m)$.

change of scale is made by dividing the x, y coordinates by K_x so that the new coordinates are x', y', the following are true:

1. The $\mathbf{G}_x(j\omega)$ plot becomes the $\mathbf{G}_x'(j\omega)$ plot.

2. The M_m circle becomes simultaneously tangent to $\mathbf{G}_x'(j\omega)$ and to the line representing $\sin \psi = 1/M_m$.

3. The $-1 + j0$ point becomes the $-1/K_x + j0$ point.

4. The radius r_0 becomes $r_0' = r_0/K_x$.

In other words, if $\mathbf{G}_x'(j\omega)$ and $\mathbf{G}_x(j\omega)$ are drawn on separate sheets of graph paper (see Fig. 9.17b), and the two graphs are superimposed so that the axes coincide, the circles and the $\mathbf{G}_x(j\omega)$ and $\mathbf{G}_x'(j\omega)$ plots also coincide. Referring to Fig. 9.17a and 9.17b, note that $oa = -1$ and $oa' = -1/K_x$.

As a consequence, it is possible to determine the required gain to achieve a desired M_m for a given system by the following graphical procedure:

Step 1. If the original system has a transfer function

$$G_x(j\omega) = \frac{K_x(1 + j\omega T_1)(1 + j\omega T_2)\cdots}{(j\omega)^m(1 + j\omega T_a)(1 + j\omega T_b)(1 + j\omega T_c)\cdots} = K_x G'_x(j\omega) \quad (9.39)$$

with an original gain K_x, only the frequency-sensitive portion $G'_x(j\omega)$ is plotted.

Step 2. Draw the straight line at the angle $\psi = \sin^{-1}(1/M_m)$, measured from the negative real axis.

Step 3. By trial and error, find a circle whose center lies on the negative real axis and is simultaneously tangent to both the $G'_x(j\omega)$ plot and the line drawn at the angle ψ.

Step 4. Having found this circle, locate the point of tangency on the ψ-angle line. Draw a line from the point of the tangency perpendicular to the real axis. Label the point where this line intersects the real axis as a'.

Step 5. For this circle to be an M circle representing M_m, the point a' must be the $-1 + j0$ point. Thus, the x', y' coordinates must be multiplied by a gain factor K_m in order to convert this plot into a plot of $G(j\omega)$. From the graphical construction the value of K_m is $1/oa'$.

Step 6. The original gain must be changed by a factor $A = K_m/K_x$.

Note that if $G_x(j\omega)$ that includes a gain K_x is already plotted, it is possible to work directly with the plot of the function $G_x(j\omega)$. Following the procedure just outlined results in the determination of the *additional* gain required to produce the specified M_m; that is, the additional gain is

$$A = \frac{K_m}{K_x} = \frac{1}{oa''}$$

Example. It is desired that the closed-loop system that has the open-loop transfer function

$$G_x(j\omega) = \frac{1.47}{j\omega(1 + j0.25\omega)(1 + j0.1\omega)}$$

have an $M_m = 1.3$. The problem is to determine the actual gain K_1 needed and the amount by which the original gain K_x must be changed to obtain this M_m. The procedure previously outlined is applied to this problem. $G'_x(j\omega)$ is plotted in Fig. 9.18 and results in $\omega_m = 2.55$ and

$$K_1 = \frac{1}{oa'} \approx \frac{1}{0.34} = 2.94\text{s}^{-1}$$

The additional gain required is

$$A = \frac{K_1}{K_x} = \frac{2.94}{1.47} = 2.0$$

In other words, the original gain must be doubled to obtain M_m of 1.3 for $\mathbf{C}(j\omega)/\mathbf{R}(j\omega)$.

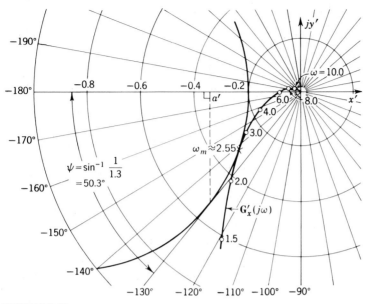

FIGURE 9.18
Gain adjustment using the procedure outlined in Sec. 9.7 for
$G'_x(j\omega) = 1/j\omega(1 + j0.25\omega)(1 + j0.1\omega)$.

9.8 CONSTANT M AND α CURVES ON THE LOG MAGNITUDE–ANGLE DIAGRAM (NICHOLS CHART)[5]

As derived earlier in this chapter, the constant M curves on the direct and inverse polar plots are circles. The transformation of these curves to the log magnitude (Lm)–angle diagram is done more easily by starting from the inverse polar plot since all the M circles have the same center. This requires a change of sign of the Lm and angle obtained, since the transformation is from the inverse transfer function on the polar plot to the direct transfer function on the Lm plot.

A constant M circle is shown in Fig. 9.19 on the inverse polar plot. The magnitude ρ and angle λ drawn to any point on this circle are shown.

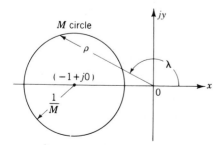

FIGURE 9.19
Constant M circle on the inverse polar plot.

The equation for this M circle is

$$y^2 + (1 + x)^2 = \frac{1}{M^2} \tag{9.40}$$

where $$x = \rho \cos \lambda \qquad y = \rho \sin \lambda \tag{9.41}$$

Combining these equations produces

$$\rho^2 M^2 + 2\rho M^2 \cos \lambda + M^2 - 1 = 0 \tag{9.42}$$

Solving for ρ and λ yields

$$\rho = -\cos \lambda \pm \left(\cos^2 \lambda - \frac{M^2 - 1}{M^2} \right)^{1/2} \tag{9.43}$$

$$\lambda = \cos^{-1} \frac{1 - M^2 - \rho^2 M^2}{2\rho M^2} \tag{9.44}$$

These equations are derived from the inverse polar plot $1/\mathbf{G}(j\omega)$. Since the Lm–angle diagram is drawn for the transfer function $\mathbf{G}(j\omega)$ and not for its reciprocal, a change in the equations must be made by substituting

$$r = \frac{1}{\rho} \qquad \phi = -\lambda$$

Since Lm $r = -$Lm ρ, it is only necessary to change the sign of Lm ρ. For any value of M a series of values of angle λ can be inserted in Eq. (9.43) to solve for ρ. This magnitude must be changed to decibels. Alternatively, for any value of M a series of values of ρ can be inserted in Eq. (9.44) to solve for the corresponding angle λ. Therefore, the constant M curve on the Lm–angle diagram can be plotted by using either of these two equations.

In a similar fashion the constant α curves can be drawn on the Lm–angle diagram. The α curves on the inverse polar plot are semi-infinite straight lines terminating on the $-1 + j0$ point and are given by

$$\tan \alpha + x \tan \alpha + y = 0 \tag{9.45}$$

Combining Eq. (9.45) with Eq. (9.41) produces

$$\tan \alpha + \rho \cos \lambda \tan \alpha + \rho \sin \lambda = 0 \tag{9.46}$$

$$\tan \alpha = \frac{-\rho \sin \lambda}{1 + \rho \cos \lambda} = \frac{-y}{1 + x} \tag{9.47}$$

For constant values of α, a series of values of λ can be inserted in this equation to solve for ρ. The constant α semi-infinite curves can then be plotted on the Lm–angle diagram.

Constant M and α curves, as shown in Fig. 9.20, repeat for every 360°. Also, there is symmetry at every 180° interval. An expanded 300° section of the constant M and α graph is shown in Fig. 9.23. This graph is commonly referred to as the Nichols chart. Note that the $M = 1$ (0dB) curve is asymptotic to $\phi = -90°$ and

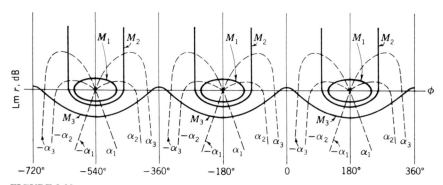

FIGURE 9.20
Constant M and α curves on the Nichols chart.

$-270°$, and the curves for $M < \frac{1}{2}$ (-6 dB) are always negative. $M = \infty$ is the point at 0 dB, $-180°$, and the curves for $M > 1$ are closed curves inside the limits $\phi = -90°$ and $\phi = -270°$. The Nichols chart should be viewed as two charts. The first has the Cartesian coordinates of dB vs. phase angle ϕ that are used to plot the open-loop frequency response, with the frequencies noted along the plot. The second is the chart for loci of constant M and α for the closed-loop transfer function that is superimposed on the open-loop frequency-response plot. These loci for constant M and α for the Nichols chart are applied only for stable unity-feedback systems. A macro, utilizing Eqs. (9.44) and (9.47), can be written to generate a computer printout of the constant M and α contours on the Nichols chart.

9.9 GENERATION OF MATLAB (1992 STUDENT VERSION) BODE AND NYQUIST PLOTS

The m-file is set up to obtain the Bode and Nyquist plots for the transfer function

$$G(s) = \frac{K}{s(s + 1)(s + 5)}$$

In using this m-file, it is important to pay attention to the selection of the appropriate scales for these plots. The smallest scale range should be selected, based upon the 0-dB crossing of the Lm plot and the value of the gain-margin frequency. For the Nyquist plots it is important to select the scale so that the horizontal axis ($-180°$ and/or $180°$) crossing points are clearly shown.

```
diary Example.out
format short e
%
%Setup transfer function (open-loop poles and gain)
p = [0 -1 -5];
num = 5;
den = poly(p);
```

```
%Look at help
%
help Bode
%
%Use Bode command to obtain plots
%
%In general, select a dB scale of +20 to -40 dB, a minimum
%angle scale, and a log w scale that includes the phase and gain
%margin frequencies
%
%Get Bode plot of interest, specify range for w
w=logspace(-1,2,200);
[mag,phase] = bode(num,den,w);
%Obtain the Bode plot
subplot(2,1,1)
semilogx(w,20*log10(mag))
%Set the axis limit
axis([0.1,100,-40,20]);
hold off
%Set the Plot labels and hold the plot on screen
title('Example'),xlabel('omega (rad/sec)'),ylabel('dB'),grid
subplot(2,1,2)
semilogx(w,phase)
%set axis
axis([0.1,100,-270,0]);
%Plot a line at phase = -180 deg
hold on
semilogx([0.1 100], [-180 -180]);
hold off
title('Example'),xlabel('omega (rad/sec)'),ylabel('Degrees'),grid
prtsc
pause
%
%Use Nyquist command to obtain plot
%
%Select the value of w = logspace( , , ) based on the
%Bode plots and the comments for the Bode plot scale
%
clg
w = logspace(log10(0.5),log10(20),200);
nyquist(num,den,w);
hold on
num = 50;
nyquist(num,dem,w);
hold off
axis([-2 0 -0.8 0.8]);
prtsc
```

The Bode plots for $K = 5$ and the Nyquist plots for $K = 5$ and 50 are shown in Figs. 9.21 and 9.22, respectively.

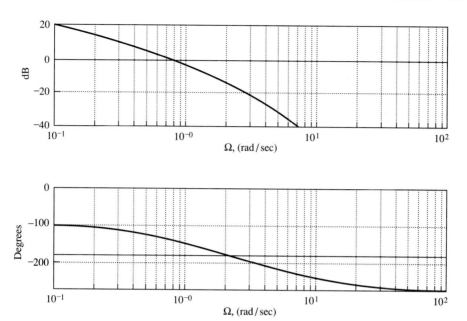

FIGURE 9.21
MATLAB Bode plots.

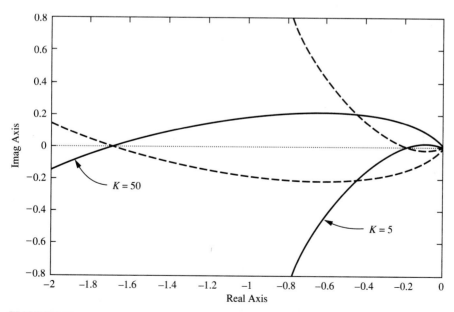

FIGURE 9.22
MATLAB Nyquist plots.

9.10 ADJUSTMENT OF GAIN BY USE OF THE LOG MAGNITUDE–ANGLE DIAGRAM

The Lm–angle diagram for

$$\mathbf{G}_x(j\omega) = \frac{2.04(1 + j2\omega/3)}{j\omega(1 + j\omega)(1 + j0.2\omega)(1 + j0.2\omega/3)} \tag{9.48}$$

is drawn as the solid curve in Fig. 9.23 on graph paper that has the constant M and α contours. (The Lm and phase-angle diagram can be drawn first to get the data for this curve.) It is convenient to use the same Lm and angle scales on both the Lm and phase diagram and the Lm–angle diagram. A pair of dividers can then be used to move the transfer function locus to the Lm–angle diagram. The $M = 1.12$ (1-dB) curve is tangent to the curve at $\omega_{m1} = 1.1$. These values are the maximum value of the control ratio M_m and the resonant frequency ω_m with the gain $K_x = 2.04$ given in Eq. (9.48).

The designer specifies that the closed-loop system be adjusted to produce an $M_m = 1.26$ (2 dB) by changing the gain. The dashed curve in Fig. 9.23 is obtained by raising the transfer function curve until it is tangent to the $M_m = 1.26$ (2-dB) curve. The resonant frequency is now equal to $\omega_{m2} = 2.09$. The curve has been raised by the amount Lm $A = 4.5$ dB, meaning that an additional gain of $A = 1.679$ must be added in cascade with $\mathbf{G}_x(j\omega)$. The gain is now equal to $K_1 = (1.679)(2.04) = 3.43$.

A simpler procedure is to make a template of the desired M_m curve that can be moved up or down until it is tangent to the transfer function curve. The amount the template is moved represents the gain change required. Once a graphical estimate of the gain is obtained by use of the template, a CAD program[2] can be used to refine the design. When the gain has been adjusted for the desired M_m, the closed-loop frequency response can be found either graphically or by use of the digital computer. Graph paper with the constant M and α curves superimposed simplifies system analysis and design. For any frequency point on the transfer function curve the values of M and α can be read from the graph. For example, with M_m made equal to 1.26 in Fig. 9.23, the value of M at $\omega = 3.4$ is -0.5 dB and α is $-110°$. The closed-loop frequency response, both Lm and angle, obtained from Fig. 9.23 is also plotted in Fig. 9.24 for both values of gain, $K_1 = 2.04$ and 3.43.

If the closed-loop response with the resultant gain adjustment has too low a resonant frequency for the desired performance, the next design step is to compensate the system to improve the frequency response. Compensation is studied in detail in Chapters 10, 11, and 12.

9.11 CORRELATION OF POLE-ZERO DIAGRAM WITH FREQUENCY AND TIME RESPONSES

Whenever the closed-loop control ratio $\mathbf{M}(j\omega)$ has the characteristic form shown in Fig. 9.5, the system may be approximated as a simple second-order system. This

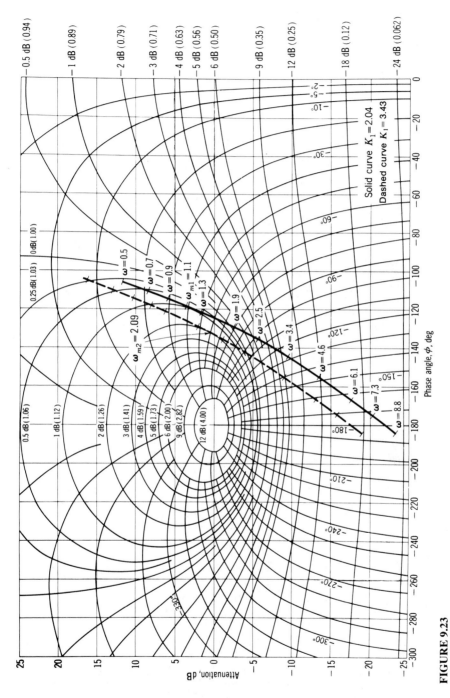

FIGURE 9.23
Log magnitude–angle diagram for Eq. (9.48).

313

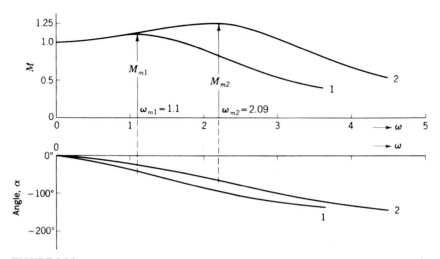

FIGURE 9.24
Control ratio vs. frequency obtained from the log magnitude–angle diagram of Fig. 9.23.

usually implies that the poles, other than the dominant complex pair, are either far to the left of the dominant complex poles or are close to zeros. When these conditions are not satisfied, the frequency response may have other shapes. This can be illustrated by considering the following three control ratios:

$$\frac{C(s)}{R(s)} = \frac{1}{s^2 + s + 1} \tag{9.49}$$

$$\frac{C(s)}{R(s)} = \frac{0.313(s + 0.8)}{(s + 0.25)(s^2 + 0.3s + 1)} \tag{9.50}$$

$$\frac{C(s)}{R(s)} = \frac{4}{(s^2 + s + 1)(s^2 + 0.4s + 4)} \tag{9.51}$$

The pole-zero diagram, the frequency response, and the time response to a step input for each of these equations are shown in Fig 9.25.

For Eq. (9.49) the following characteristics are noted from Fig. 9.25a:

1. The control ratio has only two complex dominant poles and no zeros.
2. The frequency-response curve has the following characteristics:
   (a) A single peak $M_m = 1.157$ at $\omega_m = 0.7$.
   (b) $1 < M < M_m$ in the frequency range $0 < \omega < 1$.
3. The time response has the typical waveform described in Chapter 3 for a simple second-order system. That is, the first maximum of $c(t)$ due to the oscillatory term is greater than $c(t)_{ss}$, and the $c(t)$ response after this maximum oscillates around the value of $c(t)_{ss}$.

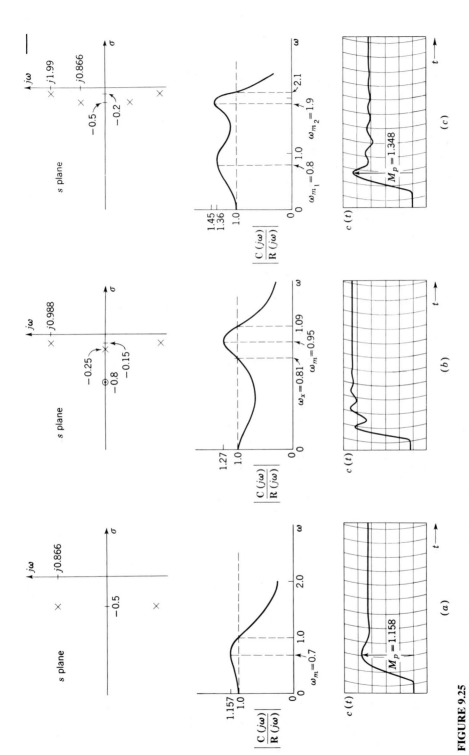

FIGURE 9.25
Comparison of frequency and time responses for three pole-zero patterns.

315

For Eq. (9.50) the following characteristics are noted from Fig. 9.25b:

1. The control ratio has two complex poles and one real pole that are all dominant, and one real zero.
2. The frequency-response curve has the following characteristics:
   (a) A single peak, $M_m = 1.27$ at $\omega_m = 0.95$.
   (b) $M < 1$ in the frequency range $0 < \omega < \omega_x$.
   (c) The peak M_m occurs at $\omega_m = 0.95 > \omega_x$.
3. The time response does not have the conventional waveform. That is, the first maximum of $c(t)$ due to the oscillatory term is less than $c(t)_{ss}$ because of the transient term $A_3 e^{-0.25t}$.

For Eq. (9.51) the following characteristics are noted from Fig. 9.25c:

1. The control ratio has four complex poles, all are dominant, and no zeros.
2. The frequency-response curve has the following characteristics:
   (a) There are two peaks, $M_{m1} = 1.36$ at $\omega_{m1} = 0.81$ and $M_{m2} = 1.45$ at $\omega_{m2} = 1.9$.
   (b) In the frequency range $0 < \omega < 2.1$, $1 < M < 1.45$.
   (c) The time response does not have the simple second-order waveform. That is, the first maximum of $c(t)$ in the oscillation is greater than $c(t)_{ss}$, and the oscillatory portion of $c(t)$ does not oscillate about a value of $c(t)_{ss}$. This time response can be predicted from the pole locations in the s plane and from the plot of M vs. ω (the two peaks).

Another example is the system represented by

$$M(s) = \frac{K}{(s - p_1)(s^2 + 2\zeta\omega_n s + \omega_n^2)} \tag{9.52}$$

When the real pole p_1 and real part of the complex poles are equal as shown in Fig. 9.26a, the corresponding frequency response is as shown in Fig. 9.26b. The magnitude at ω_m is less than unity. The corresponding time response to a step input is monotonic; i.e., there is no overshoot as shown in Fig. 9.26c. This time response may be considered as a critically damped response.

The examples discussed in this section show that the time-response waveform is closely related to the frequency response of the system. In other words, the time-response waveform of the system can be predicted from the shape of the frequency-response plot. This time-response correlation with the plot of $M = |C(j\omega)/R(j\omega)|$ vs. ω, when there is no distinct dominant complex pair of poles, is an important advantage of frequency-response analysis. Thus, as illustrated in this section, the frequency-response plot may be used as a guide in determining (or predicting) time-response characteristics.

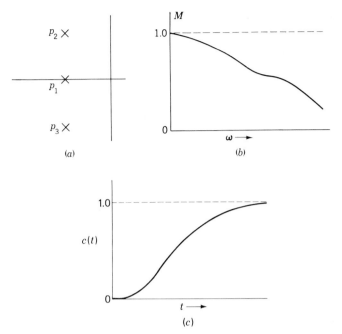

FIGURE 9.26
Form of the frequency and time responses for a particular pole pattern of
Eq. (9.52).

9.12   SUMMARY

In summary, this chapter is devoted to the correlation between the frequency and
time responses. The figures of merit M_m and ω_m, along with ω_0 and γ, are established as guideposts for evaluating a system's tracking performance. The addition of
a pole to an open-loop transfer function produces a cw shift of the direct polar plot,
which results in a larger value of M_m. The time response also suffers because ω_m
becomes smaller. The reverse is true if a zero is added to the open-loop transfer function. This agrees with the root locus analysis, which shows that the addition of a pole
or zero results in a less stable or more stable system, respectively. Thus, the qualitative correlation between the root locus and the frequency response is enhanced. The
M and α contours are developed as an aid in obtaining the closed-loop frequency response and in adjusting the gain to obtain a desired M_m. The methods described for
setting the gain for a desired M_m are based on the fact that generally the desired values of M_m are greater than 1 in order to obtain an underdamped response. When the
gain adjustment does not yield a satisfactory value of ω_m, the system must be compensated in order to increase ω_m without changing the value of M_m. Compensation
procedures are covered in the following chapters.

The procedure used in the frequency-response method is summarized as
follows:

Step 1. Derive the open-loop transfer function $G(s)H(s)$ of the system.

Step 2. Put the transfer function into the form $\mathbf{G}(j\omega)\mathbf{H}(j\omega)$.

Step 3. Arrange the various factors of the transfer function so that they are in the complex form $j\omega$, $1 + j\omega T$, and $1 + aj\omega + b(j\omega)^2$.

Step 4. Plot the Lm and phase-angle diagram for $\mathbf{G}(j\omega)\mathbf{H}(j\omega)$. The graphical methods of Chapter 8 may be used, but a CAD program[2] can provide more extensive data.

Step 5. Transfer the data from the plots in step 4 to either of the following: (*a*) Lm–angle diagram or (*b*) direct polar plot.

Step 6. Apply the Nyquist stability criterion and adjust the gain for the desired degree of stability M_m of the system. Then check the correlation to the time response for a step-input signal. This correlation reveals some qualitative information about the time response.

Step 7. If the response does not meet the desired specifications, determine the shape that the plot must have to meet these specifications. Synthesize the compensator that must be inserted into the system. The procedure is developed in the following chapters.

A digital computer is used extensively in system design. A standard CAD program[2] can be used to obtain the frequency response of both the open-loop and closed-loop transfer functions. The computer techniques are especially useful for systems containing frequency-sensitive feedback. The exact time response is then obtained to provide a correlation between the frequency and time responses. This chapter illustrates the use of the Nichols chart for analyzing the tracking performances of unity-feedback systems. The Nichols chart can also be used for analyzing the disturbance rejection performance of a nonunity-feedback system. This is discussed in detail in later chapters.

REFERENCES

1. Brown, G. S., and D. P. Campbell: *Principles of Servomechanisms*, Wiley, New York, 1948, chaps. 6 and 8.
2. A comprehensive list of CAD programs is given in App. B.
3. Higgins, T. J., and C. M. Siegel: "Determination of the Maximum Modulus, or the Specified Gain of a Servomechanism by Complex Variable Differentiation," *Trans. AIEE*, vol. 72, pt. II, pp. 467–469, January 1954.
4. Chu, Y.: "Correlation between Frequency and Transient Responses of Feedback Control Systems," *Trans. AIEE*, vol. 72, pt. II, pp. 81–92, May 1953.
5. James, H. M., N. B. Nichols, and R. S. Phillips: *Theory of Servomechanisms*, McGraw-Hill, New York, 1947, chap. 4.

CHAPTER
10

ROOT-LOCUS COMPENSATION: DESIGN

10.1 INTRODUCTION TO DESIGN

The preceding chapters deal with basic feedback control systems composed of the minimum equipment required to perform the control function and the necessary sensors and comparators to provide feedback. The designer first analyzes the performance of the basic system. This chapter presents methods of improving the design of a control system. To increase the reader's understanding of how poles and zeros affect the time response, a graphical means is presented for calculating the figures of merit of the system and the effect of additional significant nondominant poles. Typical pole-zero patterns are employed to demonstrate the correlation between the pole-zero diagram and the frequency and time responses. As a result of this analysis, the control system may be modified to achieve the desired time response. The next few chapters are devoted to achieving the necessary refinements.

Modifying a system to reshape its root locus in order to improve system performance is called *compensation* or stabilization. When the system is compensated, it is stable, has a satisfactory transient response, and has a large enough gain to ensure that the steady-state error does not exceed the specified maximum. Compensation devices may consist of electric networks or mechanical equipment containing levers, springs, dashpots, etc. The compensator (also called a filter) may be placed in cascade with the forward transfer function $G_x(s)$ (cascade or series compensation), as shown in Fig. 10.1a, or in the feedback path (feedback or parallel compensation), as shown in Fig. 10.1b and c. In Fig. 10.1b the compensation is introduced by means of a minor feedback loop. The selection of the location for inserting the compensator depends largely on the control system, the necessary physical modifications,

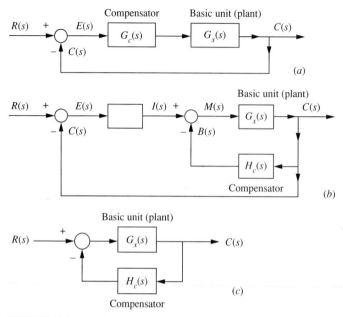

FIGURE 10.1
Block diagram showing the location of compensators: (*a*) cascade; (*b*) minor
loop feedback; and (*c*) output feedback.

and the results desired. The cascade compensator $G_c(s)$ is inserted at the low-energy
point in the forward path so that power dissipation is very small. This also requires
a high input impedance. Isolation amplifiers may be necessary to avoid loading of
or by the compensating network. The networks used for compensation are generally
called lag, lead, and lag-lead compensation. The examples used for each of these
compensators are not the only ones available but are intended primarily to show the
methods of applying compensation. Feedback compensation is used in later design
examples to improve the system's tracking of a desired input $r(t)$ or to improve the
system's rejection of a disturbance input $d(t)$.

The following factors must be considered when making a choice between cascade and feedback compensation:

1. The design procedures for a cascade compensator $G_c(s)$ are more direct than
   those for a feedback compensator $H_c(s)$. Although the design of feedback compensation is sometimes more laborious, it may be easier to implement.
2. The physical form of the control system, i.e., whether it is electrical, hydraulic,
   mechanical, etc., determines the ease of implementing a practical cascade or
   feedback compensator. A microprocessor may also be used to implement the
   required compensation.
3. The economics, size, weight, and cost of components and amplifiers are important. In the forward path the signal goes from a low to a high energy level,
   whereas the reverse is true in the feedback loop. Thus, generally an amplifier

may not be necessary in the feedback path. The cascade path generally requires an amplifier for gain and/or isolation. In applications such as aircraft and spacecraft, minimum size and weight of equipment are essential.

4. Environmental conditions in which the feedback control system is to operate affect the accuracy and stability of the controlled quantity. For example, this is a critical problem in an airplane, which is subjected to rapid changes in altitude and temperature.

5. The problem of noise within a control system may determine the choice of compensator. This is accentuated in situations in which a greater amplifier gain is required with a forward compensator than by the use of feedback networks.

6. The time of response desired for a control system is a determining factor. Often a faster time of response can be achieved by the use of feedback compensation.

7. Some systems require "tight-loop" stabilization to isolate the dynamics of one portion of a control system from other portions of the complete system. This can be accomplished by introducing an inner feedback loop around the portion to be isolated.

8. When $G_x(s)$ has a pair of dominant complex poles that yield the dominant poles of $C(s)/R(s)$, the simple first-order cascade compensators discussed in this chapter provide minimal improvement to the system's time-response characteristics. For this situation, as shown in Sec. 10.20, feedback compensation is more effective and is therefore more desirable.

9. The designer's experience and preferences influence the choice between a cascade and feedback compensator for achieving the desired performance.

10. The feedback compensator of Fig. 10.1c is used in Chapter 12 for disturbance rejection to minimize the response $c(t)$ to a disturbance input $d(t)$.

Meeting the system-performance specification often requires the inclusion of a properly designed compensator. The design process is normally performed by successively changing the compensator parameter(s) and comparing the resulting system performance. This process is expedited by the use of interactive CAD packages[14] like MATLAB, ICECAP-PC, or TOTAL-PC. The availability of a graphics capability in a CAD program permits the designer to see the changes in system performance in real time. An interactive capability allows the designer to try a number of compensator designs quickly in order to select a design that best meets the system-performance requirements. The use of an interactive CAD package is standard practice for control system designers.

10.2 TRANSIENT RESPONSE: DOMINANT COMPLEX POLES[2]

The root-locus plot permits the designer to select the best poles for the control ratio. The criteria for determining which poles are best must come from the specifications of system performance and from practical considerations. For example, the presence of nonlinear characteristics like backlash, dead zone, and Coulomb friction can

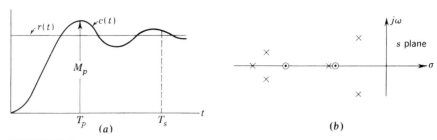

FIGURE 10.2
(a) Transient response to a step input with dominant complex poles. (b) Pole-zero pattern of
$C(s)/R(s)$ for the desired response.

produce a steady-state error with a step input, even though the linear portion of the
system is Type 1 or higher. The response obtained from the pole-zero locations of
the control ratio does not show the presence of a steady-state error caused by the
nonlinearities.[†] Frequently the system gain is adjusted so that there is a dominant
pair of complex poles, and the response to a step input has the underdamped form
shown in Fig. 10.2a. In this case the transient contributions from the other poles
must be small. From Eq. (4.68), the necessary conditions for the time response to be
dominated by one pair of complex poles require the pole-zero pattern of Fig. 10.2b
and have the following characteristics:

1. The other poles must be far to the left of the dominant poles so that the transients
   due to these other poles are small in amplitude and die out rapidly.
2. Any other pole that is not far to the left of the dominant complex poles must be
   near a zero so that the magnitude of the transient term due to that pole is small.

With the system designed so that the response to a unit step input has the underdamped form shown in Fig. 10.2a, the following transient figures of merit (described
in Secs. 3.9 and 3.10) are used to judge its performance:

1. M_p, peak overshoot, is the amplitude of the first overshoot.
2. T_p, peak time, is the time to reach the peak overshoot.
3. T_s, settling time, is the time for the response envelope first to reach and thereafter
   remain within 2 percent of the final value.
4. N is the number of oscillations in the response up to the settling time.

Consider the nonunity-feedback system shown in Fig. 10.3, using

$$G(s) = \frac{N_1}{D_1} = \frac{K_G \prod(s - z_k)}{\prod(s - p_g)} \tag{10.1}$$

[†] For a detailed study of the effect of nonlinear characteristics on system performance, the reader is
referred to the literature.[3,4]

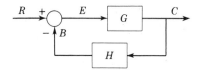

FIGURE 10.3
Feedback system.

$$H(s) = \frac{N_2}{D_2} = \frac{K_H \prod(s - z_j)}{\prod(s - p_i)} \tag{10.2}$$

$$G(s)H(s) = \frac{N_1 N_2}{D_1 D_2} = \frac{K_G K_H \prod_{h=1}^{w} (s - z_h)}{\prod_{c=1}^{n} (s - p_c)} \tag{10.3}$$

The product $K_G K_H = K$ is defined as the *static loop sensitivity*. For $G(s)H(s)$ the degree of the numerator is w and the degree of the denominator is n. These symbols are used throughout Chapter 6. The control ratio is

$$\frac{C(s)}{R(s)} = \frac{P(s)}{Q(s)} = \frac{N_1 D_2}{D_1 D_2 + N_1 N_2} = \frac{K_G \prod_{m=1}^{w'} (s - z_m)}{\prod_{k=1}^{n} (s - p_k)} \tag{10.4}$$

Note that the constant K_G in Eq. (10.4) is not the same as the *static loop sensitivity* unless the system has unity feedback. The degree n of the denominator of $C(s)/R(s)$ is the same as for $G(s)H(s)$, regardless of whether the system has unity or nonunity feedback. *The degree w' of the numerator of $C(s)/R(s)$ is equal to the sum of the degrees of N_1 and D_2.* For a unity-feedback system the degree of the numerator is $w' = w$. For a nonunity-feedback system the w' zeros of Eq. (10.4) include both the zeros of $G(s)$, i.e., N_1 and the poles of $H(s)$, i.e., D_2, as shown in Eq. (10.4).

For a unit step input the output of Fig. 10.3 is

$$C(s) = \frac{P(s)}{sQ(s)} = \frac{K_G \prod_{m=1}^{w'} (s - z_m)}{s \prod_{k=1}^{n} (s - p_k)}$$

$$= \frac{A_0}{s} + \frac{A_1}{s - p_1} + \cdots + \frac{A_k}{s - p_k} + \cdots + \frac{A_n}{s - p_n} \tag{10.5}$$

The pole $s = 0$ that comes from $R(s)$ must be included as a pole of $C(s)$, as shown in Eq. (10.5). The values of the coefficients can be obtained graphically from the pole-zero diagram by the method described in Sec. 4.11 and Eq. (4.68) or by the more practical use of a CAD program. Assume that the system represented by Eq. (10.5) has a dominant complex pole $p_1 = \sigma \pm j\omega_d$. The complete time solution is

$$c(t) = \frac{P(0)}{Q(0)} + 2 \left| \frac{K_G \prod_{m=1}^{w'} (p_1 - z_m)}{p_1 \prod_{k=2}^{n} (p_1 - p_k)} \right| e^{\sigma t} \cos\left[\omega_d t + \underline{/P(p_1)} - \underline{/p_1} - \underline{/Q'(p_1)}\right]$$

$$+ \sum_{k=3}^{n} \left[\frac{P(p_k)}{p_k Q'(p_k)}\right] e^{p_k t} \tag{10.6}$$

where
$$Q'(p_k) = \left[\frac{dQ(s)}{ds}\right]_{s=p_k} = \left[\frac{Q(s)}{s - p_k}\right]_{s=p_k} \tag{10.7}$$

By assuming the pole-zero pattern of Fig. 10.2, the last term of Eq. (10.6) may often be neglected. The time response is therefore approximated by

$$c(t) \approx \frac{P(0)}{Q(0)} + 2 \left| \frac{K_G \prod_{m=1}^{w'} (p_1 - z_m)}{p_1 \prod_{k=2}^{n} (p_1 - p_k)} \right| e^{\sigma t} \cos\left[\omega_d t + \underline{/P(p_1)} - \underline{/p_1} - \underline{/Q'(p_1)}\right]$$

$$\tag{10.8}$$

Note that, although the transient terms due to the other poles have been neglected, the effect of those poles on the amplitude and phase angle of the dominant transient has not been neglected.

The peak time T_p is obtained by setting the derivative with respect to time of Eq. (10.8) equal to zero. This gives

$$T_p = \frac{1}{\omega_d} \left[\frac{\pi}{2} - \underline{/P(p_1)} + \underline{/Q'(p_1)}\right]$$

$$= \frac{1}{\omega_d} \left\{\frac{\pi}{2} - [\text{sum of angles from zeros of } C(s)/R(s) \text{ to dominant pole } p_1]\right.$$

$$\left. + \left[\begin{array}{l}\text{sum of angles from all other poles of } C(s)/R(s) \text{ to} \\ \text{dominant pole } p_1, \text{ including conjugate pole}\end{array}\right]\right\} \tag{10.9}$$

Since the phase angle $\phi = \underline{/P(p_1)} - \underline{/p_1} - \underline{/Q'(p_1)}$ in Eq. (10.8) cannot have a value greater than 2π, the value T_p must occur within one "cycle" of the transient. Inserting this value of T_p into Eq. (10.8) gives the peak overshoot M_p. By using the value $\cos\left(\pi/2 - \underline{/p_1}\right) = \omega_d/\omega_n$, the value M_p can be expressed as

$$M_p = \frac{P(0)}{Q(0)} + \frac{2\omega_d}{\omega_n^2} \left| \frac{K_G \prod_{m=1}^{w'} (p_1 - z_m)}{\prod_{k=2}^{n} (p_1 - p_k)} \right| e^{\sigma T_p} \tag{10.10}$$

The first term in Eq. (10.10) represents the final value, and the second term represents the overshoot M_o. For a unity-feedback system that is Type 1 or higher, there is zero steady-state error when the input is a step function. Therefore, the first term on the right side of Eq. (10.8) is equal to unity. Its magnitude can also be obtained by applying the final-value theorem to $C(s)$ given by Eq. (10.5). This equality is used to solve for K_G. Note that under these conditions $K_G = K$ and $w' = w$:

$$K_G = \frac{\displaystyle\prod_{k=1}^{n}(-p_k)}{\displaystyle\prod_{m=1}^{w}(-z_m)} \tag{10.11}$$

The value of the overshoot M_o can therefore be expressed as

$$M_o = \frac{2\omega_d}{\omega_n^2}\left|\frac{\displaystyle\prod_{k=1}^{n}(-p_k)\displaystyle\prod_{m=1}^{w}(p_1-z_m)}{\displaystyle\prod_{m=1}^{w}(-z_m)\displaystyle\prod_{k=2}^{n}(p_1-p_k)}\right|e^{\sigma T_p} \tag{10.12}$$

Some terms inside the brackets in Eq. (10.12) cancel the terms in front so that M_o can be expressed in words as

$$M_o = \frac{\begin{bmatrix}\text{product of distances}\\ \text{from all poles of}\\ C(s)/R(s)\text{ to origin,}\\ \text{excluding distances of two}\\ \text{dominant poles from origin}\\ \hline \text{product of distances}\\ \text{from all other poles of}\\ C(s)/R(s)\text{ to dominant}\\ \text{pole }p_1\text{, excluding distance}\\ \text{between dominant poles}\end{bmatrix}\begin{bmatrix}\text{product of distances}\\ \text{from all zeros of}\\ C(s)/R(s)\text{ to dominant}\\ \text{pole }p_1\\ \hline \text{product of distances}\\ \text{from all zeros of}\\ C(s)/R(s)\text{ to origin}\end{bmatrix}e^{\sigma T_p} \tag{10.13}$$

The denominator terms of Eq. (10.13) clearly reveal that as a zero of Eq. (10.4) is *located closer to the imaginary axis, the magnitude of M_o increases.* If there are no finite zeros of $C(s)/R(s)$, the factors in M_o involving zeros become unity. Equation (10.13) is valid only for a unity-feedback system that is Type 1 or higher. It may be sufficiently accurate for a Type 0 system that has a large value for K_0. The value of M_o can be calculated either from the right-hand term of Eq. (10.10) or from Eq. (10.13), whichever is more convenient. The effect on M_o of other poles, which cannot be neglected, is discussed in the next section.

The values of T_s and N can be approximated from the dominant roots. Section 3.9 shows that T_s is four time constants for 2 percent error:

$$T_s = \frac{4}{|\sigma|} = \frac{4}{\zeta\omega_n} \tag{10.14}$$

$$N = \frac{\text{Settling time}}{\text{Period}} = \frac{T_s}{2\pi/\omega_d} = \frac{2\omega_d}{\pi|\sigma|} = \frac{2}{\pi} \frac{\sqrt{1-\zeta^2}}{\zeta} \tag{10.15}$$

Equation (10.9) shows that zeros of the control ratio cause a decrease in the peak time T_p, whereas poles increase the peak time. Peak time can also be decreased by shifting zeros to the right or poles (other than the dominant poles) to the left. Equation (10.13) shows that the larger the value of T_p, the smaller the value of M_o because $e^{\sigma T_p}$ decreases. M_o can also be decreased by reducing the ratios $p_k/(p_k - p_1)$ and $(z_m - p_1)/z_m$. But this reduction can have an adverse effect on T_p. The conditions on pole-zero locations that lead to a small M_o may therefore produce a large T_p. Conversely, the conditions that lead to a small T_p may be obtained at the expense of a large M_o. Thus, a compromise is required in the peak time and peak overshoot that are attainable. Some improvements can be obtained by introducing additional poles and zeros into the system and locating them appropriately. This topic is covered in the following sections on compensation.

The approximate equations, given in this section, for the response of a system to a step-function input are based on the fundamental premise that there is a pair of dominant complex poles and that the effect of other poles is small. When this premise is satisfied, a higher-order system, $n > 2$, effectively acts like a simple second-order system. These approximate equations are presented primarily to enhance the reader's understanding of how poles and zeros affect the time response. A CAD program[14] can be used to calculate and plot $c(t)$ and to yield precise values of M_p, t_p, and t_s.

10.3   ADDITIONAL SIGNIFICANT POLES[5]

When there are two dominant complex poles, the approximations developed in Sec. 10.2 give accurate results. However, there are cases where an additional pole of $C(s)/R(s)$ is significant. Figure 10.4a shows a pole-zero diagram which contains dominant complex poles and an additional real pole p_3. The control ratio, with $K = -\omega_n^2 p_3$, is given by

$$\frac{C(s)}{R(s)} = \frac{K}{(s^2 + 2\zeta\omega_n s + \omega_n^2)(s - p_3)} \tag{10.16}$$

With a unit-step input the time response is

$$c(t) = 1 + 2|A_1|e^{-\zeta\omega_n t} \sin\left(\omega_n \sqrt{1-\zeta^2}\,t + \phi\right) + A_3 e^{p_3 t} \tag{10.17}$$

The transient term due to the real pole p_3 has the form $A_3 e^{p_3 t}$, where A_3, based upon Eq. (4.68), is *always negative*. Thus, the overshoot M_p is reduced, and settling time t_s may be increased or decreased. This effect is typical of an additional real pole. The magnitude A_3 depends on the location of p_3 relative to the complex poles. The further to the left the pole p_3 is located, the smaller the magnitude of A_3, therefore the smaller its effect on the total response. A pole that is six times as far to the left as the complex poles has negligible effect on the time response. The typical time

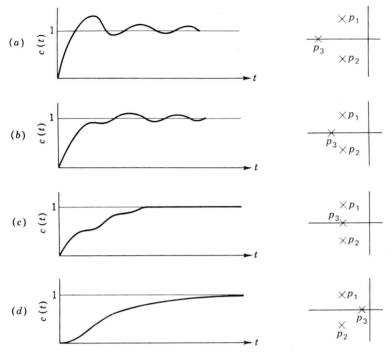

FIGURE 10.4
Typical time response as a function of the real-pole location.

response is shown in Fig. 10.4a. As the pole p_3 moves to the right, the magnitude of A_3 increases and the overshoot becomes smaller. As p_3 approaches but is still to the left of the complex poles, the first maximum in the time response is less than the final value. The largest overshoot can occur at the second or a later maximum as shown in Fig. 10.4b (see also Fig. 9.25b).

When p_3 is located at the real-axis projection of the complex poles, the response is monotonic; i.e., there is no overshoot. This represents the critically damped situation, as shown in Fig. 10.4c. The complex poles may contribute a "ripple" to the time response, as shown in the figure. When p_3 is located to the right of the complex poles, it is the dominant pole and the response is over-damped.

When the real pole p_3 is to the left of the complex poles, the peak time T_p is approximately given by Eq. (10.9). Although the effect of the *real* pole is to increase the peak time, this change is small if the real pole is fairly far to the left. A first-order correction can be made to the peak overshoot M_o given by Eq. (10.13) by adding the value of $A_3 e^{p_3 T_p}$ to the peak overshoot due to the complex poles. The effect of the real pole on the actual settling time t_s can be estimated by calculating $A_3 \exp(p_3 T_s)$ and comparing it with the size and sign of the underdamped transient at time T_s obtained from Eq. (10.14). If both are negative at T_s, then the true settling time t_s is increased. If they have opposite signs, then the value $T_s = 4T$ based on the complex roots is a good approximation for t_s.

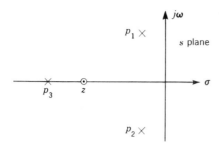

FIGURE 10.5
Pole-zero diagram of $C(s)/R(s)$ for Eq. (10.18).

The presence of a real zero in addition to the real pole further modifies the transient response. A possible pole-zero diagram for which the control ratio is

$$\frac{C(s)}{R(s)} = \frac{K(s - z)}{(s^2 + 2\zeta\omega_n s + \omega_n^2)(s - p_3)} \tag{10.18}$$

is shown in Fig. 10.5. The complete time response to a unit step-function input still has the form given in Eq. (10.17). However, the sign of A_3, based upon Eq. (4.68), depends on the relative location of the real pole and the real zero. A_3 is negative if the zero is to the left of p_3, and it is positive if the zero is to the right of p_3. Also, the magnitude of A_3 is proportional to the distance from p_3 to z [see Eq. (4.68)]. Therefore, if the zero is close to the pole, A_3 is small and the contribution of this transient term is correspondingly small. Compared with the response for Eq. (10.16), as shown in Fig. 10.4, note the following:

1. If the zero z is to the left of the real pole p_3, the response is qualitatively the same as that for a system with only complex poles; but the peak overshoot is smaller.
2. If the zero z is to the right of the real pole p_3, the peak overshoot is greater than that for a system with only complex poles.

These characteristics are evident in Fig. 10.20.

When the real pole p_3 and the real zero z are close together, the value of peak time T_p obtained from Eq. (10.9) can be considered essentially correct. Actually T_p decreases if the zero is to the right of the real pole, and vice versa. A first-order correction to M_o from Eq. (10.13) can be obtained by adding the contribution of $A_3 e^{p_3 t}$ at the time T_p. The analysis of the effect on t_s is similar to that described previously for the case of just an additional real pole.

As a summary, many control systems having an underdamped response can be approximated by one having the following characteristics: (1) two complex poles; (2) two complex poles and one real pole; and (3) two complex poles, one real pole, and one real zero. For case 1 the relations T_p, M_o, T_s, and N developed in Sec. 10.2 give an accurate representation of the time response. For case 2 these approximate values can be corrected if the real pole is far enough to the left. Then the contribution of the additional transient term is small, and the total response remains essentially the sum of a constant and an underdamped sinusoid. For case 3 the approximate values can be corrected provided the zero is near the real pole so that the amplitude of the additional transient term is small. More exact calculation of the figures of merit can be obtained by plotting the exact response as a function of time by use of a CAD program.

10.4   ROOT-LOCUS DESIGN CONSIDERATIONS

The root-locus design technique presented in Chapter 7 stresses the desired transient figures of merit. Thus, the selection of the dominant complex root can be based on a desired damping ratio ζ_D (which determines the peak overshoot), the desired damped natural frequency ω_d (which is the frequency of oscillation of the transient), or the desired real part σ_D of the dominant root (which determines the settling time). Two additional factors must be included in the design. First, the presence of additional poles will modify the peak overshoot, as described in Sec. 10.3. Second, it may be necessary to specify the value of the gain K_m. For example, in a Type 1 system it may be necessary to specify a minimum value of K_1 in order to limit the magnitude of the steady-state error e_{ss} with a ramp input, where $e(t)_{ss} = Dr(t)/K_1$. The following example incorporates these factors into a root-locus design.

Example.   Consider a unity-feedback system where

$$G_x(s) = \frac{K_x}{s(s^2 + 4.2s + 14.4)}$$

$$= \frac{K_x}{s(s + 2.1 + j3.1607)(s + 2.1 - j3.1607)} \tag{10.19}$$

and $K_1 = K_x/14.4$. For this system it is specified that the following figures of merit must be satisfied: $1 < M_p \leq 1.123$, $t_s \leq 3$ s, $e(t)_{ss} = 0$, $t_p \leq 1.6$ s, $K_1 \geq 1.5$ s^{-1}.

First Design

With the assumption that the closed-loop system can be represented by an effective simple second-order model, Eqs. (3.60), (3.61), and (10.14) can be used to determine the required values of ζ_D, ω_d, and ω_n:

$$M_p = 1.123 = 1 + \exp\frac{-\zeta\pi}{\sqrt{1 - \zeta^2}} \qquad \rightarrow \zeta_D = 0.555$$

$$t_p = \frac{\pi}{\omega_d} \qquad \rightarrow \omega_d > 1.9635$$

$$T_s = \frac{4}{\zeta_D\omega_n} \qquad \rightarrow \omega_n > 2.4024$$

Using the values $\zeta_D = 0.555$, $\omega_n = 2.4024$, and $\omega_d = 1.9984$, the required dominant complex roots are $s_{1,2} = -1.3333 \pm j1.9984$. Figure 10.6 contains the root

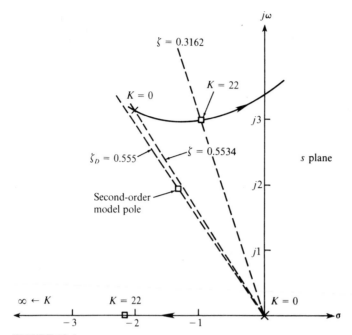

FIGURE 10.6
Root locus for Eq. (10.19).

locus for $G_x(s)$ and the line representing $\zeta_D = 0.555$. It is evident that the required dominant roots are not achievable.

Second Design

The first design does not consider the fact that there is a third root s_3. The associated transient $A_3 e^{s_3 t}$ is negative (see Sec. 10.3), and it will reduce the peak overshoot of the underdamped transient associated with the complex roots $s_{1,2}$. Therefore, it is possible to select a smaller value of ζ for the complex roots, and the effect of the third root can be used to keep M_p within the specified limits. In order simultaneously to satisfy the specification for $K_1 > 1.5$, the design is therefore based on obtaining the complex roots $s_{1,2} = -1 \pm j3$ on the root locus. The corresponding sensitivity is $K = 14.4K_1 = 22$, which yields $K_1 = 1.528$. The closed-loop transfer function is

$$\frac{C(s)}{R(s)} = \frac{22}{(s + 1 + j3)(s + 1 - j3)(s + 2.2)} \tag{10.20}$$

The plot of $|C(j\omega)/R(j\omega)|$ vs. ω very closely approximates the frequency response for a simple second-order underdamped system shown in Fig. 9.25a. A unit-step input results in the following values: $M_p \approx 1.123$, $t_p \approx 1.51$ s, and $t_s \approx 2.95$ s. Therefore, the desired specifications have been achieved.

10.5   RESHAPING THE ROOT LOCUS

The root-locus plots described in Chapter 7 show the relationship between the gain of the system and the time response. Depending on the specifications established for the system, the gain that best achieves the desired performance is selected. The performance specifications may be based on the desired damping ratio, undamped natural frequency, time constant, or steady-state error. The root locus may show that the desired performance cannot be achieved just by adjustment of the gain. In fact, in some cases the system may be unstable for all values of gain. The control-systems engineer must then investigate the methods for reshaping the root locus to meet the performance specifications.

The purpose of reshaping the root locus generally falls into one of the following categories:

1. A given system is stable and its transient response is satisfactory, but its steady-state error is too large. Thus, the gain must be increased to reduce the steady-state error (see Chapter 6). This increase must be accomplished without appreciably reducing the system stability.
2. A given system is stable, but its transient response is unsatisfactory. Thus, the root locus must be reshaped so that it is moved farther to the left, away from the imaginary axis.
3. A given system is stable, but both its transient response and its steady-state response are unsatisfactory. Thus, the locus must be moved to the left and the gain must be increased.
4. A given system is unstable for all values of gain. Thus, the root locus must be reshaped so that part of each branch falls in the left-half s plane, thereby making the system stable.

Compensation of a system by the introduction of poles and zeros is used to improve the operating performance. However, each additional compensator pole increases the number of roots of the closed-loop characteristic equation. If an underdamped response of the form shown in Fig. 10.2a is desired, the system gain must be adjusted so that there is a pair of dominant complex poles. This requires that any other pole be far to the left or near a zero so that its transient response has a small amplitude and therefore has a small effect on the total time response. The required pole-zero diagram is shown in Fig. 10.2b. The approximate values of peak overshoot M_o, peak time T_p, settling time T_s, and the number of oscillations up to settling time N can be obtained from the pole-zero pattern as described in Sec. 10.2. The effect of compensator poles and zeros on these quantities can be evaluated rapidly by use of a CAD program[14].

10.6   CAD ACCURACY CHECKS (CADAC)

The CADAC given in Secs. 3.11, 4.5, 6.8, 7.9, and 8.9 should be used, when applicable, in doing the problems associated with this chapter and the remaining chapters

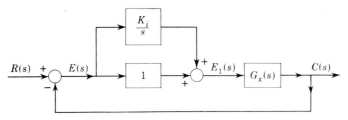

FIGURE 10.7
Ideal proportional plus integral (PI) control.

of this text. Each of these CADAC should be invoked, when applicable, at the appropriate point during the utilization of a CAD program.

10.7 IDEAL INTEGRAL CASCADE COMPENSATION (PI CONTROLLER)

When the transient response of a feedback control system is considered satisfactory but the steady-state error is too large, it is possible to eliminate the error by increasing the system type. This change must be accomplished *without appreciably changing the dominant roots of the characteristic equation.* The system type can be increased by operating on the actuating signal e to produce one that is proportional to both the magnitude and the integral of this signal. This proportional plus integral (PI) controller is shown in Fig. 10.7 where

$$E_1(s) = \left(1 + \frac{K_i}{s}\right) E(s) \tag{10.21}$$

$$G_c(s) = \frac{E_1(s)}{E(s)} = \frac{s + K_i}{s} \tag{10.22}$$

In this system the quantity $e_1(t)$ continues to increase as long as an error $e(t)$ is present. Eventually $e_1(t)$ becomes large enough to produce an output $c(t)$ equal to the input $r(t)$. The error $e(t)$ is then equal to zero. The constant K_i (generally very small) and the overall gain of the system must be selected to produce satisfactory roots of the characteristic equation. Since the system type has been increased, the corresponding error coefficient is equal to infinity. Provided that the new roots of the characteristic equation can be satisfactorily located, the transient response is still acceptable.

The locations of the pole and zero of this ideal proportional plus integral compensator are shown in Fig. 10.8. The pole alone would move the root locus to the

FIGURE 10.8
Location of the pole and zero of an ideal proportional plus integral compensator.

right, thereby slowing the time response. The zero must be near the origin in order to minimize the increase in response time of the complete system. The required integrator can be achieved by an electronic circuit. Mechanically the integral signal can be obtained by an integrating gyroscope, which is used in vehicles such as aircraft, space vehicles, ships, and submarines where the improved performance justifies the cost. Frequently, however, a passive electric network consisting of resistors and capacitors sufficiently approximates the proportional plus integral action, as illustrated in the next section.

10.8   CASCADE LAG COMPENSATION DESIGN USING PASSIVE ELEMENTS

Figure 10.9*a* shows a network that approximates a proportional plus integral output and is used as a lag compensator. Putting an amplifier of gain A in series with this network yields the transfer function

$$G_c(s) = A\frac{1 + Ts}{1 + \alpha Ts} = \frac{A\,(s + 1/T)}{\alpha\,(s + 1/\alpha T)} = \frac{A\,(s - z_c)}{\alpha\,(s - p_c)} \qquad (10.23)$$

where $\alpha = (R_1 + R_2)/R_2 > 1$ and $T = R_2C$. The pole $p_c = -1/\alpha T$ is therefore to the right of the zero $z_c = -1/T$, as shown in Fig. 10.9*b*. The locations of the pole p_c and the zero z_c of $G_c(s)$ on the s plane can be made close to those of the ideal compensator. Besides furnishing the necessary gain, the amplifier also acts as an isolating unit to prevent any loading effects between the compensator and the original system.

Assume that the original forward transfer function is

$$G_x(s) = \frac{K \prod_{h=1}^{w} (s - z_h)}{\prod_{g=1}^{n} (s - p_g)} \qquad (10.24)$$

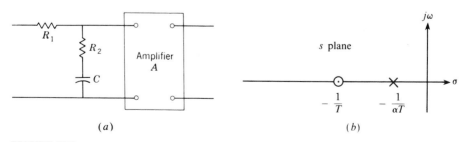

(*a*)

(*b*)

FIGURE 10.9
(*a*) Integral or lag compensator; (*b*) pole and zero location.

When the closed-loop root s_1 is selected on the root locus for the original system, the loop sensitivity K is

$$K = \frac{\prod\limits_{g=1}^{n} |s - p_g|}{\prod\limits_{h=1}^{w} |s - z_h|} \qquad (10.25)$$

With the addition of the lag compensator in cascade, the new forward transfer function is

$$G(s) = G_c(s)G_x(s) = \frac{AK}{\alpha} \frac{(s + 1/T)}{(s + 1/\alpha T)} \frac{\prod\limits_{h=1}^{w} (s - z_h)}{\prod\limits_{g=1}^{n} (s - p_g)} \qquad (10.26)$$

When the desired roots of the characteristic equation are located on the root locus, the magnitude of the loop sensitivity for the new root s' becomes

$$K' = \frac{AK}{\alpha} = \frac{|s' + 1/\alpha T|}{|s' + 1/T|} \frac{\prod\limits_{g=1}^{n} |s' - p_g|}{\prod\limits_{h=1}^{w} |s' - z_h|} \qquad (10.27)$$

As an example, a lag compensator is applied to a Type 0 system. To improve the steady-state accuracy, the error coefficient must be increased. The value of K_0 before and after the addition of the compensator is calculated by using the definition $K_0 = \lim_{s\to 0} G(s)$ from Eqs. (10.24) and (10.26), respectively,

$$K_0 = \frac{\prod\limits_{h=1}^{w} (-z_h)}{\prod\limits_{g=1}^{n} (-p_g)} K \qquad (10.28)$$

$$K_0' = \frac{\prod\limits_{h=1}^{w} (-z_h)}{\prod\limits_{g=1}^{n} (-p_g)} \alpha K' \qquad (10.29)$$

The following procedure is used to design the passive lag cascade compensator. First, the pole $p_c = -1/\alpha T$ and the zero $z_c = -1/T$ of the compensator are placed very close together. This step means that most of the original root locus remains practically unchanged. If the angle contributed by the compensator *at the original closed-loop dominant root* is less than 5°, the new locus is displaced only slightly. This 5° figure is only a guide and should not be applied arbitrarily. The new closed-loop dominant pole s' is therefore only slightly changed from the uncompensated value. This new value satisfies the restriction that the transient response must not change appreciably. As a result, the values $p'_c + 1/\alpha T$ and $z'_c + 1/T$ are almost equal, and the values K and K' in Eqs. (10.25) and (10.27) are approximately equal. The values K_0 and K'_0 in Eqs. (10.28) and (10.29) now differ only by the factor α so that $K'_0 \approx \alpha K_0$. The gain required to produce the new root s' therefore increases approximately by the factor α, which is the ratio of the compensator zero and pole. Summarizing, the necessary conditions on the compensator are that (1) the pole and zero must be close together and (2) the ratio α of the zero and pole must approximately equal the desired increase in gain. These requirements can be achieved by placing the compensator pole and zero very close to the origin. The size of α is limited by the physical parameters required in the network. A value $\alpha = 10$ is often used.

Although the preceding statements are based on a Type 0 system, the same conditions apply equally well for a Type 1 or higher system.

DESIGN EXAMPLE OF LAG COMPENSATION APPLIED TO A TYPE 1 SYSTEM.
A control system with unity feedback has a forward transfer function

$$G_x(s) = \frac{K_1}{s(1 + s)(1 + 0.2s)} = \frac{K}{s(s + 1)(s + 5)} \tag{10.30}$$

that yields the root locus shown in Fig. 10.10. For the basic control system a damping ratio $\zeta = 0.45$ yields the following pertinent data:

Dominant roots: $\qquad\qquad\qquad s_{1,2} = \quad -0.404 \pm j0.802$

Static loop sensitivity: $\qquad K = \quad |s_1| \cdot |s_1 + 1| \cdot |s_1 + 5|$

$$= \quad |-0.404 + j0.802| \cdot |0.596 + j0.802|$$

$$\cdot |4.596 + j0.802| = 4.188$$

Ramp error coefficient: $\qquad K_1 = \quad \lim_{s \to 0} sG(s) = \dfrac{K}{5} = \dfrac{4.188}{5} = 0.838 \text{ s}^{-1}$

Undamped natural frequency: $\quad \omega_n = \quad 0.898 \text{ rad/s}$

Third root: $\qquad\qquad\qquad\quad s_3 = \quad -5.192$

The values of peak time T_p, peak overshoot M_o, and settling time T_s obtained from Eqs. (10.9), (10.13), and (10.14) are $T_p = 4.12$ s, $M_o = 0.202$, and $T_s = 9.9$ s. These are approximate values, whereas the computer program gives more accurate values (see Table 10.1) including $t_s = 9.48$ s.

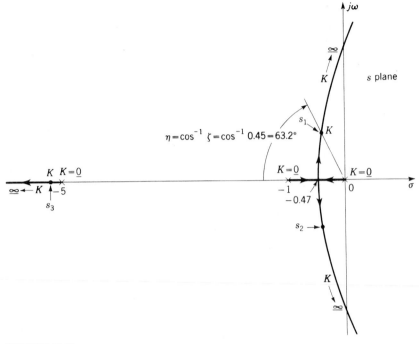

FIGURE 10.10
Root locus of $G_x(s) = K/s(s + 1)(s + 5)$

To increase the gain, a lag compensator is put in cascade with the forward transfer function. With the criteria discussed in the preceding section and $\alpha = 10$, the compensator pole is located at $p_c = -0.005$ and the zero at $z_c = -0.05$. This pole and zero are achieved for the network of Fig. 10.9 with the values $R_1 = 18$ MΩ, $R_2 = 2$ MΩ, and $C = 10$ μF. The angle of the compensator at the original dominant roots is about 2.6° and is acceptable. Since the compensator pole and zero are very close together, they make only a small change in the new root locus in the vicinity of the original roots. The compensator transfer function is

$$G_c(s) = \frac{A}{\alpha}\frac{s + 1/T}{s + 1/\alpha T} = \frac{A}{10}\frac{s + 0.05}{s + 0.005} \qquad (10.31)$$

Figure 10.11 (not to scale) shows the new root locus for $G(s) = G_x(s)G_c(s)$ as solid lines and the original locus as dashed lines. Note that for the damping ratio $\zeta = 0.45$ the new locus and the original locus are close together. The new dominant roots are $s_{1,2} = -0.384 \pm j0.763$; thus, the roots are essentially unchanged. Adjusting the gain of the compensated system to obtain roots with a damping ratio 0.45, the following results are obtained:

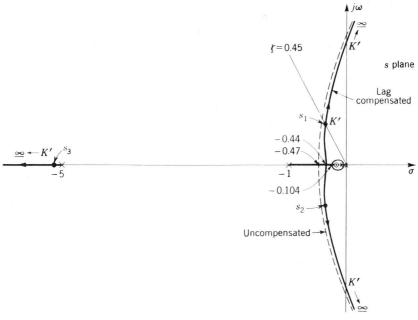

FIGURE 10.11
Root locus of

$$G(s) = K'(s + 0.05)/s(s + 1)(s + 5)(s + 0.005)$$

Dominant roots: $\quad\quad\quad\quad\quad\quad s_{1,2} = \quad -0.384 \pm j0.763$

Loop sensitivity: $\quad\quad\quad\quad\quad K' = \dfrac{|s| \cdot |s + 1| \cdot |s + 5| \cdot |s + 0.005|}{|s + 0.05|} = 4.01$

Ramp error coefficient: $\quad\quad K'_1 = \dfrac{K'\alpha}{5} = \dfrac{(4.01)(10)}{5} = 8.02 \text{ s}^{-1}$

Increase in gain: $\quad\quad\quad\quad A = \dfrac{K'_1}{K_1} = \dfrac{8.02}{0.838} = 9.57$

Undamped natural frequency: $\quad \omega_n = \quad 0.854 \text{ rad/s}$

Other roots: $\quad\quad\quad\quad\quad\quad s_3 = \quad -5.183 \quad \text{and} \quad s_4 = -0.053$

The value of T_p for the compensated system is determined by use of Eq. (10.9), with the values shown in Fig. 10.12:

$$T_p = \frac{1}{\omega_d}\left[\frac{\pi}{2} - \psi_1 + (\phi_2 + \phi_3 + \phi_4)\right]$$

$$= \frac{1}{0.763}[90° - 113.6° + (113.4° + 9.04° + 90°)]\frac{\pi \text{ rad}}{180°} = 4.32 \text{ s}$$

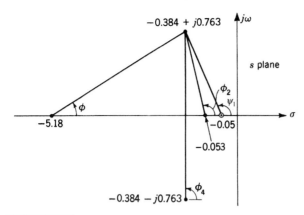

FIGURE 10.12
Angles used to evaluate T_p.

The value of M_o for the compensated system is determined by use of Eq. (10.13), with the values shown in Fig. 10.13:

$$M_o = \frac{l_1 l_2 \, (l)_1}{l_3 l_4 \, (l)_2} e^{\sigma T_p} = \frac{(5.18)(0.053)}{(4.76)(0.832)} \frac{0.834}{0.05} e^{(-0.384)(4.32)}$$

$$= 1.1563 e^{-1.659} = 0.22$$

The values obtained from a computer are $t_p = 4.30$ s and $M_o = 0.266$ and are listed in Table 10.1, Section 10.13, where a comparison is made of the performances achieved with several types of cascade compensators.

A comparison of the uncompensated system and the system with integral compensation shows that the ramp error coefficient K_1' has increased by a factor of 9.57 but the undamped natural frequency has been decreased slightly from 0.898 to 0.854 rad/s. That is, the steady-state error with a ramp input has decreased but the

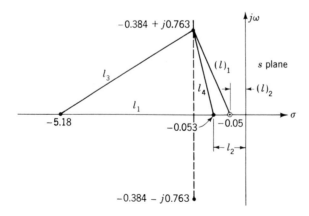

FIGURE 10.13
Lengths used to evaluate M_o.

settling time has been increased from 9.9 s to 22.0 s. Provided that the increased settling time and peak overshoot are acceptable, the system has been improved. To reduce this increase in settling time, it is necessary to include the constraint[13] that $|(s_4 - z_c)/s_4| \le -0.02$. The reader may demonstrate this by redesigning the lag compensator. Although the complete root locus is drawn in Fig. 10.11, in practice this is not necessary. Only the general shape is sketched and the particular needed points are accurately determined by using a CAD program.

10.9 IDEAL DERIVATIVE CASCADE COMPENSATION (PD CONTROLLER)

When the transient response of a feedback system must be improved, it is necessary to reshape the root locus so that it is moved farther to the left of the imaginary axis. Section 7.3 shows that introducing an additional zero in the forward transfer function produces this effect. A zero can be produced by operating on the actuating signal to produce a signal that is proportional to both the magnitude and the derivative (rate of change) of the actuating signal. This proportional plus derivative (PD) controller is shown in Fig. 10.14, where

$$E_1(s) = (1 + K_d s)E(s) \tag{10.32}$$

Physically the effect can be described as introducing anticipation into the system. The system reacts not only to the magnitude of the error but also to its rate of change. If the error is changing rapidly, then $e_1(t)$ is large and the system responds faster. The net result is to speed up the response of the system. On the root locus the introduction of a zero has the effect of shifting the curves to the left, as shown in the following example.

Example. The root locus for a Type 2 system that is unstable for all values of loop sensitivity K is shown in Fig. 10.15a. Adding a zero to the system by means of the PD compensator $G_c(s)$ has the effect of moving the locus to the left, as shown in Fig. 10.15b. The addition of a zero $s = -1/T_2$ between the origin and the pole at $-1/T_1$ stabilizes the system for all positive values of gain.

This example shows how the introduction of a zero in the forward transfer function by a proportional plus derivative compensator has modified and improved the feedback control system. However, the derivative action amplifies any spurious

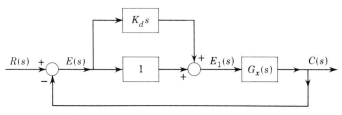

FIGURE 10.14
Ideal proportional plus derivative (PD) control.

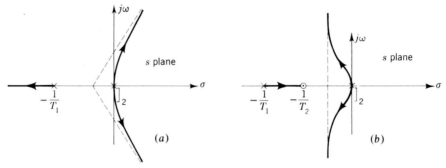

FIGURE 10.15
Root locus for

$$(a)\ G_x(s) = \frac{K}{s^2(s + 1/T_1)}; \quad (b)\ G(s) = \frac{K(s + 1/T_2)}{s^2(s + 1/T_1)}$$

signal or noise that may be present in the actuating signal. This noise amplification may saturate electronic amplifiers so that the system does not operate properly. A simple passive network is used to approximate proportional plus derivative action as shown in the next section.

10.10 LEAD COMPENSATION DESIGN USING PASSIVE ELEMENTS

Figure 10.16a shows a derivative or lead compensator made up of electrical elements. To this network an amplifier of gain A is added in cascade. The transfer function is

$$G_c(s) = A\alpha\frac{1 + Ts}{1 + \alpha Ts} = A\frac{s + 1/T}{s + 1/\alpha T} = A\frac{s - z_c}{s - p_c} \tag{10.33}$$

where $\alpha = R_2/(R_1 + R_2) < 1$ and $T = R_1C$. This lead compensator introduces a zero at $z_c = -1/T$ and a pole at $p_c = -1/\alpha T$. By making α sufficiently small, the location of the pole p_c is far to the left of the zero z_c, and therefore the pole has small effect on the dominant part of the root locus. Near the zero the net angle of the compensator is due predominantly to the zero. Figure 10.16b shows the location of the lead-compensator pole and zero and the angles contributed by each at a point s_1. It is also found that the gain of the compensated system is often increased. The maximum increases in gain and in the real part ($\zeta\omega_n$) of the dominant root of the characteristic equation do not coincide. The compensator zero location must be determined by making several trials with the aid of a CAD program for the desired optimum performance.

The loop sensitivity is proportional to the ratio of $|p_c/z_c|$. Therefore, as α decreases, the loop sensitivity increases. The minimum value of α is limited by the size of the parameters needed in the network to obtain the minimum input impedance

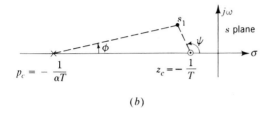

FIGURE 10.16
(a) Derivative or lead compensator. (b) Location of pole and zero of a lead compensator.

required. Note also from Eq. (10.33) that a small α requires a large value of additional gain A from the amplifier. The value $\alpha = 0.1$ is a common choice.

Design Example of Lead Compensation Applied to a Type 1 System

The same system used in Sec. 10.7 is now used with lead compensation. The locations of the pole and zero of the compensator are first selected by trial. At the conclusion of this section some rules are given to show the best location.

The forward transfer function of the original system is

$$G_x(s) = \frac{K_1}{s(1 + s)(1 + 0.2s)} = \frac{K}{s(s + 1)(s + 5)} \tag{10.34}$$

The transfer function of the compensator, using a value $\alpha = 0.1$, is

$$G_c(s) = 0.1A\frac{1 + Ts}{1 + 0.1Ts} = A\frac{s + 1/T}{s + 1/0.1T} \tag{10.35}$$

The new forward transfer function is

$$G(s) = G_c(s)G_x(s) = \frac{AK(s + 1/T)}{s(s + 1)(s + 5)(s + 1/0.1T)} \tag{10.36}$$

Three selections for the position of the zero $s = -1/T$ are made. A comparison of the results shows the relative merits of each. The three locations of the zero are $s = -0.75$, $s = -1.00$, and $s = -1.50$. Figures 10.17 to 10.19 show the resultant root loci for these three cases. In each figure the dashed lines are the locus of the original uncompensated system.

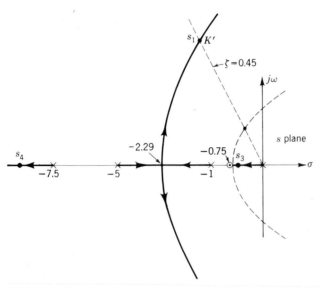

FIGURE 10.17
Root locus of

$$G(s) = \frac{AK(s + 0.75)}{s(s + 1)(s + 5)(s + 7.5)}$$

The loop sensitivity is evaluated in each case from the expression

$$K' = AK = \frac{|s| \cdot |s + 1| \cdot |s + 5| \cdot |s + 1/\alpha T|}{|s + 1/T|} \qquad (10.37)$$

and the ramp error coefficient is evaluated from the equation

$$K_1' = \lim_{s \to 0} sG(s) = \frac{K'\alpha}{5} \qquad (10.38)$$

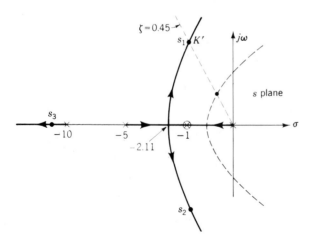

FIGURE 10.18
Root locus of

$$G(s) = \frac{AK(s + 1)}{s(s + 1)(s + 5)(s + 10)}$$

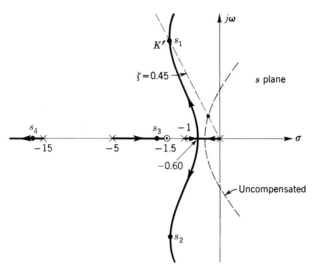

FIGURE 10.19
Root locus of

$$G(s) = \frac{AK(s + 1.5)}{s(s + 1)(s + 5)(s + 15)}$$

A summary of the results is given in Table 10.1, which shows that a large increase in ω_n has been achieved by adding the lead compensators. Since the value of ζ is held constant for all three cases, the system has a much faster response time. A comparison of the individual lead compensators shows the following.

Case A. Placing the zero of the compensator to the right of the plant pole $s = -1$ results in a root s_3 on the negative real axis close to the origin (see Fig. 10.17). From Eq. (4.68) it can be seen that the coefficient associated with this transient term is negative. Therefore, it has the desirable effect of decreasing the peak overshoot and the undesirable characteristic of having a longer settling time. If the size of this transient term is small (the magnitude is proportional to the distance from the real root to the zero), the effect on the settling time is small.

The decrease in M_o can be determined by calculating the contribution of the transient term $A_3 e^{s_3 t}$ due to the real root at time t_p, as discussed in Sec. 10.3. At the time $t = T_s$, the sign of the underdamped transient is negative and the transient due to the real root is also negative; therefore, the actual settling time is increased. The exact settling time obtained from a complete solution of the time response is $t_{sa} = 2.78$ s and is shown as plot a in Fig. 10.20.

Case B. The best settling time occurs when the zero of the compensator cancels the pole $s = -1$ of the original transfer function (see Fig. 10.18). There is no real root near the imaginary axis in this case. The actual settling time is $t_{sb} = 2.45$ s as shown in plot b in Fig. 10.20. The order of the characteristic equation is unchanged.

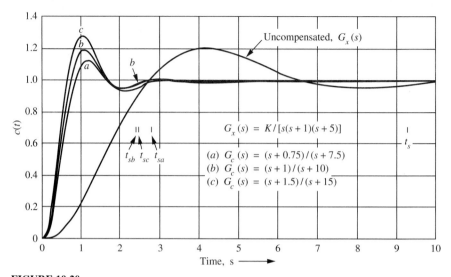

FIGURE 10.20
Time responses with a step input for the uncompensated and the lead-compensated cases of Table 10.1.

Case C. The largest gain occurs when the zero of the compensator is located at the left of the pole $s = -1$ of the original transfer function (see Fig. 10.19). Since the transient due to the root $s_3 = -1.76$ is positive, it increases the peak overshoot of the response. The value of the transient due to the real root can be added to the peak value of the underdamped sinusoid to obtain the correct value of the peak overshoot. The real root may also affect the settling time. In this case the real root decreases the peak time and increases the settling time. From plot c in Fig. 10.20 the actual values are $t_p = 1.05$ s and $t_{sc} = 2.50$ s.

From the root-locus plots it is evident that increasing the gain increases the peak overshoot and the settling time. The responses for each of the three lead compensators are shown in Fig. 10.20. The effect of the additional real root is to change the peak overshoot and the settling time. The designer must make a choice between the largest gain attainable, the smallest peak overshoot, and the shortest settling time.

Based on this example, the following guidelines are used to apply cascade lead compensators to a Type 1 or higher system:

1. If the zero $z_c = -1/T$ of the lead compensator is superimposed and cancels the largest real pole (excluding the pole at zero) of the original transfer function, a good improvement in the transient response is obtained.

2. If a larger gain is desired than that obtained by guideline 1, several trials should be made with the zero of the compensator moved to the left or right of the largest real pole of $G_x(s)$. The location of the zero that results in the desired gain and roots is selected.

For a Type 0 system it is often found that a better time response and a larger gain can be obtained by placing the compensator zero so that it cancels or is close to the second-largest real pole of the original transfer function.

10.11   GENERAL LEAD-COMPENSATOR DESIGN

Additional methods are available for the design of an appropriate lead compensator G_c that is placed in cascade with the open loop transfer function G_x, as shown in Fig. 10.1a. Figure 10.21 shows the original root locus of a control system. For a specified damping ratio ζ the dominant root of the uncompensated system is s_1. Also shown is s_2, which is the desired root of the system's characteristic equation. Selection of s_2 as a desired root is based on the performance required for the system. The design problem is to select a lead compensator that results in s_2 being a root. The first step is to find the sum of the angles at the point s_2 due to the poles and zeros of the original transfer function G_x. This angle is set equal to $180° + \phi$. For s_2 to be on the new root locus, it is necessary for the lead compensator to contribute an angle $\phi_c = -\phi$ at this point. The total angle at s_2 is then $180°$, and it is a point on the new root locus. A simple lead compensator represented by Eq. (10.33), with its zero to the right of its pole, can be used to provide the angle ϕ_c at the point s_2.

METHOD 1.   Actually, there are many possible locations of the compensator pole and zero that will produce the necessary angle ϕ_c at the point s_2. Cancellation of poles of the original open-loop transfer function by means of compensator zeros may simplify the root locus and thereby reduce the complexity of the problem. The compensator zero is simply placed over a real pole. Then the compensator pole is placed further to the left at a location that makes s_2 a point on the new root locus. The pole to be canceled depends on the system type. For a Type 1 system the largest real pole (excluding the pole at zero) should be canceled. For a Type 0 system the second-largest real pole should be canceled.

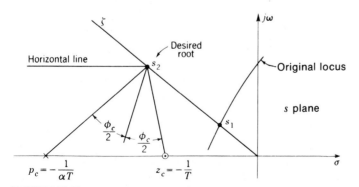

FIGURE 10.21
Graphical construction for locating the pole and zero of a simple lead compensator.

METHOD 2.   The following construction is based on obtaining the maximum value of α, which is the ratio of the compensator zero and pole. The steps in locating the lead-compensator pole and zero (see Fig. 10.21) are these:

1. Locate the desired root s_2. Draw a line from this root to the origin and a horizontal line to the left of s_2.
2. Bisect the angle between the two lines drawn in step 1.
3. Measure the angle $\phi_c/2$ on either side of the bisector drawn in step 2.
4. The intersections of these lines with the real axis locate the compensator pole p_c and zero z_c.

   The construction just outlined is shown in Fig. 10.21. It is left for the reader to show that this construction results in the largest possible value of α. Since α is less than unity and appears as the gain of the compensator [see Eq. (10.33)], the largest value of α requires the smallest additional gain A.

METHOD 3.   Changing the locations of the compensator pole and zero can produce a range of values for the gain while maintaining the desired root s_2. A specific procedure for achieving both the desired root s_2 and the system gain K_m is given in Refs. 7 and 8.

10.12   LAG-LEAD CASCADE COMPENSATION DESIGN

The preceding sections show that (1) the insertion of an integral or lag compensator in cascade results in a large increase in gain and a small reduction in the undamped natural frequency and (2) the insertion of a derivative or lead compensator in cascade results in a small increase in gain and a large increase in the undamped natural frequency. By inserting both the lag and the lead compensators in cascade with the original transfer function, the advantages of both can be realized simultaneously; i.e., a large increase in gain and a large increase in the undamped natural frequency can be obtained. Instead of using two separate networks, it is possible to use one network that acts as both a lag and a lead compensator. Such a network, shown in Fig. 10.22 with an amplifier added in cascade, is called a lag-lead compensator. The transfer function of this compensator (see Sec. 8.11) is

$$G_c(s) = A\frac{(1 + T_1 s)(1 + T_2 s)}{(1 + \alpha T_1 s)[1 + (T_2/\alpha)s]} = A\frac{(s + 1/T_1)(s + 1/T_2)}{(s + 1/\alpha T_1)(s + \alpha/T_2)} \quad (10.39)$$

where $\alpha > 1$ and $T_1 > T_2$.

   The fraction $(1 + T_1 s)/(1 + \alpha T_1 s)$ represents the lag compensator, and the fraction $(1 + T_2 s)/[1 + (T_2/\alpha)s]$ represents the lead compensator. The values T_1, T_2, and α are selected to achieve the desired improvement in system performance. The specific values of the compensator components are obtained from the relationships $T_1 = R_1 C_1, T_2 = R_2 C_2, \alpha T_1 + T_2/\alpha = R_1 C_1 + R_2 C_2 + R_1 C_2$. It may also be desirable to specify a minimum value of the input impedance over the passband frequency range.

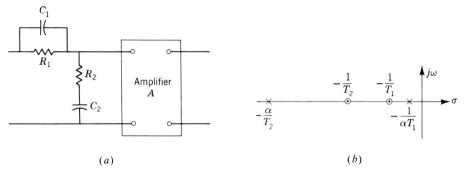

FIGURE 10.22
(*a*) Lag-lead compensator. (*b*) Pole and zero locations.

The procedure for applying the lag-lead compensator is a combination of the procedures for the individual units.

1. For the integral or lag component,

    (*a*) The zero $s = -1/T_1$ and the pole $s = -1/\alpha T_1$ are selected close together, with α set to a large value such as $\alpha = 10$.

    (*b*) The pole and zero are located to the left of and close to the origin, resulting in an increased gain.

2. For the derivative or lead component the zero $s = -1/T_2$ can be superimposed on a pole of the original system. Since α is already selected in step 1, the new pole is located at $s = -\alpha/T_2$. As a result the root locus moves to the left and therefore the undamped natural frequency increases.

The relative positions of the poles and zeros of the lag-lead compensators are shown in Fig. 10.22*b*.

DESIGN EXAMPLE OF LAG-LEAD COMPENSATION
APPLIED TO A TYPE 1 SYSTEM.
The system described in Secs. 10.8 and 10.10 is now used with a lag-lead compensator. The forward transfer function is

$$G_x(s) = \frac{K_1}{s(1 + s)(1 + 0.2s)} = \frac{K}{s(s + 1)(s + 5)} \tag{10.40}$$

The lag-lead compensator used is shown in Fig. 10.22, and the transfer function $G_c(s)$ is given by Eq. (10.39). The new complete forward transfer function is

$$G(s) = G_c(s)G_x(s) = \frac{AK(s + 1/T_1)(s + 1/T_2)}{s(s + 1)(s + 5)(s + 1/\alpha T_1)(s + \alpha/T_2)} \tag{10.41}$$

The poles and zeros of the compensator are selected in accordance with the principles outlined previously in this section. They coincide with the values used for the integral and derivative compensators when applied individually in Secs. 10.8 and 10.10. With an $\alpha = 10$, the compensator poles and zeros are

$$z_{lag} = -\frac{1}{T_1} = -0.05 \qquad p_{lag} = -\frac{1}{\alpha T_1} = -0.005$$

$$z_{lead} = -\frac{1}{T_2} = -1 \qquad p_{lead} = -\frac{\alpha}{T_2} = -10$$

The forward transfer function becomes

$$G(s) = G_c(s)G_x(s) = \frac{K'(s + 1)(s + 0.05)}{s(s + 1)(s + 0.005)(s + 5)(s + 10)} \tag{10.42}$$

The root locus (not to scale) for this system is shown in Fig. 10.23.

For a damping ratio $\zeta = 0.45$ the following pertinent data are obtained for the compensated system:

Dominant roots: $\qquad\qquad\qquad s_{1,2} = -1.571 \pm j3.119$

Loop sensitivity (evaluated at $s = s_1$):

$$K' = \frac{|s| \cdot |s + 0.005| \cdot |s + 5| \cdot |s + 10|}{|s + 0.05|} = 146.3$$

Ramp error coefficient: $\qquad K_1' = \lim_{s \to 0} sG(s) = \frac{K'(0.05)}{(0.005)(5)(10)} = 29.3 \text{ s}^{-1}$

Undamped natural frequency: $\omega_n = 3.49$ rad/s

The additional roots are $s_3 = -11.81$ and $s_4 = -0.051$. These results, when compared with those of the uncompensated system, show a large increase in both the gain and the undamped natural frequency. They indicate that the advantages of the lag-lead compensator are equivalent to the combined improvement obtained by the lag and the lead compensators. Note the practicality in the choice of $\alpha = 10$, which is based on experience. The response characteristics are shown in Table 10.1.

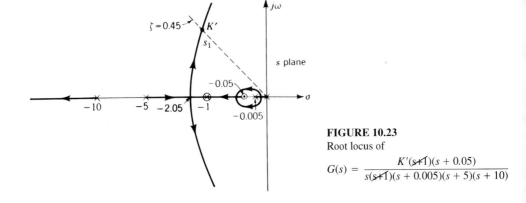

FIGURE 10.23
Root locus of
$$G(s) = \frac{K'(s+1)(s + 0.05)}{s(s+1)(s + 0.005)(s + 5)(s + 10)}$$

10.13 COMPARISON OF CASCADE COMPENSATORS

Table 10.1 shows a comparison of the system response obtained when a lag, a lead, and a lag-lead compensator are added in cascade with the same basic feedback system. These results are for the examples of Secs. 10.8, 10.10, and 10.12 and are obtained by use of a CAD program. These results are typical of the changes introduced by the use of cascade compensators, which can be summarized as follows:

1. Lag compensator
   (a) Results in a large increase in gain K_m (by a factor almost equal to α), which means a much smaller steady-state error
   (b) Decreases ω_n and therefore has the disadvantage of producing an increase in the settling time
2. Lead compensator
   (a) Results in a moderate increase in gain K_m, thereby improving steady-state accuracy
   (b) Results in a large increase in ω_n and therefore significantly reduces the settling time
   (c) The transfer function of the lead compensator, using the passive network of Fig. 10.16a, contains the gain α [see Eq. (10.33)], which is less than unity. Therefore, the additional gain A, which must be added, is larger than the increase in K_m for the system
3. Lag-lead compensator (essentially combines the desirable characteristics of the lag and the lead compensators)
   (a) Results in a large increase in gain K_m, which improves the steady-state response
   (b) Results in a large increase in ω_n, which improves the transient-response settling time

Figure 10.24 shows the frequency response for the original and compensated systems corresponding to Table 10.1. There is a close qualitative correlation between the time-response characteristics listed in Table 10.1 and the corresponding frequency-response curves shown in Fig. 10.24. The peak time response value M_p is directly proportional to the maximum value M_m, and the settling time t_s is a direct function of the passband frequency. Curves 3, 4, and 5 have a much wider passband than curves 1 and 2; therefore, these systems have a faster settling time.

For systems other than the one used as an example, these simple compensators may not produce the desired changes. *However, the basic principles described here are applicable to any system for which the open-loop poles closest to the imaginary axis are on the negative real axis.* For these systems the conventional compensators may be used, and good results are obtained.

When the complex open-loop poles are dominant, the conventional compensators produce only small improvements. A pair of complex poles near the imaginary axis often occurs in aircraft and in missile transfer functions. Effective compensation could remove the open-loop dominant complex poles and replace them with poles

TABLE 10.1
Comparison of performance of several cascade compensators for the system of Eq. (10.30) with $\zeta = 0.45$

Compensator	Dominant complex roots	Undamped natural frequency, ω_n	Other roots	Ramp error coefficients, K_1	t_p, s	M_o	t_s, s	Additional gain, required, A
Uncompensated	$-0.404 \pm j0.802$	0.898	-5.192	0.84	4.12	0.202	9.48	—
Lag: $\dfrac{s+0.05}{s+0.005}$	$-0.384 \pm j0.763$	0.854	-0.053 -5.183	8.02	4.30	0.266	22.0	9.57
Lead: $\dfrac{s+0.75}{s+7.5}$	$-1.522 \pm j3.021$	3.38	-0.689 -9.77	2.05	1.18	0.127	2.78	24.4
Lead: $\dfrac{s+1}{s+10}$	$-1.588 \pm j3.152$	3.53	-11.83	2.95	1.09	0.195	2.45	35.1
Lead: $\dfrac{s+1.5}{s+15}$	$-1.531 \pm j3.039$	3.40	-1.76 -16.18	4.40	1.05	0.280	2.50	52.5
Lag-lead: $\dfrac{(s+0.05)(s+1)}{(s+0.005)(s+10)}$	$-1.571 \pm j3.119$	3.49	-0.051 -11.81	29.26	1.10	0.213	3.33	34.9

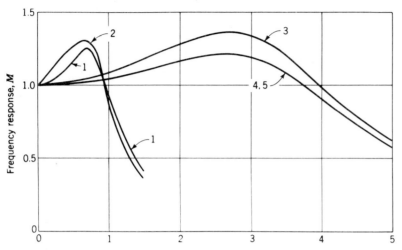

FIGURE 10.24
Frequency response for systems of Table 10.1: (1) original basic system; (2) lag-compensated; (3) lead-compensated, zero at -1.5; (4) lead-compensated, zero at -1.0; (5) lag-lead-compensated.

located farther to the left. This means that the compensator should have a transfer function of the form

$$G_c(s) = \frac{1 + Bs + As^2}{1 + Ds + Cs^2} \qquad (10.43)$$

If exact cancellation is assumed, the zeros of $G_c(s)$ would coincide with the complex poles of the original transfer function. A representative root locus, when the compensator has real poles, is shown in Fig. 10.25. The solid lines are the new locus, and the dashed lines are the original locus. Thus, the transient response has been improved.

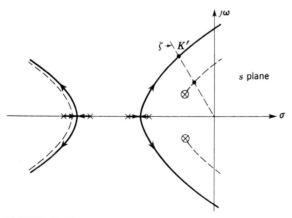

FIGURE 10.25
Root locus of

$$G_c(s)G_x(s) = \frac{K_0'}{(1 + T_1s)(1 + T_2s)(1 + T_3s)(1 + T_4s)}$$

However, the problem remaining to be solved is the synthesis of a network with the desired poles and zeros. There are references that cover network synthesis.[9,11] An alternative design method for improving the performance of plants that have dominant complex poles is the use of rate feedback compensation (see Sec. 10.21). This method does not require the assumption of exact cancellation shown in Fig. 10.25.

10.14   PID CONTROLLER

The PI and PD controller of Secs. 10.7 and 10.9, respectively, can be combined into a single controller, i.e., a proportional plus integral plus derivative (PID) controller as shown in Fig. 10.26. The transfer function of this controller is

$$G_c(s) = \frac{E_1(s)}{E(s)} = K_d s + K_p + \frac{K_i}{s} = \frac{K_d(s - z_1)(s - z_2)}{s} \tag{10.44}$$

where $(z_1 + z_2) = -K_p/K_d$ and $z_1 z_2 = K_i/K_d$. The compensator of Eq. (10.44) is placed in cascade with the basic system of Eq. (10.30), with $z_1 = -1$ and $z_2 = -0.1$, so the forward transfer function is

$$G_c(s)G_x(s) = \frac{K K_d(s + 0.1)}{s^2(s + 5)}$$

For a damping ratio $\zeta = 0.45$ the resulting control ratio is

$$\frac{C(s)}{R(s)} = \frac{30.13(s + 0.1)}{(s + 2.449 \pm j4.861)(s + 0.1017)} \tag{10.45}$$

The figures of merit for the PID-compensated system are $M_p = 1.225$, $t_p = 0.646$ s, $t_s = 2.10$ s, $K_v = \infty$, and $K_2 = 0.6026$ s^{-2}. Comparing these results with those in Table 10.1 shows that the PID controller results in the best improvement in system performance. Also note that the use of a PID controller increases the system type

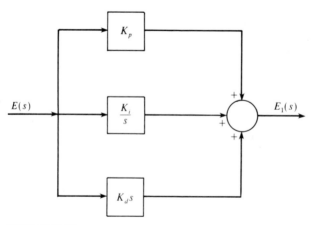

FIGURE 10.26
A PID controller.

from Type 1 to Type 2. With the advances in the state of the art of microelectronics, satisfactory derivative action may be achieved to produce the PID controller. However, to avoid saturation of the differentiated signal e due to noise that is present in the system, it is necessary to add a cascade filter to attenuate the high frequencies.

10.15   INTRODUCTION TO FEEDBACK COMPENSATION

System performance may be improved by using either cascade or feedback compensation. The factors that must be considered when making that decision are listed in Sec. 10.1. The effectiveness of cascade compensation is illustrated in the preceding sections of this chapter. The use of feedback compensation is demonstrated in the remainder of this chapter. The location of a feedback compensator depends on the complexity of the basic system, the accessibility of insertion points, the form of the signal being fed back, the signal with which it is being compared, and the desired improvement. Figure 10.27 shows various forms of feedback-compensated systems. For each case there is a unity-feedback loop (major loop) in addition to the compensation loop (minor loop) in order to maintain a direct correspondence (tracking) between the output and input.

It is often necessary to maintain the basic system type when applying feedback compensation. Care must therefore be exercised in determining the number of differentiating terms that are permitted in the numerator of $H(s)$. Table 10.2 presents the basic forms that $H(s)$ can have so that the system type remains the same after the minor feedback-compensation loop is added. $H(s)$ may have additional factors in the numerator and denominator. It can be shown by evaluating $G(s) = C(s)/E(s)$ that, to maintain the system type, the degree of the factor s in the numerator of $H(s)$

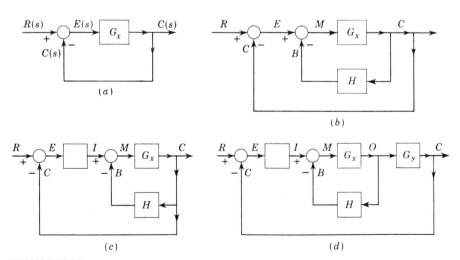

FIGURE 10.27
Various forms of feedback systems: (*a*) uncompensated system; (*b*) feedback-compensated system; (*c*) and (*d*) feedback-compensated systems with a cascade gain A or compensator G_c.

TABLE 10.2
Basic forms of $H(s)$ to be used to prevent change of system type

Basic form of $H(s)$	Type with which it can be used
K	0
Ks	0 and 1
Ks^2	0, 1, and 2
Ks^3	0, 1, 2, and 3

must be equal to or higher than the type of the forward transfer function $G_x(s)$ shown in Fig. 10.27.

The use of a direct or unity-feedback loop on any non–Type 0 control element converts it into a Type 0 element. This effect is advantageous when going from a low-energy-level input to a high-energy-level output for a given control element and also when the forms of the input and output energy are different. When $G_x(s)$ in Fig. 10.28 is Type 1 or higher, it may represent either a single control element or a group of control elements whose transfer function is of the general form

$$G_x(s) = \frac{K_m(1 + T_1 s) \cdots}{s^m(1 + T_a s)(1 + T_b s) \cdots} \tag{10.46}$$

where $m \neq 0$. The overall ratio of output to input is given by

$$G(s) = \frac{O(s)}{I(s)} = \frac{G_x(s)}{1 + G_x(s)} \tag{10.47}$$

or
$$G(s) = \frac{K_m(1 + T_1 s) \cdots}{s^m(1 + T_a s)(1 + T_b s) \cdots + K_m(1 + T_1 s) \cdots} \tag{10.48}$$

Equation (10.48) represents an equivalent forward transfer function between $O(s)$ and $I(s)$ of Fig. 10.28 and has the form of a Type 0 control element. Thus, the transformation of type has been effected. Therefore, for a step input, $O(t)_{ss} = \lim_{s \to 0}[sG(s)I(s)] = 1$. As a consequence, a given input signal produces an output signal of equal magnitude but at a higher energy level and/or with a different form of energy. The polar plots of $\mathbf{G}_x(j\omega)$ and $\mathbf{O}(j\omega)/\mathbf{I}(j\omega)$ would show that for any given frequency the phase-margin angle is greater with feedback, thus indicating a more stable system (a smaller value of M_m).

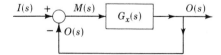

FIGURE 10.28
Transforming non–Type 0 control element to Type 0.

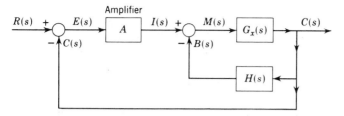

FIGURE 10.29
Block diagram for feedback compensation.

10.16   FEEDBACK COMPENSATION: DESIGN PROCEDURES

Techniques are now described for applying feedback compensation to the root-locus method of system design.[6] The system to be investigated, see Fig. 10.29, has a minor feedback loop around the original transfer function $G_x(s)$. The minor-loop feedback transfer function $H(s)$ has the form described in the following pages. Either the gain portion of $G_x(s)$ or the cascade amplifier A is adjusted to fix the closed-loop system characteristics. For more complex systems a design approach based on the knowledge obtained for this simple system can be developed.

Two methods of attack can be used. The characteristic equation of the complete system may be obtained and the root locus plotted, by use of the *partition*[1] method, as a function of a gain that appears in the system. A more general approach is first to adjust the roots of the characteristic equation of the inner loop. These roots are then the poles of the forward transfer function and are used to draw the root locus for the overall system. Both methods, using various types of feedback compensators, are illustrated in the succeeding sections. The techniques developed with cascade compensation should be kept in mind. Basically, the addition of poles and zeros changes the root locus. To improve the system time response the locus must be moved to the left, away from the imaginary axis.

10.17   SIMPLIFIED RATE FEEDBACK COMPENSATION: A DESIGN APPROACH

The feedback system shown in Fig. 10.29 provides the opportunity for varying three functions, $G_x(s)$, $H(s)$, and A. The larger the number of variables inserted into a system, the better the opportunity for achieving a specified performance. However, the design problem becomes much more complex. To develop a feel for the changes that can be introduced by feedback compensation, first a simplification is made by letting $A = 1$. The system of Fig. 10.29 then reduces to that shown in Fig. 10.30a. It can be further simplified as shown in Fig. 10.30b.

The first and simplest case of feedback compensation to be considered is the use of a transducer that has the transfer function $H(s) = K_t s$. The original forward transfer function is

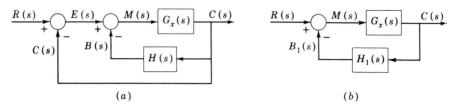

FIGURE 10.30
(a) Minor-loop compensation. (b) Simplified block diagram, $H_1(s) = 1 + H(s)$.

$$G_x(s) = \frac{K}{s(s + 1)(s + 5)} \tag{10.49}$$

The control ratio is

$$\frac{C(s)}{R(s)} = \frac{G_x(s)}{1 + G_x(s)H_1(s)} = \frac{G_x(s)}{1 + G_x(s)[1 + H(s)]} \tag{10.50}$$

The characteristic equation is

$$1 + G_x(s)[1 + H(s)] = 1 + \frac{K(1 + K_t s)}{s(s + 1)(s + 5)} = 0 \tag{10.51}$$

Equation (10.51) shows that the transducer plus the unity feedback have introduced the ideal proportional plus derivative control $(1 + K_t s)$ that can also be achieved by a cascade compensator. The results should therefore be qualitatively the same.

Partitioning the characteristic equation to obtain the general format $F(s) = -1$, which is required for obtaining a root locus, yields

$$G_x(s)H_1(s) = \frac{KK_t(s + 1/K_t)}{s(s + 1)(s + 5)} = -1 \tag{10.52}$$

The root locus is drawn as a function of KK_t. Introduction of minor-loop rate feedback results in the introduction of a zero [see Eq.(10.52)]. It therefore has the effect of moving the locus to the left and improving the time response. *This zero is not the same as introducing a cascade zero because no zero appears in the rationalized equation of the control ratio*

$$\frac{C(s)}{R(s)} = \frac{K}{s(s + 1)(s + 5) + KK_t(s + 1/K_t)} \tag{10.53}$$

The rate constant K_t must be chosen before the root locus can be drawn for Eq. (10.52). *Care must be taken in selecting the value of K_t. For example*, if $K_t = 1$ is used, cancellation of $(s + 1)$ in the numerator and denominator of Eq. (10.52) yields

$$\frac{K}{s(s + 5)} = -1 \tag{10.54}$$

and it is possible to get the impression that $C(s)/R(s)$ has only two poles. *This is not the case*, as can be seen from Eq. (10.53). Letting $K_t = 1$ provides the common factor $s + 1$ in the characteristic equation so that it becomes

$$(s + 1)[s(s + 5) + K] = 0 \qquad (10.55)$$

One closed-loop pole has therefore been made equal to $s = -1$, and the other two poles are obtained from the root locus of Eq. (10.54). If complex roots are selected from the root locus of Eq. (10.54), then the pole $s = -1$ is dominant and the response of the closed-loop system is overdamped (see Fig. 10.4d).

The proper design procedure is to select a number of trial locations for the zero $s = -1/K_t$, to tabulate or plot the system characteristics that result, and then to select the best combination. As an example, with $\zeta = 0.45$ as the criterion for the complex closed-loop poles and $K_t = 0.4$, the root locus is similar to that of Fig. 10.31 with $A = 1$. Table 10.3 gives a comparison of the original system with the rate-compensated system. From the root locus of Eq. (10.52) the roots are $s_{1,2} = -1 \pm j2$, and $s_3 = -4$, with $K = 20$. The gain K_1 must be obtained from the forward transfer function:

$$K_1 = \lim_{s \to 0} sG(s) = \lim_{s \to 0} \frac{sG_x(s)}{1 + G_x(s)H(s)} = \lim_{s \to 0} \frac{sK}{s[(s + 1)(s + 5) + KK_t]}$$

$$= \frac{K}{5 + KK_t} = 1.54 \qquad (10.56)$$

The results in Table 10.3 show that the effects of rate feedback are similar to those of a cascade lead compensator, but the gain K_1 is not so large (see Table 10.1). The control ratio of the system with rate feedback, with $K = 20$ and $K_t = 0.4$, is

$$\frac{C(s)}{R(s)} = \frac{20}{(s + 1 - j2)(s + 1 + j2)(s + 4)} \qquad (10.57)$$

TABLE 10.3
Comparison of performances

System	Dominant complex poles $\zeta = 0.45$	Other roots	Ramp error coefficient K_1, s^{-1}	t_p, s	M_o	t_s, s
Original	$-0.40 \pm j0.8$	-5.19	0.84	4.12	0.202	9.48
Cascade lead	$-1.59 \pm j3.15$	-11.83	2.95	1.09	0.195	2.45
Rate feedback	$-1.0 \pm j2.0$	-4.0	1.54	1.87	0.17	3.95

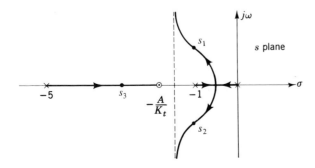

FIGURE 10.31
Root locus for Eq. (10.60).

10.18 DESIGN OF RATE FEEDBACK

The more general case of feedback compensation shown in Fig. 10.29, with $H(s) = K_t s$ and the amplifier gain A not equal to unity, is considered in this section. Two methods of attacking this problem are presented. The block diagrams of Fig. 10.32a and b are mathematically equivalent to the block diagram of Fig. 10.29 since they all yield the same closed-loop transfer function:

$$\frac{C(s)}{R(s)} = \frac{AG_x(s)}{1 + G_x(s)H(s) + AG_x(s)} = \frac{AG_x(s)}{1 + G_x(s)[H(s) + A]} \tag{10.58}$$

METHOD 1. The characteristic equation obtained from the denominator of Eq. (10.58) for the plant given by Eq. (10.49) is

$$1 + G_x(s)[H(s) + A] = 1 + \frac{K(K_t s + A)}{s(s + 1)(s + 5)} = 0 \tag{10.59}$$

Partitioning the characteristic equation to obtain the general format $F(s) = -1$, which is required for the root locus, yields

$$\frac{KK_t(s + A/K_t)}{s(s + 1)(s + 5)} = -1 \tag{10.60}$$

Since the numerator and denominator are factored, the roots can be obtained by plotting a root locus. The angle condition of $(1 + 2h)180°$ must be satisfied for the

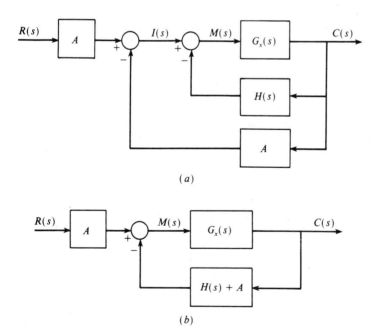

(a)

(b)

FIGURE 10.32
Block diagrams equivalent to Figure 10.29.

root locus of Eq. (10.60). It should be kept in mind, however, that Eq. (10.60) is *not* the forward transfer function of this system; it is *just an equation whose roots are the poles of the control ratio.* The three roots of this characteristic equation vary as K, K_t, and A are varied. Obviously, it is necessary to fix two of these quantities and to determine the roots as a function of the third. A sketch of the root locus as a function of K is shown in Fig. 10.31 for arbitrary values of A and K_t. Note that rate feedback results in an equivalent open-loop zero, so the system is stable for all values of gain for the value of A/K_t shown. This observation is typical of the stabilizing effect that can be achieved by derivative feedback and corresponds to the effect achieved with a proportional plus ideal derivative (PD) compensator in cascade.

The angles of the asymptotes of two of the branches of Eq. (10.60) are $\pm 90°$. For the selection A/K_t shown in Fig. 10.31, these asymptotes are in the left half of the s plane, and the system is stable for any value of K. However, if $A/K_t > 6$, the asymptotes are in the right half of the s plane and the system can become unstable if K is large enough. For the best transient response the root locus should be moved as far to the left as possible. This is determined by the location of the zero $s = -A/K_t$. To avoid having a dominant real root, the value of A/K_t is made greater than unity.

The forward transfer function obtained from Fig. 10.29 is

$$G(s) = \frac{C(s)}{E(s)} = \frac{AG_x(s)}{1 + G_x(s)H(s)} = \frac{AK}{s(s+1)(s+5) + KK_t s} \qquad (10.61)$$

The ramp error coefficient obtained from Eq. (10.61) is

$$K_1' = \frac{AK}{5 + KK_t} \qquad (10.62)$$

Selection of the zero ($s = -A/K_t$), the sensitivity KKt for the root locus drawn for Eq. (10.60), and the ramp error coefficient of Eq. (10.62) are all interrelated. The designer therefore has some flexibility in assigning the values of A, K_t, and K. The values $A = 2.0$ and $K_t = 0.8$ result in a zero at $s = -2.5$, the same as the example in Sec. 10.17. For $\zeta = 0.45$, the control ratio is given by Eq. (10.57) and the performance is the same as that listed in Table 10.3. The use of the additional amplifier A does not change the performance; instead, it permits greater flexibility in the allocation of gains. For example, K can be decreased and both A and K_t increased by a corresponding amount with no change in performance.

METHOD 2. A more general design method is to start with the inner loop, adjusting the gain to select the poles. Then successive loops are adjusted, the poles for each loop being set to desired values. For the system of Eq. (10.49), the transfer function of the inner loop is

$$\frac{C(s)}{I(s)} = \frac{K}{s(s+1)(s+5) + KK_t s} = \frac{K}{s[(s+1)(s+5) + KK_t]}$$

$$= \frac{K}{s(s - p_1)(s - p_2)} \qquad (10.63)$$

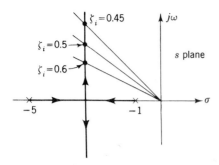

FIGURE 10.33
Poles of the inner loop for several values of ζ_i.

The equation for which the root locus is plotted to find the roots p_1 and p_2 is

$$(s + 1)(s + 5) + KK_t = 0$$

which can be rewritten as
$$\frac{KK_t}{(s + 1)(s + 5)} = -1 \qquad (10.64)$$

The locus is plotted in Fig. 10.33, where several values of the inner-loop damping ratio ζ_i for the roots are shown. Selection of a ζ_i permits evaluation of the product KK_t by use of Eq. (10.64). Upon assignment of a value to one of these quantities, the other can then be determined. The poles of the inner loop are the points on the locus that are determined by the ζ_i selected. They also are the poles of the open-loop transfer function for the overall system as represented in Eq. (10.63). It is therefore necessary that the damping ratio ζ_i of the inner loop have a higher value than the desired damping ratio of the overall system. The forward transfer function of the complete system becomes

$$G(s) = \frac{C(s)}{E(s)} = \frac{AK}{s(s^2 + 2\zeta_i\omega_{n,i}s + \omega_{n,i}^2)} \qquad (10.65)$$

The root locus of the complete system is shown in Fig. 10.34 for several values of the inner-loop damping ratio ζ_i. These loci are plotted as a function of AK.

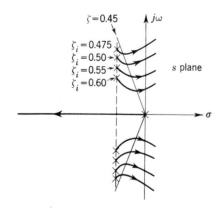

FIGURE 10.34
Root locus of complete system for several values of ζ_i.

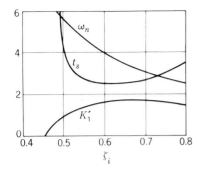

FIGURE 10.35
Variation of K_1', t_s, and ω_n with ζ_i.

The control ratio has the form

$$\frac{C(s)}{R(s)} = \frac{AK}{s(s^2 + 2\zeta_i\omega_{n,i}s + \omega_{n,i}^2) + AK}$$

$$= \frac{AK}{(s + \alpha)(s^2 + 2\zeta\omega_n s + \omega_n^2)} \qquad (10.66)$$

In this problem the damping ratio ζ of the complex poles of the control ratio is used as the basis for design. Both the undamped natural frequency ω_n and the ramp error coefficient $K_1' = AK/(5 + KK_t)$ depend on the selection of the inner-loop damping ratio ζ_i. Several trials are necessary to show how these quantities and the remaining figures of merit vary with the selection of ζ_i. This information is shown graphically in Fig. 10.35. The system performance specifications must be used to determine a suitable design. The associated value of K_1' determines the required values of A, K, and K_t. The results in Table 10.4 show that $\zeta_i = 0.6$ produces the shortest settling

TABLE 10.4
System response characteristics using method 2

ζ_i	KK_t	AK	**Roots**	t_p, s	M_o	t_s, s	K_1'
0.475	34.91	19.33	$-2.74 \pm j5.48$ -0.52	†	0	7.64	0.484
0.5	31.01	30.53	$-2.5 \pm j4.98$ -0.93	†	0	4.34	0.847
0.55	24.76	39.0	$-2.1 \pm j4.22$ -1.75	†	0	2.36	1.31
0.6	20.01	38.63	$-1.82 \pm j3.62$ -2.35	1.40	0.01	2.19	1.54
0.7	13.37	30.51	$-1.38 \pm j2.74$ -3.24	1.52	0.12	2.88	1.66
0.8	9.06	22.14	$-1.03 \pm j2.14$ -3.83	1.77	0.20	3.75	1.57

† Overdamped response.

time $t_s = 2.19$, an overshoot $M_o = 0.01$, and $K_1' = 1.54$. Table 10.4 also shows for $\zeta < 0.6$ that a real root is most dominant, which yields an overdamped system with an increased value of t_s.

10.19   DESIGN: FEEDBACK OF SECOND DERIVATIVE OF OUTPUT

To improve the system performance further, a signal proportional to the second derivative of the output may be fed back. For a position system an accelerometer will generate such a signal. An approximation to this desired signal may be generated by modifying the output of a rate transducer with a high-pass filter. An example of this technique for a position-control system is shown in Fig. 10.36.

The transfer function for the inner feedback, which consists of a tachometer in cascade with a lead compensator, is

$$H(s) = \frac{B(s)}{C(s)} = K_t s \left[\frac{s}{s + 1/RC} \right] = \frac{K_t s^2}{s + 1/RC} \tag{10.67}$$

Method 1 is used for the application of this compensator. Using the denominator of Eq. (10.58), the root locus is drawn for $G_x(s)[H(s)+A] = -1$, which yields

$$\frac{KK_t(s^2 + As/K_t + A/K_t RC)}{s(s + 1)(s + 5)(s + 1/RC)} = -1 \tag{10.68}$$

Since A/K_t and $1/RC$ must be selected in order to specify the zeros in Eq. (10.68) before the locus can be drawn, it is well to look at the expression for K_1' to determine the limits to be placed on these values. The forward transfer function is

$$G(s) = \frac{AG_x(s)}{1 + G_x(s)H(s)} = \frac{AK(s + 1/RC)}{s(s + 1)(s + 5)(s + 1/RC) + KK_t s^2} \tag{10.69}$$

Since it is a Type 1 transfer function, the gain K_1' is equal to

$$K_1' = \lim_{s \to 0} sG(s) = \frac{AK}{5} \tag{10.70}$$

Since K_1' is independent of RC, the value of RC is chosen to produce the most desirable root locus. The two zeros of $s^2 + (A/K_t)s + A/K_t RC$ may be either real or complex and are equal to

FIGURE 10.36
Tachometer and RC feedback network.

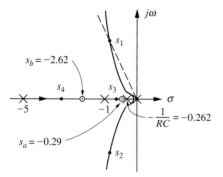

FIGURE 10.37
Root locus of Eq. (10.68).

$$s_{a,b} = -\frac{A}{2K_t} \pm \sqrt{\left(\frac{A}{2K_t}\right)^2 - \frac{A}{K_t RC}} \tag{10.71}$$

A representative sketch of the root locus is shown in Fig. 10.37 for an arbitrary selection of A, K_t, and RC. For Fig. 10.37 the values $A = 1$, $K_t = 0.344$, and $1/RC = 0.262$ give zeros at $s_a = -0.29$ and $s_b = -2.62$. Note that the zero s_a and the pole $s = -1/RC$ have the properties of a cascade lag compensator; therefore, an increase in gain can be expected. The zero s_b results in an improved time response. By specifying a damping ratio $\zeta = 0.35$ for the dominant roots, the result is $K = 52.267$, $K_1' = 10.45$, and $\omega_n = 3.75$; and the overall control ratio is

$$\frac{C(s)}{R(s)} = \frac{52.267(s + 0.262)}{(s + 1.31 + j3.51)(s + 1.31 - j3.51)(s + 3.35)(s + 0.291)} \tag{10.72}$$

Note that a passive lead network in the feedback loop acts like a passive lag network in cascade and results in a large increase in K_1'. Several trials may be necessary to determine the best selection of the closed-loop poles and ramp error coefficient. A $\zeta = 0.35$ is chosen instead of $\zeta = 0.45$ to obtain a larger value for K_1'. Since the third pole $s = -3.34$ reduces the peak overshoot, a smaller value of ζ is justifiable; however, the peak overshoot and settling time are larger than the best results obtained with just rate feedback (see Table 10.5).

TABLE 10.5
Comparison of cascade and feedback compensation using the root locus

System	K_1	ω_n	ζ	M_o	t_s,s
Uncompensated	0.84	0.89	0.45	0.202	9.48
Rate feedback	1.54	4.0	0.45	0.01	2.19
Second derivative of output	10.45	3.75	0.35	0.267	6.74
Cascade lag compensator	8.02	0.854	0.45	0.266	22.0
Cascade lead compensator	2.95	3.53	0.45	0.195	2.45
Cascade lag-lead compensator	29.26	3.49	0.45	0.213	3.33

The use of an *RC* network in the feedback path may not always produce a desired increase in gain. Also, selecting the proper location of the zeros and the corresponding values of A, K_t, and RC in Eq. (10.68) may require several trials. Therefore, a possible alternative is to use only rate feedback to improve the time response, and then to add a cascade lag compensator to increase the gain. The corresponding design steps are essentially independent.

10.20 RESULTS OF FEEDBACK COMPENSATION DESIGN

The preceding sections show the application of feedback compensation by the root-locus method. Basically, rate feedback is similar to ideal derivative control in cascade. The addition of a high-pass *RC* filter in the feedback path permits a large increase in the error coefficient. For good results the feedback compensator $H(s)$ should have a zero, $s = 0$, of higher order than the type of the original transfer function $G_x(s)$. The method of adjusting the poles of the inner loop and then adjusting successively larger loops is applicable to all multiloop systems. This method is used extensively in the design of aircraft flight control systems.[12] The results of feedback compensation applied in the preceding sections are summarized in Table 10.5. Results of cascade compensation are also listed for comparison. This tabulation shows that comparable results are obtained by cascade and feedback compensation.

10.21 RATE FEEDBACK: PLANTS WITH DOMINANT COMPLEX POLES

The simple cascade compensators of Secs. 10.8 to 10.12 yield very little system performance improvement for plants having the pole-zero pattern of Fig. 10.38*a* where

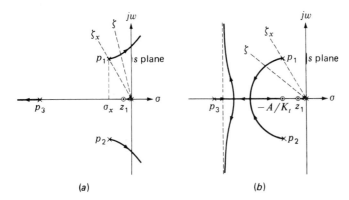

(a)   (b)

FIGURE 10.38
Root locus for the longitudinal motion of an aircraft: (*a*) uncompensated;
(*b*) compensated (rate feedback).

the plant has dominant complex poles. The following transfer function, representing the longitudinal motion of an aircraft,[12] yields such a pattern:

$$G_x(s) = \frac{\theta(s)}{e_g(s)} = \frac{K_x(s - z_1)}{s(s - p_1)(s - p_2)(s - p_3)} \tag{10.73}$$

where θ is the pitch angle and e_g is the input to the elevator servo. With unity feedback the root locus of Fig. 10.38a shows, for $K_x > 0$, that only $\zeta < \zeta_x$ is achievable for the dominant roots. However, good improvement in system performance is obtained by using a minor loop with rate feedback $H(s) = K_t s$. Using method 1 of Sec. 10.18 for the system configuration of Fig. 10.29 yields

$$\frac{C(s)}{E(s)} = \frac{AK_x(s - z_1)}{s(s - p_1)(s - p_2)(s - p_3) + K_t K_x s(s - z_1)} \tag{10.74}$$

and $$\frac{C(s)}{R(s)} = \frac{AK_x(s - z_1)}{s(s - p_1)(s - p_2)(s - p_3) + K_t K_x s(s - z_1) + AK_x(s - z_1)} \tag{10.75}$$

The characteristic equation formed from the denominator of Eq. (10.58), or from the denominator of Eq. (10.75), is partitioned to yield

$$\frac{K_t K_x(s - z_1)(s + A/K_t)}{s(s - p_1)(s - p_2)(s - p_3)} = -1 \tag{10.76}$$

A comparison of Eqs. (10.73) and (10.76) shows that rate feedback has added a zero to the root locus. One possible root locus, depending on the value of A/K_t, is shown in Fig. 10.38b. This illustrates that using rate feedback with $K_t > 0$ allows a good improvement in system performance, since it is possible to achieve $\zeta > \zeta_x$ and $\sigma < \sigma_x$ for the complex roots. The real root near the origin may appear to be dominant, but it is close to the zero z_1. Therefore, it may not make a large contribution to the transient response.

Note that the cascade compensator of Eq. (10.43), as applied in Fig. 10.25, shows exact cancellation of the pair of complex poles by the compensator zeros. In practice, exact cancellation cannot be achieved because the aircraft poles vary with different flight conditions. Thus, as illustrated by this section, using simple rate feedback is a more practical approach for achieving improved system performance.

10.22 SUMMARY

This chapter shows the applications of cascade and feedback compensation. The procedures for applying the compensators are given so that they can be used with any system. They involve a certain amount of trial and error and the exercise of judgment based on past experience. The improvements produced by each compensator are shown by application to a specific system. In cases where the conventional compensators do not produce the desired results, other methods of applying compensators are available in the literature.

It is important to emphasize that the conventional compensators cannot be applied indiscriminately. For example, the necessary improvements may not fit the four categories listed in Sec. 10.5, or the use of the conventional compensator may cause a deterioration of other performance characteristics. In such cases the designer must use some ingenuity in locating the compensator poles and zeros to achieve the desired results. The use of several compensating networks in cascade and/or in feedback may be required to achieve the desired root locations.

Realization of compensator pole-zero locations by means of passive networks requires an investigation of network synthesis techniques. Design procedures for specialized networks that produce two poles and two zeros are available.[9-11]

It may occur to the reader that the methods of compensation in this and the following chapters can become laborious because some trial and error is involved. The use of an interactive CAD program[14] to solve design problems minimizes the work involved. Using a computer means that an acceptable design of the compensator can be determined in a shorter time. It is still useful to obtain by an analytical or graphical method a few values of the solution to ensure that the problem has been properly set up and that the CAD program is operating correctly.

REFERENCES

1. D'Azzo, J. J., and C. H. Houpis: *Feedback Control System Analysis and Synthesis*, 2nd ed., McGraw-Hill, New York, 1966.
2. Chu, Y.: "Synthesis of Feedback Control System by Phase-Angle Loci," *Trans. AIEE*, vol. 71, pt. II, pp. 330–339, November 1952.
3. Gibson, J. E.: *Nonlinear Automatic Control*, McGraw-Hill, New York, 1963.
4. Thaler, G. J., and M. P. Pastel: *Analysis and Design of Nonlinear Feedback Control Systems*, McGraw-Hill, New York, 1960.
5. Elgerd, O. I., and W. C. Stephens: "Effect of Closed-Loop Transfer Function Pole and Zero Locations on the Transient Response of Linear Control Systems," *Trans. AIEE*, vol. 78, pt. II, pp. 121–127, May 1959.
6. Truxal, J. G.: *Automatic Feedback Control System Synthesis*, McGraw-Hill, New York, 1955.
7. Ross, E. R., T. C. Warren, and G. J. Thaler: "Design of Servo Compensation Based on the Root Locus Approach," *Trans. AIEE*, vol. 79, pt. II, pp. 272–277, September 1960.
8. Pena, L. Q.: "Designing Servocompensators Graphically," *Control Eng.*, vol. 82, pp. 79–81, January 1964.
9. Lazear, T. J., and A. B. Rosenstein: "On Pole-Zero Synthesis and the General Twin-T," *Trans. IEEE*, vol. 83, pt. II, pp. 389–393, November 1964.
10. Kuo, F. F.: *Network Analysis and Synthesis*, 2nd ed., Wiley, New York, 1966.
11. Mitra, S. K.: *Analysis and Synthesis of Linear Active Networks*, Wiley, New York, 1969.
12. Blakelock, J. H.: *Automatic Control of Aircraft and Missiles*, 2nd ed., Wiley, New York, 1991.
13. Ali, W., and J. H. Burghart: *Effects of Lag Controllers on Transient Response*, IEE Proc. Part D, vol. 138, no. 2, pp. 119–122, March 1991.
14. A comprehensive list of CAD programs is given in App. B.

CHAPTER
11

FREQUENCY-RESPONSE COMPENSATION DESIGN

11.1 INTRODUCTION TO FEEDBACK COMPENSATION DESIGN

Improvements in the response of feedback tracking control systems and disturbance rejection can be achieved in the frequency domain. This chapter presents a frequency-domain tracking techniques, and Chapter 12 presents a disturbance-rejection technique. The preceding chapter presents the factors used with the root-locus method for selecting compensators and the corresponding effects on the time response. The designer can place the poles and zeros of the proposed compensator on the s plane and determine the poles of the closed-loop tracking system, which in turn permits evaluation of the closed-loop time response. Similarly, new values of M_m, ω_m, and K_m can be determined by the frequency-response method, using either the polar plot or the log plot. However, the closed-loop poles are not explicitly determined. Furthermore, the correlation between the frequency-response parameters M_m and ω_m and the time response is only qualitative, as discussed in Secs. 9.3, 9.4, and 9.11. The presence of real roots near the complex dominant roots further changes the correlation between the frequency-response parameters and the time response, as discussed in Sec. 9.11.

This chapter presents the changes that can be made in the frequency-response characteristics by use of the three types of cascade compensators (lag, lead, and lag-lead) and by feedback compensation. They are only representative of the compensators that can be used. Design procedures used to obtain improvement of the system performance are described. A designer can extend or modify these design procedures, which are based on the presence of one pair of complex dominant roots for the closed-loop system, to those systems where there are other dominant roots besides the main complex pair.[1] Both the log plots and the polar plots are used in this chapter in applying the frequency-response compensation criteria to show that either type of plot can be used; the choice depends on individual preference.

The performance of a closed-loop system can be described in terms of M_m, ω_m, and the system error coefficient K_m. The value of M_m (or γ) essentially describes the damping ratio ζ and therefore the amount of overshoot in the transient response. For a specific value of M_m, the resonant frequency ω_m (or ω_ϕ) determines the undamped natural frequency ω_n, which in turn determines the response time of the system. The system error coefficient K_m is important because it determines the steady-state error with an appropriate standard input. The design procedure is usually based on selecting a value of M_m and using the methods described in Chapter 9 to find the corresponding values of ω_m and the required gain K_m. If the desired performance specifications are not met by the basic system after this is accomplished, compensation must be used. This alters the shape of the frequency-response plot in an attempt to meet the performance specifications. Also, for those systems that are unstable for all values of gain, it is mandatory that a stabilizing or compensating network be inserted in the system. The compensator may be placed in cascade or in a minor feedback loop. The reasons for reshaping the frequency-response plot generally fall into the following categories:

1. A given system is stable, and its M_m and ω_m (and therefore the transient response) are satisfactory, but its steady-state error is too large. The gain K_m must therefore be increased to reduce the steady-state error (see Chapter 6) without appreciably altering the values of M_m and ω_m. It is shown later that in this case the high-frequency portion of the frequency-response plot is satisfactory, but the low-frequency portion is not.

2. A given system is stable, but its transient settling time is unsatisfactory; i.e., the M_m is set to a satisfactory value, but the ω_m is too low. The gain may be satisfactory, or a small increase may be desirable. The high-frequency portion of the frequency-response plot must be altered in order to increase the value of ω_m.

3. A given system is stable and has a desired M_m, but both its transient response and its steady-state response are unsatisfactory. Therefore, the values of both ω_m and K_m must be increased. The portion of the frequency plot in the vicinity of the $-1 + j0$ or $(-180°, 0\text{-dB})$ point must be altered to yield the desired ω_m, and the low-frequency portion must be changed to obtain the increase in gain desired.

4. A given system is unstable for all values of gain. The frequency-response plot must be altered in the vicinity of the $-1 + j0$ or $(-180°, 0\text{-dB})$ point to produce a stable system with a desired M_m and ω_m.

Thus, the objective of compensation is to reshape the frequency-response plot of the basic system by means of a compensator in order to achieve the performance specifications. Examples that demonstrate this objective are presented in the following sections. The design methods in this chapter and the effects of compensation are illustrated by the use of frequency-response plots in order to provide a means for understanding the principles involved. It is expected, however, that a CAD program such as MATLAB, ICECAP-PC, and TOTAL-PC will be used extensively in the design process (see Sec. 10.6). Such a program provides both the frequency response and the time response of the resultant system.

11.2 SELECTION OF A CASCADE COMPENSATOR

Consider the system shown in Fig. 11.1.

$G_x(j\omega)$ = forward transfer function of a basic feedback control system

$\mathbf{G}(j\omega) = \mathbf{G}_c(j\omega)\mathbf{G}_x(j\omega)$ = forward transfer function that results in the required stability and steady-state accuracy

Dividing $\mathbf{G}(j\omega)$ by $\mathbf{G}_x(j\omega)$ gives the necessary cascade compensator

$$\mathbf{G}_c(j\omega) = \frac{\mathbf{G}(j\omega)}{\mathbf{G}_x(j\omega)} \tag{11.1}$$

A physical electrical network for the compensator described in Eq. (11.1) can be synthesized by the techniques of Foster, Cauer, Brune, and Guillemin. In the preceding and present chapters the compensators are limited to relatively simple *RC* networks. Based on the design criteria presented, the parameters of these simple networks can be evaluated in a rather direct manner. The simple networks quite often provide the desired improvement in the control system's performance. When they do not, more complex networks must be synthesized.[2,3] Physically realizable networks are limited in their performance over the whole frequency spectrum; this makes it difficult to obtain the exact transfer-function characteristic given by Eq. (11.1). However, this is not a serious limitation, because the desired performance specifications are interpreted in terms of a comparatively small bandwidth of the frequency spectrum. A compensator can be constructed to have the desired magnitude and angle characteristics over this bandwidth. Minor loop feedback compensation can also be used to achieve the desired $\mathbf{G}(j\omega)$. This may yield a simpler network than the cascade compensator given by Eq. (11.1).

The characteristics of the lag and lead compensators and their effects upon a given polar plot are shown in Fig. 11.2*a* and *b*. The cascade lag compensator results in a clockwise (cw) rotation of $\mathbf{G}(j\omega)$ and can produce a large increase in gain K_m with a small decrease in the resonant frequency ω_m. The cascade lead compensator results in a counterclockwise (ccw) rotation of $\mathbf{G}(j\omega)$ and can produce a large increase in ω_m with a small increase in K_m. When the characteristics of both the lag and the lead networks are necessary to achieve the desired specifications, a lag-lead compensator can be used. The transfer function and polar plot of this lag-lead

FIGURE 11.1
Cascade compensation.

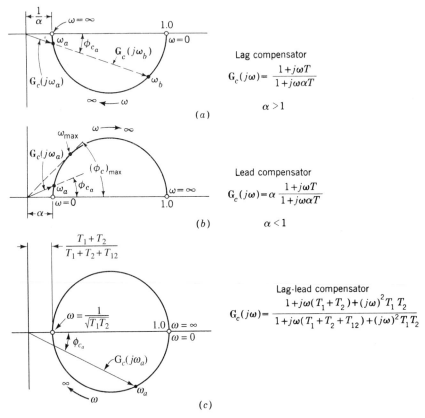

FIGURE 11.2
Polar plots: (*a*) lag compensator; (*b*) lead compensator; (*c*) lag-lead compensator.

compensator are shown in Fig. 11.2*c*. The corresponding log plots for each of these compensators are included with the design procedures later in the chapter.

The choice of a compensator depends on the characteristics of the given (or basic) feedback control system and the desired specifications. When these networks are inserted in cascade, they must be of the (1) proportional plus integral and (2) proportional plus derivative types. The plot of the lag compensator shown in Fig. 11.2*a* approximates the ideal integral plus proportional characteristics shown in Fig. 11.3*a* at high frequencies. Also, the plot of the lead compensator shown in Fig. 11.2*b* approximates the ideal derivative plus proportional characteristics shown in Fig. 11.3*b* at low frequencies. These are the frequency ranges where the proportional plus integral and proportional plus derivative controls are needed to provide the necessary compensation.

The equations of $G_c(j\omega)$ in Fig. 11.2*a* and 11.2*b*, which represent the lag and the lead compensators used in this chapter, can be expressed as

$$G_c(j\omega) = |G_c(j\omega)| \underline{/\phi_c} \tag{11.2}$$

FIGURE 11.3
(a) Ideal integral plus proportional compensator. (b) Ideal derivative plus proportional compensator.

Solving for the angle ϕ_c yields

$$\phi_c = \tan^{-1}\omega T - \tan^{-1}\omega\alpha T = \tan^{-1}\frac{\omega T - \omega\alpha T}{1 + \omega^2\alpha T^2} \qquad (11.3)$$

or

$$T^2 + \frac{\alpha - 1}{\omega\alpha\tan\phi_c}T + \frac{1}{\omega^2\alpha} = 0 \qquad (11.4)$$

where $\alpha > 1$ for the lag compensator and $\alpha < 1$ for the lead compensator. In the design procedures shown in later sections, a frequency ω is specified at which a particular value of ϕ_c is desired. For the selected values of α, ϕ_c, and ω, Eq. (11.4) is used to determine the value of T required for the compensator. For a lag compensator, a small value of ϕ_c is required, whereas a large value of ϕ_c is required for a lead compensator. For a lead compensator the value ω_{max}, at which the maximum phase shift occurs, can be determined for a given T and α by setting the derivative of ϕ_c with respect to the frequency ω equal to zero. Therefore, from Eq. (11.3) the maximum angle occurs at the frequency

$$\omega_{max} = \frac{1}{T\sqrt{\alpha}} \qquad (11.5)$$

Inserting this value of frequency into Eq. (11.3) gives the maximum phase shift

$$(\phi_c)_{max} = \sin^{-1}\frac{1 - \alpha}{1 + \alpha} \qquad (11.6)$$

Typical values for a lead compensator are

α	0	0.1	0.2	0.3	0.4	0.5
$(\phi_c)_{max}$	90°	54.9°	41.8°	32.6°	25.4°	19.5°

A knowledge of the maximum phase shift is useful in the application of lead compensators. The design procedures developed in the following sections utilize the characteristics of compensators discussed in this section.

11.3 CASCADE LAG COMPENSATOR

When M_m and ω_m, which determine the transient response of a feedback control system, are satisfactory but the steady-state error is too large, it is necessary to increase the gain without appreciably altering that portion of the given log magnitude (Lm)–angle plot in the vicinity of the $(-180°, 0\text{-dB})$ point (category 1). This can be done by introducing an integral or lag compensator in cascade with the original forward transfer function as shown in Fig. 11.1. Figure 10.9 shows a representative lag network having the transfer function

$$G_c(s) = A\frac{1 + Ts}{1 + \alpha Ts} \tag{11.7}$$

where $\alpha > 1$. As an example of its application, consider a basic Type 0 system. For an improvement in the steady-state accuracy, the system step error coefficient K_p must be increased. Its values before and after the addition of the compensator are

$$K_p = \lim_{s\to 0} G_x(s) = K_0 \tag{11.8}$$

$$K'_p = \lim_{s\to 0} G_c(s)G_x(s) = AK_0 = K'_0 \tag{11.9}$$

The compensator amplifier gain A must have a value that gives the desired increase in the value of the step error coefficient. Thus

$$A = \frac{K'_0}{K_0} \tag{11.10}$$

This increase in gain must be achieved while maintaining the same M_m and without appreciably changing the value of ω_m. However, the lag compensator has a negative angle, which moves the original Lm–angle plot to the left, or closer to the $(-180°, 0\text{-dB})$ point. This has a destabilizing effect and reduces ω_m. To limit this decrease in ω_m, the lag compensator is designed so that it introduces a small angle, generally no more than $-5°$, at the original resonant frequency ω_{m1}. The value of $-5°$ at $\omega = \omega_{m1}$ is an empirical value, determined from practical experience in applying this type of network. A slight variation in method, which gives equally good results, is to select $G_c(s)$ so that the magnitude of its angle is 5° or less at the original phase-margin frequency $\omega_{\phi1}$.

The overall characteristics of lag compensators are best visualized from the log plot. The Lm and phase-angle equations for the compensator of Eq. (11.7) are

$$\text{Lm } \mathbf{G}'_c(j\omega) = \text{Lm}(1 + j\omega T) - \text{Lm}(1 + j\omega\alpha T) \qquad (11.11)$$

$$\underline{/\mathbf{G}_c(j\omega)} = \underline{/1 + j\omega T} - \underline{/1 + j\omega\alpha T} \qquad (11.12)$$

For various values of α there is a family of curves for the log magnitude and phase-angle diagram, as shown in Fig. 11.4. These curves show that an attenuation equal to Lm α is introduced above $\omega T = 1$. Assume that a system has the original forward transfer function $\mathbf{G}_x(j\omega)$ and that the gain has been adjusted for the desired phase margin γ. The Lm and phase-angle diagram of $\mathbf{G}_x(j\omega)$ are sketched in Fig. 11.5. The addition of the lag compensator $\mathbf{G}_c(j\omega)$ reduces the log magnitude by Lm α for the frequencies above $\omega = 1/T$. The value of T of the compensator is selected so that the attenuation Lm α occurs at the original phase-margin frequency $\omega_{\phi 1}$. Notice that the phase margin angle at $\omega_{\phi 1}$ has been reduced. To maintain the specified value of γ, the phase-margin frequency has been reduced to $\omega_{\phi 2}$. It is now possible to increase the gain of $\mathbf{G}_x(j\omega)\mathbf{G}'_c(j\omega)$ by selecting the value A so that the Lm curve will have the value 0 dB at the frequency $\omega_{\phi 2}$.

The effects of the lag compensator can now be analyzed. The reduction in log magnitude at the higher frequencies due to the compensator is desirable. This permits an increase in the gain to maintain the desired γ. Unfortunately, the negative angle of the compensator lowers the angle curve. To maintain the desired γ, the

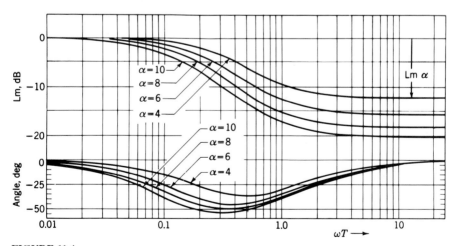

FIGURE 11.4

Log magnitude and phase-angle diagram for lag compensator

$$\mathbf{G}'_c(j\omega) = \frac{1 + j\omega T}{1 + j\omega\alpha T}$$

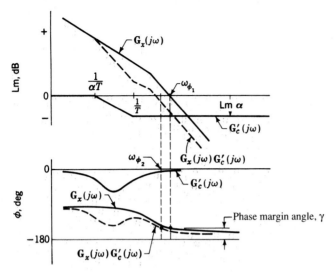

FIGURE 11.5
Original and lag-compensated log magnitude and phase-angle plots.

phase-margin frequency is reduced. To keep this reduction small, the value of $\omega_{cf} = 1/T$ should be made as small as possible.

The Lm–angle diagram permits gain adjustment for the desired M_m. The curves for $\mathbf{G}_x(j\omega)$ and $\mathbf{G}_x(j\omega)\mathbf{G}_c'(j\omega)$ are shown in Fig. 11.6. The increase in gain to obtain the same M_m is the amount that the curve of $\mathbf{G}_x(j\omega)\mathbf{G}_c'(j\omega)$ must be raised in order to be tangent to the M_m curve. Because of the negative angle introduced by the lag compensator, the new resonant frequency ω_{m2} is smaller than the original resonant frequency ω_{m1}. Provided that the decrease in ω_m is small and the increase in gain is sufficient to meet the specifications, the system is considered satisfactorily compensated.

In summary, the lag compensator is basically a low-pass filter: the low frequencies are passed and the higher frequencies are attenuated. This attenuation charac-

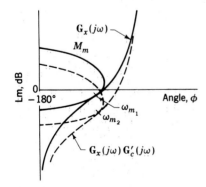

FIGURE 11.6
Original and lag-compensated Nichols plots.

teristic of the lag compensator is useful and permits an increase in the gain. The negative phase-shift characteristic is detrimental to system performance but must be tolerated. Because the predominant and useful feature of the compensator is attenuation, a more appropriate name for it is *high-frequency attenuation compensator.*

11.4 DESIGN EXAMPLE: CASCADE LAG COMPENSATION

To give a basis for comparison of all methods of compensation, the examples of Chapter 10 are used in this chapter. For the unity-feedback Type 1 system of Sec. 10.8 the forward transfer function is

$$G_x(s) = \frac{K_1}{s(1 + s)(1 + 0.2s)} \tag{11.13}$$

By the methods of Secs. 9.7 and 9.10, the gain required for this system to have an M_m of 1.26 (2 dB) is $K_1 = 0.872 \text{ s}^{-1}$, and the resulting resonant frequency is $\omega_{m1} = 0.715$ rad/s. With these values the transient response is considered to be satisfactory, but the steady-state error is too large. Putting a lag compensator in cascade with a basic system (category 1) permits an increase in gain and therefore reduces the steady-state error.

The cascade compensator to be used has the transfer function given by Eq. (11.7). Selection of an $\alpha = 10$ means that an increase in gain of almost 10 is desired. Based on the criterion discussed in the last section (a phase shift of $-5°$ at the original ω_m), the compensator time constant can be determined. Note that in Fig. 11.4, $\phi_c = -5°$ occurs for values of ωT approximately equal to 0.01 and 10. These two points correspond, respectively, to points ω_b and ω_a in Fig. 11.2a. The choice $\omega T = \omega_m T = 10$ or $\omega T = \omega_{\phi 1}T = 10$ (where $\omega_a = \omega_{\phi 1}$ or $\omega_a = \omega_{m1}$) is made because this provides the maximum attenuation at ω_{m1}. This ensures a maximum possible gain increase in $\mathbf{G}(j\omega)$ to achieve the desired M_m. Thus, for $\omega_{m1} = 0.72$, the value $T \approx 13.9$ is obtained. Note that the value of T can also be obtained from Eq. (11.4). For ease of calculation, the value of T is rounded off to 14.0, which does not appreciably affect the results.

In Figs. 11.7a and 11.7b the Lm and phase-angle diagrams for the compensator $\mathbf{G}'_c(j\omega)$ are shown, together with the curves for $\mathbf{G}'_x(j\omega)$. The sum of the curves is also shown and represents the function $\mathbf{G}'_c(j\omega)\mathbf{G}'_x(j\omega)$. The curves of $\mathbf{G}'_c(j\omega)\mathbf{G}'_x(j\omega)$ from Fig. 11.7 are used to draw the Lm–angle diagram of Fig. 11.8. By using this curve, the gain is adjusted to obtain an $M_m = 1.26$. The results are $K'_1 = 6.68$ and $\omega_m = 0.53$. Thus an improvement in K_1 has been obtained, but a penalty of a smaller ω_m results. An increase in gain of approximately $A = 7.7$ results from the use of a lag compensator with $\alpha = 10$, which is typical of the gain increase permitted by a given α. This additional gain may be obtained from an amplifier already present in the basic system if it is not already operating at maximum gain. Note that the value of gain is somewhat less than the chosen value of α. Provided that the decrease in ω_m and the increase in gain are acceptable, the system has been suitably compensated.

(a)

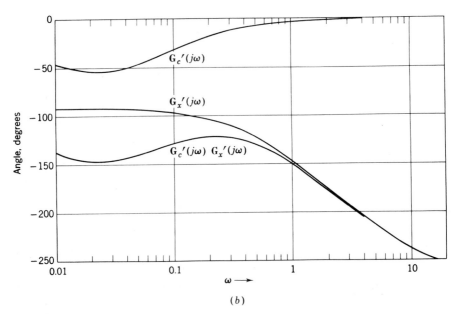

(b)

FIGURE 11.7

Log magnitude and phase-angle diagrams for

$$G'_x(j\omega) = \frac{1}{j\omega(1 + j\omega)(1 + j0.2\omega)}$$

$$G'_c(j\omega)G'_x(j\omega) = \frac{1 + j14\omega}{j\omega(1 + j\omega)(1 + j0.2\omega)(1 + j140\omega)}$$

FIGURE 11.8
Log magnitude–angle diagram: original and lag-compensated systems.

The polar plots representing the basic and compensated systems shown in Fig. 11.9 also show the effects of compensation. Both curves are tangent to the circle representing $M_m = 1.26$. Note that there is essentially no change of the high-frequency portion of $\mathbf{G}(j\omega)$ compared with $\mathbf{G}_x(j\omega)$. However, in the low-frequency region there is a cw rotation of $\mathbf{G}(j\omega)$ due to the lag compensator.

Table 11.1 presents a comparison between the root-locus method and the frequency-plot method of applying a lag compensator to a basic control system. Note the similarity of results obtained by these methods. The selection of the method depends upon individual preference. The compensators developed in the root-locus and frequency-plot methods are different, but achieve comparable results. To show this explicitly, Table 11.1 includes the compensator developed in the root-locus method as applied to the frequency-response method. In the frequency-response method, either the log or the polar plot may be used. An analytical technique for designing a lag or PI compensator, using both phase margin and a desired value of K_m, is presented in Ref. 5.

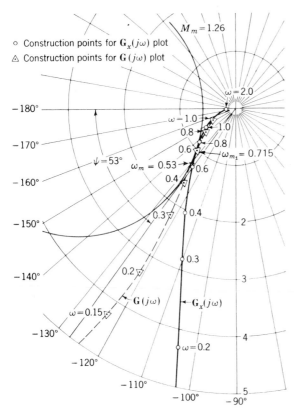

○ Construction points for $\mathbf{G}_x(j\omega)$ plot
△ Construction points for $\mathbf{G}(j\omega)$ plot

$M_m = 1.26$

FIGURE 11.9
Polar plots of a Type 1 control system, with and without lag compensation:

$$\mathbf{G}_x(j\omega) = \frac{0.872}{j\omega(1 + j\omega)(1 + j0.2\omega)}$$

$$\mathbf{G}(j\omega) = \mathbf{G}_c(j\omega)\mathbf{G}_x(j\omega) =$$

$$\frac{6.675(1 + j14\omega)}{j\omega(1 + j\omega)(1 + j0.2\omega)(1 + j140\omega)}$$

TABLE 11.1
Comparison between root-locus and frequency methods of applying a cascade lag compensator to a basic control system

$G_c'(s)$	ω_{m1}	ω_m	ω_{n1}	ω_n	K_1	K_1'	M_m	M_p	t_p, s	t_s, s
Root-locus method, ζ of dominant roots $= 0.45$										
$\dfrac{1 + 20s}{1 + 200s}$	—	0.66	0.898	0.854	0.840	8.02	1.31	1.266	4.30	22.0
Frequency-response method, $M_m = 1.26$ (effective $\zeta = 0.45$)										
$\dfrac{1 + 14s}{1 + 140s}$	0.715	0.53	—	—	0.872	6.68	1.260	1.257	4.90	22.8
$\dfrac{1 + 20s}{1 + 200s}$	0.715	0.60	—	—	0.872	7.38	1.260	1.246	4.55	23.6

11.5   LEAD COMPENSATOR

Figure 10.6*a* shows a lead compensator made up of passive elements that has the transfer function

$$\mathbf{G}'_c(j\omega) = \alpha\frac{1 + j\omega T}{1 + j\omega\alpha T} \qquad \alpha < 1 \tag{11.14a}$$

This equation is marked with a prime since it does not contain the gain A. The Lm and phase-angle equations for this compensator are

$$\text{Lm } \mathbf{G}'_c(j\omega) = \text{Lm } \alpha + \text{Lm}(1 + j\omega T) - \text{Lm}(1 + j\omega\alpha T) \tag{11.14b}$$

$$\underline{/\mathbf{G}'_c(j\omega)} = \underline{/1 + j\omega T} - \underline{/1 + j\omega\alpha T} \tag{11.14c}$$

A family of curves for various values of α is shown in Fig. 11.10. It is seen from the shape of the curves that an attenuation equal to Lm α is introduced at frequencies below $\omega T = 1$. Thus, the lead network is basically a high-pass filter: the high frequencies are passed, and the low frequencies are attenuated. Also, an appreciable lead angle is introduced in the frequency range from $\omega = 1/T$ to $\omega = 1/\alpha T$ (from $\omega T = 1$ to $\omega T = 1/\alpha$). Because of its angle characteristic a lead network can be used to increase the bandwidth for a system that falls in category 2.

Application of the lead compensator can be based on adjusting the γ and ω_ϕ. Assume that a system has the original forward transfer function $\mathbf{G}_x(j\omega)$, and that the gain has been adjusted for the desired γ or for the desired M_m. The Lm and

FIGURE 11.10
Log magnitude and phase-angle diagram for lead compensator

$$\mathbf{G}'_c(j\omega) = \alpha\frac{1 + j\omega T}{1 + j\omega\alpha T}$$

FIGURE 11.11
Original and lead-compensated log
magnitude and phase-angle plots.

phase-angle diagram of $G_x(j\omega)$ is sketched in Fig. 11.11. The purpose of the lead compensator is to increase ω_ϕ, and therefore to increase ω_m. The lead compensator introduces a positive angle over a relatively narrow bandwidth. By properly selecting the value of T, the phase-margin frequency can be increased from $\omega_{\phi 1}$ to $\omega_{\phi 2}$. Selection of T can be accomplished by physically placing the angle curve for the compensator on the same graph with the angle curve of the original system. The location of the compensator angle curve must produce the specified γ at the highest possible frequency; this location determines the value of the time constant T. The gain of $G'_c(j\omega)G_x(j\omega)$ must then be increased so that the Lm curve has the value 0 dB at the frequency $\omega_{\phi 2}$. For a given α, an analysis of the Lm curve of Fig. 11.10 shows that the farther to the right the compensator curves are placed, i.e., the smaller T is made, the larger the new gain of the system.

It is also possible to use the criterion derived in the previous chapter for the selection of T. This criterion is to select T, for Type 1 or higher systems, equal to or slightly smaller than the largest time constant of the original forward transfer function. This is the procedure used in the example of the next section. For a Type 0 system, the compensator time constant T is made equal to or slightly smaller than the second-largest time constant of the original system. Several locations of the compensator curves should be tested and the best results selected.

More accurate application of the lead compensator is based on adjusting the gain to obtain a desired M_m by use of the Lm–angle diagram. Figure 11.12 shows the original curve $G_x(j\omega)$ and the new curve $G'_c(j\omega)G_x(j\omega)$. The increase in gain to obtain the same M_m is the amount that the curve $G'_c(j\omega)G_x(j\omega)$ must be raised to be tangent to the M_m curve. Because of the positive angle introduced by the lead compensator, the new resonant frequency ω_{m2} is larger than the original resonant frequency ω_{m1}. The increase in gain is not as large as that obtained by use of the lag compensator.

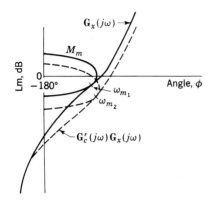

FIGURE 11.12
Original and lead-compensated log magnitude–angle curves.

11.6   DESIGN EXAMPLE: CASCADE LEAD COMPENSATION

For the unity-feedback Type 1 system of Sec. 10.7, the forward transfer function is

$$G_x(s) = \frac{K_1}{s(1 + s)(1 + 0.2s)} \tag{11.15}$$

From Sec. 11.4 for $M_m = 1.26$ it is found that $K_1 = 0.872$ s^{-1} and $\omega_{m1} = 0.715$ rad/s. Putting a lead compensator in cascade with the basic system increases the value of ω_m, and therefore improves the time response of the system. This situation falls in category 2.

The choice of values of α and T of the compensator must be such that the angle ϕ_c adds a sizeable positive phase shift at frequencies above ω_{m1} (or $\omega_{\phi1}$) of the overall forward transfer function. The nominal value of $\alpha = 0.1$ is often used; thus only the determination of the value of T remains. Two values of T selected in Sec. 10.9 are used in this example so that a comparison can be made between the root-locus and frequency-response techniques. The results are used to establish the design criteria for lead compensators.

In Fig. 11.13a and 11.13b the Lm and phase-angle diagrams for the basic system $G'_x(j\omega)$, the two compensators $G'_c(j\omega)$, and the two composite transfer functions $G_c(j\omega)G'_x(j\omega)$ are drawn. Using the Lm and phase-angle curves of $G'_c(j\omega)G'_x(j\omega)$ from Fig. 11.13, the Lm–angle diagrams are drawn in Fig. 11.14. The gain is again adjusted to obtain $M_m = 1.26$ (2 dB). The corresponding polar plots of the original system $G_x(j\omega)$ and the compensated systems $G(j\omega) = G_c(j\omega)G_x(j\omega)$ are shown in Fig. 11.15. Note that the value of ω_m is increased for both values of T.

Table 11.2 presents a comparison between the root-locus method and the frequency-response method of applying a lead compensator to a basic control system. The results show that a lead compensator increases the resonant frequency ω_m. However, a range of values of ω_m is possible, depending on the compensator time constant used. The larger the value of ω_m, the smaller the value of K_1. The characteristics selected must be based on the system specifications. The increase in gain is not as large as that obtained with a lag compensator. Note in the table the

(a)

(b)

FIGURE 11.13

Log magnitude and phase-angle diagrams of

$$\mathbf{G}'_x(j\omega) = \frac{1}{j\omega(1 + j\omega)(1 + j0.2\omega)}$$

$$\mathbf{G}'_c(j\omega)\mathbf{G}'_x(j\omega) = \frac{0.1(1 + j\omega T)}{j\omega(1 + j\omega)(1 + j0.2\omega)(1 + j0.1\omega T)}$$

382

FIGURE 11.14
Log magnitude–angle diagrams with lead compensators from Fig. 11.13a and b.

similarity of results obtained by the root-locus and frequency-response methods. Because of this similarity, the design rules of Sec. 10.9 can be interpreted in terms of the frequency-response method as follows.

RULE 1. The value of the time constant T in the numerator of $\mathbf{G}_c(j\omega)$ given by Eq. (11.14a) is made equal to the value of the largest time constant in the denominator of the original transfer function $\mathbf{G}_x(j\omega)$ for a Type 1 or higher system. This usually results in the largest increase in ω_m and ω_ϕ (best time response), and there is an increase in system gain K_m.

RULE 2. The value of the time constant T in the numerator of $\mathbf{G}_c(j\omega)$ is made slightly smaller than the largest time constant in the denominator of the original transfer function for a Type 1 or higher system. This results in achieving the largest gain increase for the system with an appreciable increase in ω_m and ω_ϕ.

Rule 1 is the simplest to apply to the design of the lead compensator. Where maximum gain increase and a good improvement in the time of response are desired, Rule 2 is applicable. For a Type 0 system, Rules 1 and 2 are modified so that the

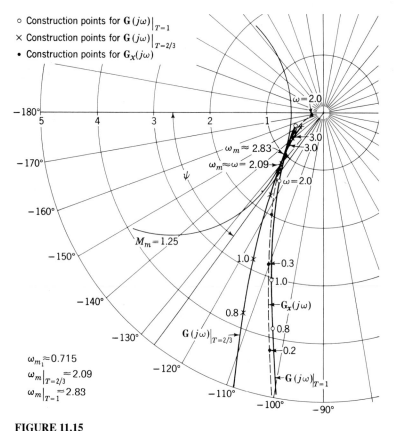

○ Construction points for $\mathbf{G}(j\omega)\big|_{T=1}$
× Construction points for $\mathbf{G}(j\omega)\big|_{T=2/3}$
• Construction points for $\mathbf{G}_x(j\omega)$

$\omega_{m_1} \approx 0.715$
$\omega_m\big|_{T=2/3} \approx 2.09$
$\omega_m\big|_{T=1} \approx 2.83$

FIGURE 11.15
Polar plots of Type 1 control system with and without lead compensation:

$$\mathbf{G}_x(j\omega) = \frac{0.872}{j\omega(1 + j\omega)(1 + j0.2\omega)}$$

$$\mathbf{G}(j\omega)\big|_{T=2/3} = \frac{3.48(1 + j2\omega/3)}{j\omega(1 + j\omega)(1 + j0.2\omega)(1 + j0.2\omega/3)}$$

$$\mathbf{G}(j\omega)\big|_{T=1} = \frac{3.13}{j\omega(1 + j0.1\omega)(1 + j0.2\omega)}$$

lead-compensator time constant is selected either equal to or slightly smaller than the second-largest time constant of the original system.

Remember that the lag and lead networks are designed for the same basic system; thus from Tables 11.1 and 11.2 it is seen that both result in a gain increase. The distinction between the two types of compensation is that the lag network gives the largest increase in gain (with the best improvement in steady-state accuracy) at the expense of increasing the response time, whereas the lead network gives an appreciable improvement in the time of response and a small improvement in steady-state accuracy. The particular problem at hand dictates the type of compen-

TABLE 11.2
Comparison between root-locus and frequency-response methods of applying a lead compensator to a basic control system

$G_c'(s)$	ω_{m1}	ω_m	ω_{n1}	ω_n	K_1	K_1'	M_m	M_p	t_p, s	t_s, s
Root-locus method, ζ of dominant roots = 0.45										
$0.1\dfrac{1+s}{1+0.1s}$	—	2.66	0.89	3.52	0.84	2.95	1.215	1.196	1.09	2.45
$0.1\dfrac{1+0.667s}{1+0.0667s}$	—	2.67	0.89	3.40	0.84	4.40	1.38	1.283	1.05	2.50
Frequency-response method, M_m = 1.26 (effective ζ = 0.45)										
$0.1\dfrac{1+s}{1+0.1s}$	0.715	2.83	—	—	0.872	3.13	1.26	1.215	1.05	2.40
$0.1\dfrac{1+0.667s}{1+0.0667s}$	0.715	2.09	—	—	0.872	3.48	1.26	1.225	1.23	2.73

sation to be used. An analytical technique for designing a lead compensator is presented in Ref. 5, using γ as the basis of design.

11.7 LAG-LEAD COMPENSATOR

The introduction of a lag compensator results in an increase in the gain, with a consequent reduction of the steady-state error in the system. Introducing a lead compensator results in an increase in the resonant frequency ω_m and a reduction in the system's settling time. If both the steady-state error and the settling time are to be reduced, a lag and a lead compensator must be used simultaneously. This improvement can be accomplished by inserting the individual lag and lead networks in cascade, but it is more effective to use a new network that has both the lag and the lead characteristics. Such a network, shown in Fig. 10.22a, is called a lag-lead compensator. Its transfer function is

$$\mathbf{G}_c(j\omega) = \frac{A(1 + j\omega T_1)(1 + j\omega T_2)}{(1 + j\omega\alpha T_1)(1 + j\omega T_2/\alpha)} \tag{11.16}$$

where $\alpha > 1$ and $T_1 > T_2$. The first half of this transfer function produces the lag effect, and the second half produces the lead effect. The Lm and phase-angle diagram for a representative lag-lead compensator is shown in Fig. 11.16, where the selection is made $T_1 = 5T_2$. A simple design method is to make T_2 equal to a time constant in the denominator of the original system. An alternative procedure is to locate the compensator curves on the log plots of the original system to produce the highest possible ω_ϕ. Figure 11.17 shows a sketch of the Lm and phase-angle diagram for the original forward transfer function $\mathbf{G}_x(j\omega)$, the compensator $\mathbf{G}_c'(j\omega)$, and the

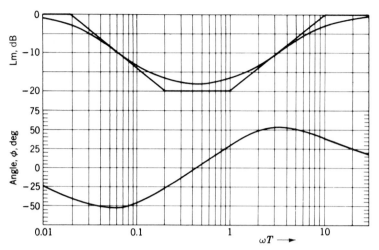

FIGURE 11.16
Log magnitude and phase-angle diagram for lag-lead compensator

$$\mathbf{G}'_c(j\omega) = \frac{(1 + j5\omega T_2)(1 + j\omega T_2)}{(1 + j50\omega T_2)(1 + j0.1\omega T_2)}$$

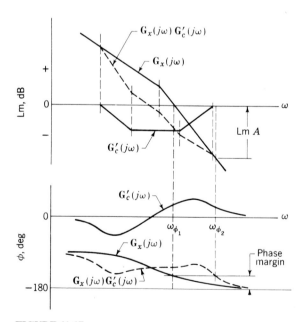

FIGURE 11.17
Original and lag-lead compensated log magnitude and phase
angle plots.

combination $\mathbf{G}_c(j\omega)\mathbf{G}'_x(j\omega)$. The phase-margin frequency is increased from $\omega_{\phi 1}$ to $\omega_{\phi 2}$, and the additional gain is labeled Lm A. The new value of M_m can be obtained from the Lm–angle diagram in the usual manner.

Another method is simply the combination of the design procedures used for the lag and lead networks, respectively. Basically, the Lm and phase-angle curves of the lag compensator can be located as described in Sec. 11.3. The lag-compensator curves are given by Fig. 11.4 and are located so that the negative angle introduced at either the original $\omega_{\phi 1}$ or the original ω_{m1} is small—of the order of $-5°$. This permits determination of the time constant T_1. The Lm and phase-angle curves of the lead compensator can be located as described in Sec. 11.5. The lead-compensator curves are given by Fig. 11.10, except that the Lm curve should be raised to have a value of 0 dB at the low frequencies. This is necessary because the lag-lead compensator transfer function does not contain α as a factor. Either the time constant T_2 of the compensator can be made equal to the largest time constant of the original system (for a Type 1 or higher system), or the angle curve is located to produce the specified phase margin angle at the highest possible frequency. In the latter case, the location chosen for the angle curve permits determination of the time constant T_2. When using the lag-lead network of Eq. (11.16), the value of α used for the lag compensation must be the reciprocal of the α used for the lead compensation. The example given in the next section implements this second method because it is more flexible.

11.8 DESIGN EXAMPLE: CASCADE LAG-LEAD COMPENSATION

For the basic system treated in Secs. 11.4 and 11.6, the forward transfer function is

$$\mathbf{G}_x(j\omega) = \frac{K_1}{j\omega(1 + j\omega)(1 + j0.2\omega)} \qquad (11.17)$$

For an $M_m = 1.26$, the result is $K_1 = 0.872 \text{ s}^{-1}$ and $\omega_{m1} = 0.715 \text{ rad/s}$. According to the second method described above, the lag-lead compensator is the combination of the individual lag and lead compensators designed in Secs. 11.4 and 11.6. Thus the compensator transfer function is

$$\mathbf{G}_c(j\omega) = A\frac{(1 + jT_1\omega)(1 + j\omega T_2)}{(1 + j\alpha T_1\omega)(1 + j0.1\omega T_2)} \qquad \text{for } \alpha = 10 \qquad (11.18)$$

The new forward transfer function with $T_1 = 14$ and $T_2 = 1$ is

$$\mathbf{G}(j\omega) = \mathbf{G}_c(j\omega)\mathbf{G}_x(j\omega) = \frac{AK_1(1 + j14\omega)}{j\omega(1 + j0.2\omega)(1 + j140\omega)(1 + j0.1\omega)} \qquad (11.19)$$

Figures 11.18a and 11.18b show the Lm and phase-angle diagrams of the basic system $\mathbf{G}'_x(j\omega)$, the compensator $\mathbf{G}'_c(j\omega)$, and the composite transfer function $\mathbf{G}'_c(j\omega)\mathbf{G}'_x(j\omega)$. From these curves, the Lm–angle diagram $\mathbf{G}'_a(j\omega)$ is shown in Fig. 11.19. Upon adjusting for an $M_m = 1.26$, the results are $K'_1 = 29.93$ and $\omega_{ma} = 2.69$. When these values are compared with those of Tables 11.1 and 11.2, it is seen that lag-lead compensation results in a value of K_1 approximately equal

FIGURE 11.18
Log magnitude and phase-angle diagrams of Eqs. (11.17) to (11.19) with $T_1 = 14$.

FIGURE 11.19
Log magnitude–angle diagram of $\mathbf{G}'(j\omega)$ from Eq. (11.19)—$\mathbf{G}_a'$ with $T_1 = 14$ and $\mathbf{G}_b'$ with $T_1 = 20$.

to the product of the K_1's of the lag- and lead-compensated systems. Also, the ω_m is only slightly less than that obtained with the lead-compensated system. Thus, a larger increase in gain is achieved than with lag compensation, and the value of ω_m is almost as large as the value obtained by using the lead-compensated system. One may therefore be inclined to use a lag-lead compensator exclusively. $\mathbf{G}_b$ in Fig. 11.19 is with $T_1 = 20$.

The resulting error coefficient (gain) due to the insertion of a lag-lead compensator is approximately equal to the product of

1. The original error coefficient K_m of the uncompensated system
2. The increase in K_m due to the insertion of a lag compensator
3. The increase in K_m due to the insertion of a lead compensator

Table 11.3 shows the results for both the root-locus and log-plot methods of applying a lag-lead compensator to a basic control system. The compensator designed from the log plots is different than the one designed by the root-locus method. To show that the difference in performance is small, Table 11.3 includes the results obtained with both compensators when applying the frequency-response method. Both plots are shown in Fig. 11.19. The plot using Eq. (10.42) with $T_1 = 20$ yields $\omega_{mb} = 2.74$, as shown in this figure.

TABLE 11.3
Comparison between root-locus and frequency-response methods of applying a lag-lead compensator to a basic control system

$G'_c(s)$	ω_{m1}	ω_m	ω_{n1}	ω_n	K_1	K'_1	M_m	M_p	t_p, s	t_s, s
Root-locus methods, ζ of dominant roots = 0.45										
$\dfrac{(1+s)(1+20s)}{(1+0.1s)(1+200s)}$	—	2.63	0.898	3.49	0.84	29.26	1.23	1.213	1.10	3.33
Frequency-response method, M_m = 1.26 (effective ζ = 0.45)										
$\dfrac{(1+s)(1+20s)}{(1+0.1s)(1+200s)}$	0.715	2.74	—	—	0.872	30.36	1.26	1.225	1.08	3.30
$\dfrac{(1+s)(1+14s)}{(1+0.1s)(1+140s)}$	0.715	2.69	—	—	0.872	29.93	1.26	1.229	1.09	3.55

11.9 FEEDBACK COMPENSATION DESIGN USING LOG PLOTS[1]

The previous sections demonstrate straightforward procedures for applying cascade compensation. When feedback compensation is applied to improve the tracking qualities of a control system, additional parameters must be determined. Besides the gain constant of the basic forward transfer function, the gain constant and frequency characteristics of the minor feedback must also be selected. Additional steps must therefore be introduced into the design procedure. The general effects of feedback compensation are demonstrated through application to the specific system that is used throughout this text. The use of a minor loop for feedback compensation is shown in Fig. 11.20. The transfer function $G_x(j\omega)$ represents the basic system, $H(j\omega)$ is the feedback compensator forming a minor loop around $G_x(j\omega)$, and A is an amplifier that is used to adjust the overall performance. The forward transfer function of this system is

$$G_1(j\omega) = \frac{C(j\omega)}{I(j\omega)} = \frac{G_x(j\omega)}{1 + G_x(j\omega)H(j\omega)} \tag{11.20}$$

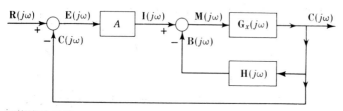

FIGURE 11.20
Block diagram for feedback compensation.

In order to apply feedback compensation, new techniques must be developed. This is done by first using some approximations with the straight-line Lm curves and then developing an exact procedure. Consider the two cases when

$$\left|\mathbf{G}_x(j\omega)\mathbf{H}(j\omega)\right| \ll 1 \qquad \text{and} \qquad \left|\mathbf{G}_x(j\omega)\mathbf{H}(j\omega)\right| \gg 1$$

The forward transfer function of Eq. (11.20) can be approximated by

$$\mathbf{G}_1(j\omega) \approx \mathbf{G}_x(j\omega) \qquad \text{for } \left|\mathbf{G}_x(j\omega)\mathbf{H}(j\omega)\right| \ll 1 \qquad (11.21)$$

and $$\mathbf{G}_1(j\omega) \approx \frac{1}{\mathbf{H}(j\omega)} \qquad \text{for } \left|\mathbf{G}_x(j\omega)\mathbf{H}(j\omega)\right| \gg 1 \qquad (11.22)$$

The condition when $\left|\mathbf{G}_x(j\omega)\mathbf{H}(j\omega)\right| \approx 1$ is still undefined, in which case neither Eq. (11.21) nor Eq. (11.22) is applicable. In the approximate procedure this condition is neglected, and Eqs. (11.21) and (11.22) are used when $\left|\mathbf{G}_x(j\omega)\mathbf{H}(j\omega)\right| < 1$ and $\left|\mathbf{G}_x(j\omega)\mathbf{H}(j\omega)\right| > 1$, respectively. This approximation allows investigation of the qualitative results to be obtained. After these results are found to be satisfactory, the refinements for an exact solution are introduced.

An example illustrates the use of these approximations. Assume that $\mathbf{G}_x(j\omega)$ represents a motor having inertia and damping. The transfer function derived in Sec. 2.13 can be represented by

$$\mathbf{G}_x(j\omega) = \frac{K_M}{j\omega(1 + j\omega T_m)} \qquad (11.23)$$

Let the feedback $\mathbf{H}(j\omega) = 1\underline{/0°}$. This problem is sufficiently simple to be solved exactly algebraically. However, use is made of the Lm curve and the approximate conditions. In Fig. 11.21 the Lm curve for $\mathbf{G}_x(j\omega)\mathbf{H}(j\omega)$ is sketched.

From Fig. 11.21 it is seen that $\left|\mathbf{G}_x(j\omega)\mathbf{H}(j\omega)\right| > 1$ for all frequencies below ω_1. Using the approximation of Eq. (11.22), $\mathbf{G}_1(j\omega)$ can be represented by $1/\mathbf{H}(j\omega)$ for frequencies below ω_1. Also, because $\left|\mathbf{G}_x(j\omega)\mathbf{H}(j\omega)\right| < 1$ for all frequencies above ω_1, the approximate of Eq. (11.21) can be used. Therefore $\mathbf{G}_1(j\omega)$ can be represented, as shown, by the line of zero slope and 0 dB for frequencies up to ω_1 and the line of -12-dB slope above ω_1. The approximate equation of $\mathbf{G}_1(j\omega)$ therefore has a quadratic in the denominator with $\omega_n = \omega_1$:

$$\mathbf{G}_1(j\omega) = \frac{1}{1 + 2\zeta j\omega/\omega_1 + (j\omega/\omega_1)^2} \qquad (11.24)$$

FIGURE 11.21
Log magnitude curves for $\mathbf{G}_x(j\omega)\mathbf{H}(j\omega) = K_1/j\omega(1 + j\omega T)$ and the approximate forward transfer function $\mathbf{G}_1(j\omega)$ for $\mathbf{H}(j\omega) = 1$.

Of course $\mathbf{G}(j\omega)$ can be obtained algebraically for this simple case from Eq. (11.20), with the result given as

$$\mathbf{G}_1(j\omega) = \frac{1}{1 + (1/K_M)j\omega + (j\omega)^2 T/K_M} \tag{11.25}$$

where $\omega_1 = \omega_n = (K_M/T)^{1/2}$ and $\zeta = 1/[2(K_M T)^{1/2}]$. Note that the approximate result is basically correct, but that some detailed information, in this case the value of ζ, is missing. The approximate angle curve can be drawn to correspond to the approximate Lm curve of $\mathbf{G}_1(j\omega)$ or to Eq. (11.24).

11.10 DESIGN EXAMPLE: FEEDBACK COMPENSATION (LOG PLOTS)

The system of Fig. 11.20 is investigated with the value of $\mathbf{G}_x(j\omega)$ given by

$$\mathbf{G}_x(j\omega) = \frac{K_x}{j\omega(1 + j\omega)(1 + j0.2\omega)} \tag{11.26}$$

The system having this transfer function has been used throughout this text for the various methods of compensation. The Lm and phase-angle diagram using the straight-line Lm curve is drawn in Fig. 11.22. A γ of 45° can be obtained, which yields $\omega_{\phi 1} = 0.8$, provided the gain is changed by –2 dB.

The object in applying compensation is to increase both the ω_ϕ and the gain. In order to select a feedback compensator $\mathbf{H}(j\omega)$, consider the following facts:

1. The system type should be maintained. In accordance with the conditions outlined in Sec. 10.15 and the results given in Sec. 10.20 for the root-locus method, it is known that good improvement is achieved by using a feedback compensator

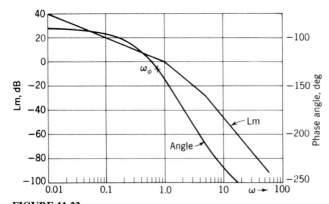

FIGURE 11.22
Log magnitude–phase-angle diagram of

$$\mathbf{G}'_x(j\omega) = \frac{1}{j\omega(1 + j\omega)(1 + j0.2\omega)}$$

$H(j\omega)$ that has a zero, $s = 0$, of order equal to (or preferably higher than) the type of the original transfer function $G_x(j\omega)$.

2. It is shown in Sec. 11.9 that the new forward transfer function can be approximated by

$$G_1(j\omega) \approx \begin{cases} G_x(j\omega) & \text{for } |G_x(j\omega)H(j\omega)| < 1 \\ \dfrac{1}{H(j\omega)} & \text{for } |G_x(j\omega)H(j\omega)| > 1 \end{cases}$$

This information is used for the proper selection of $H(j\omega)$.

In order to increase the values of ω_ϕ and the gain, a portion of the curves of $G_x(j\omega)$ for a selected range of frequencies can be replaced by the curves of $1/H(j\omega)$. This requires that the value of $H(j\omega)$ be selected such that $|G_x(j\omega)H(j\omega)| > 1$ for that range of frequencies. A feedback unit using a tachometer and an RC derivative network is considered. The circuit is shown in Fig. 10.36, and the transfer function is

$$H(j\omega T) = \frac{(K_t/T)(j\omega T)^2}{1 + j\omega T} = K_h H'(j\omega T) \tag{11.27}$$

where $K_h = K_t/T$. The Lm and phase-angle diagram for $1/H'(j\omega)$ is plotted in Fig. 11.23 with a nondimensionalized frequency scale, ωT. When this scale is converted to a dimensionalized scale ω, the curves of Fig. 11.23 shift to the left or to the right as T is increased or decreased. The angle curve $1/H(j\omega)$ shows that the desired γ can be obtained at a frequency that is a function of the compensator time constant T. This shows promise for increasing γ, provided that the magnitude $|G_x(j\omega)H(j\omega)|$ can be made greater than unity over the correct range of frequencies. Also, the larger K_t/T is made, the smaller the magnitude of $1/H(j\omega)$ at the phase-margin frequency. This permits a large value of A and therefore a large error coefficient.

Section 8.17 describes the desirability of a slope of -20 dB/decade to produce a large γ. The design objective is therefore to produce a section of the $G_1(j\omega)$ Lm curve with a slope of -20 dB/decade and with a smaller magnitude than (i.e., below) the Lm curve of $G_x(j\omega)$. This new section of the $G_1(j\omega)$ curve must occur at a higher

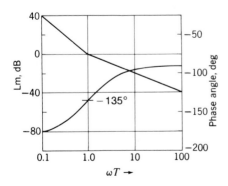

FIGURE 11.23

Log magnitude and phase-angle diagram of

$$\frac{1}{H'(j\omega T)} = \frac{1 + j\omega T}{(j\omega T)^2}$$

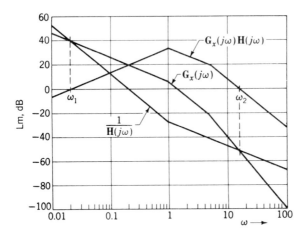

FIGURE 11.24
Log magnitude plots of $1/\mathbf{H}(j\omega)$, $\mathbf{G}_x(j\omega)$, and $\mathbf{G}_x(j\omega)\mathbf{H}(j\omega)$.

frequency than the original $\omega_{\phi 1}$. To achieve compensation, the Lm curve $1/\mathbf{H}(j\omega)$ is placed over the $\mathbf{G}_x(j\omega)$ curve so that there are one or two points of intersection. There is no reason to restrict the value of K_x, but a practical procedure is to use the value of K_x that produces the desired M_m without compensation. One such arrangement is shown in Fig. 11.24, with $K_x = 2$, $T = 1$, and $K_t = 25$. Intersections between the two curves occur at $\omega_1 = 0.021$ and $\omega_2 = 16$. The Lm plot of $\mathbf{G}_x(j\omega)\mathbf{H}(j\omega)$ is also drawn in Fig. 11.24 and shows that

$$\left|\mathbf{G}_x(j\omega)\mathbf{H}(j\omega)\right| \begin{cases} <1 & \text{for } \omega_2 < \omega < \omega_1 \\ >1 & \text{for } \omega_2 > \omega > \omega_1 \end{cases}$$

Therefore $\mathbf{G}_1(j\omega)$ can be represented approximately by

$$\mathbf{G}_1(j\omega) \approx \begin{cases} \mathbf{G}_x(j\omega) & \text{for } \omega_2 < \omega < \omega_1 \\ 1/\mathbf{H}(j\omega) & \text{for } \omega_2 > \omega > \omega_1 \end{cases}$$

The composite curve has corner frequencies at $1/\omega_1$, $1/T$, and $1/\omega_2$ and can therefore be approximately represented, when $T = 1$, by

$$\mathbf{G}_1(j\omega) = \frac{2(1 + j\omega/T)}{j\omega(1 + j\omega/\omega_1)(1 + j\omega/\omega_2)^2} = \frac{2(1 + j\omega)}{j\omega(1 + j47.5\omega)(1 + j0.0625\omega)^2}$$

$$(11.28)$$

The Lm and phase-angle diagram for Eq. (11.28) is shown in Fig. 11.25. Selecting $T = 1$, then $\gamma = 45°$ can occur at $\omega = 1$. However, the new approximate forward transfer function $\mathbf{G}_1(j\omega)$ has a second-order pole at $\omega = \omega_2$, as shown in Eq. (11.28). Therefore, there is higher frequency at which $\gamma = 45°$ can be achieved, as shown in Fig. 11.25. The higher frequency $\omega_\phi = 4.5$ is chosen in this example,

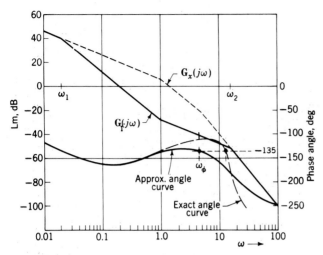

FIGURE 11.25

Log magnitude and phase-angle diagrams of

$$G_1(j\omega) = \frac{2(1 + j\omega)}{j\omega(1 + j47.5\omega)(1 + j0.0625\omega)^2}$$

which yields a gain of Lm $A = 41$ dB. This shows a considerable improvement of the system performance. determined from

The angle curve of $1/H'(j\omega T)$ in Fig. 11.23 shows that the desired values of γ and ω_ϕ are a function of the value of T. For a given value of γ the angle $\underline{/1/H'(j\omega T)} = -180° + \gamma$ is evaluated. The corresponding value of $(\omega T)_\gamma$ is then $\tan \gamma = \omega T$. Then the required value of T is $T = (\omega T)_\gamma/\omega_\phi$. It is desired that the frequency ω_2 at the upper intersection of $G_x(j\omega)$ and $1/H(j\omega)$ (see Figs. 11.24 and 11.25) be much larger than the new phase margin frequency ω_ϕ. This is achieved with $\omega_2 \geq 20\omega_\phi$, which satisfies the desired condition that $\underline{/G_1(j\omega_\phi)} = \underline{/1/H(j\omega_\phi)}$. When this condition is satisfied, a precise determination of T is possible.

An exact curve of $G_1(j\omega)$ can be obtained analytically. The functions $G_x(j\omega)$ and $H(j\omega)$ are

$$G_x(s) = \frac{2}{s(1 + s)(1 + 0.2s)} \qquad H(s) = \frac{25s^2}{1 + s} \qquad (11.29)$$

The function $G_1(j\omega)$ is obtained as follows:

$$G_1(j\omega) = \frac{G_x(j\omega)}{1 + G_x(j\omega)H(j\omega)} = \frac{2(1 + j\omega)}{j\omega(1 + j\omega)^2(1 + j0.2\omega) + 50(j\omega)^2}$$

$$= \frac{2(1 + j\omega)}{j\omega(1 + j52\omega)[0.00385(j\omega)^2 + 0.027(j\omega) + 1]} \qquad (11.30)$$

A comparison of Eqs. (11.28) and (11.30) shows that there is little difference between the approximate and the exact equations. The main difference is that the approximate equation has the term $(1 + j0.0625\omega)^2$, which assumes a damping ratio $\zeta = 1$ for this quadratic factor. The actual damping ratio is $\zeta = 0.217$ in the correct quadratic factor $[0.00385(j\omega)^2 + 0.027(j\omega) + 1]$. The corner frequency is the same for both cases, $\omega_n = 16$. As a result, the exact angle curve given in Fig. 11.25 shows a higher value of $\omega_\phi(= 13)$ than that obtained for the approximate curve ($\omega_\phi = 4.5$). If the basis of design is the value of M_m, the exact curve should be used. The data for the exact curve are readily available from a computer (see App. B).

The adjustment of the amplifier gain A is based on either γ or M_m. An approximate value for A is the gain necessary to raise the curve of Lm $G_1(j\omega)$ so that it crosses the 0-dB line at ω_ϕ. A more precise value for A is obtained by plotting $G_1(j\omega)$ on the Lm–angle diagram and adjusting the system gain for a desired value of M_m. The exact Lm $G_1(j\omega)$ curve should be used for best results.

The shape of the Nichols plot for $G_1(j\omega)$ is shown in Fig. 11.26. When adjusting the gain for a desired M_m, the objectives are to achieve a large resonant frequency ω_m and a large gain constant K_1. The tangency of the M_m curve at $\omega_c = 15.3$ requires a gain $A = 95.5$ and yields $K_1 = 191$. However, the rapid change in phase angle in the vicinity of ω_c can produce unexpected results. The M vs. ω characteristic curve c, shown in Fig. 11.27, has two peaks. This is not the form desired and must be rejected. This example serves to call attention to the fact that one point, the peak M value, is not sufficient to determine the suitability of the system response.

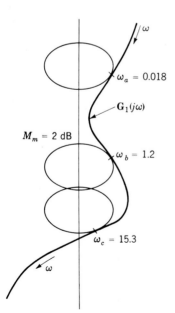

FIGURE 11.26
Nichols plot of $G_1(j\omega)$.

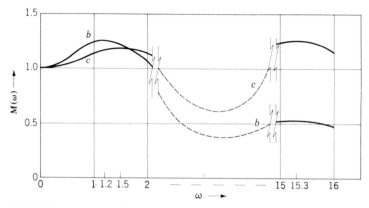

FIGURE 11.27
Closed-loop frequency response.

Tangency at $\omega_b = 1.2$ in Fig. 11.26 requires $A = 57.5$ and yields $K_1 = 115$. Curve b in Fig. 11.27 has the form desired and is acceptable. Compared with the original system, the minor-loop feedback compensation has produced a modest increase in ω_m from 0.715 to 1.2. There is also a very large increase in K_1 from 0.872 to 115. Tangency at ω_a is not considered, as it would represent a degradation of the uncompensated performance.

This example qualitatively demonstrates that feedback compensation can be used to produce a section of the Lm curve of the forward transfer function $G(j\omega)$ with a slope of -20 dB/decade. This section of the curve can be placed in the desired frequency range, with the result that the new ω_ϕ is larger than that of the original system. However, to obtain the correct quantitative results, the exact curves should be used instead of the straight-line approximations. Also, the use of the Lm–angle diagram as the last step in the design process, to adjust the gain for a specified M_m, gives a better indication of performance than the γ does.

The reader may wish to try other values of K_t and T in the feedback compensator in order to modify the system performance. Also, it is worth considering whether or not the RC network should be used as part of the feedback compensator. The results using the root-locus method (see Secs. 10.16 to 10.20) show that rate feedback alone does improve system performance. The use of second-derivative feedback produces only limited improvements.

11.11 APPLICATION GUIDELINES: BASIC MINOR-LOOP FEEDBACK COMPENSATORS

Table 11.4 presents three basic feedback compensators and the corresponding angle characteristics of $1/H(j\omega)$. One approach in designing these compensators is to determine the parameters of $H(s)$ that yield the desired value of the phase margin angle γ at a desired value of ω_ϕ.

TABLE 11.4
Basic feedback compensators

$H(s)$	$1/H(s)$	Characteristic of $\underline{/1/H(j\omega)}$ for $0 \leq \omega \leq \infty$
1. $K_t s$	$\dfrac{1}{K_t s}$	$-90°$
2. $\dfrac{K_t s}{T_s + 1}$	$\dfrac{T_s + 1}{K_t s}$	$-90°$ to $0°$
3. $\dfrac{K_t T_s^2}{T_s + 1}$	$\dfrac{T_s + 1}{K_t T s^2}$	$-180°$ to $-90°$

For compensators 1 and 2 in Table 11.4, the desired values of γ ($< 90°$) and ω_ϕ should occur in the vicinity of the intersection of the Lm[$1/\mathbf{H}(j\omega)$] and the Lm $\mathbf{G}_x(j\omega)$ plots, assuming that the angle of $\mathbf{G}_x(j\omega)$ varies from its initial value to $-90°i$ ($i = 2, 3, \ldots$) as the frequency increases from zero to infinity. Thus, to achieve an acceptable design of these compensators, an initial choice is made for the value of K_x or K_t and T. It may be necessary to lower the plot of Lm[$1/\mathbf{H}(j\omega)$] to ensure that the intersection value of ω is in the proper range, such that the desired value of ω_ϕ is attainable. The intersection of Lm[$1/\mathbf{H}(j\omega)$] with Lm $\mathbf{G}_x(j\omega)$ must occur on that part of Lm $\mathbf{G}(j\omega)$ that has a slope that will allow a $\gamma < 90°$ to be achievable in the vicinity of the intersection. By trial and error, the parameters of $1/H(s)$ are varied to achieve an acceptable value of ω_ϕ. For practical control systems there is an upper limit to an acceptable value of ω_ϕ, based upon the desire to attenuate high-frequency system noise.

For compensator 3, based upon Table 11.4 and Sec. 11.10, there are usually two intersections of Lm $\mathbf{G}_x(j\omega)$ and Lm($1/\mathbf{H}(j\omega)$). For the system of Fig. 11.20, the desired values of γ ($< 90 \deg$) and ω_ϕ for Lm $A\mathbf{G}_1$ can be made to occur on the -6-dB/octave portion of the plot of Lm[$1/\mathbf{H}(j\omega)$]. This requires that the intersections of the plots of Lm[$\mathbf{G}_x(j\omega)$] and Lm[$1/\mathbf{H}(j\omega)$] occur at least one decade away from the value of ω that yields the desired values of ω_ϕ and γ. With this separation, the forward transfer function $\mathbf{G}_1(j\omega_\phi) \approx 1/\mathbf{H}(j\omega_\phi)$.

The guidelines presented in this section for the three compensators of Table 11.4 are intended to provide some fundamental insight to the reader in developing effective feedback compensator design procedures.

11.12 SUMMARY

A basic unity-feedback control system is used throughout this and other chapters. It is therefore possible to compare the results obtained by adding each of the compensators to the system. The results obtained are consolidated in Table 11.5. The results achieved by using the frequency response are comparable to those obtained by us-

TABLE 11.5
Summary of cascade and feedback compensation of a basic control system using the frequency-response method with $M_m = 1.26$ (effective $\zeta = 0.45$)

Compensator	ω_m	K_1	Additional gain required	M_p	t_p, s	t_s, s
Uncompensated	0.715	0.872	—	1.203	4.11	9.9
Lag:						
$\dfrac{1 + 14s}{1 + 140s}$	0.53	6.68	7.66	1.257	4.90	22.8
$\dfrac{1 + 20s}{1 + 200s}$	0.60	7.38	8.46	1.246	4.55	23.6
Lead:						
$0.1\dfrac{1 + s}{1 + 0.1s}$	2.83	3.13	35.9	1.215	1.05	2.40
$0.1\dfrac{1 + 2s/3}{1 + 0.2s/3}$	2.09	3.48	39.9	1.225	1.23	2.73
Lag-lead:						
$\dfrac{(1 + s)(1 + 14s)}{(1 + 0.1s)(1 + 140s)}$	2.69	29.93	34.3	1.229	1.088	3.55
$\dfrac{(1 + s)(1 + 20s)}{(1 + 0.1s)(1 + 200s)}$	2.74	30.36	35.0	1.225	1.077	3.30
Feedback:						
$H(s) = \dfrac{25s^2}{1 + s}$	1.20	115	57.5	1.203	1.49	3.27

ing the root locus (see Chap. 10). The design methods that have been demonstrated use some trial-and-error techniques. This design process can be expedited by use of appropriate CAD packages (see App. B). The greater complexity of the feedback design procedures makes reliance on these CAD packages more valuable.

The following properties can be attributed to each compensator:

Cascade lag compensator   Results in an increase in the gain K_m and a small reduction in ω_m. This reduces steady-state error but increases transient settling time.

Cascade lead compensator   Results in an increase in ω_m, thus reducing the settling time. There may be a small increase in gain.

Cascade lag-lead compensator   Results in both an increase in gain K_m and in resonant frequency ω_m. This combines the improvements of the lag and the lead compensators.

Feedback compensator   Results in an increase in both the gain K_m and the resonant frequency ω_m. The basic improvement in ω_m is achieved by derivative feedback. This may be modified by adding a filter such as an *RC* network.

REFERENCES

1. Chestnut, H., and R. W. Mayer: *Servomechanisms and Regulating Systems Design*, 2d ed., vol. 1, Wiley, New York, 1959, chaps. 10 and 12.
2. Kuo, F. F.: *Network Analysis and Synthesis*, 2d ed., Wiley, New York, 1966.
3. Mitra, S. K.: *Analysis and Synthesis of Linear Active Networks*, Wiley, New York, 1969.
4. A comprehensive list of CAD programs is given in App. B.
5. Phillips, C. L.: "Analytical Bode Design of Controllers," *IEEE Trans. Educ.*, vol. E-28, no. 1, pp. 43–44, 1985.

CHAPTER 12

CONTROL-RATIO
MODELING

12.1 INTRODUCTION[4]

The conventional control-design techniques of the previous chapters determine closed-loop system performance based on the open-loop transfer function [$G(s)$ or $G(s)H(s)$] or the open-loop transfer function of an equivalent unity-feedback system. The frequency response or the root-locus technique yields a closed-loop response that is based on some of the closed-loop specifications. If necessary, compensation is then added in order to meet additional specifications. There are some techniques that first require the modeling of a desired control ratio that satisfies the desired figures of merit. The desired control ratio is then used to determine the necessary compensation. Formulating the desired control ratio is the main objective of this chapter. Figure 12.1 represents a multiple-input single-output (MISO) control system in which $r(t)$ is the input, $y(t)$ is the output required to track the input, $d(t)$ is an external disturbance input that must have minimal effect on the output, and $F(s)$ is a prefilter to assist in the tracking of $r(t)$ by $y(t)$. Thus, two types of control ratios need to be modeled: a tracking transfer function $M_T(s)$ and a disturbance transfer function $M_D(s)$. In Fig. 12.1 and the remainder of this text, except for Chap. 15, the output is denoted by $y(t)$ or $Y(s)$ instead of $c(t)$ or $C(s)$; this conforms with the symbols typically found in the literature for this material. For single-input single-output (SISO) systems, the symbol $c(t)$ is most often used to represent the output in most basic texts and agrees with the presentation in the earlier parts of this book.

The control systems considered in the previous chapters represent systems for which $F(s) = 1$ and $d(t) = 0$; thus, they are SISO control systems. For these SISO systems the Guillemin-Truxal method, which requires the designer to specify a

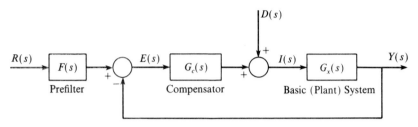

FIGURE 12.1
A MISO control system.

desired model $M_T(s)$, represents a method of designing the required compensator $G_c(s)$ of Fig. 12.1.

MISO control systems, in which both a desired input $r(t)$ and a disturbance input $d(t)$ are present, require the design of a cascade compensator $G_c(s)$ and an input prefilter $F(s)$, as shown in Fig. 12.1. The *quantitative feedback theory* (QFT) technique presented in Chapter 18 addresses the design of such systems, which have both a desired and a disturbance input. This design technique requires both types of control-ratio models, $M_T(s)$ and $M_D(s)$, in order to effect a satisfactory design. Some state-variable design techniques (Chapter 13) also requires the specification of a model tracking control ratio. Disturbance rejection while tracking a command input is discussed for multiple-input multiple-output systems in Chapter 18. The use of proportional plus integral plus derivative (PID) controllers leads naturally to the rejection of disturbance inputs and the tracking of the command inputs.

This chapter is devoted to the discussion of tracking control-ratio models, the presentation of the Guillemin-Truxal design method, the presentation of a disturbance-rejection design method for a control system having $r(t) = 0$ and a disturbance input $d(t) \neq 0$, and the discussion of disturbance-rejection control-ratio models. CAD packages (see App. B) are available to assist the designer in synthesizing desired model control ratios (see Sec. 10.6).

12.2 MODELING A DESIRED TRACKING CONTROL RATIO

The pole-zero placement method of this section requires the modeling of a desired tracking control ratio $[Y(s)/R(s)]_T = M_T(s)$ in order to satisfy the desired tracking performance specifications for M_p, t_p, and T_s (or t_s) for a step input (transient characteristic) and the gain K_m (steady-state characteristic). The model must be consistent with the degrees of the numerator and denominator, w and n respectively, of the basic plant $G_x(s)$. Before proceeding, it is suggested that the reader review Secs. 3.9, 3.10, 9.11, 10.2, and 10.3.

CASE 1. The model that yields the desired figures of merit and is simplest to synthesize is one that represents an effective simple second-order control ratio $[Y(s)/R(s)]_T$, and is independent of the numerator and denominator degrees w and n of the plant. The simple second-order closed-loop model for this case is

$$M_{T1}(s) = \frac{A_1}{s^2 + as + A_1} = \frac{\omega_n^2}{s^2 + 2\zeta\omega_n s + \omega_n^2}$$

$$= \frac{\omega_n^2}{(s - p_1)(s - p_2)} = \frac{G_{eq}(s)}{1 + G_{eq}(s)} \tag{12.1}$$

where $A_1 = \omega_n^2$ ensures that, with a step input of amplitude R_0, the output tracks the input so that $y(t)_{ss} = R_0$. The control ratio of Eq. (12.1) corresponds to a unity-feedback system with an equivalent forward transfer function

$$G_{eq}(s) = \frac{A_1}{s(s + a)} \tag{12.2}$$

Note that although there are four possible specifications, M_p, t_p, T_s, and K_m, only two adjustable parameters, ζ and ω_n, are present in Eq. (12.1). When Eqs. (3.60), (3.61), and (3.64) are analyzed for peak time t_p, peak overshoot M_p, and settling time T_s, respectively, and the appropriate value for K_m is selected, it is evident that only two of these relationships can be used to obtain the two adjustable parameters that locate the two dominant poles $p_{1,2}$ of Eq. (12.1). Therefore, it may not be possible to satisfy all four specifications. For example, specifying the minimum value of the gain K_1 (where $K_1 = A_1/a$) yields the minimum acceptable value of ω_n. Also, the required value of ζ is determined from $M_p = 1 + \exp(-\pi\zeta/\sqrt{1 - \zeta^2})$. Having determined ω_n and ζ, the values of $t_p = \pi/\omega_d$ and $T_s = 4/\zeta\omega_n$ can be evaluated. If these values of t_p and T_s are within the acceptable range of values, the desired model has been achieved. If not, an alternative procedure is to determine ζ and ω_n by using the desired M_p and T_s, or M_p and $t_p = \pi/\omega_n\sqrt{1 - \zeta^2}$.

In order to satisfy the requirement that the specifications $1 < M_p \leq \alpha_p, t_p \leq t_{\alpha_p}$, and $t_s \leq t_{\alpha_s}$, two of these values α_p, t_{α_p}, and t_{α_s} are used to determine the two unknowns ζ_α and ω_{n_α}. These values of ζ_α and ω_{n_α} yield the desired dominant poles $p_{1,2}$ shown in Fig. 12.2. Any dominant complex pole not lying in the shaded area satisfies these specifications. If an acceptable second-order $M_T(s)$, as given by Eq. (12.1), cannot be found that satisfies all the desired performance specifications, a higher-order control ratio must be used.

CASE 2. It may be possible to satisfy the specifications by use of the following third-order model:

$$M_{T2}(s) = \frac{A}{s + d}M_{T1}(s) = \frac{A_2}{(s + d)(s^2 + 2\zeta\omega_n s + \omega_n^2)} \tag{12.3}$$

The dominant complex poles of Eq. (12.3) are determined in the same manner as for case 1; i.e., the values of ζ and ω_n are determined by using the specifications for M_p, t_p, T_s, and K_m. Because the model of Eq. (12.1) cannot satisfy all four specifications, the known effects of the real pole $s = -d$ in Eq. (12.3) can be used to try to achieve these specifications. With a CAD program (see App. B), various trial values of d are used to check whether or not the desired specifications can be achieved. Note that for each value of d, the corresponding value of A_2 must be equal to $A_2 = d\omega_n^2$

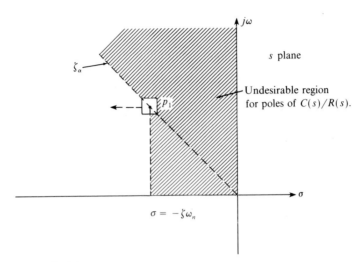

FIGURE 12.2
Location of the desired dominant pole.

in order to satisfy the requirement that $e(t)_{ss} = 0$ for a step input. As an example, the specifications $M_p = 1.125$, $t_p = 1.5\,s$, and $T_s = 3.0\,s$ cannot be satisfied by the second-order control ratio of the form of Eq. (12.1), but they can be satisfied by

$$\frac{Y(s)}{R(s)} = \frac{22}{(s + 1 \pm j3)(s + 2.2)} \tag{12.4}$$

For a third-order all-pole open-loop plant, the model of Eq. (12.3) must be used. Then, if the model of Eq. (12.1) is satisfactory, the pole $-d$ is made a nondominant pole.

CASE 3.　The third case deals with the third-order model [see Eq. (10.18)]

$$M_{T3}(s) = \frac{A(s + c)}{s + d}M_{T1}(s) = \frac{A_3(s + c)}{(s + d)(s^2 + 2\zeta\omega_n s + \omega_n^2)}$$

$$= \frac{A_3(s - z_1)}{(s - p_1)(s - p_2)(s - p_3)} \tag{12.5}$$

which has one zero. The zero can be a zero of the plant or a compensator zero that is inserted into the system to be used in conjunction with the pole $-d$ in order to satisfy the specifications. The procedure used in synthesizing Eq. (12.5) is based upon the nature of the zero; i.e., the zero (1) is not close to the origin or (2) it is close to the origin and is therefore a troublesome zero. (A stable system with a zero close to the origin may have a large transient overshoot for a step input.)

When the system has a step input, Fig. 12.3 represents the pole-zero diagram of $Y(s)$ for Eq. (12.5). Consider three possible locations of the real zero: to the right of, canceling, and to the left of the real pole $p_3 = -d$. When the zero cancels the pole p_3, the value of M_p is determined only by the complex-conjugate poles p_1 and

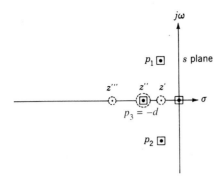

FIGURE 12.3
Pole-zero location of $Y(s)$ for Eq. (12.5), with a step input.

p_2; when the zero is to the right of p_3, the value of M_p is larger (see Sec. 10.9); and when it is to the left, the value of M_p is smaller [see Eq. (4.68) and Sec. 10.9]. The amount of this increase or decrease is determined by how close the zero is to the pole p_3. Thus the optimum location of the real pole, to the right of the zero, minimizes the effect of the troublesome zero; i.e., it reduces the peak of the transient overshoot.

In order to achieve M_T, it may be necessary to insert additional equipment in cascade and/or in feedback with $G_x(s)$. Also, in achieving the desired model for a plant containing a zero, the design method may use (1) the cancellation of the zero by a dominator pole, (2) the addition of a second zero, or (3) the replacement of the zero with one at a new location.

In synthesizing the desired control ratio, a system designer has a choice of one of the following approaches.

METHOD 1. Cancel some or all of the zeros of the control ratio by poles of the control ratio. The number of zeros α to be canceled must be equal to or less than the excess of the number of poles n minus the number of desired dominant poles β. If necessary, cascade compensators of the form $G_c(s) = 1/(s - P_i)$ can be inserted into the open-loop system in order to increase the number of poles of $Y(s)/R(s)$ so that cancellation of zeros can be accomplished. The number of compensator poles γ required must increase the order of the system sufficiently to ensure that $\alpha \leq n + \gamma - \beta$. Theoretically it is possible to insert a compensator of this form just preceding the block that contains the unwanted zero to be canceled by the pole of this compensator. However, in many systems it is not physically possible to locate the compensator in this manner. Also, setting P_i equal to the value of the unwanted zero z_h of the plant results in an unobservable state. All the states must be controllable and observable in order to achieve an *optimal response* (see Chap. 16). Thus, to achieve an optimal response and to cancel the unwanted zero, choose P_i such that $P_i \neq z_h$.

METHOD 2. Relocate some or all of the zeros of $G_x(s)$ by using cascade compensators of the form $G_c(s) = (s - z_i)/(s - P_i)$. The P_i is the value of the zero of $G_x(s)$ that is to be canceled, and the zero z_i assumes the value of the desired zero location. The number γ of these compensators to be inserted equals the number of zeros α to be relocated. The values of z_i represent the desired zeros. The values of P_i are selected to meet the condition that α *poles of the control ratio* are used to cancel the α unwanted zeros.

METHOD 3. A combination of the above two approaches can be used; i.e., canceling some zeros by closed-loop poles and replacing some of them by the zeros of the cascade compensators $G_c(s)$.

The modeling technique discussed in this section requires that the poles and zeros of $Y(s)/R(s)$ be located in a pattern that achieves the desired time response. This is called the *pole-zero placement technique*. The number of poles minus zeros of $(Y/R)_T$ must equal or exceed the number of poles minus zeros of the plant transfer function. For a plant having w zeros, w of the closed-loop poles are either located "very close" to these zeros, in order to minimize the effect of these poles on $y(t)$, or suitably placed for the desired time response to be obtained. If there are β dominant poles, the remaining $n - \beta - w$ poles are located in the nondominant region of the s plane. Another method of selecting the locations of the poles and zeros of $M_T(s)$ is given in Ref. 5, Chap. 16; it is based on a performance index that optimizes the closed-loop time responses. The specification of $Y(s)/R(s)$ is the basis for the Guillemin-Truxal method of Sec. 12.3 and for the state-feedback designs discussed in this and in the next chapter.

12.3 GUILLEMIN-TRUXAL DESIGN PROCEDURE[1]

In contrast to designing the cascade compensator $G_c(s)$ based upon the analysis of the open-loop transfer function, the Guillemin-Truxal method is based upon designing $G_c(s)$ to yield a desired control ratio $Y(s)/R(s)$. The control ratio for the unity-feedback cascade-compensated control system of Fig. 12.4 is

$$M_T(s) = \frac{Y(s)}{R(s)} = \frac{N(s)}{D(s)} = \frac{G_c(s)G_x(s)}{1 + G_c(s)G_x(s)} \tag{12.6}$$

where $N(s)$ and $D(s)$ represent the specified zeros and poles, respectively, of the desired control ratio. Solving this equation for the required cascade compensator $G_c(s)$ yields

$$G_c(s) = \frac{M_T(s)}{[1 - M_T(s)]G_x(s)} = \frac{N(s)}{[D(s) - N(s)]G_x(s)} \tag{12.7}$$

In other words, when $N(s)$ and $D(s)$ have been specified, the required transfer function for $G_c(s)$ is obtained. Network-synthesis techniques are then utilized to synthesize a network having the required transfer function of Eq. (12.7). Usually a passive network containing only resistors and capacitors is desired.

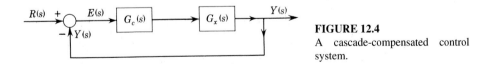

FIGURE 12.4
A cascade-compensated control system.

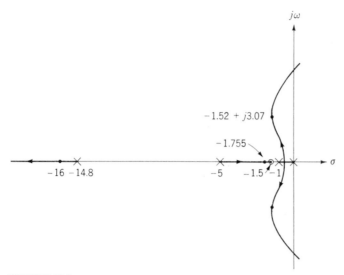

FIGURE 12.5
Root-locus plot for Example 1.

Example 1. The desired control ratio of a unity-feedback control system is

$$\frac{Y(s)}{R(s)} = \frac{N(s)}{D(s)} = \frac{210(s + 1.5)}{(s + 1.75)(s + 16)(s + 1.5 \pm j3)} \qquad (12.8)$$

The numerator constant has been selected to yield zero steady-state error with a step input and the desired figures of merit are $M_p = 1.28$, $t_p = 1.065s$, and $t_s = 2.54s$. The basic plant is

$$G_x(s) = \frac{4}{s(s + 1)(s + 5)} \qquad (12.9)$$

The required cascade compensator, determined by substituting Eqs. (12.8) and (12.9) into Eq. (12.7), is

$$G_c(s) = \frac{52.5s(s + 1)(s + 1.5)(s + 5)}{s^4 + 20.75s^3 + 92.6s^2 + 73.69s}$$

$$= \frac{52.5s(s + 1)(s + 1.5)(s + 5)}{s(s + 1.02)(s + 4.88)(s + 14.86)} \approx \frac{52.5(s + 1.5)}{s + 14.86} \qquad (12.10)$$

The approximation made in $G_c(s)$ yields a simple, physically realizable, minimum-phase lead network with an $\alpha = 0.101$. Based on a root-locus analysis of $G_c(s)G_x(s)$, the ratios of the terms $(s + 1)/(s + 1.02)$ and $(s + 5)/(s + 4.88)s$ of $G_c(s)$ have a negligible effect on the location of the desired dominant poles of $Y(s)/R(s)$. With the simplified $G_c(s)$ the root-locus plot, using $G_c(s)G_x(s)$, is shown in Fig. 12.5, and the control ratio achieved is

$$\frac{Y(s)}{R(s)} = \frac{210(s + 1.5)}{(s + 1.755)(s + 16.0)(s + 1.52 \pm j3.07)}$$

Because this equation is very close to that of Eq. (12.8), it is satisfactory. Note that the root-locus plot of $G_c(s)G_x(s) = -1$ has asymptotes of $180°$ and $\pm 60°$; therefore the system becomes unstable for high gain.

The Guillemin-Truxal design method, as illustrated by this example, involves three steps:

1. Specifying the desired zeros, poles, and numerator constant of the desired closed-loop function $Y(s)/R(s)$ in the manner discussed in Sec. 12.2.
2. Solving for the required cascade compensator $G_c(s)$ from Eq. (12.7).
3. Synthesizing a physically realizable compensator $G_c(s)$, preferably a passive network.

Using practical cascade compensators, the number of poles minus zeros of the closed-loop system must be equal to or greater than the number of poles minus zeros of the basic open-loop plant. This must include the poles and zeros introduced by the cascade compensator. The practical aspects of system synthesis using the Guillemin-Truxal method impose the limitation that the poles of $G_x(s)$ must lie in the left-half s plane. Failure to exactly cancel poles in the right-half s plane would result in an unstable closed-loop system. The state-variable-feedback method, described in the next chapter, is a compensation method that achieves the desired $Y(s)/R(s)$ without this limitation.

12.4   INTRODUCTION TO DISTURBANCE REJECTION[2,3]

The previous chapters and Sec. 12.3 deal with designing a control system whose output follows (tracks) the input signal. The design approach for the disturbance-rejection problem, by contrast, is based upon the performance specification that the output of the control system of Fig. 12.6 must not be affected by a system disturbance input $d(t)$. In other words, the steady-state output $y(t)_{ss} = 0$ for the input $d(t) \neq 0$. Consider the control ratio

$$\frac{Y(s)}{D(s)} = \frac{K(s^w + c_{w-1}s^{w-1} + \cdots + c_1 s + c_0)}{s^n + q_{n-1}s^{n-1} + \cdots + q_1 s + q_0} \tag{12.11}$$

Assuming a stable system and a step disturbance $D(s) = D_0/s$, the output is given by

$$Y(s) = \frac{K(s^w + c_{w-1}s^{w-1} + \cdots + c_1 s + c_0)}{s(s^n + q_{n-1}s^{n-1} + \cdots + q_1 s + q_0)} \tag{12.12}$$

The desired output $y(t)_{ss} = 0$ is achieved only if $c_0 = 0$. *This requires that the numerator of $Y(s)/D(s)$ have at least one zero at the origin.*

It is desired for this type of control problem that the disturbance have no effect on the steady-state output; also, the resulting transient must die out as fast as possible with a limit α_p on the maximum magnitude of the output.

Plant

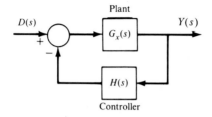

FIGURE 12.6
A simple disturbance-rejection control system.

The next three sections present a frequency-domain design technique for minimizing the effect of a disturbance input on the output of a control system. This technique is enhanced by simultaneously performing a qualitative root-locus analysis for a proposed feedback-compensator structure.

12.5 A SECOND-ORDER DISTURBANCE-REJECTION MODEL[2]

A simple second-order continuous-time disturbance-rejection model transfer function for the system of Fig. 12.6 is

$$M_D(s) = \frac{Y(s)}{D(s)} = \frac{G_x}{1 + G_x H} = \frac{K_x s}{s^2 + 2\zeta\omega_n s + \omega_n^2} \tag{12.13}$$

where $K_x > 0$. For a unit-step disturbance $D(s) = 1/s$ and $y(t)_{ss} = 0$. Eq. (12.13) yields (see Transform Pair 24a in App. A)

$$y(t) = \mathcal{L}^{-1}\left[\frac{K_x}{s^2 + 2\zeta\omega_n s + \omega_n^2}\right]$$

$$= \frac{K_x}{\omega_n\sqrt{1 - \zeta^2}}e^{-\zeta\omega_n t}\sin\omega_n\sqrt{1 - \zeta^2}\,t \tag{12.14}$$

where $a^2 + b^2 = \omega_n^2$, $b = \omega_d$, and $a = |\sigma| = \zeta\omega_n$. The time domain or frequency domain approaches can be used to determine appropriate model values of ζ and ω_n, while K_x is assumed fixed. The application of these methods is described below.

Time Domain

The time response of Eq. (12.14) is plotted vs. $\omega_n t$ in Fig. 4.14 for several values of $\zeta \le 1$. By setting the derivative of $y(t)$ with respect to time equal to zero, it can be shown that the maximum overshoot occurs at the time

$$t_p = \frac{\cos^{-1}\zeta}{\omega_n\sqrt{1 - \zeta^2}} \tag{12.15}$$

and the maximum value of $y(t)$ is

$$y(t_p) = \frac{K_x}{\omega_n}\exp\left(-\frac{\zeta\cos^{-1}\zeta}{\sqrt{1 - \zeta^2}}\right) \tag{12.16}$$

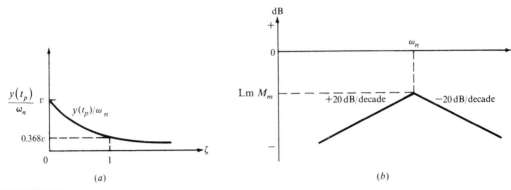

FIGURE 12.7
(a) Peak value of Eq. (12.14), using Eq. (12.17). (b) Bode plot characteristics of Eq. (12.18).

Letting $K_x = v\omega_n^2$ (with $v > 0$), Eq. (12.16) is rearranged to the form

$$\frac{y(t_p)}{\omega_n} = v \exp\left(-\frac{\zeta \cos^{-1}\zeta}{\sqrt{1 - \zeta^2}}\right) \tag{12.17}$$

A plot of Eq. (12.17) vs. ζ is shown in Fig. 12.7.

Frequency Domain

The frequency response function of Eq. (12.13) is

$$\mathbf{M}_D(j\omega) = \frac{\mathbf{Y}(j\omega)}{\mathbf{D}(j\omega)} = \frac{(K_x/\omega_n^2)(j\omega)}{[1 - (\omega/\omega_n)^2] + j(2\zeta\omega/\omega_n)} = \frac{K_x}{\omega_n^2}\mathbf{M}_D'(j\omega) \quad (12.18)$$

A straight-line approximation of the Bode plot of Lm $\mathbf{M}_D(j\omega)$ vs. log ω (drawn on semi-log paper) is shown in Fig. 12.7b for a selected value of ω_n. This plot reveals that

1. The closed-loop transfer function $\mathbf{M}_D(j\omega)$ must attenuate all frequencies present in the disturbance input $\mathbf{D}(j\omega)$. The plot of Lm $\mathbf{M}_D'(j\omega)$ vs. ω is raised or lowered by an amount equal to Lm(K_x/ω_n^2). The value of K_x/ω_n^2 therefore determines the maximum value M_m shown in Fig. 12.7b.
2. As ω_n increases, the peak value moves to the right to a higher frequency.

Selection of the desired disturbance model of Eq. (12.13) is based on both the time-response and the frequency-response characteristics. For example, selection of ω_n determines the bandwidth, as shown in Fig. 12.7b. Also, the maximum value of the permitted peak overshoot, as shown in Figs. 4.14 and 12.7a, is determined by the value of ζ. In addition, the peak time t_p and the settling time $t_s = 4/\zeta\omega_n$ of the transient may be used to determine the parameters in the model of Eq. (12.13). Some trial and error may be necessary to ensure that all specifications for the disturbance model are satisfied.

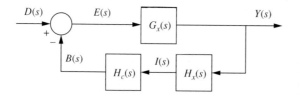

FIGURE 12.8
A control system with a disturbance input.

12.6 DISTURBANCE-REJECTION DESIGN PRINCIPLES FOR SISO SYSTEMS[2]

The desired disturbance model $M_D(s)$ of a system must be based on the specified time-domain and frequency-domain characteristics as described in Secs. 12.4 and 12.5. In order to obtain the desired characteristics of $M_D(s)$, it is necessary to design the required feedback controller $H(s)$ in Fig. 12.6. A first trial may be $H(s) = K_H/s$. If this choice of $H(s)$ does not produce the desired characteristics, then a higher-order feedback transfer function must be used. The order of the transfer functions in the system may also require a disturbance model of higher than second order.

The approximations developed in Sec. 11.9 are now restated for the disturbance-rejection system of Fig. 12.8.

$$\frac{Y(j\omega)}{D(j\omega)} = \frac{G_x(j\omega)}{1 + G_x(j\omega)H(j\omega)} \tag{12.19}$$

where $H(j\omega) = H_x(j\omega)H_c(j\omega)$. This control ratio can be approximated by

$$\frac{Y(j\omega)}{D(j\omega)} \approx \begin{cases} G_x(j\omega) & \text{for } |G_x(j\omega)H(j\omega)| \ll 1 \tag{12.20} \\[2ex] \dfrac{1}{H(j\omega)} & \text{for } |G_x(j\omega)H(j\omega)| \gg 1 \tag{12.21} \end{cases}$$

The condition when $|G_x(j\omega)H(j\omega)| \approx 1$ is still undefined, in which case neither of these conditions is applicable. In the approximate procedure the straight-line Bode plots are used and this condition is neglected. Therefore Eqs. (12.20) and (12.21) are used when $|G_x(j\omega)H(j\omega)| < 1$ and $|G_x(j\omega)H(j\omega)| > 1$, respectively. This approximation, along with a root-locus sketch of $G_x(s)H(s) = -1$, allows investigation of the qualitative results to be obtained. After these results are found to be satisfactory, the refinements for an exact solution are introduced.

The design procedure using the approximations of Eqs. (12.20) and (12.21) are

Step 1. Using the straight-line approximations, plot Lm $G_x(j\omega)$ vs. log ω.

Step 2. Assume that the feedback compensator $H_c(s)$ is put in cascade with a specified feedback unit, $H_x(s)$, in order to yield the desired system performance. To use the straight-line approximations, with $H(j\omega) = H_x(j\omega)H_c(j\omega)$, plot Lm $1/H(j\omega)$ vs. ω. Note that:

1. The condition Lm $\mathbf{G}_x(j\omega)\mathbf{H}(j\omega) = 0$ may also be expressed as

$$\text{Lm } \mathbf{G}_x(j\omega) = -\text{Lm } \mathbf{H}(j\omega) = \text{Lm } 1/\mathbf{H}(j\omega) \qquad (12.22)$$

This condition is satisfied at the intersection of the plots of $\mathbf{G}_x(j\omega)$ and $1/\mathbf{H}(j\omega)$. The frequency at the intersections represents the boundary between the two regions given by Eqs. (12.20) and (12.21). There must be at least one intersection between the plots in order for both approximations to be valid. Disturbance rejection over the frequency band of $0 \leq \omega \leq \omega_b$, the condition of Eq. (12.21), results in only one intersection.

2. For disturbance rejection it is desired to have Lm $\mathbf{Y}(j\omega)/\mathbf{D}(j\omega) < $ Lm $\mathbf{M}_D$. Also, Eq. (12.21) must hold for $0 \leq \omega \leq \omega_b$ in order to yield the desired frequency-domain specifications. To satisfy these objectives it is usually necessary for the gain constant of $H(s)$ to be very high.

Step 3. Plot Lm $\mathbf{G}_x(j\omega)\mathbf{H}(j\omega)$ vs. ω by using the straight-line approximation, and determine the slope of the plot at the 0-dB crossover point. Achieving a stable system requires (see Sec. 8.16) that this slope be greater than -40 dB/decade if the adjacent corner frequencies are not close. Also, the loop transmission frequency ω_ϕ (the 0-dB crossover or phase-margin frequency) of Lm $\mathbf{G}_x(j\omega)\mathbf{H}(j\omega)$ must be within the allowable specified value in order to attenuate high-frequency loop noise. Placing zeros of $H_c(s)$ "too close" to the imaginary axis may increase the value of ω_ϕ. It may be necessary at this stage of the design procedure to modify $H_c(s)$ in order to satisfy the stability requirement and the frequency-domain specifications.

Step 4. Once a trial $H(s)$ is found that satisfies the frequency-domain specifications, sketch the root locus $[G_x(s)H(s) = -1]$ to ascertain that (a) at least a conditionally stable system exists, and (b) it is possible to satisfy the time-domain specifications. [Section 12.5 discusses the desired locations of the poles of $Y(s)$ that achieve the desired specifications (see also Prob. 12.7).] If the sketch seems reasonable, (1) obtain the exact root-locus plots and select a value of the static loop sensitivity $K = K_G K_x K_c$ and (2) for this value of K, obtain a plot of $y(t)$ vs. t in order to obtain the figures of merit.

Step 5. If the results of step 4 do not yield satisfactory results, then use these results to select a new $H_c(s)$ and repeat steps 2 through 4.

Example 2. Consider the nonunity-feedback system of Fig. 12.8, where

$$G_x(s) = \frac{K_G}{s(s+1)} = \frac{1}{s(s+1)} \qquad (12.23)$$

and

$$H_x(s) = \frac{K_x}{s+200} = \frac{200}{s+200} \qquad (12.24)$$

Because a disturbance input is present, it is necessary to design the compensator $H_c(s) = K_c H_c'(s)$ so that the control system of Fig. 12.8 satisfies the following specifications for a unit-step disturbance input $d(t)$: $|y(t_p)| \leq 0.002$, $y(t)_{ss} = 0$, and $t_p \leq 0.2$ s. Using the empirical estimate that $M_p = |y(t_p)/d(t)| \approx M_m$ gives the specification Lm $M_m \leq$ Lm $0.002 = -54$ dB for the desired model $M_D(s)$. It is also estimated that the bandwidth of the disturbance transfer function model $\mathbf{Y}(j\omega)/\mathbf{D}(j\omega)$ is $0 \leq \omega \leq 10$ rad/s.

Trial Solution

Step 1. The plot of Lm $\mathbf{G}_x(j\omega)$ vs. ω is shown in Fig. 12.9.

Step 2. Because $G_x(s)$ does not have a zero at the origin and $Y(s)/D(s)$ does require one for disturbance rejection (so that $y(\infty)_{ss} = 0$), then $H(s)$ must have a pole at the origin. Thus the feedback transfer function must have the form

$$H(s) = H_x(s)H_c(s) = \frac{K_H(s + a_1)\cdots}{s(s + 200)(s + b_1)\cdots} \tag{12.25}$$

where the degree of the denominator is equal to or greater than the degree of the numerator. As a first trial, let

$$H_c(s) = \frac{K_c}{s} \tag{12.26}$$

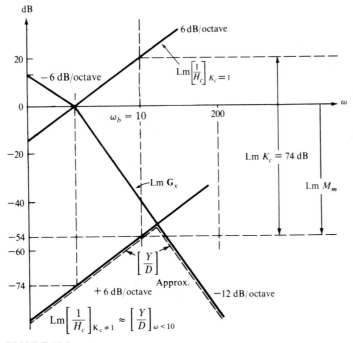

FIGURE 12.9
Log magnitude plots of Example 2.

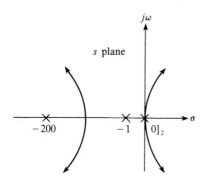

FIGURE 12.10
A root-locus sketch for Eq. (12.29).

Then $K_H = K_x K_c = 200K_c$. To determine the value of K_c required to satisfy Lm $\mathbf{Y}(j\omega)/\mathbf{D}(j\omega) \le -54$ dB for $0 \le \omega \le 10$ rad/s, initially plot the curve for [Lm $1/\mathbf{H}_c(j\omega)]_{K_c=1}$. The value of K_c is then selected so that the plot of Lm $1/\mathbf{H}_c(j\omega)$ is located such that it is on or below the -54 dB line for the frequency range $\omega \le \omega_b = 10$. Because the corner frequency of $\mathbf{H}(j\omega)$ at $\omega = 200$ lies outside the desired bandwidth, Lm $1/\mathbf{H}_c(j\omega) =$ Lm $1/\mathbf{H}(j\omega)$ is the controlling factor in the frequency range $\omega \le 10$. Next, lower this plot, as shown in the figure, until the plot of Lm$1/\mathbf{H}(j\omega)$ yields the desired value of -54 dB at $\omega_b = 10$ rad/s. Equation (12.21), for this case, satisfies the requirement $|\mathbf{M}(j\omega)| \le M_m$ over the desired bandwidth and is below -54 dB. Lm $200K_c = 74$ dB is required to lower the plot of Lm $1/\mathbf{H}(j\omega)$ to the required location. With this value of K_c the resulting first-trial feedback transfer function is

$$H(s) = \frac{5000}{s(s+200)} \tag{12.27}$$

Step 3.   An analysis of

$$\mathbf{G}_x(j\omega)\mathbf{H}(j\omega) = \frac{K_c}{(j\omega)^2(1+j\omega)\left(1 + j\dfrac{\omega}{200}\right)} \tag{12.28}$$

reveals that the plot of Lm$[\mathbf{G}_x(j\omega)\mathbf{H}(j\omega)]$ vs. ω crosses the 0-dB axis with a minimum slope of -40 dB/decade regardless of the value of K_c. This crossover characteristic is indicative of an unstable system.

Step 4.   This is confirmed by the root-locus sketch in Fig. 12.10, for

$$G_x(s)H(s) = \frac{K_H}{s^2(s+1)(s+200)} = -1 \tag{12.29}$$

which reveals a completely unstable system for the $H(s)$ of Eq. (12.27).

The next trial requires that at least one zero be added to Eq. (12.27) in order to make the system at least conditionally stable while maintaining the desired bandwidth. A final acceptable design of $H_c(s)$ that meets the system specifications requires a trial-and-error procedure. A solution to this design problem is given in Sec. 12.7.

Example 3. Consider the nonunity-feedback system of Fig. 12.9, where

$$G_x(s) = \frac{2}{s+2}$$

and
$$H(s) = H_x(s)H_c(s) = \frac{K_H}{s}$$

Determine the response $y(t)$ for a step disturbance input $d(t) = u_{-1}(t)$ as the gain K_H is increased.

The system output is

$$Y(s) = \frac{2}{s^2 + 2s + 2K_H} = \frac{2}{(s-p_1)(s-p_2)} = \frac{A_1}{(s-p_1)} + \frac{A_2}{(s-p_2)} \quad (12.30)$$

Thus
$$y(t) = A_1 e^{p_1 t} + A_2 e^{p_2 t} \quad (12.31)$$

where $p_{1,2} = -1 \pm j\sqrt{2K_H - 1}$. For $K_H > 0.5$,

$$A_1 = \frac{1}{j\sqrt{2K_H - 1}} \qquad A_2 = -A_1$$

The coefficients of the output transient, as $K_H \to \infty$, are

$$\lim_{K_H \to \infty} A_1 = \lim_{K_H \to \infty} A_2 = 0$$

For this example, as K_H is made larger, the coefficients A_1 and A_2 in Eq. (12.31) become smaller. Thus

$$\lim_{K_H \to \infty} y(t) = 0$$

which is the desired result for disturbance rejection. The output for a step disturbance input has two desired properties. First, the steady-state output is zero because the feedback contains an integrator. Second, the magnitude of the transient becomes smaller as the feedback gain K_H increases. (Note that the value of K_H is limited by practical considerations.) A similar analysis can be made for a more complicated system (see Probs. 12.8 and 12.9). In some applications it may not be necessary to achieve $y(\infty)_{ss} = 0$.

12.7 DISTURBANCE-REJECTION DESIGN EXAMPLE

In Example 2, the feedback compensator $H_c(s) = K_c/s$ results in an unstable system. An analysis of the root locus in Fig. 12.10 indicates that $H_c(s)$ must contain at least one zero near the origin in order to pull the root-locus branches into the left-half s plane. In order to minimize the effect of an unwanted disturbance on $y(t)$, it is necessary to minimize the magnitude of all the coefficients of

$$y(t) = A_1 e^{p_1 t} + \cdots + A_n e^{p_n t} \quad (12.32)$$

Problem 4.18 shows that this requires the poles of $Y(s)/D(s)$ to be located as far to the left in the s plane as possible. Figure 12.11 graphically describes the result of the analysis made in Prob. 12.7 for the desired placement of the control-ratio poles. The root-locus branch which first crosses the imaginary axis determines the

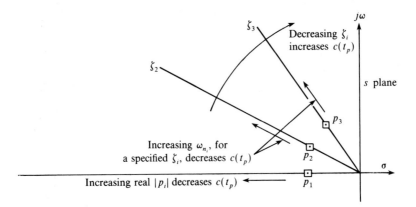

FIGURE 12.11
Effects on the peak overshoot to a disturbance input due to changes in the parameters ζ_i and ω_{n_i} of complex poles and the values of the real poles of the control ratio.

maximum value of gain for a stable system. The root selected on this branch should have $0.5 < \zeta < 0.7$, which has been determined empirically. When there is a choice of roots for a given ζ, the root selected should have the largest value of gain and ω_n. These properties are indicated in Fig. 12.11.

Example 4. The system of Example 2, Sec. 12.6, is now redesigned in order to achieve a stable system. Also, the root-locus branches must be pulled as far to the left as possible. Therefore, assume the following compensator, which contains two zeros:

$$H_c(s) = \frac{K_c(s + a_1)(s + a_2)}{s(s + b_1)(s + b_2)} \tag{12.33}$$

As a trial, let $a_1 = 4$ and $a_2 = 8$. The poles $-b_1$ and $-b_2$ must be chosen as far to the left as possible; therefore values $b_1 = b_2 = 200$ are selected. The plot of Lm $1/\mathbf{H}_x(j\omega)\mathbf{H}_c(j\omega) \leq -54$ dB, where $|\mathbf{Y}(j\omega)/\mathbf{D}(j\omega)| \approx |1/\mathbf{H}_x(j\omega)\mathbf{H}_c(j\omega)|$ over the range $0 \leq \omega \leq \omega_b = 10$ rad/s, is shown in Fig. 12.12. This requires that $200K_c \geq 50 \times 10^7$. The open-loop transfer function of Fig. 12.8 is now

$$G_x(s)H_x(s)H_c(s) = \frac{200K_c(s + 4)(s + 8)}{s^2(s + 1)(s + 200)^3}$$

$$= \frac{(K_c/1250)(0.25s + 1)(0.125s + 1)}{s^2(s + 1)(0.005s + 1)^3} \tag{12.34}$$

The poles and zeros of Eq. (12.34) are plotted in Fig. 12.13. The value of static-loop sensitivity $200K_c = 5 \times 10^8$ yields the set of roots

$$p_1 = -3.8469 \qquad p_2 = -10.079$$

$$p_{3,4} = -23.219 \pm j66.914 \qquad p_{5,6} = -270.32 \pm j95.873$$

that are shown in Fig. 12.13. This choice of gain results in a satisfactory design, because $M_p = 0.001076 \leq 0.002$ (−54 dB), $t_p = 0.155$ s (<0.2 s), $y(t)_{ss} = 0$, and

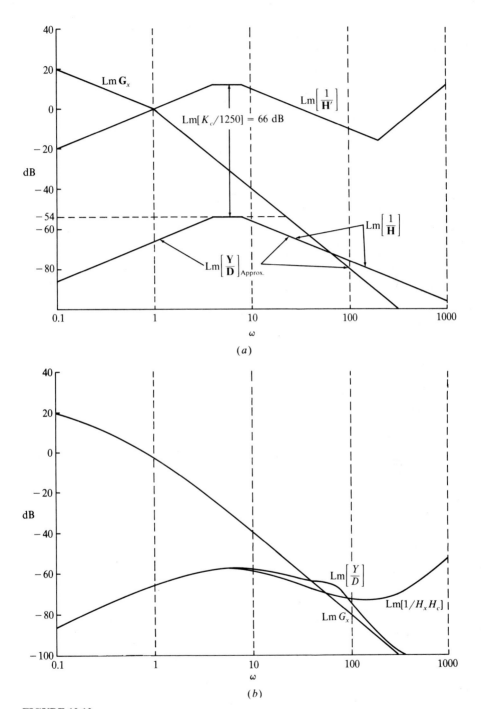

FIGURE 12.12
Bode plot for Example 4: (*a*) straight-line approximations; (*b*) exact poles.

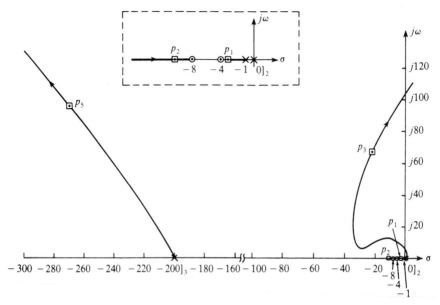

FIGURE 12.13
Sketch of root locus for Eq. (12.34). Not to scale.

$0 \leq \omega \leq \omega_b = 10$ rad/s. Note that the gain can be reduced until either $M_p = 0.002$ with $t_p \leq 0.2$ s or $t_p = 0.2$ s with $M_p \leq 0.002$ is achieved. Although not specified in this problem, the phase-margin frequency $\omega_\phi = 56.2$ rad/s must also be taken into account in the design of a practical system.

12.8 DISTURBANCE-REJECTION MODELS

The multiple-input single-output (MISO) control system of Fig. 12.14 has three inputs, $r(t)$, $d_1(t)$, and $d_2(t)$, which represent a tracking and two disturbance inputs, respectively. The previous chapters are concerned with the tracking problem alone, with $d(t) = 0$. This section uses the pole-zero placement method for the development of a desired disturbance-rejection control-ratio model $(Y/R)_D = M_D$. The spec-

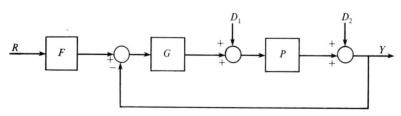

FIGURE 12.14
A multiple-input single-output (MISO) system.

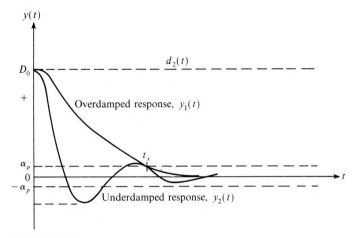

FIGURE 12.15
Disturbance-rejection time response for Fig. 12.14 with $r(t) = d_1(t) = 0$ and
$d_2(t) = D_0 u_{-1}(t)$.

ification for disturbance rejection for the control system of Fig. 12.14, with a step
disturbance input $d_2(t) = D_0 u_{-1}(t)$ and $r(t) = d_1(t) = 0$ is

$$|y(t)| \le \alpha_p \qquad \text{for } t \ge t_x \tag{12.35}$$

as shown in Fig. 12.15. Note that the initial value of the output for this condition is
$y(0) = D_0$. When $d_1(t) = D_0 u_{-1}(t)$ and $r(t) = d_2(t) = 0$, then the requirement for
disturbance rejection is

$$|y(t_p)| \le \alpha_p \tag{12.36}$$

as shown in Fig. 12.16, where $y(0) = 0$. The simplest control-ratio models for Figs.
12.15 and 12.16, respectively, are shown in Table 12.1.

With the values of α_p, t_x, and D_0 specified, the value of a for the case represented by Eqs. (12.37) and (12.39) is given by

$$a = -\frac{1}{t_x} \ln\left(\frac{\alpha_p}{D_0}\right) \tag{12.45}$$

For the case represented by Eqs. (12.38) and (12.40), then

$$a = -\frac{1}{t_x} \ln\left(\frac{\alpha_p}{D_0 \cos b t_x}\right) \tag{12.46}$$

In Eqs. (12.38), the two quantities a and b must be determined. Since Eq. (12.46)
shows that a is a function of b, a trial-and-error procedure can be used to determine
suitable values. Thus, assume a value of b and determine the corresponding value
of a. If the time-response characteristics are satisfactory, the model is completely
determined. If not, try another value for b.

TABLE 12.1
Disturbance rejection control ratio models

$r(t) = d_1(t) = 0$ and $d_2(t) = D_0 u_{-1}(t)$ **(See Fig. 12.15)**			
$M_D = \dfrac{Y_1(s)}{D_2(s)} = \dfrac{s}{s+a}$	(12.37)	$Y_1(s) = \dfrac{D_0}{s+a}$	$y_1(t) = D_0 e^{-at}$ (12.39)
$M_D = \dfrac{Y_2(s)}{D_2(s)} = \dfrac{s(s+a)}{(s+a)^2 + b^2}$	(12.38)	$Y_2(s) = \dfrac{D_0(s+a)}{(s+a)^2 + b^2}$	$y_2(t) = D_0 e^{-at} \cos bt$ (12.40)
$r(t) = d_2(t) = 0$ and $d_1(t) = D_0 u_{-1}(t)$ **(See Fig. 12.16)**			
$M_D = \dfrac{Y_3(s)}{D(s)} = \dfrac{K_D s}{(s+a)(s+b)}$	(12.41)	$Y_3(s) = \dfrac{K_D D_0}{(s+a)(s+b)}$	$y_3(t) = \dfrac{K_D D_0}{b-a}(e^{-at} - e^{-bt})$ (12.43)
$M_D = \dfrac{Y_4(s)}{D_1(s)} = \dfrac{K_D s}{(s+a)^2 + b^2}$	(12.42)	$Y_4(s) = \dfrac{K_D D_0}{(s+a)^2 + b^2}$	$y_4(t) = \dfrac{K_D D_0}{b} e^{-at} \sin bt$ (12.44)

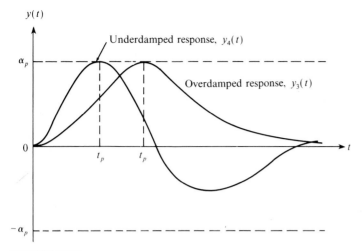

FIGURE 12.16
Disturbance-rejection time response for Fig. 12.14 with $r(t) = d_2(t) = 0$ and $d_1(t) = D_0 u_{-1}(t)$.

Consider the case where the settling time specification is given by

$$\left| y(t_s) \right| = \left| D_0 e^{-a t_s} \cos b t_s \right| = 0.02 \alpha_p = \alpha_s \qquad (12.47)$$

that is, the output decays to 2 percent of the maximum value at $t = t_s$. With t_s specified, then Eq. (12.40) can be used to obtain the value of a, i.e.,

$$a = -\frac{1}{t_s} \ln \left(\frac{\alpha_s}{D_0 \cos b t_s} \right) \qquad (12.48)$$

If all four quantities (t_p, t_s, α_p, and D_0) are specified, then the intersection of the plots of Eqs. (12.46) and (12.48), if one exists, yields the value of a that meets all the specifications.

For the case represented by Eqs. (12.41) and (12.43), assume that the values of α_p, α_s, t_p, t_s, and D_0 are specified. Then, from Eq. (12.43),

$$\alpha_p = K_\alpha (e^{-a t_p} - e^{-b t_p}) > 0 \qquad (12.49)$$

where $K_\alpha = K_D D_0/(b - a)$. Also from Eq. (12.43)

$$\frac{dy(t)}{dt} \bigg|_{t = t_p} = K_\alpha(-a e^{-a t_p} + b e^{-b t_p}) = 0 \qquad (12.50)$$

At $t = t_s$, Eq. (12.43) yields

$$\alpha_s = K_\alpha (e^{-a t_s} - e^{-b t_s}) \qquad (12.51)$$

From Eqs. (12.49) and (12.50) obtain, respectively,

$$e^{-a t_p} = \frac{\alpha_p}{K_\alpha} + e^{-b t_p} \qquad (12.52)$$

and
$$\frac{b}{a} = e^{(b-a)t_p} \tag{12.53}$$

Also, from Eq. (12.51) with $K_\alpha > 0$, obtain

$$e^{-at_s} > e^{-bt_s} \rightarrow 1 > \frac{e^{-(b/a)at_s}}{e^{-at_s}} = e^{(1-(b/a))at_s} \tag{12.54}$$

Equation (12.54) is valid only when $b/a > 1$, or

$$b > a \tag{12.55}$$

Thus, assume a value for b/a, and for various values of b (or a), obtain values for a (or b) from Eqs. (12.51) and (12.53). Obtain a plot of a vs. b for each of these equations. If these plots intersect, then the values of a or b at the intersection result in a transfer function $M_D(s)$ that satisfies the desired performance specifications. If no intersection exists, assume another value for b/a and repeat the process.

If the desired performance specifications cannot be obtained by the simple models of Table 12.1, then poles and/or zeros must be added, by a trial-and-error procedure, until a model is synthesized that is satisfactory. For either case, it may be necessary to add nondominant poles and/or zeros to $M_D(s)$ in order that the degrees of its numerator and denominator polynomials are consistent with the degrees of the numerator and denominator polynomials of the control ratio of the actual control system.

12.9 SUMMARY

This chapter presents the concepts of synthesizing a desired control-ratio model for a system's output in order to either track a desired input or to reject a disturbance input. These models are based on the known characteristics of a second-order system. Once these simple models are obtained, nondominant poles and/or zeros can be added to ensure that the final model is compatible with the degree of the numerator and denominator polynomials of the control ratio of the actual control system. The Guillemin-Truxal method is presented as one design method that requires the use of a tracking model. A disturbance-rejection frequency-domain design technique is presented. This technique is based upon the known correlation between the frequency domain, time domain, and the s plane.

REFERENCES

1. Truxal, J. G.: *Automatic Feedback Control System Synthesis*, McGraw-Hill, New York, 1955.
2. Houpis, C. H., and G. B. Lamont: *Digital Control Systems: Theory, Hardware, Software*, 2nd ed., McGraw-Hill, New York, 1992.
3. D'Azzo, J. J., and C. H. Houpis: *Feedback Control System Analysis and Synthesis*, 2nd ed., McGraw-Hill, New York, 1966.
4. Voland, G.: *Control Systems Modeling and Analysis*, Prentice-Hall, Englewood Cliffs, N. J., 1986.
5. D'Azzo, J. J., and C. H. Houpis: *Linear Control System Analysis and Design: Conventional and Modern*, 3rd ed., McGraw-Hill, New York, 1988.

CHAPTER
13

DESIGN: CLOSED-LOOP POLE-ZERO ASSIGNMENT (STATE-VARIABLE FEEDBACK)

13.1 INTRODUCTION

Chapters 10, 11, and 12 deal with the improvement of system performance in the time and frequency domains by conventional design procedures. Such procedures are based on an analysis of the open-loop transfer function and yield an improved closed-loop response as a consequence of properly designed cascade and/or feedback compensators. The design method presented in this chapter is based upon achieving a desired or model control ratio $M_T(s)$, i.e., a pole-zero placement method. Section 12.2 describes the modeling of a desired closed-loop transfer function and Sec. 12.3 presents the Guillemin-Truxal method of designing a cascade compensator to yield this closed-loop transfer function. This chapter introduces the concept of feeding back all the system states to achieve the desired improvement in system performance. The design concepts developed with conventional control theory are also used to design state-variable feedback. The state-variable-feedback concept requires that all states be accessible in a physical system, but for most systems this requirement is not met; i.e., some of the states are inaccessible. Techniques for handling systems with inaccessible states are presented in Chap. 14. The state-variable feedback design method presented in this chapter is based upon achieving a desired closed-loop state equation or a desired control ratio for a *single-input single-output* (SISO) system. Synthesis methods using matrix methods for *multiple-input multiple-output*

423

(MIMO) systems are presented in Chaps. 16 and 17. Following the convention often used with the state-variable representation of systems, the output and actuating signals are represented by y and u instead of c and e, respectively, in the remainder of this text.

The principal objective of the state-variable method in this chapter is the assignment of the closed-loop eigenvalues for SISO systems. This design process can be simplified for higher-order systems by first transforming the state equation into control canonical form $\dot{\mathbf{x}} = \mathbf{A}_c\mathbf{x} + \mathbf{b}_c\mathbf{u}$. For this representation of the state equation, the plant matrix $\mathbf{A}_c$ is in companion form, and the control matrix $\mathbf{b}_c$ contains all zeros except for the unit element in the last row. This form leads directly to the determination of the required state-feedback matrix that is unique.

Like the previous chapters, the analysis and synthesis involved in this chapter's problems relies on the use of the CAD packages like MATLAB, ICECAP-PC, or TOTAL-PC (see App. B). The utilization of such programs expedites the design process, achieving a solution, and eliminates tedious hand calculations.

13.2 CONTROLLABILITY AND OBSERVABILITY[3,4]

Before introducing the state feedback design methods, the necessary conditions and tests for *controllability* and *observability* are presented. Recall that the state and output equations representing a system have the form

$$\dot{\mathbf{x}} = \mathbf{Ax} + \mathbf{Bu} \tag{13.1}$$

$$\mathbf{y} = \mathbf{Cx} + \mathbf{Du} \tag{13.2}$$

An important objective of state variable control is the design of systems that have an optimum performance. The optimal control[6,7] is based on the optimization of some specific performance criterion, and the achievement of such optimal linear control systems is governed by the *controllability* and *observability* properties of the system. For example, in order to be able to relocate or reassign the open-loop plant poles to more desirable closed-loop locations in the s plane, it is necessary that the plant satisfy the controllability property. Further, these properties establish the conditions for complete equivalence between the state-variable and transfer-function representations. A study of these properties, presented below, provides a basis for consideration of the assignment of the closed-loop eigenstructure (eigenvalue and eigenvector spectra) in Chap. 16.

CONTROLLABILITY. A system is said to be completely *state-controllable* if, for any initial time t_0, each initial state $\mathbf{x}(t_0)$ can be transferred to any final state $\mathbf{x}(t_f)$ in a finite time, $t_f > t_0$, by means of an unconstrained control input vector $\mathbf{u}(t)$. An unconstrained control vector has no limit on the amplitudes of $\mathbf{u}(t)$. This definition implies that $\mathbf{u}(t)$ is able to affect each state variable in the state equation of Eq. (13.1). The solution of the state equation [see Sec. 3.15, Eq. (3.102)] is

$$\mathbf{x}(t) = \mathbf{\Phi}(t - t_0)\mathbf{x}(t_0) + \int_{t_0}^{t} \mathbf{\Phi}(t - \tau)\mathbf{Bu}(\tau)\,d\tau \tag{13.3}$$

OBSERVABILITY.   A system is said to be completely *observable* if every initial state $\mathbf{x}(t_0)$ can be exactly determined from the measurements of the output $\mathbf{y}(t)$ over the finite interval of time $t_0 \leq t \leq t_f$. This definition implies that every state of $\mathbf{x}(t)$ affects the output $\mathbf{y}(t)$:

$$\mathbf{y}(t) = \mathbf{C}\mathbf{x}(t) = \mathbf{C}\mathbf{\Phi}(t - t_0)\mathbf{x}(t_0) + \mathbf{C}\int_{t_0}^{t} \mathbf{\Phi}(t - \tau)\mathbf{B}\mathbf{u}(\tau)\,d\tau \qquad (13.4)$$

where the initial state $\mathbf{x}(t_0)$ is the result of control inputs prior to t_0.

The concepts of controllability and observability[5] can be illustrated graphically by the block diagram of Fig. 13.1. By the proper selection of the state variables it is possible to divide a system into the four subdivisions shown in Fig. 13.1.

S_{co} = completely controllable and completely observable subsystem

S_o = completely observable but uncontrollable subsystem

S_c = completely controllable but unobservable subsystem

S_u = uncontrollable and unobservable subsystem

This figure readily reveals that only the controllable and observable subsystem S_{co} satisfies the definition of a transfer-function matrix

$$\mathbf{G}(s)\mathbf{U}(s) = \mathbf{Y}(s) \qquad (13.5)$$

If the entire system is completely controllable and completely observable, then the state-variable and transfer-function matrix representations of a system are equivalent and accurately represent the system. In that case the resulting transfer function does carry all the information characterizing the dynamic performance of the system. As mentioned in Chap. 2, there is no unique method of selecting the state variables to be used in representing a system, but the transfer-function matrix is completely and uniquely specified once the state-variable representation of the system is known. In the transfer-function approach, the state vector is suppressed; the transfer-function method is concerned only with the system's input-output characteristics. The state-variable method also includes a description of the system's internal behavior.

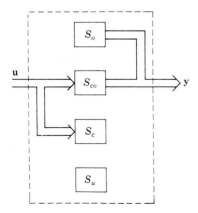

FIGURE 13.1
The four possible subdivisions of a system.

The determination of controllability and observability of a system by subdividing it into its four possible subdivisions is not always easy. A simple method for determining these system characteristics is developed by using Eq. (13.3) with $t_0 = 0$ and specifying that the final state vector $x(t_f) = 0$. This yields

$$0 = e^{At_f}x(0) + \int_0^{t_f} e^{A(t_f-\tau)}Bu(\tau)d\tau$$

or

$$x(0) = -\int_0^{t_f} e^{-A\tau}Bu(\tau)\,d\tau \tag{13.6}$$

The Cayley-Hamilton method presented in Sec. 3.14 shows that $\exp(-A\tau)$ can be expressed as a polynomial of order $n - 1$. Therefore,

$$e^{-A\tau} = \sum_{k=0}^{n-1} \alpha_k(\tau)A^k \tag{13.7}$$

Inserting this value into Eq. (13.6) yields

$$x(0) = -\sum_{k=0}^{n-1} A^k B \int_0^{t_f} \alpha_k(\tau)u(\tau)d\tau \tag{13.8}$$

With the input $u(t)$ of dimension r, the integral in Eq. (13.8) can be evaluated, and the result is

$$\int_0^{t_f} \alpha_k(\tau)u(\tau)\,d\tau = \beta_k$$

Equation (13.8) can now be expressed as

$$x(0) = -\sum_{k=0}^{n-1} A^k B\beta_k = -\left[\,B\,\middle|\,AB\,\middle|\,\cdots\,\middle|\,A^{n-1}B\,\right]\begin{bmatrix} \beta_0 \\ \hline \beta_1 \\ \hline \vdots \\ \hline \beta_{n-1} \end{bmatrix} \tag{13.9}$$

According to the definition of complete controllability, each mode $v_i e^{\lambda_i t}$ must be directly affected by the input $u(t)$. This requires that the *controllability* matrix M_c have the property

$$\text{Rank } M_c = \text{Rank }\left[\,B\,\middle|\,AB\,\middle|\,\cdots\,\middle|\,A^{n-1}B\,\right] = n \tag{13.10}$$

For a single-input system, the matrix B is the vector b, and the matrix of Eq. (13.10) has dimension $n \times n$.

When the state equation of a SISO system is in control canonical form; that is, the plant matrix A is in companion form and the control vector $b =$

$[0 \quad \cdots \quad 0 \quad K_G]^T$, then the controllability matrix $\mathbf{M}_c$ has full rank n (see Sec. 5.13). Thus, the system is completely controllable and it is not necessary to perform a controllability test.

When the matrix pair $(\mathbf{A}, \mathbf{B})$ is not completely controllable, the uncontrollable modes can be determined by transforming the state equation so that the plant matrix is in diagonal form (see Sec. 5.10). This approach is described below. An alternate method is given in Chap. 16. A simple method for determining controllability when the system has nonrepeated eigenvalues is to use the transformation $\mathbf{x} = \mathbf{Tz}$ to convert the $\mathbf{A}$ matrix into diagonal (canonical) form, as described in Sec. 5.10, so that the modes are decoupled. The resulting state equation is

$$\dot{\mathbf{z}} = \mathbf{T}^{-1}\mathbf{ATz} + \mathbf{T}^{-1}\mathbf{Bu} = \Lambda\mathbf{z} + \mathbf{B}'\mathbf{u} \tag{13.11}$$

Each transformed state z_i represents a mode and can be directly affected by the input $\mathbf{u}(t)$ only if $\mathbf{B}' = \mathbf{T}^{-1}\mathbf{B}$ has no zero row. Therefore, for nonrepeated eigenvalues, *a system is completely controllable if* $\mathbf{B}'$ *has no zero row.* For repeated eigenvalues see Refs. 3 and 5. $\mathbf{B}'$ is called the *mode-controllability* matrix. A zero row of $\mathbf{B}'$ indicates that the corresponding mode is uncontrollable. The eigenvalues associated with uncontrollable modes are called *input-decoupling zeros.*

In a similar manner, the condition for observability is derived from the homogeneous system equations

$$\dot{\mathbf{x}} = \mathbf{Ax} \tag{13.12}$$

$$\mathbf{y} = \mathbf{Cx} \tag{13.13}$$

When Eq. (13.7) is used, the output vector $\mathbf{y}(t)$ can be expressed as

$$\mathbf{y}(t) = \mathbf{C}e^{\mathbf{A}t}\mathbf{x}(0) + \sum_{k=0}^{n-1} \alpha_k(t)\mathbf{CA}^k\mathbf{x}(0) \tag{13.14}$$

For observability the output $\mathbf{y}(t)$ must be influenced by each state $\mathbf{x}_i$. This imposes restrictions on $\mathbf{CA}^k$. It can be shown that the system is completely observable if the following *observability* matrix $\mathbf{M}_o$ has the property

$$\text{Rank } \mathbf{M}_o = \text{Rank } \left[\mathbf{C}^T \, \middle| \, \mathbf{A}^T\mathbf{C}^T \, \middle| \, (\mathbf{A}^T)^2\mathbf{C}^T \, \middle| \cdots \middle| (\mathbf{A}^T)^{(n-1)}\mathbf{C}^T \right] = n \tag{13.15}$$

For a single-output system, the matrix $\mathbf{C}^T$ is a column matrix that can be represented by $\mathbf{c}$, and the matrix of Eq. (13.15) has dimension $n \times n$. When the matrix pair $(\mathbf{A}, \mathbf{C})$ is not completely observable, the unobservable modes can be determined by transforming the state and output equations so that the plant matrix is in diagonal form. This approach is described below. An alternate method is given in Chap. 16.

A simple method of determining observability that can be used when the system has nonrepeated eigenvalues is to use the transformation $\mathbf{x} = \mathbf{Tz}$ in order to put the equations in canonical form; that is, the plant matrix is transformed to the diagonal form Λ. The output equation is then

$$\mathbf{y} = \mathbf{Cx} = \mathbf{CTz} = \mathbf{C}'\mathbf{z} \tag{13.16}$$

If a column of $\mathbf{C}'$ has all zeros, then one mode is not coupled to *any* of the outputs and the system is unobservable. *The condition for observability, for the case of nonrepeated eigenvalues, is that $\mathbf{C}'$ have no zero columns.* $\mathbf{C}'$ is called the *mode-observability* matrix. The eigenvalues associated with an unobservable mode are called *output-decoupling zeros*.

Example. Figure 13.2 shows the block-diagram representation of a system whose state and output equations are

$$\dot{\mathbf{x}} = \begin{bmatrix} -2 & 0 \\ -1 & -1 \end{bmatrix} \mathbf{x} + \begin{bmatrix} 1 \\ 1 \end{bmatrix} u \tag{13.17}$$

$$y = [0 \quad 1] \mathbf{x} \tag{13.18}$$

Determine: (*a*) the eigenvalues from the state equation, (*b*) the transfer function $Y(s)/U(s)$, (*c*) whether the system is controllable and/or observable, (*d*) the transformation matrix $\mathbf{T}$ that transforms the general state-variable representation of Eqs. (13.17) and (13.18) into the canonical form representation, and (*e*) draw a simulation diagram based on the canonical representation.

(*a*)
$$|s\mathbf{I} - \mathbf{A}| = \begin{vmatrix} s+2 & 0 \\ 1 & s+1 \end{vmatrix} = (s+2)(s+1)$$

The eigenvalues are $\lambda_1 = -2$ and $\lambda_2 = -1$. Thus,

$$\Lambda = \begin{bmatrix} -2 & 0 \\ 0 & -1 \end{bmatrix}$$

(*b*)
$$\mathbf{\Phi}(s) = [s\mathbf{I} - \mathbf{A}]^{-1} = \frac{\begin{bmatrix} s+1 & 0 \\ -1 & s+2 \end{bmatrix}}{(s+1)(s+2)}$$

$$\mathbf{G}(s) = \mathbf{c}^T \mathbf{\Phi}(s)\mathbf{b} = \left([0 \quad 1] \begin{bmatrix} s+1 & 0 \\ -1 & s+2 \end{bmatrix} \begin{bmatrix} 1 \\ 1 \end{bmatrix} \right) \frac{1}{(s+1)(s+2)}$$

$$= \frac{s+1}{(s+1)(s+2)} = \frac{1}{s+2}$$

Note that there is a cancellation of the factor $s+1$, and only the mode with $\lambda_1 = -2$ is both controllable and observable.

(*c*)
$$\text{Rank} \begin{bmatrix} \mathbf{b} & \mathbf{Ab} \end{bmatrix} = \text{Rank} \begin{bmatrix} 1 & -2 \\ 1 & -2 \end{bmatrix} = 1$$

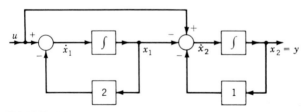

FIGURE 13.2
Simulation diagram for the system of Eq. (13.17).

The controllability matrix $\mathbf{M}_c = [\mathbf{b} \quad \mathbf{Ab}]$ is singular and has a rank deficiency of one; thus the system is *not* completely controllable and has one uncontrollable mode.

$$\text{Rank} \left[\mathbf{c} \vdots \mathbf{A}^T\mathbf{c}\right] = \text{Rank} \begin{bmatrix} 0 & -1 \\ 1 & -1 \end{bmatrix} = 2$$

The observability matrix $[\mathbf{c} \vdots \mathbf{A}^T\mathbf{c}]$ is nonsingular; thus the system is completely observable. Note that the mode with $\lambda_2 = -1$, which is canceled in the determination of the transfer function, is observable but not controllable.

(*d*) Using Eq. (5.91) of Sec. 5.10, with $\mathbf{x} = \mathbf{Tz}$, gives

$$\mathbf{AT} = \mathbf{T\Lambda}$$

$$\begin{bmatrix} -2 & 0 \\ -1 & -1 \end{bmatrix} \begin{bmatrix} v_{11} & v_{12} \\ v_{21} & v_{22} \end{bmatrix} = \begin{bmatrix} v_{11} & v_{12} \\ v_{21} & v_{22} \end{bmatrix} \begin{bmatrix} -2 & 0 \\ 0 & -1 \end{bmatrix}$$

$$\begin{bmatrix} -2v_{11} & -2v_{12} \\ -v_{11}-v_{21} & -v_{12}-v_{22} \end{bmatrix} = \begin{bmatrix} -2v_{11} & -v_{12} \\ -2v_{21} & -v_{22} \end{bmatrix}$$

Equating elements on both sides of this equation yields

$$-2v_{11} = -2v_{11} \qquad -2v_{12} = -v_{12}$$

$$-v_{11}-v_{21} = -2v_{21} \qquad -v_{12}-v_{22} = -v_{22}$$

Because there are only two independent equations, assume that $v_{11} = v_{22} = 1$. Then $v_{21} = 1$ and $v_{12} = 0$. Utilizing these values yields

$$\mathbf{b}' = \mathbf{T}^{-1}\mathbf{b} = \begin{bmatrix} 1 & 0 \\ -1 & 1 \end{bmatrix} \begin{bmatrix} 1 \\ 1 \end{bmatrix} = \begin{bmatrix} 1 \\ 0 \end{bmatrix}$$

$$\mathbf{c}'^T = \mathbf{c}^T\mathbf{T} = [0 \quad 1] \begin{bmatrix} 1 & 0 \\ 1 & 1 \end{bmatrix} = [1 \quad 1]$$

The canonical equations for this system are therefore

$$\dot{\mathbf{z}} = \begin{bmatrix} -2 & 0 \\ 0 & -1 \end{bmatrix} \mathbf{z} + \begin{bmatrix} 1 \\ 0 \end{bmatrix} u$$

$$\mathbf{y} = [1 \quad 1] \mathbf{z}$$

Because $\mathbf{b}'$ contains a zero row, the system is uncontrollable and $\lambda = -1$ is an input-decoupling zero. Because $\mathbf{c}'$ contains all nonzero elements, the system is completely observable. These results agree with the results obtained by the method used in part (*c*).

(*e*) The individual state and output equations are

$$\dot{z}_1 = -2z_1 + u \tag{13.19}$$

$$\dot{z}_2 = -z_2 \tag{13.20}$$

$$y = z_1 + z_2 \tag{13.21}$$

Equation (13.19) satisfies the requirement of controllability; for any t_0 the initial state $z_1(t_0)$ can be transferred to any final state $z_1(t_f)$ in a finite time $t_f \geq t_0$ by means

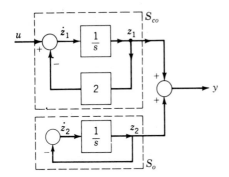

FIGURE 13.3
Simulation diagram for Eqs. (13.17) and (13.18).

of the unconstrained control vector u. Because u does not appear in Eq. (13.20), it is impossible to control the state z_2; thus $\lambda_2 = -1$ is an input-decoupling zero and the system is uncontrollable. Equation (13.21) satisfies the definition of observability, i.e., both states appear in y and, therefore, $z_1(t_0)$ and $z_2(t_0)$ can be determined from the measurements of the output $y(t)$ over the finite interval of time $t_0 \leq t \leq t_f$.

The simulation diagram of Fig. 13.3 is synthesized from Eqs. (13.19) to (13.21). The advantage of the canonical form is that the modes $\lambda_1 = -2$ and $\lambda_2 = -1$ are completely decoupled. This feature permits the determination of the existence of some or all of the four subdivisions of Fig. 13.1 and the application of the definitions of controllability and observability.

MATLAB Controllability and Observability

```
Diary
A = [-2 0; -1 -1];
B = [1; 1];
C = [0 1];
co = ctrb(A, B)
co =
   1  -2
   1  -2
rank(co)
ans =
   1
ob = obsv(A, C)
ob =
   0   1
  -1  -1
rank(ob)
ans =
   2
% Number of uncontrollable states
unco = length(A) - rank(co)
unco =
   1
```

13.3 STATE FEEDBACK FOR SISO SYSTEMS

This section covers the design of a single-input single-output (SISO) system in which the set of closed-loop eigenvalues is assigned by state feedback. The plant state and output equations have the form

$$\dot{\mathbf{x}} = \mathbf{Ax} + \mathbf{Bu} \tag{13.22}$$

$$\mathbf{y} = \mathbf{Cx} \tag{13.23}$$

The state-feedback control law (see Fig. 13.4) is

$$\mathbf{u} = \mathbf{r} + \mathbf{Kx} \tag{13.24}$$

where $\mathbf{r}$ is the input vector and $\mathbf{K}$ is the state-feedback matrix having dimension $m \times n$. The closed-loop state equation obtained by combining Eqs. (13.22) and (13.24) yields

$$\dot{\mathbf{x}} = (\mathbf{A} + \mathbf{BK})\mathbf{x} + \mathbf{Br} \tag{13.25}$$

Thus the closed-loop system matrix produced by state feedback is

$$\mathbf{A}_{cl} = \mathbf{A} + \mathbf{BK} \tag{13.26}$$

which results in the closed-loop characteristic equation

$$Q(\lambda) = |\lambda\mathbf{I} - \mathbf{A}_{cl}| = |\lambda\mathbf{I} - \mathbf{A} - \mathbf{BK}| = 0 \tag{13.27}$$

Equation (13.27) reveals that the closed-loop eigenvalues can be assigned by the proper selection of the state-feedback matrix $\mathbf{K}$. A necessary and sufficient condition for the selection of $\mathbf{K}$ is that the plant be completely controllable[7] (see Sec. 13.2).

Example 1. A second-order system is represented by the state and output equations

$$\dot{\mathbf{x}} = \begin{bmatrix} -1 & 3 \\ 0 & -2 \end{bmatrix}\mathbf{x} + \begin{bmatrix} 1 \\ 1 \end{bmatrix}\mathbf{u} \tag{13.28}$$

$$\mathbf{y} = [1 \quad 0]\mathbf{x} = x_1 \tag{13.29}$$

The open-loop eigenvalue spectrum is $\sigma(\mathbf{A}) = \{-1, -2\}$. The system is completely controllable since, from Eq. (13.10),

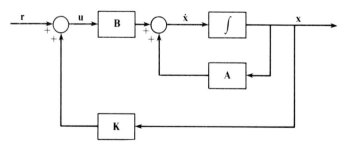

FIGURE 13.4
State feedback control system.

$$\text{Rank } \mathbf{M}_c = \text{Rank} \begin{bmatrix} 1 & 2 \\ 1 & -2 \end{bmatrix} = 2 \tag{13.30}$$

The closed-loop eigenvalues are determined from Eq. (13.27) with $\mathbf{K} = [\, k_1 \quad k_2\,]$

$$|\lambda\mathbf{I} - \mathbf{A} - \mathbf{BK}| = \left\| \begin{bmatrix} \lambda & 0 \\ 0 & \lambda \end{bmatrix} - \begin{bmatrix} -1 & 3 \\ 0 & -2 \end{bmatrix} - \begin{bmatrix} 1 \\ 1 \end{bmatrix} [\, k_1 \quad k_2\,] \right\|$$

$$= \lambda^2 + (3 - k_1 - k_2)\lambda + (2 - 5k_1 - k_2) \tag{13.31}$$

Assigning the closed-loop eigenvalues $\lambda_1 = -5$ and $\lambda_2 = -6$ requires the characteristic polynomial to be $\lambda^2 + 11\lambda + 30$. Equating the coefficients of this desired characteristic equation with those of Eq. (13.31) results in the values $k_1 = -5$ and $k_2 = -3$, or $\mathbf{K} = [-5 \quad -3\,]$. The resulting closed-loop state equation obtained by using Eq. (13.25) is

$$\dot{\mathbf{x}} = \begin{bmatrix} -6 & 0 \\ -5 & -5 \end{bmatrix} \mathbf{x} + \begin{bmatrix} 1 \\ 1 \end{bmatrix} \mathbf{r} \tag{13.32}$$

This state equation has the required eigenvalue spectrum $\sigma(\mathbf{A}_{cl}) = \{-5, -6\}$.

The following example illustrates the requirement of controllability for closed-loop eigenvalue assignment by state feedback.

Example 2. A cascade gain K is inserted in the plant of the example in Sec. 13.2, Eq. (13.17) so that its transfer function is

$$G(s) = \frac{K(s + 1)}{(s + 1)(s + 2)}$$

It is shown in that example that this plant is not controllable. Figure 13.5 shows a state-feedback system for this plant. The closed-loop control ratio is given by

$$\frac{Y(s)}{R(s)} = \frac{K(s + 1)}{s^2 + (3 + Kk_1 + Kk_2)s + (2 + Kk_2 + Kk_1)} \tag{13.33}$$

In order for the system to follow a step input with no steady-state error, the requirement obtained by applying the final value theorem is that $K = 2 + Kk_2 + Kk_1$, or

$$K = \frac{2}{1 - k_1 - k_2} \tag{13.34}$$

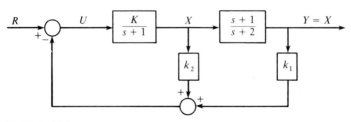

FIGURE 13.5
An uncontrollable state-feedback system.

Because this system is uncontrollable, as determined in Sec. 13.2, the pole located at $s_2 = -1$ cannot be altered by state feedback. The characteristic equation of Eq. (13.33) illustrates this property because it can be factored to the form

$$(s + 1)(s + 2 + K k_2 + K k_1) = 0 \qquad (13.35)$$

Therefore one eigenvalue, $s_1 = -2$, is altered by state feedback to the value $s_1 = -(2 + K k_2 + K k_1) = -K$, where K can be assigned as desired. Because the uncontrollable pole cancels the zero of Eq. (13.33), the control ratio becomes

$$\frac{Y(s)}{R(s)} = \frac{K(s + 1)}{(s + 1)(s + K)} = \frac{K}{s + K} \qquad (13.36)$$

This example demonstrates that the uncontrollable pole $s_1 = -1$ cannot be changed by state feedback.

When deriving the transfer function from the system differential equations, cancellation of a pole by a zero is not permitted in developing the state representation. Cancellation of a pole by a zero in the transfer function may hide some of the dynamics of the entire system. This eliminates information pertaining to S_o, S_c, and S_u; that is, the observability and controllability of the system are determined from the state equations, as illustrated in these examples.

13.4 STATE-FEEDBACK DESIGN FOR SISO SYSTEMS USING THE CONTROL CANONICAL (PHASE-VARIABLE) FORM

The examples of Sec. 13.3 are of second order, and therefore the algebraic solution for the **K** matrix is computationally simple and direct. For higher-order systems, the computations are simplified by first transforming the state equation to control canonical form, so that the plant matrix is in companion form. That procedure is presented in this section.

The model of an open-loop SISO system can be represented by a physical state variable and an output equation

$$\dot{\mathbf{x}}_p = \mathbf{A}_p \mathbf{x}_p + \mathbf{b}_p \mathbf{u} \qquad (13.37)$$

$$\mathbf{y} = \mathbf{c}_p^T \mathbf{x}_p \qquad (13.38)$$

The actuating signal **u** is formed in accordance with the feedback control law

$$\mathbf{u} = \mathbf{r} + \mathbf{k}_p^T \mathbf{x}_p \qquad (13.39)$$

where **r** is the reference input and

$$\mathbf{k}_p^T = [\, k_{p1} \quad k_{p2} \quad \cdots \quad k_{pn} \,] \qquad (13.40)$$

The closed-loop state equation obtained from Eqs. (13.37) and (13.39) is

$$\dot{\mathbf{x}}_p = [\mathbf{A}_p + \mathbf{b}_p \mathbf{k}_p^T]\mathbf{x}_p + \mathbf{b}_p \mathbf{r} = \mathbf{A}_{p\text{cl}}\mathbf{x}_p + \mathbf{b}_p \mathbf{r} \qquad (13.41)$$

The design procedure that assigns the desired set of closed-loop eigenvalues, $\sigma(\mathbf{A}_{p_{cl}}) = \{\lambda_i\}, i = 1, 2, \ldots, n$, is accomplished as follows:

1. Use the method of Sec. 5.13 to obtain the matrix **T**, which transforms the state equation (13.37), by the change of state variables $\mathbf{x}_p = \mathbf{Tx}$, into the *control canonical form* $\dot{\mathbf{x}} = \mathbf{A}_c\mathbf{x} + \mathbf{b}_c\mathbf{u}$, where $\mathbf{A}_c$ is in companion form and $\mathbf{b}_c = [0 \quad \cdots \quad 0 \quad 1]^T$.

2. Select the set of desired closed-loop eigenvalues and obtain the corresponding closed-loop characteristic polynomial

$$Q_{cl}(\lambda) = (\lambda - \lambda_1)(\lambda - \lambda_2)\cdots(\lambda - \lambda_n)$$
$$= \lambda^n + \alpha_{n-1}\lambda^{n-1} + \cdots + \alpha_1\lambda + \alpha_0 \tag{13.42}$$

3. Write the companion form of the desired closed-loop plant matrix

$$A_{C_{cl}} = \begin{bmatrix} 0 & 1 & & & & \\ & 0 & 1 & & & \\ & & 0 & \ddots & & \\ & & & \ddots & 1 & \\ & & & & 0 & 1 \\ -\alpha_0 & -\alpha_1 & & \cdots & -\alpha_{n-2} & -\alpha_{n-1} \end{bmatrix} \tag{13.43}$$

4. Use the feedback matrix

$$\mathbf{k}_c^T = [\, k_{c1} \quad k_{c2} \quad \cdots \quad k_{cn} \,] \tag{13.44}$$

to form the closed-loop matrix

$$\mathbf{A}_{C_{cl}} = \mathbf{A}_c + \mathbf{b}_c\mathbf{k}_c^T$$

$$= \begin{bmatrix} 0 & 1 & \cdots & 0 & 0 \\ 0 & 0 & \cdots & 0 & 0 \\ \hline \cdots & \cdots & & \cdots & \cdots \\ 0 & 0 & \cdots & 1 & 0 \\ 0 & 0 & \cdots & 0 & 1 \\ -a_0 + k_{c1} & -a_1 + k_{c2} & \cdots & -a_{n-2} + k_{c(n-1)} & -a_{n-1} + k_{cn} \end{bmatrix}$$

$$\tag{13.45}$$

5. Equate the matrices of Eqs. (13.43) and (13.45), and solve for the elements of the matrix $\mathbf{k}_c^T$.

6. Compute the required physical feedback matrix using

$$\mathbf{k}_p^T = \mathbf{k}_c^T\mathbf{T}^{-1} \tag{13.46}$$

Example. Using the physical system shown by Eq. (13.47), assign the set of eigenvalues $\{-1 + j1, -1 - j1, -4\}$.

$$\mathbf{A}_p = \begin{bmatrix} 1 & 6 & -3 \\ -1 & -1 & 1 \\ -2 & 2 & 0 \end{bmatrix} \quad \mathbf{b}_p = \begin{bmatrix} 1 \\ 1 \\ 1 \end{bmatrix} \tag{13.47}$$

This open-loop system is unstable because $\sigma(\mathbf{A}) = \{1, 1, -2\}$.

Step 1. As derived in the example of Sec. 5.13,

$$\mathbf{A}_c = \begin{bmatrix} 0 & 1 & 0 \\ 0 & 0 & 1 \\ -2 & 3 & 0 \end{bmatrix} \quad \mathbf{b}_c = \begin{bmatrix} 0 \\ 0 \\ 1 \end{bmatrix} \tag{13.48}$$

and

$$\mathbf{T} = \begin{bmatrix} -5 & 4 & 1 \\ -6 & -1 & 1 \\ -13 & 0 & 1 \end{bmatrix} \tag{13.49}$$

Step 2. The required closed-loop characteristic polynomial is

$$Q_{cl}(\lambda) = (\lambda + 1 - j1)(\lambda + 1 + j1)(\lambda + 4) = \lambda^3 + 6\lambda^2 + 10\lambda + 8 \tag{13.50}$$

Step 3. The closed-loop plant matrix in companion form is

$$A_{c_{cl}} = \begin{bmatrix} 0 & 1 & 0 \\ 0 & 0 & 1 \\ -8 & -10 & -6 \end{bmatrix} \tag{13.51}$$

Step 4.

$$\mathbf{A}_c + \mathbf{b}_c \mathbf{k}_c^T = \begin{bmatrix} 0 & 1 & 0 \\ 0 & 0 & 1 \\ -2 + k_{c1} & 3 + k_{c2} & k_{c3} \end{bmatrix} \tag{13.52}$$

Step 5. Equating the coefficients of $\mathbf{A}_{c_{cl}}$ and $\mathbf{A}_c + \mathbf{b}_c \mathbf{k}_c^T$ yields

$$\mathbf{k}_c^T = [-6 \quad -13 \quad -6] \tag{13.53}$$

Step 6. The feedback matrix required for the physical system is

$$\mathbf{k}_p^T = \mathbf{k}_c^T \mathbf{T}^{-1} = \frac{1}{36}[-175 \quad -232 \quad 191] \tag{13.54}$$

The reader should verify that the physical closed-loop matrix

$$\mathbf{A}_{p_{cl}} = \mathbf{A}_p + \mathbf{b}_p \mathbf{k}_p^T = \frac{1}{36} \begin{bmatrix} -139 & -16 & 83 \\ -211 & -268 & 227 \\ -247 & -160 & 191 \end{bmatrix} \tag{13.55}$$

has the required set of eigenvalues $\sigma(\mathbf{A}_{p_{cl}}) = \{-1 + j1, -1 - j1, -4\}$.

This design method provides a very straightforward procedure for computing the state feedback matrix $\mathbf{K}_p^T$. MATLAB contains procedures for assisting with this design.

13.5   STATE-VARIABLE FEEDBACK[1]
(PHYSICAL VARIABLES)

The open-loop position-control system of Fig. 13.6 is used to illustrate the effects of state-variable feedback. The dynamic equations for this basic system are

$$e_a = Ae \quad A > 0 \tag{13.56}$$

$$e_a - e_m = (R_m + L_m D)i_m \tag{13.57}$$

$$e_m = K_b \omega_m \tag{13.58}$$

$$T = K_T i_m = JD\omega_m + B\omega_m \tag{13.59}$$

$$\omega_m = D\theta_m \tag{13.60}$$

In order to obtain the maximum number of accessible (measurable) states, *physical variables are selected as the state variables*. The two variables i_m and ω_m and the desired output quantity θ_m are identified as the three state variables: $x_1 = \theta_m = y$, $x_2 = \omega_m = \dot{x}_1$, and $x_3 = i_m$, and the input is $u = e$. Figure 13.6 shows that all three state variables are accessible for measurement. The sensors are selected to produce voltages that are proportional to the state variables and to have negligible effect on the measured variables.

Figure 13.7a is the block diagram of a closed-loop position-control system utilizing the system of Fig. 13.6. Each of the states is fed back through amplifiers whose gain values k_1, k_2, and k_3 are called *feedback coefficients*. These feedback coefficients include the sensor constants and any additional gain required by the system design to yield the required values of k_1, k_2, and k_3. The summation of the three feedback quantities, where $\mathbf{k}^T = [\, k_1 \quad k_2 \quad k_3 \,]$ is the feedback matrix, is

$$\mathbf{k}^T \mathbf{X} = k_1 X_1 + k_2 X_2 + k_3 X_3 \tag{13.61}$$

Using block-diagram manipulation techniques, Fig. 13.7a is simplified to the block diagram of Fig. 13.7b. A further reduction is achieved, as shown in Fig. 13.7c. Note

FIGURE 13.6
Open-loop position-control system and measurement of states.

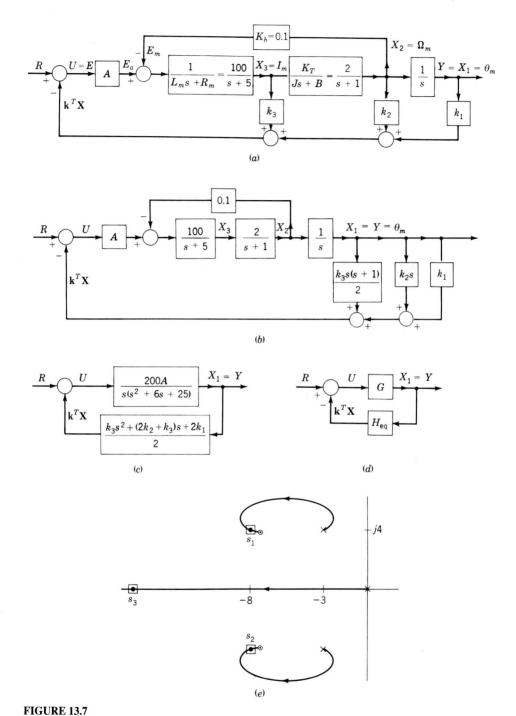

FIGURE 13.7
A position-control system with state feedback, its H_{eq} representation, and the corresponding root locus.

that the forward transfer function $G(s) = X_1(s)/U(s)$ shown in this figure is identical with that obtained from Fig. 13.7a with zero-state feedback coefficients. Thus, state-variable feedback is *equivalent* to inserting a feedback compensator $H_{eq}(s)$ in the feedback loop of a conventional feedback control system, as shown in Fig. 13.7d. The reduction of Fig. 13.7a to that of Fig. 13.7c or Fig. 13.7d is the *H-equivalent reduction*. The system control ratio is therefore

$$\frac{Y(s)}{R(s)} = \frac{G(s)}{1 + G(s)H_{eq}(s)} \tag{13.62}$$

For Figs. 13.7c and 13.7d the following equations are obtained:

$$G(s)H_{eq}(s) = \frac{100A[k_3 s^2 + (2k_2 + k_3)s + 2k_1]}{s(s^2 + 6s + 25)} \tag{13.63}$$

$$\frac{Y(s)}{R(s)} = \frac{200A}{s^3 + (6 + 100k_3 A)s^2 + (25 + 200k_2 A + 100k_3 A)s + 200k_1 A} \tag{13.64}$$

The equivalent feedback transfer function $H_{eq}(s)$ must be designed so that the system's desired figures of merit are satisfied. The following observations are noted from Eqs. (13.63) and (13.64) for this system in which $G(s)$ has no zeros; i.e., it is an all-pole plant.

1. In order to achieve zero steady-state error for a step input, $R(s) = R_0/s$, Eq. (13.64) must yield

$$y(t)_{ss} = \lim_{s \to 0} sY(s) = \frac{R_0}{k_1} = R_0$$

This requires that $k_1 = 1$.

2. The numerator of $H_{eq}(s)$ is of second order; i.e., the denominator of $G(s)$ is of degree $n = 3$ and the numerator of $H_{eq}(s)$ is of degree $n - 1 = 2$.

3. The poles of $G(s)H_{eq}(s)$ are the poles of $G(s)$.

4. The zeros of $G(s)H_{eq}(s)$ are the zeros of $H_{eq}(s)$.

5. The use of state feedback produces additional zeros in the open-loop transfer function $G(s)H_{eq}(s)$ without adding poles. This is in contrast to cascade compensation using passive networks. The zeros of $H_{eq}(s)$ are to be located to yield the desired response. It is known that the addition of zeros to the open-loop transfer function moves the root locus to the left; i.e., they make the system more stable and improve the time-response characteristics.

6. Since the root-locus plot of $G(s)H_{eq}(s) = -1$ has one asymptote, $\gamma = -180°$ for $k_3 > 0$, it is possible to pick the locations of the zeros of $H_{eq}(s)$ to ensure a completely stable system for all positive values of gain. Figure 13.7e shows a possible selection of the zeros and the resulting root locus. Note that when the conventional cascade compensators of Chap. 10 are used, the angles of the asymptotes yield branches that go, in general, into the right-half (RH) s plane. The maximum gain is therefore limited in order to maintain system stability. This restriction is removed with state feedback.

7. The desired figures of merit, for an underdamped response, are satisfied by selecting the desired locations of the poles of $Y(s)/R(s)$ of Eq. (13.64) at $s_{1,2} = -a \pm jb$ and $s_3 = -c$. For these specified pole locations the known characteristic equation has the form

$$(s + a \pm jb)(s + c) = s^3 + d_2 s^2 + d_1 s + d_0 = 0 \qquad (13.65)$$

Equating the corresponding coefficients of the denominators of Eqs. (13.64) and (13.65) yields

$$d_2 = 6 + 100k_3 A \qquad (13.66)$$

$$d_1 = 25 + 200k_2 A + 100k_3 A \qquad (13.67)$$

$$d_0 = 200k_1 A = 200A \qquad (13.68)$$

These three equations with three unknowns can be solved for the required values of A, k_2, and k_3.

It should be noted that if the basic system $G(s)$ has any zeros, they become poles of $H_{eq}(s)$. Thus, the zeros of $G(s)$ cancel these poles of $H_{eq}(s)$ in $G(s)H_{eq}(s)$. Also, the zeros of $G(s)$ become the zeros of the control ratio. The reader is able to verify this property by replacing $2/(s + 1)$ in Fig. 13.7a by $2(s + \alpha)/(s + 1)$ and performing an H-equivalent reduction. This property is also illustrated in the examples of Sec. 13.11.

13.6 GENERAL PROPERTIES OF STATE FEEDBACK (USING PHASE VARIABLES)

In order to obtain general state-variable-feedback properties that are applicable to any $G(s)$ transfer function, consider the minimum-phase transfer function

$$G(s) = \frac{Y(s)}{U(s)} = \frac{K_G(s^w + c_{w-1}s^{w-1} + \cdots + c_1 s + c_0)}{s^n + a_{n-1}s^{n-1} + \cdots + a_1 s + a_0} \qquad (13.69)$$

where $K_G > 0$, $w < n$, and the coefficients c_i are all positive. The system characteristics and properties are most readily obtained by representing this system in terms of phase variables [see Secs. 5.6 and 5.8, Eqs. (5.38) and (5.45), with $c_w = 1$ and u replaced by $K_G u$, and Sec. 13.4]:

$$\dot{\mathbf{x}} = \begin{bmatrix} 0 & 1 & 0 & \cdots & 0 \\ 0 & 0 & 1 & \cdots & 0 \\ \cdots\cdots\cdots\cdots\cdots\cdots\cdots & & & & 1 \\ -a_0 & -a_1 & -a_2 & \cdots & -a_{n-1} \end{bmatrix} \mathbf{x} + \begin{bmatrix} 0 \\ 0 \\ \cdot \\ K_G \end{bmatrix} u \qquad (13.70)$$

$$y = [\, c_0 \quad c_1 \quad \cdots \quad c_{w-1} \quad 1 \quad 0 \quad \cdots \quad 0\,] \mathbf{x} = \mathbf{c}_c^T \mathbf{x} \qquad (13.71)$$

From Fig. 13.7d and Eq. (13.71)

$$H_{eq}(s) = \frac{\mathbf{k}_c^T \mathbf{X}(s)}{Y(s)} = \frac{\mathbf{k}_c^T \mathbf{X}(s)}{\mathbf{c}_c^T \mathbf{X}(s)} = \frac{k_1 X_1(s) + k_2 X_2(s) + \cdots + k_n X_n(s)}{c_0 X_1(s) + c_1 X_2(s) + \cdots + X_{w+1}} \qquad (13.72)$$

The state-feedback matrix is $\mathbf{k}_c^T$ because the state variables are phase variables. Using the phase-variable relationship $X_j(s) = s^{j-1}X_1(s)$, Eq. (13.72) becomes

$$H_{eq}(s) = \frac{k_n s^{n-1} + k_{n-1}s^{n-2} + \cdots + k_2 s + k_1}{s^w + c_{w-1}s^{w-1} + \cdots + c_1 s + c_0} \tag{13.73}$$

Thus, from Eqs. (13.69) and (13.73), after cancellation of common terms,

$$G(s)H_{eq}(s) = \frac{K_G(k_n s^{n-1} + k_{n-1}s^{n-2} + \cdots + k_2 s + k_1)}{s^n + a_{n-1}s^{n-1} + \cdots + a_1 s + a_0} \tag{13.74}$$

Factoring k_n from the numerator polynomial gives an alternate form of this equation,

$$G(s)H_{eq}(s) = \frac{K_G k_n(s^{n-1} + \cdots + \alpha_1 s + \alpha_0)}{s^n + a_{n-1}s^{n-1} + \cdots + a_1 s + a_0} \tag{13.75}$$

where the loop sensitivity is $K_G k_n = K$. Substituting Eqs. (13.69) and (13.74) into Eq. (13.62) yields

$$\frac{Y(s)}{R(s)} = \frac{K_G(s^w + c_{w-1}s^{w-1} + \cdots + c_0)}{s^n + (a_{n-1} + K_G k_n)s^{n-1} + \cdots + (a_0 + K_G k_1)} \tag{13.76}$$

An analysis of these equations yields the following conclusions:

1. The numerator of $H_{eq}(s)$ is a polynomial of degree $n - 1$ in s; that is, it has $n - 1$ zeros. The values of the coefficients k_i of this numerator polynomial, and thus the $n - 1$ zeros of $H_{eq}(s)$, can be selected by the system designer.
2. The numerator of $G(s)$, except for K_G, is equal to the denominator of $H_{eq}(s)$. Therefore, the open-loop transfer function $G(s)H_{eq}(s)$ has the same n poles as $G(s)$ and has the $n - 1$ zeros of $H_{eq}(s)$ which are to be determined in the design process. [An exception to this occurs when physical variables are utilized and the transfer function $X_n(s)/U(s)$ contains a zero. In that case this zero does not appear as a pole of $H_{eq}(s)$, and the numerator and denominator of $G(s)H_{eq}(s)$ are both of degree n.]
3. The root-locus diagram, based on Eq. (13.75) for $k_n > 0$, reveals that $n - 1$ branches terminate on the $n - 1$ arbitrary zeros, which can be located anywhere in the s plane by properly selecting the values of the k_i. There is a single asymptote at $\gamma = -180°$, and therefore one branch terminates on the negative real axis at $s = -\infty$. System stability is ensured for high values of K_G if all the zeros of $H_{eq}(s)$ are located in the left-half (LH) s plane. For $k_n < 0$, a single asymptote exists at $\gamma = 0°$; thus the system becomes unstable for high values of K_G.
4. State-variable feedback provides the ability to select the locations of the zeros of $G(s)H_{eq}(s)$, as given by Eq. (13.75), in order to locate closed-loop system poles that yield the desired system performance. From Eq. (13.76) it is noted that the result of feedback of the states through constant gains k_i is to change the closed-loop poles and to leave the system zeros, if any, the same as those of $G(s)$. (This requires the plant to be controllable, as described in Sec. 13.2.) It may be desirable

to specify that one or more of the closed-loop poles cancel one or more unwanted zeros in $Y(s)/R(s)$.

5. In synthesizing the desired $Y(s)/R(s)$, the number of poles minus the number of zeros of $Y(s)/R(s)$ must equal the number of poles minus the number of zeros of $G(s)$.

    A thorough steady-state error analysis of Eq. (13.76) is made in Secs. 13.7 and 13.8 for the standard step, ramp, and parabolic inputs. From the conclusions obtained in that section, some or all of the feedback gains k_1, k_2, and k_3 are specified.

    Summarizing, the $H_{eq}(s)$ can be designed to yield the specified closed-loop transfer function in the manner described for the system of Fig. 13.7. A root-locus sketch of $G(s)H_{eq}(s) = \pm 1$ can be used as an aid in synthesizing the desired $Y(s)/R(s)$ function. From this sketch it is possible to determine the locations of the closed-loop poles, in relation to any system zeros, that will achieve the desired performance specifications. Note that although the zeros of $G(s)$ are *not* involved in obtaining the root locus for $G(s)H_{eq}(s)$, they do appear in $Y(s)/R(s)$ and thus affect the time function $y(t)$.

DESIGN PROCEDURE.

The design procedure is summarized as follows:

Step 1. Draw the system state-variable block diagram when applicable.

Step 2. Assume that all state variables are accessible.

Step 3. Obtain $H_{eq}(s)$ and $Y(s)/R(s)$ in terms of the k_i's.

Step 4. Determine the value of k_1 required for zero steady-state error for a unit step input, $u_{-1}(t)$; that is, $y(t)_{ss} = \lim_{s\to 0} sY(s) = 1$.

Step 5. Synthesize the desired $[Y(s)/R(s)]_T$ from the desired performance specifications for M_p, t_p, t_s, and steady-state error requirements (see Sec. 12.2). It may also be desirable to specify that one or more of the closed-loop poles should cancel unwanted zeros of $Y(s)/R(s)$.

Step 6. Set the closed-loop transfer functions obtained in steps 3 and 5 equal to each other. Equate the coefficients of like powers of s of the denominator polynomials and solve for the required values of the k_i. If phase variables are used in the design, a linear transformation must be made in order to convert the phase-variable-feedback coefficients required for the physical variables present in the control system.

    If some of the states are not accessible, suitable cascade or minor-loop compensators can be determined using the known values of the k_i. This method is discussed in Sec. 14.5.

    The procedure for designing *a state-variable-feedback system* is illustrated by means of examples using physical variables in Secs. 13.9 through 13.12. Before the examples are considered, however, the steady-state error properties are investigated in Secs. 13.7 and 13.8.

13.7 STATE VARIABLE FEEDBACK: STEADY-STATE ERROR ANALYSIS

A steady-state error analysis of a stable control system is an important design consideration (see Chap. 6). The three standard inputs are used for a steady-state analysis of a state-variable-feedback control system. The phase-variable representation is used in the following developments, in which the necessary values of the feedback coefficients are determined. The results apply only when phase variables are used and do not apply when general state variables are used.

Step Input $r(t) = R_0 u_{-1}(t), R(s) = R_0/s$

For a step input, solving Eq. (13.76), which is based upon the phase-variable representation for $Y(s)$, and applying the final-value theorem yields

$$y(t)_{ss} = \frac{K_G c_0 R_0}{a_0 + K_G k_1} \tag{13.77}$$

The system error is defined by

$$e(t) \equiv r(t) - y(t) \tag{13.78}$$

For zero steady-state error, $e(t)_{ss} = 0$, the steady-state output must be equal to the input:

$$y(t)_{ss} = r(t) = R_0 = \frac{K_G c_0 R_0}{a_0 + K_G k_1} \tag{13.79}$$

where $c_0 \neq 0$. This requires that $K_G c_0 = a_0 + K_G k_1$, or

$$k_1 = c_0 - (a_0/K_G) \tag{13.80}$$

Thus, zero steady-state error with a step input can be achieved with state-variable feedback even if $G(s)$ is Type 0. For an all-pole plant $c_0 = 1$, and for a Type 1 or higher plant, $a_0 = 0$. In that case the required value is $k_1 = 1$. *Therefore, the feedback coefficient k_1 is determined by the plant parameters once zero steady-state error for a step input is specified.* This specification is maintained for all state-variable-feedback control-system designs throughout the remainder of this text. From Eq. (13.76), the system characteristic equation is

$$s^n + (a_{n-1} + K_G k_n)s^{n-1} + \cdots + (a_2 + K_G k_3)s^2 + (a_1 + K_G k_2)s + (a_0 + K_G k_1) = 0 \tag{13.81}$$

A necessary condition for all closed-loop poles of $Y(s)/R(s)$ to be in the LH s plane is that all coefficients in Eq. (13.81) must be positive. Therefore, the constant term of the polynomial $a_0 + K_G k_1 = K_G c_0$ must be positive.

Ramp Input $r(t) = R_1 u_{-2}(t) = R_1 t u_{-1}(t), R(s) = R_1/s^2$

To perform the analysis for ramp and parabolic inputs, the error function $E(s)$ must first be determined. Solving for $Y(s)$ from Eq. (13.76) and evaluating $E(s) = R(s) - Y(s)$ yields

$$E(s) = R(s) \frac{\begin{array}{c} s^n + (a_{n-1} + K_G k_n)s^{n-1} + \cdots + (a_0 + K_G k_1) \\ -K_G(s^w + c_{w-1}s^{w-1} + \cdots + c_1 s + c_0) \end{array}}{s^n + (a_{n-1} + K_G k_n)s^{n-1} + \cdots + (a_1 + K_G k_2)s + (a_0 + K_G k_1)}$$

(13.82)

Substituting Eq. (13.80), the requirement for zero steady-state error for a step input, into Eq. (13.82) produces a cancellation of the constant terms in the numerator. Thus,

$$E(s) = R(s) \frac{\begin{array}{c} s^n + (a_{n-1} + K_G k_n)s^{n-1} + \cdots + (a_1 + K_G k_2)s \\ -K_G(s^w + c_{w-1}s^{w-1} + \cdots + c_1 s) \end{array}}{s^n + (a_{n-1} + K_G k_n)s^{n-1} + \cdots + (a_1 + K_G K_2)s + K_G c_0}$$

(13.83)

For a ramp input, the steady-state error is

$$e(t)_{ss} = \lim_{s \to 0} sE(s) = \frac{a_1 + K_G k_2 - K_G c_1}{K_G c_0} R_1$$

(13.84)

In order to obtain $e(t)_{ss} = 0$ it is necessary that $a_1 + K_G k_2 - K_G c_1 = 0$. This is achieved when

$$k_2 = c_1 - (a_1/K_G)$$

(13.85)

For an all-pole plant, $c_0 = 1$ and $c_1 = 0$; thus $k_2 = -a_1/K_G$. In that case, the coefficient of s in the system characteristic equation (13.81) is zero, and the system is unstable. *Therefore, zero steady-state error for a ramp input cannot be achieved by state-variable feedback for an all-pole stable system.* If $a_1 + K_G k_2 > 0$, in order to obtain a positive coefficient of s in Eq. (13.81), the system may be stable but a finite steady-state error exists. The error in an all-pole plant is

$$e(t)_{ss} = \left(k_2 + \frac{a_1}{K_G}\right) R_1$$

(13.86)

When a plant contains at least one zero, so that $c_1 > 0$, it is possible to specify the value of k_2 given by Eq. (13.85) and to have a stable system. In that case the system has zero steady-state error with a ramp input. Substituting Eq. (13.85) into Eq. (13.81) results in the positive coefficient $K_G c_1$ for the first power of s. *Thus, a stable state-variable-feedback system having zero steady-state error with both a step and a ramp input can be achieved only when the control ratio has at least one zero.* Note that this occurs even though $G(s)$ may be Type 0 or 1.

Parabolic Input $r(t) = R_2 u_{-3}(t) = (R_2 t^2/2)u_{-1}(t), R(s) = R_2/s^3$

The all-pole and pole-zero plants are considered separately for a parabolic input. For the all-pole plant with Eq. (13.80) satisfied, Eq. (13.82) reduces to

$$E(s) = \frac{R_2}{s^3} \left[\frac{s^n + (a_{n-1} + K_G k_n)s^{n-1} + \cdots + (a_1 + K_G k_2)s}{s^n + (a_{n-1} + K_G k_n)s^{n-1} + \cdots + (a_1 + K_G k_2)s + K_G c_0} \right] \quad (13.87)$$

Because $e(t)_{ss} = \lim_{s \to 0} sE(s) = \infty$, it is concluded that *an all-pole plant cannot follow a parabolic input.* This is true even if Eq. (13.80) is not satisfied.

Under the condition that $e(t)_{ss} = 0$ for both step and ramp inputs to a pole-zero stable system, Eqs. (13.80) and (13.85) are substituted into Eq. (13.83) to yield, for a plant containing at least one zero,

$$E(s) = \frac{R_2}{s^3} \left[\frac{\begin{matrix} s^n + (a_{n-1} + K_G k_n)s^{n-1} + \cdots + (a_2 + K_G k_3)s^2 \\ -K_G(c_w s^w + c_{w-1} s^{w-1} + \cdots + c_2 s^2) \end{matrix}}{s^n + (a_{n-1} + K_G k_n)s^{n-1} + \cdots + K_G c_1 s + K_G c_0} \right] \quad (13.88)$$

Note that s^2 is the lowest power term in the numerator. Applying the final-value theorem to this equation yields

$$e(t)_{ss} = \lim_{s \to 0} sE(s) = \frac{a_2 + K_G k_3 - K_G c_2}{K_G c_0} R_2 \quad (13.89)$$

In order to achieve $e(t)_{ss} = 0$, it is necessary that $a_2 + K_G k_3 - K_G c_2 = 0$. Thus

$$k_3 = c_2 - (a_2/K_G) \quad (13.90)$$

With this value of k_3 the s^2 term in Eq. (13.81) has the positive coefficient $K_G c_2$ only when $w \geq 2$. *Thus, a stable system having zero steady-state error with a parabolic input can be achieved, even if $G(s)$ is Type 0, only for a pole-zero system with two or more zeros.*

In a system having only one zero ($w = 1$), the value $c_2 = 0$ and $k_3 = -a_2/K_G$ produces a zero coefficient for the s^2 term in the characteristic equation (13.81), resulting in an unstable system. *Therefore, zero steady-state error with a parabolic input cannot be achieved by a state-variable-feedback pole-zero stable system with one zero.* If $k_3 \neq -a_2/K_G$ and $a_2 + K_G k_3 > 0$ in order to maintain a positive coefficient of s^2 in Eq. (13.81), a finite error exists. The system can be stable, and the steady-state error is

$$e(t)_{ss} = \frac{R_2(a_2 + K_G k_3)}{K_G c_0} \quad (13.91)$$

13.8   USE OF STEADY-STATE ERROR COEFFICIENTS

For the example represented by Fig. 13.7a the feedback coefficient blocks are manipulated to yield the H-equivalent block diagram in Fig. 13.7d. Similarly a G-equivalent unity-feedback block diagram, as shown in Fig. 13.8, can be obtained. The overall transfer functions for Figs. 13.7d and 13.8 are, respectively,

$$M(s) = \frac{Y(s)}{R(s)} = \frac{G(s)}{1 + G(s)H_{eq}(s)} \quad (13.92)$$

FIGURE 13.8
G-equivalent block diagram.

$$M(s) = \frac{Y(s)}{R(s)} = \frac{G_{eq}(s)}{1 + G_{eq}(s)} \qquad (13.93)$$

Solving for $G_{eq}(s)$ from Eq. (13.93) yields

$$G_{eq}(s) = \frac{M(s)}{1 - M(s)} \qquad (13.94)$$

Using the control ratio $M(s)$ from Eq. (13.76) in Eq. (13.94) yields

$$G_{eq}(s) = \frac{K_G(s^w + \cdots + c_1 s + c_0)}{\substack{s^n + (a_{n-1} + K_G k_n)s^{n-1} + \cdots + (a_1 + K_G k_2)s \\ +(a_0 + K_G k_1) - K_G(s^w + \cdots + c_1 s + c_0)}} \qquad (13.95)$$

This is a Type 0 transfer function and has the step-error coefficient $K_p = K_G c_0/(a_0 + K_G k_1 - K_G c_0)$. When the condition $k_1 = c_0 - a_0/K_G$ is satisfied so that $e(t)_{ss} = 0$ with a step input, Eq. (13.95) reduces to

$$G_{eq}(s) = \frac{K_G(s^w + \cdots + c_1 s + c_0)}{s^n + (a_{n-1} + K_G k_n)s^{n-1} + \cdots + (a_1 + K_G k_2)s - K_G(s^w + \cdots + c_1 s)} \qquad (13.96)$$

Since this is a Type 1 transfer function, the step error coefficient is $K_p = \lim_{s \to 0}[sG_{eq}(s)] = \infty$, and the ramp error coefficient is

$$K_v = K_G c_0/(a_1 + K_G k_2 - K_G c_1). \qquad (13.97)$$

Satisfying the conditions for $e(t)_{ss} = 0$ for both step and ramp inputs, as given by Eqs. (13.80) and (13.85), Eq. (13.96) reduces to

$$G_{eq}(s) = \frac{K_G(s^w + \cdots + c_1 s + c_0)}{s^n + (a_{n-1} + K_G k_n)s^{n-1} + \cdots + (a_2 + K_G k_3)s^2 - K_G(s^w + \cdots + c_2 s^2)} \qquad (13.98)$$

Since this is a Type 2 transfer function, there is zero steady-state error with a ramp input. It follows a parabolic input with a steady-state error. The error coefficients are $K_p = \infty$, $K_v = \infty$, and $K_a = K_G c_0/(a_2 + K_G k_3 - K_G c_2)$.

For the case when $w \ge 2$ and the conditions for $e(t)_{ss} = 0$ for all three standard inputs are satisfied [Eqs. (13.80), (13.85), and (13.96)], Eq. (13.98) reduces to

$$G_{eq}(s) = \frac{K_G(s^w + \cdots + c_1 s + c_0)}{s^n + (a_{n-1} + K_G k_n)s^{-1} + \cdots + (a_3 + K_G k_4)s^3 - K_G(s^w + \cdots + c_3 s^3)} \qquad (13.99)$$

TABLE 13.1
State-variable-feedback system: steady-state error coefficients†

State-variable-feedback system	Number of zeros	K_p	K_v	K_a
All-pole plant	$w = 0$	∞	$\dfrac{K_G}{a_1 + K_G k_2}$	0
Pole-zero plant	$w = 1$	∞	∞	$\dfrac{K_G c_0}{a_2 + K_G k_3}$
	$w \geq 2$	∞	∞	∞

†For phase-variable representation.

This is a Type 3 transfer function that has zero steady-state error with a parabolic input. The error coefficients are $K_p = K_v = K_a = \infty$. The results of the previous analysis are contained in Table 13.1. Table 13.2 summarizes the zero steady-state error requirements for a stable state-variable-feedback control system. Based on Tables 13.1 and 13.2, the specification on the type of input that the system must follow determines the number of zeros that must be included in $Y(s)/R(s)$. The number of zeros determines the ability of the system to follow a particular input with no steady-state error and to achieve a stable system.

   With state-variable feedback, any desired system type can be achieved for $G_{eq}(s)$, regardless of the type represented by $G(s)$. In order to achieve a Type m system, the requirement for $m > 0$ is that

$$a_{i-1} + K_G k_1 - K_G c_{i-1} = 0 \qquad \text{for all} \qquad i = 1, 2, 3, \ldots, m \qquad (13.100)$$

Thus, a system in which the plant is not Type m can be made to behave as a Type m system by use of state feedback. This is achieved without adding pure integrators in cascade with the plant. *In order to achieve a Type m system it is seen from Table*

TABLE 13.2
Stable state-variable-feedback system: requirements for zero steady-state error†

System	Number of zeros required	Input			$G_{eq}(s)$ type
		Step	**Ramp**	**Parabola**	
All-pole plant	$w = 0$	$k_1 = 1 - \dfrac{a_0}{K_G}$	‡	§	1
Pole-zero plant	$w = 1$	$k_1 = c_0 - \dfrac{a_0}{K_G}$	$k_2 = c_1 - \dfrac{a_1}{K_G}$	‡	2
	$w \geq 2$			$k_3 = c_1 - \dfrac{a_2}{K_G}$	3

†For phase-variable representation.   ‡A steady-state error exists.   §Cannot follow.

13.2 *that the control ratio must have a minimum of $m-1$ zeros.* This is an important characteristic of state-variable-feedback systems.

The requirement for $e(t)_{ss} = 0$ for a ramp input can also be specified in terms of the locations of the zeros of the plant with respect to the desired locations of the poles of the control ratio (see Chap. 12). Expanding

$$\frac{E(s)}{R(s)} = \frac{1}{1 + G_{eq}(s)} \tag{13.101}$$

in a Maclaurin series in s yields

$$F(s) = \frac{E(s)}{R(s)} = e_0 + e_1 s + e_2 s^2 + \cdots + e_i s^i + \cdots \tag{13.102}$$

where the coefficients $e_0, e_1, e_2, \ldots$ are called the *generalized error coefficients*[2] and are given by

$$e_i = \frac{1}{i!} \left[\frac{d^i F(s)}{ds^i} \right]_{s=0} \tag{13.103}$$

From this equation, e_0 and e_1 are:

$$e_0 = \frac{1}{1 + K_p} \tag{13.104}$$

$$e_1 = \frac{1}{K_v} \tag{13.105}$$

Note that e_0 and e_1 are expressed in terms of the system error coefficients defined in Chap. 6. When the requirement of Eq. (13.80) is satisfied, Eq. (13.96) yields $K_p = \infty$. Solving for $E(s)$ from Eq. (13.102) and substituting the result into $Y(s) = R(s) - E(s)$ yields

$$\frac{Y(s)}{R(s)} = 1 - e_0 - e_1 s - e_2 s^2 - \cdots \tag{13.106}$$

The relation between K_v and the poles and zeros of $Y(s)/R(s)$ is readily determined if the control ratio is written in factored form:

$$\frac{Y(s)}{R(s)} = \frac{K_G(s - z_1)(s - z_2) \cdots (s - z_w)}{(s - p_1)(s - p_2) \cdots (s - p_n)} \tag{13.107}$$

The first derivative of $Y(s)/R(s)$ in Eq. (13.106) with respect to s, evaluated at $s = 0$, yields $-e_1$ which is equal to $-1/K_v$.

When it is noted that $[Y(s)/R(s)]_{s=0} = 1$ for a Type 1 or higher system with unity feedback, the following equation can be written:[2]

$$\frac{1}{K_v} = -\frac{\left[\frac{d}{ds}\left(\frac{Y}{R}\right) \right]_{s=0}}{\left[\frac{Y}{R} \right]_{s=0}} = -\left[\frac{d}{ds} \ln\left(\frac{Y(s)}{R(s)}\right) \right]_{s=0} \tag{13.108}$$

Substitution of Eq. (13.107) into Eq. (13.108) yields

$$
\frac{1}{K_v} = -\left(\frac{1}{s - z_1} + \cdots + \frac{1}{s - z_w} - \frac{1}{s - p_1} - \cdots - \frac{1}{s - p_n} \right)_{s=0} = \sum_{j=1}^{n} \frac{1}{p_j} - \sum_{j=1}^{w} \frac{1}{z_j}
$$

(13.109)

Thus, the condition for $e(t)_{ss} = 0$ for a ramp input, which occurs for $K_v = \infty$, that is, $1/K_v = 0$, is satisfied when

$$
\sum_{j=1}^{n} \frac{1}{p_j} = \sum_{j=1}^{w} \frac{1}{z_j}
$$

(13.110)

Therefore, in synthesizing the desired control ratio, its poles can be located not only to yield the desired transient response characteristics but also to yield the value $K_v = \infty$. In addition, it must contain at least one zero that is not canceled by a pole.

The values of k_1, k_2, and k_3 determined for $e(t)_{ss} = 0$ for the respective input from Eqs. (13.80), (13.85), and (13.90) are based upon the phase-variable representation of the system. These values may not be compatible with the desired roots of the characteristic equation. In that case it is necessary to leave at least one root unspecified so that these feedback gains may be used. If the resulting location of the unspecified root(s) is not satisfactory, other root locations may be tried in order to achieve a satisfactory performance. Once a system is designed that satisfies both the steady-state and the transient time responses, the feedback coefficients for the physical system can be calculated. This is accomplished by equating the corresponding coefficients of the characteristic equations representing, respectively, the physical- and the phase-variable systems. It is also possible to achieve the desired performance by working with the system represented entirely by physical variables, as illustrated by the examples in the following sections.

13.9 STATE-VARIABLE FEEDBACK: ALL-POLE PLANT

The design procedure presented in Sec. 13.6 is now applied to an all-pole plant.

Design Example. For the basic plant having the transfer function shown in Fig. 13.9a, state-variable feedback is used in order to achieve a closed-loop response with $M_p = 1.043$, $t_s = 5.65$ s, and zero steady-state error for a step input $r(t) = R_0 u_{-1}(t)$.
Steps 1 to 3. See Fig. 13.9b and c, and use Eq. (13.62) to obtain

$$
\frac{Y(s)}{R(s)} = \frac{10A}{s^3 + (6 + 2Ak_3)s^2 + [5 + 10A(k_3 + k_2)]s + 10Ak_1}
$$

(13.111)

Step 4. Inserting Eq. (13.111) into $y(t)_{ss} = \lim_{s \to 0} sY(s) = R_0$, for a step input, yields the requirement that $k_1 = 1$.
Step 5. Assuming that Eq. (13.111) has complex poles that are truly dominant, Eq. (3.61) is used to determine the damping ratio.

(a)

(b)

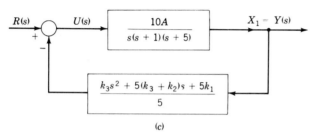

(c)

FIGURE 13.9
A state-variable-feedback system, all-pole plant: (a) plant; (b) state-variable feedback; (c) $H_{eq}(s)$.

$$M_p = 1.043 = 1 + \exp\left(-\frac{\zeta \pi}{\sqrt{1 - \zeta^2}}\right)$$

or
$$\zeta \approx 0.7079$$

From $T_s = 4/|\sigma|, |\sigma| = 4/5.65 = 0.708$. For these values of ζ and $|\sigma|$, the desired closed-loop complex-conjugate poles are $p_{1,2} = -0.708 \pm j0.7064$. Because the denominator of $G(s)$ is of third order, the control ratio also has a third-order denominator, as shown in Eq. (13.111). To ensure the dominancy of the complex poles, the third pole is made significantly nondominant and is arbitrarily selected as $p_3 = -100$. Therefore, the desired closed-loop transfer function is

$$\frac{Y(s)}{R(s)} = \frac{10A}{(s + 100)(s + 0.708 \pm j0.7064)} = \frac{10A}{s^3 + 101.4s^2 + 142.6s + 100} \quad (13.112)$$

Step 6. In order to achieve zero steady-state error, a step input requires the selection $k_1 = 1$ (see Sec. 13.5). Then, equating corresponding terms in the denominators of Eqs. (13.111) and (13.112) yields

$$
\begin{aligned}
10Ak_1 &= 100 & A &= 10 \\
6 + 2Ak_3 &= 101.4 & k_3 &= 4.77 \\
5 + 100(k_3 + k_2) &= 142.7 & k_2 &= -3.393
\end{aligned}
$$

Inserting Eq. (13.112) into Eq. (13.94) yields the G-equivalent form (see Fig. 13.8):

$$G_{eq}(s) = \frac{100}{s(s^2 + 101.4s + 142.7)} \tag{13.113}$$

Thus, $G_{eq}(s)$ is Type 1 and the ramp error coefficient is $K_1 = 0.701$.

An analysis of the root locus for the unity-feedback and the state-feedback systems, shown in Fig. 13.10a and b, respectively, reveals several characteristics.

For large values of gain the unity-feedback system is unstable, even if cascade compensation is added. For any variation in gain the corresponding change in the dominant roots changes the system response. Thus, the response is sensitive to changes of gain.

The state-variable-feedback system is stable for all $A > 0$. Because the roots are very close to the zeros of $G(s)H_{eq}(s)$, any increase in gain from the value $A = 10$ produces negligible changes in the location of the roots. In other words, for high-forward-gain (hfg) operation, $A > 10$, the system's response is essentially unaffected by variations in A, provided the feedback coefficients remain fixed. Thus, for the example represented by Fig. 13.10b, when $K = 9.54A \geq 95.4$, the change in the values of the dominant roots is small and there is little effect on the time response for a step input. The insensitivity of the system response to the variation of the gain A is related to the system's sensitivity function, which is discussed in the next chapter. *This insensitivity is due to the fact that the two desired dominant roots are close to the two zeros of $H_{eq}(s)$ and the third root is nondominant. To achieve this insensitivity to gain variation there must be β dominant roots that are located close to β zeros of $H_{eq}(s)$. The remaining $n - \beta$ roots must be nondominant. In this example $\beta = n - 1 = 2$, but in general $\beta \leq n - 1$.*

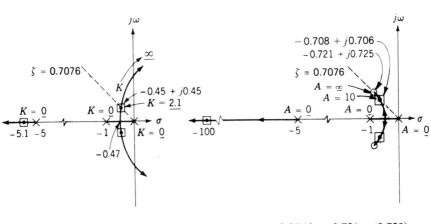

$$(a) \quad G(s) = \frac{K}{s(s + 1)(s + 5)} \qquad (b) \quad G(s)H_{eq}(s) = \frac{9.54A(s + 0.721 \pm j0.726)}{s(s + 1)(s + 5)}$$

FIGURE 13.10
Root-locus diagrams for (a) basic unity-feedback and (b) state-variable-feedback control systems.

Both the conventional and state-feedback systems can follow a step input with zero steady-state error. Also, they both follow a ramp input with a steady-state error. The state-variable-feedback system cannot be designed to produce zero steady-state error for a ramp input because $G(s)$ does not contain a zero. In order to satisfy this condition a cascade compensator that adds a zero to $G(s)$ can be utilized in the state-variable-feedback system.

In general, as discussed in Sec. 12.2, when selecting the poles and zeros of $[Y(s)/R(s)]_T$ to satisfy the desired specifications, one pair of dominant complex-conjugate poles $p_{1,2}$ is often chosen. The other $n - 2$ poles are located far to the left of the dominant pair or close to zeros in order to ensure the dominancy of $p_{1,2}$. It is also possible to have more than just a pair of dominant complex-conjugate poles to satisfy the desired specifications. For example, an additional pole of $Y(s)/R(s)$ can be located to the left of the dominant complex pair. When properly located, it can reduce the peak overshoot and may also reduce the settling time. Choosing only one nondominant root, that is, $n - 1$ dominant roots, results in lower values of K and the k_i than when a larger number of nondominant roots is chosen. *For state-variable feedback the value of the gain A is always taken as positive.*

13.10 PLANTS WITH COMPLEX POLES

This section illustrates how the state-variable representation of a plant containing complex poles is obtained. Because the system shown in Fig. 13.11a is of the fourth order, four state variables must be identified. The transfer function $O(s)/I(s)$ involves two states. If the detailed block-diagram representation of $O(s)/I(s)$ contains two explicit states, it should not be combined into the quadratic form; by combining the blocks, the identity of one of the states is lost. That situation is illustrated by the system of Fig. 13.6. As represented in Fig. 13.7a, the two states X_2 and X_3 are accessible, but in the combined mathematical form of Fig. 13.7c the physical variables are not identifiable. When one of the states is not initially identified, an equivalent model containing two states must be formed for the quadratic transfer function $O(s)/I(s)$ of Fig. 13.11a. This decomposition can be performed by the method of Sec. 5.12. The simulation diagram of Fig. 5.34 is shown in the block diagram of Fig. 13.11b. This decomposition is inserted into the complete control system that contains state-variable feedback, as shown in Fig. 13.11c. If the state X_3 is inaccessible, then, once the values of k_i are determined, the physical realization of the state feedback can be achieved by block-diagram manipulation. An alternate approach is to represent the transfer function $O(s)/I(s)$ in state-variable form, as shown in Fig. 13.12. This representation is particularly attractive when X_3 is recognizable as a physical variable. If it is not, then the pick-off for k_3 can be moved to the right by block-diagram manipulation.

13.11 COMPENSATOR CONTAINING A ZERO

When a cascade compensator of the form $G_c(s) = (s - z_i)/(s - p_i)$ is required, it can be constructed in either of the forms shown in Fig. 13.13a or b. The input signal

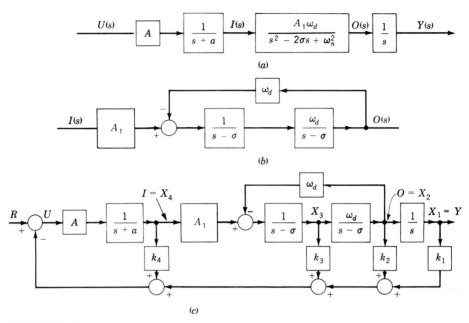

(a)

(b)

(c)

FIGURE 13.11
Representation of a plant containing complex poles.

FIGURE 13.12
State-variable representation of $O(s)/I(s)$ in Fig. 13.11a.

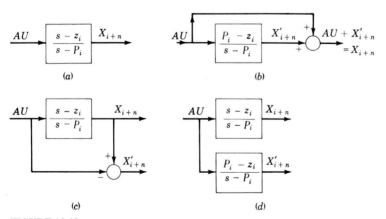

(a)   (b)

(c)   (d)

FIGURE 13.13
Cascade compensators: $G_c(s) = (s - z_i)/(s - P_i)$: ($a$) conventional; ($b$) feedforward form; ($c$) generating the state variable X'_{i+n}; (d) mathematical model for (c).

FIGURE 13.14
Forward path.

is shown as AU because, in general, this is the most accessible place to insert the compensator. The differential equations for the two forms are

For Fig. 13.13a: $\qquad \dot{x}_{i+n} = P_i x_{i+n} - z_i A u + A \dot{u}$ (13.114)

For Fig. 13.13b: $\qquad \dot{x}'_{i+n} = P_i x'_{i+n} + (P_i - z_i) A u$ (13.115)

The presence of $\dot{u}$ in Eq. (13.114) does not permit describing the system with this compensator by the state equation $\dot{\mathbf{x}} = \mathbf{A}\mathbf{x} + \mathbf{b}u$, and x_{i+n} in Fig. 13.13a is not a state variable. The presence of a zero in the first block of the forward path can also occur in a system before a compensator is inserted. The design of such a system can be treated like that of a system containing the compensator of Fig. 13.13a. The consequence of the $\dot{u}$ term in Eq. (13.114) is that the zero of $G_c(s)$ does not appear as a pole of $H_{eq}(s)$. Thus the root-locus characteristics discussed in Sec. 13.6 must be modified accordingly. There is no difficulty in expressing the system in state-variable form when the function of Fig. 13.13a is not the first frequency-sensitive term in the forward path. When the compensator of Fig. 13.13a is inserted at any place to the right of the G_1 block in the forward path shown in Fig. 13.14, the system can be described by a state equation. Thus, the root-locus characteristics discussed in Sec. 13.6 also apply for this case.

A system containing the feedforward compensator of Fig. 13.13b can be described by a state equation, because $\dot{u}$ does not appear in Eq. (13.115). Thus, the poles of $H_{eq}(s)$ continue to be the zeros of $G(s)$. Therefore, the root-locus characteristics discussed in Sec. 13.6 also apply for this case. This is a suitable method of constructing the compensator $G_c(s) = (s - z_i)/(s - P_i)$ because it produces the accessible state variable X'_{i+n}. The implementation of $G_c(s)$ consists of the lag term $K_c/(s - P_i)$, where $K_c = P_i - z_i$, with the unity-gain feedforward path shown in Fig. 13.13b.

An additional possibility is to generate a new state variable X'_{i+n} from the physical arrangement shown in Fig. 13.13c. This is equivalent to the form shown in Fig. 13.13d and is mathematically equivalent to Fig. 13.13b. Thus, X'_{i+n} is a valid state variable and may be used with a feedback coefficient k_{i+n}.

13.12 STATE-VARIABLE FEEDBACK: POLE-ZERO PLANT

Section 13.9 deals with a system having an all-pole plant; the control ratio therefore has no zeros. Locating the roots of the characteristic equation for such a system to yield the desired system response is relatively easy. This section deals with a system having a plant that contains both poles and zeros and has a control ratio of the form

$$\frac{Y(s)}{R(s)} = \frac{K_G(s - z_1)(s - z_2)\cdots(s - z_h)\cdots(s - z_w)}{(s - p_1)(s - p_2)\cdots(s - p_c)\cdots(s - p_n)} \qquad (13.116)$$

that has the same zeros as the plant. When open-loop zeros are present, the synthesis of the overall characteristic equation becomes more complex than for the all-pole plant. As discussed in Sec. 13.8 and summarized in Table 13.2, the presence of a zero in the control ratio is necessary in order to achieve $e(t)_{ss} = 0$ for a ramp input. In synthesizing the desired control ratio, Eq. (13.110) can also be used to assist in achieving this condition while simultaneously satisfying the other desired figures of merit.

If the locations of the zeros of $G_x(s)$ are satisfactory, the design procedure is the same as for the all-pole case of Sec. 13.9. That is, the coefficients of the desired characteristic equation are equated to the corresponding coefficients of the numerator polynomial of $1 + G(s)H_{eq}(s)$ in order to determine the required K_G and k_i. It is possible to achieve low sensitivity to variation in gain by requiring that

1. $Y(s)/R(s)$ have β dominant poles and *at least one nondominant pole.*
2. The β dominant poles be close to β zeros of $H_{eq}(s)$.

The latter can be achieved by a judicious choice of the compensator pole(s) and/or the nondominant poles.

Three examples are used to illustrate state feedback design for plants that contain one zero. The procedure is essentially the same for systems having more than one zero. The design technique illustrated in these examples can be modified as desired.

Design Example 1. The control ratio for the system of Fig. 13.15 is

$$\frac{Y(s)}{R(s)} = \frac{A(s + 2)}{s^3 + [6 + (k_3 + k_2)A]s^2 + [5 + (k_3 + 2k_2 + k_1)A]s + 2k_1 A} \qquad (13.117)$$

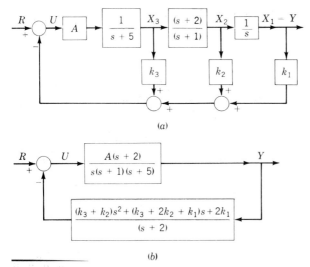

(a)

(b)

FIGURE 13.15
Closed-loop zero cancellation by state-variable feedback: (*a*) detailed block diagram; (*b*) *H*-equivalent reduction (where $K_G = A$).

A trial function for the desired control ratio for this system is specified as

$$\frac{Y(s)}{R(s)}\bigg|_{\text{desired}} = \frac{2}{s^2 + 2s + 2} \tag{13.118}$$

If Eq. (13.117) is to reduce to Eq. (13.119), then one pole of Eq. (13.117) must have the value $s = -2$, in order to cancel the zero, and $A = 2$. Equation (13.118) is modified to reflect this requirement.

$$\frac{Y(s)}{R(s)}\bigg|_{\text{desired}} = \frac{2(s + 2)}{(s^2 + 2s + 2)(s + 2)} = \frac{2(s + 2)}{s^3 + 4s^2 + 6s + 4} \tag{13.119}$$

Equating the corresponding coefficients of the denominators of Eqs. (13.117) and (13.119) yields $k_1 = 1$, $k_2 = 1/2$, and $k_3 = -3/2$. Thus

$$G(s)H_{\text{eq}}(s) = \frac{-A(s^2 - s/2 - 2)}{s(s + 1)(s + 5)} = -1 \tag{13.120}$$

whose zeros are $s_1 = -1.189$ and $s_2 = 1.686$. Canceling the minus sign shows that the $0°$ root locus is required, and it is plotted in Fig. 13.16. For $A = 2$, the roots are $p_{1,2} = -1 \pm j1$ and $p_3 = -2$, as specified by Eq. (13.119). Note that because Eq. (13.118) has no other poles in addition to the dominant poles, the necessary condition of at least one nondominant pole is not satisfied. As a result, two branches of the root locus enter the RH of the s plane, yielding a system that can become unstable for high gain. The response of this system to a unit step input has $M_p = 1.043$, $t_p = 3.1$ s, and $t_s = 4.2$ s. Because the roots are not near zeros, the system is highly sensitive to gain variation, as seen from the root locus in Fig. 13.16. Any increase in A results in a significant change in performance characteristics, and the system becomes unstable if A exceeds the value of 6. The ramp error coefficient obtained from $G_{\text{eq}}(s)$ is $K_1 = 1.0$.

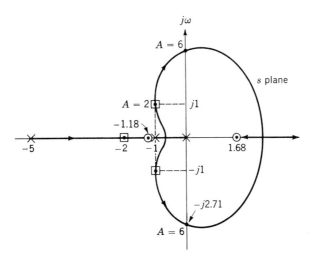

FIGURE 13.16
Root locus for Eq. (13.120).

Design Example 1 illustrates that insensitivity to gain variation and stability for all positive values of gain is not achieved. In order to satisfy these conditions, $[Y(s)/R(s)]_T$ must have β dominant poles and at least one nondominant pole. The next example illustrates this design concept.

Design Example 2. For the third-order system of Fig. 13.15, it is desired to cancel the control ratio zero at $s = -2$ and to reduce the system output sensitivity to variations in the gain A while effectively achieving the second-order response of Eq. (13.118). The time response is satisfied if $Y(s)/R(s)$ has three dominant poles $s_{1,2} = -1 \pm j1$ and $s_3 = -2$. However, in order to achieve insensitivity to gain variation, the transfer function $H_{eq}(s)$ must have three zeros. In accordance with Eq. (13.74), the number of poles of $G(s)$ must be increased to $n = 4$. Therefore, one pole must be added to $G(s)$ by using the cascade compensator $G_c(s) = 1/(s + a)$. Then $Y(s)/R(s)$ can have one nondominant pole p and the desired control ratio is given by

$$\left.\frac{Y(s)}{R(s)}\right|_{\text{desired}} = \frac{K_G(s + 2)}{(s^2 + 2s + 2)(s + 2)(s - p)} \tag{13.121}$$

The selections of the values of $K_G = A$ and p are not independent because the condition of $e(t)_{ss} = 0$ for a step input $r(t) = R_0 u_{-1}(t)$ is required. Thus

$$y(t)_{ss} = \lim_{s \to 0} sY(s) = \frac{-K_G}{2p}R_0 = R_0 \tag{13.122}$$

or

$$K_G = -2p \tag{13.123}$$

As mentioned previously, the larger the value of the loop sensitivity for the root locus of $G(s)H_{eq}(s) = -1$, the closer the roots will be to the zeros of $H_{eq}(s)$. It is this "closeness" that reduces the system's output sensitivity to gain variations. Therefore, the value of K_G should be chosen as large as possible consistent with the capabilities of the system components. This large value of K_G may have to be limited to a value that does not accentuate the system noise to an unacceptable level. For this example, selecting a value $K_G = 100$ ensures that p is a nondominant pole. This results in a value $p = -50$ so that Eq. (13.121) becomes

$$\left.\frac{Y(s)}{R(s)}\right|_{\text{desired}} = \frac{100(s + 2)}{s^4 + 54s^3 + 206s^2 + 304s + 200} \tag{13.124}$$

The detailed state feedback and H-equivalent block diagrams for the modified control system are shown in Fig. 13.17a and b, respectively.

In order to maintain a controllable and observable system the value of a can be chosen to be any value *other than* the value of the zero term of $G(s)$. Thus, assume $a = 1$. With this value of a and with $K_G = 100$, the control ratio of the system of Fig. 13.17b is

$$\frac{Y(s)}{R(s)} = \frac{100(s + 2)}{s^4 + (7 + 100k_4)s^3 + [11 + 100(6k_4 + k_3 + k_2)]s^2} \tag{13.125}$$
$$+ [5 + 100(5k_4 + k_3 + k_2 + k_1)]s + 200k_1$$

For Eq. (13.125) to yield $e(t)_{ss} = 0$ with a step input requires $k_1 = 1$. Equating the corresponding coefficients of the denominators of Eqs. (13.124) and (13.125) and solv-

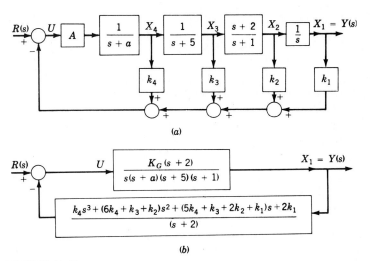

(a)

(b)

FIGURE 13.17
Closed-loop zero cancellation by state-variable feedback: (*a*) detailed
block diagram; (*b*) H-equivalent reduction (where $K_G = A$).

ing for the feedback coefficients yields $k_2 = 0.51$, $k_3 = -1.38$, and $k_4 = 0.47$. Thus,
from Fig. 13.17*b*:

$$G(s)H_{eq}(s) = \frac{0.47K_G(s^3 + 4.149s^2 + 6.362s + 4.255)}{s^4 + 7s^3 + 11s^2 + 5s} \tag{13.126}$$

The root locus shown in Fig. 13.18 reveals that the three desired dominant roots are
close to the three zeros of $H_{eq}(s)$, the fourth root is nondominant, and the root locus lies
entirely in the LH s plane. Thus, the full benefits of state-variable feedback have been

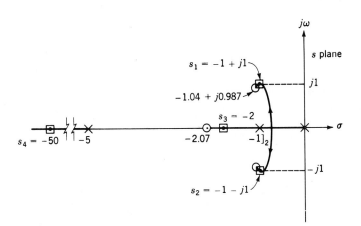

FIGURE 13.18
The root locus for Eq. (13.126).

achieved. In other words, a completely stable system with low sensitivity to parameter variation has been realized. The figures of merit of this system are $M_p = 1.043$, $t_p = 3.2$ s, $t_s = 4.2$ s, and $K_1 = 0.98\,s^{-1}$.

Design Example 3. A further improvement in the time response can be obtained by adding a pole and zero to the overall system function of Design Example 2. The control ratio to be achieved is

$$\frac{Y(s)}{R(s)} = \frac{K_G(s + 1.4)(s + 2)}{(s + 1)(s^2 + 2s + 2)(s + 2)(s - p)}$$

$$= \frac{K_G(s + 1.4)(s + 2)}{s^5 + (5 - p)s^4 + (10 - 5p)s^3 + (10 - 10p)s^2 + (4 - 10p)s - 4p}$$

$$(13.127)$$

Analysis of Eq. (13.127) shows the following features:

1. Because $Y(s)/R(s)$ has four dominant poles, the nondominant pole p yields the desired property of insensitivity to gain variation.

2. The pole-zero combination $(s + 1.4)/(s + 1)$ is added to reduce the peak overshoot in the transient response.

3. The factor $s + 2$ appears in the numerator because it is present in $G(s)$. Because it is not desired in $Y(s)/R(s)$, it is canceled by producing the same factor in the denominator.

4. In order to achieve zero steady-state error with a step input $r(t) = u_{-1}(t)$, the output is obtained by applying the final-value theorem to $Y(s)$. This yields $y(t)_{ss} = K_G \times 1.4 \times 2/(-4p) = 1$. Therefore, $p = -0.7K_G$. Assuming the large value of $K_G = 100$ to be satisfactory results in $p = -70$, which is nondominant, as required.

The required form of the modified system is shown in Fig. 13.19. The cascade compensator $1/(s + 1)$ is added to the system in order to achieve the degree of the denominator of Eq. (13.127). The cascade compensator $(s + 1.4)/(s + 3)$ is included in order to produce the desired numerator factor $s + 1.4$. The denominator is arbitrarily chosen as $s + 3$. This increases the degree of the denominator in order to permit canceling the factor $s + 2$ in $Y(s)/R(s)$.

The forward transfer function obtained from Fig. 13.19 is

$$G(s) = \frac{100(s + 1.4)(s + 2)}{s(s + 1)^2(s + 3)(s + 5)} = \frac{100(s^2 + 3.4s + 2.8)}{s^5 + 10s^4 + 32s^3 + 38s^2 + 15s} \qquad (13.128)$$

The value of $H_{eq}(s)$ is evaluated from Fig. 13.19 and is

$$H_{eq}(s) = \frac{\begin{array}{l}(k_5 + k_4)s^4 + (9k_5 + 7.4k_4 + k_3 + k_2)s^3 \\ +(23k_5 + 13.4k_4 + 2.4k_3 + 3.4k_2 + k_1)s^2 \\ +(15k_5 + 7k_4 + 1.4k_3 + 2.8k_2 + 3.4k_1)s + 2.8k_1\end{array}}{(s + 1.4)(s + 2)} \qquad (13.129)$$

The desired overall system function is obtained from Eq. (13.127), with the values $K_G = 100$ and $p = -70$, as follows.

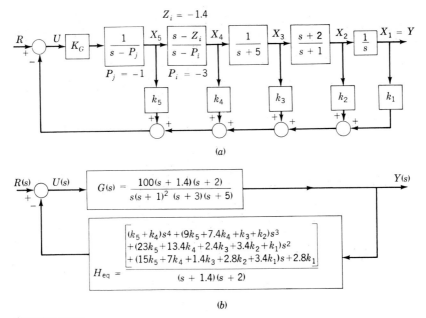

FIGURE 13.19
State feedback diagram for Design Example 3.

$$\frac{Y(s)}{R(s)} = \frac{100(s + 1.4)(s + 2)}{s^5 + 75s^4 + 360s^3 + 710s^2 + 704s + 280} \tag{13.130}$$

The system function obtained by using $G(s)$ and $H_{eq}(s)$ is

$$\frac{Y(s)}{R(s)} = \frac{100(s + 1.4)(s + 2)}{\begin{aligned} &s^5 + [10 + 100(k_5 + k_4)]s^4 \\ &+ [32 + 100(9k_5 + 7.4k_4 + k_3 + k_2)]s^3 \\ &+ [38 + 100(23k_5 + 13.4k_4 + 2.4k_3 + 3.4k_2 + k_1)]s^2 \\ &+ [15 + 100(15k_5 + 7k_4 + 1.4k_3 + 2.8k_2 + 3.4k_1)]s + 100(2.8k_1) \end{aligned}} \tag{13.131}$$

Equating the coefficients in the denominators of Eqs. (13.130) and (13.131) yields $280k_1 = 280$, or $k_1 = 1$, and

$$\begin{aligned} 1500k_5 + 700k_4 + 140k_3 + 280k_2 &= 349 \\ 2300k_5 + 1340k_4 + 240k_3 + 340k_2 &= 572 \\ 900k_5 + 740k_4 + 100k_3 + 100k_2 &= 328 \\ 100k_5 + 100k_4 &= 65 \end{aligned} \tag{13.132}$$

Solving these four simultaneous equations gives $k_2 = 1.0$, $k_3 = -2.44167$, $k_4 = 0.70528$, and $k_5 = -0.05528$.

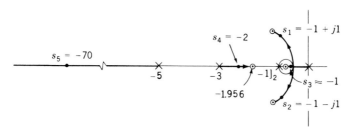

FIGURE 13.20
Root locus for Eq. (13.133).

In order to demonstrate the desirable features of this system, the root locus is drawn in Fig. 13.20 for

$$G(s)H_{eq}(s) = \frac{100(0.65s^4 + 3.28s^3 + 6.72s^2 + 6.89s + 2.8)}{s(s+1)^2(s+3)(s+5)}$$

$$= \frac{65(s + 1.04532 \pm j1.05386)(s + 0.99972)(s + 1.95565)}{s(s+1)^2(s+3)(s+5)} \quad (13.133)$$

The root locus shows that the zero of $Y(s)/R(s)$ at $s = -2$ has been canceled by the pole at $s = -2$. Because the root $s_4 = -2$ is near a zero in Fig. 13.20, its value is insensitive to increases in gain. This ensures the desired cancellation. The dominant roots are all near zeros in Fig. 13.20, thus assuring that the desired control ratio $Y(s)/R(s)$ as given by Eq. (13.127) has been achieved. The system is stable for all values of gain K_G. Also, state-variable feedback has produced a system in which $Y(s)/R(s)$ is unaffected by increases in A above 100. The time response $y(t)$ to a step input has a peak overshoot $M_p = 1.008$, a peak time $t_p = 3.92$, and a settling time $t_s = 2.94$. The ramp error coefficient evaluated for $G_{eq}(s)$ is $K_1 = 0.98$. These results show that the addition of the pole-zero combination $(s + 1.4)/(s + 1)$ to $Y(s)/R(s)$ has improved the performance. This system is insensitive to variation in gain.

The three design examples in this section are intended to provide a firm foundation in the procedures for

1. Synthesizing the desired system performance and the overall control ratio.
2. Analyzing the root locus of $G(s)H_{eq}(s) = -1$ for the desired system performance and verifying insensitivity to variations in gain.
3. Determining the values of k_i to yield the desired system performance.

These approaches are not all-inclusive, and variations may be developed.[1,6]

13.13 SUMMARY

This chapter presents the controllability and observability properties of a system, which are important in the application of many control system design techniques This is followed by the presentation of the design method using the state-variable representation of SISO systems. A model control ratio incorporating the desired system tracking performance specifications is used in this chapter to design a compensated state-variable-feedback control system. The root-locus properties and characterization, the steady-state error analysis, and the use of steady-state error coefficients, as they pertain to state-variable-feedback systems (all-pole and pole-zero plants) are discussed in detail. The presence of at least one nondominant closed-loop pole is required in order to minimize the sensitivity of the output response to gain variations. A comparison is made between the sensitivity to gain variation of unity-feedback and state-variable-feedback systems. System sensitivity for parameter variations is presented in more detail in Chap. 14. The use of CAD programs expedite the achievement of an acceptable design. These programs include the root-locus, time-response, and state-variable feedback-coefficient determination (see App. B).

REFERENCES

1. Schultz, D. G., and J. L. Melsa: *State Functions and Linear Control Systems*, McGraw-Hill, New York, 1967.
2. James, H. M., N. B. Nichols, and R. S. Phillips: *Theory of Servomechanisms*, McGraw-Hill, New York, 1947.
3. Chen, C. T., and C. A. Desoer: "Proof of Controllability of Jordan Form State Equations," *Trans. IEEE*, vol. AC-13, pp. 195–196, 1968.
4. Rugh, W. J.: *Linear System Theory*, Prentice-Hall, Englewood Cliffs, N. J., 1993.
5. Gilbert, E. G.: "Controllability and Observability in Multi-Variable Control Systems," *J. Soc. Ind. Appl. Math.*, Ser. A., Control, vol. 1, no. 2, pp. 128–151, 1963.
6. Anderson, B. D. O., and J. B. Moore: *Linear Optimal Control*, Prentice-Hall, Englewood Cliffs, N.J., 1971.
7. Kalman, R. E.: "When Is a Linear System Optimal? *"ASME J. Basic Eng.*, vol. 86, pp. 51–60, March, 1964.

CHAPTER
14

PARAMETER
SENSITIVITY
AND STATE SPACE
TRAJECTORIES

14.1 INTRODUCTION

In Chapter 13 it is shown that an advantage of a state-feedback control system operating with high forward gain (hfg) is the insensitivity of the system output to gain variations in the forward path. The design method of Chapter 13 assumes that all states are accessible. This chapter investigates in depth the insensitive property of hfg operation of a state-feedback control system. This is followed by a treatment of inaccessible states. In order for the reader to develop a better feel for the kinds of transient responses of a system, this chapter includes an introduction to state-space trajectories. This includes the determination of the steady-state, or equilibrium, values of a system response. While linear time invariant (LTI) systems have only one equilibrium value, a nonlinear system may have a number of equilibrium solutions. These can be determined from the state equations. They lead to the development of the Jacobian matrix, which is used to represent a nonlinear system by approximate linear equations in the region close to the singular points.

14.2 SENSITIVITY

The environmental conditions to which a control system is subjected affect the accuracy and stability of the system. The performance characteristics of most components are affected by their environment and by aging. Thus, any change in the component characteristics causes a change in the transfer function and therefore in the controlled quantity. The effect of a parameter change on system performance can be expressed in terms of a *sensitivity function*. This sensitivity function S_δ^M is a measure of the sensitivity of the system's response to a system parameter variation and is given by

$$S_\delta^M = \frac{d(\ln M)}{d(\ln \delta)} = \frac{d(\ln M)}{d\delta}\frac{d\delta}{d(\ln \delta)} \tag{14.1}$$

where $\ln$ = logarithm to base e, M = system's output response (or its control ratio), and δ = system parameter that varies. Now

$$\frac{d(\ln M)}{d\delta} = \frac{1}{M}\frac{dM}{d\delta} \quad \text{and} \quad \frac{d(\ln \delta)}{d\delta} = \frac{1}{\delta} \tag{14.2}$$

Accordingly, Eq. (14.1) can be written

$$S_\delta^M \bigg|_{\substack{M=M_o \\ \delta=\delta_o}} = \frac{dM/M_o}{d\delta/\delta_o}$$

$$= \frac{\delta}{M}\left(\frac{dM}{d\delta}\right)\bigg|_{\substack{M=M_o \\ \delta=\delta_o}} = \frac{\text{fractional change in output}}{\text{fractional change in system parameter}} \tag{14.3}$$

where M_o and δ_o represent the nominal values of M and δ. When M is a function of more than one parameter, say $\delta_1, \delta_2, \ldots, \delta_k$, the corresponding formulas for the sensitivity entail partial derivatives. For a small change in δ from δ_o, M changes from M_o, and the sensitivity can be written as

$$S_\delta^M \bigg|_{\substack{M=M_o \\ \delta=\delta_o}} \approx \frac{\Delta M/M_o}{\Delta\delta/\delta_o} \tag{14.4}$$

To illustrate the effect of changes in the transfer function, four cases are considered for which the input signal $r(t)$ and its transform $R(s)$ are fixed. Although the response $Y(s)$ is used in these four cases, the results are the same when $M(s)$ is the control ratio.

Case 1: Open-Loop System of Fig. 14.1a

The effect of a change in $G(s)$, for a fixed $r(t)$ and thus a fixed $R(s)$, can be determined by differentiating, with respect to $G(s)$, the output expression

$$Y_o(s) = R(s)G(s) \tag{14.5}$$

giving

$$dY_o(s) = R(s)dG(s) \tag{14.6}$$

FIGURE 14.1
Control systems: (*a*) open loop; (*b*) closed loop.

Combining these two equations gives

$$dY_o(s) = \frac{dG(s)}{G(s)}Y_o(s) \;\rightarrow\; S_{G(s)}^{Y(s)}(s) = \frac{dY_o(s)/Y_o(s)}{dG(s)/G(s)} = 1 \qquad (14.7)$$

A change in the transfer function $G(s)$ therefore causes a proportional change in the transform of the output $Y_o(s)$. This requires that the performance specifications of $G(s)$ be such that any variation still results in the degree of accuracy within the prescribed limits. In Eq. (14.7) the varying function in the system is the transfer function $G(s)$.

Case 2: Closed-Loop Unity-Feedback System of Fig. 14.1b [$H(s) = 1$]

Proceeding in the same manner as for case 1, for $G(s)$ varying, leads to

$$Y_c(s) = R(s)\frac{G(s)}{1 + G(s)} \qquad (14.8)$$

$$dY_c(s) = R(s)\frac{dG(s)}{[1 + G(s)]^2} \qquad (14.9)$$

$$dY_c(s) = \frac{dG(s)}{G(s)[1 + G(s)]}Y_c(s) = \frac{1}{1 + G(s)}\frac{dG(s)}{G(s)}Y_c(s)$$

$$S_{G(s)}^{Y(s)} = \frac{dY_c(s)/Y_c(s)}{dG(s)/G(s)} = \frac{1}{1 + G(s)} \qquad (14.10)$$

Comparing Eq. (14.10) with Eq. (14.7) readily reveals that the effect of changes of $G(s)$ upon the transform of the output of the closed-loop control is reduced by the factor $1/|1 + G(s)|$ compared to the open-loop control. *This is an important reason why feedback systems are used.*

Case 3: Closed-Loop Nonunity-Feedback System of Fig. 14.1b [Feedback Function $H(s)$ Fixed and $G(s)$ Variable]

Proceeding in the same manner as for case 1, for $G(s)$ varying, leads to

$$Y_c(s) = R(s)\frac{G(s)}{1 + G(s)H(s)} \qquad (14.11)$$

$$dY_c(s) = R(s)\frac{dG(s)}{[1 + G(s)H(s)]^2} \qquad (14.12)$$

$$dY_c(s) = \frac{dG(s)}{G(s)[1 + G(s)H(s)]}Y_c(s) = \frac{1}{1 + G(s)H(s)}\frac{dG(s)}{G(s)}Y_c(s)$$

$$S_{G(s)}^{Y(s)} = \frac{dY_c(s)/Y_c(s)}{dG(s)/G(s)} = \frac{1}{1 + G(s)H(s)} \qquad (14.13)$$

Comparing Eqs. (14.7) and (14.13) shows that the closed-loop variation is reduced by the factor $1/|1 + G(s)H(s)|$. In comparing Eqs. (14.10) and (14.13), if the term $|1 + G(s)H(s)|$ is larger than the term $|1 + G(s)|$, then there is an advantage to using a nonunity-feedback system. Further, $H(s)$ may be introduced both to provide an improvement in system performance and to reduce the effect of parameter variations within $G(s)$.

Case 4: Closed-Loop Nonunity-Feedback System of Fig. 14.1*b* [Feedback Function $H(s)$ Variable and $G(s)$ Fixed]

From Eq. (14.11),

$$dY_c(s) = R(s)\frac{-G(s)^2dH(s)}{[1 + G(s)H(s)]^2} \tag{14.14}$$

Multiplying and dividing Eq. (14.14) by $H(s)$ and also dividing by Eq. (14.11) results in

$$dY_c(s) = -\left[\frac{G(s)H(s)}{1 + G(s)H(s)}\frac{dH(s)}{H(s)}\right]Y_c(s) \approx -\frac{dH(s)}{H(s)}Y_c(s)$$

$$S_{H(s)}^{Y(s)} = \frac{dY_c(s)/Y_c(s)}{dH(s)/H(s)} \approx -1 \tag{14.15}$$

The approximation applies for those cases where $|G(s)H(s)| \gg 1$. When Eq. (14.15) is compared with Eq. (14.7), it is seen that a variation in the feedback function has approximately a direct effect upon the output, the same as for the open-loop case. Thus, the components of $H(s)$ must be selected as precision fixed elements in order to maintain the desired degree of accuracy and stability in the transform $Y(s)$.

The two situations of cases 3 and 4 serve to point out the advantage of feedback compensation from the standpoint of parameter changes. Because the use of fixed feedback compensation minimizes the effect of variations in the components of $G(s)$, prime consideration can be given to obtaining the necessary power requirement in the forward loop rather than to accuracy and stability. $H(s)$ can be designed as a precision device so that the transform $Y(s)$, or the output $y(t)$, has the desired accuracy and stability. *In other words, by use of feedback compensation the performance of the system can be made to depend more on the feedback term than on the forward term.*

Applying the sensitivity equation (14.3) to each of the four cases, where $M(s) = Y(s)/R(s)$, where $\delta = G(s)$ (for cases 1, 2, and 3), and $\delta = H(s)$ (for case 4), yields the results shown in Table 14.1. This table reveals that S_δ^M *never exceeds a magnitude of* 1, and the smaller this value, the less sensitive the system is to a variation in the transfer function. For an increase in the variable function, a positive value of the sensitivity function means that the output increases from its nominal response. Similarly, a negative value of the sensitivity function means that the output decreases from its nominal response. The results presented in this table are based upon a functional analysis; i.e., the "variations" considered are in $G(s)$ and $H(s)$. The results are easily interpreted when $G(s)$ and $H(s)$ are real numbers. When they

TABLE 14.1
Sensitivity functions

Case	System variable parameter	S_δ^M
1	$G(s)$	$\dfrac{dY_o/Y_o}{dG/G} = 1$
2	$G(s)$	$\dfrac{dY_c/Y_c}{dG/G} = \dfrac{1}{1+G}$
3	$G(s)$	$\dfrac{dY_c/Y_c}{dG/G} = \dfrac{1}{1+GH}$
4	$H(s)$	$\dfrac{dY_c/Y_c}{dH/H} \approx -1$

are not real numbers, and where δ represents the parameter that varies within $G(s)$ or $H(s)$, the interpretation can be made as a function of frequency.

The analysis in this section so far has considered variations in the transfer functions $G(s)$ and $H(s)$. Take next the case when $r(t)$ is sinusoidal. Then the input can be represented by the phasor $\mathbf{R}(j\omega)$ and the output by the phasor $\mathbf{Y}(j\omega)$. The system is now represented by the frequency transfer functions $\mathbf{G}(j\omega)$ and $\mathbf{H}(j\omega)$. All the formulas developed earlier in this section are the same in form, but the arguments are $j\omega$ instead of s. As parameters vary within $\mathbf{G}(j\omega)$ and $\mathbf{H}(j\omega)$, the magnitude of the sensitivity function can be plotted as a function of frequency. The magnitude $|S_\delta^M(j\omega)|$ does not have the limits of 0 to 1 given in Table 14.1, but can vary from 0 to any large magnitude. An example of sensitivity to parameter variation is investigated in detail in the following section. That analysis shows that the sensitivity function can be considerably reduced with appropriate feedback included in the system structure.

14.3 SENSITIVITY ANALYSIS[1,2]

This section shows how the use of complete state-variable or output feedback minimizes the sensitivity of the output response to a parameter variation. This feature of state feedback is in contrast to conventional control-system design and is illustrated in the example of Sec. 13.9 for the case of system gain variation. In order to illustrate the advantage of state-variable feedback over the conventional method of achieving the desired system performance, a system sensitivity analysis is made for both designs. The basic plant $G_x(s)$ of Fig. 13.9a is used for this comparison. The conventional control system is shown in Fig. 14.2a. The control-system design can be implemented by the complete state-feedback configuration of Fig. 14.2b or by the output feedback representation of Fig. 14.2c, which uses H_{eq} as a fixed feedback compensator. The latter configuration is shown to be best for minimizing the sensitivity of the output response with a parameter variation that occurs in the forward path, between the state x_n and the output y. For both the state-feedback configuration and its output feedback equivalent, the coefficients k_i are determined for the nominal values of the parameters and are assumed to be invariant. The analysis is based upon

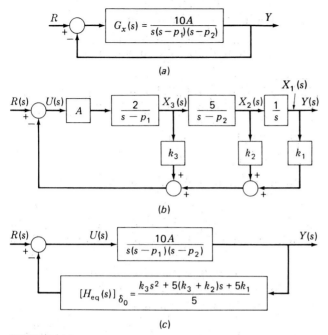

FIGURE 14.2
Control systems: (a) unity feedback; (b) state feedback; (c) state feedback, H-equivalent. The nominal values are $p_1 = -1$, $p_2 = -5$, and
$A = 10$.

determining the value of the passband frequency ω_b at $|\mathbf{M}_o(j\omega_b)| \doteq M_o = 0.707$
for the nominal system values of the conventional and state-variable-feedback
control systems. The system sensitivity function of Eq. (14.3), repeated here, is
determined for each system of Fig. 14.2 and evaluated for the frequency range
$0 \le \omega \le \omega_b$

$$S_\delta^M(s)\bigg|_{\substack{M=M_o \\ \delta=\delta_o}} = \left(\frac{\delta}{M}\frac{dM}{d\delta}\right)\bigg|_{\substack{M=M_o \\ \delta=\delta_o}} \tag{14.16}$$

Note that M is the control ratio. The passband frequency is $\omega_b = 0.642$ for the system of Fig. 14.2a with nominal values, and $\omega_b = 1.0$ for the state-feedback system
of Figs. 14.2b and 14.2c with nominal values.

The control ratios for each of the three cases to be analyzed are

Fig. 14.2a:

$$M(s) = \frac{10A}{s^3 - (p_1 + p_2)s^2 + p_1 p_2 s + 10A} \tag{14.17}$$

Fig. 14.2b:

$$M(s) = \frac{10A}{s^3 + (2Ak_3 - p_1 - p_2)s^2 + [p_1 p_2 + 2A(5k_2 - k_3 p_2)]s + 10Ak_1}$$

$$\tag{14.18}$$

Fig. 14.2c:

$$M(s) = \frac{10A}{s^3 + (2Ak_3 - p_1 - p_2)s^2 + [p_1 p_2 + 10A(k_3 + k_2)]s + 10Ak_1}$$

(14.19)

The system sensitivity function for any parameter variation is readily obtained for each of the three configurations of Fig. 14.2 by using the respective control ratios of Eqs. (14.17) to (14.19). The plots of $|\mathbf{S}_{p_1}^M(j\omega)|$, $|\mathbf{S}_{p_2}^M(j\omega)|$, and $|\mathbf{S}_A^M(j\omega)|$ vs. ω, shown in Fig. 14.3, are drawn for the nominal values of $p_1 = -1$, $p_2 = -5$, and $A = 10$ for the state-variable-feedback system. For the conventional system, the value of loop sensitivity equal to 2.1 corresponds to $\zeta = 0.7076$, as shown on the root locus in Fig. 13.10a. The same damping ratio is used in the state-variable-feedback system developed in the example in Sec. 13.9. Table 14.2 summarizes the values obtained.

The sensitivity function for each system, at $\omega = \omega_b$, is determined for a 100 percent variation of the plant poles $p_1 = -1$, $p_2 = -5$, and of the forward gain A, respectively. The term $\%|\Delta M|$ is defined as

$$\%|\Delta M| = \frac{|\mathbf{M}(j\omega_b)| - |\mathbf{M}_o(j\omega_b)|}{|\mathbf{M}_o(j\omega_b)|} 100$$

(14.20)

where $|\mathbf{M}_o(j\omega_b)|$ is the value of the control ratio with nominal values of the plant parameters, and $|\mathbf{M}(j\omega_b)|$ is the value with one parameter changed to the extreme value of its possible variation. The time-response data for systems 1 to 8 are given in Table 14.3 for a unit step input. Note that the response for the state-variable-feedback system (5 to 8, except 7a) is essentially unaffected by parameter variations.

The sensitivity function S_δ^M represents a measure of the change to be expected in the system performance for a change in a system parameter. This measure is borne out by comparing Tables 14.2 and 14.3. That is, in the frequency-domain plots of Fig. 14.3, a value $|\mathbf{S}_A^M(j\omega)| = 1.29$ at $\omega = \omega_b = 0.642$ is indicated for the conventional system, compared with 0.051 for the state-variable-feedback system. Also, the percent change $\%|\Delta M(j\omega)|$ for a doubling of the forward gain is 74 percent for the conventional system and only 1.41 percent for the state-variable-feedback system. These are consistent with the changes in the peak overshoot in the time domain of $\%\Delta M_p(t) = 13.45$ and 0.192 percent, respectively, as shown in Table 14.3. Thus, the magnitudes of $\%\Delta M(j\omega)$ and $\%\Delta M_p(t)$ are consistent with the relative magnitudes of $\%|\mathbf{S}_A^M(j\omega_b)|$. Similar results are obtained for variations of the pole p_1.

An interesting result occurs when the pole $p_2 = -5$ is subject to variations. If the state X_3 (see Fig. 14.2b) is accessible and is fed back directly through k_3, the sensitivity $|\mathbf{S}_{p_2}^M(j\omega)|$ is much larger than for the conventional system, having a value of 3.40 at $\omega_b = 1.0$. However, if the equivalent feedback is obtained from X_1 through the fixed transfer function $[H_{eq}(s)]_{\delta_o = p_2}$, Fig. 14.2c, the sensitivity is considerably reduced to $|\mathbf{S}_{p_2}^M(j1)| = 0.050$, compared with 1.28 for the conventional system. If the components of the feedback unit are selected so that the transfer function $[H_{eq}(s)]$ is invariant, then the time-response characteristic for output feedback

TABLE 14.2
System sensitivity analysis

Control system	Figure	$\|M(j\omega_b)\|$	$\% \|\Delta M(j\omega_b)\|$	$\|S^M_{p_1}(j\omega_b)\|$	$\|S^M_{p_2}(j\omega_b)\|$	$\|S^M_A(j\omega_b)\|$
		Conventional system design				
1. Nominal plant $(A = 0.21, p_1 = -1, p_2 = -5)$: $G(s) = \dfrac{2.1}{s(s+1)(s+5)}$	14.2a	0.707 $(\omega_b = 0.642)$	—	1.09	1.28	1.29
2. $p_1 = -2$ $G(s) = \dfrac{2.1}{s(s+2)(s+5)}$	14.2a	0.338	52.4	—	—	—
3. $p_2 = -10$: $G(s) = \dfrac{2.1}{s(s+1)(s+10)}$	14.2a	0.658	6.9	—	—	—
4. $A = 0.42$: $G(s) = \dfrac{4.2}{s(s+1)(s+5)}$	14.2a	1.230	74.0	—	—	—

TABLE 14.2
System sensitivity analysis

State-variable feedback system design [M is given in Eqs. (13.111), (14.18), and (14.19), respectively]

5. Nominal plant: $G(s)H_{eq}(s) = \dfrac{95.4(s^2 + 1.443s + 1.0481)}{s(s+1)(s+5)}$	14.2b, c	0.707 ($\omega_b = 1.0$)	—	0.036	0.05 for X_3 inaccessible 3.40 for X_3 accessible	0.051
6. $p_1 = -2$: $G(s)H_{eq}(s) = \dfrac{95.4(s^2 + 1.443s + 1.0481)}{s(s+2)(s+5)}$	14.2b, c	0.683	3.39	—	—	—
7a. $p_2 = -10$, X_3 accessible: $G(s)H_{eq}(s) = \dfrac{95.4(s^2 + 6.443s + 1.0481)}{s(s+1)(s+10)}$	14.2b	0.160	77.4	—	—	—
7b. $p_2 = -10$, X_3 inaccessible: $G(s)H_{eq}(s) = \dfrac{95.4(s^2 + 1.443s + 1.0481)}{s(s+1)(s+10)}$	14.2c	0.682	3.54	—	—	—
8. $A = 20$: $G(s)H_{eq}(s) = \dfrac{190.8(s^2 + 1.443s + 1.0481)}{s(s+1)(s+5)}$	14.2b, c	0.717	1.41	—	—	—

through $[H_{eq}(s)]$ is essentially unchanged when p_2 doubles in value (see Table 14.3) compared with the conventional system.

In analyzing Fig. 14.3, the values of the sensitivities must be considered over the entire passband ($0 \leq \omega \leq \omega_b$). In this passband the state-variable-feedback system has a much lower value of sensitivity than the conventional system. The performance of the state-variable-feedback systems is essentially unaffected by any variation in A, p_1, or p_2 (with restrictions) when the feedback coefficients remain fixed, provided that the forward gain satisfies the condition $K \geq K_{min}$ (see Probs. 14.2 and 14.3). *The low-sensitivity state-variable-feedback systems considered in this section satisfy the following conditions:* the system's characteristic equation must have (1) β dominant roots ($< n$) which are located close to zeros of $H_{eq}(s)$ and (2) at least one nondominant root.

One approach for determining the nominal value for the variable parameter δ for a state-variable-feedback design is to determine $\mathbf{k}$ for the minimum, midrange, and maximum value of δ. Then evaluate and plot $|\mathbf{S}_\delta^M(j\omega)|$ for the range of $0 \leq \omega \leq \omega_b$ for each of these values of δ. The value of δ that yields the minimum area under the curve of $|\mathbf{S}_\delta^M(j\omega)|$ vs. ω is chosen as the value for δ_o.

The analysis of this section reveals that for the state-variable-feedback system any parameter variation between the control U and the state variable X_n has minimal effect on system performance. In order to minimize the effect on system performance for a parameter variation that occurs between the states X_n and X_1, the feedback signals should be moved to the right of the block containing the variable parameter (see Sec. 14.4), even if all the states are accessible. The output feedback system that incorporates an invariant $H_{eq}(s)$ has a response that is more invariant to parameter variations than a conventional system.

TABLE 14.3
Time-response data

Control system	$M_p(t)$	t_p, s	t_s, s	$\%\Delta M_p(t) = \dfrac{\|M_p - M_{po}\|}{M_{po}}100$
1	1.048	7.2	9.8	—
2	1.000	—	16.4	4.8
3	1.000	—	15.0	4.8
4	1.189	4.1	10.15	13.45
5	1.044	4.45	6	—
6	1.035	4.5	6^-	0.86
7a	1.000	—	24	4.4
7b	1.046	4.6	6.5	0.181
8	1.046	4.4	5.6^+	0.192

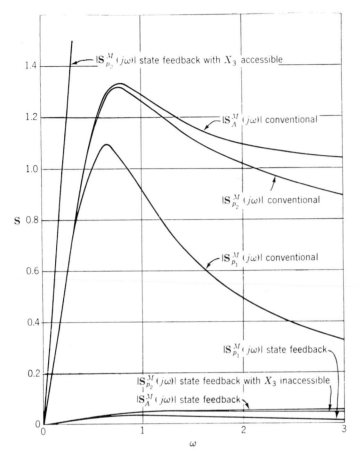

FIGURE 14.3
Sensitivity due to pole and gain variation.

14.4   PARAMETER SENSITIVITY EXAMPLES

Example 1. The $(Y/R)_T$, Eq. (13.118), for Design Example 1 of Chapter 13 contains no nondominant pole. As seen from the root locus of Fig. 13.16 for this example, the system is highly sensitive to gain variations. Using Eq. (13.117) in Eq. (14.3) with $\delta = A$ yields $|S_\delta^M(j\omega)| = 1.84$ at $\omega_b = 1.41$.

Example 2. The $(Y/R)_T$ of Design Example 1 is modified to contain one nondominant pole; see Eq. (13.121). Using Eq. (13.121), with $p = -50$ and $\delta = K_G$, in Eq. (14.3) yields $|S_\delta^M(j\omega_b)| = 0.064$ at $\omega_b = 1.41$. Thus, the presence of a nondominant pole drastically reduces $|S_\delta^M(j\omega_b)|$, which is indicative of the reduction of the sensitivity magnitude for the frequency range $0 \leq \omega \leq \omega_b$.

Example 3. The $(Y/R)_T$ of Design Example 2 is modified to improve the value of t_s; see Eq. (13.127). Using Eq. (13.127), with $p = -70$ and $\delta = K_G$, in Eq. (14.3) yields $|S_\delta^M(j\omega_b)| = 0.087$ at $\omega_b = 1.41$.

TABLE 14.4
Time-response and gain-sensitivity data

Example	$(Y/R)_D$	M_p	t_p, s	t_s, s	K_1, s^{-1}	$\left\|S^M_{K_G}(j\omega_b)\right\|$ $\omega_b = 1.41$
1	$\dfrac{2(s + 2)}{(s^2 + 2s + 2)(s + 2)}$	1.043	3.1	4.2	1.0	1.84
2	$\dfrac{100(s + 2)}{(s^2 + 2s + 2)(s + 2)(s + 50)}$	1.043	3.2	4.2	0.98	0.064
3	$\dfrac{100(s + 1.4)(s + 2)}{(s^2 + 2s + 2)(s + 1)(s + 2)(s + 70)}$	1.008	3.92	2.94	0.98	0.087

The data in Table 14.4 summarize the results of these three examples and show that the presence of at least one nondominant pole in $(Y/R)_T$ drastically reduces the system's sensitivity to gain variation while achieving the desired time-response characteristics.

14.5 INACCESSIBLE STATES[3]

In order to achieve the full benefit of a state-variable-feedback system, all the states must be accessible. Although, in general, this requirement may not be satisfied, the design procedures presented in this chapter are still valid. That is, on the basis that all states are accessible, the required value of the state-feedback gain vector **k** which achieves the desired $Y(s)/R(s)$ is computed. Then, for states that are not accessible, the corresponding k_i blocks are moved by block-diagram manipulation techniques to states that are accessible. As a result of these manipulations the full benefits of state-variable feedback may not be achieved (low sensitivity S^M_δ and a completely stable system), as described in Sec. 14.3. For the extreme case when the output $y = x_1$ is the only accessible state, the block-diagram manipulation to the G-equivalent (see Fig. 14.5c) reduces to the Guillemin-Truxal design. More sophisticated methods for reconstructing (estimating) the values of the missing states do exist (Luenberger observer theory). They permit the accessible and the reconstructed (inaccessible) states all to be fed back through the feedback coefficients, thus eliminating the need for block-diagram manipulations and maintaining the full benefits of a state-variable feedback-designed system[3]. Even when all states are theoretically accessible, design and/or cost of instrumentation may make it prohibitive to measure some of the states.

Various block diagram manipulations of Fig. 14.2b, depending on which states are accessible, are shown in Figs. 14.4 and 14.5. The availability of practical state-variable sensors determines which physical implementation is most feasible. The active minor-loop compensation of Fig. 14.4b results in the requirement for first- and second-order derivative action in the minor loops; thus $H_c(s)$ is an improper function. It is possible to add poles to $H_c(s)$ that are far to the left in order to achieve a realizable $H_c(s)$.

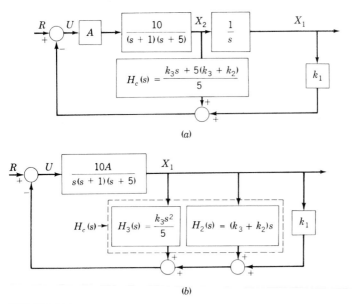

FIGURE 14.4
Minor-loop compensation (manipulation to the right) of the control system of
Fig. 14.2b with $p_1 = -1$ and $p_2 = -2$.

The compensators required for each case represented in Figs. 14.4 and 14.5, using the values of $k_1 = 1$, $k_2 = -3.393$, $k_3 = 4.77$, and $A = 10$ determined in Sec. 13.9, are given in Table 14.5, which also gives the sensitivity function $|S_{K_G}^M(j\omega_b)|$ for each case at $\omega_b = 1.0$ and $M = 0.707$. As noted in Table 14.5, the low sensitivity of the output response to gain variations is maintained as long as invariant feedback compensators $H_c(s)$ are used. In other words, the block-diagram manipulations that result in *only minor-loop* compensation, i.e., using only *feedback* compensators $H_c(s)$, have the same $H_{eq}(s)$ as before the manipulations are performed. Therefore, the same stability and sensitivity characteristics are maintained for the manipulated as for the nonmanipulated systems. When the shifting is to the left, to yield a resulting cascade compensator $G_c(s)$, the low sensitivity to parameter variation is not achieved. This sensitivity characteristic is borne out by analyzing the root locus of Figs. 13.10b and 14.6. Figure 13.10b is the root locus representing the block-diagram manipulations using only $H_c(s)$ networks. Figure 14.6 shows the root locus representing the block-diagram manipulations using cascade $G_c(s)$ networks. In Fig. 13.10b, an increase in the gain does not produce any appreciable change in the closed-loop system response. In contrast, any gain variation in the systems represented by Fig. 14.6 greatly affects the closed-loop system response.

Note that the manipulated system of Fig. 14.5d, using both cascade $G_c(s)$ and minor-loop $H_c(s) = k_2$ compensators, yields the same relative sensitivity as for the system of Fig. 14.5c which uses only cascade compensation. In general, for a system

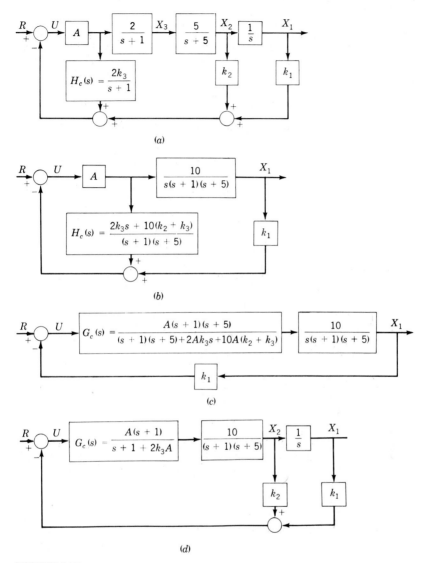

FIGURE 14.5
Manipulation of state feedback to the left for the control system of Fig. 14.2b.

using both $G_c(s)$ and $H_c(s)$ compensators, it may be possible to achieve a sensitivity that is between that of a system using only minor-loop compensation and that of one using only cascade compensation.

The technique presented in this section for implementing a system having inaccessible states is rather simple and straightforward. It is very satisfactory for systems that are noise-free and time-invariant. For example, if in Fig. 14.2b the state X_3 is

TABLE 14.5
Compensators and sensitivity functions for control systems of Figs. 14.4 and 14.5

Figure	Compensator	$\|S_A^M(j\omega_b)\|$
14.4a	$H_c(s) = 0.954s + 1.377$	0.0508
14.4b	$H_c(s) = 0.954s^2 + 1.377s$	0.0508
14.5a	$H_c(s) = \dfrac{9.54}{s+1}$	0.0508
14.5b	$H_c(s) = \dfrac{9.54(s+1.443)}{(s+1)(s+5)}$	0.0508
14.5c	$G_c(s) = \dfrac{10(s+1)(s+5)}{(s+1.43)(s+99.98)}$	0.51
14.5d	$G_c(s) = \dfrac{10(s+1)}{s+96.4}$	0.51

inaccessible and X_2 contains noise, it is best to shift the k_3 block to the left. This assumes that the amplifier output signal has minimal noise content. If all signals contain noise, the system performance is degraded from the desired performance. This technique is also satisfactory under parameter variations and/or if the poles and zeros of the transfer function $G_x(s)$ are not known exactly when operating under hfg.

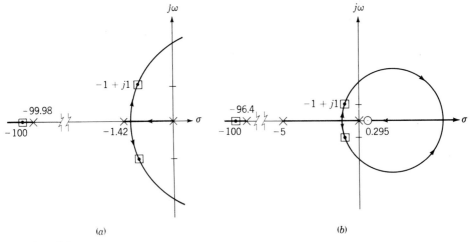

(a)   (b)

FIGURE 14.6
Root locus for (a) Fig. 14.5c and (b) Fig. 14.5d.

14.6 STATE-SPACE TRAJECTORIES[4-6]

The definition of stability for a linear time-invariant (LTI) system is an easy concept to understand. The complete response to any input contains a particular solution that has the same form as the input and a complementary solution containing terms of the form $Ae^{\lambda_i t}$. When the eigenvalues λ_i have negative real parts, the transients associated with the complementary solution decay with time, and the response is called stable. When the roots have positive real parts, the transients increase without bound and the system is called unstable. It is necessary to extend the concept of stability to nonlinear systems. This section introduces the properties of state-space trajectories, which give a visual picture of transient and steady-state performance. The kinds of singular points and their associated stability are categorized for both linear and nonlinear systems.

The *state space* is defined as the n-dimensional space in which the components of the state vector represent its coordinate axes. The unforced response of a system having n state variables, released from any initial point $\mathbf{x}(t_0)$, traces a curve or trajectory in an n-dimensional state space. Time t is an implicit function along the trajectory. When the state variables are represented by phase variables, the state space may also be called a phase space. While it is impossible to visualize a space with more than three dimensions, the concept of a state space is nonetheless very useful in systems analysis. The behavior of second-order systems is conveniently viewed in the state space because it is two-dimensional and is in fact a state (or phase) plane. It is easy to obtain a graphical or geometrical representation and interpretation of second-order system behavior in the state plane. The concepts developed for the state plane can be extrapolated to the state space of higher-order systems. Examples of a number of second-order systems are presented in this section. These concepts are extended qualitatively to third-order systems. The symmetry that results in the state-plane trajectories is demonstrated for the cases when the $\mathbf{A}$ matrix is converted to normal (canonical) or modified normal (modified canonical) form. The second-order examples that are considered include the cases where the eigenvalues are (1) negative real and unequal (overdamped), (2) negative real and equal (critically damped), (3) imaginary (oscillatory), and (4) complex with negative real parts (underdamped). The system that is studied has no forcing function and is represented by the state equation

$$\dot{\mathbf{x}} = \begin{bmatrix} a_{11} & a_{12} \\ a_{21} & a_{22} \end{bmatrix} \mathbf{x} \tag{14.21}$$

Because there is no forcing function, the response results from energy stored in the system as represented by initial conditions.

Example 1. Overdamped response. Consider the following example with the given initial conditions:

$$\dot{\mathbf{x}} = \begin{bmatrix} 0 & 3 \\ -1 & -4 \end{bmatrix} \mathbf{x} \qquad \mathbf{x}(0) = \begin{bmatrix} 0 \\ 2 \end{bmatrix} \tag{14.22}$$

Because one state equation is $\dot{x}_1 = 3x_2$, then x_2 is a phase variable. Therefore, the resulting state plane is called a *phase plane*.

The state transition matrix $\mathbf{\Phi}(t)$ is obtained by using the Sylvester expansion theorem (see Sec. 3.13) or any other convenient method. The system response is overdamped, with eigenvalues $\lambda_1 = -1$ and $\lambda_2 = -3$. For the given initial conditions the system response is

$$\mathbf{x}(t) = \mathbf{\Phi}(t)\mathbf{x}(0) = \begin{bmatrix} 3e^{-t} - 3e^{-3t} \\ -e^{-t} + 3e^{-3t} \end{bmatrix} \tag{14.23}$$

The trajectory represented by this equation can be plotted in the phase plane x_2 versus x_1. A family of trajectories in the phase plane for a number of initial conditions is drawn in Fig. 14.7. Such a family of trajectories is called a *phase portrait*. The arrows show the direction of the states along the trajectories for increasing time. Motion of the point on a trajectory is in the clockwise direction about the origin because $\dot{x}_1 = 3x_2$. Thus, when x_2 is positive, x_1 must be increasing in value, and when x_2 is negative, x_1 must be decreasing in value. The value of N in Figs. 14.7 and 14.8 is the slope of the trajectory at each point in the phase plane, i.e., $N = dx_2/dx_1$, as described in Eq. (14.24). The rate of change of x_1 is zero at the points where the trajectories cross the x_1 axis. Therefore, the trajectories always cross the x_1 axis in a perpendicular direction.

One method of drawing the phase trajectory is to insert values of time t into the solution $\mathbf{x}(t)$ and to plot the results. For the example above the solutions are $x_1 = 3e^{-t} - 3e^{-3t}$ and $x_2 = -e^{-t} + 3e^{-3t}$. Another approach is to eliminate t from these

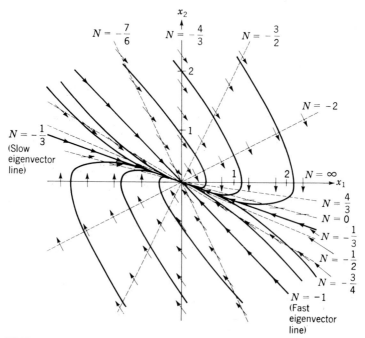

FIGURE 14.7
Phase-plane portrait for Eq. (14.22).

equations to obtain an analytical expression for the trajectory. This can be done by first solving for e^{-t} and e^{-3t}:

$$e^{-t} = \frac{x_1 + x_2}{2} \quad \text{and} \quad e^{-3t} = \frac{x_1 + 3x_2}{6}$$

Raising e^{-t} to the third power, e^{-3t} to the first power, and equating the results yields

$$\left(\frac{x_1 + x_2}{2}\right)^3 = \left(\frac{x_1 + 3x_2}{6}\right)^1$$

This equation is awkward to use and applies only for the specified initial conditions. Further, it is difficult to extend this procedure to nonlinear systems.

A practical graphical method for obtaining trajectories in the state plane is the *method of isoclines*. The previous example is used to illustrate this method. Taking the ratio of the state equations of this example yields

$$\frac{\dot{x}_2}{\dot{x}_1} = \frac{dx_2/dt}{dx_1/dt} = \frac{dx_2}{dx_1} = \frac{-x_1 - 4x_2}{3x_2} = N \tag{14.24}$$

This equation represents the slope of the trajectory passing through any point in the state plane. The integration of this equation traces the trajectory starting at any initial condition. The method of isoclines is a graphical technique for performing this integration. When the slope of the trajectory in Eq. (14.24) is fixed at any specific value N, the result is the equation of a straight line having a slope m and represented by

$$x_2 = -\frac{1}{4 + 3N} x_1 = m x_1 \tag{14.25}$$

This equation describes a family of curves that are called *isoclines*. They have the property that all trajectories crossing a particular isocline have the same slope at the crossing points. For LTI second-order systems the isoclines are all straight lines. A trajectory can be sketched by using the following procedure:

1. Draw a number of isoclines reasonably close together.
2. From any starting point A (see Fig. 14.8), draw two lines having slopes N_1 and N_2 and extending from the isocline for N_1 to the isocline for N_2. The trajectory must stay between these lines. Select point B midway in the segment of the isocline for N_2 between these lines.
3. Continue from point B using the procedure of step 2.
4. Connect the points with a smooth curve that crosses each isocline with the correct slope. An example is shown in Fig. 14.8.

The system represented by Example 1 is stable, as indicated by the negative eigenvalues. Because there is no forcing function, the response is due to the energy initially stored in the system. As this energy is dissipated, the response approaches

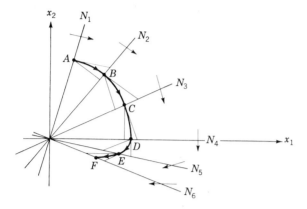

FIGURE 14.8
Construction of a trajectory using isoclines.

equilibrium at $x_1 = x_2 = 0$. All the trajectories approach and terminate at this point. The slope of the trajectory going through this equilibrium point is therefore indeterminate; that is, $N = dx_2/dx_1 = 0/0$, representing a *singularity* point. This indeterminacy can be used to locate the equilibrium point by letting the slope of the trajectory in the state plane be $N = dx_2/dx_1 = 0/0$. Applying this condition to Eq. (14.24) yields the two equations $x_2 = 0$ and $-x_1 - 4x_2 = 0$. This confirms that the equilibrium point is $x_1 = x_2 = 0$. For the overdamped response this equilibrium point is called a *node*. The procedure can be extended to all systems of any order, even when they are nonlinear. This is done by solving the vector equation $\dot{\mathbf{x}} = \mathbf{0}$ for the equilibrium point(s).

It is interesting to note the symmetry that results in the phase portrait when the states are transformed to canonical form in which the plant matrix $\mathbf{A}$ is diagonal [see Eq. (5.70)].[8] The methods for accomplishing this transformation are described in Sec. 5.10. The resulting state equations in canonical form for a second-order system are

$$\dot{z}_1 = \lambda_1 z_1 \quad (14.26) \qquad\qquad \dot{z}_2 = \lambda_2 z_2 \quad (14.27)$$

When the eigenvalues $\lambda_1 \neq \lambda_2$, the ratio of these equations yields

$$\frac{dz_1}{dz_2} = \frac{\lambda_1 z_1}{\lambda_2 z_2} \tag{14.28}$$

The variables in this equation are separable, and the resulting equation can be integrated to obtain the equation of the state-plane trajectory

$$z_2 = C z_1^{\lambda_2/\lambda_1} \tag{14.29}$$

For a stable system with λ_1 and λ_2 both negative and real, where $\lambda_2 < \lambda_1 < 0$, the state-plane portraits are parabolic, as shown in Fig. 14.9a and b. All the trajectories are tangent to the z_1 axis. When returning to the x_2, x_1 plane, the trajectories are tangent to the line labeled z_1 in Fig. 14.9c and d. This line is the transformation of the z_1 axis into the x_2, x_1 plane. The slope of this line can be obtained from the

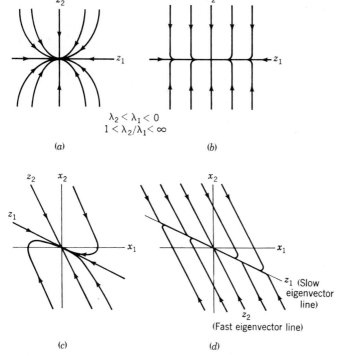

FIGURE 14.9
Stable node: (*a*) normal form: roots of similar magnitudes; (*b*) normal form: roots of widely different magnitudes; (*c*) x_2 vs. x_1: roots with similar magnitudes; (*d*) x_2 vs. x_1: roots with widely different magnitudes (see Ref. 8).

modal matrix $\mathbf{T}$ (see Sec. 5.10), which transforms the $\mathbf{z}$ plane into the $\mathbf{x}$ plane. Using $\mathbf{T} = [v_{ij}]$ and $\mathbf{z} = \mathbf{T}^{-1}\mathbf{x}$ yields

$$z_1 = \frac{1}{\Delta_T}(v_{22}x_1 - v_{12}x_2) \tag{14.30}$$

$$z_2 = \frac{1}{\Delta_T}(-v_{21}x_1 + v_{11}x_2) \tag{14.31}$$

The z_1 axis is defined by $z_2 = 0$. Therefore, from Eq. (14.31), the slope of the z_1 axis after transformation to the x_2, x_1 plane is

$$m_1 = \frac{x_2}{x_1} = \frac{v_{21}}{v_{11}} \tag{14.32}$$

Thus the slope is determined by the eigenvector $\mathbf{v}_1$ that is the first column of the modal matrix $\mathbf{T}$. The ratio v_{21}/v_{11} can be evaluated, using method 2 of Sec. 5.10, by equating the elements of $\mathbf{AT} = \mathbf{T\Lambda}$ [see Eq. (5.91)], which yields, for real eigenvalues,

$$m_1 = \frac{v_{21}}{v_{11}} = \frac{\lambda_1 - a_{11}}{a_{12}} = \frac{a_{21}}{\lambda_1 - a_{22}} \tag{14.33}$$

Because λ_1 yields the transient mode with the longest settling time, the line z_1 in Fig. 14.9c and d is labeled the "slow eigenvector line."

The trajectories in Fig. 14.9a and b are parallel to the z_2 axis at large distances from the origin. In the x_2, x_1 plane (see Fig. 14.9c and d), the trajectories retain this property except that they are parallel to the line labeled z_2. This line is the transformation of the z_2 axis into the x_2, x_1 plane. The z_2 axis is defined by the condition $z_1 = 0$. Using Eq. (14.30), the slope of this axis transformed into the **x** plane is

$$m_2 = \frac{x_2}{x_1} = \frac{v_{22}}{v_{12}} \tag{14.34}$$

This slope is therefore determined by the eigenvector, the second column of **T**. From the coefficients of $\mathbf{AT} = \mathbf{T\Lambda}$ [see Eq. (5.91)], this slope is obtained as a function of the real eigenvalue λ_2:

$$m_2 = \frac{v_{22}}{v_{12}} = \frac{\lambda_2 - a_{11}}{a_{12}} = \frac{a_{21}}{\lambda_2 - a_{22}} \tag{14.35}$$

Because λ_2 yields the transient mode with the shortest settling time, the line z_2 in Fig. 14.9c and d is labeled the "fast eigenvector line."

For the system of Example 1 the eigenvectors are $\mathbf{v}_1 = [\,1 \quad -3\,]^T$ and $\mathbf{v}_2 = [\,1 \quad -1\,]^T$. Thus, the slopes are $m_1 = -1/3$ and $m_2 = -1$. The lines going through the equilibrium point with these slopes are drawn in Fig. 14.7. The transient term $e^{\lambda_2 t} = e^{-3t}$ decays faster than the term $e^{\lambda_1 t} = e^{-t}$. Therefore, near the origin the trajectories are tangent to the line having the slope $m_1 = -1/3$ determined by the slow eigenvalue $\lambda_1 = -1$.

When the real eigenvalues are equal and negative, the phase portrait is similar to that in Fig. 14.7, except that there is only one eigenvector line.

When any real eigenvalue is positive, the system response is unstable. For the second-order system with two positive eigenvalues the equilibrium point is an *unstable node*. In this case the trajectories depart from the equilibrium point and go to infinity.

Example 2. Underdamped response. The underdamped system considered here is

$$\dot{\mathbf{x}} = \begin{bmatrix} 0 & 1 \\ -4 & -2 \end{bmatrix} \mathbf{x} \tag{14.36}$$

The characteristic equation is $\lambda^2 + 2\lambda + 4 = 0$, and the eigenvalues are $\lambda_{1,2} = -1 \pm j\sqrt{3}$, representing a damping ratio $\zeta = 0.5$. The equilibrium point is determined from the equation of the slope of the trajectory

$$\frac{dx_2}{dx_1} = \frac{-4x_1 - 2x_2}{x_2} = N \tag{14.37}$$

The singularity condition $N = 0/0$ determines the equilibrium point $x_1 = x_2 = 0$.

Using Eq. (14.37), the isoclines are the straight lines represented by

$$x_2 = \frac{-4}{2 + N} x_1 \tag{14.38}$$

Because the matrix $\mathbf{A}$ in Eq. (14.36) is in companion form, the states are phase variables and the state plane is a phase plane. Several phase-plane trajectories are shown in Fig. 14.10. The trajectories are modified logarithmic spirals.

The equilibrium point for an underdamped response is called a *focus*. For stable responses the motion of the point on the spiral trajectories is clockwise (cw) toward the focus. Thus, when $x_2 = \dot{x}_1$ is positive, the trajectory moves to the right in the direction of increasing values of x_1. Also, when $x_2 = \dot{x}_1$ is negative, the trajectory moves to the left in the direction of decreasing values of x_1.

When the state variables are transformed to canonical form, the equation representing the trajectories is best represented in polar form, because the trajectories are known to be spiral. When the eigenvalues are complex conjugates, $\lambda_{1,2} = \sigma \pm j\omega_d$, a simplified form of the trajectory equation is obtained by assuming that the canonical state variables are also complex conjugates:

$$z_1 = w_1 - jw_2 \qquad z_2 = w_1 + jw_2 \tag{14.39}$$

Inserting these values into the canonical state equations $\dot{z}_1 = \lambda_1 z_1$ and $\dot{z}_2 = \lambda_2 z_2$ yields

$$\dot{w}_1 - j\dot{w}_2 = (\sigma + j\omega_d)(w_1 - jw_2) \qquad \dot{w}_1 + j\dot{w}_2 = (\sigma - j\omega_d)(w_1 + jw_2) \tag{14.40}$$

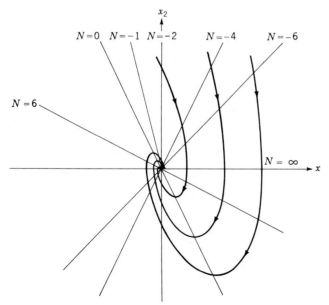

FIGURE 14.10
Underdamped phase-plane trajectories.

Separating the real and imaginary components of these equations produces

$$\dot{w}_1 = \sigma w_1 + \omega_d w_2 \qquad \dot{w}_2 = -\omega_d w_1 + \sigma w_2 \tag{14.41}$$

These equations can be recognized as being in the modified normal form described in Sec. 5.12. The slope of the trajectories in the w_1, w_2 plane, obtained from Eq. (14.41), is

$$\frac{dw_2}{dw_1} = \frac{-\omega_d w_1 + \sigma w_2}{\sigma w_1 + \omega_d w_2} \tag{14.42}$$

This equation is converted from the rectangular coordinates w_1 and w_2 to the polar coordinates r and ϕ by using $w_1 = r\cos\phi$ and $w_2 = r\sin\phi$. Then,

$$dw_1 = \cos\phi\, dr - r\sin\phi\, d\phi \qquad \text{and} \qquad dw_2 = \sin\phi\, dr + r\cos\phi\, d\phi$$

The polar magnitude r and its angle ϕ are related to the rectangular coordinates w_1 and w_2 by

$$r(t) = \sqrt{w_1^2(t) + w_2^2(t)} \qquad \text{and} \qquad \phi(t) = \tan^{-1}[w_2(t)/w_1(t)] \tag{14.43}$$

These values apply for all values of time. After simplification of Eq. (14.42), the resulting equation representing the slope of the trajectory in the w_1, w_2 plane is

$$\frac{dr}{d\phi} = -\frac{\sigma}{\omega_d} r$$

The variables can be separated, and integration of the equation yields

$$r = Ce^{-(\sigma/\omega_d)\phi} = Ce^{\left(\zeta/\sqrt{1-\zeta^2}\right)\phi} \tag{14.44}$$

For any initial values, at time $t = 0$, of $w_1(0)$ and $w_2(0)$ (see Fig. 14.11d), the corresponding values r_0 and ϕ_0 are obtained from Eq. (14.43); i.e., $\tan\beta = |w_2(0)|/|w_1(0)|$ and $\phi_0 = -360° + \beta$. Then the value of C in Eq. (14.44) is given by

$$C = r_0/e^{\left(\zeta/\sqrt{1-\zeta^2}\right)\phi_0}$$

The trajectories represented by this equation are logarithmic spirals. Motion along the trajectories is clockwise for increasing time. For positive values of damping ratio, $1 > \zeta > 0$, the system response is stable, and the trajectories approach the equilibrium point (*stable focus*). Figure 14.11a and b shows the shapes of the trajectories in the **w** and **x** planes for a stable response. For negative damping ratio, $-1 < \zeta < 0$, the response is unstable, and the trajectories diverge from the equilibrium point (*unstable focus*). Figure 14.11c shows the trajectories for an unstable response in the **x** plane.

For the state equation (14.36) the transformation matrix $\mathbf{T}_m$ from the **x** plane to the **w** plane [see Eq. (5.130)] is

$$\mathbf{T}_m = \begin{bmatrix} 1 & 0 \\ -1 & \sqrt{3} \end{bmatrix} \tag{14.45}$$

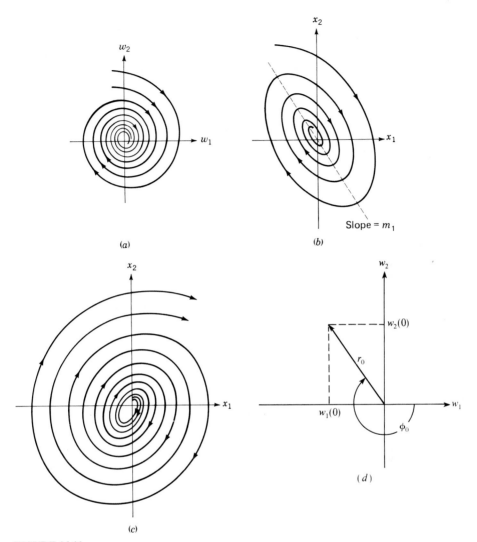

FIGURE 14.11
Underdamped trajectories: (a) stable, modified canonical form; (b) stable, general form; (c) unstable, general form; (d) polar coordinates.

Thus, when Eq. (14.32) is applied, the w_1 axis becomes the straight line having the slope $m_1 = -1$ in the x_2, x_1 plane. The trajectories are perpendicular to this line at the crossing points.

For the undamped case, $\zeta = 0$, the canonical and modified canonical coefficient matrices have the form

$$\mathbf{\Lambda} = \begin{bmatrix} j\omega_d & 0 \\ 0 & -j\omega_d \end{bmatrix} \quad \text{and} \quad \mathbf{\Lambda}_m = \begin{bmatrix} 0 & \omega_d \\ -\omega_d & 0 \end{bmatrix} \qquad (14.46)$$

The trajectories obtained from Eq. (14.44) are circles having constant radii in the **w** plane. In the **x** plane they become ellipses. The equilibrium point is called a *center* or *vortex*. The portraits for the **w** and **x** planes are illustrated in Fig. 14.12. The principal axis of the ellipses in the **x** plane is a function of the **A** matrix. As an example, the matrix

$$\mathbf{A} = \begin{bmatrix} 1 & 1 \\ -2 & -1 \end{bmatrix} \tag{14.47}$$

has the eigenvalues $\lambda_{1,2} = \pm j1$, and the transformation matrix $\mathbf{T}_m$ from the **x** plane to the **w** plane [see Eq. (5.130)] is

$$\mathbf{T}_m = \begin{bmatrix} 1 & 0 \\ -1 & 1 \end{bmatrix} \tag{14.48}$$

Thus, when Eq. (14.32) is applied, the w_1 axis becomes the straight line having the slope $m_1 = -1$ in the x_2, x_1 plane. The principal axes of the ellipses in the x_2, x_1 plane are shown in Fig. 14.12b.

Example 3. One positive and one negative real root. When the roots are real but of opposite sign, the equilibrium point is called a *saddle point* and represents an unstable equilibrium. The trajectories in canonical form are represented by Eq. (14.29), which can be rewritten as

$$z_2 = C z_1^m \tag{14.49}$$

where $m = \lambda_2/\lambda_1$. The trajectories are hyperbolas that are asymptotic to the z_1 and z_2 axes, as shown in Fig. 14.13a. When transformed back to the **x** plane, the trajectories are rotated and modified in shape as shown in Fig. 14.13b. The transformation of the z axes to the **x** plane yields straight lines with the slopes given by Eqs. (14.32)

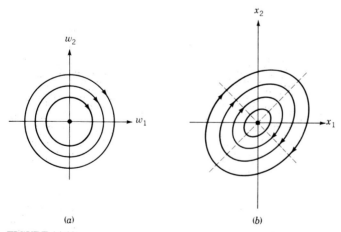

(a)　　　　　　　　　　　(b)

FIGURE 14.12
Center and trajectories with no damping in (a) normal form and (b) general form.

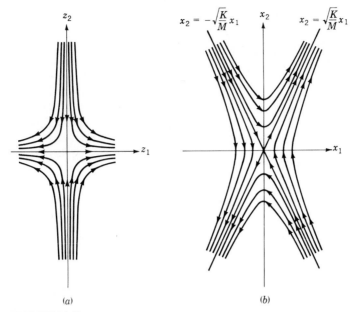

FIGURE 14.13
Saddle point and trajectories in (*a*) normal form and (*b*) general form.

to (14.35). These lines are called *separatrices* since they divide the state plane into separate regions.

The mechanical system containing mass, a "repulsive" spring force, and no damping is represented by the differential equation

$$M\ddot{x} - Kx = 0 \tag{14.50}$$

This equation describes the motion of a simple pendulum with no damping in the neighborhood of the upper unstable equilibrium position. In terms of phase variables the state equations are

$$\begin{bmatrix} \dot{x}_1 \\ \dot{x}_2 \end{bmatrix} = \begin{bmatrix} 0 & 1 \\ \dfrac{K}{M} & 0 \end{bmatrix} \begin{bmatrix} x_1 \\ x_2 \end{bmatrix} \tag{14.51}$$

The eigenvalues are $\lambda_{1,2} = \pm\sqrt{K/M}$. The slopes of the z_1 and z_2 axes transformed to the x_1, x_2 plane are obtained from Eqs. (14.33) and (14.35).

$$m_1 = \sqrt{\dfrac{K}{M}} = \lambda_1 \qquad m_2 = -\sqrt{\dfrac{K}{M}} = \lambda_2 \tag{14.52}$$

The trajectories are shown in Fig. 14.13*b*. If the initial value $\mathbf{x}(t_0)$ lies on the separatrix in the second or fourth quadrant, the trajectory terminates at the origin. That is, $\mathbf{x}(t_f) = \mathbf{0}$ satisfies the equation $M\dot{x}_2 = Kx_1$ and implies that the pendulum stops at the top equilibrium position where all the system energy is potential energy.

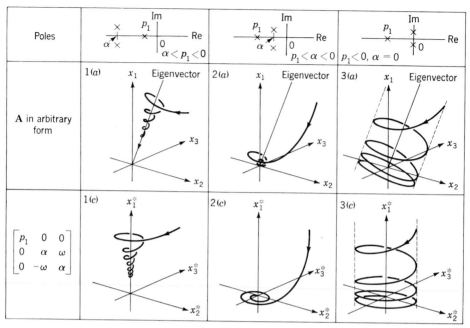

FIGURE 14.14

Trajectory patterns of third-order oscillatory systems. (*Reproduced by permission from Y. Takahashi, M. J. Rabins, and D. J. Anslander,* Control and Dynamic Systems, *Addison-Wesley Publishing Company, Reading, Mass., 1970.*)

Third-order systems require a three-dimensional state space. At least one of the eigenvalues must be real. Figure 14.14 contains examples for third-order systems with a pair of complex-conjugate poles. It is observed that transforming the states to modified canonical form results in a symmetry of the trajectories with respect to the principal axes. The eigenvector for the real pole coincides with one of the axes.

The properties of trajectories in the state plane are developed in this section. The type of equilibrium points and trajectories can be extended conceptually to higher-order systems, even when it is not possible to draw them in a higher-dimensional space. It is also possible to show the effect of nonlinearities on the shape of the trajectories, and an example is presented in Sec. 14.7. An important consideration in the stability analysis of such systems is that the system response in the neighborhood of the singularities can be determined from the linearized equations. Use is made of this principle in the next section.

14.7 LINEARIZATION (JACOBIAN MATRIX)[4,7]

A linear system with no forcing function (an autonomous system) and with $|A| \neq 0$ has only one equilibrium point, x_0. All the examples in Sec. 14.6 have the equilib-

FIGURE 14.15
A nonlinear feedback control system.

rium point at the origin $\mathbf{x}_0 = \mathbf{0}$. A nonlinear system, on the other hand, may have more than one equilibrium point. This is easily illustrated by considering the unity-feedback angular position-control system shown in Fig. 14.15. The feedback action is provided by synchros that generate the actuating signal $e = \sin(\theta_i - \theta_o)$. With no input, $\theta_i = 0$, the differential equation of the system is

$$\ddot{\theta}_o + a\dot{\theta}_o + K \sin \theta_o = 0 \tag{14.53}$$

This is obviously a nonlinear differential equation because of the term $\sin \theta_o$. With the phase variables $x_1 = \theta_o$ and $x_2 = \dot{x}_1 = \dot{\theta}_o$, the corresponding state equations are

$$\dot{x}_1 = x_2 \qquad \dot{x}_2 = -K \sin x_1 - ax_2 \tag{14.54}$$

The slope of the trajectories in the phase plane is obtained from

$$N = \frac{\dot{x}_2}{\dot{x}_1} = \frac{-K \sin x_1 - ax_2}{x_2} \tag{14.55}$$

In the general case the unforced nonlinear state equation is

$$\dot{\mathbf{x}} = \mathbf{f}(\mathbf{x}) \tag{14.56}$$

Because equilibrium points $\mathbf{x}_0$ exist at $\dot{\mathbf{x}} = \mathbf{f}(\mathbf{x}_0) = \mathbf{0}$, the singularities are $x_2 = 0$ and $x_1 = k\pi$, where k is an integer. The system therefore has multiple equilibrium points.

In a small neighborhood about each of the equilibrium points, a nonlinear system behaves like a linear system. The states can therefore be written as $\mathbf{x} = \mathbf{x}_0 + \mathbf{x}^*$, where $\mathbf{x}^*$ represents the perturbation or *state deviation* from the equilibrium point $\mathbf{x}_0$. Each of the elements of $\mathbf{f}(\mathbf{x})$ can be expanded in a Taylor series about one of the equilibrium points $\mathbf{x}_0$. Assuming that $\mathbf{x}^*$ is restricted to a small neighborhood of the equilibrium point, the higher-order terms in the Taylor series may be neglected. Thus, the resulting linear variational state equation is

$$\dot{\mathbf{x}}^* = \begin{bmatrix} \dfrac{\partial f_1}{\partial x_1} & \dfrac{\partial f_1}{\partial x_2} & \cdots & \dfrac{\partial f_1}{\partial x_n} \\[2mm] \dfrac{\partial f_2}{\partial x_1} & \dfrac{\partial f_2}{\partial x_2} & \cdots & \dfrac{\partial f_2}{\partial x_n} \\ \hdotsfor{4} \\ \dfrac{\partial f_n}{\partial x_1} & \dfrac{\partial f_n}{\partial x_2} & \cdots & \dfrac{\partial f_n}{\partial x_n} \end{bmatrix}_{\mathbf{x}=\mathbf{x}_0} \mathbf{x}^* = \mathbf{J}_x \mathbf{x}^* \tag{14.57}$$

where f_i is the ith row of $\mathbf{f(x)}$ and $\mathbf{J}_x = \partial\mathbf{f}/\partial\mathbf{x}^T$ is called the *Jacobian matrix* and is evaluated at $\mathbf{x}_0$.

For the system of Eq. (14.54), the motion about the equilibrium point $x_1 = x_2 = 0$ is represented by

$$\dot{\mathbf{x}}^* = \begin{bmatrix} \dot{x}_1^* \\ \dot{x}_2^* \end{bmatrix} = \begin{bmatrix} 0 & 1 \\ -K & -a \end{bmatrix} \begin{bmatrix} x_1^* \\ x_2^* \end{bmatrix} = \mathbf{J}_x\mathbf{x}^* \tag{14.58}$$

For these linearized equations the eigenvalues are $\lambda_{1,2} = -a/2 \pm \sqrt{(a/2)^2 - K}$. This equilibrium point is stable and is either a node or a focus, depending upon the magnitudes of a and K.

For motion about the equilibrium point $x_1 = \pi, x_2 = 0$, the state equations are

$$\dot{\mathbf{x}}^* = \begin{bmatrix} 0 & 1 \\ K & -a \end{bmatrix} \mathbf{x}^* \tag{14.59}$$

The eigenvalues of this $\mathbf{J}_x$ are $\lambda_{1,2} = -a/2 \pm \sqrt{(a/2)^2 + K}$. Thus, one eigenvalue is positive and the other is negative, and the equilibrium point represents an unstable saddle point. The motion around the saddle point is considered unstable because every point on all trajectories, except on the two separatrices, moves away from this equilibrium point. A phase-plane portrait for this system can be obtained by the method of isoclines (see Fig. 14.8 and Sec. 14.6). From Eq. (14.55), the isocline equation is

$$x_2 = \frac{-K \sin x_1}{N + a} \tag{14.60}$$

The phase portrait is shown in Fig. 14.16. Note that the linearized equations are applicable only in the neighborhood of the singular points. Thus, they describe stability *in the small*.

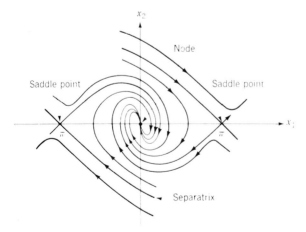

FIGURE 14.16
Phase portrait for Eq. (14.53).

When an input $\mathbf{u}$ is present, the nonlinear state equation is

$$\dot{\mathbf{x}} = \mathbf{f}(\mathbf{x}, \mathbf{u}) \tag{14.61}$$

Its equilibrium point $\mathbf{x}_0$, when $\mathbf{u} = \mathbf{u}_0$ is a constant, is determined by letting $\dot{\mathbf{x}} = \mathbf{f}(\mathbf{x}_0, \mathbf{u}_0) = \mathbf{0}$. Then the linearized variational state equation describing the variation from the equilibrium point is obtained by using only the linear terms from the Taylor series expansion. For variations in $\mathbf{u}$, appearing as $\mathbf{u}^*$ in the input $\mathbf{u} = \mathbf{u}_0 + \mathbf{u}^*$, the linearized state equation is

$$\dot{\mathbf{x}}^* = \begin{bmatrix} \dfrac{\partial f_1(\mathbf{x},\mathbf{u})}{\partial x_1} & \dfrac{\partial f_1(\mathbf{x},\mathbf{u})}{\partial x_2} & \cdots & \dfrac{\partial f_1(\mathbf{x},\mathbf{u})}{\partial x_n} \\[2mm] \dfrac{\partial f_2(\mathbf{x},\mathbf{u})}{\partial x_1} & \dfrac{\partial f_2(\mathbf{x},\mathbf{u})}{\partial x_2} & \cdots & \dfrac{\partial f_2(\mathbf{x},\mathbf{u})}{\partial x_n} \\[2mm] \cdots & \cdots & \cdots & \cdots \\[2mm] \dfrac{\partial f_n(\mathbf{x},\mathbf{u})}{\partial x_1} & \dfrac{\partial f_n(\mathbf{x},\mathbf{u})}{\partial x_2} & \cdots & \dfrac{\partial f_n(\mathbf{x},\mathbf{u})}{\partial x_n} \end{bmatrix}_{\substack{\mathbf{x}=\mathbf{x}_0 \\ \mathbf{u}=\mathbf{u}_0}} \mathbf{x}^*$$

$$+ \begin{bmatrix} \dfrac{\partial f_1(\mathbf{x},\mathbf{u})}{\partial u_1} & \cdots & \dfrac{\partial f_1(\mathbf{x},\mathbf{u})}{\partial u_r} \\[2mm] \cdots & \cdots & \cdots \\[2mm] \dfrac{\partial f_n(\mathbf{x},\mathbf{u})}{\partial u_1} & \cdots & \dfrac{\partial f_n(\mathbf{x},\mathbf{u})}{\partial u_r} \end{bmatrix}_{\substack{\mathbf{x}=\mathbf{x}_0 \\ \mathbf{u}=\mathbf{u}_0}} \mathbf{u}^* = \mathbf{J}_x \mathbf{x}^* + \mathbf{J}_u \mathbf{u}^* \tag{14.62}$$

In analyzing the system performance in the vicinity of equilibrium, it is usually convenient to translate the origin of the state space to that point. This is done by inserting $\mathbf{x} = \mathbf{x}_0 + \mathbf{x}^*$ into the original equations. With the origin at $\mathbf{x}_0$, the variations from this point are described by $\mathbf{x}^*$.

The characteristic equation describing the motion about the equilibrium point can be obtained from the linearized state equations. For a second-order system it has the form

$$s^2 + ps + q = (s - \lambda_1)(s - \lambda_2) = 0 \tag{14.63}$$

The eigenvalues λ_1 and λ_2 determine whether the singular point is a node, a focus, or a saddle. For a node or focus, the eigenvalues also determine whether the equilibrium point is stable or unstable. Figure 14.17 shows the type of equilibrium point, based on the coefficients p and q in Eq. (14.63).

Phase-plane portraits have been applied to provide a visual picture of the responses of many linear and nonlinear systems. They are described extensively in the literature. The purpose in presenting them here is to provide some insight into trajectories and their relationships to the singular points. This understanding of phase-plane trajectories should provide the basis for a clearer understanding of the stability analysis for higher-order systems.

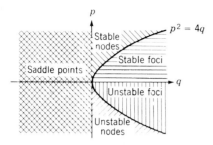

FIGURE 14.17
Type of equilibrium point for the characteristic equation $s^2 + ps + q = 0$.

14.8 SUMMARY

This chapter presents an analysis of the sensitivity of the system output to variations of system parameters. This is accomplished by defining a sensitivity function S_δ^M, where M is the control ratio and δ is the parameter that varies. A sensitivity analysis for a specific example system shows that a state-variable-feedback system has a much lower sensitivity than a conventional output feedback design. When there are inaccessible states, the states can be regenerated from the output, provided that the system is completely observable. For SISO systems this requires derivative action, accentuating noise that may be present in the output. This can be avoided by designing an observer.[9–11] The properties of state-space trajectories and singular points are introduced. Most of the examples used are second-order systems because they clearly and easily demonstrate the important characteristics, such as stability and the kinds of singular points. Both linear and nonlinear differential equations are considered. The Jacobian matrix is used to represent nonlinear systems by approximate linear equations in the vicinity of their singular points. The important property is demonstrated that a linear equation has only one singular point, whereas a nonlinear equation may have more than one singular point.

REFERENCES

1. "Sensitivity and Modal Response for Single-Loop and Multiloop Systems," *Flight Control Lab. ASD, AFSC Tech. Doc. Rep.* ASD-TDR-62-812, Wright-Patterson Air Force Base, Ohio, January 1963.
2. Cruz, J. B., Jr.: *Feedback Systems,* McGraw-Hill, New York, 1972.
3. D'Azzo, J. J., and C. H. Houpis: *Linear Control System Analysis & Design: Conventional and Modern,* McGraw-Hill, New York, 3rd ed., 1988.
4. Andronow, A. A., et al.: *Theory of Oscillations,* Addison-Wesley, Reading, Mass., 1966.
5. Thaler, G. J., and M. P. Pastel: *Analysis and Design of Nonlinear Feedback Control Systems,* McGraw-Hill, New York, 1962.
6. Takahashi, Y., et al.: *Control and Dynamic Systems,* Addison-Wesley, Reading, Mass., 1970.
7. Csaki, F.: *Modern Control Theories,* Akademiai, Kiado, Budapest, 1972.
8. Gibson, J. E.: *Nonlinear Automatic Control,* McGraw-Hill, New York, 1963.
9. Luenberger, D. G.: "Observing the State of a Linear System," *IEEE Trans. Mil. Electron.,* vol. 8, pp. 74–80, 1964.
10. Luenberger, D. G.: "Observers for Multivariable Systems," *IEEE Trans. Autom. Control,* vol. AC-11, pp. 190–197, 1966.
11. Luenberger, D. G.: "An Introduction to Observer," *IEEE Trans. Autom. Control,* vol. AC-16, pp. 569-602, 1971.

CHAPTER
15

DIGITAL
CONTROL
SYSTEMS

15.1 INTRODUCTION[1-6]

The methods presented in the previous chapters deal with the fundamental properties of continuous systems. However, digital computers are available not only for the design of control systems, but also to perform the control function. Digital computers and microprocessors are used in many control systems, and the size and cost advantages associated with the microcomputer make its use as a controller both economical and practical. Using such a digital processor with a plant that operates in the continuous-time domain requires that its input signal be discrete, which requires the sampling of the signals used by the controller. Such sampling may be an inherent characteristic of some systems. For example, a radar tracking system supplies information on an airplane's position and motion to a digital processor at discrete periods of time. This information is therefore available as a succession of data points. When there is no inherent sampling, an analog-to-digital (A/D) converter must be incorporated in a digital or sampled-data (SD) control system. The sampling process can be performed at a constant rate or at a variable rate, or it may be random. The discussion in this chapter is based on a constant-rate sampling. The output of the controller must then be converted from discrete form into an analog signal by a digital-to-analog (D/A) converter. A functional block diagram of such a system is shown in Fig. 15.1, in which the signals e^* and u^* are in discrete form, u is piecewise-continuous, and the remaining signals are continuous. The notation e^* and u^* denotes that these signals are sampled at specified time intervals and are therefore in discrete form. Systems which include a digital computer are known as *digital control systems*. The synthesis techniques for such systems are based on *representing the entire system* as an equivalent sampled, discrete, or *pseudo-continuous-time* (PCT) system.

493

FIGURE 15.1
Control system incorporating a digital computer.

There are various approaches that may be used in analyzing the stability and time-response characteristics of sampled-data systems. These approaches may be divided into two distinct categories: (1) *direct* (DIR) and (2) *digitization* (DIG) or *discrete* digital control analysis techniques. In the first category, the analysis is performed entirely in the discrete domain (z plane). In the second category, the analysis and synthesis of the sampled-data system is carried out entirely by transformation to the w' plane or, by use of the Padé approximation (see Ref. 1, App. C), entirely in the s plane. The use of the Padé approximation, along with the information provided by a Fourier analysis of the sampled signal, results in the modeling of a sampled-data control system by a PCT control system. The w'-plane analysis is not covered in this text, but the reader is referred to Ref. 1 where an overview of the DIR and DIG methods is presented.

The analysis and design of digital control systems, as discussed in this chapter, is expedited by the use of CAD packages such as MATLAB, ICECAP-PC, or TOTAL-PC (see Sec. 10.6 and App. B).

15.2 SAMPLING

Sampling may occur at one or more places in a system. The sampling operation is represented in a block diagram by the symbol for a switch. Figure 15.2 shows a sampled-data control system with sampling of the actuating signal. Note that the output $c(t)$ is a continuous function of time. The sampling process can be considered a modulation process in which a pulse train $p(t)$, with magnitude $1/\gamma$ and period T, multiplies a continuous-time function $f(t)$ and produces the sampled function $f_p^*(t)$. This is represented by

$$f_p^*(t) = p(t)f(t) \tag{15.1}$$

FIGURE 15.2
Block diagram of a sampled-data control system containing sampling of the actuating signal.

These quantities are shown in Fig. 15.3. A Fourier-series expansion of $p(t)$ is

$$p(t) = \frac{1}{\gamma} \sum_{n=-\infty}^{+\infty} C_n e^{jn\omega_s t} \tag{15.2}$$

where the sampling frequency is $\omega_s = 2\pi/T$ and the Fourier coefficients C_n are given by

$$C_n = \frac{1}{T} \int_0^T p(t) e^{-jn\omega_s t} dt = \frac{1}{T} \frac{\sin(n\omega_s \gamma/2)}{n\omega_s \gamma/2} e^{-jn\omega_s \gamma/2} \tag{15.3}$$

The sampled function is therefore

$$f_p^*(t) = \sum_{n=-\infty}^{+\infty} C_n f(t) e^{jn\omega_s t} \tag{15.4}$$

If the Fourier transform of the continuous function $f(t)$ is $\mathbf{F}(j\omega)$, the Fourier transform of the sampled function $f_p^*(t)$ is[2,3,6]

$$\mathbf{F}_p^*(j\omega) = \sum_{n=-\infty}^{+\infty} C_n \mathbf{F}(j\omega + jn\omega_s) \tag{15.5}$$

A comparison of the Fourier spectra of the continuous and sampled functions is shown in Fig. 15.4. It is seen that the sampling process produces a fundamental spectrum similar in shape to that of the continuous function. It also produces a succession of spurious complementary spectra which are shifted periodically by a frequency separation $n\omega_s$.

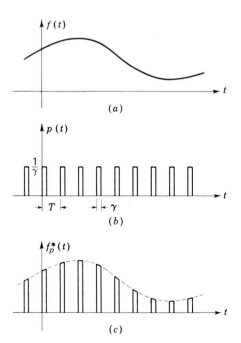

FIGURE 15.3
(a) Continuous function $f(t)$; (b) sampling pulse train $p(t)$; (c) sampled function $f_p^*(t)$.

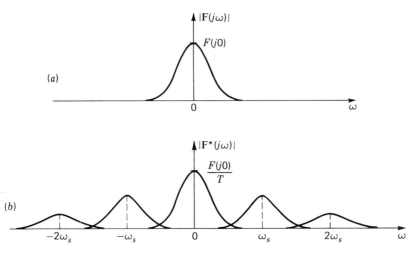

FIGURE 15.4
Frequency spectra for (*a*) a continuous function $f(t)$ and (*b*) a pulse-sampled function $f_p^*(t)$.

If the sampling frequency is sufficiently high, there is very little overlap between the fundamental and complementary frequency spectra. In that case a low-pass filter could extract the spectrum of the continuous input signal by attenuating the spurious higher-frequency spectra. The forward transfer function of a control system generally has a low-pass characteristic, so the system responds with more or less accuracy to the continuous signal.

15.3 IDEAL SAMPLING

To simplify the mathematical analysis of sampled-data control systems, the concept of an ideal sampler (impulse sampling) is introduced. If the duration γ of the sampling pulse is much less than the sampling time T and much smaller than the smallest time constant of $e(t)$, then the pulse train $p(t)$ can be represented by an ideal sampler. Because the area of each pulse of $p(t)$ has unit value, the impulse train also has a magnitude of unity. The ideal sampler produces the impulse train $\delta_T(t)$, which is shown in Fig. 15.5 and is represented by

$$\delta_T(t) = \sum_{k=-\infty}^{+\infty} \delta(t - kT) \tag{15.6}$$

where $\delta(t-kT)$ is the unit impulse that occurs at $t = kT$. The frequency spectrum of the function that is sampled by the ideal impulse train is shown in Fig. 15.6. It is sim-

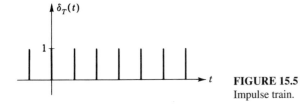

FIGURE 15.5
Impulse train.

ilar to that shown in Fig. 15.4, except that the complementary spectra have the same amplitude as the fundamental spectrum.[1] Because the forward transfer function of a control system attenuates the higher frequencies, the overall system response is essentially the same with the idealized impulse sampling as with the actual pulse sampling. Note in Figs. 15.4 and 15.6 that sampling reduces the amplitude of the spectrum by the factor $1/T$. The use of impulse sampling simplifies the mathematical analysis of sampled systems and is therefore used extensively to represent the sampling process.

When $f(t) = 0$ for $t < 0$, the impulse sequence is

$$f^*(t) = f(t)\delta_T(t) = \sum_{k=0}^{+\infty} f(t)\delta_T(t - kT) \qquad (15.7)$$

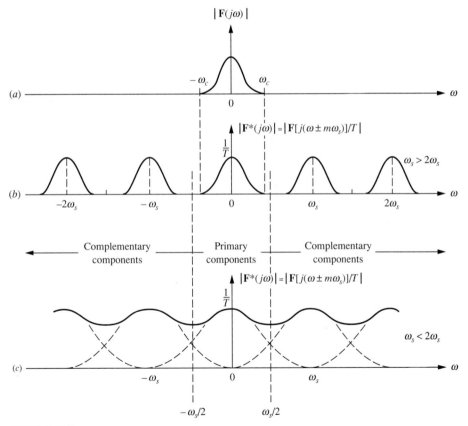

FIGURE 15.6
Frequency spectra for (a) a continuous function $f(t)$; (b) an impulse-sampled function $f^*(t)$, when $\omega_s > 2\omega_c$; (c) an impulse-sampled function $f^*(t)$, when $\omega < 2\omega_c$. (*From* Houpis, C. H., and G. B. Lamont: *Digital Control Systems, Theory, Hardware, Software,* 2nd ed., McGraw-Hill, New York, 1992, p. 82. Reprinted by permission of the publisher.)

and the Laplace transform is given by the infinite series

$$F^*(s) = \sum_{k=0}^{+\infty} f(kT)e^{-kTs} \tag{15.8}$$

where $f(kT)$ represents the function $f(t)$ at the sampling times kT. Because the expression $F^*(s)$ contains the term e^{Ts}, it is not an algebraic expression but a transcendental one. Therefore a change of variable is made:

$$z \equiv e^{Ts} \tag{15.9}$$

where $s = (1/T)\ln z$. Equation (15.8) can now be written as

$$[F^*(s)]_{s=(1/T)\ln z} = F(z) = \sum_{k=0}^{+\infty} f(kT)z^{-k} \tag{15.10}$$

For functions $f(t)$ which have zero value for $t < 0$, the one-sided z transform is the infinite series

$$F(z) = f(0) + f(T)z^{-1} + f(2T)z^{-2} + \cdots \tag{15.11}$$

If the Laplace transform of $f(t)$ is a rational function, it is possible to write $F^*(s)$ in closed form. When the degree of the denominator of $F(s)$ is at least 2 higher than the degree of the numerator, the closed form can be obtained from

$$F^*(s) = \sum_{\text{at poles of } F(p)} \text{residues of } \left[F(p)\frac{1}{1 - e^{-(s-p)T}} \right] \tag{15.12}$$

where $F(p)$ is the Laplace transform of $f(t)$ with s replaced by p. The z transform in closed form may be obtained from Eq. (15.12) and is written as[4]

$$F(z) = \hat{F}(z) + \beta = \sum_{\text{at poles of } F(p)} \text{residues of } \left[F(p)\frac{1}{1 - e^{pT}z^{-1}} \right] + \beta \tag{15.13}$$

where

$$\beta = \lim_{s \to \infty} sF(s) - \lim_{z \to \infty} \hat{F}(z) \tag{15.14}$$

The value of β given by Eq. (15.14) ensures that the initial value $f(0)$ represented by $F(s)$ and $F(z)$ are identical. $F(z)$ *is called the z transform of* $f^*(t)$. The starred and z forms of the *impulse response transfer function* are easily obtained for the ordinary Laplace transfer function.

Example 1. Consider the transfer function

$$G(s) = \frac{K}{s(s + a)} \tag{15.15}$$

The corresponding impulse transfer function in the s domain is

$$G^*(s) = \frac{Ke^{-sT}(1 - e^{-aT})}{a(1 - e^{-sT})(1 - e^{-(s+a)T})} \tag{15.16}$$

and in the z domain

$$G(z) = \frac{(1 - e^{-aT})Kz^{-1}}{a(1 - z^{-1})(1 - e^{-aT}z^{-1})} \tag{15.17}$$

The poles of $G^*(s)$ in Eq. (15.16) are infinite in number. These poles exist at $s = jk\omega_s$ and $s = -a + jk\omega_s$ for all values of $-\infty < k < +\infty$. Note that they are uniformly spaced with respect to their imaginary parts with a separation $j\omega_s$. By contrast, there are just two poles of Eq. (15.17), located at $z = 1$ and $z = e^{-aT}$. The root-locus method can therefore be easily applied in the z plane, whereas for sampled functions it is not very convenient to use in the s plane because of the infinite number of poles.

The transformation of the s plane into the z plane is investigated by inserting $s = \sigma + j\omega$ into

$$z = e^{Ts} = e^{\sigma T} e^{j\omega T} = e^{\sigma T} e^{j2\pi\omega/\omega_s} \tag{15.18}$$

1. Lines of constant σ in the s plane map into circles of radius equal to $e^{\sigma T}$ in the z plane. Specifically, the segment of the imaginary axis in the s plane of width ω_s maps into the circle of unit radius in the z plane; successive segments on the imaginary axis map into overlapping circles. This fact shows that the *proper* consideration of sampled-data systems in the z plane requires the use of a multiple-sheeted surface, i.e., a Riemann surface. But by virtue of the uniform repetition of the roots of the characteristic equation, the condition for stability is that all roots of the characteristic equation contained in the principal branch lie within the unit circle in the principal sheet of the z plane.

2. Lines of constant ω in the s plane map into radial rays drawn at the angle ωT in the z plane. The portion of the constant ω line in the left half of the s plane becomes the radial ray within the unit circle in the z plane. The negative part of the real axis $-\infty < \sigma \leq 0$ in the s plane is mapped on the segment of the real axis defined by $0 < z \leq 1$.

3. The constant-damping-ratio ray in the s plane is defined by the equation

$$s = \sigma + j\omega = -\omega \cot \eta + j\omega$$

where $\eta = \cos^{-1} \zeta$. Therefore

$$z = e^{sT} = e^{-\omega T \cot \eta + j\omega T} = e^{-\omega T \cot \eta} \underline{/\omega T} \tag{15.19}$$

The corresponding map describes a logarithmic spiral in the z plane.

The corresponding paths, as discussed above, are shown in Fig. 15.7.

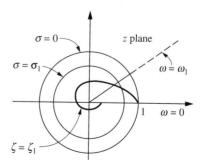

FIGURE 15.7
Transformation from the s plane to the z plane.

15.4 z-TRANSFORM THEOREMS

A short list of z transforms is given in Table 15.1. Several simple properties of the one-sided z transform permit the extension of the table and facilitate its use in solving difference equations. These properties are presented as theorems. They apply when the z transform of $f^*(t)$, denoted by $\mathscr{L}[f^*(t)]$, is $F(z)$ and the sampling time is T.

> **Theorem 1. Translation in time (time shift).** (a) Shifting to the right (delay) yields
>
> $$\mathscr{L}[f^*(t - pT)] = z^{-p}F(z) \tag{15.20}$$
>
> (b) Shifting to the left (advance) yields
>
> $$\mathscr{L}[f^*(t + pT)] = z^p F(z) - \sum_{i=0}^{p-1} f(iT)z^{p-i} \tag{15.21}$$

> **Theorem 2. Final value.** If $F(z)$ converges for $|z| > 1$ and all poles of $(1 - z)F(z)$ are inside the unit circle, then
>
> $$\lim_{k \to \infty} f(kT) = \lim_{z \to 1}[(1 - z^{-1})F(z)] \tag{15.22}$$

> **Theorem 3. Initial value.** If $\lim_{z \to \infty} F(z)$ exists, then
>
> $$\lim_{k \to 0} f(kT) = \lim_{z \to \infty} F(z) \tag{15.23}$$

15.5 SYNTHESIS IN THE z DOMAIN (DIRECT METHOD)

For the block diagram of Fig. 15.2, the system equations are

$$C(s) = G(s)E^*(s) \tag{15.24}$$

$$E(s) = R(s) - B(s) = R(s) - G(s)H(s)E^*(s) \tag{15.25}$$

The starred transform of Eq. (15.25) is

$$E^*(s) = R^*(s) - GH^*(s)E^*(s) \tag{15.26}$$

where $GH^*(s) \equiv [G(s)H(s)]^*$. Note that in general, $GH^*(s) \neq G^*(s)H^*(s)$.

Solving for $C(s)$ from Eqs. (15.24) and (15.26) gives

$$C(s) = \frac{G(s)R^*(s)}{1 + GH^*(s)} \tag{15.27}$$

The starred transform $C^*(s)$ obtained from Eq. (15.27) is

$$C^*(s) = \frac{G^*(s)R^*(s)}{1 + GH^*(s)} \tag{15.28}$$

The z transform is obtained by replacing each starred transform by the corresponding function of z:

$$C(z) = \frac{G(z)R(z)}{1 + GH(z)} \tag{15.29}$$

TABLE 15.1
Table of z transforms

Continuous time function	Laplace transform	z transform of $f^*(t)$ $F(z)$	Discrete time function $f(kT)$ for $k \geq 0$
1. $\delta_0(t)$	1	1	$\delta_0(0)$
2. $\delta(t - kT)$ k is any integer	e^{-kTs}	z^{-1}	$\delta(t - kT)$
3. $u_{-1}(t)$	$\dfrac{1}{s}$	$\dfrac{z}{z - 1}$	$\delta(t - kT)$
4. $tu_{-1}(t)$	$\dfrac{1}{s^2}$	$\dfrac{Tz}{(z - 1)^2}$	$kT\,\delta(t - kT)$
5. $\dfrac{t^2}{2}u_{-1}(t)$	$\dfrac{1}{s^3}$	$\dfrac{T^2 z(z + 1)}{2(z - 1)^3}$	$kT^2 \delta(t - kT)$
6. e^{-at}	$\dfrac{1}{s + a}$	$\dfrac{z}{z - e^{-aT}}$	e^{-akT}
7. $\dfrac{e^{-bt} - e^{-at}}{a - b}$	$\dfrac{1}{(s + a)(s + b)}$	$\dfrac{1}{a - b}\left[\dfrac{z}{z - e^{-bT}} - \dfrac{z}{z - e^{-aT}}\right]$	$\dfrac{1}{a - b}(e^{-bkT} - e^{-akT})$
8. $u_{-1}(t) - e^{-at}$	$\dfrac{a}{s(s + a)}$	$a\dfrac{(1 - e^{-aT})z}{(z - 1)(z - e^{-aT})}$	$\delta(t - kT)e^{-akT}$
9. $t - \dfrac{1 - e^{-aT}}{a}$	$\dfrac{a}{s^2(s + a)}$	$\dfrac{Tz}{(z - 1)^2} - \dfrac{(1 - e^{-aT})z}{a(z - 1)(z - e^{-aT})}$	$kT - \dfrac{1 - e^{-akT}}{a}$
10. $\sin at$	$\dfrac{a}{s^2 + a^2}$	$\dfrac{z \sin aT}{z^2 - 2z \cos aT + 1}$	$\sin akT$
11. $\cos at$	$\dfrac{s}{s^2 + a^2}$	$\dfrac{z(z - \cos aT)}{z^2 - 2z \cos aT + 1}$	$\cos akT$
12. $e^{-at} \sin bt$	$\dfrac{b}{(s + a)^2 + b^2}$	$\dfrac{ze^{-aT} \sin bT}{z^2 - 2ze^{-aT} \cos bT + e^{-2aT}}$	$e^{-akT} \sin bkT$
13. $e^{-at} \cos bt$	$\dfrac{s + a}{(s + a)^2 + b^2}$	$\dfrac{z^2 - ze^{-aT} \cos bT}{z^2 - 2ze^{-aT} \cos bT + e^{-2aT}}$	$e^{-akT} \cos bkT$
14. te^{-at}	$\dfrac{1}{(s + a)^2}$	$\dfrac{Tze^{-aT}}{(z - e^{-aT})^2}$	$kT e^{-akT}$

Although the output $c(t)$ is continuous, the inverse of $C(z)$ in Eq. (15.29) yields only the set of values $c(kT)$, $k = 0, 1, 2, \dots$, corresponding to the values at the sampling instants. Thus, $c(kT)$ is the set of impulses from an ideal sampler located at the output that operates in synchronism with the sampler of the actuating signal.

Analysis of the performance of the sampled system, corresponding to the response given by Eq. (15.29), can be performed by the frequency-response or root-locus method. However, it should be kept in mind that the geometric boundary for stability in the z plane is the unit circle. As an example of the root-locus method, consider the sampled feedback system represented by Fig. 15.2 with $H(s) = 1$. Using $G(s)$ given by Eq. (15.15) with $a = 1$ and a sampling time $T = 1$, the equation of $G(z)$ is

$$G(z) = GH(z) = \frac{0.632Kz}{(z - 1)(z - 0.368)} \tag{15.30}$$

The characteristic equation is $1 + GH(z) = 0$ or

$$GH(z) = -1 \tag{15.31}$$

The usual root-locus techniques can be used to obtain a plot of the roots of Eq. (15.31) as a function of the static loop sensitivity K. The root locus is drawn in Fig. 15.8. The maximum value of K for stability is obtained from the magnitude condition and yields $K_{max} = 2.73$. This occurs at the crossing of the root locus and the unit circle. The selection of the desired roots can be based on the damping ratio ζ desired. For the specified value of ζ, the spiral given by Eq. (15.19) and shown in Fig. 15.7 must be drawn. The intersection of this curve with the root locus determines the roots. Alternatively, it is possible to specify the settling time. This determines the value of σ in the s plane, so the circle of radius $e^{\sigma T}$ can be drawn in the z plane. The intersection of this circle and the root locus determines the roots. For the value $\zeta = 0.48$ the roots are $z = 0.368 \pm j0.482$, shown on the root locus in Fig. 15.8, and the value of $K = 1$. The control ratio is therefore

$$\frac{C(z)}{R(z)} = \frac{0.632z}{z - 0.368 \pm j0.482} = \frac{0.632z}{z^2 - 0.736z + 0.368} \tag{15.32}$$

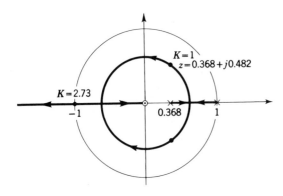

FIGURE 15.8
Root locus for Eqs. (15.30) and (15.31).

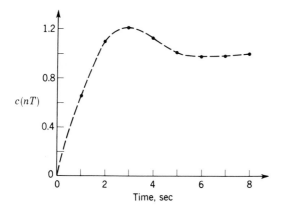

FIGURE 15.9
Plot of $c(nT)$ for Eq. (15.36).

For a unit-step input the value of $R(z)$ is

$$R(z) = \frac{z}{z - 1} \tag{15.33}$$

so that
$$C(z) = \frac{0.632z^2}{(z - 1)(z^2 - 0.736z + 0.368)} \tag{15.34}$$

The expression for $C(z)$ can be expanded by dividing its denominator into its numerator to get a power series in z^{-1}.

$$C(z) = 0.632z^{-1} + 1.096z^{-2} + 1.205z^{-3} + 1.120z^{-4}$$
$$+ 1.014z^{-5} + 0.98z^{-6} + \cdots \tag{15.35}$$

The inverse transform of $C(z)$ is

$$c(kT) = 0\delta(t) + 0.632\delta(t - T) + 1.096\delta(t - 2T) + \cdots \tag{15.36}$$

Hence, comparing Eq. (15.35) with Eq. (15.11), the values of $c(kT)$ at the sampling instants are the coefficients of the terms in the series of Eq. (15.36) at the corresponding sampling instant. A plot of the values of $c(kT)$ is shown in Fig. 15.9. The curve of $c(t)$ is drawn as a smooth curve through these plotted points.

This section presents several basic characteristics of the analysis of sampled-data systems. Additional topics of importance include the reconstruction of the continuous signal from the sampled signal by use of *hold* circuits, and the problem of compensation to improve performance. These topics are presented in the following sections.

15.6   THE INVERSE z TRANSFORM

In the example of Sec. 15.5 the inverse of $C(z)$ is obtained by expanding this function into an infinite series in terms of z^{-k}. This is referred to as the *power-series method*. Then the coefficients of z^{-k} are the values of $c(kT)$. The division required to obtain $C(z)$ as an infinite series (open form) is easily performed on a digital computer. However, if an analytical expression in closed form for $c^*(t)$ is desired, $C(z)$ can be expanded into partial fractions, which appear in Table 15.1. Because the z transforms in Table 15.1 contain z in the numerator, the Heaviside partial fraction is first performed on $C(z)/z$, which must be in proper form.

Example 2.   For $C(z)$ given by Eq. (15.34) with $T = 1$

$$\frac{C(z)}{z} = \frac{A}{z - 1} + \frac{Bz + C}{z^2 - 0.736z + 0.368} \tag{15.37}$$

The coefficients are evaluated by the usual Heaviside partial-fraction methods and yield $A = 1, B = -1$, and $C = 0.368$. Inserting these values into Eq. (15.37) and using the form of entries 3 and 13 of Table 15.1, the response transform $C(z)$ is

$$C(z) = \frac{z}{z - 1} - \frac{z(z - e^{-a}\cos b)}{z^2 - 2ze^{-a}\cos b + e^{-2a}} \tag{15.38}$$

where $a = 0.5$ and $b = 1$. Thus, from Table 15.1, the system output of Eq. (15.38) in closed form is

$$c(kT) = 1 - e^{-0.5k}\cos k \tag{15.39}$$

The open form introduces an error that is propagated in the division process due to roundoff. This does not occur with the closed form. Remember that $C(z)$ represents the sampled values $c^*(t)$ of $c(t)$ at the sampling instants kT. Thus, the inversion from the z domain to the time domain yields $c^*(t)$ or the values $c(kT)$. Thus the inversion process yields values of $c(t)$ only at the sampling instants. In other words, $C(z)$ does not contain any information about the values of $c(t)$ in between the sampling instants. This is an inherent limitation of the z-transform method.

15.7   ZERO-ORDER HOLD

The function of a zero-order hold (ZOH) is to reconstruct a piecewise-continuous signal from the sampled function $f^*(t)$. It holds the output amplitude constant at the value of the impulse for the duration of the sampling period T. Thus, it has the transfer function

$$G_{\text{ZOH}}(s) = \frac{1 - e^{-Ts}}{s} \tag{15.40}$$

The action of the ZOH is shown in Fig. 15.10. When the sampling time T is small or the signal is slowly varying, the output of the ZOH is frequently used in digital control systems. It is used to convert the discrete signal obtained from a digital computer into a piecewise-continuous signal which is the input to the plant. The ZOH is a low-pass filter that , together with the basic plant, attenuates the complementary frequency spectra introduced by the sampling process. The ZOH does introduce a lag angle into the system, and therefore it can affect the stability and time-response characteristics of the system.

FIGURE 15.10
Input and output signals for a zero-order hold: (*a*) continuous input signal $e(t)$ and the sampled signal $e^*(t)$; (*b*) continuous signal $e(t)$ and the piecewise-constant output $m(t)$ of the zero-order hold.

Example 3. Figure 15.11 shows a unity-feedback system in which a zero-order hold is placed after the sampler. The z transform of the forward transfer function is

$$G(z) = \frac{C(z)}{E(z)} = \mathscr{L}\left[\frac{1 - e^{-Ts}}{s}\frac{K}{s(s+1)}\right] \tag{15.41}$$

Because $\mathscr{L}(1 - e^{-Ts}) = 1 - z^{-1}$, this term can be factored from Eq. (15.41) to give, for $T = 1$,

$$G(z) = (1 - z^{-1})\mathscr{L}\left[\frac{K}{s^2(s+1)}\right] = \frac{K(ze^{-1} + 1 - 2e^{-1})}{(z-1)(z-e^{-1})} = \frac{0.368K(z+0.717)}{(z-1)(z-0.368)} \tag{15.42}$$

A comparison with the z transform without the ZOH as given by Eq. (15.30) shows that the zero is moved from the origin to $z = -0.717$.

The root locus for the system of Fig. 15.11 is shown in Fig. 15.12 and may be compared with the root locus without the ZOH in Fig. 15.8. The maximum value of the gain K for stability is $K_{max} = 0.8824$. With $\zeta = 0.707$, the closed-loop control ratio is

$$\frac{C(z)}{R(z)} = \frac{0.118(z+0.717)}{(z-0.625+j0.249)(z-0.625-j0.249)} = \frac{0.118(z+0.717)}{z^2 - 1.25z + 0.453} \tag{15.43}$$

For this control ratio, $e^{\sigma T} = 0.673$ so that the corresponding real part of the closed-loop poles in the s-domain is $\sigma = -0.396$ and the approximate settling time is $T_s = 4/0.396 = 10.1$ s.

This example illustrates that a second-order sampled system becomes unstable for large values of gain. This is in contrast to a continuous second-order system, which is stable for all positive values of gain. The advantage of a discrete control system is the greater flexibility of compensation that can be achieved with a digital compensator.

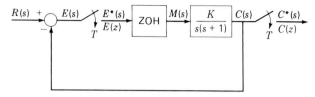

FIGURE 15.11
Sampled system containing a ZOH.

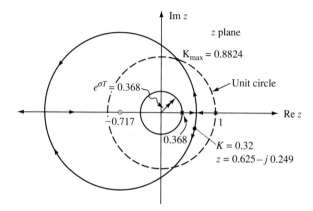

FIGURE 15.12
Root locus for the system of
Fig. 15.11.

15.8 LIMITATIONS

Like any method of analysis, there are limitations to the use of the z-transform method. In spite of these limitations, however, this method is a very useful and convenient analytical tool because of its simplicity. Nevertheless, the following limitations must be kept in mind when applying and interpreting the results of the z-transform method:

1. The use of an ideal sampler in the analysis of a discrete-data system is based upon a model in which the strengths (area) of the impulses in the impulse train, $c^*(t)$, are equal to the corresponding values of the input signal $c(t)$ at the sampling instants kT. For this mathematical model to be "close" to correct, it is necessary for the sampling duration (or pulse width) γ to be very small in comparison to the smallest time constant of the system. It must also be very much smaller than the sampling time T.

2. While the inverse z transform $z^{-1}[C(z)]$ yields only $c(kT)$ and not a unique $c(t)$ function, the *modified $\mathscr{L}$-transform method*,[6] z_m, removes this limitation.

3. The impulse responses of rational $C(s)$ functions (*exclusive of a hold device*) experience a jump behavior at $t = 0$ [at $t = kT$ for $c^*(t)$] unless these functions have at least two more poles than zeros. For these functions,

$$\lim_{t \to kT^+} c(t) = L_+ \neq \lim_{t \to kT^-} c(t) = L_-$$

i.e., the value of $c(kT)$, as t approaches the value of kT from the low side, is not identical to the value obtained when approaching kT from the high side. In other words, a discontinuity occurs in the value of $c(kT)$ at $t = kT$. Fortunately, the $C(z)$ encountered in the analysis of most practical control systems have at least two more poles than zeros ($n \geq w+2$), and this limitation does not apply for such systems.

15.9   TUSTIN TRANSFORMATION[1,5,6]

The transformation from the s-plane to the z-plane given by $z = e^{Ts}$ is a transcendental relationship. A linear approximation for this transformation is obtained by use of the Tustin algorithm. Solving for s in terms of z yields

$$s = \frac{1}{T} \ln z \tag{15.44}$$

The natural logarithm $\ln z$ may be expanded into the series

$$\ln z = 2\left(x + \frac{1}{3}x^3 + \frac{1}{5}x^5 + \cdots\right) \tag{15.45}$$

where
$$x = \frac{1 - z^{-1}}{1 + z^{-1}} \tag{15.46}$$

Using only the first term in the infinite series of Eq. (15.45) yields the Tustin transformation

$$s = \frac{2}{T}\left[\frac{1 - z^{-1}}{1 + z^{-1}}\right] \tag{15.47}$$

The expression for s in Eq. (15.47) can be inserted into a function such as $G(s)$, which represents a continuous-time function. Rationalizing the expression yields the function $G(z)$, which represents a discrete-time function. The root-locus and frequency response design methods, developed for continuous-time systems, can now be applied to the design of discrete-time systems.

The response of the Tustin z-domain transfer function has good accuracy when compared with the response of the exact z-domain transfer function; however, the accuracy decreases as the frequency increases.

Ideally, the z-domain transfer function obtained by digitizing an s-domain transfer function should retain all the properties of the s-domain transfer function. This is not achieved completely. The transformation has four properties that are of particular interest to the designer. These four properties are the cascading property, stability, dc gain, and the impulse response. The first three properties are maintained by the Tustin transformation; i.e., the three properties are invariant. The Tustin algorithm maintains the cascading property of the s-domain transfer function. Thus, cascading two z-domain Tustin transfer functions yields the same result as cascading the s-domain counterparts and Tustin-transforming the result to the z domain. This cascading property implies that no ZOH device is required when cascading

$$\frac{E(s)}{e(t)} \qquad \frac{E^*(s)}{e^*(t)} \longrightarrow$$

FIGURE 15.13
A sampled function.

s-domain transfer functions. The Tustin transformation of a stable s-domain transfer function is also stable. The dc gains for the s-domain and the z-domain Tustin transfer functions are identical; that is, $F(s)|_{s=0} = F(z)|_{z=1}$. However, the dc gains for the exact $\mathscr{L}$ transfer functions are not identical.

Example 4. This example illustrates the properties of the Tustin transformation of a function $e(t)$ that is sampled as shown in Fig. 15.13. Given $E(s) = E_x(s) = 1/[(s + 1)(s + 2)]$, *with no ZOH involved*, and $T = 0.1$ s, the exact $\mathscr{L}$ transform yields

$$E(z) = \mathscr{L}[e^*(t)] = \frac{0.086107z}{(z - 0.81873)(z - 0.90484)}$$

The Tustin transformation, indicated by the subscript T, is obtained by inserting s from Eq. (15.47) into $E(s)$ and simplifying the resulting equation

$$[E(z)]_T = \frac{0.0021645(z + 1)^2}{(z - 0.81818)(z - 0.90476)}$$

For each transform, the initial- and final-value theorems yield, respectively, the results shown in the following table:

	$E(s)$	$E(z)$	$[E(z)]_T$	$\frac{1}{T}[E(z)]_T$
Initial value	0	0	0.0021645	0.021646
Final value	0	0	0	0
dc gain	0.5	4.9918	0.499985	4.99985
n	2	2	2	2
w	0	1	2	2

This example demonstrates that the Tustin transformation results in a $\mathscr{L}$ transform for which, in general, the degrees of the numerator and denominator are the same, i.e., $w = n$. Thus, because $n = w$, the initial value of $[E(z)]_T$ is always different from zero; that is,

$$\lim_{z \to \infty} E(z) \neq \lim_{z \to \infty} [E(z)]_T$$

Based on the dc gain, it is evident that the values of $E(z)$ and $[E(z)]_T$ differ by a factor of $1/T$; that is,

$$E(z) = \mathscr{L}[e^*(t)] \approx \frac{1}{T}[E(z)]_T \qquad (15.48)$$

The reason for this difference is that the Tustin transformation does not take into account the attenuation factor $1/T$ introduced by the sampling process, whereas $E(z) = \mathscr{L}[e^*(t)]$ does take this attenuation factor into account. This factor must be included when the DIG technique is used for the PCT system analysis and design.

15.10 TUSTIN TRANSFORMATION PROPERTIES

A comparison is now made of the properties of the s-plane with the properties of the z-plane that results from the Tustin transformation. Solving Eq. (15.47) for z yields

$$z = \frac{1 + sT/2}{1 - sT/2} \tag{15.49}$$

Letting $s = j\hat{\omega}_{sp}$ gives

$$z = \frac{1 + j\hat{\omega}_{sp}T/2}{1 - j\hat{\omega}_{sp}T/2} \tag{15.50}$$

The exact $\mathscr{L}$ transform yields $z = e^{j\omega_{sp}T}$, where ω_{sp} is an equivalent s-plane frequency. Equation (15.49) can be put in the form (see Prob. 15.15)

$$e^{j\omega_{sp}T} = \exp\left(j2\tan^{-1}\frac{\hat{\omega}_{sp}T}{2}\right) \tag{15.51}$$

Thus $\omega_{sp}T/2 = \tan^{-1}(\hat{\omega}_{sp}T/2)$, or

$$\tan\frac{\omega_{sp}T}{2} = \frac{\hat{\omega}_{sp}T}{2} \tag{15.52}$$

When $\omega_{sp}T/2 < 17°$, or about 0.30 rad, then

$$\omega_{sp} \approx \hat{\omega}_{sp} \tag{15.53}$$

This means that in the frequency domain the Tustin approximation is good for small values of $\omega_{sp}T/2$.

Using Eq. (15.51), it is easy to realize that the imaginary axis of the s plane is mapped as the unit-circle in the z plane as shown in Fig. 15.14. The left-half s plane is mapped into the unit circle. Therefore, the stability region, inside the unit circle in the z-plane, is the same for both the exact $\mathscr{L}$ transform and the Tustin approximation. Also, in this approximation, the entire imaginary axis of the s plane is mapped once, and only once, onto the unit circle. The Tustin approximation prevents pole and zero *aliasing* because the folding phenomenon does not occur with this method. However, the warping of Eq. (15.52) is depicted by the arrow in Fig. 15.15b. Compensation can be accomplished by prewarping of ω_{sp} by using Eq. (15.52) to generate $\hat{\omega}_{sp}$. The continuous controller is mapped into the z plane by means of Eq. (15.47), using the prewarped frequency $\hat{\omega}_{sp}$. The digital compensator (controller) must be tuned (i.e., its numerical coefficients adjusted) to finalize the design because approximations have been employed. As seen from Fig. 15.15a, $\omega_{sp} \approx \hat{\omega}_{sp}$ is a good approximation when $\omega_{sp}T/2$ and $\hat{\omega}_{sp}T/2$ are both less than 0.3 rad.

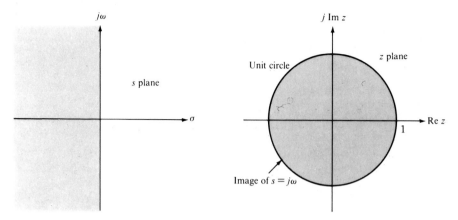

FIGURE 15.14
Tustin approximation s- to z-plane mapping.

The prewarping approach for the Tustin approximation takes the imaginary axis s plane and folds it back to $\pi/2$ to $-\pi/2$, as seen from Fig. 15.15a. The spectrum of the input must also be taken into consideration when selecting an approximation procedure with or without prewarping. In the previous discussion, only the frequency has been prewarped due to the interest in the controller frequency response.

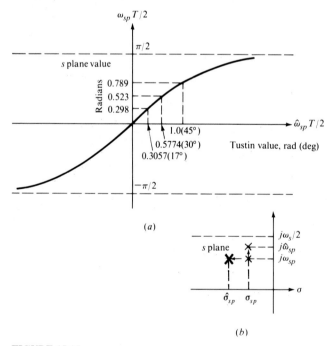

FIGURE 15.15
Map of $\hat{\omega}_{sp} = 2(\tan \omega_{sp} T/2)/T$. ($a$) Plot of Eq. (15.51) and (b) warping effect.

The real part of the s-plane pole influences such characteristics as rise time, peak overshoot, and settling time. Consideration of the warping of the real pole component is now analyzed. Substituting $z = e^{\sigma_{sp}T}$ and $s = \hat{\sigma}_{sp}$ into Eq. (15.49) yields

$$e^{\sigma_{sp}T} = \frac{1 + \hat{\sigma}_{sp}T/2}{1 - \hat{\sigma}_{sp}T/2} \tag{15.54}$$

Replacing $e^{\sigma_{sp}T}$ by its exponential series and dividing the numerator by the denominator in Eq. (15.54) results in an expression that can be put in closed form by forming the sum of the infinite series:

$$1 + \sigma_{sp}T + \frac{(\sigma_{sp}T)^2}{2} + \cdots = 1 + \frac{\hat{\sigma}_{sp}T}{1 - \hat{\sigma}_{sp}T/2} \tag{15.55}$$

If $|\sigma_{sp}T| \gg (\sigma_{sp}T)^2/2$ (or $1 \gg |\hat{\sigma}_{sp}T/2|$) and $1 \gg |\hat{\sigma}_{sp}T/2|$, then

$$|\hat{\sigma}_{sp}| \approx |\sigma_{sp}| \ll \frac{2}{T} \tag{15.56}$$

Thus when Eqs. (15.53) and (15.56) are satisfied, the Tustin approximation in the s domain is good for small magnitudes of the real and imaginary components of the variable s. The cross-hatched area in Fig. 15.16 represents the allowable location of the poles and zeros in the s plane for a good Tustin approximation. Because of the mapping properties and its ease of use, the Tustin transformation is employed for the DIG technique in the next section.

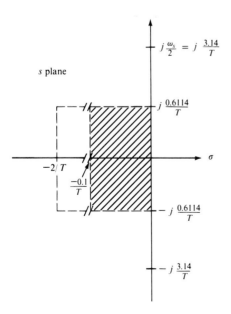

FIGURE 15.16
Allowable location (cross-hatched area) of dominant poles and zeros in s plane for a good Tustin approximation.

15.11 PSEUDO-CONTINUOUS-TIME (PCT) CONTROL SYSTEM (DIG METHOD)

The DIG method of designing a sampled-data system is performed in the complex-frequency s plane. First a satisfactory pseudo-continuous-time (PCT) model of the sampled-data system must be obtained. Therefore, for the sampled-data system of Fig. 15.17, the sampler and ZOH must be approximated by a linear continuous-time unit $G_A(s)$, as shown in Fig. 15.18c. The DIG method requires that the dominant poles and zeros of the PCT model should lie in the shaded area of Fig. 15.16 for a high level of correlation with the response of the sampled-data system. To determine $G_A(s)$, first consider the frequency component of $E^*(j\omega)$ representing the continuous-time signal $E(j\omega)$, where all its sidebands are multiplied by $1/T$ (see Sec. 15.2).[1] Because of the low-pass filtering characteristics of the plant in a sampled-data system, only the primary component is considered in the analysis of the system. Therefore, the PCT approximation of the sampler of Fig. 15.17 is equal to $1/T$, as shown in Fig. 15.18b.

Using the first-order Padé approximation,[1] the transfer function of the ZOH, when the value of T is small enough, is approximated as follows:

$$G_{zo}(s) = \frac{1 - e^{-Ts}}{s} \approx \frac{2T}{Ts + 2} = G_{pa}(s) \qquad (15.57)$$

Thus the Padé approximation $G_{pa}(s)$ is used to replace $G_{zo}(s)$, as shown in Fig. 15.18a and b. This approximation is good for $\omega_c \leq \omega_s/10$, whereas the second-order approximation is good for $\omega_c \leq \omega_s/3$.[1] Therefore the sampler and ZOH of a sampled-data system are approximated in the PCT system of Fig. 15.18c by the transfer function

$$G_A(s) = \frac{1}{T}G_{pa}(s) = \frac{2}{Ts + 2} \qquad (15.58)$$

Because $\lim_{T \to 0} G_A(s) = 1$, Eq. (15.58) is an accurate PCT representation of the sampler and ZOH units, because it satisfies the requirement that the output of $G_A(s)$ must equal its input as $T \to 0$. Further note that in the frequency domain as $\omega_s \to \infty$

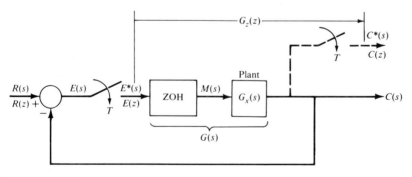

FIGURE 15.17
The uncompensated sampled-data control system.

FIGURE 15.18
(*a*) Sampler and ZOH. (*b*) Approximations of the sampler and ZOH. (*c*) The approximate continuous-time control system equivalent of Fig. 15.17. (*d*) A cascade compensated digital control system. (*e*) The PCT equivalent system.

($T \to 0$), the primary strip becomes the entire frequency-spectrum domain that is the representation for the continuous-time system.

    Note that in obtaining the PCT model for the compensated digital control system of Fig. 15.18d, the factor $1/T$ *replaces only the sampler that is sampling the continuous-time signal.* This multiplier of $1/T$ is reflected in the Fourier series analysis of $e^*(t)$ for ideal impulse sampling, because the fundamental frequency

TABLE 15.2
Analysis of a PCT system representing a digital control system for $\zeta = 0.45$

Method	T, s	Domain	K_x	M_p	t_p, s	t_s, s
DIR	0.01	z	4.147	1.202	4.16	9.53
DIG		s	4.215	1.206	4.11	9.478
DIR	0.1	z	3.892	1.202	4.2–4.3	9.8–9.9
DIG		s	3.906	1.203	4.33$^-$	9.90
DIR	1	z	2.4393	1.199	6	13–14
DIG		s	2.496	1.200	6.18	13.76

spectrum of the sampled signal and all its harmonics are attenuated by $1/T$. *The sampler at the output of the digital controller of Fig. 15.18d is replaced by a factor of 1.* The PCT representation of Fig. 15.18d is shown in Fig. 15.18e to illustrate the effect of the value of T on the validity of the results obtained by the DIG method. Consider the sampled-data closed-loop control system of Fig. 15.17 for the $G_x(s)$ of Sec. 10.8 with $K_x = 4.2$. The closed-loop system performances for three values of T and $\zeta = 0.45$ are determined in both the s and z domains, i.e., using both the DIG and DIR design methods, respectively. Table 15.2 presents the required value of K_x and time-response characteristics for each value of T. Note that there is a high level of correlation between the DIG and DIR models. (The designer must specify what is considered to be "good correlation" for a given application.)

15.12 ANALYSIS OF A BASIC (UNCOMPENSATED) SYSTEM

The lag characteristic of $G_{zo}(s)$ reduces the degree of system stability, as mentioned in Sec. 15.7. This degradation is illustrated in some of the examples in this chapter. Two design methods for improving the performance of a basic system are illustrated here.

Case 1.

Figure 15.17 represents a basic or uncompensated digital control system where

$$G_x(s) = \frac{K_x}{s(s + 1)} \tag{15.59}$$

PCT Control-System Model

One approach for designing a unity-feedback digital control system is to first obtain a closed-loop model $[C(s)/R(s)]_M$ for the PCT unity-feedback control system of Fig. 15.18c, utilizing the plant of the digital control system without compensation. This model is then used as a guide for selecting an acceptable $C(z)/R(z)$.

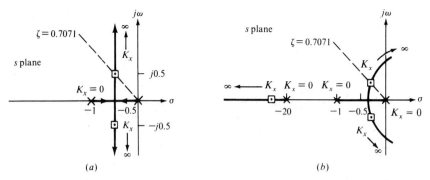

FIGURE 15.19
Root locus for (*a*) Case 1, Eq. (15.59); (*b*) Case 2, Eq. (15.60) with $T = 0.1$s. Not to scale.

Case 2.

For the plant of Eq. (15.59) and the ZOH Padé model of Eq. (15.58),

$$G_{PC}(s) = G_A(s)G_x(s) = \frac{2K_x/T}{s(s+1)(s+2/T)} \tag{15.60}$$

For $T = 0.1$ s, the closed-loop transfer function is

$$\left[\frac{C(s)}{R(s)}\right]_M = \frac{G_{PC}(s)}{1 + G_{PC}(s)} = \frac{20K_x}{s^3 + 21s^2 + 20s + 20K_x} \tag{15.61}$$

The root locus for $G_{PC}(s) = -1$ is shown in Fig. 15.19*b*. For comparison purposes the root-locus plot for $G_x(s) = -1$ is shown in Fig. 15.19*a*. These figures illustrate that the effect of sampling yields the same results as a lag network in cascade in an analog feedback control system; i.e., it transforms a completely stable system into a conditionally stable system. Thus, for a given value of ζ, the values of t_p and t_s (and T_s) are increased. Therefore, the ZOH degrades the degree of system stability.

For the model of Eq. (15.61) the desired value of the damping ratio ζ for the dominant roots is 0.7071. Thus, for a unit-step input, with $K_x = 0.4767$,

$$[C(s)]_M = \frac{9.534}{s(s^3 + 21s^2 + 20s + 9.534)}$$

$$= \frac{9.534}{s(s + 0.4875 \pm j0.4883)(s + 20.03)} \tag{15.62}$$

The real and imaginary parts of the desired poles of Eq. (15.62), with $T = 0.1$ s, lie in the acceptable region of Fig. 15.16 for a good Tustin approximation.

Sampled-Data Control System

The determination of the time-domain performance of the sampled-data system of Fig. 15.17 may be achieved either by obtaining the exact expression for $C(z)$ or by applying the Tustin transformation to Eq. (15.61) to obtain the approximate expression for $C(z)$. The exact approach first requires the z-transfer function of the forward path of Fig. 15.17. For the plant transfer function of Eq. (15.59), using transform pair 9 in Table 15.1,

$$G_z(z) = \mathscr{L}\left[\frac{K_x(1 - e^{-sT})}{s^2(s + 1)}\right] = (1 - z^{-1})\mathscr{L}\left[\frac{K_x}{s^2(s + 1)}\right]$$

$$= \frac{K_x[(T - 1 + e^{-T})z + (1 - Te^{-T} - e^{-T})]}{z^2 - (1 + e^{-T} + K_x - TK_x - K_xe^{-T})z + e^{-T} + K_x - K_x(T + 1)e^{-T}} \tag{15.63}$$

Case 3.

Thus, for $T = 0.1$ s and $K_x = 0.4767$,

$$G_z(z) = \frac{0.002306(z + 0.9672)}{(z - 1)(z - 0.9048)} \tag{15.64}$$

or

$$\frac{C(z)}{R(z)} = \frac{0.002306(z + 0.9672)}{(z - 0.9513 \pm j0.04649)} \tag{15.65}$$

The DIG technique requires that the s-domain model control ratio be transformed into a z-domain model. Applying the Tustin transformation to

$$\left[\frac{C(s)}{R(s)}\right]_M = \frac{G_{PC}(s)}{1 + G_{PC}(s)} = F_M(s) \tag{15.66}$$

yields

$$\frac{[C(z)]_T}{[R(z)]_T} = [F(z)]_T \tag{15.67}$$

This equation is rearranged to

$$[C(z)]_T = [F(z)]_T[R(z)]_T \tag{15.68}$$

As stated in Sec. 15.9,

$$R(z) = \mathscr{L}[r^*(t)] = \frac{1}{T}[R(z)]_T \tag{15.69}$$

$$C(z) = \mathscr{L}[c^*(t)] = \frac{1}{T}[C(z)]_T \tag{15.70}$$

Substituting from Eqs. (15.69) and (15.70) into Eq. (15.67) yields

$$\frac{C(z)}{R(z)} = [F(z)]_T \tag{15.71}$$

Substituting from Eq. (15.70) into Eq. (15.68) and rearranging yields

$$C(z) = \frac{1}{T}[F(z)]_T[R(z)]_T = \frac{1}{T}[\text{Tustin of } F_M(s)R(s)] \tag{15.72}$$

Case 4.

Based upon Eq. (15.71), the Tustin transformation applied to Eq. (15.61), with $K_x = 0.4767$, results in a Tustin model of the control ratio:

$$\frac{C(z)}{R(z)} = \left[\frac{C(z)}{R(z)}\right]_T$$

$$= \frac{5.672 \times 10^{-4}(z + 1)^3}{(z - 0.9513 \pm j0.04651)(z + 6.252 \times 10^{-4})} \tag{15.73}$$

Note that the dominant poles of Eq. (15.73) are essentially the same as those of Eq. (15.65) because the value of T resulted in the dominant roots lying in the good Tustin region of Fig. 15.16. In using the exact $\mathscr{L}$ transformation, the degree of the numerator polynomial of $C(z)/R(z)$, Eq. (15.65), is one less than the degree of its denominator polynomial. When using the Tustin transformation, the degree of the numerator polynomial of the resulting $[C(z)/R(z)]_T$ is in general equal to the degree of its corresponding denominator polynomial [see Eq. (15.73)]. Thus $[C(z)]_T$ results in a value of $c^*(t) \neq 0$ at $t = 0$, which is in error, based upon the zero initial conditions. Table 15.3 illustrates the effect of this characteristic of the Tustin transformation on

TABLE 15.3
Comparison of time responses between $C(z)$ and $[C(z)]_T$ for a unit-step input and $T = 0.1$ s

	$c(kT)$				$c(kT)$	
k	Case 3 (exact), $C(z)$	Case 4 (Tustin), $[C(z)]_T$	k	Case 3 (exact), $C(z)$	Case 4 (Tustin), $[C(z)]_T$
0	0.	0.5672E - 03	42	0.9460	0.9452
2	0.8924E - 02	0.9823E - 02	44	0.9668	0.9660
4	0.3340E - 01	0.3403E - 01	46	0.9844	0.9836
6	0.7024E - 01	0.7064E - 01	48	0.9992	0.9984
8	0.1166	0.1168	50	1.011	1.011
10	0.1701	0.1701	52	1.021	1.020
12	0.2284	0.2283	54	1.029	1.028
14	0.2897	0.2894	56	1.035	1.034
16	0.3525	0.3521	58	1.039	1.038
18	0.4153	0.4148	60	1.042	1.041
20	0.4771	0.4766	61	1.042	1.042
22	0.5370	0.5364	62	1.043	1.043
24	0.5943	0.5937	63	1.043	1.043
26	0.6485	0.6478	64	1.043 $\Big\}\,M_p$	1.043 $\Big\}\,M_p$
28	0.6992	0.6984	65	1.043	1.043
30	0.7461	0.7453	66	1.043	1.043
32	0.7890	0.7883	67	1.043	1.043
34	0.8281	0.8273	68	1.042	1.042
36	0.8632	0.8624	85	1.022	1.022
38	0.8944	0.8936	86	1.021 $\Big\}\,t_s$	1.021 $\Big\}\,t_s$
40	0.9220	0.9212	87	1.019	1.019

the time response due to a unit-step forcing function. The difference in the time response by use of the Tustin transformation is minimal; i.e., the figures of merit are in close agreement with those obtained by the use of the exact $\mathscr{L}$ transformation. Therefore, the Tustin transformation is a valid design tool when the dominant zeros and poles of $[C(s)/R(s)]_M$ lie in the acceptable region of Fig. 15.16.

Table 15.4 summarizes the time-response characteristics for a unit-step forcing function of (1) the continuous-time system of Fig. 15.18 for the two cases of $G(s) = G_x(s)$ and $G(s) = G_A(s)G_x(s)$ with the continuous-time representation of the sampler and ZOH included, and (2) the sampled-data system of Fig. 15.17 based upon the exact and Tustin expressions for $C(z)$.

Steady-State Error Coefficients

The steady-state error coefficients have the same meaning and importance for sampled-data systems as for continuous-time systems; i.e., how well the system output can follow a given type of input forcing function in the steady state. *The following definitions, given without proof,[1] of the error coefficients are independent of the system type. They apply to any system type and are defined for specific forms of the input. These error coefficients are useful only for stable unity-feedback systems.*

Step input, $R(z) = R_0 z/(z - 1)$.
The step error coefficient K_p is defined as

$$K_p \equiv \frac{c^*(\infty)}{e^*(\infty)} = \lim_{z \to 1} G(z) \tag{15.74}$$

which applies only for a step input, $r(t) = R_0 u_{-1}(t)$.

Ramp input, $R(z) = R_1 T z/(z - 1)^2$.
The ramp error coefficient K_v is defined as

$$K_v \equiv \frac{\text{steady-state value of first derivative of output}}{e^*(\infty)}$$

$$= \frac{1}{T} \lim_{z \to 1} \left[\frac{z - 1}{z} G(z) \right] \tag{15.75}$$

which applies only for a ramp input, $r(t) = R_1 t u_{-1}(t)$.

Parabolic input, $R(z) = R_2 T^2 z(z + 1)/2(z - 1)^3$.
The parabolic error coefficient K_a is defined as

$$K_a \equiv \frac{\text{steady-state value of second derivative of output}}{e^*(\infty)}$$

$$= \frac{1}{T^2} \lim_{z \to 1} \left[\frac{(z - 1)^2}{z^2} G(z) \right] \tag{15.76}$$

which applies only for a parabolic input, $r(t) = R_2 t^2 u_{-1}(t)/2$.

TABLE 15.4
Time-response characteristics of the closed-loop system

Case		M_p	t_p, s	t_s, s	K_1
		Continuous-time system			
1	$G(s) = G_x(s)$	1.0335	6.3	8.43	0.5005
2	$G_M(s) = G_A(s)G_x(s)$	1.04342	6.48	8.69	0.4767
		Sampled-data system			
3	$C(z)$	1.043	6.45	8.65	0.4765
4	$[C(z)]_T$	1.043	6.45	8.6	0.4765

The ramp error coefficients for Cases 2 to 4 are obtained by applying Eq. (15.75). Table 15.4 reveals that

1. In converting a continuous-time system (Case 1) into a sampled-data system (Case 2) the time-response characteristics are degraded.
2. The time-response characteristics of the sampled-data system (Case 3), using the value of gain obtained from the continuous-time model, agree favorably with those of the continuous-time model of the discrete system. As may be expected, there is some variation in the values obtained when utilizing the exact $C(z)$ and $[C(z)]_T$ (Case 4).

The results of this section demonstrate that when the Tustin approximation is valid, the PCT approximation of the sampled-data system is a practical design method. When a cascade compensator $D(s)$ (see Fig. 15.18e) is designed to yield the desired performance specifications in the pseudo-continuous-time domain, the discrete cascade compensator $D(z)$ (see Fig. 15.18d where the controller does not include a ZOH) is accurately obtained by using the Tustin transformation.

15.13 DESIGN OF DIGITAL CONTROL SYSTEMS[1]

The previous sections present the DIR and DIG techniques for analyzing and achieving the desired performance of the basic system. Generally, gain adjustment alone is not sufficient to achieve the desired system performance specifications; i.e., the figures of merit (FOM) (see Secs. 3.9 and 3.10). Thus, a cascade and/or feedback compensator (controller) can be used to achieve the design objectives. This chapter discusses the design objective of satisfying the specified values of the conventional control theory FOM by using cascade compensators. The reader is referred to Ref. 1 for the design of feedback compensators.

The analysis of a system's performance may be carried out in the time or the frequency domain. If the performance of the basic system is unsatisfactory, then, based on the results of this analysis, a compensator can be designed. Two approaches may be used to design the digital cascade compensator of Fig. 15.20—the DIG and DIR techniques. *Both techniques rely heavily on a trial-and-error approach that*

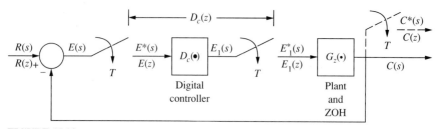

FIGURE 15.20
A compensated digital control system.

requires a firm understanding of the fundamentals of compensation. To facilitate this trial-and-error approach, a computer-aided-design (CAD) package (see App. B), such as MATLAB, ICECAP-PC, and TOTAL-PC, is used. The main advantage of the DIR design is that the performance specification can be met with less stringent requirements on the controller parameters.

In the DIG method, the controller is first designed in the s domain (see Ref. 1 for the w domain approach), and then the bilinear Tustin transformation is employed to transform the controller into its equivalent z-domain discrete controller. The advantage of this technique is that tried and proven continuous-time domain methods are used for designing an acceptable $D_c(s)$ or $D_c(w)$ controller. However, if T is not small enough, the poles and zeros of the controller lie outside the good bilinear transformation region, which may result in not achieving the desired performance.

After $D_c(z)$ is included in the system model, as shown in Fig. 15.20, a system analysis is again performed. If the modified system does not meet the desired specifications, the controller is modified accordingly. To illustrate this design process, the basic system of Sec. 15.12 is used for the design of lead, lag, and lag-lead or lead-lag cascade compensators. The design is done in both the s and z domains.

15.14   DIRECT (DIR) DESIGN TECHNIQUE

Considering the DIR technique for cascade compensation, the standard lead and lag s domain transfer function is transformed to the z-domain where the specific design is developed. The simple lead ($\alpha < 1$) and lag ($\alpha > 1$) compensators in the s domain have the form

$$D_c(s) = \frac{K_{sc}(s - z_s)}{s - p_s} \tag{15.77}$$

where $p_s = z_s/\alpha$. These s-plane zeros and poles are transformed into the z-domain zeros and poles as follows:

$$z_z = e^{sT}\big|_{s=z_s} = e^{z_s T} \tag{15.78}$$

$$p_z = e^{sT}\big|_{s=p_s} = e^{p_s T} = e^{z_s T/\alpha} \tag{15.79}$$

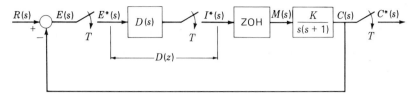

FIGURE 15.21
Sampled-data control system containing a cascade digital controller $D(z)$.

Thus, the corresponding first-order z-domain compensator (digital filter) is

$$D_c(z) = \frac{K_{zc}(z - z_z)}{z - p_z} = \frac{K_{zc}(z - z_z)}{z - z_z/\beta} \tag{15.80}$$

where $p_z \equiv z_z/\beta$.

A relationship between α and β is obtained by taking the natural log of Eqs. (15.78) and (15.79) as follows:

$$z_s T = \ln z_z \tag{15.81}$$

$$\frac{z_s T}{\alpha} = \ln p_z \tag{15.82}$$

Taking the ratio of these equations and rearranging yields $\alpha \ln p_z = \ln z_z$. Thus,

$$p_z^\alpha = z_z \tag{15.83}$$

and

$$\beta = \frac{z_z}{p_z} = p_z^{\alpha - 1} \tag{15.84}$$

For a lead compensator, $\alpha < 1$ and $p_z < 1$. *Therefore, for a lead digital filter, $\beta > 1$ from Eq. (15.84). For a lag digital filter, $\beta < 1$.* (Note that the magnitude condition on β is just the opposite of that on α.)

In the z-domain $p_z > z_z > 0$ for a lag controller and $z_z > p_z$ for a lead controller. Thus, the standard lag-lead controller is of the form

$$D_c(z) = K_c \left[\frac{z - z_1}{z - p_1} \right] \left[\frac{z - z_2}{z - p_2} \right] \tag{15.85}$$

where $p_1 > z_1 > 0$ (the lag portion) and $z_2 > p_2 > 0$ (the lead portion). The cascade controller design rules given in Chap. 10 are used for the design of these z-domain controllers.

15.15 LEAD CONTROLLER (COMPENSATOR): DIR DESIGN METHOD

A damping ratio of $\zeta = 0.707$ and a settling time of $T_s \approx 4$ s are desired for the system of Fig. 15.11. The settling time requires that the roots be located inside the circle of radius $e^{\sigma T}$. Because $T_s = 4/|\sigma|$, the requirement is that $\sigma = -1$. The circle of radius $e^{-1} = 0.368$, for $T = 1$ s, is drawn in Fig. 15.12. It is evident that the root locus does not cross this circle and therefore the settling time cannot be achieved.

A digital lead cascade controller $D(z)$ can be added to the system to improve the performance, and is shown in Fig. 15.21. The digital controller must be selected to bring the root locus within the circle of radius $e^{\sigma T} = 0.368$. Using the cancellation rule for a lead controller, where the zero z_z of $D_c(z)$ cancels the pole of $G(z)$ at $z = 0.368$ [see Eq. (15.42)], a possible controller is:

$$D(z) = \frac{I(z)}{E(z)} = \frac{z - 0.368}{z + 0.230} \tag{15.86}$$

Thus, the forward transfer function becomes:

$$D(z)G(z) = \frac{0.368K(z + 0.717)}{(z - 1)(z + 0.230)} \tag{15.87}$$

The new root locus is shown in Fig. 15.22. Because a portion of the root locus is within the circle of radius equal to 0.368, achieving the desired settling time depends on the damping ratio. For $\zeta = 0.707$ the closed-loop sampled control ratio is

$$\frac{C(z)}{R(z)} = \frac{0.473(z + 0.717)}{(z - 0.149 + j0.295)(z - 0.149 - j0.295)}$$

$$= \frac{0.473(z + 0.717)}{z^2 + 0.298z + 0.109} \tag{15.88}$$

which yields $\sigma = -1.1$. Based on continuous-time criteria, the transient settling time is $T_s = 4/|\sigma| = 3.64$ s. The use of the cascade digital controller has reduced the settling time from approximately 10.1 to 3.64 s and meets the specifications. The responses of both the compensated and uncompensated systems with a unit-step input are shown in Fig. 15.23. The characteristic values for the basic system are $M_p = 1.037, t_p = 8$ s, and $t_s = 10$–11 s, and $K_1 = 0.325$ s^{-1}. For the compensated system the values are $M_p = 1.044, t_p = 3$ s, $t_s = 4$ s, and $K_1 = 0.66$ s^{-1}. Thus, the digital controller has reduced the settling time, as desired.

The digital controller given by Eq. (15.86) is expressed in the discrete-time domain by the difference equation

$$i(kT) = e(kT) - 0.368e[(k - 1)T] - 0.230i[(k - 1)T] \tag{15.89}$$

This equation is very easily implemented on a digital computer.

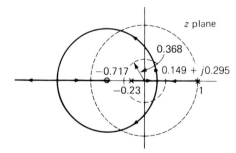

FIGURE 15.22
Root locus for the system of Fig. 15.21, where $D(z)G(z)$ is given by Eq. (15.87).

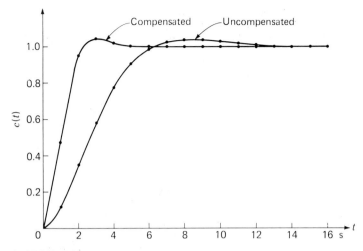

FIGURE 15.23
Step input responses for the uncompensated system of Eq. (15.43) and the compensated system of Eq. (15.88).

15.16 LAG AND LAG-LEAD CONTROLLERS: DIR DESIGN METHOD

A cascade lag controller can be used to increase the gain. The selection of $p_z = 0.99$ and $z_z = 0.98$, for the lag controller of Eq. (15.80), introduces an angular contribution of less than 5° to the dominant closed-loop pole of the basic system. The resulting control ratio for the compensated system is

$$\frac{C(z)}{R(z)} = \frac{0.12859(z + 0.717)(z - 0.98)}{(z - 0.625 \pm j0.2713)(z - 0.9794)} \tag{15.90}$$

The resulting figures of merit are: $M_p = 1.081$, $t_p = 8$ s, $t_s = 23^+$ s, and $K_1 = 0.6986$ s^{-1}. Thus, the gain K_1 is increased, but the penalty is that the peak overshoot and the settling time are increased. When the zeros and poles of the lag and lead controllers are used as the zeros and poles of the lag-lead controller of Eq. (15.85), the resulting control ratio for the lag-lead compensated system for $\zeta = 0.7071$ is

$$\frac{C(z)}{R(z)} = \frac{0.19313(z + 0.717)(z - 0.98)}{(z - 0.52368 \pm j0.31106)(z - 0.9795)} \tag{15.91}$$

The resulting figures of merit are $M_p = 1.0786$, $t_p = 6$ s, $t_s = 10^+$ s, and $K_1 = 0.8613$ s^{-1}. The results obtained for all three types of controllers are similar to those obtained for analog compensated systems.

15.17 DIGITIZATION (DIG) DESIGN TECHNIQUE

Figure 15.24 represents the trial-and-error design philosophy in applying the DIG technique in the design of a cascade compensator via the analysis of the open-loop

FIGURE 15.24
Digitization (DIG) design philosophy for Fig. 15.20.

system. An important tool for any design technique is the utilization of a CAD package like MATLAB (see App. B). This presentation of the design procedure is intended to establish firmly the manner in which the DIG technique is applied to the design of the compensator $D_s(\cdot) = (\text{gain})D'_c(\cdot)$. If path A does not satisfy the specifications set forth by the digital control system of Fig. 15.21, then path B is used to try to determine a satisfactory value of K_{zc}. That is:

1. Follow path A if the dominant poles and zeros of $C(\cdot)/R(\cdot)$ lie in the shaded area of Fig. 15.16 (Tustin approximation is good!), that is, when the degree of warping is deemed not to affect negatively the achievement of the desired design results. If the desired results are not achieved, try path B.

2. Follow path B when severe warping exists.
   The DIG design procedure is as follows:

Step 1. Convert the basic sampled-data control system to an s-domain PCT control system.

Step 2. Use the standard s-domain cascade compensation procedures of Chap. 10 to design the compensator. Thus, by means of a root-locus analysis, determine $D_c(s) = K_{sc}D'_c(s)$.

Step 3. Obtain the control ratio of the compensated system and the corresponding time response for the desired forcing function. If the desired FOM are not achieved, repeat step 2 by selecting a different value of ζ, σ, ω_d, etc.

Step 4. When an acceptable $D_c(s)$ has been achieved, transform the compensator, via the Tustin transformation, into the z domain. *Note*: In order for real poles (or zeros) $[p_i$ (or $z_i)]$ that lie on the negative real axis of the s plane to lie between 0 and 1 in the z plane, *it is necessary that $|p_i| \leq 2/T$ (or $|z_i| \leq 2/T$) be satisfied when utilizing the Tustin transformation of Eq. (15.47).* This requirement should be kept in mind when the controller is to be implemented by an equivalent passive analog compensator (RC components). The input to this analog compensator must be the output of a zero-order hold (ZOH) unit.

Step 5. Obtain the z-domain control ratio of the compensated system and the corresponding time response for the desired forcing function. If the FOM for the digital control system have been achieved via path A or B, then the design of the controller is complete. If not, return to Step 2 and repeat the steps with a new controller design or proceed to the DIR technique.

   It is left to the reader to repeat the design examples of Secs. 15.15 and 15.16 by applying the DIG technique, and to compare the results obtained by the DIR and the DIG designs.

15.18 SUMMARY

The availability of inexpensive microprocessors has led to their extensive use in digital control systems. Although there is a degradation of performance by the introduction of sampling, this is more than overcome by the greater range of digital controllers that can be achieved. The flexibility that is provided by microprocessors in synthesizing digital controllers yields greatly improved closed-loop performance.

Two approaches for analyzing a digital control system are presented: the DIG (digitization) and DIR (direct) techniques. The former requires the use of the Padé approximation and the Tustin transformation for an initial design in the s plane (PCT control-system configuration). If the Tustin approximation criteria of Secs. 15.9 and 15.10 are satisfied, the correlation between the two planes is very good. If the approximations are not valid, the analysis and design of the digital control system should be done by the DIR technique.

REFERENCES

1. Houpis, C. H., and G. B. Lamont: *Digital Control Systems, Theory, Hardware, Software*, 2nd ed., McGraw-Hill, New York, 1992.
2. Franklin, G. F., Powell, J. D., and M. L. Workman: *Digital Control of Dynamic Systems*, 2nd ed., Addison-Wesley, Reading, Mass., 1990.
3. Philips, C. L., and H. T. Nagle: *Digital Control System Analysis and Design*, Prentice-Hall, Englewood Cliffs, N. J., 1984.
4. Ogata, K.: *Discrete-Time Computer Control Systems*, Prentice-Hall, Englewood Cliffs, N. J., 1987.
5. Tustin, A.: "A Method of Analyzing the Behavior of Linear Systems in Terms of Time Series," *JIEE* (Lond.) , vol. 94, IIA, 1947.
6. Houpis, C. H., and G. B. Lamont: *Digital Control Systems, Theory, Hardware, Software*, 1st ed., McGraw-Hill, New York, 1988.

CHAPTER 16

ENTIRE EIGENSTRUCTURE ASSIGNMENT FOR MULTIVARIABLE SYSTEMS

16.1 INTRODUCTION

The foundations for control-system synthesis presented in this text apply to multi-variable systems, i.e., systems with multiple-inputs and multiple-outputs (MIMO). While the emphasis of the root-locus, frequency-response, and state-variable methods in the preceding chapters is on single-input single-output (SISO) systems, the synthesis methods developed in Chapters 16 and 17 are based upon the time-domain representation of MIMO systems in terms of vector state and output equations. Matrix properties are therefore used extensively. The principal objective of the state-feedback methods of Chapter 13 is the assignment of the closed-loop eigenvalues for SISO systems. The assignment of the closed-loop eigenvalue spectrum for MIMO systems is covered in Ref. 1, Chapters 18 and 19, where the state equation is first transformed into generalized control canonical form. In this representation the plant matrix is block diagonal, where each block is in companion form, and readily leads to the determination of a state-feedback matrix that assigns the desired closed-loop eigenvalues.

In this chapter the influence on the time response of the eigenvectors that are associated with the eigenvalues is first presented in Sec. 16.2. In the MIMO case there are an infinite number of feedback matrices that can assign a specified set of closed-loop eigenvalues. With each feedback matrix there results a different set of

527

associated eigenvectors. Because the eigenvectors also affect the time response of the system (see Sec. 16.2), it is important to assign both the eigenvalues and the associated eigenvectors. Thus, the principles that are the basis of *entire eigenstructure assignment,* consisting of both the eigenvalues and the associated eigenvectors, are presented in this chapter.

The power and versatility of entire eigenstructure assignment by state feedback is demonstrated in this chapter for the synthesis of tracking systems. Tracking systems are required to track a command input vector containing polynomial signals. In these tracking systems, specified outputs y_i are required to follow corresponding inputs r_i. This objective is achieved in a straightforward manner for SISO systems by adding integrators to increase the system type. Because each input in a MIMO system is required to control only a corresponding output without affecting the other system outputs, a vector integrator is required. The method of entire eigenstructure assignment is illustrated by the design of a tracker that follows piecewise-constant input commands. When a system has uncontrollable modes, the eigenvalues associated with such modes cannot be changed by state feedback. However, the eigenvectors associated with such uncontrollable modes can be changed; this property is demonstrated by an example.

The analysis and synthesis involved in the problems associated with this chapter, like the previous chapters, rely on the use of the CAD packages such as MATLAB or ICECAP-PC (see Sec. 10.6 and App. B). The use of such programs expedites the achievement of a solution and eliminates the tedious hand calculations.

16.2 EFFECT OF EIGENSTRUCTURE ON TIME RESPONSE[2]

For the open-loop system represented by the state and output equations

$$\dot{\mathbf{x}} = \mathbf{A}\mathbf{x} + \mathbf{B}\mathbf{u} \qquad \mathbf{y} = \mathbf{C}\mathbf{x} \tag{16.1}$$

the eigenvalue spectrum $\boldsymbol{\sigma}(\mathbf{A})$ is the set of roots of the characteristic equation which is formed from

$$Q(\lambda) = |\lambda\mathbf{I} - \mathbf{A}| = \lambda^n + a_{n-1}\lambda^{n-1} + \cdots + a_1\lambda + a_0 = 0 \tag{16.2}$$

In Sec. 5.10 it is shown that when all the eigenvalues of $\mathbf{A}$ are distinct, the *modal matrix* $\mathbf{T}$ can be determined such that matrix $\mathbf{A}$ is transformed to a diagonal matrix $\boldsymbol{\Lambda}$:

$$\mathbf{T}^{-1}\mathbf{A}\mathbf{T} = \boldsymbol{\Lambda} \tag{16.3}$$

The matrix $\boldsymbol{\Lambda}$ is a diagonal matrix in which the eigenvalues appear on the diagonal [see Eq. (5.82)]. The eigenvectors $\mathbf{v}_i$ are the columns of $\mathbf{T}$ and satisfy the equation

$$[\lambda_i\mathbf{I} - \mathbf{A}]\mathbf{v}_i = \mathbf{0} \tag{16.4}$$

The rows of $\mathbf{T}^{-1}$ are the row vectors $\mathbf{w}_i^T$, which are called the reciprocal or left eigenvectors and satisfy the equation

$$\mathbf{w}_i^T [\lambda_i \mathbf{I} - \mathbf{A}] = \mathbf{0} \tag{16.5}$$

Thus
$$\mathbf{T} = [\mathbf{v}_1 \quad \mathbf{v}_2 \quad \cdots \quad \mathbf{v}_n] \qquad \mathbf{T}^{-1} = \begin{bmatrix} \mathbf{w}_1^T \\ \mathbf{w}_2^T \\ \vdots \\ \mathbf{w}_n^T \end{bmatrix} \tag{16.6}$$

It is evident from Eqs. (16.6), because the matrix product $\mathbf{TT}^{-1} = \mathbf{I}$, that the sets of eigenvectors $\mathbf{v}_i$ and reciprocal eigenvectors $\mathbf{w}_i^T$ are orthogonal, i.e.,

$$\mathbf{w}_i^T \mathbf{v}_j = \begin{cases} 1 & \text{for } i = j \\ 0 & \text{for } i \neq j \end{cases} \tag{16.7}$$

Solving for $\mathbf{A}$ in Eq. (16.3) yields $\mathbf{A} = \mathbf{T}\mathbf{\Lambda}\mathbf{T}^{-1}$, which can be substituted into the solution of the state equation (see Sec. 3.15, Eq. (3.103))

$$\mathbf{x}(t) = e^{\mathbf{A}t}\mathbf{x}(0) + \int_0^t e^{\mathbf{A}\tau}\mathbf{B}\mathbf{u}(t - \tau)d\tau \tag{16.8}$$

Thus, the state transition matrix $e^{\mathbf{A}t}$ can be expressed in terms of the eigenvectors and reciprocal eigenvectors. Using the representation of $e^{\mathbf{A}t}$ given in Eq. (5.115) yields

$$e^{\mathbf{A}t} = \mathbf{T}e^{\mathbf{\Lambda}t}\mathbf{T}^{-1} \tag{16.9}$$

The matrix $e^{\mathbf{\Lambda}t}$ has the diagonal form shown in Eq. (5.116). Therefore, the state transition matrix can be written as

$$e^{\mathbf{A}t} = [\mathbf{v}_1 \quad \mathbf{v}_2 \quad \cdots \quad \mathbf{v}_n] \begin{bmatrix} e^{\lambda_1 t} & & & 0 \\ & e^{\lambda_2 t} & & \\ & & \ddots & \\ 0 & & & e^{\lambda_2 t} \end{bmatrix} \begin{bmatrix} \mathbf{w}_1^T \\ \mathbf{w}_2^T \\ \vdots \\ \mathbf{w}_n^T \end{bmatrix}$$

$$= \mathbf{v}_1 e^{\lambda_1 t}\mathbf{w}_1^T + \mathbf{v}_2 e^{\lambda_2 t}\mathbf{w}_2^T + \cdots + \mathbf{v}_n e^{\lambda_n t}\mathbf{w}_n^T = \sum_{i=1}^n \mathbf{v}_i e^{\lambda_i t}\mathbf{w}_i^T \tag{16.10}$$

When Eqs. (16.1), (16.8), and (16.10) are used, the output equation, when the dimension of the input $\mathbf{u}$ is m, is given by

$$\mathbf{y}(t) = \sum_{i=1}^n (\mathbf{C}\mathbf{v}_i)e^{\lambda_i t}\mathbf{w}_i^T \mathbf{x}(0) + \sum_{j=1}^m \sum_{i=1}^n (\mathbf{C}\mathbf{v}_i)(\mathbf{w}_i^T \mathbf{b}_j)\int_0^t e^{\lambda_i \tau}u_j(t - \tau)d\tau \tag{16.11}$$

The transient response of the system is therefore a linear combination of n functions of the form

$$\mathbf{v}_i e^{\lambda_i t} \quad i = 1, 2, \ldots, n \tag{16.12}$$

which describes the *dynamical modes*[9] of the system. (This is an extension of the description in Sec. 3.5, which defines a mode as $e^{\lambda_i t}$.) The *shape* of a mode is therefore determined by the eigenvector v_i, and its time-domain characteristic is determined by the eigenvalue λ_i. Equation (16.11) shows that the entire eigenstructure determines the time response of the system; i.e., the eigenvalues λ_i, the associated eigenvectors v_i, and the reciprocal eigenvectors w_i all contribute to the time response. *Therefore, the design of a state-feedback control law should be based upon the entire eigenstructure in order to be most effective in determining the time response.* The terms $c_k^T v_i$, $w_i^T x(0)$, and $w_i^T b_j$, which are present in Eq. (16.11), are scalars and determine the magnitude of the modal responses $e^{\lambda_i t}$. The ability to select v_i and w_i^T provides the potential for adjusting the magnitude of each mode that appears in each of the outputs. The selection of the eigenvectors as well as the eigenvalues is a logical extension of the methods previously available, which assigned only the eigenvalues (or poles). The constraints on the selection of possible closed-loop eigenvectors are described in the next section.

16.3   ENTIRE EIGENSTRUCTURE ASSIGNMENT[3,4-7]

A linear time-invariant MIMO open-loop system is represented in the continuous-time domain by the matrix state equation

$$\dot{x} = Ax + Bu \tag{16.13}$$

The control law for applying state-variable feedback is

$$u = r + Kx \tag{16.14}$$

and leads to the closed-loop equation

$$\dot{x} = [A + BK]x + Br = A_{cl}x + Br \tag{16.15}$$

The purpose in applying state feedback is to assign both a closed-loop eigenvalue spectrum

$$\sigma(A + BK) = \{\lambda_1, \lambda_2, \ldots, \lambda_n\} \tag{16.16}$$

and an associated set of eigenvectors

$$v(A + BK) = \{v_1, v_2, \ldots, v_n\} \tag{16.17}$$

which are selected in order to achieve the desired time-response characteristics. The closed-loop eigenvalues and eigenvectors are related by the equation

$$[A + BK]v_i = \lambda_i v_i \tag{16.18}$$

This equation can be put in the form

$$[A - \lambda_i I \quad B]\begin{bmatrix} v_i \\ q_i \end{bmatrix} = 0 \quad \text{for } i = 1, 2, \ldots, n \tag{16.19}$$

where $\mathbf{v}_i$ is the eigenvector and

$$\mathbf{q}_i = \mathbf{K}\mathbf{v}_i \tag{16.20}$$

In order to satisfy Eq. (16.19), the vector $[\, \mathbf{v}_i^T \quad \mathbf{q}_i^T \,]^T$ must lie in the *kernel* or *null space* of the matrix

$$\mathbf{S}(\lambda_i) = [\mathbf{A} - \lambda_i \mathbf{I} \quad \mathbf{B}] \qquad \text{for } i = 1, 2, \ldots, n \tag{16.21}$$

The notation $\ker \mathbf{S}(\lambda_i)$ is used to define the null space that contains all the vectors $[\mathbf{v}_i^T \quad \mathbf{q}_i^T]^T$ for which Eq. (16.19) is satisfied. Equation (16.20) can be used to form the matrix equality

$$[\mathbf{q}_1 \quad \mathbf{q}_2 \quad \cdots \quad \mathbf{q}_n] = [\mathbf{K}\mathbf{v}_1 \quad \mathbf{K}\mathbf{v}_2 \cdots \mathbf{K}\mathbf{v}_n] \tag{16.22}$$

Because $\mathbf{K}$ can be factored from the right-hand matrix of Eq. (16.22),

$$\mathbf{K} = [\mathbf{q}_1 \quad \mathbf{q}_2 \quad \cdots \quad \mathbf{q}_n][\mathbf{v}_1 \quad \mathbf{v}_2 \quad \cdots \quad \mathbf{v}_n]^{-1} = \mathbf{Q}\mathbf{V}^{-1} \tag{16.23}$$

Therefore, if the desired eigenvalue spectrum of Eq. (16.16) is specified and the associated eigenvectors are selected to satisfy Eq. (16.19), then Eq. (16.23) specifies the required state-feedback matrix $\mathbf{K}$. The eigenvalues λ_i are real or appear in complex-conjugate sets, and the associated vectors $\mathbf{v}_i$ and $\mathbf{q}_i$ are also real or appear in complex-conjugate sets. The sets of eigenvalues and eigenvectors are therefore described as being *self-conjugate*. Thus, $\mathbf{K}$ is a real matrix.

The ker $\mathbf{S}(\lambda_i)$ imposes constraints on the selection of the eigenvector $\mathbf{v}_i$ that is associated with the assigned eigenvalue λ_i. The ker $\mathbf{S}(\lambda_i)$ identifies a specific subspace, and the selected eigenvector $\mathbf{v}_i$ must be located within this subspace. In addition, the selected eigenvectors must be linearly independent (see Sec. 4.16) so that the inverse matrix $\mathbf{V}^{-1}$ in Eq. (16.22) exists. Also, in order that $\mathbf{K}$ be a real matrix, the selected set of eigenvectors must be self-conjugate; i.e., they must be complex conjugates for pairs of eigenvalues that are complex conjugates. The procedure is illustrated by the examples in the following sections.

16.4 EXAMPLES OF ENTIRE EIGENSTRUCTURE ASSIGNMENT FOR REGULATORS

This section contains two examples illustrating the synthesis of regulators by the method of entire eigenstructure assignment. When the system is a regulator, the command input vector is $\mathbf{r} = \mathbf{0}$. In Example 1 the closed-loop eigenvalues are real, while in Example 2 some of the eigenvalues are complex. The eigenvector subspace corresponding to a real eigenvalue can be obtained by either of the methods illustrated in these examples. For a complex eigenvalue, the corresponding subspace is more readily obtained by the method of Example 2.

In these examples, ker $\mathbf{S}(\lambda_i)$ is evaluated for each λ_i. However, it may be simpler to obtain ker $\mathbf{S}(\lambda)$ without inserting the specific values of λ. This minimizes the work because the matrix column operations need be performed only once. The desired values of λ to be assigned can then be inserted in the expression for ker $\mathbf{S}(\lambda)$. (The answers to selected problems are given in this form.)

Example 1. Real closed-loop eigenvalues. A system whose plant, control, and output matrices are

$$\mathbf{A} = \begin{bmatrix} 0 & 1 & 0 \\ 0 & 0 & 1 \\ 4 & 4 & -1 \end{bmatrix} \quad \mathbf{B} = \begin{bmatrix} 1 & 0 \\ 0 & 0 \\ 0 & 1 \end{bmatrix} \quad \mathbf{C} = \begin{bmatrix} 1 & 0 & 0 \\ 0 & 1 & 1 \end{bmatrix} \quad (16.24)$$

has the open-loop eigenvalue spectrum $\sigma(\mathbf{A}) = \{-1, -2, 2\}$. Because the open-loop system is unstable, the state feedback control law, $\mathbf{u} = r + \mathbf{Kx}$, is used to assign the closed-loop eigenvalue spectrum

$$\sigma(\mathbf{A} + \mathbf{BK}) = \{-2, -3, -4\} \quad (16.25)$$

This is possible because the use of Eq. (13.10) shows that $(\mathbf{A}, \mathbf{B})$ is a controllable pair. The required steps for determining the spaces in which the eigenvectors must be located are

1. The matrix $\mathbf{S}(\lambda_i)$ is formed in accordance with Eq. (16.21)

$$\mathbf{S}(\lambda_i) = [\mathbf{A} - \lambda_i \mathbf{I} \quad \mathbf{B}] = \begin{bmatrix} -\lambda_i & 1 & 0 & 1 & 0 \\ 0 & -\lambda_i & 1 & 0 & 0 \\ 4 & 4 & -\lambda_i - 1 & 0 & 1 \end{bmatrix} \quad (16.26)$$

2. The matrix is put into Hermite normal form (HNF) by means of elementary row operations (see Sec. 4.16).[14]

3. This matrix is then augmented with sufficient rows of zeros to form a square matrix.

4. The rows are interchanged, if necessary, so that the leading ones in each row are on the principal diagonal.

5. Wherever a 0 appears on the principal diagonal,[13] it is replaced with -1 and the matrix is designated as $\hat{\mathbf{S}}(\lambda_i)$.

6. The columns of this matrix $\hat{\mathbf{S}}(\lambda_i)$ that contain -1 on the principal diagonal are vectors that span the null space of $\mathbf{S}(\lambda_i)$. They can be multiplied by arbitrary constants. There are m such columns, where m is the dimension of the input vector $\mathbf{u}$.

7. As verification that the correct ker $\mathbf{S}(\lambda_i)$ has been achieved, check to see that

$$[\mathbf{A} - \lambda_i \mathbf{I} \quad \mathbf{B}] \cdot \ker \mathbf{S}(\lambda_i) = \mathbf{0}.$$

8. Select the desired vector $[\mathbf{v}_i^T \quad \mathbf{q}_i^T]^T$ from ker $\mathbf{S}(\lambda_i)$.

Inserting $\lambda_1 = -2$ in Eq. (16.26) and following these steps yields

$$\hat{\mathbf{S}}(-2) = \begin{bmatrix} 1 & 0 & -\frac{1}{4} & 0 & \frac{1}{4} \\ 0 & 1 & \frac{1}{2} & 0 & 0 \\ 0 & 0 & -1 & 0 & 0 \\ 0 & 0 & 0 & 1 & -\frac{1}{2} \\ 0 & 0 & 0 & 0 & -1 \end{bmatrix} \quad (16.27)$$

Because there are two columns in this matrix with -1 on the principal diagonal, the null space consists of all vectors formed from combinations of these columns. The two columns are multiplied by -4 and 4, respectively, in order to eliminate fractions and to make the first element positive. Thus, the null space is indicated by the notation

$$
\ker \mathbf{S}(-2) = \text{span} \left\{ \begin{bmatrix} 1 \\ -2 \\ 4 \\ 0 \\ 0 \end{bmatrix}, \begin{bmatrix} 1 \\ 0 \\ 0 \\ -2 \\ -4 \end{bmatrix} \right\} = \text{span} \{\mathbf{s}_1(-2), \mathbf{s}_2(-2)\} \tag{16.28}
$$

Similarly, for $\lambda_2 = -3$

$$
\hat{\mathbf{S}}(-3) = \begin{bmatrix} 1 & 0 & 0 & \frac{1}{5} & \frac{1}{10} \\ 0 & 1 & 0 & \frac{2}{5} & -\frac{3}{10} \\ 0 & 0 & 1 & -\frac{6}{5} & \frac{9}{10} \\ 0 & 0 & 0 & -1 & 0 \\ 0 & 0 & 0 & 0 & -1 \end{bmatrix} \tag{16.29}
$$

$$
\ker \mathbf{S}(-3) = \text{span} \left\{ \begin{bmatrix} 1 \\ 2 \\ -6 \\ -5 \\ 0 \end{bmatrix}, \begin{bmatrix} 1 \\ -3 \\ 9 \\ 0 \\ -10 \end{bmatrix} \right\} = \text{span} \{\mathbf{s}_1(-3), \mathbf{s}_2(-3)\} \tag{16.30}
$$

Also, for $\lambda_3 = -4$

$$
\hat{\mathbf{S}}(-4) = \begin{bmatrix} 1 & 0 & 0 & \frac{2}{9} & \frac{1}{36} \\ 0 & 1 & 0 & \frac{1}{9} & -\frac{4}{36} \\ 0 & 0 & 1 & -\frac{4}{9} & \frac{16}{36} \\ 0 & 0 & 0 & -1 & 0 \\ 0 & 0 & 0 & 0 & -1 \end{bmatrix} \tag{16.31}
$$

$$
\ker \mathbf{S}(-4) = \text{span} \left\{ \begin{bmatrix} 2 \\ 1 \\ -4 \\ -9 \\ 0 \end{bmatrix}, \begin{bmatrix} 1 \\ -4 \\ 16 \\ 0 \\ -36 \end{bmatrix} \right\} = \text{span} \{\mathbf{s}_1(-4), \mathbf{s}_2(-4)\} \tag{16.32}
$$

The null-space vectors are selected by taking any desired combination of the two vectors in $\ker \mathbf{S}(\lambda_i)$, using the form

$$
\begin{bmatrix} \mathbf{v}_i \\ \mathbf{q}_i \end{bmatrix} = \alpha_1 \mathbf{s}_1(\lambda_i) + \alpha_2 \mathbf{s}_2(\lambda_i) + \cdots + \alpha_m \mathbf{s}_m(\lambda_i) \tag{16.33}
$$

where $\alpha_1, \alpha_2, \ldots, \alpha_m$ are arbitrary constants. For this example, the selected null-space vectors are

$$
\lambda_1 = -2: \qquad \begin{bmatrix} \mathbf{v}_1 \\ \mathbf{q}_1 \end{bmatrix} = \begin{bmatrix} 0 \\ 1 \\ -2 \\ --- \\ -1 \\ -2 \end{bmatrix} \qquad \alpha_1 = \tfrac{1}{2}, \quad \alpha_2 = \tfrac{1}{2}
$$

$$\lambda_2 = -3: \qquad \begin{bmatrix} \mathbf{v}_2 \\ \hline \mathbf{q}_2 \end{bmatrix} = \begin{bmatrix} 1 \\ 0 \\ 0 \\ \hline -3 \\ -4 \end{bmatrix} \qquad \alpha_1 = \tfrac{3}{5}, \quad \alpha_2 = \tfrac{2}{5} \tag{16.34}$$

$$\lambda_3 = -4: \qquad \begin{bmatrix} \mathbf{v}_3 \\ \hline \mathbf{q}_3 \end{bmatrix} = \begin{bmatrix} 0 \\ 1 \\ -4 \\ \hline -1 \\ -8 \end{bmatrix} \qquad \alpha_1 = \tfrac{1}{9}, \quad \alpha_2 = -\tfrac{2}{9}$$

The values of α_1 and α_2 used to form the vectors selected in Eq. (16.34) are shown in this equation. The vectors are also partitioned into the vector segments corresponding to $\mathbf{v}_i$ and $\mathbf{q}_i$, respectively, as shown in Eq. (16.19). This selection of eigenvectors produces state responses for which $x_1(t)$ contains only the mode e^{-3t}. Also, $x_2(t)$ and $x_3(t)$ contain only the modes e^{-2t} and e^{-4t}. The state-feedback matrix required to assign the closed-loop eigenvalue spectrum of Eq. (16.25) and the associated eigenvectors of Eq. (18.34) is obtained from Eq. (16.23) as

$$\mathbf{K} = \begin{bmatrix} -1 & -3 & -1 \\ -2 & -4 & 8 \end{bmatrix} \begin{bmatrix} 0 & 1 & 0 \\ 1 & 0 & 1 \\ -2 & 0 & -4 \end{bmatrix}^{-1}$$

$$= \begin{bmatrix} -1 & -3 & -1 \\ -2 & -4 & 8 \end{bmatrix} \begin{bmatrix} 0 & 2 & \tfrac{1}{2} \\ 1 & 0 & 0 \\ 0 & -1 & -\tfrac{1}{2} \end{bmatrix} = \begin{bmatrix} -3 & -1 & 0 \\ -4 & -12 & -5 \end{bmatrix} \tag{16.35}$$

The closed-loop matrix is

$$\mathbf{A}_{cl} = \mathbf{A} + \mathbf{BK} = \begin{bmatrix} -3 & 0 & 0 \\ 0 & 0 & 1 \\ 0 & -8 & -6 \end{bmatrix} \tag{16.36}$$

This closed-loop matrix has the assigned eigenvalue spectrum of Eq. (16.25) and the eigenvectors of Eq. (16.34), as required. It should be noted that one open-loop eigenvalue $\lambda_1 = -2$ has been retained in the closed-loop plant matrix, but its eigenvector has been changed from $[1 \quad -2 \quad 4]^T$ to $[0 \quad 1 \quad -2]^T$. Therefore, because the first element of this eigenvector is zero, the mode e^{-2t} does not appear in $x_1(t)$.

The state response of the closed-loop system computed for the initial condition $\mathbf{x}(0) = [1 \quad 1 \quad 2]^T$ and $\mathbf{r}(t) = 0$ is $\mathbf{x}(t) = \mathcal{L}^{-1}[s\mathbf{I} - \mathbf{A}_{cl}]^{-1}\mathbf{x}(0)$:

$$\begin{bmatrix} x_1(t) \\ x_2(t) \\ x_3(t) \end{bmatrix} = \begin{bmatrix} e^{-3t} & 0 & 0 \\ 0 & 2e^{-2t} - e^{-4t} & 0.5e^{-2t} - 0.5e^{-4t} \\ 0 & -4e^{-2t} + 4e^{-4t} & -e^{-2t} + 2e^{-4t} \end{bmatrix} \begin{bmatrix} 1 \\ 1 \\ 2 \end{bmatrix}$$

$$= \begin{bmatrix} e^{-3t} \\ 3e^{-2t} - 2e^{-4t} \\ -6e^{-2t} + 8e^{-4t} \end{bmatrix} \tag{16.37}$$

Note that the selection of the eigenvectors in Eq. (16.34) produces the result that only the mode e^{-3t} appears in $x_1(t)$. Also, only the modes e^{-2t} and e^{-4t} appear in $x_2(t)$ and $x_3(t)$. The same separation of the modes appears in the output response for this example obtained from $\mathbf{y} = \mathbf{Cx}$.

$$\begin{bmatrix} y_1(t) \\ y_2(t) \end{bmatrix} = \begin{bmatrix} e^{-3t} \\ -3e^{-2t} + 6e^{-4t} \end{bmatrix} \qquad (16.38)$$

Example 1 illustrates how particular modes can be included or excluded from particular state responses. This is analyzed by reference to Eq. (16.38) which relates the eigenvector values to the state responses.

$$\mathbf{v} = [\mathbf{v}_1 \quad \mathbf{v}_2 \quad \mathbf{v}_3] = \begin{matrix} & \mathbf{v}_1 & \mathbf{v}_2 & \mathbf{v}_3 \\ & \downarrow & \downarrow & \downarrow \\ & \begin{bmatrix} a & b & c \\ d & e & f \\ g & h & i \end{bmatrix} & \begin{matrix} \leftarrow x_1(t) \\ \leftarrow x_2(t) \\ \leftarrow x_3(t) \end{matrix} \\ & \uparrow & \uparrow & \uparrow \\ & e^{\lambda_1} & e^{\lambda_2} & e^{\lambda_3} \end{matrix} \qquad (16.39)$$

If the modes due to λ_1 and λ_3 are not to appear in $x_1(t)$, then the eigenvector element values must satisfy the requirement that $a = c = 0$ and $b \neq 0$. In this example, the mode due to λ_2 must not appear in $x_2(t)$ and $x_3(t)$, which requires that $e = h = 0$, while d, f, g, and i are not equal to zero. The above method shows that particular modes can be eliminated in $x_i(t)$, while maintaining the necessary condition that the matrix $\mathbf{V}$ must have full rank. This is required in order to compute the state-feedback matrix $\mathbf{K}$, as shown in Eq. (16.35). Equation (16.39) can be expanded for higher-order systems.

Example 2. Complex closed-loop eigenvalues.
For the system of Example 1 the desired closed-loop eigenvalue spectrum is changed to

$$\sigma(\mathbf{A}_{cl}) = \sigma(\mathbf{A} + \mathbf{BK}) = \{-2, -1 + j1, -1 - j1\} \qquad (16.40)$$

The purpose of this example is to determine the feedback matrix $\mathbf{K}$, the state response $\mathbf{x}(t)$ when $\sigma(\mathbf{A}_{cl})$ has complex eigenvalues, and a more general method of obtaining $\ker \mathbf{S}(\lambda_i)$. The rules[6,7] for obtaining $\ker \mathbf{S}(\lambda_i)$ are

Rule 1. Form the augmented matrix

$$\begin{bmatrix} \mathbf{A} - \lambda_i \mathbf{I}_n & \mathbf{B} \\ \hline \mathbf{I}_{n+m} \end{bmatrix} \qquad (16.41)$$

Rule 2. Use elementary column operations on this matrix to obtain m zero columns of $[\mathbf{A} - \lambda_i \mathbf{I} \quad \mathbf{B}]$. (Note: Do not divide columns by a function of λ.)

Rule 3. The columns below these zero columns span $\ker \mathbf{S}(\lambda_i)$. This can be verified by checking that $[\mathbf{A} - \lambda_i \mathbf{I} \quad \mathbf{B}] \cdot \ker \mathbf{S}(\lambda_i) = \mathbf{0}$. Forming the matrix of Eq. (16.41) with the plant matrices of Eq. (16.24) gives

$$
\left[
\begin{array}{ccccc}
-\lambda_i & 1 & 0 & 1 & 0 \\
0 & -\lambda_i & 1 & 0 & 0 \\
4 & 4 & -1-\lambda_i & 0 & 1 \\
\hline
1 & 0 & 0 & 0 & 0 \\
0 & 1 & 0 & 0 & 0 \\
0 & 0 & 1 & 0 & 0 \\
0 & 0 & 0 & 1 & 0 \\
0 & 0 & 0 & 0 & 1
\end{array}
\right]
\tag{16.42}
$$

$\operatorname{Ker} S(\lambda_1)$ for $\lambda_1 = -2$ is given in Example 1, Eq. (16.28). Thus, to obtain $\ker S(\lambda_2)$, insert $\lambda_2 = -1 + j1$ in the matrix of (16.42) and use column operations as specified by rule 2; this yields

$$
\left[
\begin{array}{ccccc}
0 & 0 & 0 & 1 & 0 \\
0 & 0 & 1 & 0 & 0 \\
0 & 0 & -j1 & 0 & 1 \\
\hline
1 & 0 & 0 & 0 & 0 \\
0 & 1 & 0 & 0 & 0 \\
0 & -1+j1 & 1 & 0 & 0 \\
-1+j1 & -1 & 0 & 1 & 0 \\
-4 & -5-j1 & 0 & 0 & 1
\end{array}
\right]
\tag{16.43}
$$

In accordance with rule 3, because the top portion of the first two columns is $[0 \quad 0 \quad 0]^T$, $\ker S(\lambda_2)$ is

$$
\ker S(-1+j1) = \operatorname{span}\left\{
\begin{bmatrix} 1 \\ 0 \\ 0 \\ -1+j1 \\ -4 \end{bmatrix}
\begin{bmatrix} 0 \\ 1 \\ -1+j1 \\ -1 \\ -5-j1 \end{bmatrix}
\right\}
$$

$$
= \operatorname{span}\{s_1(-1+j1), s_2(-1+j1)\}
\tag{16.44}
$$

It is not necessary to compute $\ker S(\lambda_3) = \ker S(-1 - j1)$ because its spanning vectors are the complex conjugates of the spanning vectors of $\ker S(-1 + j1)$. From Eqs. (16.28), (16.33), and (16.44) the selected null-space vectors are

$$
\left[\frac{\mathbf{v}_1}{\mathbf{q}_1}\right] =
\begin{bmatrix} 1 \\ -2 \\ 4 \\ \hline 0 \\ 0 \end{bmatrix}
\qquad
\left[\frac{\mathbf{v}_2}{\mathbf{q}_2}\right] =
\begin{bmatrix} 0 \\ 1 \\ -1+j1 \\ \hline -1 \\ -5-j1 \end{bmatrix}
\qquad
\left[\frac{\mathbf{v}_3}{\mathbf{q}_3}\right] =
\begin{bmatrix} 0 \\ 1 \\ -1-j1 \\ \hline -1 \\ -5+j1 \end{bmatrix}
\tag{16.45}
$$

Equation (16.23), which is used to compute the state-feedback matrix $\mathbf{K}$, can be modified[4] when the eigenvalues are complex conjugates. The following form for $\mathbf{K}$, where $n = 3$, uses only real vectors and therefore avoids operations with complex matrices[4]

$$
\mathbf{K} = [\mathbf{q}_1 \quad \operatorname{Re}(\mathbf{q}_2) \quad \operatorname{Im}(\mathbf{q}_2)][\mathbf{v}_1 \quad \operatorname{Re}(\mathbf{v}_2) \quad \operatorname{Im}(\mathbf{v}_2)]^{-1}
\tag{16.46}
$$

This format for computing $\mathbf{K}$ can be extended for the case when there is more than one set of complex eigenvalues and eigenvectors. Applying Eq. (16.46) with the values of Eq. (16.45) yields

$$\mathbf{K} = \begin{bmatrix} 0 & -1 & 0 \\ 0 & -5 & -1 \end{bmatrix} \begin{bmatrix} 1 & 0 & 0 \\ -2 & 1 & 0 \\ 4 & -1 & 1 \end{bmatrix}^{-1}$$

$$= \begin{bmatrix} 0 & -1 & 0 \\ 0 & -5 & -1 \end{bmatrix} \begin{bmatrix} 1 & 0 & 0 \\ 2 & 1 & 0 \\ -2 & 1 & 1 \end{bmatrix} = \begin{bmatrix} -2 & -1 & 0 \\ -8 & -6 & -1 \end{bmatrix} \tag{16.47}$$

The closed-loop matrix is

$$\mathbf{A}_{\mathrm{cl}} = \mathbf{A} + \mathbf{BK} = \begin{bmatrix} -2 & 0 & 0 \\ 0 & 0 & 1 \\ -4 & -2 & -2 \end{bmatrix} \tag{16.48}$$

The reader can verify that this matrix has the desired eigenvalue spectrum of Eq. (16.40) and the eigenvectors of Eq. (16.45). It should be noted that one open-loop eigenvalue, $\lambda_1 = -2$, and its associated eigenvector have been retained in the closed-loop matrix $\mathbf{A}_{\mathrm{cl}}$.

The state response of the closed-loop system can be obtained for the initial condition $\mathbf{x}(0) = \begin{bmatrix} 1 & 1 & 2 \end{bmatrix}$ and $\mathbf{r}(t) = \mathbf{0}$. Using the Laplace method yields

$$\mathbf{x}(t) = \mathcal{L}^{-1}[s\mathbf{I} - \mathbf{A}_{\mathrm{cl}}]^{-1}\mathbf{x}(0) = \mathcal{L}^{-1} \begin{bmatrix} \dfrac{1}{s+2} \\[2mm] \dfrac{s^2 + 5s + 2}{(s+2)(s^2 + 2s + 2)} \\[2mm] \dfrac{s^2 - 4s - 4}{(s+2)(s^2 + 2s + 2)} \end{bmatrix}$$

$$= \begin{bmatrix} e^{-2t} \\ -2e^{-2t} + 3e^{-t}\cos t \\ 4^{-2t} - 3e^{-t}\cos t \end{bmatrix} \tag{16.49}$$

Note that the selection of the eigenvectors in Eq. (16.45) produces the result that only e^{-2t} appears in $x_1(t)$ and all the modes appear in $x_2(t)$ and $x_3(t)$. (A suggested exercise for the reader is to assign the eigenvector $\begin{bmatrix} 1 & 0 & 0 \end{bmatrix}^T$ for $\lambda_1 = -2$ and to compare the results with those shown above.)

16.5   MATLAB EIGENVECTORS

The results of using the MATLAB toolbox to determine the eigenvectors for the real closed-loop eigenvalues of Example 1 of Sec. 16.4 (a script file) is as follows:

```
A   =   [0 1 0; 0 0 1; 4 4 -1];
B   =   [1 0;0 0;0 1];
S1  =   [A + 2*eye(3) B; zeros(2,5)]              % For λ₁ = -2
```

```
S1  =
       2    1    0    1    0
       0    2    1    0    0
       4    4    1    0    1
       0    0    0    0    0
       0    0    0    0    0

S2  =  [A + 3*eye(3) B; zeros(2,5)]              % For λ₂ = -3

S2  =
       3    1    0    1    0
       0    3    1    0    0
       4    4    2    0    1
       0    0    0    0    0
       0    0    0    0    0

S3  =  [A + 4*eye(3) B; zeros(2,5)]              % For λ₃ = -4

S3  =
       4    1    0    1    0
       0    4    1    0    0
       4    4    3    0    1
       0    0    0    0    0
       0    0    0    0    0

rref(S1)                                         % S1 in HNF

ans  =
    1.0000   0         -0.2500   0          0.2500
    0        1.0000     0.5000   0          0
    0        0          0        1.0000    -0.5000
    0        0          0        0          0
    0        0          0        0          0

rref(S2)                                         % S2 in HNF

ans  =
    1.0000   0          0        0.2000     0.1000
    0        1.0000     0        0.4000    -0.3000
    0        0          1.0000  -1.2000     0.9000
    0        0          0        0          0
    0        0          0        0          0

rref(S3)                                         % S3 in HNF

ans  =
    1.0000   0          0        0.2222     0.0278
    0        1.0000     0        0.1111    -0.1111
    0        0          1.0000  -0.4444     0.4444
    0        0          0        0          0
    0        0          0        0          0

diary off
```

The third and fourth rows of rref(S1) are interchanged in order to have the $+1$ on the principal diagonal. Inserting the -1's wherever a 0 appears on the principal diagonal and multiplying columns 3 and 4 by 4 gives the results of Eq. (16.28). For rref(S2) and rref(S3) no rows need be interchanged in order to achieve the desired format. The last two columns of rref(S2) are multiplied by 5 and 10, respectively, and the last two columns of rref(S3) are multiplied by 9 and 36, respectively, to yield essentially the results of Eqs. (16.30) and (16.32), respectively.

16.6   UNCONTROLLABLE SYSTEMS

A system is completely controllable (see Sec. 13.2) when

$$\text{Rank } \mathbf{M}_c = \text{Rank}[\mathbf{B} \quad \mathbf{AB} \quad \mathbf{A}^2\mathbf{B} \quad \cdots \quad \mathbf{A}^{n-m}\mathbf{B}] = n \qquad (16.50)$$

Complete controllability implies that all eigenvalues can be assigned by the use of state feedback. When Eq. (16.50) is not satisfied, the rank deficiency of $\mathbf{M}_c$ is equal to the number of modes for which the eigenvalues cannot be changed by the use of state feedback; i.e., these modes are uncontrollable. For distinct eigenvalues these uncontrollable modes have the property[9]

$$\text{Rank } \mathbf{S}(\lambda_u) = \text{Rank}[\mathbf{A} - \lambda_u\mathbf{I} \quad \mathbf{B}] = g < n \qquad (16.51)$$

The eigenvalues λ_u associated with uncontrollable modes are defined[8] as *input-decoupling zeros*. While the eigenvalue λ_u associated with an uncontrollable mode cannot be changed, it *is* possible to change the associated eigenvector by the use of state feedback. For controllable modes the null space of $\mathbf{S}(\lambda_i)$ has dimension m, which is equal to the dimension of $\mathbf{B}$. For uncontrollable modes, the dimension of $\ker \mathbf{S}(\lambda_u)$ is $n + m - g > m$. The larger dimension of this null space provides greater flexibility[4] in the selection permitted for the associated eigenvector. This is demonstrated in the following example.

Example.   A system is represented by a state equation having the plant and control matrices

$$\mathbf{A} = \begin{bmatrix} -1 & 0 & 0 \\ -2 & -2 & -2 \\ -1 & 0 & -3 \end{bmatrix} \quad \mathbf{B} = \begin{bmatrix} 1 & 0 \\ 0 & 2 \\ 0 & 1 \end{bmatrix} \qquad (16.52)$$

and the open-loop eigenvalue spectrum is

$$\sigma(\mathbf{A}) = \{-1, -2, -3\} \qquad (16.53)$$

For this system

$$\text{Rank } \mathbf{M}_c = \text{Rank}[\mathbf{B} \quad \mathbf{AB}] = \text{Rank} \begin{bmatrix} 1 & 0 & -1 & 0 \\ 0 & 2 & -2 & -6 \\ 0 & 1 & -1 & -3 \end{bmatrix} = 2 \qquad (16.54)$$

Because rank $\mathbf{M}_c < n$, the system is not completely controllable. The mode associated with $\lambda = -2$ is uncontrollable because, from Eq. (16.51),

$$\text{Rank } \mathbf{S}(-2) = \text{Rank} \begin{bmatrix} 1 & 0 & 0 & 1 & 0 \\ -2 & 0 & -2 & 0 & 2 \\ -1 & 0 & -1 & 0 & 1 \end{bmatrix} = 2 < n \qquad (16.55)$$

The desired closed-loop eigenvalue spectrum *must* include $\lambda = -2$ and is selected as

$$\sigma(\mathbf{A} + \mathbf{BK}) = \{-2, -4, -5\} \qquad (16.56)$$

Using the method of Example 2 in Sec. 16.4 with $\lambda_1 = -2$, $\lambda_2 = -4$, and $\lambda_3 = -5$ yields

$$\ker \mathbf{S}(\lambda_1) = \text{span} \left\{ \begin{bmatrix} 1 \\ 0 \\ 0 \\ -1 \\ 1 \end{bmatrix}, \begin{bmatrix} 0 \\ 1 \\ 0 \\ 0 \\ 0 \end{bmatrix}, \begin{bmatrix} 0 \\ 0 \\ 1 \\ 0 \\ 1 \end{bmatrix} \right\} = \text{span} \{s_1(\lambda_1), s_2(\lambda_1), s_3(\lambda_1)\} \quad (16.57)$$

$$\ker \mathbf{S}(\lambda_2) = \text{span} \left\{ \begin{bmatrix} 1 \\ 0 \\ 0 \\ -3 \\ 1 \end{bmatrix}, \begin{bmatrix} 0 \\ 2 \\ 1 \\ 0 \\ -1 \end{bmatrix} \right\} = \text{span} \{s_1(\lambda_2), s_2(\lambda_2)\} \qquad (16.58)$$

$$\ker \mathbf{S}(\lambda_3) = \text{span} \left\{ \begin{bmatrix} 1 \\ 0 \\ 0 \\ -4 \\ 1 \end{bmatrix}, \begin{bmatrix} 0 \\ 2 \\ 1 \\ 0 \\ -2 \end{bmatrix} \right\} = \text{span} \{s_1(\lambda_3), s_2(\lambda_3)\} \qquad (16.59)$$

The selected eigenvectors must be linearly independent. It is also noted that any desired eigenvector can be associated with $\lambda_1 = -2$. The linearly independent vectors selected from these subspaces for this example are

$$\begin{bmatrix} \mathbf{v}_1 \\ \mathbf{q}_1 \end{bmatrix} = \begin{bmatrix} 0 \\ 0 \\ 1 \\ \text{---} \\ 0 \\ 1 \end{bmatrix} \quad \begin{bmatrix} \mathbf{v}_2 \\ \mathbf{q}_2 \end{bmatrix} = \begin{bmatrix} 1 \\ 0 \\ 0 \\ \text{---} \\ -3 \\ 1 \end{bmatrix} \quad \begin{bmatrix} \mathbf{v}_3 \\ \mathbf{q}_3 \end{bmatrix} = \begin{bmatrix} 0 \\ 2 \\ 1 \\ \text{---} \\ 0 \\ -2 \end{bmatrix} \qquad (16.60)$$

The required feedback matrix is computed using Eq. (16.23) as

$$\mathbf{K} = \begin{bmatrix} 0 & -3 & 0 \\ 1 & 1 & -2 \end{bmatrix} \begin{bmatrix} 0 & 1 & 0 \\ 0 & 0 & 2 \\ 1 & 0 & 1 \end{bmatrix}^{-1}$$

$$= \begin{bmatrix} 0 & -3 & 0 \\ 1 & 1 & -2 \end{bmatrix} \begin{bmatrix} 0 & -0.5 & 1 \\ 1 & 0 & 0 \\ 0 & 0.5 & 0 \end{bmatrix} = \begin{bmatrix} -3 & 0 & 0 \\ 1 & -1.5 & 1 \end{bmatrix} \qquad (16.61)$$

The closed-loop matrix is

$$\mathbf{A_{cl}} = \mathbf{A} + \mathbf{BK} = \begin{bmatrix} -4 & 0 & 0 \\ 0 & -5 & 0 \\ 0 & -1.5 & -2 \end{bmatrix} \tag{16.62}$$

This closed-loop matrix has the specified eigenvalue spectrum of Eq. (16.56) and the eigenvectors of Eq. (16.60). For the initial condition $\mathbf{x}(0) = [1 \quad 1 \quad 2]^T$ and $\mathbf{r}(t) = 0$, the state response is

$$\begin{bmatrix} x_1(t) \\ x_2(t) \\ x_3(t) \end{bmatrix} = \begin{bmatrix} e^{-4t} & 0 & 0 \\ 0 & e^{-5t} & 0 \\ 0 & 0.5e^{-2t} - 0.5e^{-5t} & e^{-2t} \end{bmatrix} \mathbf{x}(0) = \begin{bmatrix} e^{-4t} \\ e^{-5t} \\ 2.5e^{-2t} - 0.5e^{-5t} \end{bmatrix} \tag{16.63}$$

The selection of the eigenvectors of Eq. (16.60) results in the presence of only the mode e^{-4t} in $x_1(t)$ and only the mode e^{-5t} in $x_2(t)$. Also, the uncontrollable mode e^{-2t} appears only in $x_3(t)$. This is verified by the response given in Eq. (16.63).

16.7 TRACKING SYSTEMS

A controllable open-loop system is represented by the nth-order state and ℓth-order output equations of the form

$$\dot{\mathbf{x}} = \mathbf{Ax} + \mathbf{Bu} \tag{16.64}$$

$$\mathbf{y} = \mathbf{Cx} = \begin{bmatrix} \mathbf{E} \\ \mathbf{F} \end{bmatrix} \mathbf{x} \tag{16.65}$$

where $\mathbf{y}$ is an $\ell \times 1$ vector and $\mathbf{w} = \mathbf{Ex}$ is a $p \times 1$ vector representing the outputs that are required to follow a $p \times 1$ input vector $\mathbf{r}$. The outputs y_i that must be controlled, $i = 1, 2, \ldots, p$, must be the first p elements in the output vector $\mathbf{y}$. The output vector may be reordered, if necessary, so that $\mathbf{w} = \mathbf{Ex}$ contains all the controlled outputs. In order to maintain controllability, the number of outputs p that can be tracked cannot exceed the number of control inputs m as described below ($\ell \geq p$).

The feedback controller is required to cause the output vector $\mathbf{w}$ to track the command input $\mathbf{r}$ in the sense that the steady-state response is

$$\lim_{t \to \infty} \mathbf{w}(t) = \mathbf{r}(t) \tag{16.66}$$

when $\mathbf{r}$ consists of piecewise-constant command inputs. The design method[4,10] consists of the addition of a vector comparator and integrator that satisfies the equation

$$\dot{\mathbf{z}} = \mathbf{r} - \mathbf{w} = \mathbf{r} - \mathbf{Ex} \tag{16.67}$$

The composite open-loop system is therefore governed by the augmented state and output equations formed from Eqs. (16.64) to (16.67):

$$\begin{bmatrix} \dot{\mathbf{x}} \\ \dot{\mathbf{z}} \end{bmatrix} = \begin{bmatrix} \mathbf{A} & \mathbf{0} \\ -\mathbf{E} & \mathbf{0} \end{bmatrix} \begin{bmatrix} \mathbf{x} \\ \mathbf{z} \end{bmatrix} + \begin{bmatrix} \mathbf{B} \\ \mathbf{0} \end{bmatrix} \mathbf{u} + \begin{bmatrix} \mathbf{0} \\ \mathbf{I} \end{bmatrix} \mathbf{r} = \bar{\mathbf{A}}\mathbf{x}' + \bar{\mathbf{B}}\mathbf{u} + \bar{\mathbf{B}}'\mathbf{r} \tag{16.68}$$

and

$$\mathbf{y} = [\mathbf{C} \quad \mathbf{0}] \begin{bmatrix} \mathbf{x} \\ \mathbf{z} \end{bmatrix} = \bar{\mathbf{C}}\mathbf{x}' \tag{16.69}$$

where
$$\bar{\mathbf{A}} = \begin{bmatrix} \mathbf{A} & \mathbf{0} \\ -\mathbf{E} & \mathbf{0} \end{bmatrix} \qquad \bar{\mathbf{B}} = \begin{bmatrix} \mathbf{B} \\ \mathbf{0} \end{bmatrix} \qquad \mathbf{B}' = \begin{bmatrix} \mathbf{0} \\ \mathbf{I} \end{bmatrix} \qquad \bar{\mathbf{C}} = [\mathbf{C} \quad \mathbf{0}]$$

(16.70)

The state-feedback control law to be used is

$$\mathbf{u} = \mathbf{K}_1 \mathbf{x} + \mathbf{K}_2 \mathbf{z} = [\mathbf{K}_1 \quad \mathbf{K}_2] \begin{bmatrix} \mathbf{x} \\ \mathbf{z} \end{bmatrix} \qquad (16.71)$$

where
$$\bar{\mathbf{K}} = [\mathbf{K}_1 \quad \mathbf{K}_2] \qquad (16.72)$$

A block diagram representing the feedback-control system, consisting of the plant state and output equations given by Eqs. (16.64) and (16.65) and the control law given by Eq. (16.71), is shown in Fig. 16.1. The dimensions of all matrices are shown in the figure. This control law assigns the desired closed-loop eigenvalue spectrum if and only if the augmented plant and control matrix pair $(\bar{\mathbf{A}}, \bar{\mathbf{B}})$ is controllable. It has been shown[10,11] that this condition is satisfied if $(\mathbf{A}, \mathbf{B})$ is a controllable pair and, in addition,

$$\text{Rank} \begin{bmatrix} \mathbf{B} & \mathbf{A} \\ \mathbf{0} & -\mathbf{E} \end{bmatrix} = n + p \qquad (16.73)$$

This condition is much easier to use than the determination of the rank of the con-trollability matrix for $(\bar{\mathbf{A}}, \bar{\mathbf{B}})$. *Equation (16.73) can be satisfied only if the number of outputs p that are required to track the input* $\mathbf{r}$ *is less than or equal to the number of control inputs m.* Satisfaction of the condition of Eq. (16.73) guarantees that a control law of the form of Eq. (16.71) can be synthesized such that the closed-loop output tracks the command input. In that case the closed-loop state equation is

$$\dot{\mathbf{x}}' = \begin{bmatrix} \dot{\mathbf{x}} \\ \dot{\mathbf{z}} \end{bmatrix} = \begin{bmatrix} \mathbf{A} + \mathbf{B}\mathbf{K}_1 & \mathbf{B}\mathbf{K}_2 \\ -\mathbf{E} & \mathbf{0} \end{bmatrix} \begin{bmatrix} \mathbf{x} \\ \mathbf{z} \end{bmatrix} + \begin{bmatrix} \mathbf{0} \\ \mathbf{I} \end{bmatrix} \mathbf{r} = \mathbf{A}'_{\text{cl}} \mathbf{x}' + \mathbf{B}' \mathbf{r} \qquad (16.74)$$

The feedback matrix of Eq. (16.72) can be selected so that the eigenvalues are in the left-half plane for the closed-loop plant matrix of Eq. (16.74). Then, because the inputs are constants, the steady-state values of the states $\mathbf{x}$ and $\mathbf{z}$ are also constants. Therefore, $\dot{\mathbf{z}} = 0$ in the steady state and Eq. (16.67) indicates that Eq. (16.66) is satisfied. Thus, the outputs $\mathbf{w}(t)$ track the piecewise constant command vector $\mathbf{r}(t)$ in the steady state.

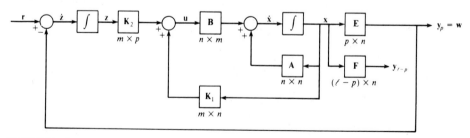

FIGURE 16.1
Tracking system.

16.8   TRACKING-SYSTEM DESIGN EXAMPLE

Let the control system represented by the state and output equations of Eqs. (16.64) and (16.65) have the following plant, control, and output matrices

$$
\mathbf{A} = \begin{bmatrix} 0 & 1 & 0 \\ 0 & 0 & 1 \\ 4 & 4 & -1 \end{bmatrix} \quad
\mathbf{B} = \begin{bmatrix} 1 & 0 \\ 0 & 0 \\ 0 & 1 \end{bmatrix} \quad
\mathbf{C} = \begin{bmatrix} 1 & 0 & 0 \\ 0 & 1 & 1 \end{bmatrix}
\tag{16.75}
$$

Because $\sigma(\mathbf{A}) = \{-1, -2, 2\}$, the open-loop system is unstable. The purpose of the design is to achieve a stable system in which the output vector $\mathbf{y}(t)$ can track the command $\mathbf{r}(t)$, where

$$
\mathbf{r}(t) = \begin{bmatrix} 1 \\ -2 \end{bmatrix}
\tag{16.76}
$$

The system is augmented by the comparator and integrator given in Eq. (16.67). Because there are two outputs ($\ell = 2$) and two control inputs ($m = 2$), tracking of all the inputs $\mathbf{r}$ by the outputs $\mathbf{y}$ is possible. Thus, $\mathbf{E} = \mathbf{C}$. Then the composite open-loop state and output equations are represented by Eqs. (16.68) and (16.69). The composite system matrices using Eqs. (16.70) are

$$
\bar{\mathbf{A}} = \begin{bmatrix} 0 & 1 & 0 & 0 & 0 \\ 0 & 0 & 1 & 0 & 0 \\ 4 & 4 & -1 & 0 & 0 \\ -1 & 0 & 0 & 0 & 0 \\ 0 & -1 & -1 & 0 & 0 \end{bmatrix}
\tag{16.77}
$$

$$
\bar{\mathbf{B}} = \begin{bmatrix} 1 & 0 \\ 0 & 0 \\ 0 & 1 \\ 0 & 0 \\ 0 & 0 \end{bmatrix}
\tag{16.78}
$$

$$
\bar{\mathbf{C}} = \begin{bmatrix} 1 & 0 & 0 & 0 & 0 \\ 0 & 1 & 1 & 0 & 0 \end{bmatrix}
\tag{16.79}
$$

The augmented system is controllable because $(\mathbf{A}, \mathbf{B})$ is a controllable pair and the necessary condition of Eq. (16.73) is satisfied, that is, $n + p = 5$ and

$$
\text{Rank} \begin{bmatrix} \mathbf{B} & \mathbf{A} \\ \mathbf{0} & -\mathbf{C} \end{bmatrix} = \text{Rank} \begin{bmatrix} 1 & 0 & 0 & 1 & 0 \\ 0 & 0 & 0 & 0 & 1 \\ 0 & 1 & 4 & 4 & -1 \\ 0 & 0 & -1 & 0 & 0 \\ 0 & 0 & 0 & -1 & -1 \end{bmatrix} = 5
\tag{16.80}
$$

The selected eigenvalue spectrum of the closed-loop plant matrix in Eq. (16.74) is

$$
\sigma(\bar{\mathbf{A}} + \bar{\mathbf{B}}\mathbf{K}) = \{-2, -3, -4, -5, -6\}
\tag{16.81}
$$

The associated eigenvectors must lie in the null spaces of

$$\bar{\mathbf{S}}(\lambda_i) = \begin{bmatrix} \bar{\mathbf{A}} - \lambda_i \mathbf{I} & \bar{\mathbf{B}} \end{bmatrix}$$

$$= \begin{bmatrix} -\lambda_i & 1 & 0 & 0 & 0 & 1 & 0 \\ 0 & -\lambda_i & 1 & 0 & 0 & 0 & 0 \\ 4 & 4 & -1-\lambda_i & 0 & 0 & 0 & 1 \\ -1 & 0 & 0 & -\lambda_i & 0 & 0 & 0 \\ 0 & -1 & -1 & 0 & -\lambda_i & 0 & 0 \end{bmatrix} \qquad (16.82)$$

Using the method[6,7] shown in Sec. 16.4, Example 2, these null spaces are

$$\ker \bar{\mathbf{S}}(-2) = \operatorname{span}\left\{ \begin{bmatrix} 2 \\ 0 \\ 0 \\ 1 \\ 0 \\ -4 \\ -8 \end{bmatrix}, \begin{bmatrix} 0 \\ 2 \\ -4 \\ 0 \\ -1 \\ -2 \\ -4 \end{bmatrix} \right\} \qquad (16.83a)$$

$$\ker \bar{\mathbf{S}}(-3) = \operatorname{span}\left\{ \begin{bmatrix} 3 \\ 0 \\ 0 \\ 1 \\ 0 \\ -9 \\ -12 \end{bmatrix}, \begin{bmatrix} 0 \\ 3 \\ -9 \\ 0 \\ -2 \\ -3 \\ 6 \end{bmatrix} \right\} \qquad (16.83b)$$

$$\ker \bar{\mathbf{S}}(-4) = \operatorname{span}\left\{ \begin{bmatrix} 4 \\ 0 \\ 0 \\ 1 \\ 0 \\ -16 \\ -16 \end{bmatrix}, \begin{bmatrix} 0 \\ 4 \\ -16 \\ 0 \\ -3 \\ -4 \\ 32 \end{bmatrix} \right\} \qquad (16.83c)$$

$$\ker \bar{\mathbf{S}}(-5) = \operatorname{span}\left\{ \begin{bmatrix} 5 \\ 0 \\ 0 \\ 1 \\ 0 \\ -25 \\ -20 \end{bmatrix}, \begin{bmatrix} 0 \\ 5 \\ -25 \\ 0 \\ -4 \\ -5 \\ 80 \end{bmatrix} \right\} \qquad (16.83d)$$

$$\ker \bar{\mathbf{S}}(-6) = \operatorname{span}\left\{ \begin{bmatrix} 6 \\ 0 \\ 0 \\ 1 \\ 0 \\ -36 \\ -24 \end{bmatrix}, \begin{bmatrix} 0 \\ 6 \\ -36 \\ 0 \\ -5 \\ -6 \\ 156 \end{bmatrix} \right\} \qquad (16.83e)$$

The following vectors are selected from $\ker \mathbf{S}(\lambda_i)$ in Eqs. (16.83):

$$
\begin{bmatrix} \mathbf{v}_1 \\ \mathbf{q}_1 \end{bmatrix} = \begin{bmatrix} 2 \\ 0 \\ 0 \\ 1 \\ 0 \\ \hline -4 \\ -8 \end{bmatrix} \qquad
\begin{bmatrix} \mathbf{v}_2 \\ \mathbf{q}_2 \end{bmatrix} = \begin{bmatrix} 0 \\ 3 \\ -9 \\ 0 \\ -2 \\ \hline -3 \\ 6 \end{bmatrix} \qquad
\begin{bmatrix} \mathbf{v}_3 \\ \mathbf{q}_3 \end{bmatrix} = \begin{bmatrix} 0 \\ 4 \\ -16 \\ 0 \\ -3 \\ \hline -4 \\ 32 \end{bmatrix} \qquad (16.84)
$$

$$
\begin{bmatrix} \mathbf{v}_4 \\ \mathbf{q}_4 \end{bmatrix} = \begin{bmatrix} 0 \\ 5 \\ -25 \\ 0 \\ -4 \\ \hline -5 \\ 80 \end{bmatrix} \qquad
\begin{bmatrix} \mathbf{v}_5 \\ \mathbf{q}_5 \end{bmatrix} = \begin{bmatrix} 6 \\ 0 \\ 0 \\ 1 \\ 0 \\ \hline -36 \\ -24 \end{bmatrix}
$$

The eigenvectors $\mathbf{v}_i (i = 1, \ldots, 5)$ in Eqs. (16.84) are linearly independent, and therefore the feedback matrix $\bar{\mathbf{K}}$ can be computed by use of Eq. (16.23) as

$$
\bar{\mathbf{K}} = \begin{bmatrix} -4 & -3 & -4 & -5 & -36 \\ -8 & 6 & 32 & 80 & -24 \end{bmatrix} \begin{bmatrix} 2 & 0 & 0 & 0 & 6 \\ 0 & 3 & 4 & 5 & 0 \\ 0 & -9 & -16 & -25 & 0 \\ 1 & 0 & 0 & 0 & 1 \\ 0 & -2 & -3 & -4 & 0 \end{bmatrix}^{-1}
$$

$$
= \begin{bmatrix} -4 & -3 & -4 & -5 & -36 \\ -8 & 6 & 32 & 80 & -24 \end{bmatrix} \left(\frac{1}{4}\right) \begin{bmatrix} -1 & 0 & 0 & 6 & 0 \\ 0 & 22 & -2 & 0 & 40 \\ 0 & -28 & 4 & 0 & -60 \\ 0 & 10 & -2 & 0 & 24 \\ 1 & 0 & 0 & -2 & 0 \end{bmatrix}
$$

$$
= \begin{bmatrix} -8 & -1 & 0 & 12 & 0 \\ -4 & 9 & -11 & 0 & 60 \end{bmatrix} \qquad (16.85)
$$

The closed-loop composite plant matrix is

$$
\bar{\mathbf{A}} + \bar{\mathbf{B}}\bar{\mathbf{K}} = \begin{bmatrix} -8 & 0 & 0 & 12 & 0 \\ 0 & 0 & 1 & 0 & 0 \\ 0 & 13 & -12 & 0 & 60 \\ -1 & 0 & 0 & 0 & 0 \\ 0 & -1 & -1 & 0 & 0 \end{bmatrix} \qquad (16.86)
$$

This closed-loop plant matrix has the desired eigenvalue spectrum of Eq. (16.81) and the eigenvectors of Eq. (16.84), as required. It should be noted that the selection of the eigenvectors in Eq. (16.84) produces the result that only the modes e^{-2t} and e^{-6t} appear in the states $x_1(t)$ and $z_1(t)$. Also, only the modes e^{-3t}, e^{-4t}, and e^{-5t} appear in the states $x_2(t)$, $x_3(t)$, and $z_2(t)$. This is verified by computing the state response from Eq. (16.74) for an initially quiescent system [$\mathbf{x}(0) = \mathbf{0}, \mathbf{z}(0) = \mathbf{0}$] and the input of Eq. (16.76). The Laplace transform of Eqs. (16.74) and (16.69) yields

$$\begin{bmatrix} \mathbf{X}(s) \\ \mathbf{Z}(s) \end{bmatrix} = \begin{bmatrix} \dfrac{12}{s(s+2)(s+6)} \\[2ex] \dfrac{-120}{s(s+3)(s+4)(s+5)} \\[2ex] \dfrac{-120}{(s+3)(s+4)(s+5)} \\[2ex] \dfrac{s+8}{s(s+2)(s+6)} \\[2ex] \dfrac{-2(s^2+12s-13)}{s(s+3)(s+4)(s+5)} \end{bmatrix} \tag{16.87}$$

$$\mathbf{Y}(s) = \begin{bmatrix} \dfrac{12}{s(s+2)(s+6)} \\[2ex] \dfrac{-120(s+1)}{s(s+3)(s+4)(s+5)} \end{bmatrix} \tag{16.88}$$

Thus, the output time response is

$$\mathbf{y}(t) = \begin{bmatrix} 1 - 1.5e^{-2t} + 0.5e^{-6t} \\ -2 - 40e^{-3t} + 90e^{-4t} - 48e^{-5t} \end{bmatrix} \tag{16.89}$$

and its final value is
$$\mathbf{y}(t)_{ss} = \begin{bmatrix} 1 \\ -2 \end{bmatrix} \tag{16.90}$$

Therefore $\mathbf{y}(t)_{ss}$ is equal to the command input given in Eq. (16.76), as required. This example demonstrates that a tracker can be synthesized that results in each output following a specified constant input command; that is, $y_1(t)$ tracks $r_1(t)$, and $y_2(t)$ tracks $r_2(t)$. Also, the use of the method of entire eigenstructure assignment permits the selection of the eigenvectors associated with each mode. The result of this selection may produce designs such that only certain modes appear in each state and in each output. This feature is a function of the spaces [see Eqs. (16.83)] from which the closed-loop eigenvectors must be selected.

16.9 MATLAB EXAMPLE OF TRACKER DESIGN IN SEC. 16.8

```
Diary
% Alternate eigenvector selection used
% This method is good only for real eigenvalues
A  =  [0 1 0; 0 0 1; 4 4 -1];       % Eq. (16.75)
B  =  [1 0 ; 0 0; 0 1];   C  =  [1 0 0; 0 1 1];
Mc  =  [B A];
rank (Mc )
```

```
ans  =  3
D   =  [0 0; 0 0];
Mcc  =  [B A;D -C]                    % Eqs.  (16.80)
rank(Mcc)
ans  =  5
E   =  [0 0; 0 0; 0 0];
Abar  =  [A E; -C D];                 % Eqs.  (16.70) and (16.77)
Bbar  =  [B;D];                       % Eqs.  (16.70) and (16.78)
% The Si are formed for Eq.  (16.26) for each eigenvalue
S1  =  [Abar  +  2*eye(5) Bbar];      % For λ 1 = -2
S2  =  [Abar  +  3*eye(5) Bbar];      % For λ 2 = -3
S3  =  [Abar  +  4*eye(5) Bbar];      % For λ 3 = -4
S4  =  [Abar  +  5*eye(5) Bbar];      % For λ 4 = -5
S5  =  [Abar  +  6*eye(5) Bbar];      % For λ 5 = -6
% The Si are put in HNF
rref(S1)
ans  =
   1.0000        0        0        0    -1.0000        0    0.2500
        0   1.0000        0        0     2.0000        0         0
        0        0   1.0000        0    -4.0000        0         0
        0        0        0   1.0000    -0.5000        0    0.1250
        0        0        0        0         0   1.0000   -0.5000
rref(S2)
ans  =
   1.0000        0        0        0         0   0.2000    0.1000
        0   1.0000        0        0         0   0.4000   -0.3000
        0        0   1.0000        0         0  -1.2000    0.9000
        0        0        0   1.0000         0   0.0667    0.0333
        0        0        0        0    1.0000  -0.2667    0.2000
rref(S3)
ans  =
   1.0000        0        0        0         0   0.2222    0.0278
        0   1.0000        0        0         0   0.1111   -0.1111
        0        0   1.0000        0         0  -0.4444    0.4444
        0        0        0   1.0000         0   0.0556    0.0069
        0        0        0        0    1.0000  -0.0833    0.0833
rref(S4)
ans  =
   1.0000        0        0        0         0   0.1905    0.0119
        0   1.0000        0        0         0   0.0476   -0.0595
        0        0   1.0000        0         0  -0.2381    0.2976
        0        0        0   1.0000         0   0.0381    0.0024
        0        0        0        0    1.0000  -0.0381    0.0476
rref(S5)
ans  =
   1.0000        0        0        0         0   0.1625    0.0063
        0   1.0000        0        0         0   0.0250   -0.0375
        0        0   1.0000        0         0  -0.1500    0.2250
        0        0        0   1.0000         0   0.0271    0.0010
        0        0        0        0    1.0000  -0.0208    0.0313
```

```
a  =  rref(S1);  b  =  rref(S2);  c  =  rref(S3);  d  =  rref(S4);
e  =  rref(S5);
% Use Step 3 of Ex. 1, Sec. 16.4, to insert two zero rows so
% the leading element in each row is on the principal diagonal
a1  =  [a(1:4,1:7);zeros(1,7);a(5,1:7);zeros(1,7)];
% Use Step 5 of Ex. 1, Sec. 16.4, to insert -1 for the zeros
% in columns 5 and 7 on the principal diagonal of a1
a1(5,5)  =  -1,  a1(7,7)  =  -1,  a1
a1  =
```

1.0000	0	0	0	-1.0000	0	0.2500
0	1.0000	0	0	2.0000	0	0
0	0	1.0000	0	-4.0000	0	0
0	0	0	1.0000	-0.5000	0	0.1250
0	0	0	0	-1.0000	0	0
0	0	0	0	0	1.0000	-0.5000
0	0	0	0	0	0	-1.0000

```
b1  =  [b(1:5,1:7);zeros(1,7);zeros(1,7)];
% Use Step 5 of Ex. 1, Sec. 16.4 to insert -1 in columns
% 6 and 7 on the principal diagonals of b1, c1, d1, and e1
b1(6,6)  =  -1;  b1(7,7)  =  -1;  b1
b1  =
```

1.0000	0	0	0	0	0.2000	0.1000
0	1.0000	0	0	0	0.4000	-0.3000
0	0	1.0000	0	0	-1.2000	0.9000
0	0	0	1.0000	0	0.0667	0.0333
0	0	0	0	1.0000	-0.2667	0.2000
0	0	0	0	0	-1.0000	0
0	0	0	0	0	0	-1.0000

```
c1  =  [c(1:5,1:7);zeros(1,7);zeros(1,7)];
c1(6,6)  =  -1;  c1(7,7)  =  -1;  c1
c1  =
```

1.0000	0	0	0	0	0.2222	0.0278
0	1.0000	0	0	0	0.1111	-0.1111
0	0	1.0000	0	0	-0.4444	0.4444
0	0	0	1.0000	0	0.0556	0.0069
0	0	0	0	1.0000	-0.0833	0.0833
0	0	0	0	0	-1.0000	0
0	0	0	0	0	0	-1.0000

```
d1  =  [d(1:5,1:7);0*ones(1,7);0*ones(1,7)];
d1(6,6)  =  -1,  d1(7,7)  =  -1;  d1
d1  =
```

1.0000	0	0	0	0	0.1905	0.0119
0	1.0000	0	0	0	0.0476	-0.0595
0	0	1.0000	0	0	-0.2381	0.2976
0	0	0	1.0000	0	0.0381	0.0024
0	0	0	0	1.0000	-0.0381	0.0476
0	0	0	0	0	-1.0000	0
0	0	0	0	0	0	-1.0000

```
e1  =  [e(1:5,1:7);0*ones(1,7);0*ones(1,7)];
```

```
e1(6,6)  =  -1;  e1(7,7)  =  -1;  e1
e1  =
  1.0000        0        0        0        0   0.1625   0.0063
       0   1.0000        0        0        0   0.0250  -0.0375
       0        0   1.0000        0        0  -0.1500   0.2250
       0        0        0   1.0000        0   0.0271   0.0010
       0        0        0        0   1.0000  -0.0208   0.0313
       0        0        0        0        0  -1.0000        0
       0        0        0        0        0        0  -1.0000
```
% The following procedure is used to first establish the
% dimensions and then to form the V and Q matrices from column
% 7 of a1, b1, c1, d1, and e1
```
V   =   zeros(5,5);
Q   =   zeros(2,5);
V(:,1)  =  a1(1:5,7);  V(:,2)  =  b1(1:5,7);  V(:,3)  =  c1(1:5,7);
V(:,4)  =  d1(1:5,7);  V(:,5)  =  e1(1:5,7)
V   =   0.2500   0.1000   0.0278   0.0119   0.0063
             0  -0.3000  -0.1111  -0.0595  -0.0375
             0   0.9000   0.4444   0.2976   0.2250
        0.1250   0.0333   0.0069   0.0024   0.0010
             0   0.2000   0.0833   0.0476   0.0313
Q(:,1)  =  a1(6:7,7);  Q(:,2)  =  b1(6:7,7);  Q(:,3)  =  c1(6:7,7);
Q(:,4)  =  d1(6:7,7);  Q(:,5)  =  e1(6:7,7)
Q   =
      -0.5000        0        0        0        0
      -1.0000  -1.0000  -1.0000  -1.0000  -1.0000
Kbar  =  Q*inv(V);              % See Eq. (16.85)
Kbar
Kbar  =
      -2.0000   -2.0000    0.0000    0.0000   -2.0000
    -184.0000   39.0000  -17.0000  360.0000  162.0000
```
% Form the closed loop matrices [see Eq. (16.86)]
```
Acl = [Abar  +  Bbar*Kbar]
Acl  =
      -2.0000   -1.0000    0.0000    0.0000   -2.0000
            0         0    1.0000         0         0
    -180.0000   43.0000  -18.0000  360.0000  162.0000
      -1.0000        0        0        0        0
            0   -1.0000   -1.0000        0        0
```
% Accuracy check (CADAC)
```
eig(Acl)
ans  =    -6.0000  -5.0000   -4.0000   -3.0000   -2.0000
Bcl  =  [0 0; 0 0; 0 0;1 0;0 -2];  % Bbar  =  B' of Eq. (16.68)
Ccl  =  [1 0 0 0 0;  0 1 1 0 0];   Dcl  =  [ 0 0;  0 0];
[tfn,tfd]  =  ss2tf(Acl,Bcl,Ccl,Dcl,1);  % See Eq. (16.88);  tfn
tfn  =
    1.0e+003 *
         0   0.0000   0.0000   0.0000   0.3600   0.7200
         0   0.0000   0.3600   1.0800   0.7200   0.0000
```

```
tfd  =
  1.0e+003 *
    0.0010       0.0200       0.1550       0.5800       1.0440       0.7200
[tfn,tfd] = ss2tf(Acl,Bcl,Ccl,Dcl,2); % See Eq. (16.88); tfn
tfn  =
  1.0e+003 *
         0       0.0000       0.0040       0.0720       0.1520       0.0000
         0       0.0000      -0.3240      -1.6920      -2.8080      -1.4400
tfd  =
  1.0e+003 *
    0.0010       0.0200       0.1550       0.5800       1.0440       0.7200
% The output time responses of Eq. (16.89) follow:
[yt,xt]   =   step(Acl,Bcl,Ccl,Dcl,1);
[yt,xt]   =   step(Acl,Bcl,Ccl,Dcl,2);
     num1  =  tfn(1,:); num2  =  tfn(2,:);   den  =  tfd;
     t  =  0:.1:5;
     % To obtain plot of output 1
     step(num1,den)
     print;       % The plot for y1(t) is shown in Fig. 16.2
     % To obtain plot of output 2
     step(num2,den)
     print;       % The plot for y2(t) is shown in Fig. 16.3
     diary off
```

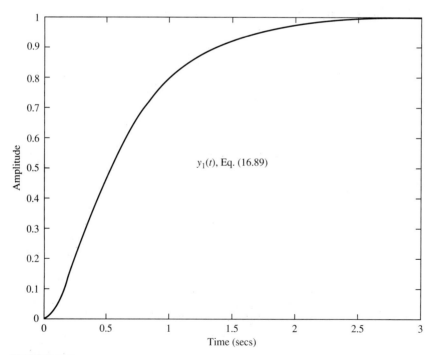

FIGURE 16.2
MATLAB plot of $y_1(t)$, Eq. (16.89).

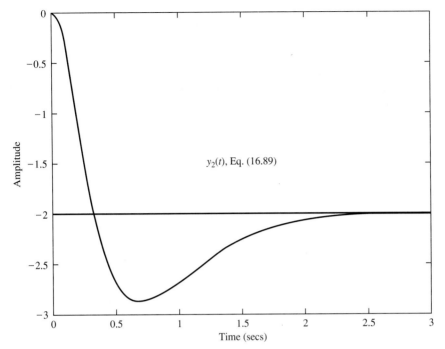

FIGURE 16.3
MATLAB plot of $y_2(t)$, Eq. (16.89).

Caution: In order to maintain the desired accuracy of the final MATLAB results it is important to stay in MATLAB *while accomplishing the entire solution*. For example, one cannot log out of MATLAB right after obtaining ker $\mathbf{S}(\lambda_i)$ and then, some time later, reenter MATLAB in order to complete the solution of the problem by entering just the $\mathbf{V}$ and $\mathbf{Q}$ matrices based upon the printed 4-digit values of ker $\mathbf{S}(\lambda_i)$. Doing so, the eigenvalues of $\mathbf{A}_{cl}$ will be in error.

16.10 SUMMARY

The use of a state-feedback control law permits the assignment of the entire eigenstructure, consisting of the eigenvalues and the associated eigenvectors. The closed-loop eigenvalues can be assigned arbitrarily, provided that the matrix pair $(\mathbf{A}, \mathbf{B})$ is controllable. The associated eigenvectors must be located within specific subspaces, which are determined by ker $[\mathbf{A} - \lambda\mathbf{I} \quad \mathbf{B}]$. A simple algorithm that is readily implemented on a computer for obtaining these subspaces for real eigenvalues is presented. The use of this algorithm, the selection of the eigenvectors, and the synthesis of the feedback matrix are illustrated by means of examples.

This chapter demonstrates the successful application of entire eigenstructure assignment to the synthesis of MIMO regulators and tracking systems. Such tracking systems achieve the objective of causing all output components to track specified inputs. Additional techniques need to be developed on the proper guidelines for selection of eigenvectors to satisfy specifications on output response. In Ref. 12, guidelines are presented for the selection of eigenvectors for the longitudinal and lateral directional modes of aircraft flight control systems.

The assignment of the eigenvalue spectrum alone does not lead to a unique feedback matrix because the feedback matrix also assigns the eigenvectors associated with the eigenvalues. The assignment of the entire eigenstructure, consisting of both the eigenvalues and the eigenvectors, leads to a unique feedback matrix. The method of entire eigenstructure assignment using state feedback presented in this chapter identifies the subspaces in which the eigenvectors associated with each assigned eigenvalue must be located. The ability to select from many possible eigenvectors for a given eigenvalue provides the means for adjusting the magnitude of each mode that appears in the output. This method of shaping the output provides the potential for achieving the "best possible" response characteristics. Following the selection of the eigenvectors, a simple formula is used to compute the required feedback gain matrix. These methods are easily implemented on a digital computer.[7]

The method of entire eigenstructure assignment is useful in the case of systems that are not completely controllable. In such systems some of the eigenvalues cannot be changed, but the associated eigenvectors can be changed. The eigenvector determines the magnitude of the uncontrollable mode in each of the states and is adjustable.

The designer must be aware of the fact that speeding up the output response by use of feedback requires an increase in plant effort. This means that the signal $u(t)$ must increase in order to increase the speed of response of $y(t)$. The input $u(t)$ to the plant may have maximum limits and rates of change that cannot be exceeded. An example of this is the aircraft in which the control surfaces have maximum displacements. Increasing the signal inputs to the servo actuators that drive the control surfaces may cause these surfaces to reach their mechanical maximum limits. Further increases in the inputs cause a saturation of the control surface deflections, and the system becomes nonlinear. It is therefore necessary in the design process to obtain a simulation of the complete system performance. Bounds on the maximum inputs for which the system remains linear can be determined from the simulation. Stability of the system in the presence of saturated control signals must be carefully investigated.

An important area of control theory is the use of digital controllers to achieve the desired system response. In such cases a discrete domain model representation of the plant may be necessary.[14] The method of entire eigenstructure assignment using state feedback is readily applied to such systems.[4,6,7] Chapter 17 provides an introduction to the design of output feedback control systems. The use of output feedback eliminates the requirement either to measure all of the states or to use an observer to generate an estimate of the state vector.

REFERENCES

1. D'Azzo, J. J., and C. H. Houpis: *Linear Control System Analysis & Design: Conventional and Modern*, 3rd ed., McGraw-Hill, New York, 1988.
2. Porter, B.: *Synthesis of Dynamical Systems*, Nelson, London, 1969.

3. Porter, B., and J. J. D'Azzo: "Algorithm for the Synthesis of State-Feedback Regulators by Entire Eigenstructure Assignment," *Electron. Lett.*, vol. 13, pp. 230–231, 1977.
4. D'Azzo, J. J.: "Synthesis of Linear Multivariable Sampled-Data Feedback Control Systems by Entire Eigenstructure Assignment," Ph.D. thesis, University of Salford, England, 1978.
5. Moore, B. C.: "On the Flexibility Offered by State Feedback in Multivariable Systems Beyond Closed-Loop Eigenvalue Assignment," *IEEE Trans. Autom. Control*, vol. AC-21, pp. 689–692, 1976.
6. Porter, B., and J. J. D'Azzo: "Closed-Loop Eigenstructure Assignment by State Feedback in Multivariable Linear Systems," *Int. J. Control*, vol. 27, pp. 487–492, 1978.
7. Porter, B., and J. J. D'Azzo: "Algorithm for Closed-Loop Eigenstructure Assignment by State Feedback in Multivariable Linear Systems," *Int. J. Control*, vol. 27, pp. 943–947, 1978.
8. Rosenbrock, H. H.: *State Space and Multivariable Theory*, Nelson, London, 1970.
9. Hautus, M. L. J.: "Controllability and Observability Conditions of Linear Autonomous Systems," *Proc. Ned. Acad.*, ser. A, vol. 72, pp. 443–448, 1969.
10. Porter, B., and A. Bradshaw: "Design of Linear Multivariable Continuous-Time Tracking Systems," *Int. J. Syst. Sci.*, vol. 5, pp. 1155–1164, 1974.
11. Davison, E. J., and H. W. Smith: "Pole Assignment in Linear Time-Invariant Multivariable Systems with Constant Disturbances," *Automatica*, vol. 7, pp. 489–498, 1971.
12. Shapiro, E. Y., A. N. Andry, and J. C. Chung: "On Eigenstructure Assignment for Linear Systems," *IEEE Trans. Aerosp. Electron. Syst.*, vol. 19, pp. 711–729, 1983.
13. Reid, J. G.: *Linear System Fundamentals: Continuous and Discrete, Classic and Modern*, McGraw-Hill, New York, 1983.
14. Kwakernaak, H., and R. Sivan: *Linear Optimal Control Systems*, Wiley-Interscience, New York, 1972.

CHAPTER
17

DESIGN OF TRACKING SYSTEMS USING OUTPUT FEEDBACK

17.1 INTRODUCTION

In the MIMO tracking systems described in Chapter 16, the system outputs $\mathbf{y}(t)$ effectively follow constant input commands $\mathbf{r}(t)$ in the steady state. The outputs are also decoupled and noninteracting so that each output follows only its respective input; that is, $y_1(t)$ tracks $r_1(t)$, $y_2(t)$ tracks $r_2(t)$, etc. There is only a transient interaction from other inputs, and the amplitudes of those transients are very small. The design method of Chapter 16 requires that all the states must be fed back as shown in Fig. 16.1. Depending upon the dimension of the state vector, it may not be feasible to physically include sensors that will measure all of the states. Therefore, it is often necessary to use a state estimator or observer to reconstruct the states, and then these estimated states are fed back.

The control system configuration of Fig. 17.1 uses only output feedback to generate an error vector $\mathbf{e}(t) = \mathbf{r}(t) - \mathbf{y}(t)$. The direct use of the output measurements to produce the input signal to the controller avoids the requirement for measuring or reconstructing the entire state vector. Further, based upon a high-gain approach, the closed-loop system exhibits a distinctive asymptotic structure in which there are *slow* and *fast* modes. The slow modes are asymptotically uncontrollable or unobservable, and the output response is therefore dominated by the fast modes. This leads to very fast tracking of the command input by the output. This chapter presents a method for designing robust control systems, as shown in Fig. 17.1.

The design is dependent upon the first Markov parameter that is equal to the matrix product **CB**. When **CB** has maximal rank the plant is described as *regular*. When **CB** does not have maximal rank, then the plant is described as *irregular*. In the case of regular plants the controller implements a proportional plus integral control law as shown in Fig. 17.1. In the case of irregular plants, the proportional plus integral control is augmented with an inner-loop that provides extra measurements for control purposes. (See Sec. 17.7.)

The analysis and design involved in the tracking problems associated with this chapter, like the previous chapters, rely on the use of the CAD packages such as MATLAB (see Sec. 10.6 and App. B). The design of output feedback tracking systems generally requires design iterations that facilitate achieving an acceptable solution.

17.2   OUTPUT FEEDBACK TRACKING SYSTEM

The state and output equations of a MIMO plant are

$$\dot{\mathbf{x}}(t) = \mathbf{A}\mathbf{x}(t) + \mathbf{B}\mathbf{u}(t) \tag{17.1}$$

$$\mathbf{y}(t) = \mathbf{C}\mathbf{x}(t) \tag{17.2}$$

where the dimension of **A** is $n \times n$, **B** is $n \times m$, **C** is $\ell \times n$, and the rank of **B** is m. These equations are often available or may be transformed to the partitioned form

$$\begin{bmatrix} \dot{\mathbf{x}}_1(t) \\ \dot{\mathbf{x}}_2(t) \end{bmatrix} = \begin{bmatrix} \mathbf{A}_{11} & \mathbf{A}_{12} \\ \mathbf{A}_{21} & \mathbf{A}_{22} \end{bmatrix} \begin{bmatrix} \mathbf{x}_1(t) \\ \mathbf{x}_2(t) \end{bmatrix} + \begin{bmatrix} \mathbf{0} \\ \mathbf{B}_2 \end{bmatrix} \mathbf{u}(t) \tag{17.3}$$

$$\mathbf{y}(t) = [\, \mathbf{C}_1 \quad \mathbf{C}_2 \,] \begin{bmatrix} \mathbf{x}_1(t) \\ \mathbf{x}_2(t) \end{bmatrix} \tag{17.4}$$

where $\mathbf{x}_2$ is $m \times 1$, $\mathbf{B}_2$ is $m \times m$ and has rank m, $\mathbf{C}_2$ is $m \times m$ and has rank m, and the remaining elements in these equations have appropriate dimensions. The design method for the tracker presented in this chapter requires that the number of controlled outputs $\mathbf{y}(t)$ must be equal to the number of controls $\mathbf{u}(t)$. Thus, $\ell = m$. A high-gain controller implements a proportional plus integral (PI) control law represented by

$$\mathbf{u}(t) = g\{\mathbf{K}_1\mathbf{e}(t) + \mathbf{K}_2\mathbf{z}(t)\} \tag{17.5}$$

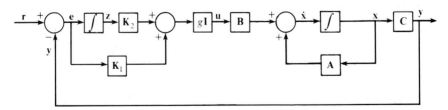

FIGURE 17.1
Output feedback tracking system containing a proportional plus integral (PI) controller.

where g is a scalar gain. The error vector between the constant command input $\mathbf{r}(t)$ and the output $\mathbf{y}(t)$ is $\mathbf{e}(t) = \mathbf{r}(t) - \mathbf{y}(t)$. The integral of the error is the vector $\mathbf{z}(t) = \int_0^t \mathbf{e}(t)\,dt$, which therefore satisfies the equation

$$\dot{\mathbf{z}}(t) = \mathbf{r}(t) - \mathbf{y}(t) \tag{17.6}$$

The closed-loop tracking system of Fig. 17.1 is represented by Eqs. (17.3) through (17.6). Combining these equations, the composite closed-loop state and output equations have the respective forms

$$\begin{bmatrix} \dot{\mathbf{z}}(t) \\ \dot{\mathbf{x}}_1(t) \\ \dot{\mathbf{x}}_2(t) \end{bmatrix} = \left[\begin{array}{cc:c} 0 & -\mathbf{C}_1 & -\mathbf{C}_2 \\ 0 & \mathbf{A}_{11} & \mathbf{A}_{12} \\ \hdashline g\mathbf{B}_2\mathbf{K}_2 & \mathbf{A}_{21} - g\mathbf{B}_2\mathbf{K}_1\mathbf{C}_1 & \mathbf{A}_{22} - g\mathbf{B}_2\mathbf{K}_1\mathbf{C}_2 \end{array} \right] \begin{bmatrix} \mathbf{z}(t) \\ \mathbf{x}_1(t) \\ \mathbf{x}_2(t) \end{bmatrix}$$

$$+ \left[\begin{array}{c} \mathbf{I}_\ell \\ 0 \\ \hdashline g\mathbf{B}_2\mathbf{K}_1 \end{array} \right] \mathbf{r}(t) \tag{17.7}$$

and
$$\mathbf{y}(t) = [\, \mathbf{0} \quad \mathbf{C}_1 \; \vdots \; \mathbf{C}_2\,] \begin{bmatrix} \mathbf{z}(t) \\ \mathbf{x}_1(t) \\ \mathbf{x}_2(t) \end{bmatrix} \tag{17.8}$$

The partitioning of the matrices in these equations, indicated by the dashed lines, is used in the following section. These closed-loop equations may be represented in these forms, respectively,

$$\dot{\bar{\mathbf{x}}}(t) = \begin{bmatrix} \bar{\mathbf{A}}_1 & \bar{\mathbf{A}}_2 \\ \bar{\mathbf{A}}_3 & \bar{\mathbf{A}}_4 \end{bmatrix} \bar{\mathbf{x}}(t) + \begin{bmatrix} \bar{\mathbf{B}}_1 \\ \bar{\mathbf{B}}_2 \end{bmatrix} \mathbf{r}(t) = \bar{\mathbf{A}}\bar{\mathbf{x}}(t) + \bar{\mathbf{B}}\mathbf{r}(t) \tag{17.9}$$

$$\mathbf{y}(t) = [\,\bar{\mathbf{C}}_1 \quad \bar{\mathbf{C}}_2\,]\bar{\mathbf{x}}(t) = \bar{\mathbf{C}}\bar{\mathbf{x}}(t) \tag{17.10}$$

where the dimension of $\bar{\mathbf{A}}_1$ is $n \times n$, $\bar{\mathbf{A}}_4$ is $m \times m$, $\bar{\mathbf{B}}_2$ is $m \times m$ and has rank m, and the remaining elements have corresponding dimensions. The closed-loop transfer function for Eqs. (17.7) and (17.8) is

$$\mathbf{G}(\lambda) = [\, \mathbf{0} \quad \mathbf{C}_1 \quad \mathbf{C}_2\,]$$

$$\times \left[\begin{array}{ccc} \lambda\mathbf{I}_\ell & \mathbf{C}_1 & \mathbf{C}_2 \\ 0 & \lambda\mathbf{I}_{n-\lambda} - \mathbf{A}_{11} & -\mathbf{A}_{12} \\ -g\mathbf{B}_2\mathbf{K}_2 & -\mathbf{A}_{21} + g\mathbf{B}_2\mathbf{K}_1\mathbf{C}_1 & \lambda\mathbf{I}_\ell - \mathbf{A}_{22} + g\mathbf{B}_2\mathbf{K}_1\mathbf{C}_2 \end{array} \right]^{-1} \left[\begin{array}{c} \mathbf{I}_\ell \\ 0 \\ g\mathbf{B}_2\mathbf{K}_1 \end{array} \right]$$

$$\tag{17.11}$$

The form of the **B** matrix in Eq. (17.3) and the corresponding partitioning in Eqs. (17.3) and (17.4) are introduced in order to develop the properties of the closed-loop system under high-gain operation. This is achieved for the control law of Eq. (17.5) when g becomes large and, in the limit, asymptotically approaches infinity: $g \to \infty$. These properties are derived by first using singular perturbation methods in Sec. 17.3 to block diagonalize the closed-loop plant matrix in Eq. (17.7). The resulting block diagonalization is presented in Sec. 17.3, and the analysis in Sec. 17.4 leads to definitions of the asymptotic slow and fast modes of the closed-loop system as $g \to \infty$. This further leads to a design procedure which is outlined in Sec. 17.5. Finally, a design example is presented in Sec. 17.6.

17.3 BLOCK DIAGONALIZATION

In this section the closed-loop system properties are determined when the scalar gain g in the control law of Eq. (17.5) has large values approaching infinity. In order to obtain these properties the state and output equations given in Eqs. (17.7) and (17.8) are transformed to the block diagonal form[1-6] given by

$$\begin{bmatrix} \dot{x}_s(t) \\ \dot{x}_f(t) \end{bmatrix} = \begin{bmatrix} A_s & 0 \\ 0 & A_f \end{bmatrix} \begin{bmatrix} x_s(t) \\ x_f(t) \end{bmatrix} + \begin{bmatrix} B_s \\ B_f \end{bmatrix} r(t) \tag{17.12}$$

$$y(t) = \begin{bmatrix} C_s & C_f \end{bmatrix} \begin{bmatrix} x_s(t) \\ x_f(t) \end{bmatrix} \tag{17.13}$$

The transformation is achieved by using

$$\bar{x}(t) = T \begin{bmatrix} x_s(t) \\ x_f(t) \end{bmatrix} \tag{17.14}$$

The transformation matrix T is selected[2-5] as

$$T = \begin{bmatrix} I_1 & M \\ -L & I_2 - LM \end{bmatrix} \tag{17.15}$$

where I_1 and I_2 are identity matrices of appropriate dimensions. The inverse of this matrix is equal to

$$T^{-1} = \begin{bmatrix} I_1 - ML & -M \\ L & I_2 \end{bmatrix} \tag{17.16}$$

Inserting the transformation of Eqs. (17.14) and (17.15) into Eq. (17.9) yields the new plant matrix

$$A_T = T^{-1}\bar{A}T = \begin{bmatrix} I_1 - ML & -M \\ L & I_2 \end{bmatrix} \begin{bmatrix} \bar{A}_1 & \bar{A}_2 \\ \bar{A}_3 & \bar{A}_4 \end{bmatrix} \begin{bmatrix} I_1 & M \\ -L & I_2 - LM \end{bmatrix} = \begin{bmatrix} A_s & A_x \\ A_y & A_f \end{bmatrix} \tag{17.17}$$

where
$$\begin{aligned} A_s &= \bar{A}_1 - ML\bar{A}_1 - M\bar{A}_3 - \bar{A}_2L + ML\bar{A}_2L + M\bar{A}_4L \\ &= \bar{A}_1 - \bar{A}_2L - MA_y \end{aligned} \tag{17.18}$$

$$\begin{aligned} A_x &= \bar{A}_1M - ML\bar{A}_1M - M\bar{A}_3M + \bar{A}_2 - ML\bar{A}_2 \\ &\quad - M\bar{A}_4 - \bar{A}_2LM + ML\bar{A}_2LM + M\bar{A}_4LM \\ &= (\bar{A}_1 - \bar{A}_2L)M + \bar{A}_2 - M(L\bar{A}_2 + \bar{A}_4) - MA_yM \end{aligned} \tag{17.19}$$

$$\mathbf{A}_y = \mathbf{L}\bar{\mathbf{A}}_1 + \bar{\mathbf{A}}_3 - \mathbf{L}\bar{\mathbf{A}}_2\mathbf{L} - \bar{\mathbf{A}}_4\mathbf{L} \tag{17.20}$$

$$\mathbf{A}_f = \mathbf{L}\bar{\mathbf{A}}_1\mathbf{M} + \bar{\mathbf{A}}_3\mathbf{M} + \mathbf{L}\bar{\mathbf{A}}_2 + \bar{\mathbf{A}}_4 - \mathbf{L}\bar{\mathbf{A}}_2\mathbf{L}\mathbf{M} - \bar{\mathbf{A}}_4\mathbf{L}\mathbf{M}$$

$$= \mathbf{A}_y\mathbf{M} + \mathbf{L}\bar{\mathbf{A}}_2 + \bar{\mathbf{A}}_4 \tag{17.21}$$

In order for the resulting matrix of Eq. (17.17) to have the block diagonal form shown in Eq. (17.12), it is necessary that $\mathbf{A}_x = \mathbf{A}_y = \mathbf{0}$. Inserting these conditions into Eqs. (17.18) and (17.21) yields

$$\mathbf{A}_s = \bar{\mathbf{A}}_1 - \bar{\mathbf{A}}_2\mathbf{L} \tag{17.22}$$

$$\mathbf{A}_f = \mathbf{L}\bar{\mathbf{A}}_2 + \bar{\mathbf{A}}_4 \tag{17.23}$$

The value of the matrix $\mathbf{L}$ is determined by using the series form

$$\mathbf{L} = \mathbf{L}_0 + g^{-1}\mathbf{L}_1 + g^{-2}\mathbf{L}_2 + \cdots \tag{17.24}$$

Inserting Eq. (17.24) into the equation $\mathbf{A}_y = \mathbf{0}$ and taking the limit as $g \to \infty$ yields (see Prob. 17.7)

$$\mathbf{L}_0 = [\, -\mathbf{C}_2^{-1}\mathbf{K}_1^{-1}\mathbf{K}_2 \quad \mathbf{C}_2^{-1}\mathbf{C}_1 \,] \tag{17.25}$$

The first-order approximation of $\mathbf{L}$ as $g \to \infty$ is therefore

$$\mathbf{L} = \mathbf{L}_0 = [\, -\mathbf{C}_2^{-1}\mathbf{K}_1^{-1}\mathbf{K}_2 \quad \mathbf{C}_2^{-1}\mathbf{C}_1 \,] \tag{17.26}$$

The evaluation of $\mathbf{M}$ is determined by using the series form

$$\mathbf{M} = \mathbf{M}_0 + g^{-1}\mathbf{M}_1 + \cdots \tag{17.27}$$

The necessary conditions for obtaining the block diagonal plant matrix of Eq. (17.12) require that $\mathbf{A}_x = \mathbf{0}$ and $\mathbf{A}_y = \mathbf{0}$. Including these equations in the values of $\mathbf{A}_x$ and $\mathbf{A}_y$ given in Eqs. (17.19) and (17.20) yields

$$\mathbf{A}_x = (\bar{\mathbf{A}}_1 - \bar{\mathbf{A}}_2\mathbf{L})\mathbf{M} - \mathbf{M}(\mathbf{L}\bar{\mathbf{A}}_2 + \bar{\mathbf{A}}_4) + \bar{\mathbf{A}}_2 = \mathbf{0} \tag{17.28}$$

The expansion of the left side of Eq. (17.28), in terms of $\mathbf{L}$ and $\mathbf{M}$ given by Eqs. (17.24) and (17.27), produces the term containing the highest power of g as $g\mathbf{M}_0\mathbf{B}_2\mathbf{K}_1\mathbf{C}_2 = \mathbf{0}$. Then, as $g \to \infty$, because $\mathbf{B}_2\mathbf{K}_1\mathbf{C}_2$ is nonsingular, the necessary condition is that $\mathbf{M}_0 = \mathbf{0}$. Therefore, a first-order approximation of $\mathbf{M}$ is $\mathbf{M} = g^{-1}\mathbf{M}_1$. Inserting this value into Eq. (17.28) and taking the limit as $g \to \infty$, the nonzero terms remaining are

$$\mathbf{M}_1\mathbf{B}_2\mathbf{K}_1\mathbf{C}_2 + \begin{bmatrix} -\mathbf{C}_2 \\ \mathbf{A}_{12} \end{bmatrix} = \mathbf{0} \tag{17.29}$$

The first-order approximation of $\mathbf{M}$ is therefore

$$\mathbf{M} = g^{-1}\mathbf{M}_1 = \begin{bmatrix} g^{-1}\mathbf{K}_1^{-1}\mathbf{B}_2^{-1} \\ g^{-1}\mathbf{A}_{12}\mathbf{C}_2^{-1}\mathbf{K}_1^{-1}\mathbf{B}_2^{-1} \end{bmatrix} \tag{17.30}$$

With the values of $\mathbf{L}$ and $\mathbf{M}$ determined by Eqs. (17.26) and (17.30), respectively, the transformation matrix $\mathbf{T}$ of Eq. (17.15) can now be obtained.

As $g \to \infty$, using the value of $\mathbf{L}$ in Eq. (17.26), the matrix $\mathbf{A}_s$ of Eq. (17.22) therefore becomes

$$\mathbf{A}_s = \begin{bmatrix} -\mathbf{K}_1^{-1}\mathbf{K}_2 & \mathbf{0} \\ \mathbf{A}_{12}\mathbf{C}_2^{-1}\mathbf{K}_1^{-1}\mathbf{K}_2 & \mathbf{A}_{11} - \mathbf{A}_{12}\mathbf{C}_2^{-1}\mathbf{C}_1 \end{bmatrix} \tag{17.31}$$

Similarly, as $g \to \infty$, Eq. (17.21) yields

$$\mathbf{A}_f = \mathbf{L}\bar{\mathbf{A}}_2 + \bar{\mathbf{A}}_4 = -g\mathbf{B}_2\mathbf{K}_1\mathbf{C}_2 \tag{17.32}$$

Inserting the transformation of Eqs. (17.14) into Eqs. (17.7) and (17.9) yields the new control matrix

$$\begin{bmatrix} \mathbf{B}_s \\ \mathbf{B}_f \end{bmatrix} = \mathbf{T}^{-1}\bar{\mathbf{B}} = \begin{bmatrix} \mathbf{I}_1 - \mathbf{ML} & -\mathbf{M} \\ \mathbf{L} & \mathbf{I}_2 \end{bmatrix}\begin{bmatrix} \bar{\mathbf{B}}_1 \\ \bar{\mathbf{B}}_2 \end{bmatrix} = \begin{bmatrix} \mathbf{I}_1 & -g^{-1}\mathbf{M}_1 \\ \mathbf{L}_0 & \mathbf{I}_2 \end{bmatrix}\begin{bmatrix} \mathbf{I}_\ell \\ \mathbf{0} \\ \hline g\mathbf{B}_2\mathbf{K}_1 \end{bmatrix} \tag{17.33}$$

Thus,

$$\mathbf{B}_s = \begin{bmatrix} \mathbf{0} \\ \mathbf{A}_{12}\mathbf{C}_2^{-1} \end{bmatrix} \tag{17.34}$$

and, as $g \to \infty$,

$$\mathbf{B}_f = g\mathbf{B}_2\mathbf{K}_1 \tag{17.35}$$

The output equation is also transformed to

$$\mathbf{y}(t) = \bar{\mathbf{C}}\bar{\mathbf{x}}(t) = \bar{\mathbf{C}}\mathbf{T}\begin{bmatrix} \mathbf{x}_s(t) \\ \mathbf{x}_f(t) \end{bmatrix} = [\bar{\mathbf{C}}_1 \quad \bar{\mathbf{C}}_2]\begin{bmatrix} \mathbf{I}_1 & g^{-1}\mathbf{M}_1 \\ -\mathbf{L}_0 & \mathbf{I}_2 - \mathbf{L}_0\mathbf{M}_1 \end{bmatrix}\begin{bmatrix} \mathbf{x}_s(t) \\ \mathbf{x}_f(t) \end{bmatrix}$$

$$= [\mathbf{C}_s \quad \mathbf{C}_f]\begin{bmatrix} \mathbf{x}_s(t) \\ \mathbf{x}_f(t) \end{bmatrix} \tag{17.36}$$

Using the values obtained from Eqs. (17.8), (17.25), and (17.30), the elements of the output matrix, as $g \to \infty$, are

$$\mathbf{C}_s = [\mathbf{K}_1^{-1}\mathbf{K}_2 \quad \mathbf{0}] \tag{17.37}$$

$$\mathbf{C}_f = \mathbf{C}_2 \tag{17.38}$$

17.4 ANALYSIS OF CLOSED-LOOP SYSTEM PERFORMANCE

The system of Fig. 17.1 contains a proportional plus integral controller that generates the input vector $\mathbf{u}$ to the plant. The signal $\mathbf{u}(t)$ is formed by the control law of Eq. (17.5). The composite closed-loop system is represented by the state and output equations (17.7) and (17.8), and the exact overall transfer function is given by Eq. (17.11).

In order to analyze the closed-loop system performance, the transformation of Eq. (17.14) is introduced in Sec. 17.3. By properly selecting the transformation matrix $\mathbf{T}$ given by Eq. (17.15), the closed-loop state and output equations are then given by Eqs. (17.12) and (17.13) for the case when the scalar gain $g \rightarrow \infty$. In Eq. (17.12) the plant matrix is in block diagonal form, and the subblocks $\mathbf{A}_s$ and $\mathbf{A}_f$ are represented by Eqs. (17.31) and (17.32), respectively. These submatrices are first analyzed in order to determine the nature of the closed-loop eigenvalues. They are then used to synthesize the gain matrices $\mathbf{K}_1$ and $\mathbf{K}_2$ in order to implement the control law of Eq. (17.5).

The closed-loop characteristic equation when $g \rightarrow \infty$, obtained from the plant matrix of Eq. (17.12), is

$$\begin{vmatrix} \lambda\mathbf{I}_n - \mathbf{A}_s & \mathbf{0} \\ \mathbf{0} & \lambda\mathbf{I}_m - \mathbf{A}_f \end{vmatrix} = \left|\lambda\mathbf{I}_n - \mathbf{A}_s\right| \cdot \left|\lambda\mathbf{I}_m - \mathbf{A}_f\right| = 0 \qquad (17.39)$$

Using the matrices for $\mathbf{A}_s$ and $\mathbf{A}_f$ given in Eqs. (17.31) and (17.32), the closed-loop characteristic equation is

$$\left|\lambda\mathbf{I}_m + \mathbf{K}_1^{-1}\mathbf{K}_2\right| \cdot \left|\lambda\mathbf{I}_{n-m} - (\mathbf{A}_{11} - \mathbf{A}_{12}\mathbf{C}_2^{-1}\mathbf{C}_1)\right| \cdot \left|\lambda\mathbf{I}_m + g\mathbf{B}_2\mathbf{K}_1\mathbf{C}_2\right| = 0 \quad (17.40)$$

The components of Eq. (17.40) lead to three classes of eigenvalues.[1]

Eigenvalue Set S_1. The first class of m eigenvalues is designated by the set S_1 and is determined by

$$\left|\lambda\mathbf{I}_m + \mathbf{K}_1^{-1}\mathbf{K}_2\right| = 0 \qquad (17.41)$$

These eigenvalues lead to *slow* modes in the output. The designation of *slow* is based on the fact that, although $g \rightarrow \infty$, these eigenvalues have finite values that are determined by the values of $\mathbf{K}_1$ and $\mathbf{K}_2$. These control law matrices are selected by the designer, and therefore the set of eigenvalues S_1 can be assigned as desired. Because $\mathbf{A}_s$ is lower triangular and $\mathbf{B}_s$ has an upper zero block [see Eqs. (17.31) and (17.34)], the modes associated with the set of eigenvalues S_1 are uncontrollable.

Eigenvalue Set S_2. The second class of $n - m$ eigenvalues is designated by the set S_2 and is determined by

$$\left|\lambda\mathbf{I}_{n-m} - \left[\mathbf{A}_{11} - \mathbf{A}_{12}\mathbf{C}_2^{-1}\mathbf{C}_1\right]\right| = 0 \qquad (17.42)$$

Because this equation is a function only of the plant [see Eqs. (17.3) and (17.4)], the set of eigenvalues S_2 cannot be changed by selection of the control law matrices $\mathbf{K}_1$ and $\mathbf{K}_2$. These eigenvalues have finite values, and the associated modes are therefore also designated as *slow* modes. Because these eigenvalues cannot be changed by the control law, it is essential that they be located in the stable region. These eigenvalues are also identified in the technical literature as *transmission zeros*.[1,7,8,10] Because $\mathbf{A}_s$

is lower triangular and C_s has a right-hand zero block [see Eqs. (17.31) and (17.37)], the modes associated with the set of eigenvalues S_2 are unobservable. Therefore, they will not appear in the output.

Eigenvalue Set S_3. The third class of m eigenvalues is designated by the set S_3 and is determined by

$$\left| \lambda \mathbf{I}_m + g\mathbf{B}_2\mathbf{K}_1\mathbf{C}_2 \right| = 0 \tag{17.43}$$

Because the matrix $\mathbf{C}_2$ is a square matrix having full rank, Eq. (17.43) can be shown to be equivalent to (see Prob. 17.8)

$$\left| \lambda \mathbf{I}_m + g\mathbf{C}_2\mathbf{B}_2\mathbf{K}_1 \right| = 0 \tag{17.44}$$

The set S_3 therefore consists of m eigenvalues which are all located at infinity because $g \to \infty$. Consequently, there are m *fast* modes in the output. The designation of *fast* is based on the fact that they can be located in the stable region at infinity by the selection of the proportional matrix $\mathbf{K}_1$.

Using Eqs. (17.12) and (17.13), the asymptotic transfer function as $g \to \infty$ is obtained as

$$\mathbf{\Gamma}(\lambda) = \mathbf{\Gamma}_s(\lambda) + \mathbf{\Gamma}_f(\lambda) = [\, \mathbf{C}_s \quad \mathbf{C}_f \,] \begin{bmatrix} \lambda \mathbf{I}_n - \mathbf{A}_s & \mathbf{0} \\ \mathbf{0} & \lambda \mathbf{I}_m - \mathbf{A}_f \end{bmatrix}^{-1} \begin{bmatrix} \mathbf{B}_s \\ \mathbf{B}_f \end{bmatrix} \tag{17.45}$$

where $\mathbf{\Gamma}_s(\lambda)$ is the slow transfer function and $\mathbf{\Gamma}_f(\lambda)$ is the fast transfer function. The slow transfer function is computed from Eq. (17.45). Using the matrices for $\mathbf{A}_s$, $\mathbf{B}_s$, and $\mathbf{C}_s$ given in Eqs. (17.31), (17.34), and (17.37) yields

$$\mathbf{\Gamma}_s(\lambda) = \mathbf{C}_s[\lambda \mathbf{I}_n - \mathbf{A}_s]^{-1}\mathbf{B}_s = \mathbf{0} \tag{17.46}$$

This is a consequence of the uncontrollability of the modes associated with the set of eigenvalues S_1 and the unobservability of the modes associated with the set of eigenvalues S_2. Therefore, the output $\mathbf{y}(t)$ contains no slow modes.

The fast transfer function is computed from Eq. (17.45). Using the matrices for $\mathbf{A}_f$, $\mathbf{B}_f$, and $\mathbf{C}_f$ given in Eqs. (17.32), (17.35), and (17.38) yields

$$\mathbf{\Gamma}_f(\lambda) = \mathbf{C}_f[\lambda \mathbf{I}_m - \mathbf{A}_f]^{-1}\mathbf{B}_f = \mathbf{C}_2[\lambda \mathbf{I}_m + g\mathbf{B}_2\mathbf{K}_1\mathbf{C}_2]^{-1}g\mathbf{B}_2\mathbf{K}_1 \tag{17.47}$$

By matrix manipulation (see Prob. 17.8), Eq. (17.47) may be rewritten as

$$\mathbf{\Gamma}_f(\lambda) = [\lambda \mathbf{I}_m + g\mathbf{C}_2\mathbf{B}_2\mathbf{K}_1]^{-1}g\mathbf{C}_2\mathbf{B}_2\mathbf{K}_1 \tag{17.48}$$

In view of Eqs. (17.46) and (17.48), the exact transfer function $\mathbf{G}(\lambda)$ represented by Eq. (17.11) has the asymptotic form, as $g \to \infty$, given by

$$\mathbf{\Gamma}(\lambda) = [\lambda \mathbf{I}_m + g\mathbf{C}_2\mathbf{B}_2\mathbf{K}_1]^{-1}g\mathbf{C}_2\mathbf{B}_2\mathbf{K}_1 \tag{17.49a}$$

The output therefore contains only the fast modes corresponding to the set of eigenvalues S_3. Because a finite value must be assigned to the scalar gain g, the output will also contain slow modes. The dominance of the fast modes is determined by the magnitude of the scalar gain g.

17.5   DESIGN PROCEDURE FOR REGULAR PLANTS

The design procedure is based on the requirement that all of the closed-loop eigen-values must be located in the LH plane so that the system is stable. In addition, non-interacting or decoupled control is desirable. This means that each output tracks its respective command input; that is, $y_1(t)$ follows $r_1(t)$, $y_2(t)$ follows $r_2(t)$, etc. Also, the amplitudes of any transients due to other inputs must be very small. The design process consists of the following steps:

1. Check the controllability and observability of the plant given by the state and output equations given by Eqs. (17.1) and (17.2), respectively.
2. Check the controllability of the augmented plant containing the PI control. This requires, in addition to controllability of the plant (step 1), that the following matrix have full rank.[9]

$$\text{Rank}\begin{bmatrix} \mathbf{B} & \mathbf{A} \\ \mathbf{0} & \mathbf{C} \end{bmatrix} = n + \ell \tag{17.50}$$

3. Determine the transmission zeros[10] that comprise set S_2 of the closed-loop eigenvalues. These are determined from

$$|\lambda \mathbf{I}_{n-m} - \mathbf{A}_{11} + \mathbf{A}_{12}\mathbf{C}_2^{-1}\mathbf{C}_1| = 0 \tag{17.51}$$

The transmission zeros must be located in the LH plane or the design cannot proceed. In the case of a controllable and observable plant, the transmission zeros can also be computed from the determinant of the system matrix which is de-fined[7] by

$$\begin{vmatrix} s\mathbf{I}_n - \mathbf{A} & \mathbf{B} \\ -\mathbf{C} & \mathbf{0} \end{vmatrix} = 0 \tag{17.52}$$

The transmission zeros are obtained from Eq. (17.52) even when the **B** matrix of the state equation is not in the partitioned form shown in Eq. (17.3). The number of transmission zeros[8] for a system having the same number of outputs and inputs, i.e., for $\ell = m$, is equal to $n - m - d$, where d is the rank deficiency of the matrix product **CB**.

4. Assign the fast eigenvalue set S_3 by use of Eq. (17.44). Because decoupling of the outputs is desired, an arbitrary diagonal matrix $\mathbf{\Sigma}$ is selected and used in

$$\mathbf{C}_2\mathbf{B}_2\mathbf{K}_1 = \mathbf{\Sigma} = \begin{bmatrix} \sigma_1 & & & \mathbf{0} \\ & \sigma_2 & & \\ & & \ddots & \\ \mathbf{0} & & & \sigma_m \end{bmatrix} \tag{17.53}$$

The values $-g\sigma_1, -g\sigma_2, \ldots, -g\sigma_m$ comprise the fast eigenvalue set S_3. From Eq. (17.53), the value of the proportional matrix $\mathbf{K}_1$ is computed from

$$\mathbf{K}_1 = (\mathbf{C}_2\mathbf{B}_2)^{-1}\mathbf{\Sigma} \tag{17.54a}$$

It is evident from Eq. (17.54) that C_2B_2 must have full rank in order for the inverse to exist. When this is true, the plant is described as regular. (Irregular plants for which C_2B_2 does not have full rank are covered in Sec. 17.7.)

The development of Eq. (17.54a) is based on the state and output equations given in Eqs. (17.3) and (17.4), in which the B matrix has the form $B = [0 \; B_2^T]^T$. When the B matrix is not in this form, it can be achieved by introducing the transformation $x = Tx'$ so that the state and output equations take the form

$$\dot{x}' = [T^{-1}AT]x' + [T^{-1}B]u \qquad y = [CT]x'$$

where the new control matrix has the form $T^{-1}B = [0 \; B_2^T]^T$ and the new output matrix is $CT = [C_1 \; C_2]$. It is now evident that the product of these matrices yields

$$[CT][T^{-1}B] = CB = [C_1 \; C_2]\begin{bmatrix} 0 \\ B_2 \end{bmatrix} = C_2B_2$$

Therefore, the matrix product CB, known as the first *Markov parameter*, is invariant with change of the state space. It is a measure only of the input-output characteristics of the system. Equations (17.49a) and (17.54a) can therefore be written, respectively, in the alternate forms as

$$\Gamma(\lambda) = [\lambda I_m + gCBK_1]^{-1}gCBK_1 \qquad (17.49b)$$

$$K_1 = [CB]^{-1}\Sigma \qquad (17.54b)$$

It is *not* necessary to determine the transformation matrix T. The matrices C and B can be used directly in Eq. (17.54b) to compute K_1.

5. Select the slow eigenvalue set S_1 and determine the integral matrix K_2 by use of Eq. (17.41). A simple and often effective selection of these eigenvalues is to assign them all to have the same value λ_o. This is achieved by using $K_1^{-1}K_2 = -\lambda_o I$, which requires that

$$K_2 = -\lambda_o K_1 \qquad (17.55)$$

6. Completion of the design process produces a diagonal asymptotic closed-loop transfer function given by Eqs. (17.49a, b) and (17.53) as

$$\Gamma(\lambda) = \text{diag}\left\{ \frac{g\sigma_1}{\lambda + g\sigma_1} \quad \frac{g\sigma_2}{\lambda + g\sigma_2} \cdots \frac{g\sigma_m}{\lambda + g\sigma_m} \right\} \qquad (17.56)$$

The value of g that produces satisfactory performance is determined from a simulation of the system response. The values of the σ_i are selected individually in the simulation in order to achieve the desired performance of each output.

17.6 REGULAR SYSTEM DESIGN EXAMPLE

The design procedure outlined in Sec. 17.5 is applied to the development of a controller for an unstable chemical reactor.[1,11] The state and output equations are given by

$$\begin{bmatrix} \dot{x}(t) \\ \dot{x}_2(t) \\ \dot{x}_3(t) \\ \dot{x}_4(t) \end{bmatrix} = \begin{bmatrix} 1.38 & -0.2077 & 6.715 & -5.676 \\ -0.5814 & -4.29 & 0 & 0.675 \\ 1.067 & 4.273 & -6.654 & 5.893 \\ 0.048 & 4.273 & 1.343 & -2.104 \end{bmatrix} \begin{bmatrix} x_1(t) \\ x_2(t) \\ x_3(t) \\ x_4(t) \end{bmatrix}$$

$$+ \begin{bmatrix} 0 & 0 \\ 5.679 & 0 \\ 1.136 & -3.146 \\ 1.136 & 0 \end{bmatrix} \begin{bmatrix} u_1(t) \\ u_2(t) \end{bmatrix} \tag{17.57}$$

and

$$\begin{bmatrix} y_1(t) \\ y_2(t) \end{bmatrix} = \begin{bmatrix} 1 & 0 & 0 & -1 \\ 0 & 1 & 0 & 0 \end{bmatrix} \begin{bmatrix} x_1(t) \\ x_2(t) \\ x_3(t) \\ x_4(t) \end{bmatrix} \tag{17.58}$$

The plant is unstable, as shown by the eigenvalue spectrum

$$\boldsymbol{\sigma}(\mathbf{A}) = \{0.06351, 1.991, -5.057, -8.666\} \tag{17.59}$$

The state equation given in Eq. (17.57) is not in the partitioned form shown in Eq. (17.3). Therefore, the procedure uses Eq. (17.54b) to take this into account.

Step 1. The open-loop plant is controllable because the controllability matrix M_c, given in Eq. (13.10), has full rank $n = 4$. The open-loop plant is observable, because the observability matrix $\mathbf{M}_o$, given in Eq. (13.15), has full rank $n = 4$.

Step 2. The augmented system with the PI controller is controllable, because the rank of the matrix formed in accordance with Eq. (17.50) has full rank $n + m = 6$.

Step 3. The transmission zeros cannot be obtained from Eq. (17.51) because the **B** matrix in Eq. (17.57) is not in the partitioned form of Eq. (17.3). Therefore, Eq. (17.52) is used and yields the transmission zeros

$$S_2 = \{-1.192, -5.039\} \tag{17.60}$$

Because these transmission zeros are in the stable region, the design can proceed.

Step 4. The fast eigenvalues in the set S_3 are all assigned the values of $-g$, thus $\boldsymbol{\Sigma} = \mathbf{I}$ is used in Eq. (17.54) to compute the proportional matrix $\mathbf{K}_1$. Because the state equation of Eq. (17.57) does not have the **B** matrix in the partitioned form of Eq. (17.3), the computation of $\mathbf{K}_1$ is obtained by using Eq. (17.54b).

$$\mathbf{K}_1 = [\mathbf{CB}]^{-1}\boldsymbol{\Sigma} = \left\{ \begin{bmatrix} 1 & 0 & 1 & -1 \\ 0 & 1 & 0 & 0 \end{bmatrix} \begin{bmatrix} 0 & 0 \\ 5.679 & 0 \\ 1.136 & -3.146 \\ 1.136 & 0 \end{bmatrix} \right\}^{-1} \begin{bmatrix} 1 & 0 \\ 0 & 1 \end{bmatrix}$$

$$= \begin{bmatrix} 0 & 0.1761 \\ -0.3179 & 0 \end{bmatrix} \tag{17.61}$$

Step 5. The two slow eigenvalues contained in the set S_1 are assigned the values $\lambda_o = -2$. Thus, from Eq. (17.55)

$$\mathbf{K}_2 = 2\mathbf{K}_1 = \begin{bmatrix} 0 & 0.3522 \\ -0.6358 & 0 \end{bmatrix} \tag{17.62}$$

Step 6. The closed-loop eigenvalues, as $g \to \infty$, are asymptotic to the values $S_1 = \{-2, -2\}$, $S_2 = \{-1.192, -5.039\}$, and $S_3 = \{-g, -g\}$. Only the fast eigenvalues from the set S_3 appear in the asymptotic transfer function given by Eq. (17.56):

$$\boldsymbol{\Gamma}(\lambda) = \begin{bmatrix} \dfrac{g}{\lambda + g} & 0 \\ 0 & \dfrac{g}{\lambda + g} \end{bmatrix} \tag{17.63}$$

A suitable value of g is determined from a computer simulation of the system. The value of g that produces acceptable values of overshoot and settling time is selected. As the value of g increases, the outputs become increasingly noninteractive and the fast modes dominate the responses. Figures 17.2 and 17.3 show the responses with the commands $\mathbf{r}(t) = [1 \ 0]^T$ and $\mathbf{r}(t) = [0 \ 1]^T$, respectively, for the value $g = 50$. It is further necessary to determine the values of $u_1(t)$ and $u_2(t)$. The system must be capable of providing this required control effort while the linear system model of Eqs. (17.57) and (17.58) remains valid. Plots of $u_1(t)$ and $u_2(t)$ must be obtained from a computer simulation to be sure that they remain within the linear operating range.

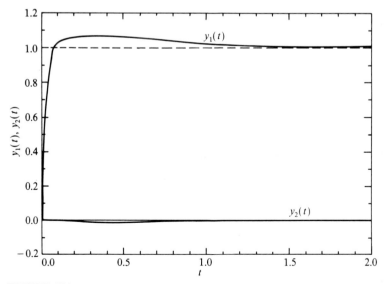

FIGURE 17.2
Step response $\mathbf{y}(t) = \{y_1(t), y_2(t)\}$ for $\mathbf{r}(t) = [1 \ 0]^T$.

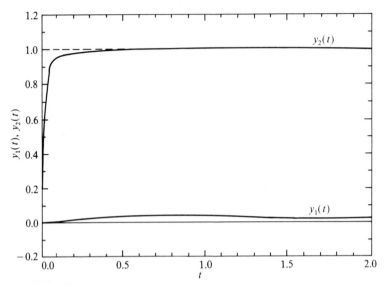

FIGURE 17.3
Step response $\mathbf{y}(t) = \{y_1(t), y_2(t)\}$ for $\mathbf{r}(t) = [0 \quad 1]^T$.

17.7   IRREGULAR PLANT CHARACTERISTICS

The fast eigenvalues for the output feedback high-gain system with PI control are determined from Eq. (17.54b) which states that $\mathbf{K}_1 = (\mathbf{CB})^{-1}\mathbf{\Sigma}$. In the case where $\mathbf{CB}$ does not have full rank, the inverse $(\mathbf{CB})^{-1}$ does not exist and $\mathbf{K}_1$ cannot be determined. When $\mathbf{CB}$ does not have full rank the system is described as irregular. Such systems can be controlled by introducing[12] the minor loop feedback shown in Fig. 17.4. This figure shows a new output described by

$$\mathbf{w}(t) = \mathbf{y}(t) + \mathbf{M}\dot{\mathbf{x}}_1 \tag{17.64}$$

When the command input to the system $\mathbf{r}(t)$ consists of step functions, steady-state values of the states for a stable system also are step functions. Because the state

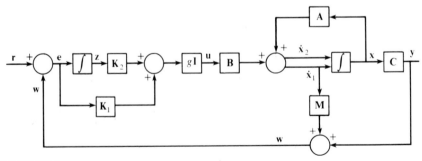

FIGURE 17.4.
A tracking system utilizing a measurement matrix (for an irregular plant).

equations for $\dot{\mathbf{x}}_1(t)$ in Eq. (17.3) have zero forcing function, the steady-state value of $\dot{\mathbf{x}}_1(t)$ is therefore

$$\lim_{t \to \infty} \dot{\mathbf{x}}_1(t) = 0 \tag{17.65}$$

Thus Eq. (17.64) shows that $\mathbf{w}(t)$ equals $\mathbf{y}(t)$ in the steady state:

$$\mathbf{w}(t)_{ss} = \mathbf{y}(t)_{ss} \tag{17.66}$$

Inserting the values obtained from Eqs. (17.3) and (17.4) into Eq. (17.64) yields the new output equation

$$\mathbf{w}(t) = [\, \mathbf{C}_1 \quad \mathbf{C}_2 \,] \begin{bmatrix} \mathbf{x}_1 \\ \mathbf{x}_2 \end{bmatrix} + \mathbf{M}[\, \mathbf{A}_{11} \quad \mathbf{A}_{12} \,] \begin{bmatrix} \mathbf{x}_1 \\ \mathbf{x}_2 \end{bmatrix}$$

$$= [(\mathbf{C}_1 + \mathbf{M}\mathbf{A}_{11}) \quad (\mathbf{C}_2 + \mathbf{M}\mathbf{A}_{12})] \begin{bmatrix} \mathbf{x}_1 \\ \mathbf{x}_2 \end{bmatrix} = [\, \mathbf{F}_1 \quad \mathbf{F}_2 \,] \begin{bmatrix} \mathbf{x}_1 \\ \mathbf{x}_2 \end{bmatrix} \tag{17.67}$$

By the proper selection[13] of the measurement matrix $\mathbf{M}$, the matrix $\mathbf{F}_2$ in Eq. (17.67) will have full rank m.

The composite continuous-time tracking system of Fig. 17.4, consisting of the plant of Eq. (17.3), the control law of Eq. (17.5), and the output of Eq. (17.67), has the state and output equations given by

$$\begin{bmatrix} \dot{\mathbf{z}}(t) \\ \dot{\mathbf{x}}_1(t) \\ \dot{\mathbf{x}}_2(t) \end{bmatrix} = \begin{bmatrix} 0 & -\mathbf{F}_1 & -\mathbf{F}_2 \\ 0 & \mathbf{A}_{11} & \mathbf{A}_{12} \\ g\mathbf{B}_2\mathbf{K}_2 & \mathbf{A}_{21} - g\mathbf{B}_2\mathbf{K}_1\mathbf{F}_1 & \mathbf{A}_{22} - g\mathbf{B}_2\mathbf{K}_1\mathbf{F}_2 \end{bmatrix} \begin{bmatrix} \mathbf{z}(t) \\ \mathbf{x}_1(t) \\ \mathbf{x}_2(t) \end{bmatrix} + \begin{bmatrix} \mathbf{I}_m \\ 0 \\ g\mathbf{B}_2\mathbf{K}_1 \end{bmatrix} \mathbf{r}(t) \tag{17.68}$$

and

$$\mathbf{y}(t) = [\, \mathbf{0} \quad \mathbf{C}_1 \mathbin{\vdots} \mathbf{C}_2 \,] \begin{bmatrix} \mathbf{z}(t) \\ \mathbf{x}_1(t) \\ \mathbf{x}_2(t) \end{bmatrix} \tag{17.69}$$

The closed-loop transfer function for Eqs. (17.68) and (17.69) is

$$\mathbf{G}(\lambda) = [\, \mathbf{0} \quad \mathbf{C}_1 \quad \mathbf{C}_2 \,] \begin{bmatrix} \lambda \mathbf{I}_\ell & \mathbf{F}_1 & \mathbf{F}_2 \\ \mathbf{0} & \lambda \mathbf{I}_{n-\ell} - \mathbf{A}_{11} & -\mathbf{A}_{12} \\ -g\mathbf{B}_2\mathbf{K}_2 & -\mathbf{A}_{21} + g\mathbf{B}_2\mathbf{K}_1\mathbf{F}_1 & \lambda \mathbf{I}_\ell - \mathbf{A}_{22} + g\mathbf{B}_2\mathbf{K}_1\mathbf{F}_2 \end{bmatrix}^{-1}$$

$$\times \begin{bmatrix} \mathbf{I}_\ell \\ \mathbf{0} \\ g\mathbf{B}_2\mathbf{K}_1 \end{bmatrix} \tag{17.70}$$

Using the procedure illustrated in Sec. 17.3, the state and output equations given in Eqs. (17.3) and (17.68) are transformed to the forms of Eqs. (17.12) and (17.13) where the composite plant matrix is block diagonal. Taking the limit as $g \to \infty$ yields the components

$$\mathbf{A}_s = \begin{bmatrix} -\mathbf{K}_1^{-1}\mathbf{K}_2 & \mathbf{0} \\ \mathbf{A}_{12}\mathbf{F}_2^{-1}\mathbf{K}_1^{-1}\mathbf{K}_2 & \mathbf{A}_{11} - \mathbf{A}_{12}\mathbf{F}_2^{-1}\mathbf{F}_1 \end{bmatrix} \tag{17.71}$$

$$\mathbf{B}_s = \begin{bmatrix} \mathbf{0} \\ \mathbf{A}_{12}\mathbf{F}_2^{-1} \end{bmatrix} \tag{17.72}$$

$$\mathbf{C}_s = [\, \mathbf{C}_2\mathbf{F}_2^{-1}\mathbf{K}_1^{-1}\mathbf{K}_2 \qquad \mathbf{C}_1 - \mathbf{C}_2\mathbf{F}_2^{-1}\mathbf{F}_1 \,] \tag{17.73}$$

$$\mathbf{A}_f = -g\mathbf{B}_2\mathbf{K}_1\mathbf{F}_2 \tag{17.74}$$

$$\mathbf{B}_f = g\mathbf{B}_2\mathbf{K}_1 \tag{17.75}$$

$$\mathbf{C}_f = \mathbf{C}_2 \tag{17.76}$$

17.8   IRREGULAR SYSTEM PERFORMANCE

The system of Fig. 17.4 shows a tracker for a plant that is irregular; that is, $\mathbf{CB}$ does not have full rank. A new output $\mathbf{w}(t)$ is formed by

$$\mathbf{w}(t) = \mathbf{y}(t) + \mathbf{M}\dot{\mathbf{x}}_1(t) = [\, \mathbf{F}_1 \quad \mathbf{F}_2 \,]\begin{bmatrix} \mathbf{x}_1 \\ \mathbf{x}_2 \end{bmatrix} \tag{17.77}$$

where
$$\mathbf{F}_1 = \mathbf{C}_1 + \mathbf{MA}_{11} \tag{17.78}$$

$$\mathbf{F}_2 = \mathbf{C}_2 + \mathbf{MA}_{12} \tag{17.79}$$

The measurement matrix $\mathbf{M}$ in Eqs. (17.78) and (17.79) is selected[13] to be sparse, containing only sufficient elements to ensure that $\mathbf{F}_2$ has full rank. This is essential to ensure that the matrix product $\mathbf{F}_2\mathbf{B}_2$ has rank m.

Eigenvalue Set S_1. An evaluation similar to that presented in Sec. 17.4 leads to the determination of the closed-loop eigenvalues. The slow eigenvalue set S_1 is determined by

$$|\lambda\mathbf{I}_m + \mathbf{K}_1^{-1}\mathbf{K}_2| = 0 \tag{17.80}$$

These finite eigenvalues are associated with uncontrollable modes because of the form of $\mathbf{B}_s$.

Eigenvalue Set S_2. The slow eigenvalue set S_2 is determined by

$$|\lambda\mathbf{I}_{n-m} - [\mathbf{A}_{11} - \mathbf{A}_{12}\mathbf{F}_2^{-1}\mathbf{F}_1]| = 0 \tag{17.81}$$

These finite eigenvalues are the transmission zeros of the plant with the addition of minor-loop feedback. This adds transmission zeros to those of the plant. Because of the form of the matrix $\mathbf{C}_s$ [see Eq. (17.73)], the associated modes are observable in the output. This observability characteristic of the set S_2 is a major difference from the case of a regular plant. It is a consequence of the minor loop containing the measurement matrix $\mathbf{M}$ which has been added to the system.

Eigenvalue Set S_3. The fast eigenvalue set S_3 is determined by

$$|\lambda\mathbf{I}_m + g\mathbf{B}_2\mathbf{K}_1\mathbf{F}_2| = 0 \tag{17.82}$$

This characteristic equation yields the same eigenvalues as

$$|\lambda \mathbf{I}_m + g\mathbf{F}_2\mathbf{B}_2\mathbf{K}_1| = 0 \tag{17.83}$$

The modes associated with the fast eigenvalues of the set S_3 are observable in the output.

The slow transfer function determined from Eqs. (17.71), (17.72), and (17.73) is

$$\mathbf{\Gamma}_s(\lambda) = [\mathbf{C}_1 - \mathbf{C}_2\mathbf{F}_2^{-1}\mathbf{F}_1][\lambda \mathbf{I}_{n-\ell} - \mathbf{A}_{11} + \mathbf{A}_{12}\mathbf{F}_2^{-1}\mathbf{F}_1]^{-1}\mathbf{A}_{12}\mathbf{F}_2^{-1} \tag{17.84}$$

This transfer function contains only the transmission zeros.

The fast transfer function determined from Eqs. (17.74), (17.75), and (17.76) is

$$\mathbf{\Gamma}_f(\lambda) = \mathbf{C}_2\mathbf{F}_2^{-1}[\lambda \mathbf{I}_\ell + g\mathbf{F}_2\mathbf{B}_2\mathbf{K}_1]^{-1}g\mathbf{F}_2\mathbf{B}_2\mathbf{K}_1 \tag{17.85}$$

The total asymptotic transfer function as $g \to \infty$, obtained from Eq. (17.45), is the sum of the slow and fast transfer functions given by Eqs. (17.84) and (17.85):

$$\mathbf{\Gamma}(\lambda) = \mathbf{\Gamma}_s(\lambda) + \mathbf{\Gamma}_f(\lambda) = [\mathbf{C}_1 - \mathbf{C}_2\mathbf{F}_2^{-1}\mathbf{F}_1][\lambda \mathbf{I}_{n-\ell} - \mathbf{A}_{11} + \mathbf{A}_{12}\mathbf{F}_2^{-1}\mathbf{F}_1]^{-1}\mathbf{A}_{12}\mathbf{F}_2^{-1}$$
$$+ \mathbf{C}_2\mathbf{F}_2^{-1}[\lambda \mathbf{I}_\ell + g\mathbf{F}_2\mathbf{B}_2\mathbf{K}_1]^{-1}g\mathbf{F}_2\mathbf{B}_2\mathbf{K}_1 \tag{17.86}$$

This transfer function contains the slow eigenvalues (transmission zeros) of the set S_2 and the fast eigenvalues of the set S_3. Because the eigenvalue set S_1 is associated with asymptotically uncontrollable modes as $g \to \infty$, these eigenvalues do not appear in the transfer function of Eq. (17.86).

It is important to note the difference between the characteristics of the system containing a regular plant and the system containing an irregular plant. The transfer function of the tracking system containing the regular plant, given in Eq. (17.49a, b), contains only fast eigenvalues. The transfer function of the tracking system containing the irregular plant, given in Eq. (17.86), contains both the transmission zeros and the fast eigenvalues. The transmission zeros include those of the plant and those determined by the measurement matrix $\mathbf{M}$. Therefore, the presence of the associated slow modes in the output limits the speed of response that is achievable. This is shown in the example in Sec. 17.10.

17.9 DESIGN OF THE MEASUREMENT MATRIX M

The design of the control law for the tracking system may include the desirable requirement that the outputs be decoupled, which minimizes the interaction between the outputs. In order to achieve this decoupling, the component transfer functions $\mathbf{\Gamma}_s(\lambda)$ and $\mathbf{\Gamma}_f(\lambda)$ must each be diagonal. It may be possible to make these transfer functions diagonal by selecting the measurement matrix $\mathbf{M}$, but there is no guarantee that this is achievable. The procedure presented in Ref. 13 provides a method for accomplishing the selection of the measurement matrix.

The design procedure is modified from that used for regular plants because it is necessary to select the measurement matrix $\mathbf{M}$ and to achieve, if possible, a diagonal transfer function $\boldsymbol{\Gamma}(\lambda)$. The design process consists of the following steps:

1. Check the controllability and observability of the plant given by the state and output equations of Eqs. (17.1) and (17.2).
2. Check that the rank of the following matrix is satisfied:

$$\text{Rank}\begin{bmatrix} \mathbf{B} & \mathbf{A} \\ \mathbf{0} & \mathbf{C} \end{bmatrix} = n + \ell \tag{17.87}$$

3. Determine the rank of $\mathbf{C}_2\mathbf{B}_2$. If this matrix is rank deficient, select the most sparse matrix $\mathbf{M}$ which produces a matrix $\mathbf{F}_2 = \mathbf{C}_2 + \mathbf{MA}_{12}$ which has full rank. Simultaneously, the matrix product $\mathbf{C}_2\mathbf{F}_2^{-1}$ must be made diagonal, if possible. This is accomplished[13,17] by the following procedure:
   (a) First form the matrix

$$\mathbf{B}^* = \begin{bmatrix} \mathbf{c}_1^T \mathbf{A}_{11}^{d_1} \mathbf{A}_{12} \\ \vdots \\ \mathbf{c}_m^T \mathbf{A}_{11}^{d_m} \mathbf{A}_{12} \end{bmatrix} \tag{17.88}$$

   where m is the number of control inputs, $\mathbf{c}_i^T$ is the ith row of $\mathbf{C}_1$, and

$$d_i = \min\{j : \mathbf{c}_i^T \mathbf{A}_{11}^j \mathbf{A}_{12} \neq \mathbf{0}, j = 0, 1, \ldots, n-1\} \tag{17.89}$$

   Equation (17.89) specifies that d_i is the smallest value of j for which $\mathbf{c}_i^T \mathbf{A}_{11}^j \mathbf{A}_{12} \neq \mathbf{0}$. The permissible values of j are $0, 1, \ldots, n-m-1$, where n is the dimension of $\mathbf{A}_{11}$. If all of these values of j result in $\mathbf{c}_i^T \mathbf{A}_{11}^j \mathbf{A}_{11} = \mathbf{0}$, then use $d_i = n - m - 1$.
   (b) Form $\mathbf{F}_2 = \mathbf{C}_2 + \mathbf{MA}_{12}$ using a general form of the $m \times (n-m)$ matrix $\mathbf{M} = [m_{ij}]$. The elements m_{ij} appearing in $\mathbf{F}_2$ are permitted nonzero values only if $\mathbf{B}^*$ has a nonzero element in a corresponding position. Set all other elements of $\mathbf{M}$ to zero.
   (c) Form the matrix $\mathbf{C}_2\mathbf{F}_2^{-1}$, where $\mathbf{F}_2$ is obtained in step b. Make the matrix $\mathbf{C}_2\mathbf{F}_2^{-1}$ diagonal, if possible, by setting off-diagonal elements to zero. This is accomplished by setting selected values of m_{ij} to zero.
   (d) Make sure that $\mathbf{F}_2$ has enough elements m_{ij} so that it has full rank. The nonzero elements of $\mathbf{M}$ are now all identified.
4. Determine the set of eigenvalues S_2 (the transmission zeros) using Eq. (17.81) to assure that they are in the LH plane (stable).
5. Select the matrix $\mathbf{K}_1$ which makes the fast transfer function $\boldsymbol{\Gamma}_f(\lambda)$ of Eq. (17.85) diagonal. This is accomplished by selecting a diagonal matrix $\boldsymbol{\Sigma}$ such that

$$\mathbf{F}_2\mathbf{B}_2\mathbf{K}_1 = \boldsymbol{\Sigma} = \begin{bmatrix} \sigma_1 & & & \mathbf{0} \\ & \sigma_2 & & \\ & & \ddots & \\ \mathbf{0} & & & \sigma_\ell \end{bmatrix} \tag{17.90}$$

The proportional matrix is computed as

$$\mathbf{K}_1 = (\mathbf{F}_2\mathbf{B}_2)^{-1}\mathbf{\Sigma} \tag{17.91}$$

Then the fast eigenvalues are

$$S_3 = \{-\sigma_1 g, -\sigma_2 g, \dots, -\sigma_\ell g\} \tag{17.92}$$

6. Assign the integral matrix using

$$\mathbf{K}_2 = -\lambda_o\mathbf{K}_1 \tag{17.93}$$

where λ_o is adjusted to give good transient performance in the simulation.

7. Determine the asymptotic transfer function $\mathbf{\Gamma}(\lambda)$.

17.10 IRREGULAR SYSTEM DESIGN EXAMPLE

The design procedure of Sec. 17.9 is applied to the development of a controller for an aircraft flight control system.[13,14] The model represents the longitudinal dynamics of an aircraft and the variables are perturbations from a steady-state straight and level trim condition. The cartesian coordinate system shown in Fig. 17.5 has three perpendicular axes labeled X, Y, and Z, and the rotation rates around these axes are labeled p, q, and r, respectively. Motion that is confined to the X-Z plane is described

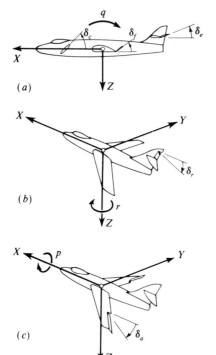

FIGURE 17.5
Airplane control surfaces. (*a*) Elevator, flaperon, and canard deflections produce pitching velocity q. (*b*) Rudder deflection produces yawing velocity r. (*c*) Aileron deflection produces rolling velocity p.

as longitudinal motion. The longitudinal dynamics are represented by the state and output equations that are partitioned as shown in Eqs. (17.3) and (17.4).

$$
\begin{bmatrix} \dot{\theta} \\ \dot{u} \\ \dot{w} \\ \dot{q} \end{bmatrix} = \left[\begin{array}{c:ccc} 0 & 0 & 0 & 1.0 \\ \hdashline -0.56025 & -0.08220 & 0.0472 & -0.34383 \\ -0.01230 & -0.05580 & -1.68 & 15.67306 \\ 0 & -0.18163 & 1.73606 & -0.253 \end{array} \right] \begin{bmatrix} \theta \\ u \\ w \\ q \end{bmatrix}
$$

$$
+ \left[\begin{array}{ccc} 0 & 0 & 0 \\ \hdashline 0.359 & -0.0479 & 0.218 \times 10^{-2} \\ -0.634 & -2.0070 & -0.744 \times 10^{-4} \\ 30.1949 & -3.7242 & -0.277 \times 10^{-3} \end{array} \right] \begin{bmatrix} \delta_c \\ \delta_f \\ \delta_t \end{bmatrix} \tag{17.94}
$$

$$
\begin{bmatrix} \theta \\ \gamma \\ v \end{bmatrix} = \left[\begin{array}{c:ccc} 1 & 0 & 0 & 0 \\ 1 & 1.4037 \times 10^{-3} & -6.41713 \times 10^{-2} & 0 \\ 0 & 1 & 0.0219 & 0 \end{array} \right] \begin{bmatrix} \theta \\ u \\ w \\ q \end{bmatrix} \tag{17.95}
$$

where the units are: θ = deg, u = ft/sec, w = ft/sec, q = deg/sec, γ = deg, v = ft/sec, δ_c = deg, δ_f = deg, and δ_t = lb. In the state equations, $\theta(t)$ is the pitch angle around the Y axis, $q(t)$ is the pitch rate around the Y axis, $u(t)$ is the perturbation velocity in the X direction, and $w(t)$ is the velocity in the Z direction. The three outputs that are to be controlled are the pitch angle $\theta(t)$, the flight path angle $\gamma(t)$, and the total aircraft velocity $v(t)$. This unconventional aircraft contains canards that are forward of the wing, and, as a result, it is able to perform uncommon maneuvers. It has the three longitudinal inputs described by δ_c (canards), δ_f (flaperons), and δ_t (thrust). The equations are partitioned in accordance with Eqs. (17.3) and (17.4). The design steps follow:

Step 1. The controllability matrix $\mathbf{M}_c$ and the observability matrix $\mathbf{M}_o$ each have full rank equal to 4, and thus the plant is completely controllable and observable.

Step 2. The matrix of Eq. (17.87) has rank $n + m = 7$, and thus the complete feedback control system containing the PI control is controllable.

Step 3. The matrix product $\mathbf{CB}$ is

$$
\mathbf{CB} = \begin{bmatrix} 0 & 0 & 0 \\ 0.0412 & 0.1287 & 7.835 \times 10^{-6} \\ 3.4512 & -0.0918 & 0.0022 \end{bmatrix} \tag{17.96}
$$

This matrix is rank deficient, having only rank 2, and thus the plant is irregular.

(a) Because $\mathbf{A}_{11} = [0]$, Eq. (17.89) yields the values $d_1 = d_2 = d_3 = 0$. Then $\mathbf{B}^*$ is formed in accordance with Eq. (17.88):

$$
\mathbf{B}^* = \begin{bmatrix} 0 & 0 & 1 \\ 0 & 0 & 1 \\ 0 & 0 & 0 \end{bmatrix} \tag{17.97}
$$

(b) In the matrix $\mathbf{F}_2 = \mathbf{C}_2 + \mathbf{MA}_{12}$, the dimension of $\mathbf{C}_2$ is 3×3. Because the dimension of $\mathbf{A}_{12}$ is 1×3, the necessary dimension of $\mathbf{M}$ is 3×1. Thus, using

$$\mathbf{M} = \begin{bmatrix} m_1 \\ m_2 \\ m_3 \end{bmatrix} \tag{17.98}$$

the matrix $\mathbf{F}_2$ is

$$\mathbf{F}_2 = \begin{bmatrix} 0 & 0 & m_1 \\ 1.4037 \times 10^{-3} & -6.41713 \times 10^{-2} & m_2 \\ 1 & 0.0219 & m_3 \end{bmatrix} \tag{17.99}$$

The assignable elements of $\mathbf{F}_2$ are permitted nonzero values only if $\mathbf{B}^*$ has a nonzero element in the same position. Thus it is required that $m_3 = 0$.

(c) The matrix $\mathbf{C}_2\mathbf{F}_2^{-1}$ is now obtained.

$$\mathbf{C}_2\mathbf{F}_2^{-1} = \begin{bmatrix} 0 & 0 & 0 \\ -m_2/m_1 & 1 & 0 \\ 0 & 0 & 1 \end{bmatrix} \tag{17.100}$$

Because $\mathbf{C}_2\mathbf{F}_2^{-1}$ must be diagonal, assign the value $m_2 = 0$.

(d) The resulting matrices $\mathbf{F}_1 = \mathbf{C}_1 + \mathbf{MA}_{11}$ and $\mathbf{F}_2 = \mathbf{C}_2 + \mathbf{MA}_{12}$ are therefore

$$\mathbf{F}_1 = \begin{bmatrix} 1 \\ 1 \\ 0 \end{bmatrix} \quad \mathbf{F}_2 = \begin{bmatrix} 0 & 0 & m_1 \\ 1.4037 \times 10^{-3} & -6.41713 \times 10^{-2} & 0 \\ 1 & 0.0219 & 0 \end{bmatrix} \tag{17.101}$$

The determination of the value of $m_1 = 0.25$ is discussed in steps 4 and 5 of this design.

Step 4. Because the $\mathbf{B}$ matrix of Eq. (17.94) is in the form $[\mathbf{0} \quad \mathbf{B}_2^T]^T$, Eq. (17.81) can be used to compute the transmission zeros. There is one transmission zero $S_2 = \{-1/m_1\}$. A stable system is therefore achieved for a positive value of m_1.

Step 5. The proportional matrix, in accordance with Eq. (17.91), is

$$\mathbf{K}_1 = (\mathbf{F}_2\mathbf{B}_2)^{-1}\boldsymbol{\Sigma} = (\mathbf{F}_2\mathbf{B}_2)^{-1}\begin{bmatrix} \sigma_1 & & \\ & \sigma_2 & \\ & & \sigma_3 \end{bmatrix}$$

At this point in the design, a complete simulation of the system is obtained, starting with assumed nominal initial values for m_1, g and $\boldsymbol{\Sigma} = \text{diag}\{\sigma_1 \quad \sigma_2 \quad \sigma_3\}$. By trial and error, values of m_1, g, and σ_i that yield acceptable performance are determined. Based on the simulation, the values $m_1 = 0.25$, $\boldsymbol{\Sigma} = \text{diag}\{1 \quad 0.05 \quad 0.2\}$, and $g = 75$ are selected. This yields

$$\mathbf{K}_1 = \begin{bmatrix} 0.1274 & 0.0461 & 0.0001 \\ -0.0394 & 0.3732 & -0.0056 \\ -21.8479 & 8.4307 & 91.5514 \end{bmatrix} \tag{17.102}$$

Step 6. A simulation of the complete system is now obtained. The value $\mathbf{K}_2 = \mathbf{K}_1$ yields good responses.

Step 7. The asymptotic closed-loop transfer function is computed using Eq. (17.86) and is shown in Eq. (17.103). This diagonal transfer function shows that the outputs are decoupled. It is also noted that the element m_1 in the measurement matrix determines the time response of the first output $\theta(t)$.

$$
\boldsymbol{\Gamma}(\lambda) = \boldsymbol{\Gamma}_s(\lambda) + \boldsymbol{\Gamma}_f(\lambda) =
\begin{bmatrix}
\dfrac{1/m_1}{\lambda + 1/m_1} & 0 & 0 \\
0 & 0 & 0 \\
0 & 0 & 0
\end{bmatrix}
+
\begin{bmatrix}
0 & 0 & 0 \\
0 & \dfrac{g\sigma_2}{\lambda + g\sigma_2} & 0 \\
0 & 0 & \dfrac{g\sigma_3}{\lambda + g\sigma_3}
\end{bmatrix}
$$

$$
=
\begin{bmatrix}
\dfrac{1/m_1}{\lambda + 1/m_1} & 0 & 0 \\
0 & \dfrac{g\sigma_2}{\lambda + g\sigma_2} & 0 \\
0 & 0 & \dfrac{g\sigma_3}{\lambda + g\sigma_3}
\end{bmatrix}
\qquad (17.103)
$$

17.11   TRACKER SIMULATION

The performance of the tracking system must be verified from a simulation of the complete system. The simulation is also used to adjust the individual parameters until the responses are satisfactory and meet the performance specifications as mentioned in step 5 of the design of Sec. 17.10. The control inputs are also obtained so that their magnitudes and rates of change are within acceptable values. For this design, the aircraft is required to perform three maneuvers which are identified as pitch pointing, vertical translation, and straight climb. The control system containing the controller designed by this method has very rapid responses with step inputs. As a result, the inputs to the plant must have the large values required to achieve the fast response. In the case of aircraft, the control surface deflections are constrained by physical limitations. For example, the flaperons in this aircraft are limited to maximum deflections of $45°$. Also, there are rate limits on the motion of the control surfaces that are necessary in order to limit the structural stresses. In addition, the size of the servo actuators which must move the control surfaces limits the speed of these control surfaces.

After the design is completed, the computer simulation must include the dynamics of the servo actuators which move the aircraft control surfaces. For this aircraft the servo transfer functions are

$$
G_s(s) = \frac{20}{s + 20} \qquad (17.104)
$$

For the engine thrust control the transfer function is $G(s) = 10/(s + 10)$.

In order to avoid accelerations of the responses which exceed the structural aircraft limitations, it is common to insert a prefilter in the input channels. Such a prefilter prevents the plant outputs from being too fast. This also keeps the control surface deflections within allowable limits. The prefilters used in this simulation have the transfer function

$$G_{PF}(s) = \frac{24}{(s + 3)(s + 8)} \tag{17.105}$$

Figure 17.6 is a block diagram showing the complete control system, including the servos and prefilters.

In a pitch pointing maneuver the aircraft is flying straight and level at a given velocity. Then a pitch up or pitch down of the angle θ is commanded, while the flight-path angle γ and velocity v are simultaneously commanded to remain unchanged. This is an unconventional maneuver for an aircraft, and can be achieved only be-cause of the addition of canards, which are not present on most aircraft. For the step command $\mathbf{r}(t) = [\theta \quad \gamma \quad v]^T = [1 \quad 0 \quad 0]^T$, the responses are shown in Fig. 17.7 and the required controls are shown in Fig. 17.8. There is a maximum flight-path angle perturbation γ of only 0.2° and the maximum velocity perturbation v is only 1.1 ft/s from a nominal velocity of 881 ft/s.

In a vertical translation command the pitch angle and forward velocity remain constant while the aircraft rises vertically. The flight-path angle shows a climb or a descent without any pitch change of the aircraft. In order to climb, the aircraft without canards must pitch up; i.e., the angle θ must increase. The ability to keep a

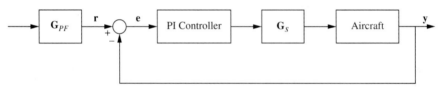

FIGURE 17.6
Complete feedback control system.

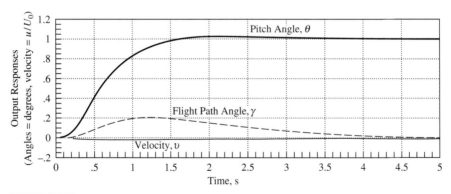

FIGURE 17.7
Pitch pointing maneuver, with actuators and prefilter.

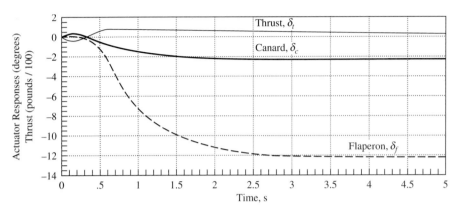

FIGURE 17.8
Controls for pitch pointing.

horizontal attitude while climbing is possible because of the presence of the canards. For the step command $\mathbf{r}(t) = [\theta \quad \gamma \quad v]^T = [0 \quad 1 \quad 0]^T$ the responses are shown in Fig. 17.9 and the controls are shown in Fig. 17.10. The response is excellent, as the maximum pitch angle and velocity perturbations are only 0.05° and 0.4 ft/s, respectively.

The straight climb maneuver consists of simultaneous pitch angle and flight-path angle commands, while holding the velocity constant. For the command $\mathbf{r}(t) = [\theta \quad \gamma \quad v]^T = [1 \quad 1 \quad 0]^T$, the responses are shown in Fig. 17.11 and the controls are shown in Fig. 17.12. Excellent responses are indicated. This response shows that the flaperons are required to have a large initial deflection θ_f in order to pitch up the aircraft.

The aircraft is able to perform all three maneuvers with a fixed control law. Effective decoupling is achieved for all commands. The control surface and thrust commands have reasonable values in all cases. Thus, this design yields a satisfactory flight control system.

FIGURE 17.9
Vertical translation maneuver, with actuators and prefilter.

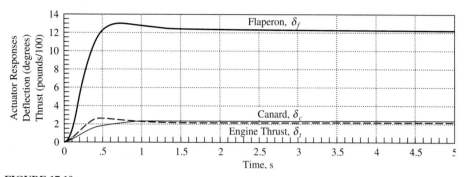

FIGURE 17.10
Controls for vertical translation.

FIGURE 17.11
Straight climb maneuver, with actuators and prefilter.

FIGURE 17.12
Controls for straight climb.

17.12 SUMMARY

A design method is presented in this chapter for output feedback systems that must function as trackers; i.e, the system outputs $\mathbf{y}(t)$ effectively follow the step command inputs $\mathbf{r}(t)$. The outputs are decoupled so that each input $r_i(t)$ commands only one output $y_i(t)$, and any transient perturbations in the other outputs are very small. The basis for the high-gain PI control law is presented, and the asymptotic transfer function as the gain $g \to \infty$ is developed. The asymptotic closed-loop eigenvalues approach three sets. The set S_1 consists of finite slow eigenvalues, the set S_2 consists of transmission zeros that are determined only by the plant and not by the control law, and the set S_3 consists of eigenvalues of the form $-g\sigma_i$, which approach infinity as $g \to \infty$. The designer selects a diagonal matrix $\mathbf{\Sigma}$ which assigns the fast eigenvalue set S_3, a location λ_o for the members of the slow eigenvalue set S_1, and a value of the scalar gain g.

In the case of regular plants for which the matrix product $\mathbf{CB}$ has full rank, the set S_1 becomes uncontrollable and the set S_2 becomes unobservable; thus these eigenvalues do not appear in the asymptotic transfer functions. Therefore, rapid response is achieved because only the fast eigenvalues affect the output response.

In the case of irregular plants for which the matrix product $\mathbf{CB}$ does not have full rank, it is necessary to add additional measurements through a measurement matrix $\mathbf{M}$. A procedure for selecting a sparse matrix $\mathbf{M}$ is illustrated. When the measurement matrix $\mathbf{M}$ is added to the system, the eigenvalue set S_2 is observable and the associated modes appear in the output.

Excellent closed-loop system performance is demonstrated by means of two examples. One example is an unstable chemical plant that has a regular plant model. The second example is a flight control system for an aircraft that is required to perform a number of maneuvers. The Markov parameter $\mathbf{CB}$ is rank deficient, and therefore a minor loop feedback containing a measurement matrix $\mathbf{M}$ is required. The design is achieved with relative ease.

The methods presented in this chapter apply to continuous control systems. They have the desirable features that they are extendable to a direct digital design[15,16] for the controller. A dual of the analog design leads to a very similar formulation for the digital design. By designing directly in the discrete domain, one avoids the common process of first designing an analog controller and then digitizing the resulting design.

REFERENCES

1. Porter, B., and A. Bradshaw: "Singular Perturbation Methods in the Design of Tracking Systems Incorporating High-Gain Error-Actuated Controllers," *Int. J. Syst. Sci.,* vol. 1, pp. 1169–1179, 1981.
2. Porter, B., and A. T. Shenton: "Singular Perturbation Analysis of the Transfer Function Matrices of a Class of Multivariable Linear Systems," *Int. J. Control,* vol. 21, pp. 655–660, 1975.
3. Phillips, R. G.: "A Two-State Design of Linear Feedback Controls," *IEEE Trans. Autom. Control,* vol. AC-25, pp. 1220–1223, 1980.
4. Tran, M. T., and M. E. Sawan: "Reduced Order Discrete-Time Models," *Int. J. Syst. Sci.,* vol. 14, pp. 745–752, 1983.

5. Mahmond, M. S., Y. Chen, and M. G. Singh: "On Eigenvalue Assignment in Discrete Systems with Fast and Slow Modes," *Int. J. Syst. Sci.,* vol. 16, pp. 61–70, 1985.
6. Kokotovic, P. V., and A. H. Haddad: "Controllability and Time-Optimal Control of Systems with Slow and Fast Modes," *IEEE Trans Autom. Control,* vol. AC-20, pp. 111–113, 1975.
7. Rosenbrock, H. H.: "The Zeros of a System," *Int. J. Control,* vol. 18, pp. 297–299, 1973 and "Correction to 'The Zeros of a System'," *Int. J. Control,* vol. 20, pp. 525–527, 1974.
8. MacFarlane, A. G. J., and N. Karcanias: "Poles and Zeros of Linear Multivariable Systems: A Survey of the Algebraic, Geometric and Complex-Variable Theory," *Int. J. Control,* vol. 24, pp. 33–74, 1976.
9. Porter, B., and A. Bradshaw: "Design of Linear Multivariable Continuous-Time Tracking Systems," *Int. J. Syst. Sci.,* vol. 5, pp. 1155–1164, 1974.
10. Porter, B., and J. J. D'Azzo, "Transmission Zeros of Linear Multivariable Continuous-Time Systems," *Electron. Lett.,* vol. 13, pp. 753–754, 1977.
11. MacFarlane, A. G. J., and B. Kouvaritakis: "Geometric Approach to Analysis and Synthesis of Systems Zeros, Part 1. Square System," *Int. J. Control,* vol. 23, pp. 149–166, 1976.
12. Porter, B., and A. Bradshaw: "Singular Perturbation Methods in the Design of Tracking Systems Incorporating Inner-Loop Compensators and High-Gain Error-Actuated Controllers," *Int. J. Syst. Sci.,* vol. 12, pp. 1193–1205, 1981.
13. Ridgely, D. B., S. S. Banda, and J. J. D'Azzo: "Decoupling of High Gain Multivariable Tracking Systems," *AIAA J. Guidance, Control, Dynamics,* vol. 8, pp. 44–49, 1985.
14. Smyth, J. S.: "Digital Flight Control System Design Using Singular Perturbation Methods," M.S. thesis, AFIT/EE/GE/81D-55, Air Force Institute of Technology, Wright-Patterson Air Force Base, Ohio, 1981.
15. Bradshaw, A., and B. Porter: "Singular Perturbation Methods in the Design of Tracking Systems Incorporating Fast-Sampling Error-Actuated Controllers," *Int. J. Syst. Sci.,* vol. 12, pp. 1181–1191, 1981.
16. Bradshaw, A., and B. Porter: "Singular Perturbation Methods in the Design of Tracking Systems Incorporating Inner-Loop Compensators and Fast-Sampling Error-Actuated Controllers," *Int. J. Syst. Sci.,* vol. 12, pp. 1207–1220, 1981.
17. Falb, P. L., and W. A. Wolovich: "Decoupling in the Design and Synthesis of Multivariable Control Systems," *IEEE Trans. Autom. Control,* vol. AC-12, pp. 651–659, 1967.

CHAPTER
18

QUANTITATIVE FEEDBACK THEORY (QFT) TECHNIQUE

18.1 INTRODUCTION[1−3,5]

Quantitative feedback theory (QFT) is a unified theory that emphasizes the use of feedback for achieving the desired robust system performance tolerances despite structured plant uncertainty and plant disturbances. QFT *quantitatively* formulates these two factors in the form of (*a*) sets $\mathcal{T}_R = \{T_R\}$ of acceptable command or tracking input-output relationships and $\mathcal{T}_D = \{T_D\}$ of acceptable disturbance input-output relationships, and (*b*) a set $\mathcal{P} = \{P\}$ of possible plants which include the uncertainties. The objective is to guarantee that the control ratio $T_R = Y/R$ is a member of $\mathcal{T}_R$ and $T_D = Y/D$ is a member of $\mathcal{T}_D$, for all P in $\mathcal{P}$. QFT has been developed for linear and nonlinear, time-invariant and time-varying, continuous and sampled-data, uncertain MISO and MIMO plants, and for both output and internal variable feedback. It has also been extended to some classes of uncertain distributed systems (in which the plant is described by partial differential equations) and the feedback and specifications are also distributed.

The representation of a MIMO plant with m inputs and ℓ outputs is shown in Fig. 18.1. The QFT synthesis technique for highly uncertain linear time-invariant MIMO plants has the following features[11].

1. The MIMO synthesis problem is converted into a number of single-loop feedback problems in which parameter uncertainty, external disturbances, and performance tolerances are derived from the original MIMO problem. The solutions to these

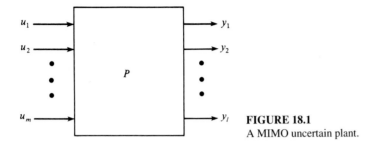

FIGURE 18.1
A MIMO uncertain plant.

single-loop *problems* represent a solution to the MIMO plant. It is not necessary to consider the complete system characteristic equation.

2. The design is tuned to the extent of the uncertainty and the performance tolerances.

This design technique is applicable to the following problem classes:

1. SISO linear time-invariant (LTI) systems.
2. SISO nonlinear systems. These are rigorously converted to equivalent class 1 systems whose solutions are guaranteed to work for a large problem class.
3. MIMO LTI systems. The performance specifications for each individual closed-loop system transfer function and for all the closed-loop disturbance response functions must be specified.
4. MIMO nonlinear systems. They are rigorously converted to equivalent class 3 systems whose solutions are guaranteed to work for a large class.
5. Distributed systems.
6. Sampled-data systems as well as continuous systems for all of the preceding.

QFT is a frequency-domain technique that is applied in this text to MISO systems. Problem classes 3 and 4 are converted into equivalent sets of MISO systems, to which the QFT design technique is applied. The objective is to solve the MISO problems, i.e., to find compensation functions that guarantee that the performance tolerances for each MISO problem are satisfied for all P in $\mathcal{P}$. The amount of feedback designed into the system is then tuned to the desired performance sets $\mathcal{T}_R$ and $\mathcal{T}_D$ and the given structured plant uncertainty set $\mathcal{P}$. Also, time-varying and nonlinear uncertain plant sets can be converted into equivalent MISO LTI plant problems, to which the MISO frequency-domain technique can be readily applied and where the fundamental tradeoffs are highly visible.[9]

This chapter is devoted to presenting a basic understanding and appreciation of the power of the QFT technique. This is accomplished by first presenting an introduction of the QFT technique using the frequency-response method as applied to a single-loop system. This is followed by a discussion of minimum-phase (m.p.) performance specifications. The next few sections are devoted to the MIMO system and the development of a suitable mapping that permits the analysis and synthesis of a MIMO control system by a set of equivalent single-loop control systems. The remaining portion of the chapter is devoted to an in-depth development of the QFT technique as applied to the design of robust single-loop control systems having two inputs (a tracking and a disturbance input, respectively) and a single output (a MISO system). For a more complete and thorough presentation of this technique as applied to MIMO control systems, especially to the design of an aircraft reconfigurable flight control system with control surface failures, for both m.p. and nonminimum-phase (n.m.p.) plants, the reader is referred to the literature.[1-17]

18.2 FREQUENCY RESPONSES WITH PARAMETER VARIATIONS

Insight into the QFT technique is provided by considering the unity-feedback cascade-compensated control system of Fig. 18.2, where G is a compensator and P is the plant, in which the plant parameters vary over some known range or there is plant parameter uncertainty. The loop transmission L is defined as

$$L = GP \qquad (18.1)$$

and the control ratio of the unity-feedback system of Fig. 18.2 is

$$T = \frac{Y}{R} = \frac{L}{1 + L} \qquad (18.2)$$

The plant with nominal plant parameters is denoted as P_o; thus, $L_o = GP_o$. For a given $\mathbf{G}(j\omega)$ and $\mathbf{P}_o(j\omega)$ a plot of $Lm\,\mathbf{L}_o(j\omega)$ vs. $\underline{/\mathbf{L}_o}$ on the Nichols chart (NC) is shown in Fig. 18.3. From this plot the closed-loop frequency-response data can be obtained for plotting M_o and α vs. ω, where

$$M_o(j\omega)\underline{/\alpha(j\omega)} = \frac{\mathbf{Y}(j\omega)}{\mathbf{R}(j\omega)} = \frac{\mathbf{L}_o(j\omega)}{1 + \mathbf{L}_o(j\omega)} \qquad (18.3)$$

Note that for the nominal plant $\mathbf{P}_o(j\omega)$, the nominal loop transmission is

$$Lm\,\mathbf{L}_o = Lm\,\mathbf{GP}_o = Lm\,\mathbf{G} + Lm\,\mathbf{P}_o \qquad (18.4)$$

whereas for all other plants $\mathbf{P}(j\omega)$

$$Lm\,\mathbf{L} = Lm\,\mathbf{GP} = Lm\,\mathbf{G} + Lm\,\mathbf{P} \qquad (18.5)$$

Thus, for $\omega = \omega_i$, the variation $\delta_p(j\omega_i)$ in $Lm\,\mathbf{L}(j\omega_i)$ is given by

$$\delta_p(j\omega_i) = Lm\,\mathbf{L}(j\omega_i) - Lm\,\mathbf{L}_o(j\omega_i) = Lm\,\mathbf{P}(j\omega_i) - Lm\,\mathbf{P}_o(j\omega_i) \qquad (18.6)$$

FIGURE 18.2
Cascade compensation.

and

$$\underline{/\Delta P(j\omega_i)} = \underline{/L} - \underline{/L_o} = \left(\underline{/G} + \underline{/P}\right) - \left(\underline{/G} + \underline{/P_o}\right) = \underline{/P} - \underline{/P_o} \quad (18.7)$$

Consider an example of Eq. (18.1), with $\mathbf{G} = 1\underline{/0°}$. If point A in Fig. 18.3 represents $\mathrm{Lm}\,\mathbf{P}_o$ vs. $\underline{/P_o}$, then a variation in $\mathbf{P}$ results in a horizontal translation in the angle of $\mathbf{P}$, given by Eq. (18.7), and a vertical translation in the log magnitude value of $\mathbf{P}$, given by Eq. (18.6). These translations are shown in Fig. 18.3 at the points B, C, and D.

The variation $\delta_p(j\omega_i)$ of the plant, and in turn $\mathbf{L}(j\omega_i)$, from the nominal value for a given frequency ω_i over the range of plant parameter variation is described by the template $\mathcal{T}\mathbf{P}(j\omega_i)$ shown in Fig. 18.3 (see Sec. 18.6 for a specific numerical example). Consider that point A on the template represents the nominal plant $\mathbf{P} = \mathbf{P}_o(j\omega_i)$ and that its corresponding closed-loop frequency response is $\mathrm{Lm}\,M_A = \mathrm{Lm}\,\beta = -6\,\mathrm{dB}$. When $\mathbf{P} = \mathbf{P}_C(j\omega)$, represented by point C in Fig. 18.3, the corresponding closed-loop frequency response is $\mathrm{Lm}\,M_C = \mathrm{Lm}\,\alpha = -2\,\mathrm{dB}$. These values are plotted in Fig. 18.4. Note that point A represents the minimum value of $\mathrm{Lm}\,M(j\omega_i)$ at $\omega = \omega_i$, i.e., $\mathrm{Lm}\,\mathbf{T}_{\min} = [\mathrm{Lm}\,M(j\omega_i)]_{\min} = \mathrm{Lm}\,\beta$. Point C

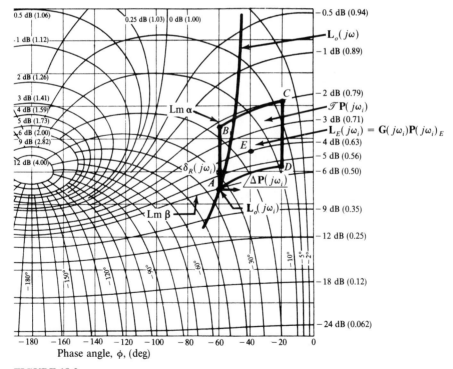

FIGURE 18.3
Nominal loop transmission plot with plant parameter area of uncertainty.

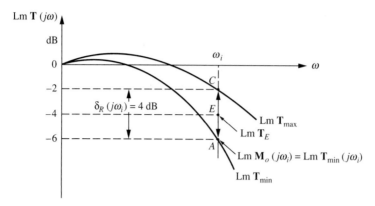

FIGURE 18.4
Closed-loop responses obtained from Fig. 18.3.

represents the largest value of $\operatorname{Lm} M(j\omega_i)$ at $\omega = \omega_i$, i.e., $\operatorname{Lm} \mathbf{T}_{max} = [\operatorname{Lm} M(j\omega_i)]_{max} = \operatorname{Lm} \alpha$, with the range of plant parameter variation described by the template. Thus, the maximum variation in $\operatorname{Lm} M$, denoted by $\delta_R(j\omega_i)$, for this example is

$$\delta_R(j\omega_i) = \operatorname{Lm} \alpha - \operatorname{Lm} \beta = -2 - (-6) = 4 \text{ dB}$$

Also shown in Fig. 18.4 is point E, which is midway between points A and C; it corresponds to point E in Fig. 18.3 which is within the variation template $\mathcal{T}P(j\omega_i)$. The procedure is repeated to obtain the maximum variation $\delta_R(j\omega)$ at each frequency, as shown in Fig. 18.4. From this figure it is possible to determine the variation in the control system's figures of merit due to the plant uncertainty. This graphical description of the effect of plant uncertainty is the basis of the QFT technique.

18.3   INTRODUCTION TO THE QFT METHOD (SINGLE-LOOP SYSTEM)

Figure 18.5 represents the feedback control system in which P is the transfer function of the plant, which may contain parameter uncertainty; G is the cascade compen-

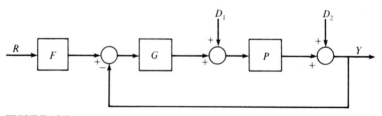

FIGURE 18.5
A feedback structure.

sator; and F is an input filter transfer function. The output $y(t)$ is required to track the command input $r(t)$ and to reject the disturbances $d_1(t)$ and $d_2(t)$. The compensator G in Fig. 18.5 is to be designed so that the variation of $y(t)$ resulting from the uncertainty in the plant P is within allowable tolerances and the effects of the disturbances $d_1(t)$ and $d_2(s)$ on $y(t)$ are acceptably small. Also, the filter properties of $F(s)$ must be designed to achieve the desired tracking by the output $y(t)$ of the input $r(t)$. Since the control system in Fig. 18.5 has two measurable quantities, $r(t)$ and $y(t)$, it is referred to as a two degree-of-freedom (DOF) feedback structure. If the two disturbance inputs are measurable, then it represents a four-DOF structure. The actual design is closely related to the extent of the uncertainty and to the narrowness of the performance tolerances. The uncertainty of the plant transfer function is denoted by the set

$$\mathcal{P} \equiv \{\mathbf{P}\} \tag{18.8}$$

and is illustrated by Example 1.

Example 1. The plant transfer function is

$$P(s) = \frac{K}{s(s + a)} \tag{18.9}$$

where the value of K is in the range $[1, 10]$ and a is in the range $[-2, 2]$. For example, an aircraft operates over a wide range of speed, altitude, and weight and it has a different transfer function for each flight condition. The region of parameter uncertainty of the plant is illustrated by Fig. 18.6.
    The uncertainty of the disturbance is denoted by

$$\mathcal{D} \equiv \{\mathbf{D}\} \tag{18.10}$$

and the acceptable closed-loop transmittances (control ratios) are denoted by

$$\mathcal{T} \equiv \{\mathbf{T}\} \tag{18.11}$$

The design objective is to guarantee that $T_R(s) = Y(s)/R(s)$ and $T_D(s) = Y(s)/D(s)$ are members of the sets of acceptable $\mathcal{T}_R$ and $\mathcal{T}_D$, respectively, for all $\mathbf{P} \in \mathcal{P}$ and all $\mathbf{D} \in \mathcal{D}$. In a feedback system, the principal challenge in the control system design is to relate the specifications to the requirements on the loop transmission function $L(s) = G(s)P(s)$ in order to achieve the desired benefits of feedback, i.e., the desired reduction in sensitivity to plant uncertainty and desired disturbance attenuation. The

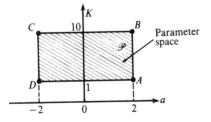

FIGURE 18.6
Plant parameter uncertainty.

advantage of the frequency domain is that $L(s) = G(s)P(s)$ is simply the multiplication of complex numbers. In the frequency domain it is possible to evaluate $\mathbf{L}(j\omega)$ at every ω_i separately, and thus, at each ω_i, the optimal bounds on $\mathbf{L}(j\omega_i)$ can be determined.

18.4   MINIMUM-PHASE SYSTEM PERFORMANCE SPECIFICATIONS[8,9,16]

In order to apply the QFT technique it is necessary to synthesize the desired or model control ratio, based upon the system's performance specifications in the time domain. For the structure of Fig. 18.5 the control ratios are

$$\frac{\text{Tracking}}{(D_1 = D_2 = 0)} : T_R(s) = \frac{F(s)G(s)P(s)}{1 + G(s)P(s)} = \frac{FL}{1 + L} \tag{18.12}$$

$$\frac{\text{Disturbance}}{(R = 0)} : T_{D_1}(s) = \frac{P(s)}{1 + G(s)P(s)} = \frac{P}{1 + L} \qquad \text{with } D_2 = 0 \tag{18.13}$$

$$T_{D_2}(s) = \frac{1}{1 + G(s)P(s)} = \frac{1}{1 + L} \qquad \text{with } D_1 = 0 \tag{18.14}$$

Note that for T_{D_1}, the specified maximum value $|y(t_p)| = \alpha_p$, due to $d_1(t) = u_{-1}(t)$, is often used as the disturbance model specification, i.e., $\text{Lm}\,T_{D_1} = \text{Lm}\,\alpha_p$.

Tracking Models

The tracking thumbprint specifications for a simple second-order system are based upon satisfying some or all of the step forcing function figures of merit for underdamped (M_p, t_p, t_s, t_r, K_m) and overdamped (t_s, t_r, K_m) responses, respectively. They are graphically depicted in Fig. 18.7. The time responses $y(t)_U$ and $y(t)_L$ in this figure represent the upper and lower bounds, respectively, of the thumbprint specifications; that is, an acceptable response $y(t)$ must lie between these bounds. The Bode plots of the upper bound B_U and lower bound B_L for $\text{Lm}\,\mathbf{T}_R(j\omega)$ vs. ω are shown in Fig. 18.8. Note that for m.p. plants, only the tolerance on $|\mathbf{T}_R(j\omega_i)|$ needs to be satisfied for a satisfactory design. For n.m.p. plants, tolerances on $\underline{/\mathbf{T}_R(j\omega_i)}$ must also be specified and satisfied in the design process. The modeling of a desired transmittance $T(s)$ is discussed in Chapter 12. It is desirable to synthesize the control

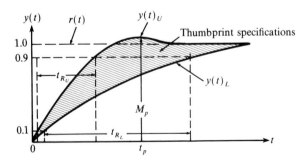

FIGURE 18.7
The tracking thumbprint specifications.

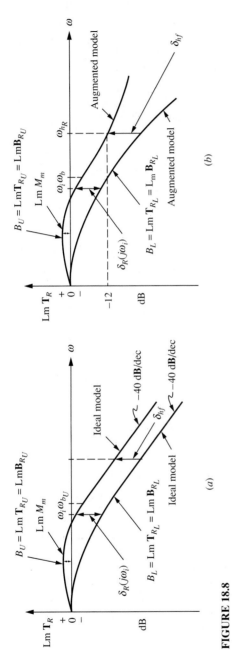

FIGURE 18.8
Bode plots of T_R: (a) Ideal simple second-order models; (b) The augmented models.

587

ratios corresponding to the upper and lower bounds T_{R_U} and T_{R_L}, respectively, so that $\delta_R(j\omega_i)$ increases as ω_i increases above the 0-dB crossing frequency of T_{R_U}. This characteristic of $\delta_R(j\omega_i)$ simplifies the process of synthesizing a loop transmission $L_o(s) = G(s)P_o(s)$, as discussed in Sec. 18.12. To synthesize $L_o(s)$ it is necessary to determine the tracking bounds $\mathbf{B}_R(j\omega_i)$ (see Sec. 18.8), which are obtained based on $\delta_R(j\omega_i)$. This characteristic of $\delta_R(j\omega_i)$ ensures that the tracking bounds $\mathbf{B}_R(j\omega)$ decrease as ω_i increases.

The QFT technique requires that the desired control ratios be modeled in the frequency domain to satisfy the required gain K_m and the desired time-domain performance specifications for a step input. Thus, M_p, t_p, t_s, and t_r must be satisfied for T_{R_U}, and T_s (or t_s) and t_r must be satisfied for T_{R_L}. The modeling of the desired control ratio T_{R_U} is described in Sec. 12.2. It may have the form

$$T_{R_U}(s) = \frac{\omega_n^2}{s^2 + 2\zeta\omega_n s + \omega_n^2} = \frac{\omega_n^2}{(s - p_1)(s - p_2)} \tag{18.15}$$

where $\omega_n^2 = p_1 p_2$ and $t_s \approx T_s = 4/\zeta\omega = 4/|\sigma_D|$ (the desired settling time). The control ratio $T_{R_U}(s)$ of Eq. (18.15) can be represented by an equivalent unity-feedback system so that

$$T_{R_U}(s) = \frac{G_{eq}(s)}{1 + G_{eq}(s)} \tag{18.16}$$

where

$$G_{eq}(s) = \frac{\omega_n^2}{s(s + 2\zeta\omega_n)} \tag{18.17}$$

The gain constant of this equivalent Type 1 transfer function $G_{eq}(s)$ is $K_1 = \lim_{s \to 0}[sG_{eq}(s)] = \omega_n/2\zeta$. Equation (18.15) satisfies the requirement that $y(t)_{ss} = R_o u_{-1}(t)$ for $r(t) = R_o u_{-1}(t)$. The frequency ω_b for which $|\mathbf{T}_{R_U}(j\omega_b)| = 0.7071$ is defined as the system bandwidth frequency ω_{b_U}.

The simplest overdamped model for $T_{R_L}(s)$ is of the form

$$T_{R_L}(s) = \frac{Y(s)}{R(s)} = \frac{K}{(s - \sigma_1)(s - \sigma_2)} = \frac{G_{eq}(s)}{1 + G_{eq}(s)} \tag{18.18}$$

where

$$G_{eq}(s) = \frac{\sigma_1\sigma_2}{s[s - (\sigma_1 + \sigma_2)]}$$

and $K_1 = -\sigma_1\sigma_2/(\sigma_1 + \sigma_2)$. For this system $y(t)_{ss} = R_o$ for $r(t) = R_o u_{-1}(t)$. Selection of the parameters σ_1 and σ_2 is used to meet the specifications for t_s and K_1. The -12 dB line determines the model response bandwidth (BW) ω_{h_R}.

Once the ideal models $\mathbf{T}_{R_U}(j\omega)$ and $\mathbf{T}_{R_L}(j\omega)$ are determined, the time and frequency response plots of Figs. 18.7 and 18.8a, respectively, can then be drawn. Because the models for T_{R_U} and T_{R_L} are both second order, the high-frequency asymptotes in Fig. 18.8a have the same slope. The high-frequency range in Fig. 18.8a is defined as $\omega \geq \omega_h$ where ω_h is the model BW frequency of B_U. An increasing spread between B_U and B_L is required in the high-frequency (hf) range; that is,

$$\delta_{hf} = B_U - B_L \tag{18.19}$$

must increase with increasing frequency. This is achieved by changing B_U and B_L, without violating the desired time-response characteristics, by augmenting T_{R_L} with a negative real pole that is as close to the origin as possible but far enough away not to affect the time response significantly. This additional pole lowers B_L in the high-frequency range. If necessary, the spread can be further increased by augmenting T_{R_U} with a zero as close to the origin as possible without significantly affecting the time response. This additional zero raises the curve B_U in the high-frequency range. In order to minimize the iteration process in achieving acceptable models for $T_{R_U}(s)$ and $T_{R_L}(s)$ which have an increasing $\delta_R(j\omega)$, the following procedure may expedite the design process: (a) first synthesize a second-order $T_{R_U}(s)$ containing the zero $|z_1| \geq \omega_n$ that meets the desired figures of merit; and (b) then, as a first trial, select all three poles of T_{R_L} to have the value of $\omega_3 = \omega_n > \omega_2 = |p_i| = \omega_1 > |\sigma_D|$. For succeding trials, if necessary, one or more of these poles are moved right or left until the desired specifications are satisfied. As illustrated by the slopes of the straight-line Bode plots in Fig. 18.8b, selecting the values of all three poles in the range specified above ensures an increasing δ_R. Other possibilities are as follows: (c) the specified values of t_p and t_s for T_{R_L} may be such that a pair of complex poles and a real pole need to be chosen for the model response. For this situation, the real pole *must* be more dominant than the complex poles. (d) Depending on the performance specifications (see the paragraph below Eq. (18.20)), $T_{R_U}(s)$ may require two real poles and a zero "close" to the origin, i.e., select $|z_1|$ very much less than $|p_1|$ and $|p_2|$, in order to effectively have an underdamped response.

At high frequencies δ_{hf} must be larger than the actual variation in the plant, ΔP. This characteristic is the result of the Bode theorem which states that $\int_0^\infty \mathrm{Lm}\, \mathbf{S}_P^T d\omega = 0$, i.e., the reduction in sensitivity $\mathbf{S}_P^T$ at the lower frequencies must be compensated by an increase in sensitivity at the higher frequencies. At some high frequency $\omega_i \geq \omega_h$ (see Fig. 18.8b), since $|\mathbf{T}_{R_L}| \approx 0$, then

$$\lim_{\omega \to \infty} \delta_R(j\omega_i) \approx B_U - (-\infty) = \infty \text{ dB} \tag{18.20}$$

For the case where $y(t)$, corresponding to T_{R_U}, is to have an allowable "large" overshoot followed by a small tolerable undershoot, a dominant complex-pole pair is not suitable for T_{R_U}. An acceptable overshoot with no undershoot for T_{R_U} can be achieved by T_{R_U} having two real dominant poles $p_1 > p_2$, a dominant real zero $(z_1 > p_1)$ "close" to p_1, and a far-off pole $p_4 << p_2$. The closeness of the zero dictates the value of M_p. Thus, a designer selects a pole-zero combination to yield the form of the desired time-domain response. This also applies for synthesizing a desired T_D.

Disturbance-Rejection Models[6]

A detailed discussion on synthesizing disturbance-rejection models, based upon desired performance specifications, is given in Secs. 12.5 and 12.8. The reader is urged to review these sections before proceeding in this chapter. The Bode plots of Eqs. (12.37), (12.38), (12.41), and (12.42) and of $\mathrm{Lm}\, \boldsymbol{\alpha}_p$ are shown in Fig. 18.9. As

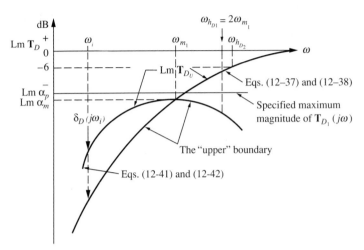

FIGURE 18.9
Bode plots of disturbance models for $\mathbf{T}_D(j\omega)$.

seen from this figure, there is only an upper boundary for $\mathrm{Lm}\,\mathbf{T}_D$. Peak time for the disturbance model is specified only to establish a disturbance control ratio. The essential specification is to equate the maximum magnitude allowable for $T_{D_1}(s)$ in Eq. (18.13) to the value of α_p (see Fig. 12.17).

18.5   MULTIPLE-INPUT MULTIPLE-OUTPUT (MIMO) UNCERTAIN PLANTS

The design for a MIMO system involves the design of an equivalent set of single-loop MISO systems. In general, an $\ell \times m$ open-loop MIMO plant can be represented in matrix notation as

$$\mathbf{Y}(s) = \mathbf{P}\mathbf{U}(s) \qquad (18.21)$$

where $\mathbf{Y}(s) = \ell$-dimensional plant output vector
   $\mathbf{U}(s) = m$-dimensional plant input vector
   $\mathbf{P} = \ell \times m$ plant transfer-function matrix relating $\mathbf{U}(s)$ to $\mathbf{Y}(s)$

Consider the MIMO plant of Fig. 18.1, for which $\mathbf{P}$ is a member of $\mathcal{P}$ (i.e., $\mathbf{P} \in \mathcal{P}$), the set of all possible plant transfer functions relating each input to each output. When a system has variable parameters, known or unknown, the transfer-function matrix of the plant may be represented by an associated set of transfer-function matrices $\{\mathbf{P}_1, \mathbf{P}_2, \ldots\}$ which are contained in $\mathcal{P}$. The QFT design method requires that the uncertainty in $\mathbf{P}$ be known or at least bounded. In any MIMO system with m inputs there are at most m outputs that can be independently controlled[3]; therefore, the dimensions of both the input and output vectors are equal, i.e., $\ell = m$. If the model defines an unequal number of inputs and outputs, the first step is to modify the model so that the dimensions of the input and output vectors are the same.

This requires a judicious selection of inputs if $m > \ell$. The system is then defined as being of order $\ell \times \ell$. Using fixed-point theory[12,14,18] (see Sec. 18.6), the MIMO problem for an $\ell \times \ell$ system can be separated into ℓ equivalent single-loop MISO systems and ℓ^2 prefilter disturbance problems. Each MISO system design is based upon the specifications relating its output and all of its inputs.

Figure 18.10 represents an $\ell \times \ell$ MIMO closed-loop system in which **F, G**, and **P** are each $\ell \times \ell$ matrices, and $\mathcal{P} = \{\mathbf{P}\}$ is a set of matrices due to plant uncertainty. There are ℓ^2 closed-loop system transfer functions (transmissions) $t_{ij}(s)$ contained within its system transmission matrix, i.e., $\mathbf{T}(s) = \{t_{ij}(s)\}$, relating the outputs $y_i(s)$ to the inputs $r_j(s)$, e.g., $y_i(s) = t_{ij}(s)r_j(s)$. In a quantitative problem statement there are tolerance bounds on each $t_{ij}(s)$, giving a set of ℓ^2 acceptable regions $\tau_{ij}(s)$, which are to be specified in the design; thus $t_{ij}(s) \in \tau_{ij}(s)$ and $\mathcal{T}(s) = \{\tau_{ij}(s)\}$. The QFT design technique for 2×2 and 3×3 systems has been developed and applied to the design of flight control systems.[10,13,15]

From Fig. 18.10 the following equations can be written:

$$\mathbf{y} = \mathbf{Px} \qquad \mathbf{x} = \mathbf{Gu} \qquad \mathbf{u} = \mathbf{v} - \mathbf{y} \qquad \mathbf{v} = \mathbf{Fr}$$

In these equations $\mathbf{P}(s) = [p_{ij}(s)]$ is the matrix of plant transfer functions. $\mathbf{G}(s)$ is the matrix of compensator transfer functions and is often simplified so that it is diagonal, that is, $\mathbf{G}(s) = \text{diag}\{g_i(s)\}$. $\mathbf{F}(s) = [f_{ij}(s)]$ is the matrix of prefilter transfer functions, which may also be a diagonal matrix. The combination of these equations yields

$$\mathbf{y} = [\mathbf{I} + \mathbf{PG}]^{-1}\mathbf{PGFr} \tag{18.22}$$

where the system control ratio relating $\mathbf{r}$ to $\mathbf{y}$ is

$$\mathbf{T} = \{t_{ij}(s)\} = [\mathbf{I} + \mathbf{PG}]^{-1}\mathbf{PGF} \tag{18.23}$$

The QFT design procedure systematizes and simplifies the manner of achieving a satisfactory system design. The objective is to design a system that behaves as desired for the entire range of structured plant uncertainty. This requires finding, for a 3×3 system, nine $f_{ij}(s)$ and three $g_i(s)$ such that each $t_{ij}(s)$ stays within its

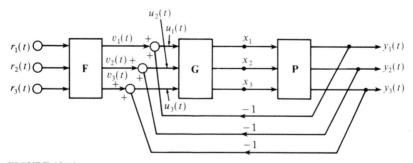

FIGURE 18.10
Three-input three-output MIMO feedback structure.

acceptable region $\tau_{ij}(s)$, no matter how the $p_{ij}(s)$ may vary. Clearly, this is a very difficult problem but is overcome by converting this MIMO system into an equivalent set of single-loop systems, and then the MISO design method for MIMO systems[16] is used. The basic MIMO compensation structure for a 2×2 MIMO system is shown in Fig. 18.11. The structure for a 3×3 MIMO system is shown in Fig. 18.10. They consist of the uncertain plant matrix **P**, the *diagonal compensation matrix* **G**, and the prefilter matrix **F**. Generally, a diagonal **G** matrix is considered, although the use of a nondiagonal **G** matrix may allow the designer more design flexibility. Reference 11 provides the justification for representing MIMO systems by ℓ^2 equivalent MISO systems. Note: there is no need to consider crossfeeding of the y_1 and y_2 outputs in Fig. 18.11 to the u_2 and u_1 inputs respectively.

Figure 18.12 shows the four effective MISO loops (in the boxed area) resulting from a 2×2 system and the nine effective MISO loops resulting from a 3×3 system. The c_{ij} input of each MISO system represents the cross-coupling effect of the MIMO system. This input is akin to the disturbance input D_1 of Fig. 18.5. In each of the nine structures of Fig. 18.12 it is necessary that the control ratio $t_{ij}(s)$ must be a member of the acceptable $t_{ij} \in \tau_{ij}(s)$. All the $g_i(s)$ and $f_{ij}(s)$ appearing in Fig. 18.12 must be chosen to ensure that this condition is satisfied, thus constituting nine MISO design problems. If all of these MISO problems are solved, then this means that there exist nine t_{ij} and t_{kj}, each in its acceptable set, which is a solution of the control problem illustrated in Fig. 18.11. If each element satisfies the Schauder fixed point theorem[12,14,18] and is 1:1, then this solution must be unique. A formal and detailed treatment is given in Ref. 12. Only the design of a MISO system is covered in the remainder of this chapter. The reader is referred to the literature listed previously for design examples of MIMO systems.

18.6   PLANT TEMPLATES OF $P(s)$, $\mathcal{T}P(j\omega_i)$

The tracking control ratio for Fig. 18.5 with $D_1 = D_2 = 0$ is

$$T_R(s) = F\left[\frac{L}{1 + L}\right] \tag{18.24}$$

where $L = GP$. From Eq. (18.24)

$$\operatorname{Lm} T_R = \operatorname{Lm} F + \operatorname{Lm}\left[\frac{L}{1 + L}\right] \tag{18.25}$$

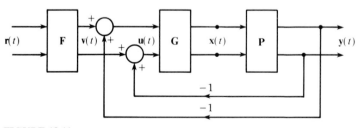

FIGURE 18.11
MIMO control structure (2×2 system).

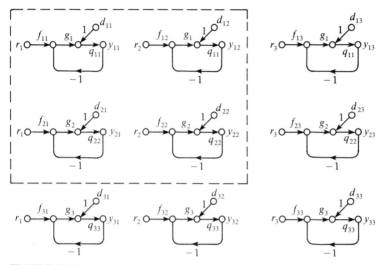

FIGURE 18.12
Effective MISO loops 2×2 (boxed-in loops) and 3×3 (all nine loops).

The change in T_R due to the uncertainty in P is

$$\Delta(\mathrm{Lm}\, \mathbf{T}_R) = \mathrm{Lm}\, \mathbf{T}_R - \mathrm{Lm}\, \mathbf{F} = \mathrm{Lm}\left[\frac{\mathbf{L}}{1+\mathbf{L}}\right] \qquad (18.26)$$

The proper design of $L = L_o$ and F must restrict this change in T_R so that the actual value of $\mathrm{Lm}\, \mathbf{T}_R$ always lies between B_U and B_L of Fig. 18.8. The first step in synthesizing an L_o is to make templates that characterize the variation of the plant uncertainty for various values of ω_i over a frequency range $\omega_x \leq \omega_i \leq \omega_h$, where ω_h is the largest value obtained from Figs. 18.8 (ω_{h_R}) and 18.9 (ω_{h_D}). The guideline for selecting the frequency range for the templates is to select three frequency values below and above the 0 dB crossing frequency, no less than an octave apart. The frequency range should extend up to approximately the -12 dB value of the $\mathbf{B}_U$ plot in Fig. 18.8. In addition, for a Type 0 plant, select $\omega_x = 0$, and for Type 1 or higher-order plants select $\omega_x \neq 0$.

The simple plant

$$P(s) = \frac{Ka}{s(s+a)} \qquad (18.27)$$

where

$$K \in \{1, 10\} \qquad a \in \{1, 10\} \qquad (18.28)$$

is used to illustrate how the templates are obtained for a plant with variable parameters. The region of plant uncertainty for the plant of Eq. (18.28) is depicted in Fig. 18.13. The boundary of the plant template on the Nichols chart (NC) can be obtained by mapping the boundary of the plant parameter uncertainty region. A number of points on the perimeter of $ABCD$ of Fig. 18.13 are selected, and values of $\mathrm{Lm}\, \mathbf{P}(j\omega_i)$ and $\underline{/\mathbf{P}(j\omega_i)}$ are obtained at each point. These data, for each

FIGURE 18.13
Region of plant uncertainty for Eqs. (18.27) and (18.28).

frequency $\omega = \omega_i$, are plotted on an NC. A curve is drawn through these points and becomes the template $\mathfrak{T}\mathbf{P}(j\omega_i)$ at the frequency ω_i. A sufficient number of points must be selected so that the contour of $\mathfrak{T}\mathbf{P}(j\omega_i)$ accurately reflects the region of structured plant uncertainty. In addition to the points A, B, C, and D marked in Fig. 18.13, it may also be necessary to include additional points on the perimeter. For the points A, B, C, and D, at the frequency $\omega = 1$, the data obtained from Eq. (18.27) is $\mathbf{P}_A(j1) = \sqrt{2}/2\underline{/-135°}$, $\mathbf{P}_B(j1) = 5\sqrt{2}\underline{/-135°}$, $\mathbf{P}_C(j1) = 100/\sqrt{101}\underline{/-95.7°}$, and $\mathbf{P}_D(j1) = 10/\sqrt{101}\underline{/-95.7°}$.

These data are plotted on the Nichols chart shown in Fig. 18.14. A curve is drawn through the points A, B, C, and D, and the shaded area is labeled $\mathfrak{T}\mathbf{P}(j1)$. The contour A, B, C, and D in Fig. 18.14 may be drawn on a plastic sheet (preferably colored) so that a plastic template for $\mathfrak{T}\mathbf{P}(j1)$ can be cut and labeled. The templates for other values of ω_i are obtained in a similar manner. For an aircraft, each point A, B, C, D, etc. in Fig. 18.14 represents a given flight condition (FC) at $\omega = \omega_i$, i.e., $[\mathbf{P}_{A/C}(j\omega_i)]_{FC_j}$. One of the flight conditions may be identified as the nominal plant P_o. For the plant of Eq. (18.27) the values $K = a = 1$ represent the lowest point of each of the templates $\mathfrak{T}\mathbf{P}(j\omega_i)$ and may be selected as the nominal plant $\mathbf{P}_o$ for all frequencies. However, any plant in $\mathcal{P}$ can be chosen as the nominal plant.[11] With $\mathbf{L} = \mathbf{GP}$ and $\mathbf{L}_o = \mathbf{GP}_o$, as given in Eqs. (18.6) and (18.7):

$$\delta_P(j\omega_i) = \mathrm{Lm}\,\mathbf{L} - \mathrm{Lm}\,\mathbf{L}_o = (\mathrm{Lm}\,\mathbf{G} + \mathrm{Lm}\,\mathbf{P}) - (\mathrm{Lm}\,\mathbf{G} + \mathrm{Lm}\,\mathbf{P}_o)$$

$$= (\mathrm{Lm}\,\mathbf{P} - \mathrm{Lm}\,\mathbf{P}_o)\ \mathrm{dB}$$

and

$$\underline{/\Delta\mathbf{P}} = \underline{/\mathbf{P}} - \underline{/\mathbf{P}_o}$$

Thus, if point A in Fig. 18.14 represents $\mathrm{Lm}\,\mathbf{P}_o$, a variation in $\mathbf{P}$ results in a horizontal translation in the angle of $\mathbf{P}$ and a vertical translation in the log magnitude value of $\mathbf{P}$. When $\mathbf{G}(j\omega)$ represents a specific transfer function, the template of Fig. 18.14 can be converted into a template of $\mathbf{L}(j\omega_i)$ by translating it vertically by $\mathrm{Lm}\,\mathbf{G}(j\omega_i)$ and horizontally by $\underline{/\mathbf{G}(j\omega_i)}$. For the template of $\mathbf{L}(j\omega_i)$, the values of the M-contours at the intersections with the template are the values of the control ratio $\mathrm{Lm}\,\mathbf{T}(j\omega_i) = \mathrm{Lm}\,[\mathbf{L}(j\omega_i)/(1 + \mathbf{L}(j\omega_i))]$. The range of values of $\mathbf{T}(j\omega_i)$ for the entire range of parameter variation given in Eq. (18.28) can therefore be determined.

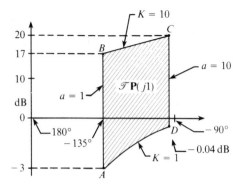

FIGURE 18.14
NC characterizing Eq. (18.27) over the region of uncertainty.

18.7 U-CONTOUR[2,12]

The specifications on system performance in the time domain (see Fig. 18.7) identify a minimum damping ratio ζ for the dominant roots of the closed-loop system. In the frequency domain (see Fig. 18.8) this becomes a bound on the value of M_m. On the NC this bound on M_m establishes a region that must not be penetrated by the plot of $L(j\omega)$ for all ω. The boundary of this region is referred to as the universal high-frequency boundary (UHFB), the U-contour, because this becomes the dominating constraint on $L(j\omega)$. The formation of the U-contour is described in this section.

For the two cases of disturbance rejection depicted in Fig. 18.5 the control ratios are, respectively, as given in Eqs. (18.13) and (18.14):

$$T_{D_2} = \frac{1}{1+L} \quad \text{and} \quad T_{D_1} = \frac{P}{1+L}$$

Thus it is necessary to synthesize an $L_o(s)$ so that the disturbances are properly attenuated. For the present, only one aspect of this disturbance-response problem is considered, namely a constraint is placed on the damping ratio ζ of the dominant complex-pole pair of poles of T_D nearest the $j\omega$-axis.[6] This damping ratio is related to the peak value of

$$\left| \mathbf{T}(j\omega) \right| = \left| \frac{\mathbf{L}(j\omega)}{1 + \mathbf{L}(j\omega)} \right| \tag{18.29}$$

For example, consider the case for which the dominant complex-pole pair of Eq. (18.29) results in a peak of $\text{Lm}\,\mathbf{T} = 8$ dB for $\zeta = 0.2$ and 2.7 dB for $\zeta = 0.4$, etc. Although this large peak does not appear in T_R due to the design of the filter $F(s)$, it does affect the response for T_D. If $d(t)$ is very small, a peak of T_D due to $\zeta = 0.2$ can be "very large," and it may be difficult to achieve the restriction on the peak overshoot α_p of the time response [see Eqs. (12.35) and (12.36)]:

$$\left| c(t) \right| \leq \alpha_p$$

Therefore, it is reasonable to add the requirement

$$\left| \mathbf{T} \right| = \left| \frac{\mathbf{L}}{1+\mathbf{L}} \right| \leq M_L \tag{18.30}$$

where M_L is a constant for all ω and over the whole range of P parameter values. This results in a constraint on ζ of the dominant complex-pole pair of T_D. This constraint can therefore be translated into a constraint on the maximum value T_m of Eq. (18.29). For example, for Lm $M_m = 2$ dB, the oval $agefa$ in Fig. 18.15 corresponds to the 2-dB M-contour on the NC. This results in limiting the peak of the disturbance response. A value of M_L can be selected to correspond to the maximum value of T_R. Therefore, the top portion, efa, of the M-contour on the NC, which corresponds to the selected value of M_L, becomes part of the U-contour.

For a large class of problems, as $\omega \to \infty$, the limiting value of the plant transfer function approaches

$$\lim_{\omega \to \infty} [P(j\omega)] = K/\omega^\lambda$$

where λ represents the excess of poles over zeros of $P(s)$. The plant template, for this problem class, approaches a vertical line, of length equal to

$$\Delta \, \mathrm{Lm} \, \mathbf{P} \equiv \lim_{\omega \to \infty} [\mathrm{Lm} \, \mathbf{P}_{\max} - \mathrm{Lm} \, \mathbf{P}_{\min}] = \mathrm{Lm} \, K_{\max} - \mathrm{Lm} \, K_{\min} = V \text{ dB} \quad (18.31)$$

If the nominal plant is chosen at $K = K_{\min}$, then the constraint M_L gives a boundary that approaches the U-contour $abcdefa$ of Fig. 18.15. (*Note:* For a MIMO plant, as $\omega \to \infty$, the templates may not approach a vertical line if the λ_{ij} are not the same for all p_{ij} elements of the plant matrix $\mathbf{P}$. When the λ_{ij} are different, then the widths of the templates are a multiple of 90°.)

For a plant of Eqs. (18.27) and (18.28), applying the limiting condition $\omega \to \infty$ to Eq. (18.31) yields

$$V = \Delta \, \mathrm{Lm} \, \mathbf{P}$$

$$= \lim_{\omega \to \infty} \left[\left\{ \mathrm{Lm} \, (Ka)_{\max} - \mathrm{Lm} \, (j\omega)^2 \right\} - \left\{ \mathrm{Lm} \, (Ka)_{\min} - \mathrm{Lm} \, (j\omega)^2 \right\} \right]$$

$$= \mathrm{Lm} \, (Ka)_{\max} - \mathrm{Lm} \, (Ka)_{\min} = \mathrm{Lm} \, 100 - \mathrm{Lm} \, 1 = 40 \text{ dB} \quad (18.32)$$

For the m.p. plant of Eq. (18.27), where the poles are real, the plant templates have the typical shape of Fig. 18.14.

The high-frequency boundary B_h, the bcd portion of the U-contour in Fig. 18.15, is obtained by measuring down V dB from the ega portion of the M-contour, as illustrated in this figure. V is determined by Eq. (18.32) and for this example is 40 dB. The remaining portions of the U-contour, portions ab and de, not necessarily straight

FIGURE 18.15
U-contour construction.

lines, are determined by satisfying the requirement of Eq. (18.30) and $\delta_R(j\omega_1)$. The $\mathcal{T}\mathbf{P}(j\omega_1)$ is used to determine the corresponding tracking bounds $\mathbf{B}_R(j\omega_i)$ on the NC in the manner described in Sec. 18.8.

18.8 TRACKING BOUNDS $\mathrm{Lm}\,\mathbf{B}_R(j\omega)$ ON THE NC

As an introduction to this section, the procedure for adjusting the gain of a unity-feedback system to achieve a desired value of M_m by use of the NC is reviewed. Consider the plot of $\mathrm{Lm}\,\mathbf{P}(j\omega)$ vs. $\underline{/\mathbf{P}(j\omega)}$ for a plant, shown in Fig. 18.16 (the solid curve). With $G(s) = A = 1$ and $F(s) = 1$ in Fig. 18.5, $\mathbf{L} = \mathbf{P}$. The plot of $\mathrm{Lm}\,\mathbf{L}(j\omega)$ vs. $\underline{/\mathbf{L}(j\omega)}$ is tangent to the $M = 1$ dB curve with a resonant frequency $\omega_{m_1} = 1.1$. If $\mathrm{Lm}\,M_m = 2$ dB is specified for $\mathrm{Lm}\,\mathbf{T}_R$, the gain A is increased, raising $\mathrm{Lm}\,\mathbf{L}(j\omega)$, until it is tangent to the 2 dB M-curve. For this example the curve is raised by $\mathrm{Lm}\,A = 4.5$ dB ($G = A = 1.679$), and the resonant frequency is $\omega_m = 2.09$.

Now consider that the plant uncertainty involves only the variation in gain A between the values of 1 and 1.679. It is desired to find a cascade compensator $G(j\omega)$, in Fig. 18.5, such that the specification 1 dB $\leq \mathrm{Lm}\,M_m \leq 2$ dB is always maintained for this plant gain variation while the resonant frequency ω_m remains constant. This requires that the loop transmission $\mathbf{L}(j\omega) = G(j\omega)\mathbf{P}(j\omega)$ must be tangent to an M-contour in the range of $1 \leq \mathrm{Lm}\,M \leq 2$ dB for the entire range of $1 \leq A \leq 1.679$, at the same resonant frequency $\omega_m = 2.09$. The manner of achieving this and other time-response specifications is the subject of the remaining portion of this chapter.

It is assumed for Eq. (18.26) that the compensators F and G are fixed, that is, they have negligible uncertainty. Thus, only the uncertainty in P contributes

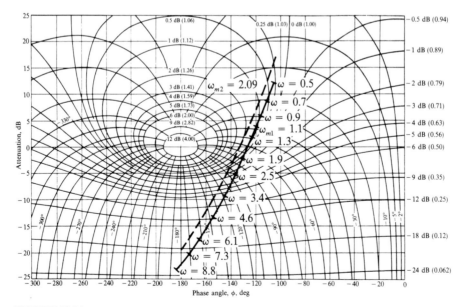

FIGURE 18.16
Log magnitude–angle diagram.

to the change in T_R given by Eq. (18.26). The solution requires that the actual $\Delta \operatorname{Lm} T_R(j\omega_i) \leq \delta_R(j\omega_i)$ dB in Fig. 18.8. Thus it is necessary to determine the resulting constraint, or bound $\mathbf{B}_R(j\omega_i)$, on $\mathbf{L}(j\omega_i)$. The procedure is to pick a nominal plant $P_o(s)$ and to derive the bounds on the resulting nominal loop transfer function $L_o(s) = G(s)P_o(s)$.

As an illustration, consider the plot of $\operatorname{Lm} \mathbf{P}(j2)$ vs. $\underline{/\mathbf{P}(j2)}$ for the plant of Eqs. (18.27) and (18.28). As shown in Fig. 18.17, the plant's region of uncertainty $\mathfrak{T}\mathbf{P}(j2)$ is given by the contour $ABCD$, i.e., $\operatorname{Lm} \mathbf{P}(j2)$ lies on or within the boundary of this contour. The nominal plant transfer function, with $K_o = 1$ and $a_o = 1$, is

$$P_o = \frac{1}{s(s+1)} \tag{18.33}$$

and is represented in Fig. 18.17 by point A for $\omega = 2[-13.0 \text{ dB}, -153.4°]$. Note, once a nominal plant is chosen, it must be used for determining all the bounds $\mathbf{B}_R(j\omega_i)$. Since $\operatorname{Lm} \mathbf{L}(j2) = \operatorname{Lm} \mathbf{G}(j2) + \operatorname{Lm} \mathbf{P}(j2)$, then $\mathfrak{T}\mathbf{P}(j2)$ is translated on the Nichols chart vertically by the value of $\operatorname{Lm} \mathbf{G}(j2)$ and horizontally by the angle $\underline{/\mathbf{G}(j2)}$. The templates $\mathfrak{T}\mathbf{P}(j\omega_1)$ are relocated to find the position of $\mathbf{L}_o(j\omega)$ which satisfies the specifications in Fig. 18.8 of $\delta_R(j\omega_1)$. For example, if a trial design of $\mathbf{L}(j2)$ requires sliding $\mathfrak{T}\mathbf{P}(j2)$ to the position $A'B'C'D'$ in Fig. 18.17, then

$$|\operatorname{Lm} \mathbf{G}(j2)| = \left| |\operatorname{Lm} \mathbf{L}(j2)|_{A'} - |\operatorname{Lm} \mathbf{P}(j2)|_A \right| = \left| |-2| - |-13| \right| = 11 \text{ dB} \quad (18.34)$$

$$\underline{/\mathbf{G}(j2)} = \underline{/\mathbf{L}(j2)_{A'}} - \underline{/\mathbf{P}(j2)_A} = -60° - (-153.4°) = 93.4° \quad (18.35)$$

Using the contours of constant $\operatorname{Lm} M = \operatorname{Lm} [\mathbf{L}/(1+\mathbf{L})]$ on the NC in Fig. 18.17, the maximum occurs at point $C'(M = -0.49 \text{ dB})$ and the minimum at point

FIGURE 18.17
Derivation of bounds $\mathbf{B}_R(j\omega_i)$ on $\mathbf{L}_o(j\omega)$ for $\omega = 2$.

$A'(M = -6 \text{ dB})$ so that the maximum change in Lm **T** is, in this case, $(-0.49) - (-6) = 5.51$ dB. If the specifications tolerate a change of 6.5 dB at $\omega = 2$, the above trial position of Lm $\mathbf{L}_o(j2)$ is well within the permissible tolerance. Lowering the template on the NC to $A''B''C''D''$, where the extreme values of Lm $[L/(1 + L)]$ are at $C''(-0.7 \text{ dB})$ and $A''(-7.2 \text{ dB})$, yields Lm $\mathbf{L}(j2)_{C''} - $ Lm $\mathbf{L}(j2)_{A''} = -0.7 - (-7.2) = 6.5 \text{ dB} = \delta_R(j2)$. Thus, if $\underline{/\mathbf{L}_o(j2)} = -60°$, then -4.2 dB is the smallest or minimum value of Lm $\mathbf{L}_o(j2)$ which satisfies the 6.5 dB specifications for $\delta_R(j\omega_i)$. Any smaller magnitude is satisfactory but represents overdesign at that frequency. The manipulation of the $\omega = 2$ template, for ease of the design process, is repeated along a new angle (vertical) line, and a corresponding new minimum of $\mathbf{L}_o(j2)$ is found. Sufficient points are obtained in this manner to permit drawing a continuous curve of the bound $\mathbf{B}_R(j2)$ on $\mathbf{L}_o(j2)$, as shown in Fig. 18.17. The above procedure is repeated at other frequencies, resulting in a family of boundaries $\mathbf{B}_R(j\omega_i)$ of permissible $\mathbf{L}_o(j\omega)$. The procedure for determining the boundaries $\mathbf{B}_R(j\omega_i)$ is summarized as follows:

1. From Fig. 18.8 obtain values of $\delta_R(j\omega_i)$ for a range of values of $\omega_i(\omega_1, \omega_2, \ldots, \omega_i, \ldots, \omega_h)$, preferably an octave apart, over the specified bandwidth. [The selection of ω_h results in the bound $\mathbf{B}_R(j\omega_h)$ passing under the U-contour.]

2. Place the template $\mathfrak{T}\mathbf{P}(j\omega_i)$ on the NC containing the U-contour to determine the bound $\mathbf{B}_R(j\omega_i)$ as follows:

    a. Use major angle divisions of the NC for lining up the $\mathfrak{T}\mathbf{P}(j\omega_i)$.

    b. Select P_o to represent, in general, the lowest point of $\mathfrak{T}\mathbf{P}(j\omega_i)$. For the design example shown in this chapter:

    (1) Select point A in Fig. 18.18 as P_o.

    (2) Line up side A-B of $\mathfrak{T}\mathbf{P}(j\omega_i)$ on the $-90°$ line, as shown in Fig. 18.18. Move the template up or down until the difference ΔLm $T_R(j\omega_i)$ between the values of two adjacent M-contours (Lm α and Lm β, respectively) is equal to the value of $\delta_R(j\omega_i)$ obtained from Fig. 18.8. Thus, in Fig. 18.18, determine the locations of the template on the $-90°$ line where

$$\Delta \text{ Lm } T_R(j\omega_i) = \text{Lm } \alpha - \text{Lm } \beta = \delta_R(j\omega_i) \qquad (18.36)$$

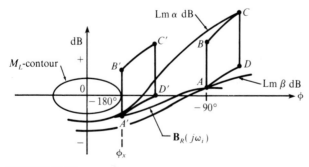

FIGURE 18.18
Graphical determination of $\mathbf{B}_R(j\omega)$.

When Eq. (18.36) is satisfied, then point A, on the M-contour $=$ Lm β, lies on the bound $\mathbf{B}_R(j\omega_i)$. Mark this point on the NC. *For other plants the shape of the template may be such that if point A represents* $\mathbf{P}_o$, *another point of the template may be the lowest point that satisfies Eq. (18.36). When this equation is satisfied, point A still yields points of* $\mathbf{B}_R(j\omega_i)$.

3. Repeat step 2 on the lines $-100°$, $-110°$, etc., up to $-180°$ or until a point of the template becomes tangent to the M_L-contour. No intersection of the M_L-contour by a template is permissible. For example, in moving the template from the $-90°$ line to the left, the template may eventually become tangent to the M_L-contour at some angle ϕ_x, as illustrated in Fig. 18.18. If the template is moved further to the left, it will intersect the M_L contour and permit a peak of $T(j\omega)$ greater than M_L. In order to satisfy the requirement of Eq. (18.30), point A' on the $\phi = \phi_x$ line becomes the left boundary or terminating point for the $\mathbf{B}_R(j\omega_i)$ contour and is a point on the U-contour (a point on ab of Fig. 18.15).[2] Draw a curve through all the points to obtain the contour for $\mathbf{B}_R(j\omega_i)$. For the plant of this example the U-contour is symmetrical about the $-180°$ axis. Note that obtaining the bounds $\mathbf{B}_R(j\omega_i)$ as described in step 2b(2) guarantees only that the difference between the upper bound Lm $\mathbf{T}_U$ and the lower bound Lm $\mathbf{T}_L$ for Lm $\mathbf{T} =$ Lm $[\mathbf{L}/(1 + \mathbf{L})]$ will be

$$\delta_R(j\omega_i) = \text{Lm}\,\mathbf{T}_{R_U}(j\omega_i) - \text{Lm}\,\mathbf{T}_{R_L}(j\omega_i)$$

$$= \text{Lm}\,\mathbf{T}_U(j\omega_i) - \text{Lm}\,\mathbf{T}_L(j\omega_i) \tag{18.37}$$

4. Repeat steps 2 and 3 over the range of $\omega_x \leq \omega_i \leq \omega_h$, generally octaves apart, until the highest bound $\mathbf{B}_R(j\omega_i)$ and lowest bound $\mathbf{B}_R(j\omega_h)$ on the NC clear the U-contour. For reasonably damped plants ($\zeta > 0.6$), over the entire region of plant uncertainty, the magnitudes of the bounds $\mathbf{B}_R(j\omega_i)$ usually decrease as ω increases. Thus, for this type of plant, it is desirable to have $\delta_R(j\omega_i)$ increasing as ω_i increases, as discussed in Sec. 18.4. When this characteristic of $\delta_R(j\omega_i)$ is not observed, it is possible to have $|\mathbf{B}_R(j\omega_j)| > |\mathbf{B}_R(j\omega_i)|$ for $\omega_j > \omega_i$. For a plant that is highly underdamped ($\zeta \leq 0.6$), over some portion of the region of plant uncertainty, avoid selecting an underdamped nominal plant $P_o(s)$. For this latter situation, it is desirable, but not necessary, to synthesize T_{R_U} and T_{R_L} based upon an overdamped $P_o(s)$ model.

18.9   DISTURBANCE BOUNDS $\mathbf{B}_D(j\omega_i)$:
CASE 1 $[d_2(t) = D_o u_{-1}(t),\ d_1(t) = 0]$

Two disturbance inputs are shown in Fig. 18.5. It is assumed that only one disturbance input exists at a time. Both cases are analyzed.

CONTROL RATIO.   From Fig. 18.5 the disturbance control ratio for input $d_2(t)$ is

$$T_D(s) = \frac{1}{1 + L} \tag{18.38}$$

Substituting $L = 1/\ell$ into Eq. (18.38) yields

$$T_D(s) = \frac{\ell}{1 + \ell} \tag{18.39}$$

This equation has the mathematical format required to use the NC. Over the specified bandwidth it is desired that $|T_D(j\omega)| \ll 1$, which results in the requirement, from Eq. (18.39), that $|L(j\omega)| \gg 1$ (or $|\ell(j\omega)| \ll 1$), i.e.,

$$|T_D(j\omega)| \approx \frac{1}{|L(j\omega)|} = |\ell(j\omega)| \tag{18.40}$$

DISTURBANCE RESPONSE CHARACTERISTIC. A time-domain tracking response characteristic based upon $r(t) = u_{-1}(t)$ often specifies a maximum allowable peak overshoot M_p. In the frequency domain this specification may be approximated by

$$|M_R(j\omega)| = |T_R(j\omega)| = \left| \frac{Y(j\omega)}{R(j\omega)} \right| \leq M_m \approx M_P \tag{18.41}$$

The corresponding time- and frequency-domain response characteristics, based upon a step disturbance forcing function are, respectively,

$$|M_D(t)| = \left| \frac{y(t)}{d(t)} \right| \leq \alpha_p \quad \text{for } t \geq t_x \tag{18.42}$$

and

$$|M_D(j\omega)| = |T_D(j\omega)| = \left| \frac{Y(j\omega)}{D(j\omega)} \right| \leq \alpha_m \approx \alpha_p \tag{18.43}$$

APPLICATION. Let $L = KL'$ in the tracking ratio $T = L/(1 + L)$, where K is an unspecified gain, and the specification on system performance is $M_m = 1.12$ (1 dB). By means of a NC determine the value of K required and obtain the data to plot $|M(j\omega)|$ vs. ω. The plot of Lm L' vs. $\angle L'$ on the NC is tangent to the $M = 1$ dB contour, resulting in the plot of Lm $L(j\omega)$ vs. $\angle L(j\omega)$ in Fig. 18.19. Intersections of Lm $L(j\omega)$ with the M-contours provide the data to plot the tracking control ratio $|M(j\omega)|$ vs. ω.

Now consider the corresponding disturbance control ratio for the same control system. The disturbance transfer function is $T_D = 1/(1+L) = \ell/(1 + \ell)$, having the desired BW $0 \leq \omega \leq \omega_2$ for which $|L| \gg 1$ and $|\ell| \ll 1$. Thus Eq. (18.4) applies within the BW region. Table 18.1 contains data for two points on the Nichols plot of Fig. 18.19. The plot of Lm ℓ vs. $\angle \ell$ for these two points is also shown in this figure. The NC of Fig. 18.19 can be rotated 180° and is redrawn in Fig. 18.20. Since Lm $\ell(j\omega)$ = Lm $(1/L(j\omega))$ = $-$Lm $L(j\omega)$, then a negative value of Lm ℓ yields a positive value for Lm L as shown in Fig. 18.20. Since $L = KL' = 1/\ell$, then

$$\ell(j\omega) = K^{-1}\ell'(j\omega) \tag{18.44}$$

TABLE 18.1
Data points for a Nichols plot

ω	Lm **L**	$\angle$**L**	Lm ℓ	$\angle\ell$
ω_1	21	$-96°$	-21	$96°$ (or $-264°$)
ω_2	15	$-98°$	-15	$98°$ (or $-262°$)

If $\ell'(j\omega)$ is given and it is required to determine K^{-1} to satisfy Eq. (18.38), then the plot Lm $\ell'(j\omega)$ vs. $\angle\ell'(j\omega)$ must be raised or lowered until it is tangent to the Lm α_p-contour ($\|\mathbf{T}_D\| \max = \alpha_m$). The amount Δ by which this plot is raised or lowered yields the value of K, i.e., Lm $K^{-1} = \Delta$. Note that this is the same procedure used for the tracking example of Fig. 18.19, except that the adjustment in Lm $\ell(j\omega)$ is K^{-1}.

TEMPLATES. For a given plant **P** having uncertain parameters, consider that its template $\mathcal{T}\mathbf{P}(j\omega_i)$ for a given ω_i has equal dB differences along its A-B and C-D boundaries, i.e., for a given $\angle\mathbf{P}(j\omega_i)$,

$$\Delta(\text{Lm }\mathbf{P}_B - \text{Lm }\mathbf{P}_A) = \Delta(\text{Lm }\mathbf{P}_C - \text{Lm }\mathbf{P}_D) = 10 \text{ dB}$$

This template is arbitrarily placed on the NC as shown in Fig. 18.19. Data corresponding to the template location shown in Fig. 18.19 are given in Table 18.2.

The template of the reciprocal, Lm $[1/\mathbf{P}(j\omega_i)]$, is arbitrarily located on the NC as shown in Fig. 18.19 for the same frequency as for the template of Lm $\mathbf{P}(j\omega_i)$ and for the angles of Table 18.2. Note that the template of Lm $[1/\mathbf{P}(j\omega_i)]$ is the same as the template of Lm $\mathbf{P}(j\omega_i)$ but is rotated by $180°$. Thus it is located by first reflecting

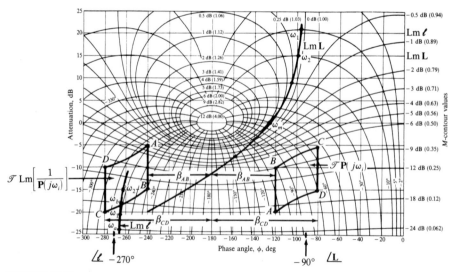

FIGURE 18.19
Regular Nichols chart.

TABLE 18.2
Data points for the templates of Fig. 18.19

Points	$\angle \mathbf{P}$	$\angle 1/\mathbf{P}$
A, B	$-120°$	$120°$ (or $-240°$)
C, D	$-80°$	$80°$ (or $-280°$)

the template of $\text{Lm }\mathbf{P}(j\omega_i)$ about the $-180°$ axis, "flipping it over" vertically, and then moving it up or down so that it lies between -5 and -20 dB. For the *arbitrary* location of the template of $\text{Lm}\,[1/\mathbf{P}(j\omega_i)]$, note that:

(a)
$$\beta_{AB} = 180° + \angle\mathbf{P} = 180° - 120° = 60°$$

$$\beta_{CD} = 180° + \angle\mathbf{P} = 180° - 80° = 100°$$

For $1/\mathbf{P}$, the corresponding angles are

$$\angle 1/\mathbf{P}_{AB} = -180° - \beta_{AB} = -180° - 60° = -240°$$

$$\angle 1/\mathbf{P}_{CD} = -180° - \beta_{CD} = -180° - 100° = -280°$$

(b) The templates of $\text{Lm }\mathbf{P}(j\omega_i)$ are used for the tracker case $T = L/(1 + L)$ and the templates of $\text{Lm}\,[1/\mathbf{P}(j\omega_i)]$ are used for the disturbance-rejection case of $T_D = 1/(1 + L) = \ell/(1 + \ell)$.

Since Lm L is desired for the disturbance-rejection problem, Eq. (18.6), the disturbance boundary $\mathbf{B}_D(j\omega_i)$ for $\mathbf{L}(j\omega_i) = 1/\ell(j\omega_i)$ is best determined on the rotated NC of Fig. 18.20. Thus, the NC of Fig. 18.19 is rotated cw by $180°$, where the rotation of the $\text{Lm}\,[1/\mathbf{P}(j\omega_i)]$ template $ABCD$ is reflected in Fig. 18.20. The rotated NC is used to determine directly the boundaries $\mathbf{B}_D(j\omega_i)$ for $\mathbf{L}_D(j\omega_i)$. Point A for the simple plant of this design example corresponds to the nominal plant parameters and is the lowest point of the template $\mathcal{T}\mathbf{P}(j\omega_i)$. This point is *again used* to determine the disturbance bounds $\mathbf{B}_D(j\omega_i)$. The lowest point of the template must be used to determine the bounds and, in general, may or may not be the point corresponding to the nominal plant parameters.

Based upon Eqs. (18.38) and (18.39) in Fig. 18.9

$$-\text{Lm }T_D = \text{Lm}\,[1 + L] \geq -\text{Lm}\,\alpha(j\omega_i) > 0 \text{ dB} \qquad (18.45)$$

where $\alpha(j\omega_i) < 0$. Since $|\mathbf{L}| \gg 1$ in the BW, then

$$-\text{Lm }T_D \cong \text{Lm L} \geq -\text{Lm}\,\alpha(j\omega_i) = -\delta_D(j\omega_i) \qquad (18.46)$$

In terms of $\mathbf{L}(j\omega_i)$, the constant M-contours of the NC can be used to obtain the disturbance performance T_D. This requires the change of sign of the vertical axis in dB and the M-contours, as shown in Fig. 18.20.

FIGURE 18.20
Rotated Nichols chart.

BOUNDS $\mathbf{B}_D(j\omega_l)$. The procedure for determining the boundaries $\mathbf{B}_D(j\omega_i)$ is:

1. From Fig. 18.9 obtain values of $\delta_D(j\omega_i)$ from the curve representing Eqs. (12.37) and (12.38) for the same values of frequency as for the tracking boundary $\mathbf{B}_R(j\omega_i)$.

2. Select the lowest point of $\mathcal{T}\mathbf{P}(j\omega_i)$ to represent the nominal plant P_o in Fig. 18.20. For the design example used in this chapter select point A in Fig. 18.20 as $\mathbf{P}_o$. The same nominal point must be used in obtaining the tracking and disturbance bounds.

3. Use major angle divisions of the NC for lining up the $\mathcal{T}\mathbf{P}(j\omega_i)$. Line up side A-B of $\mathcal{T}\mathbf{P}(j\omega_i)$, for example, on the $-280°$ line for ℓ (or the $-80°$ line for $\mathbf{L}$). Move the template up or down until point A lies on the M-contour that represents $\delta_D(j\omega_i)$. Mark this point on the NC.

4. Repeat step 3 on the vertical lines for $-100°$, $-120°$, etc., up to the $-180°$ line or the U-contour. Draw a curve through all the points to obtain the contour for $\mathbf{B}_D(j\omega_i)$.

5. Repeat steps 3 and 4 over the desired frequency range $\omega_x \leq \omega_i \leq \omega_h$.

6. Transcribe these $\mathbf{B}_D(j\omega_i)$ onto the NC that contains the bounds $\mathbf{B}_R(j\omega_i)$.

Note that when $|\mathbf{L}| \gg 1$, then, from Eq. (18.45), $|\mathbf{L}(j\omega_i)| \gg \alpha(j\omega_i)$. Thus the M-contour corresponding to $\mathbf{T}_D(j\omega_i)$ becomes the boundary $\mathbf{B}_D(j\omega_i) = -\text{Lm}\,M$

for $\mathbf{L}(j\omega_i)$. For example, if $\alpha(j\omega_i) = 0.12$, then $|\mathbf{L}(j\omega_i)| = 8.33$, and thus Lm $\alpha(j\omega_i) = -18.4$ dB and Lm $\mathbf{L}(j\omega_i) = 18.4$ dB.

18.10 DISTURBANCE BOUNDS $\mathbf{B}_D(j\omega_i)$:
CASE 2 $[d_1(t) = D_o u_{-1}(t), d_2(t) = 0]$

CONTROL RATIO. From Fig. 18.5 the disturbance control ratio for the input $d_1(t)$ is

$$\mathbf{T}_D = \frac{\mathbf{P}}{1 + \mathbf{GP}} \tag{18.47}$$

Assuming point A of the template represents the nominal plant P_o, Eq. (18.47) is rearranged to

$$\mathbf{T}_D = \frac{\mathbf{P}_o}{\mathbf{P}_o}\left[\frac{1}{\dfrac{1}{\mathbf{P}} + \mathbf{G}}\right] = \frac{\mathbf{P}_o}{\dfrac{\mathbf{P}_o}{\mathbf{P}} + \mathbf{GP}_o} = \frac{\mathbf{P}_o}{\dfrac{\mathbf{P}_o}{\mathbf{P}} + \mathbf{L}_o} = \frac{\mathbf{P}_o}{\mathbf{W}} \tag{18.48}$$

where
$$\mathbf{W} = (\mathbf{P}_o/\mathbf{P}) + \mathbf{L}_o \tag{18.49}$$

Thus, Eq. (18.48) with Lm $\mathbf{T}_D = \delta_D$ yields

$$\text{Lm } \mathbf{W} = \text{Lm } \mathbf{P}_o - \delta_D \tag{18.50}$$

DISTURBANCE RESPONSE CHARACTERISTICS. Based on Eq. (18.41), the time- and frequency-domain response characteristics, for a unit-step disturbance forcing function, are given, respectively, by

$$\left|M_D(t)\right| = \left|\frac{y(t_p)}{d(t)}\right| = \left|y(t_p)\right| \leq \alpha_p \tag{18.51}$$

and
$$\left|\mathbf{M}_D(j\omega)\right| = \left|\mathbf{T}_D(j\omega)\right| = \left|\frac{\mathbf{Y}(j\omega)}{\mathbf{D}(j\omega)}\right| \leq \alpha_m \equiv \alpha_p \tag{18.52}$$

where t_p is the peak time.

BOUNDS $\mathbf{B}_D(j\omega_l)$ The procedure for determining the boundaries $\mathbf{B}_D(j\omega_i)$ is:

1. From Fig. 18.9 obtain values of $\delta_D(j\omega_i)$ from the plot representing a model in Table 12.1 or $T_D = \alpha_p$ for the same values of frequency as for the tracker boundaries $\mathbf{B}_R(j\omega_i)$.
2. *Evaluate* in tabular form for each value of ω_i the following items in the order given:

$$\text{Lm } P_o(j\omega_i) \qquad \delta_D(j\omega_i) \qquad \text{Lm } \mathbf{W}(j\omega_i) \qquad |\mathbf{W}(j\omega_i)| \qquad \frac{P_o(j\omega_i)}{P(j\omega_i)}$$

The ratio $\mathbf{P}_o/\mathbf{P}$ is evaluated at each of the four points of Fig. 18.13 for each value of ω_i. It may be necessary to evaluate this ratio at additional points around the perimeter of the $ABCD$ contour.

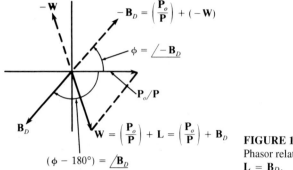

FIGURE 18.21
Phasor relationship of Eq. (18.49) with
$L = B_D$.

3. Before presenting the procedure for the graphical determination of $B_D(j\omega_i)$ it is necessary to first review, graphically, the phasor relationship between B_D, P_o/P, and W for $\omega = \omega_i$. Equation (18.49), with L_o replaced by its bound B_D, is rearranged to the form

$$-B_D = \frac{P_o}{P} + (-W) \tag{18.53}$$

For arbitrary values of $P_o(j\omega_i)/P(j\omega_i)$ and $W(j\omega_i)$, Fig. 18.21 presents the phasor relationship of Eq. (18.53). Since the values of $P_o(j\omega_i)/P(j\omega_i)$ for $P \in \mathcal{P}$ and $|W(j\omega_i)|$ are known, the following procedure can be used to evaluate $B_D(j\omega_i)$:

(a) On polar or rectangular graph paper, draw $\mathcal{T}[P_o/P]$ for each ω_i, as shown in Fig. 18.22, where point A is the nominal point.

(b) Based on Fig. 18.21 and the location of the phasor $-W(j\omega)$, the solution for $-B_D$ is obtained from

$$-B_D = \mathcal{T}[P_o/P] - W \tag{18.54}$$

For one value of P_o/P shown in Fig. 18.21, the value of $-W$ is plotted and $-B_D = |-B_D|/\underline{\phi}$ is obtained. This graphical evaluation of $B_D = |B_D|/\underline{\phi} - 180°$ is performed for various points around the perimeter of $\mathcal{T}[P_o/P]$ in Fig. 18.22. A simple graphical evaluation yields a more restrictive bound (the worst case). Use a compass to mark off arcs with a radius equal to the distance $|W(j\omega_i)|$ at a number of points on the perimeter of $\mathcal{T}[(P_o(j\omega_i))/(P(j\omega_i))]$. Draw a curve that is tangent to these arcs to form the first-quadrant portion of the Q-contour shown in Fig. 18.23. Depending

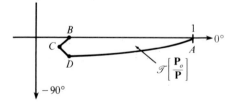

FIGURE 18.22
Template in polar coordinates.

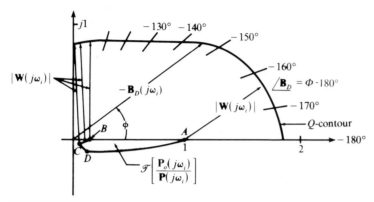

FIGURE 18.23
Graphical evaluation of $\mathbf{B}_D(j\omega_i)$.

TABLE 18.3
Data points for the boundary $\mathbf{B}_D(j\omega)$

$\angle(\mathbf{B}_D(j\omega_i))$	$|\mathbf{B}_D(j\omega_i)|$	$\mathbf{Lm}\,\mathbf{B}_D(j\omega_i)$
$-180°$		
$-170°$		
$-160°$		
⋮		

on the plant type desired for **L**, it may be necessary to extend this contour into the second and fourth quadrants.

4. Based upon Eq. (18.54) and Fig. 18.22, the phasor from the origin of Fig. 18.23 to the Q-contour represents $-\mathbf{B}_D(j\omega_i)$. This contour includes the plant uncertainty as represented by $\mathcal{T}[\mathbf{P}_o(j\omega_i)/\mathbf{P}(j\omega_i)]$. In the frequency range $\omega_x \leq \omega_i \leq \omega_h$, if $|\mathbf{W}(j\omega_i)| \gg |\mathbf{P}_o(j\omega_i)/\mathbf{P}(j\omega_i)|$, then the Q-contour is essentially a circle about the origin with radius $|\mathbf{W}(j\omega_i)| \equiv |\mathbf{B}_D(j\omega_i)|$.

5. Assuming the partial Q-contour of Fig. 18.23 is sufficient, measure from the graph the length $\mathbf{B}_D(j\omega_i)$ for every 10° of $\mathbf{B}_D(j\omega_i)$ and create Table 18.3 for each value of ω_i.

6. Plot the values from Table 18.3 for each value of ω_i, on the same NC as $\mathbf{B}_R(j\omega_i)$.

18.11   THE COMPOSITE BOUNDARY $\mathbf{B}_o(j\omega_i)$

The composite bound $\mathbf{B}_o(j\omega_i)$, which is used to synthesize the desired loop transmission transfer function $L_o(s)$ is obtained in the manner shown in Fig. 18.24. The composite bound $\mathbf{B}_o(j\omega_i)$ for each value of ω_i is composed of those portions of each respective bound $\mathbf{B}_R(j\omega_i)$ and $\mathbf{B}_D(j\omega_i)$ that are the most restrictive. For the case shown in Fig. 18.24(a) the bound $\mathbf{B}_o(j\omega_i)$ is composed of those portions of each respective bound $\mathbf{B}_R(j\omega_i)$ and $\mathbf{B}_D(j\omega_i)$ that have the largest values. For the situation of

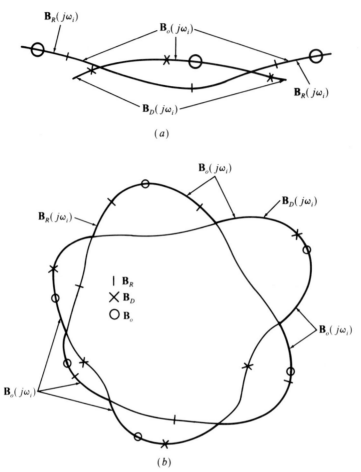

FIGURE 18.24
Composite $\mathbf{B}_o(j\omega_i)$.

Fig. 18.24(*b*), the outermost of the two boundaries $\mathbf{B}_R(j\omega_i)$ and $\mathbf{B}_D(j\omega_i)$ becomes the perimeter of $\mathbf{B}_o(j\omega_i)$. The situations of Fig. 18.24 occur when the two bounds have one or more intersections. If there are no intersections, the bound with the largest value or with the outermost boundary dominates. The synthesized $\mathbf{L}_o(j\omega_i)$, for the situation of Fig. 18.24(*a*), must lie on or just above the bound $\mathbf{B}_o(j\omega_i)$. For the situation of Fig. 18.24(*b*) the synthesized $\mathbf{L}_o(j\omega_i)$ *must not lie in the interior* of the $\mathbf{B}_o(j\omega_i)$ contour.

18.12   SHAPING OF $\mathbf{L}_o(j\omega)$[1,2,16]

A realistic definition of optimum[9,10] in an LTI system is the minimization of the high-frequency loop gain K. This gain affects the high-frequency response, because $\lim_{\omega\to\infty}[\mathbf{L}(j\omega)] = K(j\omega)^{-\lambda}$, where λ is the excess of poles over zeros assigned to

$\mathbf{L}(j\omega)$. Thus, only the gain K has a significant effect on the high-frequency response, and the effect of the other parameter uncertainty is negligible. It has been shown that the optimum $\mathbf{L}(j\omega)$ exists; it lies on the boundary $\mathbf{B}_o(j\omega_i)$ at all ω_i, and it is unique.[7,8]

Previous sections describe how tolerances on the closed-loop system frequency response, in combination with plant uncertainty templates, are translated into bounds on a nominal loop transmission function $\mathbf{L}(j\omega)$. In Fig. 18.25 the template $\mathcal{T}\mathbf{P}(j\omega_i)$ is located on the NC so that point A is on the constant M-curve Lm β, and point C is on the constant M-curve Lm α, such that

$$\delta_R(j\omega_i) = \mathrm{Lm}\,\alpha - \mathrm{Lm}\,\beta = 4\,\mathrm{dB} \tag{18.55}$$

where $\alpha = T_{\max}$, $\beta = T_{\min}$, and $\delta_R(j\omega_i) = 4$ dB are obtained from Fig. 18.4. Repeat this procedure for the required frequency range ω to determine the entire curve for $\mathbf{L}_o(j\omega_i)$ shown in Fig. 18.25.

For the set of plant parameters that correspond to point E within $\mathcal{T}\mathbf{P}(j\omega_i)$, as shown in Fig. 18.25, the open-loop transfer function is $\mathbf{L}_E = \mathbf{GP}_E$. The control ratio

$$\mathbf{T}_E = \frac{\mathbf{L}_E}{1 + \mathbf{L}_E} \tag{18.56}$$

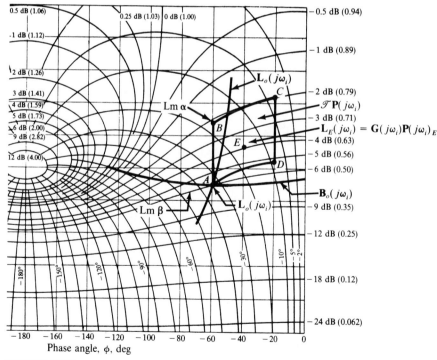

FIGURE 18.25
Graphical determination of Lm $\mathbf{T}(j\omega_i)$ for $\mathbf{P} \in \mathcal{P}$.

obtained from the constant M-contours on the NC has the value $\mathrm{Lm}\,\mathbf{T}_E = -4$ dB. Thus, any value of $\mathbf{P}$ within $\mathfrak{I}\mathbf{P}(j\omega_i)$ yields a value of $\mathrm{Lm}\,\mathbf{T}$ between points A and C in Fig. 18.25. Therefore, any value of $\mathbf{P}$ that lies within the region of uncertainty (see Fig. 18.13 for the example of this chapter) yields a maximum variation in $\mathbf{T}(j\omega_i)$ that satisfies the requirement

$$\mathrm{Lm}\,\mathbf{T}_{\max} - \mathrm{Lm}\,\mathbf{T}_{\min} \leq \delta_R(j\omega_i) \tag{18.57}$$

Thus, proper design of the prefilter F (see Fig. 18.5) yields a tracking control ratio $\mathbf{T}_R$ that lies between $\mathrm{Lm}\,\mathbf{T}_U$ and $\mathrm{Lm}\,\mathbf{T}_L$ in Fig. 18.8(b).

For the plant of Eqs. (18.27) and (18.28), the shaping of $\mathbf{L}_o(j\omega)$ is shown by the dashed curve in Fig. 18.26. A point such as $\mathrm{Lm}\,\mathbf{L}_o(j2)$ must be on or above the curve labeled $\mathbf{B}_o(j2)$. Further, in order to satisfy the specifications, $\mathbf{L}_o(j\omega)$ cannot violate the U-contour. In this example a reasonable $\mathbf{L}_o(j\omega)$ closely follows the U-contour up to $\omega = 40$ and must then stay below it, as shown in Fig. 18.26. Additional specifications are that $\lambda = 4$ and that $L_o(s)$ must be Type 1 (one pole at the origin).

A representative procedure for choosing a rational function $L_o(s)$ that satisfies the above specifications is now described. It involves building up the function

$$\mathbf{L}_o(j\omega) = \mathbf{L}_{ok}(j\omega) = \mathbf{P}_o(j\omega) \prod_{k=0}^{w} [K_k \mathbf{G}_k(j\omega)]$$

where for $k = 0$, $\mathbf{G}_0 = 1\underline{/0°}$, and $K = \prod_{k=0}^{w} K_k$. $\mathbf{L}_o(j\omega)$ is built up term by term in order to stay just outside the U-contour in the NC of Fig. 18.26. The first step is to find the $\mathbf{B}_o(j\omega)$ that "dominates" $\mathbf{L}_o(j\omega)$. For example, suppose $L_{o0}(j4) = K_o \mathbf{P}'_o(j4) = 0$ dB $\underline{/-135°}$ (point A in Fig. 18.26) and that at $\omega = 1$ the required $\mathrm{Lm}\,\mathbf{L}_{o0}(j1)$ is approximately 27 dB. In order for $\mathrm{Lm}\,\mathbf{L}_o(j\omega)$ to decrease from 27 to about 0 dB in two octaves, the slope of $\mathbf{L}_o(j\omega)$ must be about -14 dB/octave, with $\underline{/\mathbf{L}_o} < -180°$ (see Fig. 18.27). Since absolute stability is required, $\mathbf{L}_o(j\omega)$ must have a phase margin of 40° over the entire frequency range for which $\mathbf{L}_o(j\omega)$ follows the vertical right-hand side of the U-contour and not just at the 0 dB crossover. Hence $\mathbf{B}_o(j1)$ dominates $\mathbf{L}_o(j\omega)$ more than does $\mathbf{B}_o(j4)$. In the same way it is seen that $\mathbf{B}_o(j1)$ dominates all other $\mathbf{B}_o(j\omega)$ in Fig. 18.26.

The $\mathbf{B}_o(j\omega)$ curves for $\omega < 1$ are not shown in Fig. 18.26, because it is assumed that a slope of -6 dB/octave for $\omega < 1$ suffices (additional values are 33 dB at $\omega = 0.5$, 39 dB at $\omega = 0.25$, etc.). To maintain $-140°$ for $\omega \geq 1$, the first denominator factor of $\mathbf{L}(j\omega)$ has a corner frequency at $\omega = 1$ (i.e., a pole at -1 is chosen), and the value $\mathrm{Lm}\,\mathbf{L}_o(j1)$ on the straight-line approximation is selected at 30 dB (to allow for the -3 dB correction at the corner frequency). The function $L_o(s)$ determined so far is

$$L_{o0}(s) = 31.6/s(s+1)$$

whose angle $\underline{/\mathbf{L}_{o0}(j\omega)}$ is sketched in Fig. 18.27.

$\mathbf{L}_{o0}(j\omega)$ violates the $-140°$ bound at $\omega \geq 1.2$, so a numerator term $(1 + j\omega T_2)$ must be added. At $\omega = 5$, $\underline{/\mathbf{L}_{o0}(j5)} = -169°$ (see Fig. 18.27), and therefore a lead angle of 29° is needed at this frequency. Since a second denominator term

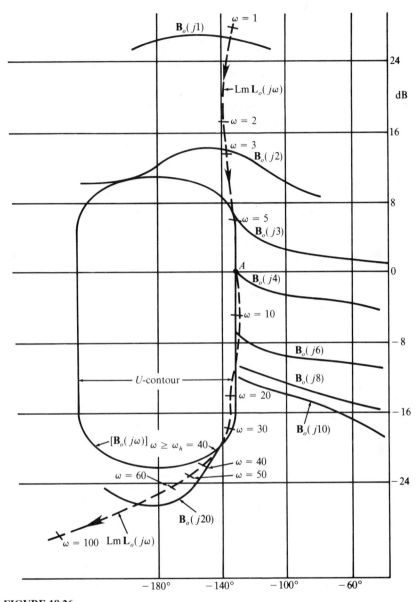

FIGURE 18.26
Shaping of $\mathbf{L}_o(j\omega)$ on the Nichols chart for the plant of Eqs. (18.27) and (18.28).

$(1 + j\omega T_3)$ will be needed, allow an additional 15° for this factor, giving a total of $15° + 29° = 45°$ lead angle required at $\omega = 5$. This is achieved by selecting $T_2 = 0.2$, i.e., a zero at -5. This results in the composite value

$$L_{o1}(s) = \frac{31.6(1 + s/5)}{s(1 + s)}$$

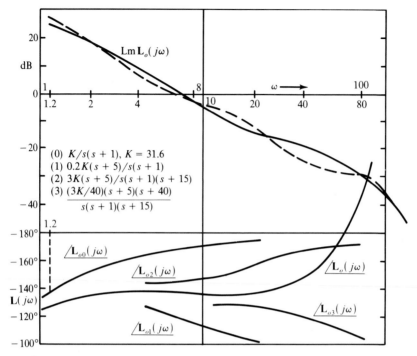

FIGURE 18.27
Shaping of $\mathbf{L}_o(j\omega)$ on the Bode plot.

whose phase angle $\angle\mathbf{L}_{o1}(j\omega)$ is sketched in Fig. 18.27. In the NC of Fig. 18.26, $\omega \geq 5$ is the region where the maximum phase *lag* allowed is $-135°$ (i.e., $\angle\mathbf{L}_o(j\omega)$ must be $\geq -135°$). At $\omega = 10$, since $\angle\mathbf{L}_{o1}(j10) = -112°$, then $135° - 112° = 23°$ more lag angle is needed, and it is provided with an additional denominator term $(1 + j\omega T_3)$. However, this is to be followed by an additional numerator term $(1 + j\omega T_4)$, so allow about $10°$ for it, giving $23° + 10° = 33°$ more lag allowable from $(1 + j\omega T_3)$. This requires selecting the corner frequency at $\omega = 15$ (or $T_3 = 1/15$), giving

$$L_{o2}(s) = \frac{31.6(1 + s/5)}{s(1 + s)(1 + s/15)}$$

A sketch of $\angle\mathbf{L}_{o2}(j\omega)$ is shown in Fig. 18.27.

Looking ahead at $\omega = 40$, $\text{Lm}\,\mathbf{L}_{o2}(j40) = -20$ dB, thus $\mathbf{L}_o(j\omega)$ can make its asymptotic left turn *under* the U-contour. The plan is to add two more numerator factors, $(1 + j\omega T_4)$ and $(1 + j\omega T_5)$, and finally two complex-pole pairs, in order to have an excess $\lambda = 4$ of poles over zeros and to minimize the *BW*. A zero is assigned at $\omega = 40$ with $T_4 = 1/40$, giving

$$L_{o3}(s) = \frac{31.6(1 + s/5)(1 + s/40)}{s(1 + s)(1 + s/15)}$$

A sketch of $\angle\mathbf{L}_{o3}(j\omega)$ is shown in Fig. 18.27.

In order to achieve (an asymptotic) horizontal segment for $\text{Lm}\,\mathbf{L}_o(j\omega)$ before the final -24 dB/octave slope is achieved, a final zero obtained from $(1 + j\omega T_5)$ is needed. Since the bottom of the U-contour (see Fig. 18.26) is at -22.5 dB, allow a 2 dB safety margin, a 3 dB correction due to $(1 + j\omega T_5)$, and 1.5 dB for the effect of $(1 + j\omega T_4)$, giving a total of $-(22.5 + 2 + 3 + 1.5) = -29$ dB. A damping ratio of $\zeta = 0.5$ is selected for the two complex-pole pairs, so no correction is needed for them. Thus, the final corner frequency for the straight-line curve of $\text{Lm}\,\mathbf{L}_{o3}(j\omega)$ is at -29 dB. Since $\text{Lm}\,\mathbf{L}_{o3}(j\omega)$ achieves this at $\omega = 60$, the last corner frequency is at $\omega = 60$, i.e., $T_5 = 1/60$. The resulting phase angle, due to L_{o3} and $(1 + j\omega/60)$, is $-66°$. An angle of $-180°$ could be selected at this point, but an additional $15°$ margin is allowed (this is a matter of judgment that depends on the problem, which may include the presence of higher-order modes, etc.). This means that $100°$ phase lag is permitted: $50°$ due to each complex-pole pair ($180° - 66° - 15° \approx 100°$). Thus, a different value of damping ratio, with the appropriate dB correction applied to the log magnitude plot, is chosen, i.e., $\zeta = 0.6$. This locates the corner frequency at $\omega = 100$. Thus

$$L_o(s) = \frac{31.6\,(1 + s/5)\,(1 + s/40)\,(1 + s/60)}{s(1 + s)(1 + s/15)\left(1 + 1.2s/100 + s^2/10^4\right)^2} = KP_o(s)\prod_{k=1}^{4} G_k(s)$$

(18.58)

The optimal loop transfer function $\mathbf{L}_o(j\omega)$ is sketched in Fig. 18.26 and is shown by the solid curves in Fig. 18.27. A well designed, i.e., an "economical," $\mathbf{L}_o(j\omega)$ is close to the $\mathbf{B}_o(j\omega)$ boundary at each ω_i. The vertical line at $-140°$ in Fig. 18.26 is the dominating vertical boundary for $\mathbf{L}_o(j\omega)$ for $\omega < 5 = \omega_x$, and the right side of the U-contour line at $-135°$ is the vertical boundary effectively for $\omega_x \approx 5 < \omega < 30 \approx \omega_y$. The final $\mathbf{L}_o(j\omega)$ is good in this respect, since it is close to these boundaries. Although there is a "slight" infringement of the U- contour for $\omega > 15$ because of the inherent overdesign, no reshaping of $L_o(s)$ is done unless the simulation reveals that the specifications are not met.

There is a trade-off between the complexity of $L_o(s)$ (the number of its poles and zeros) and its final cut-off corner frequency, which is $\omega = 100$, and the phase margin frequency $\omega_\phi = 7$. There is some phase to spare between $\mathbf{L}_o(j\omega)$ and the boundaries. Use of more poles and zeros in $L_o(s)$ permits this cut-off frequency to be reduced a bit below 100, but not by much. On the other hand, if it is desired to reduce the number of poles and zeros of $L_o(s)$, the price in achieving this is a larger cut-off frequency. It is possible to economize significantly by allowing more phase lag in the low-frequency range. If $-180°$ is permitted at $\omega = 1$, then a decrease of Lm $\mathbf{L}_o(j\omega)$ at a rate of 12 dB/octave can be achieved, e.g., with $\text{Lm}\,\mathbf{L}_o(j1) = 25$ dB, then it will be 13 dB at $\omega = 2$ (instead of the present 18 dB). Even with no more savings, this 5-dB difference allows a cut-off frequency of about $\omega = 70$ instead of 100.

Figure 18.26 reveals immediately, without any reshaping of $\mathbf{L}_o(j\omega)$ required, that reduction (i.e., easing) of the specifications at $\omega = 1$ to about 21 dB (instead of about 26 dB) has the same effect as the above. How badly the specifications are compromised by such easing can easily be checked. The design technique is thus highly "transparent" in revealing the trade-offs between performance tolerances, complexity of the compensation, stability margins, and the "cost of feedback" in bandwidth.

18.13 GUIDELINES FOR SHAPING $L_o(j\omega)$

Some general guidelines for the shaping of $\mathbf{L}_o(j\omega)$ are:

1. For the beginner it is best not to use a CAD program (see App. B) until step 6. Use the straight-line approximations on the Bode diagram for the log magnitude at the start of the design problem.
2. On the graph paper for the Bode diagram plot the points representing Lm $\mathbf{B}_o(j\omega_i)$ and the angles corresponding to the right side of the U-contour (the desired phase margin angle γ) for the frequency range of $\omega_x \leq \omega \leq \omega_y$. For the example illustrated in Fig. 18.26 the phase margin angle $\gamma = -45°$ must be maintained for the frequency range $\omega_x = 5 \leq \omega \leq \omega_y \approx 30$.
3. Do the shaping of Lm $\mathbf{L}_o(j\omega)$ on the Bode plot using straight-line approximations for Lm $\mathbf{L}_o(j\omega)$, with the plotted information of step 2, and employ the shaping discussion that follows Eq. (18.58) for the frequency range $\omega < \omega_x$ as guidelines in achieving Eq. (18.58).
4. Use frequencies an octave above and below and a decade above and below a corner frequency for both first- and second-order terms,[3,6] while maintaining the phase-margin angle corresponding to the right side of the U-contour in shaping $\mathbf{L}_o(j\omega)$.
5. The last two poles that are added to $\mathbf{L}_o(j\omega)$ are generally a complex pair (the nominal range is $0.5 < \zeta < 0.7$), which tends to minimize the BW.
6. Once $\mathbf{L}_o(j\omega)$ has been shaped, determine $F(s)$ (see Sec. 18.14) and then verify that $\mathbf{L}_o(j\omega)$ does meet the design objectives by use of a CAD program. If the synthesized $\mathbf{L}_o(j\omega)$ yields the desired performance, then the required compensator is given by

$$G(s) = \frac{L_o(s)}{P_o(s)}$$

Specific guidelines for shaping $\mathbf{L}_o(j\omega)$ are:

1. An optimum design of $\mathbf{L}_o(j\omega)$ requires that $\mathbf{L}_o(j\omega_i)$ be on the corresponding bound. In practice, place $\mathbf{L}_o(j\omega_i)$ as close as possible to the bound $\mathbf{B}_o(j\omega_i)$, but above it, in order to keep the BW of $\mathbf{L}_o/[1 + \mathbf{L}_o]$ to a minimum.
2. Since exact cancellation of a pole by a zero is rarely possible, any RH plane poles and/or zeros of $P_o(s)$ should be included in $L_o(s)$. A good starting $L_o(s)$ is $L_{oo}(s) = K_0 P_o(s)$. If it is desired that $y(\infty) = 0$ for $d(t) = u_{-1}(t)$, it is necessary to ensure that $T_D(s)$ has a zero at the origin. For this situation a possible starting point is $L_{oo}(s) = K_0 P_o(s)/s$.
3. If $P(s)$ has an excess of poles over zeros, $\#p\text{'s} - \#z\text{'s} = \lambda$, which is denoted by e^λ, then, in general, the *final form* of $L_o(s)$ must have an excess of poles over zeros of at least $e^{\lambda+i}$, where $i \geq 1$. If the BW is too large, then increase the value

of i. Experience shows that a value of $\lambda + i$ of 3 or more for $L(s)$ yields satisfactory results.[2,8]

4. Generally the BW of $L_o(s)/[1 + L_o(s)]$ is larger than required for an acceptable rise time t_R for the tracking of $r(t)$ by $y(t)$. An acceptable rise time can be achieved by the proper design of the prefilter $F(s)$ (see Fig. 18.18).

5. Generally it is desirable to first find the bounds $\mathbf{B}_D(j\omega_i)$ and then the bounds $\mathbf{B}_R(j\omega_i)$. After finding the first $\mathbf{B}_R(j\omega_i)$, it may be evident that all or some of the $\mathbf{B}_D(j\omega_i)$ are completely dominant compared to $\mathbf{B}_R(j\omega_i)$. In that case the $\mathbf{B}_D(j\omega_i)$ boundaries are the optimal boundaries, i.e., $\mathbf{B}_D(j\omega_i) = \mathbf{B}_o(j\omega_i)$.

6. If $\delta_R(j\omega_i)$ is not continuously increasing as ω_i increases, then it will be necessary to utilize complex poles and/or zeros in $G(s)$ in order to achieve an optimal $L_o(s)$. This assumes that the tracking model is not designed to yield this increasing characteristic for $\delta_R(j\omega_i)$.

7. The ability to shape the nominal loop transmission $L_o(s)$ is an art developed by the designer after much practice and patience. It is expedited by the use of an interactive CAD package.

18.14 DESIGN OF THE PREFILTER $F(s)$[1,10]

Design of a proper $L_o(s)$ guarantees only that the variation in $|\mathbf{T}_R(j\omega)|$ is less than or equal to that allowed. The purpose of the prefilter is to position $\mathrm{Lm}\,\mathbf{T}(j\omega)$ within the frequency domain specifications. For the example of this chapter the magnitude of the frequency response must lie within the bounds B_U and B_L shown in Fig. 18.8, which are redrawn in Fig. 18.28. A method for determining the bounds on $F(s)$ is as follows: Place the nominal point on the ω_i plant template on the $L_o(j\omega_i)$ point on the $L_o(j\omega)$ curve on the NC (see Fig. 18.29). Traversing the template, determine the maximum $\mathrm{Lm}\,\mathbf{T}_{max}$ and minimum $\mathrm{Lm}\,\mathbf{T}_{min}$ values of

$$\mathrm{Lm}\,\mathbf{T}(j\omega_i) = \frac{\mathbf{L}(j\omega_i)}{1 + \mathbf{L}(j\omega_i)} \tag{18.59}$$

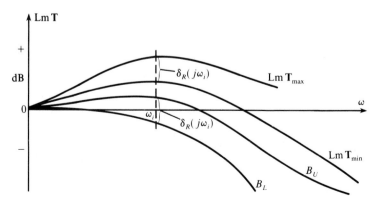

FIGURE 18.28
Requirements on $F(s)$.

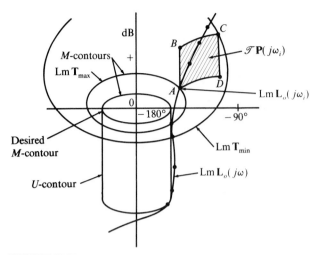

FIGURE 18.29
Prefilter determination.

obtained from the M-contours. These values are plotted as shown in Fig. 18.28. The tracking control ratio is $T_R = FL/[1 + L]$, and

$$\text{Lm}\,\mathbf{T}_R(j\omega_i) = \text{Lm}\,\mathbf{F}(j\omega_i) + \text{Lm}\,\mathbf{T}(j\omega_i) \tag{18.60}$$

The variations in Eqs. (18.59) and (18.60) are both due to the variation in P; thus

$$\delta_p(j\omega_i) = \text{Lm}\,\mathbf{T}_{max} - \text{Lm}\,\mathbf{T}_{min} \le \delta_R = B_U - B_L \tag{18.61}$$

If values of $\mathbf{L}_o(j\omega_i)$, for each value of ω_i, lie exactly on the tracking bounds $\mathbf{B}_R(j\omega_i)$, then $\delta_p = \delta_R$. Therefore, based upon Eq. (18.60), it is necessary to determine the range in dB that $\text{Lm}\,\mathbf{T}(j\omega_i)$ must be raised or lowered to fit within the bounds of the specifications by use of the prefilter $\mathbf{F}(j\omega_i)$. The process is repeated for each frequency corresponding to the templates used in the design of $\mathbf{L}_o(j\omega)$. Therefore, in Fig. 18.30 the difference between the $\text{Lm}\,\mathbf{T}_{R_U} - \text{Lm}\,\mathbf{T}_{max}$ and the $\text{Lm}\,\mathbf{T}_{R_L} - \text{Lm}\,\mathbf{T}_{min}$ curves yields the requirement for $\text{Lm}\,\mathbf{F}(j\omega)$, i.e., from Eq. (18.60)

$$\text{Lm}\,\mathbf{F}(j\omega) = \text{Lm}\,\mathbf{T}_R(j\omega) - \text{Lm}\,\mathbf{T}(j\omega) \tag{18.62}$$

The procedure for designing $F(s)$ is summarized as follows:

1. Use templates in conjunction with the $\mathbf{L}_o(j\omega)$ plot on the NC to determine T_{max} and T_{min} for each ω_i. This is done by placing $\mathscr{T}\mathbf{P}(j\omega_i)$ with its nominal point on the point $\text{Lm}\,\mathbf{L}_o(j\omega_i)$. Then use the M-contours to determine $\mathbf{T}_{max}(j\omega_i)$ and $\mathbf{T}_{min}(j\omega_i)$ (see Fig. 18.29).
2. Obtain the values of $\text{Lm}\,\mathbf{T}_{R_U}$ and $\text{Lm}\,\mathbf{T}_{R_L}$ for various values of ω_i from Fig. 18.8.

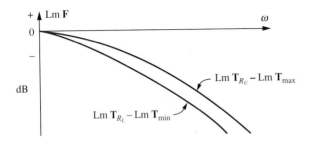

FIGURE 18.30
Frequency bounds on the prefilter $F(s)$.

3. From the values obtained in steps 1 and 2, plot

$$[\text{Lm}\,\mathbf{T}_{R_U} - \text{Lm}\,\mathbf{T}_{\max}] \text{ and } [\text{Lm}\,\mathbf{T}_{R_L} - \text{Lm}\,\mathbf{T}_{\min}] \text{ vs. } \omega$$

as shown in Fig. 18.30.
4. Use straight-line approximations to synthesize an $F(s)$ so that $\text{Lm}\,\mathbf{F}(j\omega)$ lies within the plots of step 3. For step forcing functions the resulting $F(s)$ must satisfy

$$\lim_{s \to 0}[F(s)] = 1 \tag{18.63}$$

18.15 BASIC DESIGN PROCEDURE FOR A MISO SYSTEM

The basic concepts of the QFT technique are explained by means of a design example. The system configuration shown in Fig. 18.5 contains three inputs. Parameter uncertainty for the plant of Eq. (18.9) is shown in Fig. 18.6. The first objectives are to track a step input $r(t) = u_{-1}(t)$ with no steady-state error and to satisfy the thumb print specifications of Fig. 18.7. An additional objective is to attenuate the system response due to step disturbance inputs $d_1(t)$ and $d_2(t)$, as described in Sec. 18.7 and shown in Figs. 12.15 and 12.16. An outline of the basic design procedure for the QFT technique, as applied to a m.p. plant, is as follows:

1. Synthesize the tracking model control ratio $T_R(s)$ in the manner described in Sec. 18.4, based upon the desired tracking specifications (see Fig. 18.8).
2. Synthesize the disturbance-rejection model control ratios $T_D(s)$ in the manner described in Secs. 12.5 and 12.8, based upon the disturbance-rejection specifications.
3. Obtain templates of $\mathbf{P}(j\omega_i)$ that pictorially describe the plant uncertainty on the Nichols chart (NC) for the desired passband frequency range.
4. Select a nominal plant from the set of Eq. (18.8) and denote it as $P_o(s)$.
5. Determine the U-contour based upon the specified values of $\delta_R(j\omega_i)$ for tracking, M_L for disturbance rejection, and V for the universal high-frequency boundary (UHFB) B_h in conjunction with steps 6 through 8.
6. Use the data of steps 2 and 3 and the values of $\delta_D(j\omega_i)$ (see Fig. 18.9) to determine the disturbance bound $\mathbf{B}_D(j\omega_i)$ on the loop transmission $\mathbf{L}_D(j\omega_i) = G(j\omega_i)\mathbf{P}(j\omega_i)$. For m.p. systems this requires that the synthesized loop

transmission $\text{Lm}\,\mathbf{L}_D(j\omega_i)$ must be on or above the curve for $\text{Lm}\,\mathbf{B}_D(j\omega_i)$ on the Nichols diagram (see Fig. 18.26 assuming $\mathbf{B}_D = \mathbf{B}_o$).

7. Determine the tracking bound $\mathbf{B}_R(j\omega_i)$ on the nominal transmission $\mathbf{L}_o(j\omega_i) = \mathbf{G}(j\omega_i)\mathbf{P}(j\omega_i)$, using the tracking model (step 1), the templates $\mathbf{P}(j\omega_i)$ (step 3), the values of $\delta_R(j\omega_i)$ (see Fig. 18.8), and M_L [see Eq. (18.32)]. For m.p. systems this requires that the synthesized loop transmission must satisfy the requirement that $\text{Lm}\,\mathbf{L}_R(j\omega_i)$ is on or above the curve for $\text{Lm}\,\mathbf{B}_R(j\omega_i)$ on the Nichols diagram.

8. Plot curves of $\text{Lm}\,\mathbf{B}_R(j\omega_i)$ versus $\phi_R = \angle\mathbf{B}_R(j\omega_i)$ and $\text{Lm}\,\mathbf{B}_D(j\omega_i)$ versus $\phi_D = \angle\mathbf{B}_D(j\omega_i)$ on the same NC. For a given value of ω_i at various values of the angle ϕ, select the value of $\text{Lm}\,\mathbf{B}_D(j\omega_i)$ or $\text{Lm}\,\mathbf{B}_R(j\omega_i)$, whichever is the largest value (termed the "worst" or "most severe" boundary). Draw a curve through these points. The resulting plot defines the overall boundary $\text{Lm}\,\mathbf{B}_o(j\omega_i)$ vs. ϕ. Repeat this procedure for sufficient values of ω_i.

9. Design $\mathbf{L}_o(j\omega_i)$ to be as close as possible to the boundary value $\mathbf{B}_o(j\omega_i)$ by selecting an appropriate compensator transfer function $\mathbf{G}(j\omega)$. Synthesize an $\mathbf{L}_o(j\omega) = \mathbf{G}(j\omega)\mathbf{P}(j\omega)$ using the $\text{Lm}\,\mathbf{B}_o(j\omega_i)$ boundaries and U-contour so that $\text{Lm}\,\mathbf{L}_o(j\omega_i)$ is on or above the curve for $\text{Lm}\,\mathbf{B}_o(j\omega_i)$ on the Nichols diagram. This procedure achieves the lowest possible value of the loop transmission frequency (phase margin frequency ω_ϕ). Note that $\mathbf{L}_o(j\omega_i) \geq \mathbf{B}_o(j\omega_i)$ represents the loop transfer function that satisfies the most severe boundary $\mathbf{B}_R$ and $\mathbf{B}_D$.

10. Based upon the information available from steps 1 and 9, synthesize an $F(s)$ that results in a $\text{Lm}\,\mathbf{T}_R$ [Eq. (18.12)] vs. ω that lies between B_U and B_L of Fig. 18.8.

11. Obtain the time-response data for $y(t)$: (a) with $d(t) = u_{-1}(t)$ and $r(t) = 0$ and (b) with $r(t) = u_{-1}(t)$ and $d(t) = 0$ for sufficient points around the parameter space describing the plant uncertainty (see Fig. 18.15).

For the $\mathbf{L}_o(j\omega)$ obtained, the plot of $\text{Lm}\,(\mathbf{L}(j\omega)/[1 + \mathbf{L}(j\omega)])$ may be larger or smaller than $\text{Lm}\,\mathbf{T}_{R_U}$ or $\text{Lm}\,\mathbf{T}_{R_L}$ of Fig. 18.8, but $\delta_R(j\omega_i)$ is satisfied. By the proper design of the input filter $F(s)$, $\text{Lm}\,\mathbf{T}_R = \text{Lm}\,\mathbf{FL}/[1 + \mathbf{L}]$ will lie within the bounds of $\text{Lm}\,\mathbf{T}_{R_U}$ and $\text{Lm}\,\mathbf{T}_{R_L}$. In problems with very large uncertainty and in disturbance rejection requiring a very large $|\mathbf{L}(j\omega)|$ over a "large" BW, then

$$\frac{\mathbf{L}}{1 + \mathbf{L}} \approx 1 \tag{18.64}$$

and $T \approx F$. For these situations design $F(s)$ in the same manner as for the tracking models of Secs. 12.2 and 18.4.

The simple plant of Eq. (18.9) is used in the following sections to illustrate the details in applying this QFT design procedure. The CAD package TOTAL-PC facilitates executing these design procedures (see Sec. 10.6 and App. B).

18.16   DESIGN EXAMPLE 1

This design example is for the control system of Fig. 18.5 with $r(t) = d_2(t) = u_{-1}(t)$ and $d_1(t) = 0$. The plant transfer function

$$P(s) = \frac{Ka}{s(s + a)} \qquad 1 \le K \le 10 \qquad 1 \le a \le 10 \qquad (18.65)$$

has the nominal values: $a = K = 1$.

Step 1. Modeling the tracking control ratio $T_R(s) = Y(s)/R(s)$:

$$T_{R_U} : M_p = 1.2, \; t_s = 2s$$
$$T_{R_L} : \text{Overdamped}, \; t_s = 2s$$

(a) A tracking model for the upper bound, based on the given desired performance specifications, is tentatively identified by

$$T_{R_U}(s) = \frac{19.753}{(s + 2 \pm j3.969)} \qquad (18.66)$$

A zero is inserted in Eq. (18.66) that does not affect the desired performance specifications but widens δ_{hf} between T_{R_U} and T_{R_L} in the high-frequency range. Thus

$$T_{R_U}(s) = \frac{0.6584(s + 30)}{(s + 2 \pm j3.969)} \qquad (18.67)$$

The figures of merit for this transfer function with a step input are $M_p = 1.2078$, $t_R = 0.342$ s, $t_p = 0.766$ s, $t_s = 1.84$ s, Lm $M_m = 1.95$ dB, and $\omega_m = 3.4$ r/s.

(b) A tracking model (see Chapter 12) for the lower bound, based upon the desired performance specifications, is tentatively identified by

$$T_{R_L}(s) = \frac{120}{(s + 3)(s + 4)(s + 10)} \qquad (18.68)$$

and is modified with the addition of a pole to yield

$$T_{R_L}(s) = \frac{8400}{(s + 3)(s + 4)(s + 10)(s + 70)} \qquad (18.69)$$

The pole at $s = -70$ is inserted in Eq. (18.69) to further widen δ_{hf} between T_{R_U} and T_{R_L} at high frequencies. The figures of merit for Eq. (18.69) are $M_p = 1$, $t_R = 1.02$ s, and $t_s = 1.849$ s. In practice, these synthesized models should yield figures of merit within 1 percent of the specified values.

(c) Determination of $\delta_R(j\omega_i)$. From the data for the log magnitude plots of Eqs. (18.67) and (18.69), the following values are obtained for $\delta_R(j\omega_i)$:

ω	$\delta_R(j\omega_i)$, dB
0.5	0.25
1	1.06
2	3.728
5	10.57
10	9.851
15	11.522
100	40

These values are sufficient to meet the disturbance-rejection specifications (see step 6). Normally, values of ω an octave apart up to ω_h should be obtained.

Step 2. Modeling the disturbance control ratio $T_{D_U}(s) = Y(s)/D_2(s)$.

Since $y(0) = 1$, it is desired that $y(t)$ decay "as fast as possible," i.e., $|y(t)| \le 0.01$ for $t \ge t_x = 60$ ms. Let the model disturbance control ratio be of the form

$$T_{D_U}(s) = \frac{Y(s)}{D_2(s)} = \frac{s(s+g)}{(s+g)^2 + h^2} \tag{18.70}$$

For $d_2(t) = u_{-1}(t)$ the output response has the form $y(t) = e^{-gt}\cos ht$. The desired specifications (see Fig. 18.9) are satisfied by choosing

$$T_{D_U}(s) = \frac{s(s+70)}{s+70 \pm j18} = \frac{1}{1+L_D} \tag{18.71}$$

For Eq. (18.71) the output decays rapidly so that $|y(t)| \le 0.01$ at $t_x \approx 0.0565$ s.

Steps 3 and 4. Forming the templates of $P(j\omega_i)$.

Analysis of $P(s)$ is based on the nominal values $K = a = 1$ (point A of Fig. 18.31). The template for each frequency is based on the data listed in Table 18.4 which is computed using Eq. (18.65). The values in Table 18.4 are used to draw the templates $\mathcal{T}P(j\omega_i)$ shown in Fig. 18.32.

Step 5. Determination of the U-contour.

(a) Determination of V [see Eq. (18.32)] for the plant of Eq. (18.65) yields

$$\lim_{s \to \infty} [P(s)] = \left[\frac{Ka}{s^2}\right] \quad \text{and} \quad \Delta \text{Lm } [\mathbf{P}]_{\max} = \text{Lm } [Ka]_{\max} = 40 \text{ dB} \tag{18.72}$$

To determine the value of ω_h, where $\Delta \text{Lm } \mathbf{P} \approx V = 40$ dB, select various values of ω_i, as shown below. For $\omega = 100$:

Point	Lm $P(j\omega_i)$, dB
A	−80
B	−60
C	−40
D	−60

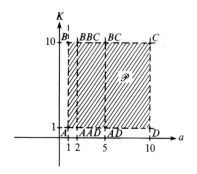

FIGURE 18.31
Template points.

TABLE 18.4
Data points for $\mathcal{T}P(j\omega_i)$

Point ↓	$\omega_i \rightarrow$	0.5	1	2	5	10	15
A	Lm P	5	−3	−13	−28.1	−40	−47
	$\angle$ P	−116.6°	−135°	−153.5°	−168.7°	−174.3°	−176.2°
B	Lm P	25	17	7	−8.1	−20	−27
	$\angle$ P	−116.6°	−135°	−153.5°	−168.7°	−174.3°	−176.2°
BBC	Lm P	25.8	19	11	−2.6	−14.2	−21.1
	$\angle$ P	−104.1°	−116.6°	−135°	−158.2°	−168.7°	−171.4°
BC	Lm P	26	19.8	13.3	3	−7	−13.5
	$\angle$ P	−95.7°	−101.3°	−111.8°	−135°	−153.5°	−161.6°
C	Lm P	26	20	13.8	5	−3	−8.64
	$\angle$ P	−93°	−95.7°	−101.3°	−116.6°	−13.5°	−146.4°
D	Lm P	6	−0.04	−6.2	−14.95	−23	−28.6
	$\angle$ P	−93°	−95.7°	−101.3°	−116.6°	−135°	−146.4°
AD	Lm P	6	−0.17	6.67	−17	−27	−33.5
	$\angle$ P	−95.7°	−101.3°	−111.8°	−135°	−153.5°	−116.6°
ADD	Lm P	5.65	0.97	−9	−22.6	−34.2	−41.1
	$\angle$ P	−104.1°	−116.6°	−135°	−158.2°	−168.7°	−172.4°

Note: Values of Lm are in dB.

From this data it is seen that the maximum value of $\Delta \text{Lm } P(j\omega_i)$ occurs between points A and C. For other values of ω_i the maximum values of $\Delta \text{Lm } P(j\omega_i)$ are:

Frequency ω_i	max $\Delta \text{Lm } P(j\omega_i)$ dB
15	38.4
16	38.6
20	39.0
40	39.65

Therefore it can be seen, for this example, that V is achieved essentially at $\omega_h \approx 40$.

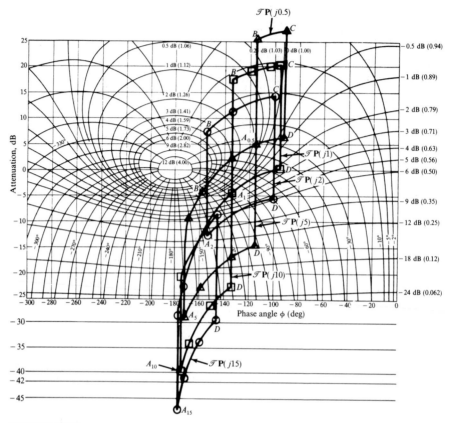

FIGURE 18.32
Construction of the plant templates.

(*b*) Determination of the B_h boundary. Select the M-contour that represents the desired value of M_L for T (see Sec. 18.8). At various points around the lower half of this contour measure down 40 dB and draw the B_h boundary in the manner shown in Fig. 18.21.

Step 6. Determination of bounds $\mathbf{B}_D(j\omega_i)$ for

$$T_{D_U} = \frac{Y}{D} = \frac{1}{1+L} = \frac{\ell}{1+\ell} \tag{18.73}$$

where $\ell = 1/L = 1/PG$ and $Y = DT_{D_U}$. For the disturbance-rejection case:

$$\Delta\operatorname{Lm}\mathbf{Y} = \Delta\operatorname{Lm}\left[\frac{\ell}{1+\ell}\right] = \Delta\operatorname{Lm}\mathbf{T}_{D_U} \tag{18.74}$$

Note that there are no lower bounds on the response $y(t)$. Thus, choose

$$\mathbf{L} \geq \mathbf{B}_D(j\omega_i)$$

to satisfy

$$\text{Lm}\left[\frac{\mathbf{Y}}{\mathbf{D}}\right] = \text{Lm}\left[\frac{\ell}{1+\ell}\right] \leq \text{Lm}[\mathbf{T}_{D_U}] \qquad (18.75)$$

for each frequency ω_i of $\mathbf{T}_D(j\omega_i)$ over the frequency range $0 \leq \omega \leq 15$ r/s. The data of Table 18.5 are obtained from Eq. (18.71). Point A of the templates of step 3 yields the maximum value for $\ell = 1/L$. For $0 \leq \omega \leq 15$ the contours of Lm $\mathbf{T}_D(j\omega_i)$ (see Fig. 18.33) become the contours for $\mathbf{B}_D$, i.e., Lm $\mathbf{T}_D$ = Lm $\mathbf{B}_D$. For $\omega \geq 20$, the values of Lm $\mathbf{L}_D$ in Table 18.5, obtained by use of Eq. (18.71), are points of $\mathbf{B}_D(j\omega_i)$ on and in the vicinity of the U-contour. For this example, as shown in the next step, the $\mathbf{B}_D$ bounds completely dominate over the $\mathbf{B}_R$ bounds. Thus, $\mathbf{B}_D = \mathbf{B}_o$ as shown in Fig. 18.33.

Step 7. Determination of the bounds $\mathbf{B}_R(j\omega_i)$. By use of the templates, the values of $\delta_R(j\omega_i)$ given in step 2, the M_L contour, and the B_h boundary, the bounds $\mathbf{B}_R(j\omega_i)$ and the U-contour are determined and are drawn in Fig. 18.34. From this figure it is seen that the phase margin angle $\gamma \approx 58°$.

Step 8. Determination of the composite bounds $\mathbf{B}_o(j\omega_i)$. An analysis of Figs. 18.33 and 18.34 reveals that the bounds $\mathbf{B}_D(j\omega_i)$ for this example all lie above the tracking contours $\mathbf{B}_R(j\omega_i)$. Thus the $\mathbf{B}_D$ contours of Fig. 18.33 are more severe and become the $\mathbf{B}_o(j\omega_i)$ contours for the overall system of Fig. 18.35.

Step 9. Synthesizing or shaping of $L_o(s)$.
   (*a*) Minimum structure. Since $P(s)$ is Type 1, then $L_o(s)$ must be at least Type 1 in order to maintain the Type 1 tracking characteristic for $L/(1+L)$. Thus, the initial $L_o(s)$ has the form

$$L_{o0}(s) = \frac{K_{o1}}{s^m} \qquad (18.76)$$

TABLE 18.5
Frequency data for Eq. (18.71)

ω	Lm T_D, dB	$\angle T_D$, degrees	Lm L_D
0.5	−43.5	89.65	43.5
1	−37.5	89.3	37.5
2	−31.44	88.6	31.4
5	−23.5	86.4	23.5
10	−17.52	82.85	17.5
15	−14.1	79.31	14.1
20	−11.67	75.81	11.4
50	−4.84	56.8	4.46
100	−1.67	36.17	−2.91
200	−0.453	19.55	−9.05
500	−0.0737	7.99	−17.1
1000	−0.01845	4.007	−21.6
2000	−0.002954	2.005	−31.0
4000	−0.001154	1.003	−35.2

FIGURE 18.33
Construction of the $\mathbf{B}_D(j\omega_i)$ contours.

where $m = 1$ for this example. (Note: In general, select $L_{o0}(s) = P(s)$. In addition, ensure that the desired system type is achieved.) For the case where $T_D = P/(1+L)$, it is desired that T_D have a zero at the origin so that $y(\infty) = 0$ for $d_1(t) = u_{-1}(t)$. Therefore $G(s)$ must have a pole at the origin. This requirement may place a severe restriction on trying to synthesize an L_o.

   (b) Synthesis of $L_o(s)$. Using four-cycle semilog graph paper and the data in Table 18.6, construct or shape $L_o(s)$ so that it comes as close as possible to the U-contour. $L_o(j\omega_i)$ must be as close as possible to $\mathbf{B}_o(j\omega_i)$ but never below it. For the approximate range $\omega_x \approx 15 \le \omega_i \le \omega_y$, the angle $\angle L_o(j\omega_i) \ge -123°$ must be satisfied. For this example the frequency ω_y is the value of $\mathbf{L}_o(j\omega_y)$ that results in $\text{Lm}\,\mathbf{L}_o(j\omega_i) \approx -36$ dB. These restrictions can be satisfied by assuming the format of the optimal transfer function as

$$
\mathbf{L}_o(j\omega) = \mathbf{L}_{o0}(j\omega) \left[\frac{(j\omega - z_1)(j\omega - z_2) \cdots}{(j\omega - p_3)(j\omega - p_4) \cdots \left[\left(1 - \dfrac{\omega^2}{\omega_n^2}\right) + j2\zeta \dfrac{\omega}{\omega_n}\right]} \right] \qquad (18.77)
$$

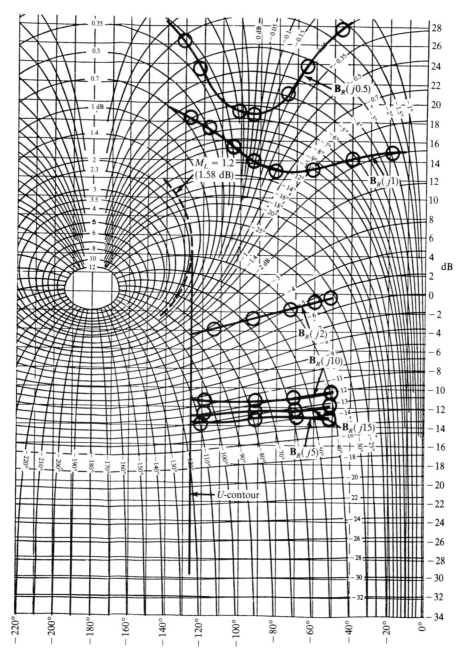

FIGURE 18.34
Construction of the $\mathbf{B}_R(j\omega_i)$ boundaries.

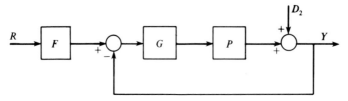

FIGURE 18.35
MISO system.

which has w real zeros, $n = w + \lambda - 2$ real poles, and a pair of complex-conjugate poles. For this example a table having the format shown in Table 18.6 is used to assist in obtaining the desired $L_o(s)$. The resulting loop transfer function is

$$L_o(s) = \frac{4.11 \times 10^{10}(s + 0.6)^2(s + 176)}{s^2(s + 1)(s + 200)(s + 14700 \pm j15000)} \tag{18.78}$$

where $\zeta = 0.7$, $\omega_n = 1000$ r/s, and $\omega_\phi \approx 200$ r/s. $L_o(j\omega)$ is drawn on the Bode plot of Fig. 18.36. It is also plotted on the NC of Fig. 18.40. Since L_o is Type 2, then $T_D = 1/(1 + L)$ has two zeros at the origin.

(c) Determination of $G(s)$. Use $L_o(s)$ and $P_o(s)$, where $P_o(s) = 1/s(s + 1)$ is the nominal plant, to obtain the transfer function $G(s) = L_o(s)/P_o(s)$.

Step 10. The input filter $F(s)$ is synthesized to yield the desired tracking of the input by the output $y(t)$ in the manner described in Sec. 18.14. The curves of $\mathbf{B}_U$, $\mathbf{B}_L$, Lm $T_{\min}$, and Lm $T_{\max}$ are plotted in Fig. 18.37. Since, for this example $|L_o(j\omega)| \gg 1$ over the desired BW, then $F(s) = T_R(s)$. Thus, the following $F(s)$ is synthesized to lie between $\mathbf{B}_U$ and $\mathbf{B}_L$ as shown in Fig. 18.37.

TABLE 18.6
Data for loop shaping

ω	$\angle \mathbf{L}_{ol}$	Angle contribution of pole (p) and zero (z) in degrees				$\angle L_o$
		z_1	p_2	z_2	$p_{3,4}$	
1	$-180°$	32.0	-2.20	1.27	0	-149.0
5	$-180°$	72.26	-10.89	6.34	-0.40	-112.7
10	$-180°$	80.90	-21.04	12.53	-0.80	-108.4
15	$-180°$	83.91	-29.98	18.43	-1.20	-108.8
50	$-180°$	88.17	-62.53	48.01	-4.01	-110.4
100	$-180°$	89.08	-75.43	65.77	-8.05	-108.6
200	$-180°$	89.54	-82.59	77.32	-16.26	-111.99
1000	$-180°$	89.90	-88.51	87.42	-90.0	-181.2
2000	$-180°$	89.95	-89.26	88.71	-136.98	-227.6

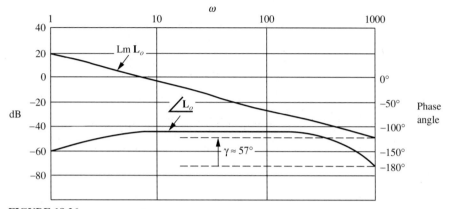

FIGURE 18.36
Bode plot of $\mathbf{L}_o(j\omega)$ of Eq. (18.78).

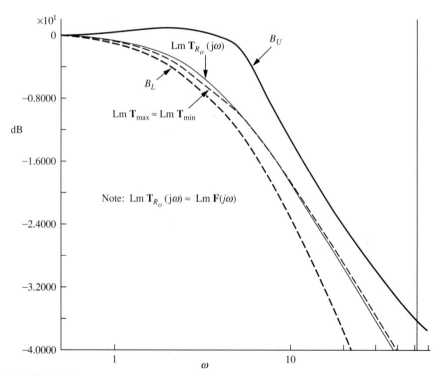

FIGURE 18.37
Requirements on $F(s)$ for design example 1.

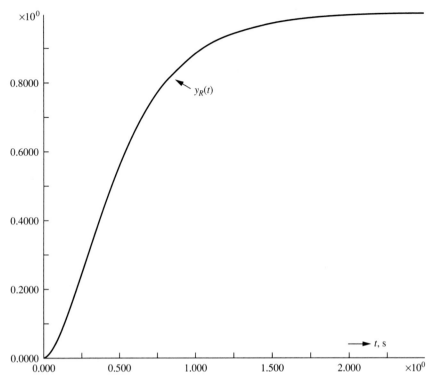

FIGURE 18.38
Time responses to unit step input.

$$F(s) = \frac{18.85}{(s + 2.9)(s + 6.5)} \tag{18.79}$$

The output responses for this example with a unit step input are all identical, as shown in Fig. 18.38 for the entire region of plant parameter uncertainty, where $M_p = 1.002$ and $t_s = 1.606$ s.

Step 11. Time responses for a disturbance input: $d(t) = u_{-1}(t)$. For each point of Fig. 18.31, substitute the corresponding data into

$$Y_D(s) = \frac{1}{1 + P(s)G(s)} D(s) \tag{18.80}$$

and determine $y_D(t)$ for each point. As indicated in Table 18.7, the disturbance time response characteristics for points A, B, C, and D satisfy the specification that $|y(t_x)| \le 0.01$ for $t_x \ge 60$ ms. The disturbance response plots, for all eight cases, are shown in Fig. 18.39. As shown in the figure, although the initial peak overshoots exceed the magnitude of 0.01, the value of $|y_D(t)| \le 0.01$ for $t_x \ge 60$ ms is achieved.

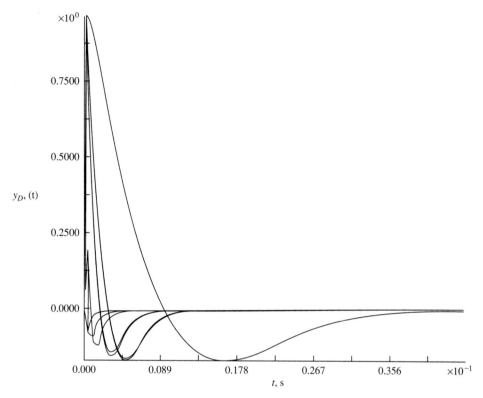

FIGURE 18.39
Disturbance response plots.

18.17 DESIGN EXAMPLE 2

The plant and tracking specifications of Example 1 are used for this example, with $r(t) = d_1(t) = u_{-1}(t)$, $d_2(t) = 0$, and $\alpha_p = 0.1$. Steps 1, 3 through 5, and 7 are the same as for Example 1. For Step 2, $M_p \approx M_m = 0.1$; thus $\operatorname{Lm} T_{D_1} = \operatorname{Lm} M_{m_D} = -20$ dB. Yaniv's MISO QFT CAD package[17] is used to obtain Fig. 18.40, which shows the U-contour, $\mathbf{B}_o$ bounds, and $\operatorname{Lm} \mathbf{L}_o(j\omega)$. The synthesized $L_o(s)$ is

$$L_o(s) = \frac{4.58 \times 10^6 (s + 1.6)(s + 45)}{s^2(s + 1)(s + 26)(s + 700 \pm j714.1)} \qquad (18.81)$$

TABLE 18.7
Time response characteristics for Eq. (18.80)

Point	$y_D(t_x)$	Specified t_x	Actual t_x for $\|y_D(t)\| = 0.01$	$y(\infty)$
A	0.00998	0.06	0.058	0
B	0.0090	0.06	0.0074	0
C	0.00004	0.06	0.00226	0
D	0.000399	0.06	0.007	0

FIGURE 18.40

U-contour, $\mathbf{B}_o$ bounds, and $\mathrm{Lm}\,\mathbf{L}_o(j\omega_i)$ for obtaining $L_o(s)$.

Equations (18.67) and (18.69) and Fig. 18.40 are used to obtain the data for plotting $\mathbf{B}_R$, $\mathbf{B}_L$, $\mathrm{Lm}\,T_{\min}$, and $\mathrm{Lm}\,T_{\max}$ in Fig. 18.41. Since the tracking specifications and models are identical for both examples, the prefilter for this example is the same as given by Eq. (18.79) and is plotted in Fig. 18.41. The simulation results are shown in Figs. 18.42 and 18.43. These figures reveal that the largest disturbance response is $|y_D(t)_{\max}| = 0.03956$, and all eight plant cases satisfy the specification $|y_D(t)_{\max}| \leq 0.1$. Also, the tracking responses all lie in the range of $1 < M_p < 1.02$ and $t_s \leq 1.89$ s. Thus, the desired performance specifications have all been met, and a robust design has been achieved.

18.18 TEMPLATE GENERATION FOR UNSTABLE PLANTS

In the generation of templates for unstable plants, proper care must be exercised in analyzing the angular variation of $\mathbf{P}(j\omega)$ as the frequency is varied from zero to infinity. As an example, consider the plant

$$\mathbf{P}(j\omega) = \frac{K(j\omega - z_1)}{j\omega(j\omega - p_1)(j\omega - p_2)(j\omega - p_3)} \tag{18.82}$$

where p_1 and p_2 are a complex-conjugate pair and

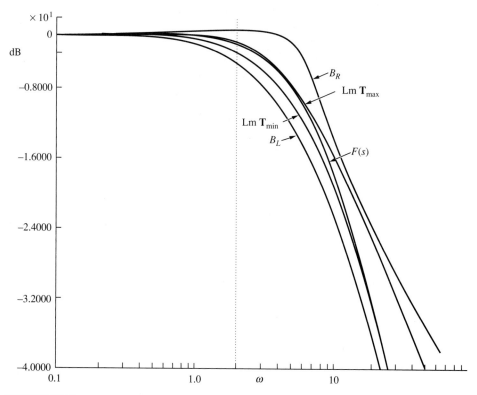

FIGURE 18.41
Requirements and resulting prefilter for design examples.

$$\underline{/\lim_{\omega \to \infty} \mathbf{P}(j\omega)} = \underline{/K/(j\omega)^3} = -270° \tag{18.83}$$

In Eq. (18.83) the convention used is that, in the limit as $\omega \to \infty$, each $j\omega$ term contributes an angle of $+90°$. Two cases are analyzed for Eq. (18.82) as follows:

Case 1. The poles and zeros of Eq. (18.82) for a stable plant are plotted in Fig. 18.44a. In Eq. (18.83) the angle of $(j\omega_i - p_1)$ is negative (cw) for $\omega_i < \omega_{d_1}$ and is positive (ccw) for $\omega_i > \omega_{d_1}$. All other angles are positive (ccw) for all $0 \le \omega \le \infty$ as illustrated in Fig. 18.44a, and in Table 18.8.

Case 2. The poles and zeros of Eq. (18.82) for an unstable plant are plotted in Fig. 18.44b. For Eq. (18.82), all angles of first-order factors $(j\omega_i - p_i)$ are positive (ccw), as illustrated in Fig. 18.44b and in Table 18.8.

  Note that the difference in the angle of $\mathbf{P}(j0)$ between the two cases is $360°$ and is necessary to account for the RH plane poles. Thus, because all angles are measured ccw, the angle of $\mathbf{P}(j\omega)$ varies continuously in a given direction as the frequency is varied between zero and infinity. This feature is very important in obtaining a

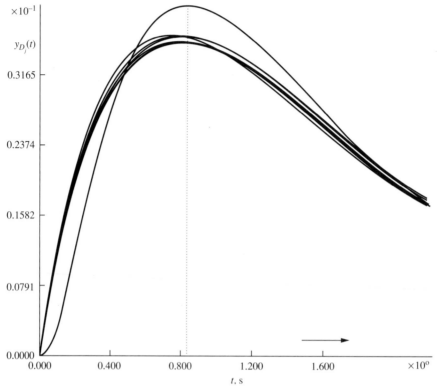

FIGURE 18.42
Disturbance responses of system designed in example 2.

template where $\mathscr{P}$ contains both stable and unstable plants. The following guidelines should be used in the angular determination of $\underline{/\mathbf{P}(j\omega_i)}$:

1 For stable plants, all angular directions are taken so that their values always lie within the range of $-90°$ to $+90°$.
2 For plants with complex RH plane poles and zeros: (*a*) the angular directions for LH plane poles and zeros of $P(s)$ are taken in the same manner as for the stable plant; (*b*) the angular directions of all RH plane poles and zeros are all taken ccw.

18.19 SUMMARY

A general introduction to the QFT technique is presented in this chapter. It is a frequency-response design method applied to the design of a MIMO control system with an uncertain MIMO plant **P**. The method of representing a MIMO system by an equivalent set of MISO systems is presented, using $\mathbf{P}^{-1}$. Two design approaches are available in which the equivalent MISO loops are designed according to the MISO design method presented in this chapter. This design is based upon

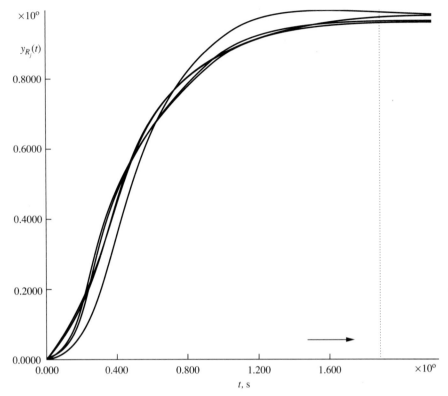

FIGURE 18.43
Tracking responses of system designed in example 2.

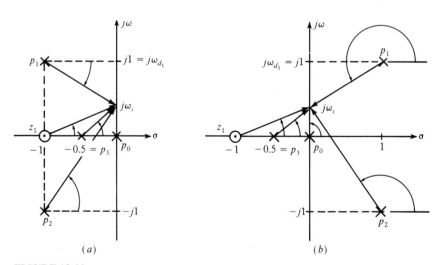

FIGURE 18.44
Poles and zeros in the s plane for (a) a stable plant and (b) an unstable plant.

TABLE 18.8
Angular variations of the minimum-phase stable and unstable plants $P(j\omega)$ [see Eq. (18.80)] shown in Fig. 18.44

Case	ω_i	Angle					
		$\angle p_0$	$\angle p_1$	$\angle p_2$	$\angle p_3$	$\angle z_1$	$\angle P(j\omega_i)$
1	0	$+90°$	$-45°$	$+45°$	$0°$	$0°$	$-90°$
	1	$+90°$	$0°$	$+63.5°$	$+63.5°$	$+45°$	$-172°$
	∞	$+90°$	$+90°$	$+90°$	$+90°$	$+90°$	$-270°$
2	0	$+90°$	$+225°$	$+135°$	$0°$	$0°$	$-450°$
	1	$+90°$	$+180°$	$+116.5°$	$+63.5°$	$+45°$	$-405°$
	∞	$+90°$	$+90°$	$+90°$	$+90°$	$+90°$	$-270°$

1 Specifying the tolerance in the ω-domain by means of the sets of plant transfer functions and closed-loop control ratios, $\mathscr{P} = \{P(j\omega)\}$ and $\mathscr{T}(j\omega) = \{T(j\omega)\}$, respectively.

2 Finding the resulting bounds on the loop transfer functions L_i and input filter transfer functions F_i of Figs. 18.2 and 18.17.

The robust design technique of this chapter permits the design of an analog control system that satisfies the desired performance specifications within the specified range of structured plant parameter variation. Since the QFT technique is based on the design of the loop transmission function $L(s)$ of the MISO control system of Fig. 18.2, it is referred to as a SISO design technique. Although this chapter deals with m.p. plants, the QFT technique can be applied to n.m.p. plants and to digital control systems.[4,13,45]

REFERENCES

1. Horowitz, I. M.: "Advanced Control Theory and Applications," unpublished notes, The Weizmann Institute of Science, Rehovot, Israel, 1982.
2. Horowitz, I. M., and M. Sidi: "Synthesis of Feedback Systems with Large Plant Ignorance for Prescribed Time Domain Tolerances," *Int. J. Control*, vol. 16, pp. 287–309, 1972.
3. Horowitz, I. M.: *Synthesis of Feedback Systems*, Academic, New York, 1963.
4. Schneider, D. L.: "QFT Digital Flight Control Design as Applied to the AFTI/F-16," M.S. Thesis, AFIT/GE/EE(86D-4), School of Engineering, Air Force Institute of Technology, Wright-Patterson AFB, Ohio, December 1986.
5. Bode, H. W.: *Network Analysis and Feedback Amplifier Design*, Van Nostrand, New York, 1945.
6. Houpis, C. H., and G. B. Lamont: *Digital Control Systems, Theory, Hardware, Software*, 2nd ed., McGraw-Hill, New York, 1992.
7. Horowitz, I. M., and M. Sidi: "Optimum Synthesis of Nonminimum-Phase Feedback Systems with Parameter Uncertainty," *Int. J. Control*, vol. 27, pp. 361–386, 1978.
8. Horowitz, I. M.: "Optimum Loop Transfer Function in Single-Loop Minimum Phase Feedback Systems," *Int. J. Control*, vol. 22, pp. 97–113, 1973.
9. Horowitz, I. M.: "Synthesis of Feedback Systems with Non-Linear Time Uncertain Plants to Satisfy Quantitative Performance Specifications," *IEEE Proc.*, vol. 64, pp. 123–130, 1976.

10. Betzold, R. W.: "Multiple Input-Multiple Output Flight Control Design with Highly Uncertain Parameters, Application to the C-135 Aircraft," M.S. Thesis, AFIT/GE/EE(83D-11), School of Engineering, Air Force Institute of Technology, Wright-Patterson AFB, Ohio, December 1983.
11. Horowitz, I. M., and C. Loecher: "Design of a 3×3 Multivariable Feedback System with Large Plant Uncertainty," *Int. J. Control*, vol. 33, pp. 677–699, 1981.
12. Horowitz, I. M.: "Quantitative Synthesis of Uncertain Multiple Input-Output Feedback Systems," *Int. J. Control*, vol. 30, pp. 81–106, 1979.
13. Horowitz, I. M., et al.: "Multivariable Flight Control Design with Uncertain Parameters (YF16CCV)," AFWAL-TR-83-3036, Air Force Wright Aeronautical Laboratories, Wright-Patterson AFB, Ohio, 1982.
14. Horowitz, I. M.: "A Synthesis Theory for Linear Time-Varying Feedback Systems with Plant Uncertainty," *IEEE Trans. Autom. Control*, vol. AC-20, pp. 454–463, 1975.
15. Horowitz, I. M.: "A Quantitative Inherent Reconfiguration Theory for a Class of Systems," *Int. J. Syst. Sci.*, vol. 16, pp. 1377–1390, 1985.
16. Houpis, C. H.: "Quantitative Feedback Theory (QFT): Technique for Designing Multivariable Control Systems," AFWAL-TR-86-3107, Air Force Wright Aeronautical Laboratories, Wright-Patterson AFB, Ohio, January 1987. (Available from Defense Technical Information Center, Cameron Station, Alexandria, VA 22314, document number AD-A176883.)
17. Yaniv, O.:"Multiple-Input Single-Output (MISO) QFT CAD Package," Tel Aviv University, Tel Aviv 69 978, Israel, 1992.
18. D'Azzo, J. J., and C. H. Houpis: *Linear Control System Analysis & Synthesis: Conventional and Modern,* 3rd ed., McGraw-Hill, New York, 1988.

APPENDIX

A

TABLE OF LAPLACE TRANSFORM PAIRS

$F(s)$	$f(t) \quad 0 \le t$
1. $\quad 1$	$u_0(t) \qquad$ unit impulse at $t = 0$
2. $\quad \dfrac{1}{s}$	1 or $u_{-1}(t) \qquad$ unit step starting at $t = 0$
3. $\quad \dfrac{1}{s^2}$	$tu_{-1}(t) \qquad$ ramp function
4. $\quad \dfrac{1}{s^n}$	$\dfrac{1}{(n-1)!} t^{n-1} \qquad n = $ positive integer
5. $\quad \dfrac{1}{s} e^{-as}$	$u_{-1}(t - a) \qquad$ unit step starting at $t = a$
6. $\quad \dfrac{1}{s}(1 - e^{-as})$	$u_{-1}(t) - u_{-1}(t - a) \qquad$ rectangular pulse
7. $\quad \dfrac{1}{s + a}$	$e^{-at} \qquad$ exponential decay
8. $\quad \dfrac{1}{(s + a)^n}$	$\dfrac{1}{(n-1)!} t^{n-1} e^{-at} \qquad n = $ positive integer
9. $\quad \dfrac{1}{s(s + a)}$	$\dfrac{1}{a}(1 - e^{-at})$
10. $\quad \dfrac{1}{s(s + a)(s + b)}$	$\dfrac{1}{ab}\left(1 - \dfrac{b}{b - a} e^{-at} + \dfrac{a}{b - a} e^{-bt}\right)$

$F(s)$	$f(t) \qquad 0 \le t$
11. $\dfrac{s + \alpha}{s(s + a)(s + b)}$	$\dfrac{1}{ab}\left[\alpha - \dfrac{b(\alpha - a)}{b - a}e^{-at} + \dfrac{a(\alpha - b)}{b - a}e^{-bt}\right]$
12. $\dfrac{1}{(s + a)(s + b)}$	$\dfrac{1}{b - a}(e^{-at} - e^{-bt})$
13. $\dfrac{s}{(s + a)(s + b)}$	$\dfrac{1}{a - b}(ae^{-at} - be^{-bt})$
14. $\dfrac{s + \alpha}{(s + a)(s + b)}$	$\dfrac{1}{b - a}[(\alpha - a)e^{-at} - (\alpha - b)e^{-bt}]$
15. $\dfrac{1}{(s + a)(s + b)(s + c)}$	$\dfrac{e^{-at}}{(b - a)(c - a)} + \dfrac{e^{-bt}}{(c - b)(a - b)} + \dfrac{e^{-ct}}{(a - c)(b - c)}$
16. $\dfrac{s + \alpha}{(s + a)(s + b)(s + c)}$	$\dfrac{(\alpha - a)e^{-at}}{(b - a)(c - a)} + \dfrac{(\alpha - b)e^{-bt}}{(c - b)(a - b)} + \dfrac{(\alpha - c)e^{-ct}}{(a - c)(b - c)}$
17. $\dfrac{\omega}{s^2 + \omega^2}$	$\sin \omega t$
18. $\dfrac{s}{s^2 + \omega^2}$	$\cos \omega t$
19. $\dfrac{s + \alpha}{s^2 + \omega^2}$	$\dfrac{\sqrt{\alpha^2 + \omega^2}}{\omega}\sin(\omega t + \phi) \qquad \phi = \tan^{-1}\dfrac{\omega}{\alpha}$
20. $\dfrac{s \sin\theta + \omega \cos\theta}{s^2 + \omega^2}$	$\sin(\omega t + \theta)$
21. $\dfrac{1}{s(s^2 + \omega^2)}$	$\dfrac{1}{\omega^2}(1 - \cos \omega t)$
22. $\dfrac{s + \alpha}{s(s^2 + \omega^2)}$	$\dfrac{\alpha}{\omega^2} - \dfrac{\sqrt{\alpha^2 + \omega^2}}{\omega^2}\cos(\omega t + \phi) \qquad \phi = \tan^{-1}\dfrac{\omega}{\alpha}$
23. $\dfrac{1}{(s + a)(s^2 + \omega^2)}$	$\dfrac{e^{-at}}{a^2 + \omega^2} + \dfrac{1}{\omega\sqrt{a^2 + \omega^2}}\sin(\omega t - \phi) \qquad \phi = \tan^{-1}\dfrac{\omega}{a}$
24. $\dfrac{1}{(s + a)^2 + b^2}$	$\dfrac{1}{b}e^{-at}\sin bt$
24a. $\dfrac{1}{s^2 + 2\zeta\omega_n s + \omega_n^2}$	$\dfrac{1}{\omega_n\sqrt{1 - \zeta^2}}e^{-\zeta\omega_n t}\sin\omega_n\sqrt{1 - \zeta^2}t$
25. $\dfrac{s + a}{(s + a)^2 + b^2}$	$e^{-at}\cos bt$

$F(s)$	$f(t) \qquad 0 \le t$
26. $\dfrac{s + \alpha}{(s + a)^2 + b^2}$	$\dfrac{\sqrt{(\alpha - a)^2 + b^2}}{b} e^{-at} \sin(bt + \phi) \qquad \phi = \tan^{-1}\dfrac{b}{\alpha - a}$
27. $\dfrac{1}{s[(s + a)^2 + b^2]}$	$\dfrac{1}{a^2 + b^2} + \dfrac{1}{b\sqrt{a^2 + b^2}} e^{-at} \sin(bt - \phi) \qquad \phi = \tan^{-1}\dfrac{b}{-a}$
27a. $\dfrac{1}{s(s^2 + 2\zeta\omega_n s + \omega_n^2)}$	$\dfrac{1}{\omega_n^2} - \dfrac{1}{\omega_n^2\sqrt{1 - \zeta^2}} e^{-\zeta\omega_n t} \sin(\omega_n\sqrt{1 - \zeta^2}\,t + \phi)$ $\phi = \cos^{-1}\zeta$
28. $\dfrac{s + \alpha}{s[(s + a)^2 + b^2]}$	$\dfrac{\alpha}{a^2 + b^2} + \dfrac{1}{b}\sqrt{\dfrac{(\alpha - a)^2 + b^2}{a^2 + b^2}} e^{-at} \sin(bt + \phi)$ $\phi = \tan^{-1}\dfrac{b}{\alpha - a} - \tan^{-1}\dfrac{b}{-a}$
29. $\dfrac{1}{(s + c)[(s + a)^2 + b^2]}$	$\dfrac{e^{-ct}}{(c - a)^2 + b^2} + \dfrac{e^{-at}\sin(bt - \phi)}{b\sqrt{(c - a)^2 + b^2}} \qquad \phi = \tan^{-1}\dfrac{b}{c - a}$
30. $\dfrac{1}{s(s + c)[(s + a)^2 + b^2]}$	$\dfrac{1}{c(a^2 + b^2)} - \dfrac{e^{-ct}}{c[(c - a)^2 + b^2]}$ $+ \dfrac{e^{-at}\sin(bt - \phi)}{b\sqrt{a^2 + b^2}\sqrt{(c - a)^2 + b^2}}$ $\phi = \tan^{-1}\dfrac{b}{-a} + \tan^{-1}\dfrac{b}{c - a}$
31. $\dfrac{s + \alpha}{s(s + c)[(s + a)^2 + b^2]}$	$\dfrac{\alpha}{c(a^2 + b^2)} + \dfrac{(c - \alpha)e^{-ct}}{c[(c - a)^2 + b^2]}$ $+ \dfrac{\sqrt{(\alpha - a)^2 + b^2}}{b\sqrt{a^2 + b^2}\sqrt{(c - a)^2 + b^2}} e^{-at} \sin(bt + \phi)$ $\phi = \tan^{-1}\dfrac{b}{\alpha - a} - \tan^{-1}\dfrac{b}{-a} - \tan^{-1}\dfrac{b}{c - a}$
32. $\dfrac{1}{s^2(s + a)}$	$\dfrac{1}{a^2}(at - 1 + e^{-at})$
33. $\dfrac{1}{s(s + a)^2}$	$\dfrac{1}{a^2}(1 - e^{-at} - ate^{-at})$
34. $\dfrac{s + \alpha}{s(s + a)^2}$	$\dfrac{1}{a^2}[\alpha - \alpha e^{-at} + a(a - \alpha)te^{-at}]$
35. $\dfrac{s^2 + \alpha_1 s + \alpha_0}{s(s + a)(s + b)}$	$\dfrac{\alpha_0}{ab} + \dfrac{a^2 - \alpha_1 a + \alpha_0}{a(a - b)} e^{-at} - \dfrac{b^2 - \alpha_1 b + \alpha_0}{b(a - b)} e^{-bt}$

$F(s)$	$f(t)$ $0 \le t$
36. $\dfrac{s^2 + \alpha_1 s + \alpha_0}{s[(s+a)^2 + b^2]}$	$\dfrac{\alpha_0}{c^2} + \dfrac{1}{bc}[(a^2 - b^2 - \alpha_1 a + \alpha_0)^2$ $+\, b^2(\alpha_1 - 2a)^2]^{1/2} e^{-at} \sin(bt + \phi)$ $\phi = \tan^{-1} \dfrac{b(\alpha_1 - 2a)}{a^2 - b^2 - \alpha_1 a + \alpha_0} - \tan^{-1} \dfrac{b}{-a}$ $c^2 = a^2 + b^2$
37. $\dfrac{1}{(s^2 + \omega_2)[(s+a)^2 + b^2]}$	$\dfrac{(1/\omega)\sin(\omega t + \phi_1) + (1/b)e^{-at}\sin(bt + \phi_2)}{[4a^2\omega^2 + (a^2 + b^2 - \omega^2)^2]^{1/2}}$ $\phi_1 = \tan^{-1} \dfrac{-2a\omega}{a^2 + b^2 - \omega^2} \qquad \phi_2 = \tan^{-1} \dfrac{2ab}{a^2 - b^2 + \omega^2}$
38. $\dfrac{s + \alpha}{(s^2 + \omega^2)[(s+a)^2 + b^2]}$	$\dfrac{1}{\omega}\left(\dfrac{\alpha^2 + \omega^2}{c}\right)^{1/2} \sin(\omega t + \phi_1)$ $+\, \dfrac{1}{b}\left[\dfrac{(\alpha - a)^2 + b^2}{c}\right]^{1/2} e^{-at}\sin(bt + \phi_2)$ $c = (2a\omega)^2 + (a^2 + b^2 - \omega^2)^2$ $\phi_1 = \tan^{-1} \dfrac{\omega}{a} - \tan^{-1} \dfrac{2a\omega}{a^2 + b^2 + \omega^2}$ $\phi_2 = \tan^{-1} \dfrac{b}{\alpha - a} + \tan^{-1} \dfrac{2ab}{a^2 - b^2 + \omega^2}$
39. $\dfrac{s + \alpha}{s^2[(s+a)^2 + b^2]}$	$\dfrac{1}{c}\left(\alpha t + 1 - \dfrac{2\alpha a}{c}\right) + \dfrac{[b^2 + (\alpha - a)^2]^{1/2}}{bc}e^{-at}\sin(bt + \phi)$ $c = a^2 + b^2$ $\phi = 2\tan^{-1}\left(\dfrac{b}{a}\right) + \tan^{-1}\dfrac{b}{\alpha - a}$
40. $\dfrac{s^2 + \alpha_1 s + \alpha_0}{s^2(s+a)(s+b)}$	$\dfrac{\alpha_1 + \alpha_0 t}{ab} - \dfrac{\alpha_0(a+b)}{(ab)^2} - \dfrac{1}{a-b}\left(1 - \dfrac{\alpha_1}{a} + \dfrac{\alpha_0}{a^2}\right)e^{-at}$ $-\, \dfrac{1}{1-b}\left(1 - \dfrac{\alpha_1}{b} + \dfrac{\alpha_0}{b^2}\right)e^{-bt}$

APPENDIX
B

INTERACTIVE COMPUTER-AIDED-DESIGN PROGRAMS FOR DIGITAL AND CONTINUOUS CONTROL-SYSTEM ANALYSIS AND SYNTHESIS

B.1 INTRODUCTION

Computer-aided-design (CAD) software packages[B.1–B.12] are available for the development of continuous- and discrete-time control systems. These packages can assist the control engineer in performing time-consuming calculations for both scalar and multivariable models. Some of the available CAD packages provide the user with an integrated set of design tools and perform a broad range of calculations applicable to control system analysis and synthesis. Each of these CAD packages has various excellent features. Some are ideally suited as educational tools, while others are more suited for the practicing engineer. In general, most packages are command-driven interactive programs accessed through the keyboard or a file designation.

Some commercial software products include extensive system-building capabilities, very high-order plant modeling, nonlinear system construction, and a multitude of data displays. Some of these packages require extensive knowledge to use their full capabilities, whereas others are not comprehensive in their capabilities.

The computer programs called *ICECAP-PC,*[†] *TOTAL-PC,*[†] and *MATLAB®* are discussed in this appendix as examples of specific CAD packages. These CAD packages are the ones employed in the text examples.

Support materials, which are also very important, include user's manuals, programmer's manuals, source listings, tutorials, and on-line help. These materials assist users to learn the operation of the package efficiently and, as appropriate, to modify it for a given environment. Other concerns may focus on real-time applications and portability of the package. Simulation capabilities, which are embedded in some packages, are very useful in analyzing and synthesizing various control systems. An appropriate software computation and visualization package is *MATLAB®*, developed by The MathWorks, Inc. *SIMULINK®* and the *Control System Toolbox*, which run with *MATLAB*, permit simulation and viewing of the dynamic system's behavior in real time.

B.2 OVERVIEW OF ICECAP-PC AND TOTAL-PC

TOTAL[B.1] is an option-number package that reflects a hand calculator or line-terminal environment for interface speed and agility. This FORTRAN package was developed at the Air Force Institute of Technology (AFIT) in the late 1970s and provides an extensive set of control system analysis and design CAD capabilities. In the early 1980s the improved version of TOTAL, designated as Interactive Control Engineering Computer-Aided Design Package (ICECAP), enhanced the analysis and design of control systems. An outgrowth of TOTAL and ICECAP are two PC CAD packages: ICECAP-PC[B.2] and TOTAL-PC[B.3].

ICECAP-PC (version 11, 1994) reflects an object-oriented software engineering design based upon Borland Pascal and Turbo Vision. The interface is through a window system with extensive mouse and hot-key interaction. The general-purpose polynomial and matrix data structures and efficient numerical algorithms support the required variety of commands for continuous and discrete scalar and state-space model manipulation. Graphical displays and command files, along with toolbox extensions and on-line help, provide an effective computer-aided control system design pedagogical environment.

TOTAL-PC,[B.3] a version of TOTAL,[B.1] contains over 140 commands (see Table B.1), which are divided according to general functional categories. It divides each of its options (commands) into various groups. It contains both conventional

---

[†]Portions of ICECAP-PC and TOTAL-PC were developed as individual programs (FREQR, PARTL, and ROOTL) in the 1960s at the Air Force Institute of Technology.

TABLE B.1
Some TOTAL-PC Options

No.	Option
	Transfer-function input options
0	List options
1	Recover all data from file memory
2	Polynomial form—GTF (forward transfer function)
3	Polynomial form—HTF (feedback transfer function)
4	Polynomial form—OLTF (open-loop transfer function)
5	Polynomial form—CLTF (closed-loop transfer function)
6	Factored form—GTF
7	Factored form—HTF
8	Factored form—OLTF
9	Factored form—CLTF
	Matrix input options for state equations
10	List options
11	AMAT—Continuous plant matrix
12	BMAT—Continuous input matrix
13	CMAT—Output matrix
14	DMAT—Direct transmission matrix
15	KMAT—State variable feedback matrix
16	FMAT—Discrete plant matrix
17	GMAT—Discrete input matrix
18	Set up state space model of system
19	Explain use of above matrices
	Block diagram manipulation and state-space options
20	List options
21	Form OLTF = GTF∗HTF (in cascade)
22	Form CLTF = (GAIN∗GTF)/(1 + GAIN∗GTF∗HTF)
23	Form CLTF = (GAIN∗OLTF)/(1 + GAIN∗OLTF)
24	Form CLTF = GTF + HTF (in parallel)
25	GTF(s) and HTF(s) from continuous state space model
26	GTF(z) and HTF(z) from discrete state space model
27	Write adjoint(SI-AMAT) to ANSWER file
28	Find HTF from CLTF & GTF for CLTF = GTF∗HTF/(1 + GTF∗HTF)
29	Find HTF from CLTF & GTF for CLTF = GTF/(1 + GTF∗HTF)
	Time-response options, continuous $F(t)$ and discrete $F(kT)$[†]
30	List options
31	Tabular listing of $F(t)$ or $F(kT)$
32	Plot $F(t)$ or $F(kT)$ at user's terminal
33	Printer plot (written to file answer)
34	Create GNUPlot (written to file plot)
35	Print time or difference equation [$F(t)$ or $F(kT)$]
36	Partial fraction expansion of CLTF (or OLTF)
37	List T-PEAK, T-RISE, T-SETTLING, T-DUP, M-PEAK, final value
38	Quick sketch at user's terminal
39	Select input: step, ramp, pulse, impulse, sin ωT

[†] Option 37 is not available for the discrete domain.

TABLE B.1
Some TOTAL-PC Options—continued

No.	Option
	Root-locus options
40	List options
41	General root locus
42	Root locus with a gain of interest
43	Root locus with a zeta (damping ratio) of interest
44	List N Points on a branch of interest
45	List all points on a branch of interest
46	List locus roots at a gain of interest
47	List locus roots at a zeta of interest
48	Plot root locus at user's terminal
49	List current values of all root locus variables
	Frequency-response options
50	List options
51	Tabular listing
52	Two cycle scan of magnitude (or dB)
53	Two cycle scan of phase (degrees or radians)
54	Plot $F(j\omega)$ at user's terminal
55	Create GNUPlot—Frequency plot
56	Create GNUPlot—Root locus plot
57	Tabulate points of interest: peaks, breaks, etc.
58	Create GNUPlot—Nichols Log magnitude/angle plot
59	Chalk Pitch Axis HQ Criterion Analysis
	Polynomial operations
60	List options
61	Factor polynomial (POLYA)
62	Add polynomials (POLYC = POLYA + POLYB)
63	Subtract polynomials (POLYC = POLYA − POLYB)
64	Multiply polys (POLYC = POLYA∗POLYB)
65	Divide polys (POLYC + REM = POLYA/POLYB)
66	Store Polynomial (POLY_) into POLYD
67	Expand roots into a polynomial
68	$(s + a)^n$ Expansion into a polynomial
69	Activate polynomial calculator
	Matrix operations
70	List options
71	ROOTA = eigenvalues of AMAT
72	CMAT = AMAT + BMAT
73	CMAT = AMAT − BMAT
74	CMAT = AMAT∗BMAT
75	CMAT = AMAT inverse
76	CMAT = AMAT transposed
77	CMAT = identity matrix **I**
78	DMAT = zero matrix **0**
79	Copy one matrix to another

TABLE B.1
Some TOTAL-PC Options—continued

No.	Option
	Digitization options
80	List options
82	CLTF(s) to CLTF(z) by first difference approximation
83	CLTF(s) to CLTF(z) by tustin transformation
84	CLTF(z) to CLTF(s) by impulse invariance
85	CLTF(z) to CLTF(s) by inverse first difference
86	CLTF(z) to CLTF(s) by inverse tustin
87	Find FMAT and GMAT from AMAT and BMAT
88	FMAT from AMAT (FMAT = dexp(AMAT*T))
89	CLTF(X) to CLTF(Y) by X = ALPHA*(Y + A)/(Y + B)
	Miscellaneous options
90	List options
91	Rewind and update MEMORY file with current data
93	List current switch settings (ECHO, ANSWER, etc.)
96	List special commands allowed in option mode
97	List variable name directory
98	List main options of TOTAL
99	Print new features bulletin
119	Augment AMAT:[AMAT] = [AMAT]−GAIN*[BMAT]*[KMAT]
121	Form OLTF = CLK*CLNPOLY/[CLDPOLY−CLK*CLNPOLY] (G-EQUIVALENT)
129	(Integral of (CLTF SQUARED))/2PI =
	Double-precision discrete transform options
140	List options
141	CLTF(s) to CLTF(z) (impulse variance)
142	CLTF(s) to CLTF(w) W = (Z−1)/(Z + 1)
143	CLTF(s) to CLTF(w') W' = (2/T)(Z−1)/(Z + 1)
144	Hi-rate CLTF(z) to lo-rate CLTF(z)
145	Hi-rate CLTF(w) to lo-rate CLTF(w)
146	Hi-rate CLTF(w') to lo-rate CLTF(w')
147	Option 144 (avoiding internal factoring)
148	Option 144 (all calculations in z plane)
	Quantitative feedback technique (QFT) options
150	List options
151	Define—Plant
152	Select—Nominal plant: (1-10)
153	Define—TLTF: (Lower Tracking Ratio)
154	Define—TUTF: (Upper Tracking Ratio)
155	Define—TDTF: (Disturbance)
156	Define—FTF: (Pre-filter function)
157	Define—LOTF: (Nominal loop function)
158	Display Plant, TLTF, TUTF, TDTF, FTF, LOTF
159	QFT design/report macro

analog and discrete control system analysis and design options. It also contains the Quantitative Feedback Theory (QFT) CAD design routines under the Option 150 series for MISO systems.

Both CAD packages, ICECAP-PC and TOTAL-PC, are available by contacting the individuals in Refs. B.2 and B.3, respectively.

B.3 OVERVIEW OF MATLAB®

MATLAB computation and visualization software, the Control System Toolbox, and the *Student Edition of MATLAB*, all developed by The MathWorks, Inc., are used for some of the examples and problems contained in this text. MATLAB's basic data structures are matrices and vectors. The *capabilities of MATLAB*[B.4–B.6, B.14] are shown in Table B.2. MATLAB program printouts for solution of text examples are shown in Chaps. 4, 7, 9, 13, and 16 in order to demonstrate the capabilities of this CAD package. References B.5, B.6, and B.13 contain numerous additional examples demonstrating the use of the MATLAB program. This CAD package provides the ability to run this program either by use of a list of commands, stored in *M-files*, or interactively. It also has the ability to produce plots with simple instructions. Selected functions from the professional Control System and Signal Processing Toolboxes have been consolidated into a Signals and Systems Toolbox, which is included with the Student Edition of MATLAB. The MathWorks offers additional toolboxes for Neural Networks, System Identification, Nonlinear Control Design, Optimization, Robust Control, Mu-Analysis, Model Predictive Control, and other areas for use with the professional version of MATLAB. The functions in the Systems and Signals Toolbox are listed in Table B.2.

TABLE B.2
The Student Edition of MATLAB: available functions from the Signals and Systems Toolbox

Model building

append	Append system dynamics
augstate	Augment states as outputs
blkbuild	Build state-space system from block diagram
cloop	Close loops of system
connect	Block diagram modeling
conv	Convolution of two polynomials
dreg	Form discrete controller/estimator from gain matrices
drmodel	Generate random discrete model
feedback	Feedback system connection
ord2	Generate A, B, C, D for a second-order system
pade	Pade approximation to time delay
parallel	Parallel system connection
rmodel	Generate random continuous model
series	Series system connection
ssdelete	Delete inputs, outputs, or states from model
ssselect	Select subsystem from larger system

TABLE B.2
The Student Edition of MATLAB: available functions from the Signals and Systems Toolbox—continued

Model conversions

c2d	Continuous to discrete-time conversion
c2dm	Continuous to discrete-time conversion with method
c2dt	Continuous to discrete conversion with delay
d2c	Discrete to continuous-time conversion
d2cm	Discrete to continuous-time conversion with method
poly	Roots to polynomial conversion
residue	Partial fraction expansion
ss2tf	State-space to transfer function conversion
ss2zp	State-space to zero-pole conversion
tf2ss	Transfer function to state-space conversion
tf2zp	Transfer function to zero-pole conversion
zp2tf	Zero-pole to transfer function conversion
zp2ss	Zero-pole to state-space conversion

Model reduction

minreal	Minimal realization and pole-zero cancellation
modred	Model order reduction

Model realizations

canon	Canonical form
ss2ss	Apply similarity transform

Model properties

ctrb	Controllability matrix
damp	Damping factors and natural frequencies
dcgain	Continuous steady-state (dc) gain
ddamp	Discrete damping factors and natural frequencies
ddcgain	Discrete steady-state (dc) gain
eig	Eigenvalues and eigenvectors
esort	Sort continuous eigenvalues by real part
obsv	Observability matrix
printsys	Display system in formatted form
roots	Polynomial roots
tzero	Transmission zeros
tzero2	Transmission zeros using random perturbation method

Time response

dimpulse	Discrete unit sample response
dinitial	Discrete initial condition response
dlsim	Discrete simulation to arbitrary inputs
dstep	Discrete step response
filter	SISO Z-transform simulation
impulse	Impulse response
initial	Continuous initial condition response
lsim	Continuous simulation to arbitrary inputs
ltitr	Low-level time response function
step	Step response
stepfun	Step function

TABLE B.2
The Student Edition of MATLAB: available functions
from the Signals and Systems Toolbox—continued

Frequency response	
bode	Bode plot (frequency response)
dbode	Discrete Bode plot (frequency response)
dnichols	Discrete Nichols plot
dnyquist	Discrete Nyquist plot
freqs	Laplace-transform frequency response
freqz	Z-transform frequency response
ltifr	Low-level frequency response function
margin	Gain and phase margins
nichols	Nichols plot
ngrid	Draw grid lines for Nichols plot
nyquist	Nyquist plot

Root locus	
pzmap	Pole-zero map
rlocfind	Interactive root locus gain determination
rlocus	Evans root locus
sgrid	Draw continuous root locus ω_n, ζ grid
zgrid	Draw discrete root locus ω_n, ζ grid

Gain selection	
acker	SISO pole placement
place	Pole placement

Demonstrations	
ctrldemo	Introduction to the Control Toolbox
jetdemo	Classical design of jet transport yaw damper
diskdemo	Digital control design of hard disk controller

B.4   QFT CAD PACKAGES

Up to about 1986 there were essentially no Quantitative Feedback Theory (QFT) CAD packages designed specifically to assist in doing a complete QFT design for analog and digital control systems. The ICECAP/QFT CAD package was designed in 1986 at the Air Force Institute of Technology. This package was designed as an educational tool, as illustrated in Fig. B.1. Since 1986 other QFT CAD packages have become available that are more attuned to the practicing engineer who is familiar with QFT.[B.10,B.15]

B.5   COMPUTER-AIDED DESIGN ACCURACY CHECKS (CADAC)

As stressed in the Preface, the 21st century requires an individual to be computer-literate and competent. Thus, throughout the text, in order to help the student to

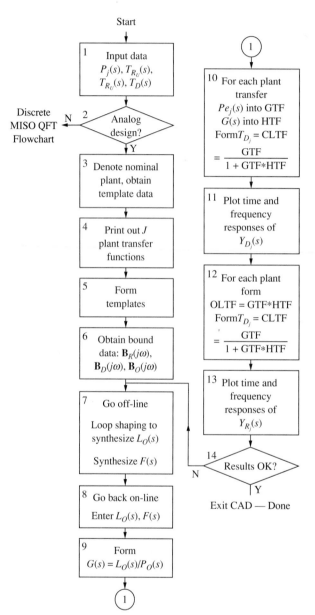

FIGURE B.1
MISO/QFT design option using TOTAL-PC.

develop this competence, CAD accuracy checks (CADAC) are given. Their purpose is to convey to the reader the importance of using these necessary, but not sufficient, conditions to assist in attesting to the viability of the computer solution. An individual should augment these CADAC with others that are appropriate with the CAD package being used.

B.6 OTHER COMPUTER-AIDED DESIGN PACKAGES

There are many other computer-aided-design packages available to assist a control engineer in the analysis, design, and simulation of a control system design. Some of these packages are CTRL-C by Systems Control Technology, Inc.; Program CC by Systems Technology, Inc.[B.11]; Simnon[B.12] program by SSPA Systems; MATRIX$_x$ program by Integrated Systems, Inc.; and MACSYMA program by Math Lab Group, Laboratory for Computer Science, M.I.T. The latter two programs are very powerful programs: the MATRIX$_x$ program is a data analysis, system identification, control design, and simulation package, and the MACSYMA program is a large computer program that performs symbolic mathematical computations. Other programs available are Mathematica, L-A-S, LOGALPHA, and Mathcad, to mention just a few.[B.8,B.9]

The program MULTI[B.7] assists in the design of high-gain multiple-input multiple-output digital tracking systems similar to those described in Chap. 17. This is an output feedback system with a proportional plus integral controller. The controller gains are adjusted in the simulation of the closed-loop system, providing state, control input, and output responses.

REFERENCES

1. Larimer, S. J.: "An Interactive Computer-Aided Design Program for Digital and Continuous System Analysis and Synthesis (TOTAL)," M.S. thesis, GE/GGC/EE/78-2, Graduate School of Engineering, Air Force Institute of Technology, Wright-Patterson AFB, OH., 1978.
2. "ICECAP-PC," G. B. Lamont, Graduate School of Engineering, Air Force Institute of Technology, AFIT/ENG, Bldg 642, 2950 P Street, Wright-Patterson AFB, OH 45433-7765, 1993.
3. "TOTAL-PC," R. Ewing, Graduate School of Engineering, Air Force Institute of Technology, AFIT/ENG, Bldg 642, 2950 P Street, Wright-Patterson AFB, OH 45433-7765, 1992.
4. "MATLAB® Version 4," The MathWorks, Inc., Natick, MA 01760-1500, 1993.
5. Saadat, H.: *Computational Aids in Control Systems Using MATLAB*, McGraw-Hill, New York, 1993.
6. The MathWorks, Inc.: *The Student Edition of MATLAB*, Prentice Hall, Englewood Cliffs, NJ, 1992.
7. "MULTI," J. J. D'Azzo, Graduate School of Engineering, Air Force Institute of Technology, AFIT/ENG, Bldg 642, 2950 P Street, Wright-Patterson AFB, OH 45433-7765, Version 4, 1993.
8. Putz, P., and M. J. Wozny: "A New Computer Graphics Approach to Parameter Space Design of Control Systems," *IEEE Trans. AC*, vol. AC-32, pp. 294–302, April 1987.
9. Wolfram, Stephen, et al.: *Mathematica: A System for Doing Mathematics by Computer*, Addison-Wesley, Reading, MA, 1991.
10. "ICECAP/QFT," G. B. Lamont, Graduate School of Engineering, Air Force Institute of Technology, AFIT/ENG, Bldg 642, 2950 P Street, Wright-Patterson AFB, OH 45433-7765, 1986.
11. Thompson, P. M.: "USER'S Guide to Program, Version 3," System Technology, Inc., Hawthorne, CA, March 1985.
12. "Simnon", SSPA Systems, PO Box 24001, S-400 22 Göteborg, Sweden, 1993.

13. Strum, R. D., and D. E. Kirk: *Contemporary Linear Systems,* PWS Publishing Company, Boston, MA, 1994.
14. Sigmon, K.: "MATLAB Primer," Department of Mathematics, University of Florida, Gainesville, FL, 1992.
15. "MIMO QFT CAD," C.H. Houpis, Graduate School of Engineering, Air Force Institute of Technology, AFIT/ENG, Bldg 642, 2950 P Street, Wright-Patterson AFB, OH 45433-7765, 1993.

Chapter 2

2.1. Write the (a) loop, (b) node, and (c) state equations for the circuit shown after the switch is closed. Let $u = e$, $y_1 = v_C$, and $y_2 = v_{R_2}$. (d) Determine the transfer function y_1/e and $y_2/u = G_2$.

2.2. Write the (a) loop, (b) node, and (c) state equations for the circuit shown after the switch S is closed.

2.3. The circuits shown are in the steady state with the switch S closed. At time $t = 0$, S is opened. (a) Write the necessary differential equations for determining $i_1(t)$. (b) Write the state equations.

Circuit 1 · Circuit 2

2.4. Write all the necessary equations to determine v_o. (*a*) Use nodal equations. (*b*) Use loop equations. (*c*) Write the state and output equations.

2.5. Derive the state equations. Note that there are only two independent state variables.

2.6. (*a*) Derive the differential equation relating the position $y(t)$ and the force $f(t)$. (*b*) Draw the mechanical network. (*c*) Determine the transfer function $G(D) = y/f$. (*d*) Identify a suitable set of independent state variables. Write the state equation in matrix form.

2.7. (*a*) Draw the mechanical network for the mechanical system shown. (*b*) Write the differential equations of performance. (*c*) Draw the analogous electric circuit in which force is analogous to current. (*d*) Write the state equations.

2.8. (*a*) Write the differential equations describing the motion of the following system, assuming small displacements. (*b*) Write the state equations.

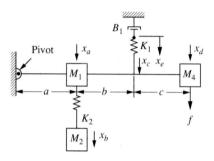

2.9. A simplified model for the vertical suspension of an automobile is shown in the figure. (*a*) Draw the mechanical network; (*b*) write the differential equations of performance; (*c*) derive the state equations; (*d*) determine the transfer function $x_1/\hat{u}_2$, where $\hat{u}_2 = (B_2 D + K_2) u_2$ is the force exerted by the road.

2.10. An electromagnetic actuator contains a solenoid, which produces a magnetic force proportional to the current in the coil, $f = K_i i$. The coil has resistance and inductance. (*a*) Write the differential equations of performance. (*b*) Write the state equations.

2.11. A warehouse transportation system has a motor drive moving a trolley on a rail. Write the equation of motion.

2.12. For the system shown, the torque T is transmitted through a noncompliant shaft and a hydraulic clutch to a pulley #1, which has a moment of inertia J_1. The clutch is modeled by the damping coefficient B_1 (it is assumed to be massless). A bearing between the clutch and the pulley #1 has a damping coefficient B_2. Pulley #1 is connected to pulley #2 with a slipless belt drive. Pulley #2, which has a moment of inertia J_2, is firmly connected to a wall with a compliant shaft that has a spring constant K. The desired outputs are: $y_1 = \theta_b$, the angular position of the J_1, and $y_2 = \dot{\theta}_a$, the angular velocity of the input shaft, before the clutch. (*a*) Draw the mechanical network; (*b*) write the differential equations of performance; (*c*) write the state equations; (*d*) determine the transfer functions y_1/u and y_2/u.

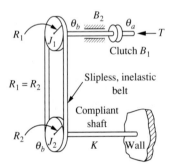

2.13. (*a*) Write the equations of motion for this system. (*b*) Using the physical energy variables, write the matrix state equation.

2.14. The figure represents a cylinder of inertia J_1 inside a sleeve of inertia J_2. There is viscous damping B_1 between the cylinder and the sleeve. The springs K_1 and K_2

are fastened to the inner cylinder. (*a*) Draw the mechanical network. (*b*) Write the system equations. (*c*) Draw the analogous electric circuit. (*d*) Write the state and output equations. The outputs are θ_1 and θ_2. (*e*) Determine the transfer function $G = \theta_2/T$.

2.15. The two gear trains have an identical net reduction, have identical inertias at each stage, and are driven by the same motor. The number of teeth on each gear is indicated on the figures. At the instant of starting, the motor develops a torque T. Which system has the higher initial load acceleration? (*a*) Let $J_1 = J_2 = J_3$; (*b*) let $J_1 = 40J_2 = 4J_3$.

2.16. In the mechanical system shown, r_1 is the radius of the drum. (*a*) Write the necessary differential equations of performance for this system. (*b*) Obtain a differential equation expressing the relationship of the output x_2 in terms of the input $T(t)$ and the corresponding transfer function. (*c*) Write the state equations with $T(t)$ as the input.

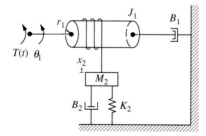

2.17. Write the state and system output equations for the rotational mechanical system of Fig. 2.16. (*a*) With T as the input, use $x_1 = \theta_3$, $x_2 = D\theta_3$, $x_3 = D\theta_2$ and (*b*) with θ_1 as the input, use $x_1 = \theta_3$, $x_2 = D\theta_3$, $x_3 = \theta_2$, $x_4 = D\theta_2$.

2.18. For the hydraulic preamplifier shown, write the differential equations of performance relating x_1 to y_1. (*a*) Neglect the load reaction. (*b*) Do not neglect load reaction. There is only one fixed pivot, as shown in the figure.

2.19. The mechanical load on the hydraulic translational actuator in Sec. 2.10, Fig. 2.22, is changed to that in the accompanying figure. Determine the new state equations and compare with Eq. (2.111).

2.20. The sewage system leading to a treatment plant is shown. The variables q_A and q_B are input flow rates into tanks 1 and 2, respectively. Pipes 1, 2, and 3 have resistances as shown. Derive the state equations.

2.21. Most control systems require some type of motive power. One of the most commonly used units is the electric motor. Write the differential equation for the angular displacement of a moment of inertia, with damping, connected directly to a dc motor shaft when a voltage is suddenly applied to the armature terminals with the field separately energized.

2.22. A sketch of a moving-coil microphone is shown. The diaphragm has the spring elastance K, mass M, and damping B. Fastened to the diaphragm is a coil, which moves in the magnetic field produced by the permanent magnets. (*a*) Derive the differential equations of the system, considering changes from the equilibrium conditions. (*b*) Draw an analogous electric circuit.

2.23. The damped simple pendulum shown in the figure is suspended in a uniform gravitational field g. There is a horizontal force f and a rotational viscous friction B. (*a*) Determine the equation of motion (note: $J = M\ell^2$). (*b*) Linearize the equation of part (*a*) by assuming $\cos\theta \approx 1$ and $\sin\theta \approx \theta$ for small values of θ. (*c*) Write the state and output equations of the system using $x_1 = \theta$, $x_2 = \dot{\theta}$, and $y = x_1$.

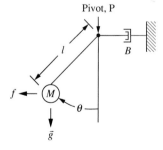

2.24. Given the mechanical system shown, the load J_2 is coupled to the rotor J_1 of a motor through a reduction gear and springs K_1 and K_2. The gear teeth ratio is $1 : N$. The torque T is the mechanical torque generated by the motor. B_1 and B_2 are damping coefficients, and $\theta_1, \theta_2, \theta_3$, and θ_4 are angular displacements. (*a*) Draw a mechanical network for this system. (*b*) Write the necessary differential equations in order to solve for θ_4.

2.25. In some applications the motor shaft of Prob. 2.21 is of sufficient length so that its elastance K must be taken into account. Consider that $v_2(t)$ is constant and that the velocity $\omega_m(t) = D\theta_m(t)$ is being controlled. (*a*) Write the necessary differential equation for determining the velocity of the load $\omega_J(t) = D\theta_J$. (*b*) Determine the transfer function $\omega_J(t)/v_1(t)$.

Chapter 3

3.1. (*a*) With $v_c(0^-) = 0$, what are the initial values of current in all elements when the switch is closed? (*b*) Write the state equations. (*c*) Solve for the voltage across C as a function of time from the state equations. (*d*) Find M_p, t_p, and t_s.

$$R_1 = R_2 = 2\,\text{k}\Omega \qquad C = 50\,\mu\text{F} \qquad L = 1\,\text{H} \qquad E = 100\,\text{V}$$

3.2. In Prob. 2.12, the parameters have the following values:

$$J_1 = 1.0\,\text{lb} \cdot \text{ft} \cdot \text{s}^2 \qquad B_1 = 0.5\,\text{lb} \cdot \text{ft}/(\text{rad/s}) \qquad K = 0.5\,\text{lb} \cdot \text{ft/rad}$$
$$J_2 = 1.0\,\text{slug} \cdot \text{ft}^2 \qquad B_2 = 3.35\,\text{oz} \cdot \text{in}/(\text{deg/s})$$

Solve for $\theta_b(t)$ if $T(t) = tu_{-1}(t)$.

3.3. In Prob. 2.10, the parameters have the following values:

$$M_1 = M_2 = 0.05\,\text{slug} \qquad L = 1\,\text{H} \qquad l_1 = 10\,\text{in.} \qquad K_1 = K_2 = 1.2\,\text{lb/in.}$$
$$R = 10\,\Omega \qquad l_2 = 20\,\text{in.} \qquad B_1 = B_2 = 15\,\text{oz}/(\text{in./s}) \qquad K_i = 24\,\text{oz/A}$$

With $e(t) = u_{-1}(t)$ and the system initially at rest, solve for $x_b(t)$.

3.4. In part (*a*) of Prob. 2.18, the parameters have the following values:

$$a = b = 12 \text{ in.} \qquad C_1 = 9.0 \text{ (in./s)/in.} \qquad c = 10 \text{ in.} \qquad d = 2 \text{ in.}$$

Solve for $y_1(t)$ with $x_1(t) = 0.1u_{-1}(t)$ in. and zero initial conditions.

3.5. (*a*) $r(t) = (D^3 + 6D^2 + 11D + 6)c(t)$. With $r(t) = \sin t$ and zero initial conditions, determine $c(t)$. (*b*) $r(t) = [(D + 1)(D^2 + 4D + 5)]c(t)$. With $r(t) = tu_{-1}(t)$ and all initial conditions zero, determine the complete solution with all constants evaluated. (*c*) $D^3c + 10D^2c + 32Dc + 32c = 10r$. Find $c(t)$ for $r(t) = u_{-1}(t)$, $D^2c(0) = Dc(0) = 0$, and $c(0) = -2$. One of the eigenvalues is $\lambda = -2$. (*d*) For each equation determine $C(j\omega)/R(j\omega)$ and $c(t)_{ss}$ for $r(t) = 10 \sin 5t$.

3.6. The system shown is initially at rest. At time $t = 0$ the string connecting M to W is severed at X. Measure $x(t)$ from the initial position. Find $x(t)$.

3.7. Switch S_1 is open, and there is no energy stored in the circuit. (*a*) Write the state equations. (*b*) Find $v_o(t)$ after switch S_1 is closed. (*c*) After the circuit reaches steady state, the switch S_1 is opened. Find $v_o(t)$.

3.8. Solve the following differential equations. Assume zero initial conditions. Sketch the solutions.

(*a*) $D^2x + 16x = 1$ (*b*) $D^2x + 4Dx + 3x = 9$

(*c*) $D^2x + Dx + 4.25x = t + 1$ (*d*) $D^3x + 3D^2x + 4Dx + 2x = 10 \sin 10t$

3.9. For Prob. 2.2, solve for $i_2(t)$, where

$$E = 10 \text{ V} \qquad L = 1 \text{ H} \qquad C_1 = C_2 = 0.001 \text{ F} \qquad R_1 = 10 \ \Omega \qquad R_2 = 15 \ \Omega$$

3.10. For the circuit of Prob. 2.2: (*a*) obtain the differential equation relating $i_2(t)$ to the input E; (*b*) write the state equations using the physical energy variables; (*c*) solve the state and output equations and compare with the solution of Prob. 3.9; (*d*) write the state and output equations using the phase variables, starting with the differential equation from part (*a*).

3.11. In Prob. 2.3, the parameters have the values $R_1 = R_2 = R_3 = 10\,\Omega$, $C_1 = C_2 = 1\,\mu F$, and $L_1 = L_2 = 50\,$H. If $E = 100\,$V, (a) find $i_1(t)$; (b) sketch $i_1(t)$ and compute M_o, t_p, and t_s; (c) determine the value of T_s (± 2 percent of $i(0)$); (d) solve for $i_1(t)$ from the state equation.

3.12. For $L = 0.4$, $C = 0.5$, and (1) $R = 1$, (2) $R = 0.5$: (a) show that the differential equation is $(RCD^2 + D + R/L)v_a = De$; (b) find $v_a(t)_{ss}$ for $e(t) = u_{-1}(t)$; (c) find the roots m_1 and m_2; (d) find $v_a(0^+)$; (e) find $Dv_a(0^+)$ [use $i_c(0^+)$]; (f) determine the complete solution $v_a(t)$; (g) sketch and dimension the plot of $v_a(t)$.

3.13. Solve the following equations. Show explicitly $\mathbf{\Phi}(t)$, $\mathbf{x}(t)$, and $\mathbf{y}(t)$.

$$\dot{\mathbf{x}} = \begin{bmatrix} -6 & 4 \\ -2 & 0 \end{bmatrix}\mathbf{x} + \begin{bmatrix} 0 \\ 1 \end{bmatrix}u \qquad \mathbf{y} = [1 \quad 0]\mathbf{x}$$

$$x_1(0) = 2 \qquad x_2(0) = 0 \qquad u(t) = u_{-1}(t)$$

3.14. A single-degree-of-freedom representation of the rolling dynamics of an aircraft, together with a first-order representation of the aileron servomotor, is given by

$$p(t) = \dot{\phi}(t) \qquad J_x\dot{p}(t) = L_{\delta_A}\delta_A(t) + L_p p(t)$$

$$T\dot{\delta}_A(t) = (\delta_A)_{\text{comm}}(t) - \delta_A(t)$$

(a) Derive a state-variable representation of the aircraft and draw a block diagram of the control system with

$$(\delta_A)_{\text{comm}}(t) = AV_{\text{ref}} - \mathbf{P}\mathbf{x}(t)$$

where $\quad\mathbf{x}(t) =$ state vector of the aircraft system $= [\phi \quad p \quad \delta_A]^T$

$\mathbf{P} =$ matrix of correct dimensions and with constant nonzero elements p_{ij}

$A. =$ scalar gain

$V_{\text{ref}} =$ the reference input

$(\delta_A)_{\text{comm}} =$ the input to the servo

(b) Determine the aircraft state equations in response to the reference input V_{ref}.

3.15. Use the Sylvester expansion of Sec. 3.14 to obtain $\mathbf{A}^{10}$ for

$$\mathbf{A} = \begin{bmatrix} 0 & 1 & 0 \\ 0 & 0 & 1 \\ -6 & -11 & -6 \end{bmatrix}$$

3.16. For the mechanical system of Fig. 2.11a the state equation for Example 2, Sec. 2.6, is given in *phase-variable form*. With the input as $u = x_a$, the state variables are $x_1 = x_b$ and $x_2 = \dot{x}_b$. Use $M = 5$, $K = 10$, and $B = 15$. The initial conditions are $x_b(0) = 1$ and $\dot{x}_b(0) = -2$. (*a*) Find the homogeneous solution for $\mathbf{x}(t)$. (*b*) Find the complete solution with $u(t) = u_{-1}(t)$.

3.17. For the autonomous system

$$\dot{\mathbf{x}} = \begin{bmatrix} 0 & 1 & 0 \\ 0 & 0 & 1 \\ -5 & -7 & -3 \end{bmatrix} \mathbf{x} \qquad \mathbf{y} = [1 \quad 0 \quad 0]\mathbf{x}$$

(*a*) Find the system eigenvalues. (*b*) Evaluate the state transmission matrix $\mathbf{\Phi}(t)$ by means of Sylvester's expansion theorem.

Chapter 4

4.1. Determine the inverse Laplace transform by use of partial fraction expansion:

(*a*) $F(s) = \dfrac{9}{(s + 1)^2(s^2 + 2s + 10)}$ (*b*) $F(s) = \dfrac{90}{s(s^2 + 4s + 13)(s + 5)}$

(*c*) $F(s) = \dfrac{10(s + 10)}{(s + 1)^2(s^2 + 4s + 5)}$

4.2. Find $x(t)$ for

(*a*) $X(s) = \dfrac{4}{s^3 + 6s^2 + 16s + 16}$ (*b*) $X(s) = \dfrac{20}{s^4 + 13s^3 + 64s^2 + 142s + 120}$

(*c*) $X(s) = \dfrac{0.969426(s^2 + 4.2s + 13.44)}{s(s + 1)(s^2 + 4s + 13)}$ (*d*) Sketch $x(t)$ for parts (*a*) to (*c*).

4.3. Given an ac servomotor with inertia load, find $\omega(t)$ by (*a*) the classical method; (*b*) the Laplace-transform method.

$$\omega(0) = 10^4 \text{ rad/s} \qquad K_\omega = -4 \times 10^{-3} \text{ oz} \cdot \text{in./(rad/s)} \qquad K_c = 1.2 \text{ oz} \cdot \text{in./V}$$

$$e(t) = 5u_{-1}(t)\text{V} \qquad J = 15.36 \text{ oz} \cdot \text{in.}^2$$

4.4. Repeat the following problems using the Laplace transform:

(*a*) Prob 3.1 (*b*) Prob 3.3 (*c*) Prob 3.5
(*d*) Prob 3.6 (*e*) Prob 3.10 (*f*) Prob 3.11(*d*)
(*g*) Prob 3.13 (*g*) Prob. 3.14

4.5. Find the partial-fraction expansions of the following:

(*a*) $F(s) = \dfrac{7}{(s + 1)(s + 7)}$ (*b*) $F(s) = \dfrac{12}{s(s + 2)(s + 6)}$

(*c*) $F(s) = \dfrac{68}{s(s^2 + 6s + 34)}$ (*d*) $F(s) = \dfrac{0.5(s + 6)}{s^2(s + 1)(s + 3)}$

(*e*) $F(s) = \dfrac{10(s + 5)}{s^2(s^2 + 6s + 25)}$ (*f*) $F(s) = \dfrac{4(s + 3)}{(s + 1)(s^2 + 8s + 20)}$

(g) $F(s) = \dfrac{30}{s^2(s+15)(s^2+6s+10)}$ (h) $F(s) = \dfrac{15(s+2)}{s(s+3)(s^2+6s+10)}$

(i) $F(s) = \dfrac{10(s+2.02)}{s(s+2)(s^2+2s+10)}$ (j) $F(s) = \dfrac{10(s^2+4.1s+5.305)}{s(s+1)(s^2+4s+5)}$

NOTE: For parts (g) through (j) use a CAD program.

4.6. Solve the differential equations of Prob. 3.8 by means of the Laplace transform.

4.7. Write the Laplace transforms of the following equations and solve for $x(t)$; the initial conditions are given to the right.

(a) $Dx + 8x = 0$ $x(0) = -2$
(b) $D^2x + 2.8Dx + 4x = 10$ $x(0) = 2, Dx(0) = 3$
(c) $D^2x + 4Dx + 3x = t$ $x(0) = 0, Dx(0) = -2$
(d) $D^3x + 4D^2x + 9Dx + 10x = \sin 5t$ $x(0) = -4,\ Dx(0) = 1,\ D^2x(0) = 0$

4.8. Determine the final value for:

(a) Prob. 4.1 (b) Prob. 4.2 (c) Prob. 4.5

4.9. Determine the initial value for:

(a) Prob. 4.1 (b) Prob. 4.2 (c) Prob. 4.5

4.10. For the functions of Prob. 4.5, plot M vs. ω and α vs. ω.

4.11. Find the complete solution of $x(t)$ with zero initial conditions for

(1) $(D^2 + 2D + 2)(D + 5)x = (D + 3)f(t)$
(2) $(D^2 + 3D + 2)(D + 5)x = (D + 6)f(t)$

The forcing function $f(t)$ is

(a) $u_0(t)$ (b) $10u_{-1}(t)$ (c) $tu_{-1}(t)$ (d) $2t^2u_{-1}(t)$

4.12. A linear system is described by

(1) $\dot{\mathbf{x}} = \begin{bmatrix} -2 & 1 \\ 2 & -3 \end{bmatrix}\mathbf{x} + \begin{bmatrix} 0 \\ 1 \end{bmatrix}u$ $y = [1 \quad 0]\mathbf{x}$

(2) $\dot{\mathbf{x}} = \begin{bmatrix} 0 & 3 \\ -5 & -8 \end{bmatrix}\mathbf{x} + \begin{bmatrix} 1 \\ 0 \end{bmatrix}u$ $y = [1 \quad 1]\mathbf{x}$

where $u(t) = u_{-1}(t)$ and the initial conditions are $x_1(0) = 0, x_2(0) = 1$. (a) Using Laplace transforms, find $\mathbf{X}(s)$. Put the elements of this vector over a common denominator. (b) Find the transfer function $G(s)$. (c) Find $y(t)$.

4.13. A linear system is represented by

(1) $\dot{\mathbf{x}} = \begin{bmatrix} -3 & -2 \\ -2 & -3 \end{bmatrix}\mathbf{x} + \begin{bmatrix} 1 \\ 1 \end{bmatrix}u$ $y = \begin{bmatrix} 1 & 0 \\ 1 & 1 \end{bmatrix}\mathbf{x}$ $u = u_{-1}(t)$

(2) $\dot{\mathbf{x}} = \begin{bmatrix} 0 & 3 \\ -5 & -8 \end{bmatrix}\mathbf{x} + \begin{bmatrix} 1 & 1 \\ 0 & -1 \end{bmatrix}u$ $y = \begin{bmatrix} 1 & 0 \\ 1 & 2 \end{bmatrix}\mathbf{x}$ $\mathbf{u} = \begin{bmatrix} 1 \\ 1 \end{bmatrix}u_{-1}(t)$

(3) $\dot{\mathbf{x}} = \begin{bmatrix} 0 & 4 \\ 0 & -5 \end{bmatrix}\mathbf{x} + \begin{bmatrix} 0 \\ 1 \end{bmatrix}u$ $y = \begin{bmatrix} 0 & 1 \\ 1 & 1 \end{bmatrix}\mathbf{x}$ $u = u_{-1}(t)$

(*a*) Find the complete solution for $\mathbf{y}(t)$ when $x_1(0) = 1$ and $x_2(0) = 0$. (*b*) Determine the transfer functions. (*c*) Draw a block diagram representing the system.

4.14. A system is described by

$$(1) \quad \dot{\mathbf{x}} = \begin{bmatrix} -6 & -1 \\ 5 & 0 \end{bmatrix} \mathbf{x} + \begin{bmatrix} 1 \\ 0 \end{bmatrix} \mathbf{u} \qquad y = x_1$$

$$(2) \quad \dot{\mathbf{x}} = \begin{bmatrix} 0 & 1 \\ -2 & -3 \end{bmatrix} \mathbf{x} + \begin{bmatrix} 1 \\ 1 \end{bmatrix} \mathbf{u} \qquad y = x_1$$

(*a*) Find $\mathbf{x}(t)$ with $\mathbf{x}(0) = 0$ and $u = u_{-1}(t)$. (*b*) Determine the transfer function $G(s) = Y(s)/U(s)$.

4.15. Repeat Prob. 3.17 using the Laplace transform.

4.16. Show that Eqs. (4.52) and (4.53) are equivalent expressions.

4.17. Given

$$Y(s) = \frac{K(s + a_1)\cdots(s + a_w)}{s(s + \zeta_1\omega_{n_1} \pm j\omega_{d_1})(s + \zeta_3\omega_{n_3} \pm j\omega_{d_3})(s + b_5)\cdots(s + b_n)}$$

all the a_i's and b_k's are positive and real and $\zeta_1\omega_{n_1} = \zeta_3\omega_{n_3}$ with $\omega_{n_3} > \omega_{n_1}$. Determine the effect on the Heaviside partial-fraction coefficients associated with the pole $p_3 = -\zeta_3\omega_{n_3} + j\omega_{d_3}$ when ω_{n_3} is allowed to approach infinity. *Hint:* Analyze

$$\lim_{\omega \to \infty} A_3 = \lim_{\omega \to \infty} [(s - p_k)Y(s)]_{s = p_3 = -\zeta_3\omega_{n_3} + j\omega_{d_3}}$$

K has the value required for $y(t)_{ss} = 1$ for a unit step input. Consider the cases (1) $n = w$, (2) $n > w$.

4.18. Repeat Prob. 4.17 with $a_1 = 0$, $w = 2$, $n = 6$, and K is a fixed value.

4.19. Use the method of Sec. 4.16, Eqs. (4.103) and (4.104), to determine: (*a*) the characteristic polynomial (*Hint:* For (1) use $\mathbf{b} = [1 \quad 0 \quad 1]^T$ and for (2) use $\mathbf{b} = [1 \quad 0 \quad 0]^T$); (*b*) the matrix inverse $\mathbf{A}^{-1}$.

$$(1) \quad \mathbf{A} = \begin{bmatrix} -1 & 0 & 0 \\ -1 & -2 & -1 \\ -1 & 0 & -3 \end{bmatrix} \qquad (2) \quad \mathbf{A} = \begin{bmatrix} 0 & 2 & -4 \\ 1 & 3 & 1 \\ 2 & 0 & 1 \end{bmatrix}$$

4.20. Repeat Prob. 4.19 with

$$\mathbf{A} = \begin{bmatrix} 0 & 1 & 0 & 0 \\ 1 & 0 & 1 & -2 \\ 0 & 0 & 0 & 2 \\ 1 & 1 & 0 & 0 \end{bmatrix}$$

Hint: Use $\mathbf{b} = [0 \quad 1 \quad 0 \quad 1]^T$.

4.21. For the linearized system of Prob. 2.23 determine (*a*) the transfer function and (*b*) the value of B that makes this system critically damped.

4.22. Determine the transfer function and draw a block diagram.

$$\dot{\mathbf{x}} = \begin{bmatrix} -2 & 0 \\ 3 & -1 \end{bmatrix} \mathbf{x} + \begin{bmatrix} 1 & 0 \\ 1 & 2 \end{bmatrix} \mathbf{u} \qquad \mathbf{y} = \begin{bmatrix} 0 & 1 \\ 2 & 0 \end{bmatrix} \mathbf{x}$$

Chapter 5

5.1. For the temperature-control system of Fig. 5.1, some of the pertinent equations are $b = K_b\theta$, $f_s = K_s i_s$, $q = K_q x$, $\theta = K_c q/(D + a)$. The solenoid coil has resistance R and inductance L. The solenoid armature and valve have mass M, damping B, and a restraining spring K. (a) Determine the transfer function of each block in Fig. 5.1b. (b) Determine the forward transfer function $G(s)$. (c) Write the state and output equations in matrix form. Use $x_1 = \theta$, $x_2 = x$, $x_3 = \dot{x}$, $x_4 = i_s$.

5.2. Find an example of a practical closed-loop control system not covered in this book. Briefly describe the system and show a block diagram.

5.3. A satellite-tracking system is shown in the accompanying schematic diagram. The transfer function of the tracking receiver is

$$\frac{V_1}{\Theta_{comm} - \Theta_L} = \frac{5}{1 + s/40\pi}$$

The following parameters apply:

$$K_a = \text{gain of servoamplifier} = 60$$
$$K_t = \text{tachometer constant} = 0.04 \text{ V} \cdot \text{s}$$
$$K_T = \text{motor torque constant} = 0.5 \text{ N} \cdot \text{m/A}$$
$$K_b = \text{motor back-emf constant} = 0.75 \text{ V} \cdot \text{s}$$
$$J_L = \text{antenna inertia} = 2{,}880 \text{ kg} \cdot \text{m}^2$$
$$J_m = \text{motor inertia} = 7 \times 10^{-3} \text{ kg} \cdot \text{m}^2$$
$$1 : N = \text{gearbox stepdown ratio} = 1 : 12{,}000$$

(a) Draw a detailed block diagram showing all the variables. (b) Derive the transfer function $\Theta_L(s)/\Theta_{comm}(s)$.

5.4. A photographic control system is shown in simplified form in the diagram. The aperture slide moves to admit the light from the high-intensity lamp to the sensitive plate, the illumination of which is a linear function of the exposed area of the aperture. The maximum area of the aperture is 4 m². The plate is illuminated by 1 candela (cd) for every square meter of aperture area. This luminosity is detected by the photocell, which provides an output voltage of 1 V/cd. The motor torque constant is 2 N·m/A, the motor inertia is 0.2 kg·m², and the motor back-emf constant is 1 V·s. The viscous friction is 0.2 N·m·s, and the amplifier gain is 50 V/V. (a) Draw the block diagram of the system with the appropriate transfer function inserted in each block. (b) Derive the overall transfer function.

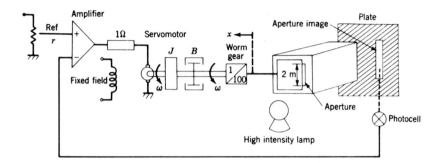

5.5. The longitudinal motion of an aircraft is represented by the vector differential equation

$$\dot{\mathbf{x}} = \begin{bmatrix} -0.09 & 1 & -0.02 \\ -8.0 & -0.06 & -6.0 \\ 0 & -0 & -10 \end{bmatrix} \mathbf{x} + \begin{bmatrix} 0 \\ 0 \\ 10 \end{bmatrix} \delta_E$$

where x_1 = angle of attack

x_2 = rate of change of pitch angle

x_3 = incremental elevator angle

δ_E = control input into the elevator actuator

Derive the transfer function relating the system output, rate of change of pitch angle, to the control input into the elevator actuator, x_2/δ_E.

5.6. (a) Use the hydraulic valve and power piston in a closed-loop position control system, and derive the transfer function of the system. (b) Draw a diagram of a control system for the elevators on an airplane. Use a hydraulic actuator.

5.7. In the diagram $[q_c]$ represents compressibility flow, and the pressure p is the same in both cylinders. Assume there is no leakage flow around the pistons. Draw a block diagram that relates the output $Y(s)$ to the input $X(s)$.

5.8. The dynamic equations that describe an aircraft in the landing configuration are as follows:

$$\dot{h}(t) = b_{33}h(t) + b_{32}\theta(t)$$

$$\ddot{\theta}(t) = b_{11}\dot{\theta}(t) + b_{12}\theta(t) + b_{13}h(t) + c_{11}\delta_e(t)$$

where the coefficients are defined as follows:

$$b_{11} = \frac{1}{T_s} - 2\zeta\omega_s, \quad b_{13} = \frac{1}{VT_s^2} - \frac{2\zeta\omega_s}{VT_s} + \frac{\omega_s^2}{V}, \quad b_{33} = -\frac{1}{T_s}$$

$$b_{12} = \frac{2\zeta\omega_s}{T_s} - \omega_s^2 - \frac{1}{T_s^2}, \quad b_{32} = \frac{V}{T_s}, \quad c_{11} = \omega_s^2 K_s T_s$$

These coefficients are constant for a given flight condition and aircraft configuration. The parameters are defined as follows:

K_s = short-period aircraft gain T_s = flight path time constant
ω_s = short-period natural frequency h = altitude
ζ = short-period damping ratio θ = pitch angle

(*a*) Draw a detailed block diagram that explicitly shows each time-varying quantity and the constant coefficients (left in terms of the *b*'s and *c*'s). The input is $\delta_e(t)$ and the output is $h(t)$. (*b*) Determine the transfer function $G(s) = H(s)/\Delta_e(s)$.

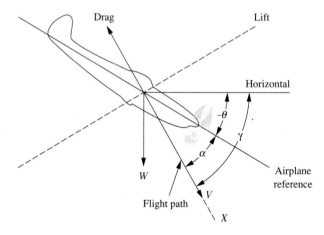

5.9. A MIMO digital robot arm control system (see Fig. 1.5) may be decomposed into sets of equivalent MISO control systems by the technique shown in Sec. 18.5. Figure 18.5 represents one of these equivalent MISO systems, where $r(t)$ is the input that must be tracked, $d_1(t)$ is the unwanted disturbance input, $d_2(t) = 0$, and

$$F(s) = \frac{1.020833(s + 12)}{s^2 + 6.3s + 12.25}$$

$$G(s) = \frac{823768.2(s + 8)(s + 444)}{(s + 100)(s + 140 \pm j279.1)}$$

$$P(s) = \frac{124.498}{s(s + 444)(s - 2.9811)}$$

The specifications, based on unit-step tracking and disturbance inputs, are: for the tracking input $y_r(t_p) \leq 1.1$ and $t_s \leq 1.5s$, for the disturbance input $|y_d(t_p)| \leq 0.1$, and for the total output $y(t_p) \leq 1.1$ when both inputs are applied at the same time. (a) Derive the transfer functions $Y_r(s)/R(s)$ and $Y_d(s)/D(s)$; (b) by use of a CAD program, obtain the time responses $y_r(t)$, $y_d(t)$, and $y(t)$ for unit-step forcing functions. Have the specifications been satisfied?

5.10. For the control system shown, (1) the force of attraction on the solenoid is given by $f_c = K_c i_c$, where K_c has the units pounds per ampere. (2) The voltage appearing across the generator field is given by $e_f = K_x x$, where K_x has the units volts per inch and x is in inches. (3) When the voltage across the solenoid coil is zero, the spring K_s is unstretched and $x = 0$. (a) Derive all necessary equations relating all the variables in the system. (b) Draw a block diagram for the control system. The diagram should include enough blocks to indicate specifically the variables $I_c(s)$, $X(s)$, $I_f(s)$, $E_g(s)$, and $T(s)$. Give the transfer function for each block in the diagram. (c) Draw an SFG for this system. (d) Determine the overall transfer function. (e) Write the system state and output equations in matrix form.

5.11. Given

(1) $D^3 y + 5D^2 y + 12Dy + 8y = 2D^3 u + 6D^2 u + 12Du + 6u$
(2) $D^4 y + 14D^3 y + 71D^2 y + 154Dy + 120y = 4D^2 u + 8u$
(3) $D^3 y - 11D^2 y + 38Dy - 40y = 2D^2 u + 6Du + u$

Obtain state and output equations using (a) phase variables, (b) canonical variables.

5.12. Given

$$G(s) = \frac{b_1 s + b_0}{(s - \sigma)^2 + \omega_d^2}$$

write the state equation with the **A** matrix in modified canonical form so that all elements are real. Draw the SFG.

5.13. The state and output equations of a system are as follows:

$$(1)\ \dot{\mathbf{x}} = \begin{bmatrix} 0 & 1 & 0 \\ 0 & 0 & 1 \\ -24 & -26 & -9 \end{bmatrix}\mathbf{x} + \begin{bmatrix} 0 \\ 2 \\ -10 \end{bmatrix}\mathbf{u} \qquad \mathbf{y} = [1\ \ 0\ \ 0]\mathbf{x}$$

$$(2)\ \dot{\mathbf{x}} = \begin{bmatrix} -4 & 1 & 0 \\ 0 & -3 & 0 \\ 0 & 0 & -1 \end{bmatrix}\mathbf{x} + \begin{bmatrix} 0 \\ 2 \\ 2 \end{bmatrix}\mathbf{u} \qquad \mathbf{y} = [1\ \ 2\ \ 0]\mathbf{x}$$

(a) Draw a state transition SFG for these equations. (b) Determine the eigenvalues. (c) Determine the modal matrix **T** that uncouples the states. (d) Write the new state and output equations in matrix form. (e) Draw a simulation diagram for the new equations, and determine the uncontrollable or unobservable mode. (f) Determine the transfer function $Y(s)/U(s)$.

5.14. For the system of Prob. 4.14 (a) draw an SFG; (b) from the transfer function determine a new state equation and draw the SFG using phase variables; (c) change the state equation to normal form and draw the SFG.

5.15. Use the state and output equations of Prob. 3.13, (a) draw a simulation diagram for the given equations. (b) Determine the modal matrix **T**. Show that $\mathbf{T}^{-1}\mathbf{AT} = \mathbf{\Lambda}$. (c) Convert the equations using the similarity transformation $\mathbf{x} = \mathbf{Tz}$. (d) Draw the simulation diagram for (c).

5.16. A system has a state-variable equation $\dot{\mathbf{x}} = \mathbf{A}_p\mathbf{x} + \mathbf{B}_p u$, where **x** is the 3×1 state vector and u is the scalar control input.

$$\mathbf{A}_p = \begin{bmatrix} 0 & 2 & 0 \\ 0 & -1 & -1 \\ 0 & 0 & -4 \end{bmatrix} \qquad \mathbf{B}_p = \begin{bmatrix} 0 \\ 4 \\ 4 \end{bmatrix}$$

(a) Find a nonsingular matrix **T** that transforms the state variables into phase variables. Be sure to satisfy the $\mathbf{B}_c$ matrix for the phase-variable representation. (This will lead to a unique matrix **T**.) (b) Write the new state equation of the system in terms of the phase variables. (c) Draw a state-variable diagram representing the system.

5.17. Given

$$(1)\ \dot{\mathbf{x}} = \begin{bmatrix} 0 & 1 & 0 \\ 0 & 0 & 1 \\ -15 & -11 & -5 \end{bmatrix}\mathbf{x} + \begin{bmatrix} 0 \\ 0 \\ 1 \end{bmatrix}\mathbf{u}$$

$$(2)\ \dot{\mathbf{x}} = \begin{bmatrix} -2 & 0 & 2 \\ 0 & -1 & 0 \\ -2 & 0 & -2 \end{bmatrix}\mathbf{x} + \begin{bmatrix} 0 \\ 0 \\ 1 \end{bmatrix}\mathbf{u}$$

(a) Find a modified matrix $\mathbf{\Lambda}_m$ so that its elements are all real. (b) Determine the modified modal matrix $\mathbf{T}_m$. (c) Draw a simulation diagram for the transformed equations.

5.18. For the system described by the state equation

$$(1)\ \dot{\mathbf{x}} = \begin{bmatrix} 0 & 2 \\ 0 & -4 \end{bmatrix}\mathbf{x} + \begin{bmatrix} 1 \\ 2 \end{bmatrix}\mathbf{u}$$

(2) $\dot{\mathbf{x}} = \begin{bmatrix} -2 & -1 \\ 0 & 3 \end{bmatrix} \mathbf{x} + \begin{bmatrix} 0 \\ 10 \end{bmatrix} \mathbf{u}$

(a) Derive the system transfer function if $y = x_1$. (b) Draw an appropriate state-variable diagram. (c) For zero initial state and a unit-step input evaluate $x_1(t)$ and $x_2(t)$.

5.19. For $\dot{\mathbf{x}} = \mathbf{Ax} + \mathbf{Bu}$

(1) $\mathbf{A} = \begin{bmatrix} -2 & 0 & 0 \\ 1 & -3 & 0 \\ -1 & -2 & -5 \end{bmatrix}$ $\quad \mathbf{B} = \begin{bmatrix} 1 \\ 0 \\ 1 \end{bmatrix}$

(2) $\mathbf{A} = \begin{bmatrix} 0 & 1 & 0 \\ -3 & -3 & 0 \\ 0 & -3 & -1 \end{bmatrix}$ $\quad \mathbf{B} = \begin{bmatrix} 0 \\ 1 \\ 2 \end{bmatrix}$

Determine the matrix $\mathbf{T}$ so that the transformation $\mathbf{x} = \mathbf{Tz}$ produces (a) a state equation in control canonical form and (b) a state equation with a diagonalized plant matrix Λ.

5.20. Given the real symmetric matrix

$$\mathbf{A} = \begin{bmatrix} 3 & \sqrt{6} & \sqrt{6} \\ \sqrt{6} & -2 & 1 \\ \sqrt{6} & 1 & -2 \end{bmatrix}$$

Determine (a) the eigenvalues, (b) the eigenvectors, and (c) the modal matrix $\mathbf{T}$. Show that $\mathbf{T}^{-1}\mathbf{AT} = \Lambda$. Note that this matrix can be diagonalized even though there are repeated eigenvalues.

5.21. Repeat Prob. 5.20 for

(1) $\mathbf{A} = \begin{bmatrix} 0 & 1 & 0 \\ 0 & 0 & 1 \\ 9 & 9 & -1 \end{bmatrix}$

5.22. Determine the modal matrix $\mathbf{T}$ that will uncouple the states

(1) $\dot{\mathbf{x}} = \begin{bmatrix} -4 & -2 & 0 \\ 0 & 0 & 1 \\ 0 & -2 & -3 \end{bmatrix} \mathbf{x} + \begin{bmatrix} 1 \\ 0 \\ 1 \end{bmatrix} \mathbf{u}$

Find $\mathbf{x}(t)$ for $u(t) = u_{-1}(t)$. Use the method of Sec. 5.11 where $\mathbf{x}(0) = \mathbf{0}$.

5.23. Draw a simulation diagram for the state equations of Prob. 3.7 using physical variables.

5.24. Given the following system:

(*a*) Draw an equivalent signal flow graph. (*b*) Apply Mason's gain rule to find $C(s)/R(s)$. (*c*) Write the four differential and three algebraic equations that form the math model depicted by the block diagram above. Use only the signals $r(t)$, $e(t)$, $x(t)$, $y(t)$, $z(t)$, and $c(t)$ in these equations.

5.25. For the following system,

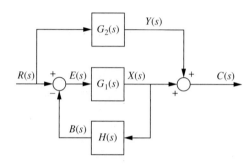

(*a*) Draw a signal flow graph. (*b*) Derive transfer functions for $E(s)/R(s)$, $C(s)/R(s)$, $X(s)/R(s)$, $B(s)/R(s)$, and $Y(s)/R(s)$.

5.26. For Prob. 5.25, given $G_1 = 1/(s + 1)$, $G_2 = 1/s$, and $H = s/(s^2 + 2s + 2)$, write the differential equations (*a*) relating x to e, (*b*) relating y to r, and (*c*) relating b to c. Then write the algebraic equations (*d*) relating e to r and b and (*e*) relating c to x and y.

5.27. The pump in the system shown below supplies the pressure $p_1(t) = 4u_{-1}(t)$. Calculate the resulting velocity $v_2(t)$ of the mass M_2. Include the effect of the mass of the power piston. (*a*) Neglect load reaction. (*b*) Include load reaction.

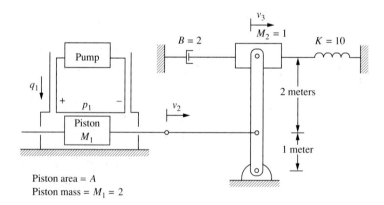

Piston area = A
Piston mass = $M_1 = 2$

Chapter 6

6.1. For Prob. 5.3, (*a*) use the Routh criterion to check the stability of the system. (*b*) Compute the steady-state tracking error of the system in response to a command input which is a ramp of 0.0317°/s. (c) Repeat (b) with the gain of the tracking receiver increased from 5 to 40.

6.2. For each of the following cases, determine the range of values of K for which the response $c(t)$ is stable, where the driving function is a step function. Determine the roots on the imaginary axis that yield sustained oscillations.

(a) $C(s) = \dfrac{K}{s[s(s + 2)(s^2 + 4s + 20) + K]}$

(b) $C(s) = \dfrac{K(s + 2)}{s[s^2(s^2 + 2s + 2) + K(s + 2)]}$

(c) $C(s) = \dfrac{20K}{s[s(s + 1)(s + 5) + 20K]}$

(d) $C(s) = \dfrac{K(s + 4)}{s[s(s + 1)(s + 2)(s + 5) + K(s + 4)]}$

(e) For Prob. 5.24.

6.3. Factor the following equations:

(a) $s^3 + 6.4s^2 + 18.48s + 19.36 = 0$
(b) $s^4 + 2s^2 + 9 = 0$
(c) $s^3 + 6s^2 + 15.25s + 18.75 = 0$
(d) $s^3 + 9s^2 + 26s + 24 = 0$
(e) $s^4 + 10s^3 + 50s^2 + 120s + 144 = 0$
(f) $s^4 + 3s^3 - 15s^2 - 19s + 30 = 0$

(g) $s^5 + 3s^4 + 28s^3 + 226s^2 + 600s + 400 = 0$

6.4. Use Routh's criterion to determine the number of roots in the right-half s plane for the equations of Prob. 6.3.

6.5. A unity-feedback system has the forward transfer function

$$G(s) = \frac{K(0.2s + 1)}{s(s + 2)(s + 4)}$$

In order to obtain the best possible ramp error coefficient, the highest possible gain K is desirable. Do stability requirements limit this choice of K?

6.6. The equation relating the output $y(t)$ of a control system to its input is

(a) $[s^5 + 2s^4 + 3s^3 + 2s^2 + (K + 1)s + (K + 2)]Y(s) = 5X(s)$
(b) $[s^3 - 7s^2 + 10s + 2 + K(s + 2)]Y(s) = X(s)$

Determine the range of K for stable operation of the system. Consider both positive and negative values of K.

6.7. The output of a control system is related to its input $r(t)$ by

$$[s^4 + 4s^3 + 2s^2 + (2 + K)s + K]C(s) = K(s + 1)R(s)$$

Determine the range of K for stable operation of the system. Consider both positive and negative values of K. Find K that yields imaginary roots.

6.8.

$$F(s) = \frac{1}{s[(s^4 + 0.5s^3 + 4.5s^2 + s + 2) + K(s + 1)]}$$

K is real but may be positive or negative. (*a*) Find the range of values of *K* for which the time response is stable. (*b*) Select a value of *K* that will produce imaginary poles for $F(s)$. Find these poles. What is the physical significance of imaginary poles as far as the time response is concerned?

6.9. For the system shown, (*a*) find $C(s)/R(s)$. (*b*) What type of system does $C(s)/E(s)$ represent? (*c*) Find the step-, ramp-, and parabolic-error coefficients. (*d*) Find the steady-state value of $c(t)$, $e(t)$, and $m(t)$ if $r(t) = 4u_{-1}(t)$, $K = 1$, and $A = 1$. (*e*) A steady-state error less than 0.5 is required. Determine the values for *A* and *K* that meet this condition. What effect does *K* have on the steady-state error? (*f*) For parts (*d*) and (*e*), is the closed-loop system stable?

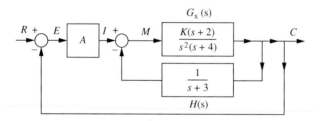

6.10. For a unity-feedback control system:

$$(1)\ \ G(s) = \frac{15(s + 4)}{s(s + 1)(s + 2)(s^2 + 2s + 10)} \qquad (2)\ \ G(s) = \frac{800(s + 1)}{s^2(s + 10)(s^2 + 10s + 50)}$$

(*a*) Determine the step-, ramp-, and parabolic-error coefficients (K_p, K_v, and K_a) for this system. (*b*) Using the appropriate error coefficients from part (*a*), determine the steady-state actuating signal $e(t)_{ss}$ for $r_1(t) = (16 + 2t)u_{-1}(t)$ and for $r_2(t) = 5t^2 u_{-1}(t)$. (*c*) Is the closed-loop system stable? (*d*) Use the results of part (*b*) to determine $c(t)_{ss}$.

6.11. (*a*) Determine the open-loop transfer function $G(s)H(s)$ of Fig. (*a*). (*b*) Determine the overall transfer function. (*c*) Figure (*b*) is an equivalent block diagram of Fig. (*a*). What must the transfer function $H_x(s)$ be in order for Fig. (*b*) to be equivalent to Fig. (*a*)? (*d*) Figure (*b*) represents what type of system? (*e*) Determine the system error coefficients of Fig. (*b*). (*f*) If $r(t) = u_{-1}(t)$, determine the final value of $c(t)$. (*g*) What are the values of $e(t)_{ss}$ and $m(t)_{ss}$?

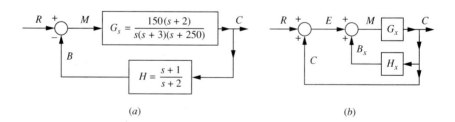

(*a*)   (*b*)

6.12. Find the step-, ramp-, and parabolic-error coefficients for unity-feedback systems that have the following forward transfer functions:

(a) $G(s) = \dfrac{20}{(0.2s + 1)(0.5s + 1)}$

(b) $G(s) = \dfrac{200}{s^2(s^2 + 4s + 4)(s^2 + 10s + 25)}$

(c) $G(s) = \dfrac{64(s + 4)}{s(s^2 + 4s + 8)}$

(d) $G(s) = \dfrac{48(s + 3)}{s(s + 6)(s^2 + 4s + 4)}$

(e) $G(s) = \dfrac{20(s + 4)}{(s + 2)(s^2 + 4s + 20)}$

6.13. For Prob. 6.12, find $e(\infty)$ by use of the error coefficients, with the following inputs:

  (a) $r(t) = 5$ (b) $r(t) = 2t$ (c) $r(t) = t^2$

6.14. A unity-feedback control system has

  (1) $G(s) = \dfrac{20K}{s[(s + 1)(s + 5) + 20]}$ (2) $G(s) = \dfrac{K(s + 2.5)}{s(s + 2)(s + 5)(s + 8)}$

  where $r(t) = 2t$. (a) If $K = 1.5$, determine $e(t)_{ss}$. (b) It is desired that for a ramp input $e(t)_{ss} \leq 1.5$. What minimum value must K_1 have for this condition to be satisfied? (c) For the value of K_1 determined in part (b), is the system stable?

6.15. (a) Derive the ratio $G(s) = C(s)/E(s)$. (b) Based upon $G(s)$, the figure has the characteristic of what type of system? (c) Derive the control ratio for this control system. (d) Repeat with the minor loop feedback replaced by K_h. (e) What conclusion can be reached about the effect of minor loop feedback on system type? (f) For (c) and (d) determine $G_{eq}(s)$.

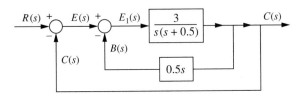

6.16. A unity-feedback system has the forward transfer function

$$G(s) = \dfrac{K}{s(s + 2)(s + 2.5)}$$

The input $r(t) = 6 + 8t$ is applied to the system. (a) It is desired that the steady-state value of the error be equal to or less than 1.6 for the given input function. Determine the minimum value that K_1 must have to satisfy this requirement. (b) By use of Routh's stability criterion, determine whether the system is stable for the minimum value of K_1 determined in part (a).

6.17. The angular position θ of a radar antenna is required to follow a command signal θ_c. The command signal e_c is proportional to θ_c with a proportionality constant K_θ V/deg. The positioning torque is applied by an ac motor that is activated by a position error signal e. The shaft of the electric motor is geared to the antenna with a gear ratio n. A potentiometer having a proportionality constant K_θ V/deg produces a feedback signal e_θ proportional to θ. The actuating signal $e = e_c - e_\theta$ is amplified with gain K_a V/V to produce the voltage e_α that drives the motor. The amplifier output e_a is an ac voltage. With the motor shaft clamped, the motor stall torque measured at the motor shaft is proportional to e_a with a proportionality constant K_c in lb/V. The motor torque decreases linearly with increasing motor speed $\dot{\theta}_m$ and is zero at the no-load speed $\dot{\theta}_0$. The no-load speed is proportional to e_a with proportionality constant K_0. The moment of inertia of the motor, gearing, and antenna, referred to the output θ, is J and the viscous friction B is negligible. The motor torque is therefore $T = J\ddot{\theta} + T_L$, where T_L is a wind torque. (a) Draw a completely labeled detailed block diagram for this system. (b) Neglecting T_L, find the open-loop transfer function $\theta(s)/E(s)$ and the closed-loop transfer function $\theta(s)/\theta_c(s)$. (c) For a step input θ_c, describe the response characteristics as the gain K_a is increased. (d) For a constant wind torque T_L and $\theta_c(t) = 0$ find the steady-state position error as a function of K_a.

6.18. The components of a single-axis stable platform with direct drive can be described by the following equations:

Controlled platform: $T = JD\omega + B\omega + T_L$

Integrating gyro: $De_{ig} = K_{ig}(\omega_c - \omega + \omega_d + \omega_b)$

Servomotor: $T = (K_a/K_{ig})e_{ig}$

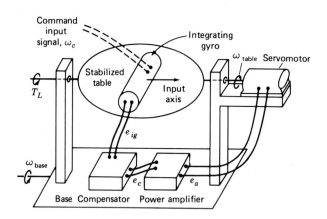

where J = moment of platform and servomotor

   B = damping

   ω = angular velocity of platform with respect to inertial frame of reference

T = servomotor torque
T_L = interfering torque
e_{ig} = gyro output voltage
ω_c = command angular velocity
ω_d = drift rate of gyro
ω_b = angular velocity of the base
K_{ig} = gyro gain
K_a/K_{ig} = amplifier gain

Consider that the base has a given orientation and that $\omega_b = 0$. (a) Show that the overall equation of performance for the stable platform is

$$(JD^2 + BD + K_a)\omega = K_a(\omega_c + \omega_d) - DT_L$$

(b) Draw a detailed block diagram representing the system. (c) Using e_{ig} and ω as state variables, write the state equations of the system. (d) With $\omega_c = 0$ and $T_L = 0$ determine the effect of a constant drift rate ω_d. (e) With $\omega_c = 0$ and $\omega_d = 0$, determine the effect of a constant interfering torque T_L on ω.

6.19. In the feedback system shown, K_h is adjustable and

$$G(s) = \frac{K_G(s^w + \cdots + c_0)}{s^2 + \cdots + a_1 s + a_0}$$

The output is required to follow a step input with no steady-state error. Determine the necessary conditions on K_h for a stable system.

6.20. For Prob. 2.25, (a) derive the transfer function $G_x(s) = \Omega(s)/V_1(s)$. Figure (a) shows a velocity control system, where K_x is the velocity sensor coefficient. The equivalent unity-feedback system is shown in Fig. (b), where $\Omega(s)/R(s) = K_x G_x(s)/[1 + K_x G_x(s)]$. (b) What system type does $G(s) = K_x G_x(s)$ represent? (c) Determine $\omega(t)_{ss}$, when $r(t) = R_0 u_{-1}(t)$. Assume the system is stable.

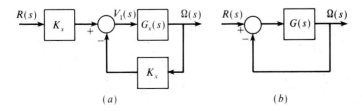

(a)   (b)

6.21. Given the following system, find (*a*) $C(s)/R(s)$, (*b*) the range of values of K for which the system is stable; (*c*) determine $G_{eq}(s)$, and (*d*) the equivalent system type.

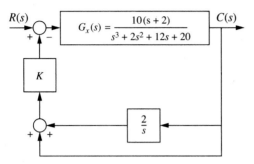

6.22. Given the following system, find $C(s)/R(s)$. (*a*) Find the range of values of K for which the system is stable. (*b*) Find the *positive* value of K that yields pure oscillations in the homogeneous system and find the frequency associated with these oscillations.

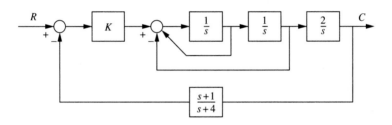

6.23. Based on this figure, do the following:

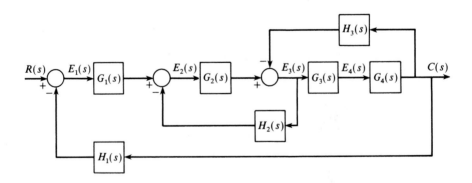

(*a*) By block diagram reduction find $G(s)$ and $H(s)$ such that the system can be represented in this form:

(*b*) For this system find $G_{eq}(s)$ for an equivalent unity-feedback system of this form:

6.24. By block diagram manipulations or the use of SFG simplify the following systems to the form of Fig. 6.1.

(1)

(2)

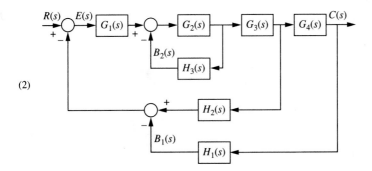

6.25. A unity-feedback system has the forward transfer function

$$G(s) = \frac{20K}{s(s^2 + 10s + 14 + K)}$$

(a) What is the maximum value of K for stable operation of the system? (b) For K equal to one-half of the maximum value, what is the $e(t)_{ss}$ for the ramp input $r(t) = 2tu_{-1}(t)$?

6.26. A unity-feedback system has the forward transfer function

$$G(s) = \frac{K(s + 30)}{(s + 1)(s^2 + 20s + 116)}$$

(a) What is the maximum value of K for stable operation of the system? (b) For a unit-step input, $e(t)_{ss}$ must be less than 0.1 for a stable system. Find K_m to satisfy this condition. (c) For the value of K_m of part (b), determine $c(t)$ and the figures of merit by use of a CAD package.

Chapter 7

7.1. Determine the pertinent geometrical properties and *sketch* the root-locus for the following transfer functions for both positive and negative values of K:

(a) $G(s)H(s) = \dfrac{K}{s(s + 1)^2}$

(b) $G(s)H(s) = \dfrac{K(s - 1)}{(s + 1)(s^2 + 8s + 20)}$

(c) $G(s)H(s) = \dfrac{K(s + 4)}{s(s^2 + 4s + 29)}$

(d) $G(s)H(s) = \dfrac{K}{s^2(s + 2)(s + 3)}$

(e) $G(s)H(s) = \dfrac{K_0}{(1 + 0.5s)(1 + 0.2s)(1 + 0.125s)^2}$

(f) $G(s)H(s) = \dfrac{K(s + 2)^2}{s(s^2 + 4s + 20)}$

(g) $G(s)H(s) = \dfrac{K(s^2 + 6s + 34)}{(s + 2)(s^2 + 2s + 2)}$

Determine the range of values of K for which the closed-loop system is stable.

7.2. (CAD Problem) Obtain the root locus of the following for a unity-feedback system with $K \geq 0$:

(a) $G(s) = \dfrac{K}{(1 + s)(0.25 + s)^2}$

(b) $G(s) = \dfrac{K}{s(s^2 + s + 0.24)(s + 4)}$

(c) $G(s) = \dfrac{K}{(s^2 + 4)(s^2 + 8s + 20)}$

(d) $G(s) = \dfrac{K(s + 2)}{s^2(s^2 + 2s + 2)}$

(e) $G(s) = \dfrac{K}{s(s^2 + 2s + 2)(s^2 + 6s + 13)}$

(f) $G(s) = \dfrac{K(s + 15)}{(s + 2)(s^2 + 12s + 100)(s + 10)}$

(g) $G(s) = \dfrac{K}{s(s^2 + 10s + 29)}$

Does the root locus cross any of the asymptotes?

7.3. A system has the following transfer functions:

$$G(s) = \dfrac{K}{s(s^2 + 4s + 8)} \qquad H(s) = 1 \qquad K > 0$$

(a) Plot the root locus. (b) A damping ratio of 0.58 is required for the dominant roots. Find $C(s)/R(s)$. The denominator should be in factored form. (c) With a unit step input, find $c(t)$. (d) Plot the magnitude and angle of $\mathbf{C}(j\omega)/\mathbf{R}(j\omega)$ vs. ω.

7.4. (CAD Problem) A feedback control system with unity feedback has a transfer function

$$G(s) = \dfrac{100K(s + 10)}{s(s^2 + 10s + 29)}$$

(a) Plot the locus of the roots of $1 + G(s) = 0$ as the loop sensitivity $K(> 0)$ is varied. (b) The desired figures of merit for this system are $M_p = 1.0432, t_p \leq 1.8s$, and $t_s \leq 2.27s$. Based on these values, determine the dominant closed-loop poles that the control ratio must have in order to be able to achieve these figures of merit. Are these closed-loop poles achievable? If so, determine the value of K. (c) For the value of K_1 found in part (b), determine $e(t)$ for $r(t) = u_{-1}(t)$ and determine the actual values of the figures of merit that are achieved. (d) Plot the magnitude and angle of $M(j\omega)$ vs. ω for the closed-loop system.

7.5. (a) Sketch the root locus for the control system having the following open-loop transfer function. (b) Calculate the value of K_1 that causes instability. (c) Determine $C(s)/R(s)$ for $\zeta = 0.3$. (d) Use a CAD package to determine $c(t)$ for $r(t) = u_{-1}(t)$.

(1) $\quad G(s) = \dfrac{K_1(1 + 0.05s)}{s(1 + 0.0714s)(1 + 0.1s + 0.0125s^2)} \qquad$ (2) $\quad G(s) = \dfrac{K(s - 5)(s + 4)}{s(s + 1)(s + 3)}$

7.6. (a) Sketch the root locus for a control system having the following forward and feedback transfer functions:

$$G(s) = \dfrac{K_2(1 + s/5)}{s^2(1 + s/12)} \qquad H(s) = K_H(1 + s/15)$$

(b) Express $C(s)/R(s)$ in terms of the unknown gains K_2 and K_H. (c) Determine the value of K_H that will yield $c(\infty) = 1$ for $r(t) = u_{-1}(t)$. (d) Choose closed-loop pole locations for a $\zeta = 0.5$. (e) Write the factored form of $C(s)/R(s)$. (f) Determine the figures of merit M_p, t_p, t_s, and K_m.

7.7. A unity-feedback control system has the transfer function

$$G(s) = \frac{K}{(s + 1)(s + 7)(s^2 + 8s + 25)}$$

(a) Sketch the root locus for positive and negative values of K. (b) For what values of K does the system become unstable? (c) Determine the value of K for which all the roots are equal. (d) Determine the value of K from the root locus that makes the closed-loop system a perfect oscillator.

7.8. (CAD Problem) A unity-feedback control system has the transfer function

$$(1) \ G(s) = \frac{K_1(1 - 0.5s)}{(1 + 0.5s)(1 + 0.2s)(1 + 0.1s)}$$

$$(2) \ G(s) = \frac{K_1}{s(1 + 0.1s)(1 + 0.08s + 0.008s^2)}$$

(a) Obtain the root locus for positive and negative values of K_1. (b) What range of values of K_1 makes the system unstable?

7.9. (CAD problem) For positive values of gain, obtain the root locus for unity-feedback control systems having the following open-loop transfer functions. For what value or values of gain does the system become unstable in each case?

$$(a) \ G(s) = \frac{K(s^2 + 4s + 13)}{s(s + 1)(s + 5)^2}$$

$$(b) \ G(s) = \frac{K(s^2 + 4s + 8)}{s^2(s + 1)(s + 4)}$$

$$(c) \ G(s) = \frac{K(s + 1)^2}{s(s^2 - 3s + 4.5)}$$

7.10. (CAD Problem) For each system of Prob. 7.9 a stable system with $M_p = 1.14$ is desired. If this value of M_p is achievable, determine the control ratio $C(s)/R(s)$. Select the *best* roots in each case.

7.11. A nonunity-feedback control system has the transfer functions

$$G(s) = \frac{K_G(1 + s/3.2)}{(1 + s/5)[1 + 2(0.7)s/23 + (s/23)^2][1 + 2(0.49)s/7.6 + (s/7.6)^2]}$$

$$H(s) = \frac{K_H(1 + s/5)}{1 + 2(0.89)s/42.7 + (s/42.7)^2}$$

(a) Sketch the root locus for the system, using as few trial points as possible. Determine the angles and locations of the asymptotes and the angles of departure of the branches from the open-loop poles. (b) Determine the gain $K_G K_H$ at which the system becomes unstable, and determine the approximate locations of all the closed-loop poles for this value of gain. (c) Using the closed-loop configuration determined in part (*b*), write the expression for the closed-loop transfer function of the system.

7.12. (CAD Problem) A nonunity-feedback control system has the transfer functions

$$(1)\ G(s) = \frac{10A(s^2 + 8s + 20)}{s(s + 4)} \qquad H(s) = \frac{0.2}{s + 2}$$

$$(2)\ G(s) = \frac{K(s^2 + 6s + 13)}{s(s + 3)} \qquad H(s) = \frac{1}{s + 1}$$

For system (1): (*a*) determine the value of the amplifier gain A that will produce complex roots having the *minimum* possible value of ζ. (*b*) Express $C(s)/R(s)$ in terms of its poles, zeros, and constant term.

For system (2) an $M_p = 1.0948$ is specified: (*a*) determine the values of K that will produce complex roots for a ζ corresponding to this value of M_p. Note that there are two intersections of the ζ-line. (*b*) Express $C(s)/R(s)$ in terms of its poles, zeros, and constant term for each intersection. (*c*) For each control ratio determine the figures of merit. Compare these values. Note the differences, even though the pair of complex poles in each case has the same value of ζ.

7.13. For the figure shown, only the forward and feedback gains, K_x and K_h, are adjustable. (*a*) Express $C(s)/R(s)$ in terms of the poles and zeros of $G_x(s)$ and $H(s)$. (*b*) Determine the value K_h must have in order that $c(\infty) = 1.0$ for a unit step input. (*c*) Determine the value of K_xK_h that will produce complex roots having a $\zeta = 0.65$. (*d*) Determine the corresponding $C(s)/R(s)$. (*e*) Determine $G_{eq}(s)$ for a unity feedback system that is equivalent to this nonunity feedback system. (*f*) Determine the figures of merit M_p, t_p, t_s, and K_m.

$$G_x = \frac{K_x}{s(s + 1)(s + 5)} \qquad H_s = \frac{K_h(s + 5)}{s + 2}$$

7.14. Yaw rate feedback and sideslip feedback have been added to an aircraft for turn coordination. The transfer functions of the resulting aircraft and the aileron servo are, respectively,

$$\frac{\dot{\phi}(s)}{\delta_a(s)} = \frac{15}{s + 2} \quad \text{and} \quad \frac{\delta_a(s)}{e_{\delta_a}(s)} = \frac{20}{s + 20}$$

For the bank-angle control system shown in the figure, design a suitable system by drawing the root-locus plots on which your design is based. (*a*) Use $\zeta = 0.8$ for the inner loop and determine K_{rrg}. (*b*) Use $\zeta = 0.7$ for the outer loop and determine K_{vg}. (*c*) Determine $\phi(s)/\phi_{comm}(s)$.

7.15. A heading control system using the bank-angle control system is shown in the following figure, where g is the acceleration of gravity and V_T is the true airspeed. Use the design of Prob. 7.14 for $\phi(s)/\phi_{comm}(s)$. (*a*) Draw a root locus for this system and use $\zeta = 0.6$ to determine K_{dg}. (*b*) Determine $\psi(s)/\psi_{comm}(s)$ and the time response with a step input for $V_T = 500$ ft/s. (The design can be repeated using $\zeta = 1.0$ for obtaining K_{rrg}.)

7.16. (CAD Problem) The simplified open-loop transfer function for a T-38 aircraft at 35,000 feet and 0.9 Mach is

$$G(s) = \frac{\theta(s)}{\delta_e(s)} = \frac{22.3K_x(0.995s + 1)}{(s + 50)(s^2 + 1.76s + 7.93)}$$

The open-loop transfer function contains the simplified longitudinal aircraft dynamics (a second-order system relating pitch rate to elevator deflection). The $K_x/(s + 50)$ term models the hydraulic system between the pilot's control and the elevator. Obtain a plot of the root locus for this system. Select characteristic roots that have a damping ratio of 0.4. Find the gain value K_x that yields these roots. Determine the system response to an impulse. What is the minimum value of damping ratio available at this flight condition?

Chapter 8

8.1. For each of the transfer functions

$$(1)\ G(s) = \frac{1}{s(1 + 0.5s)(1 + 0.4s)}$$

$$(2)\ G(s) = \frac{2}{s(1 + 0.1s)(1 + 0.024s + 0.0016s^2)}$$

$$(3)\ G(s) = \frac{32(s + 2)}{s(s^2 + 4s + 16)}$$

$$(4)\ G(s) = \frac{40(1 - 0.2s)}{s(1 + s)(1 + 0.2s)}$$

(*a*) draw the log magnitude (exact and asymptotic) and phase diagrams; (*b*) determine ω_ϕ and ω_c for each transfer function.

8.2. For each of the transfer functions

$$(1)\ G(s)\ =\ \frac{4(1 + 0.5s)}{s^2(1 + 0.2s)(1 + 0.05s)} \qquad (2)\ G(s)\ =\ \frac{10}{(1 + s)(1 + 4s)^2}$$

$$(3)\ G(s)\ =\ \frac{2}{s^2(1 + 0.3s)(1 + 0.4s)} \qquad (4)\ G(s)\ =\ \frac{10(1 + 0.125s)}{s(1 + 0.5s)(1 + 0.25s)}$$

(*a*) Draw the log magnitude (exact and asymptotic) and phase diagrams.

8.3. (CAD Problem) Plot to scale the log magnitude and angle vs. log ω curves for $\mathbf{G}(j\omega)$. Is it an integral or a derivative compensating network?

$$(a)\ \mathbf{G}(j\omega)\ =\ \frac{10(1 + j\omega)}{1 + j4\omega} \qquad (b)\ \mathbf{G}(j\omega)\ =\ \frac{10(1 + j12\omega)}{1 + j2\omega}$$

8.4. A control system with unity feedback has the forward transfer function, with $K_m = 1$,

$$(1)\ G(s)\ =\ \frac{K_1(1 + 0.2s)}{s(1 + 0.1s)(1 + s/10\sqrt{2} + s^2/400)}$$

$$(2)\ G(s)\ =\ \frac{K_0(1 + 0.4s)}{(1 + s)(1 + 0.2s)(1 + 0.5s)^2}$$

(*a*) Draw the log magnitude and angle diagrams. (*b*) Draw the polar plot of $\mathbf{G}'(j\omega)$. Determine all key points of the curve.

8.5. A system has

$$G(s)\ =\ \frac{5(1 + T_2s)(1 + T_3s)}{s^2(1 + T_1s)^2(1 + T_4s)}$$

where $T_1 = 2$, $T_2 = 0.5$, $T_3 = 0.25$, $T_4 = 0.125$. (*a*) Draw the asymptotes of Lm $\mathbf{G}(j\omega)$ on a decibel vs. log ω plot. Label the corner frequencies on the graph. (*b*) What is the *total* correction from the asymptotes at $\omega = 2$?

8.6. (*a*) What characteristic must the plot of magnitude in decibels vs. log ω possess if a velocity servo system (ramp input) is to have no *steady-state velocity error* for a constant velocity input $dr(t)/dt$? (*b*) What is true of the corresponding phase-angle characteristic?

8.7. Explain why the phase-angle curve cannot be calculated from the plot of $|\mathbf{G}(j\omega)|$ in decibels vs. log ω if some of the factors are not minimum phase.

8.8. The asymptotic gain vs. frequency curve of the open-loop minimum-phase transfer function is shown for a unity-feedback control system. (*a*) Evaluate the open-loop transfer function. Assume only first-order factors. (*b*) What is the frequency at which $|\mathbf{G}(j\omega)|$ is unity? What is the phase angle at this frequency? (*c*) Draw the polar diagram of the open-loop control system.

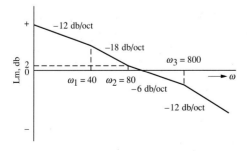

8.9. For each plot shown, (*a*) evaluate the transfer function; (*b*) find the correction that should be applied to the straight-line curve at $\omega = 4$; (*c*) determine K_m for Cases I and II. Assume only first-order factors.

Case I

Case II

Case III

8.10. Determine the value of the error coefficient from parts (*a*) and (*b*) of the figure.

(a)   (b)

8.11. An experimental transfer function gave, respectively, the following results:

Plant 1

ω	Lm G($j\omega$), dB	Angle, deg
0.20	−18.57	−91.4
0.60	−27.97	−94.1
1.0	−32.11	−97.1
2.0	−36.65	−107.1
3.0	−37.34	−126.9
4.0	−38.13	−172.9
4.2	−38.93	−184.4
4.4	−40.02	−195.0
4.8	−42.68	−212.2
5.0	−44.09	−218.7
5.4	−46.83	−228.4
6.0	−50.55	−237.7
7.0	−55.72	−246.4
9.0	−63.50	−254.3
14.0	−76.04	−261.1
20.0	−85.69	−264.0
30.0	−96.44	−266.1
49.0	−109.74	−267.7

Plant 2

ω	Lm G($j\omega$), dB	Angle, deg
0.20	1.97	−177.7
0.40	−9.98	−175.5
1.00	−25.33	−170.3
2.40	−38.75	−167.6
3.40	−43.87	−170.5
3.80	−45.52	−172.0
4.00	−46.29	−172.9
4.40	−47.74	−174.6
4.80	−49.09	−176.3
5.00	−49.73	−177.2
6.00	−52.64	−181.4
8.00	−57.47	−189.3
10.00	−61.44	−196.1
15.00	−69.20	−209.5
20.00	−75.18	−219.4
40.00	−91.10	−240.6
50.00	−96.58	−245.9
59.80	−101.06	−249.6

(*a*) Determine the transfer function represented by the above data. (*b*) What type of system does it represent?

8.12. Determine the transfer function by using the straight-line asymptotic log plot shown and the fact that the correct angle is −167.3° at $\omega = 0.8$. Assume a minimum-phase transfer function.

8.13. Determine whether each system shown is stable or unstable in the absolute sense by sketching the *complete* Nyquist diagrams. $H(s) = 1$.

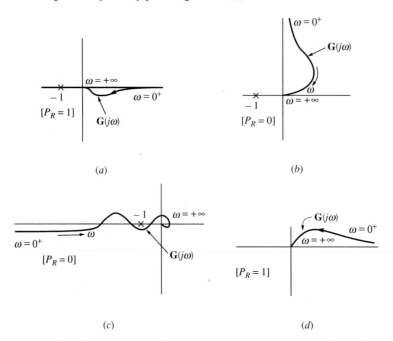

(a)

(b)

(c)

(d)

8.14. Use the Nyquist stability criterion and the polar plot to determine, with the aid of a CAD package, the range of values K (positive or negative) for which the closed-loop system is stable.

$$(a)\ G(s)H(s) = \frac{K(s + 1)^2(s + 2)}{s^3(s + 10)}$$

$$(b)\ G(s)H(s) = \frac{K}{s^2(-1 + 5s)(5 + s)}$$

$$(c)\ G(s)H(s) = \frac{K(4 - s)}{s^2(1 - 0.5s)(1 + 0.1s)}$$

$$(d)\ G(s)H(s) = \frac{K(s + 1)}{s^2(s + 15)(s^2 + 6s + 10)}$$

$$(e)\ G(s)H(s) = \frac{K}{s(s^2 + 2s + 16)}$$

8.15. For the following transfer functions, sketch a direct Nyquist locus to determine the closed-loop stability. Determine, with the aid of a CAD package, the range of values of K that produce stable closed-loop operation and those which produce unstable closed-loop operation. $H(s) = 1$.

$$(a)\ G(s) = \frac{K}{2 + s} \qquad (b)\ G(s) = \frac{K}{s^2(s - 2)}$$

$(c)\ G(s)\ =\ \dfrac{K(1+0.5s)}{s^2(8+s)}$ $\qquad$ $(d)\ G(s)\ =\ \dfrac{-K(s+1)}{s(1-s)(6+s)}$

$(e)\ G(s)\ =\ \dfrac{K}{s(s+3)(s-1)(s+6)}$

8.16. (CAD Problem) For the control systems having the transfer functions

$(1)\ G(s)\ =\ \dfrac{K_1}{s(1+0.02s)(1+0.05s)(1+0.10s)}$

$(2)\ G(s)\ =\ \dfrac{K(1+0.25s)}{(1+0.1s)(2+3s+s^2)}$

$(3)\ G(s)\ =\ \dfrac{K_0(1+0.333s)}{s(1+0.1s+0.001s^2)}$

$(4)\ G(s)\ =\ \dfrac{K(s+1)}{(s+2)(s^2+4s+5)}$

determine from the logarithmic curves the required value of K_m and the phase-margin frequency so that each system will have (a) a positive phase margin angle of 45°; (b) a positive phase margin angle of 60°. (c) From these curves determine the maximum permissible value of K_m for stability.

8.17. A system has the transfer functions

$$G(s)\ =\ \frac{K}{s(s+4)(s^2+12s+136)} \qquad H(s)\ =\ 1$$

(a) Draw the log magnitude and phase diagram of $G'(j\omega)$. Draw both the straight-line and the corrected log magnitude curves. (b) Draw the log magnitude–angle diagram. (c) Determine the maximum value of K_1 for stability.

8.18. What gain values would just make the systems in Prob. 8.2 unstable?

8.19. Use the Nyquist criterion to determine the maximum value of K_m for stability of the closed-loop systems having the following transfer functions:

$(a)\ G(s)H(s)\ =\ \dfrac{K_1 e^{-0.5s}}{s(1+s)(1+0.2s)}$ $\qquad$ $(b)\ G(s)H(s)\ =\ \dfrac{Ke^{-3s}}{(s+2)^2(s+1)}$

(c) The transfer functions of Prob. 8.16 with the transport lag $e^{-1.5s}$ included in the numerator.

8.20. Use the Nyquist criterion to determine the maximum value of T for stability of the closed-loop system which has the open-loop transfer function

$$G(s)H(s)\ =\ \frac{2e^{-Ts}}{s(s^2+6s+25)}$$

Hint: Determine the frequency ω_x for which $|G(j\omega)H(j\omega)| = 1$. What additional angle ωT, due to the transport lag, will make the angle of $G(j\omega_x)H(j\omega_x)$ equal to $-180°$?

8.21. A remotely controlled control system includes a transport lag in transmitting the actuating signal to the system and for the feedback signal to return. Tabulated data for the open-loop frequency response of the system without transport is:

ω	Magnitude	Degrees	ω	Magnitude	Degrees
0.20	2.988	−18.17	1.90	1.060	−120.3
0.40	2.796	−35.47	2.00	0.992	−123.9
0.60	2.533	−51.30	3.00	0.531	−152.7
0.80	2.247	−65.49	4.00	0.307	−172.6
1.00	1.970	−83.88	5.00	0.190	−187.1
1.20	1.717	−89.33	6.00	0.124	−199.0
1.40	1.494	−99.35	7.00	0.085	−206.7
1.60	1.301	−108.36	8.00	0.060	−213.5
1.80	1.135	−116.50	10.00	0.033	−223.7

(a) Determine the gain margin, the phase margin frequency, and the phase margin angle without the transport lag. (b) Determine the gain margin, the phase margin angle, and the phase margin frequency with a transport lag of 0.3 s.

Chapter 9

9.1. For Prob. 8.1, find K_1, M_m, ω_c, ω_ϕ, and ω_m for a $\gamma = 45°$ by

(a) the direct-polar-plot method;

(b) the log magnitude–angle diagram;

(c) a computer program.

(d) Repeat (a) through (c) by determining the value of K_1 for an $M_m = 1.26$.

(e) Compare the values of ω_m, ω_c, ω_ϕ, and K_1 obtained in (d) with those obtained in (a) and (b).

9.2. (CAD Problem) For the feedback control system of Prob. 7.4,

(a) determine, by use of the polar-plot method, the value of K_1 that just makes the system unstable;

(b) determine the value of K_1 that makes $M_m = 1.16$;

(c) for the value of K_1 found in part (b), find $c(t)$ for $r(t) = u_{-1}(t)$;

(d) obtain the data for plotting the curve of M vs. ω for the closed-loop system and plot this curve.

(e) Why do the results of this problem differ from those obtained in Prob. 7.4? [See Eq. (9.16).]

9.3. (CAD Problem) Determine the value K_1 must have for an $M_m = 1.13$.

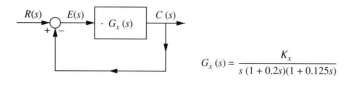

$$G_x(s) = \frac{K_x}{s(1 + 0.2s)(1 + 0.125s)}$$

9.4. Using the plot of $\mathbf{G}'(j\omega)$ shown, determine the number of values of gain K_m which produce the same value of M_m. Which of these values yields the best system performance? Give the reasons for your answer.

$$G(s) = \frac{K_1(1 + 0.15s)^2}{s(1 + s)^2(1 + 0.004s)^3}$$

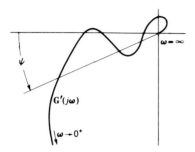

9.5. (CAD Problem) (*a*) Determine the values of M_m and ω_m for each of the transfer functions of Prob. 8.16 with $K_m = 2$.

    (*b*) Repeat with the gain-constant values obtained in part (*a*) of Prob. 8.16;

    (*c*) part (*b*) of Prob. 8.16.

    (*d*) For part (*c*), plot M vs. ω and obtain $c(t)$ for a step input.

9.6. (CAD Problem) (*a*) In Prob. 8.17, adjust the gain for Lm $M_m = 2$ dB and determine ω_m. For this value of gain find the phase margin angle, and plot M vs. ω and α vs. ω.

    (*b*) Repeat for Prob. 8.4.

9.7. For

$$G_x(s) = \frac{K}{s(s + 20)(s + 50)} \qquad \text{and} \qquad H(s) = 1$$

determine the required gain to achieve an $M_m = 1.26$. What are the values of ω_m, w_ϕ, and γ for this value of M_m?

    (*a*) Use log plots.

    (*b*) Use polar plots.

    (*c*) Use a computer.

9.8. (*a*) For Prob. 8.2, determine, by use of the Nichols chart, the values of M_m and ω_m.

    (*b*) What must the gain be in order to achieve an $M_m = 1.12$?

    (*c*) Refine the values of parts (*a*) and (*b*) by use of a CAD package.

    (*d*) What is the value of ω_m and the phase margin angle for this value of M_m for each case?

9.9. (CAD Problem) Refer to Prob. 8.1.

    (*a*) Determine the values of K_1, ω_m, and γ corresponding to the following values of M_m: 1.05, 1.1, 1.2, 1.4, 1.6, 1.8, and 2.0. For each value of M_m, determine ζ_{eff} from Eq. (9.16) and plot M_m vs. ζ_{eff}.

    (*b*) For each value of K_1 determined in part (*a*), calculate the value of M_p. For each value of M_p, determine ζ_{eff} from Fig 3.7 and plot M_p vs. ζ_{eff}. (See the last paragraph of Sec. 7.10.)

    (*c*) What is the correlation between M_m vs. ζ_{eff} and M_p vs. ζ_{eff}? What is the effect of a third real root that is also dominant?

(*d*) What is the correlation between γ and both M_m and M_p?

9.10. For Prob. 8.19, determine the value of K_m for Lm $M_m = 3$ dB with $H(s) = 1$. Hint: If the CAD program does not include dead zone, the second-order Padé approximation can be used:

$$e^{-\tau s} \approx \frac{1 - (\tau/2)s + (\tau^2/8)s^2}{1 + (\tau/2)s + (\tau^2/8)s^2}$$

9.11. For the given transfer function:

(*a*) With the aid of the Nichols chart, specify the value of K_2 that will make the peak M_m in the frequency response as small as possible.

(*b*) At what frequency does this peak occur?

(*c*) What value does $\mathbf{C}(j\omega)/\mathbf{R}(j\omega)$ have at the peak?

$$(1) \quad G(s) = \frac{K_2(1 + 2s)}{s^2(1 + 0.2s)(1 + 0.05s)}$$

$$(2) \quad G(s) = \frac{K(s + 0.05)}{s^2(s + 1.5)}$$

9.12. By use of the Nichols chart for

$$G(s) = \frac{K}{s(s + 1)(s + 2)}$$

(*a*) What is the maximum value of the unity-feedback closed-loop frequency response magnitude M_m with $K = 1$?

(*b*) Determine the value of K for a phase margin angle of $\gamma = 45°$?

(*c*) What is the corresponding gain margin a for this value of K?

9.13. Evaluate the transfer function for the frequency-response plot shown below. The plot is not drawn to scale.

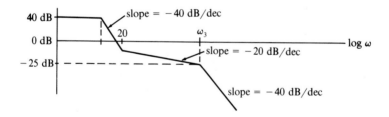

9.14. For a unity-feedback system, $G(s)$ is

$$G(s) = \frac{K_1(1 + 0.125s)}{s(1 + 0.25s)\left(1 + \dfrac{s}{25} + \dfrac{s^2}{225}\right)}$$

(*a*) Sketch the normalized Nyquist polar plot and determine the range of K_1 for stability by using the Nyquist stability criterion.

(b) For $K_1 = 40$, plot the log magnitude and phase-angle curves (Bode diagrams). Then find the gain margin and phase margin angle. Is the system stable for this gain?

Chapter 10

For the design of all problems of Chapter 10, obtain the figures of merit (M_p, t_p, t_s, and K_m) for the uncompensated and compensated systems. Where appropriate, use a CAD package in solving these problems.

10.1. It is desired that a control system have a damping ratio of 0.5 for the dominant complex roots. Using the root-locus method, (a) add a lag compensator, with $\alpha = 10$, so that this value of ζ can be obtained; (b) add a lead compensator with $\alpha = 0.1$; (c) add a lag-lead compensator with $\alpha = 10$. Indicate the time constants of the compensator in each case. Obtain $c(t)$ with a step input in each case. Compare the results obtained by the use of each type of compensator with respect to the error coefficient K_m, ω_d, T_s, and M_o.

$$(1) \quad G(s) = \frac{K}{(s + 1)(s + 2)(s + 5)} \qquad (2) \quad G(s) = \frac{K}{s(s + 1)(s + 4)(s + 5)}$$

Note: For plant (1), the 5° rule for a lag compensator does not apply. By trial and error determine the best location of the compensator's zero and pole.

10.2. A control system has the forward transfer function

$$G_x(s) = \frac{K_2}{s^2(1 + 0.1s)}$$

The closed-loop system is to be made stable by adding a compensator $G_c(s)$ and an amplifier A in cascade with $G_x(s)$. A ζ of 0.55 is desired with a value of $\omega_n \approx 2.0$ rad/s. By use of the root-locus method, determine the following: (a) What kind of compensator is needed? (b) Select an appropriate α and T for the compensator. (c) Determine the value of the error coefficient K_2. (d) Plot the compensated root locus. (e) Plot M vs. ω for the compensated closed-loop system. (f) From the plot of part (e) determine the values of M_m and ω_m. With this value of M_m determine the effective ζ of the system by the use of $M_m = (2\zeta \sqrt{1 - \zeta^2})^{-1}$. Compare the effective ζ and ω_m with the values obtained from the dominant pair of complex roots. *Note:* The effective $\omega_m = \omega_n \sqrt{1 - 2\zeta^2}$. (g) Obtain $c(t)$ for a unit step input.

10.3. For a unity feedback system whose basic plant is

$$(1) \quad G_x = \frac{K}{s(s^2 + 8s + 20)}$$

$$(2) \quad G_x = \frac{K(s + 10)}{s(s + 4)(s^2 + 12s + 100)}$$

Use $\zeta = 0.5$ for the dominant roots. Determine the figures of merit M_p, t_p, t_s, and K_m for each of the following cases: (a) original system; (b) lag compensator added, $\alpha = 10$; (c) lead compensator added, $\alpha = 0.1$; (d) lag-lead compensator added, $\alpha = 10$. *Note:* Except for the original system, it is not necessary to obtain the complete root locus for each type of compensation. When the lead and lag-lead compensator are added to the original system, there may be other dominant roots in addition to the complex pair. When a real root is also dominant, a $\zeta = 0.3$ may produce the desired improvement (see Sec. 10.3).

10.4. A unity-feedback system has the forward transfer function

$$G_x(s) = \frac{K_2}{s^2}$$

(*a*) Design a cascade compensator that will produce a stable system and that meets the following requirements without reducing the system type: (1) The dominant poles of the closed-loop control ratio are to have a damping ratio $\zeta = 0.5$. (2) The settling time is to be $T_s = 4$ s. Is it a lag or lead compensator? (*b*) Using the cascade compensator, determine the control ratio $C(s)/R(s)$. (*c*) Find $c(t)$ for a step input. What is the effect of the real pole of $C(s)/R(s)$ on the transient response?

10.5. Using the root-locus plot of Prob. 7.7, adjust the damping ratio to $\zeta = 0.5$ for the dominant roots of the system. Find $K_0, \omega_n, M_0, T_p, T_s, N, C(s)/R(s)$ for (*a*) the original system and (*b*) the original system with a cascade lag compensator using $\alpha = 10$; (*c*) design a cascade compensator that will improve the response time, i.e., will move the dominant branch to the left in the s plane.

10.6. The plant in a unity feedback system is

$$G_x(s) = \frac{K}{(s + 1)(s + 2)(s + 4)(s + 6)}$$

(*a*) Determine the roots of the characteristic equation of this system for $\zeta = 0.5$. (*b*) Determine the ratio $\sigma_{3,4}/\sigma_{1,2}$. (*c*) It is desired that this ratio be increased to at least 10 with no increase in T_s for $\zeta = 0.5$. A compensator that may accomplish this has the transfer function

$$G_c = A\frac{s + a}{s + b}$$

Determine appropriate positive values of a and b that can be achieved with a practical value of α.

10.7. A unity-feedback system has the transfer function $G_x(s)$. The closed-loop roots must satisfy the specifications $\zeta = 0.707$ and $T_s = 2$ s. A suggested compensator $G_c(s)$ must maintain the same degree as the characteristic equation for the basic system:

$$G_x(s) = \frac{K(s + 6)}{s(s + 4)} \qquad G_c(s) = \frac{A(s + a)}{s + b}$$

(*a*) Determine the values of a and b. (*b*) Determine the value of α. (*c*) Is this a lag or a lead compensator?

10.8. A unity-feedback control system contains the forward transfer function shown below. The system specifications are $\omega_n = 4$ and $T_s \leq 2$ s, which are to be achieved by the proposed cascade compensator which has the form indicated.

$$G_x = \frac{K(s + 6)}{s(s + 2)(s + 5)} \qquad G_c = A\frac{s + a}{s + b}$$

(*a*) For the uncompensated system, based upon the given performance specifications, determine the value of ζ which yields the desired dominant complex-conjugate poles.

For this value of ζ determine the value of K and the corresponding figures of merit. (b) Determine the values of a and b such that the desired complex-conjugate poles of $C(s)/R(s)$, for the compensated system, are truly dominant and the degree of the compensated system is 3. (c) Determine the values of A and α for this compensator. (d) Is this $G_c(s)$ a lag or lead compensator? (e) Determine the control ratio and the figures of merit for the compensated system.

10.9. For Prob. 10.6: (a) Sketch the root locus. (b) For $\zeta \geq 0.7071$ the system's performance specifications are: $T_s \leq 2.5$ s and $K_0 \geq 0.5$. Identify the region on the root-locus that meets these specifications. (c) Determine the roots and the value of K_x for $\zeta = 0.71$. (d) Add a lead compensator that cancels the pole at $s = -1$. (e) Add a lead compensator that cancels the pole at $s = -2$. (f) Compare the results of parts (d) and (e). Establish a "rule" for adding a lead compensator to a Type 0 system.

10.10. For the system of Prob. 7.4, the desired dominant roots are $-2.62 \pm j4.54$. (a) Design a compensator that will achieve these characteristics. (b) Determine $c(t)$ with a unit step input. (c) Are the desired complex roots dominant? (d) Compare the results of the basic system for the same value of ζ with those of the compensated system.

10.11. A unity-feedback system has the transfer function

$$G_x(s) = \frac{K}{s(s + 2)(s + 12)}$$

The poles of the closed-loop system must be $s = -1 \pm j2$. (a) Design a lead compensator with the maximum possible value of α that will produce these roots. (b) Determine the control ratio for the compensated system. (c) Add a compensator to increase the gain without increasing the settling time.

10.12. A system has the transfer equation

$$G_x(s) = \frac{K}{(s^2 - 4s + 4)(s + 20)}$$

The roots of the closed-loop system must have $\zeta = 0.5$. Add a lead compensator, with two zeros and two poles, to produce a stable system. Determine $C(s)/R(s)$. Is this a practical compensator? Explain.

10.13. In the figure the servomotor has inertia but no viscous friction. The feedback through the accelerometer is proportional to the acceleration of the output shaft. (a) When $H(s) = 1$, is the servo system stable? (b) Is the system stable if $K_x = 15$, $K_y = 2$, and $H(s) = 1/s$? (c) Use $H(s) = 1/(1 + Ts)$ and show that it produces a stable system. Show that the system is stable with this value of $H(s)$ by obtaining the transfer function $C(j\omega)/E(j\omega)$ and sketching the root locus.

10.14. The accompanying schematic shows a method for maintaining a constant rate of discharge from a water tank by regulating the level of the water in the tank. The relationships governing the dynamics of the flow into and out of the tank are:

$$Q_1 - Q_2 = 16\frac{dh}{dt} \qquad Q_2 = 4h \qquad Q_1 = 10\theta$$

where Q_1 = volumetric flow into tank
  Q_2 = volumetric flow out of tank
  h = pressure head in tank
  θ = angular rotation of control valve

The motor damping is much smaller than the inertia. The system shown is stable only for very small values of system gain. Therefore, feedback from the control valve to the amplifier is proposed. Show conclusively which of the following feedback functions would produce the best results: (*a*) feedback signal proportional to control-valve position; (*b*) feedback signal proportional to rate of change of control-valve position; (*c*) feedback signal comprising a component proportional to valve position and a component proportional to rate of change of valve position.

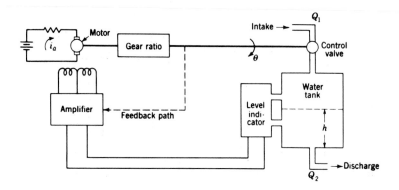

For Probs. 10.15 through 10.18, the block diagram shows a simplified form of roll control for an airplane. Overall system specifications with a step input are $t_s \leq 1.0$ s and $1 < M_p \leq 1.3$. A_h is the gain of an amplifier in the H_2 feedback loop. $G_2 = A_2$ is an amplifier of adjustable gain with a maximum value of 100. Restrict b to values between 15 and 50. With $G_1(s) = 1$, derive general expressions for $G_y(s)$ and $\Phi(s)/R(s)$ before solving these problems.

$$G_3(s) = \frac{1.7}{(1 + 0.2s)(1 + s)} \qquad H_2(s) = 0.2A_h\frac{s + a}{s + b} \qquad G_4(s) = \frac{1}{s}$$

10.15. (*a*) By use of the root locus method 1, with $G_1(s) = 1$, determine a set of parameters A_h, a, and b in $H_2(s)$ to meet the overall system specifications. [*Note:* Prob. 10.17 requires the design of the cascade compensator $G_1(s)$ to improve K_1 by a factor of 5. When a lag compensator is used to meet this requirement, it is known that it results in an increase in the values of M_p and t_s. Therefore, modify the specifications for this problem to $M_p < 1.2$ and $0.8s < t_s < 0.95$ s in order to allow for the degradation in transient performance produced by this compensator.] (*b*) Determine $G_y(s) = s\Phi(s)/E_1(s)$ and $G(s) = A_1 G_y(s)G_4(s)$ in factored form. Then with $A_1 = 1$ compute K_1. (*c*) Determine the figures of merit (M_p, t_p, t_s, and K_1) for the closed-loop system. (*d*) With $e_1(t) = u_1(t)$, determine the values of M_p and t_s for the inner loop represented by $G_y(s)$.

10.16. (*a*) By use of the root locus method 2, determine the parameters A_h, a, and b to meet the inner-loop specifications $M_p = 1.05$ and $T_s = 1.0$ s. [The same note given in Prob. 10.15(*a*) also applies to this problem.] Select A_2 so that the steady-state value of $\dot{\phi}(t)$ follows a step input $e_1(t) = u_1(t)$. Determine the necessary values of A_h and A_2. (*b*) Determine $G_y(s)$ in factored form. (*c*) Determine the figures of merit for $G_y(s)$ and compare them with those of problem 10.15(*d*).

10.17. (*a*) Using the values determined in Prob. 10.15, derive the expression $G_z(s) = G_y(s)G_4(s)$. (*b*) Design $G_1(s)$ to increase the value of K_1 of Prob. 10.15 by a factor of 5 while maintaining the desired overall system specifications. (*c*) Determine the resulting $\Phi(s)/R(s)$ in factored form. (*d*) Determine the figures of merit for a step input, and compare them with those of Prob. 10.15.

10.18. Repeat Prob. 10.17 with the values determined in Prob. 10.16. In tabular form, summarize the results of Probs. 10.15 through 10.18.

10.19. For the feedback-compensated system of Fig. 10.29,

$$G_x(s) = \frac{2}{s(s + 1)} \qquad H(s) = K_h \frac{s(s + a)}{s + 4}$$

(*a*) Use method 1 to determine $C(s)/R(s)$ using $T_s = 2$ s (2 percent criterion) and $\zeta = 0.707$ for the dominant roots. (*b*) Design a cascade compensator that yields the same dominant roots as part (*a*). (*c*) Compare the gains of the two systems.

10.20. A unity-feedback system has

$$G_x(s) = \frac{K}{s(s + 5)}$$

(*a*) For $\zeta = 0.5$, determine K and $C(s)/R(s)$. (*b*) Minor-loop and cascade compensation are to be added (see Fig. 10.29) with the amplifier replaced by $G_c(s)$:

$$G_c(s) = \frac{A(s + c)}{s + b} \qquad H(s) = \frac{K_h s}{s + a}$$

Use the value of K determined in part (*a*). By method 1 of the root-locus design procedure, determine values of a, b, K_h, and A to produce dominant roots having $\zeta = 0.5$, $\sigma = -7$, and $c(t)_{ss} = 1.0$. Only positive values of K_h and A are acceptable. To minimize the effect of the third real root, it should be located near a zero of $C(s)/R(s)$. (*c*) Determine $C(s)/E(s)$ and $C(s)/R(s)$ in factored form.

10.21. For $G(s)$ given in Prob. 10.3, add feedback rate (tachometer) compensation. Use root-locus method 1 or 2 as assigned. (*a*) Design the tachometer so that an improvement in system performance is achieved while maintaining a $\zeta = 0.5$ for the dominant roots.

(b) Repeat part (a) for a $\zeta = 0.3$. Compare with the results of Prob. 10.3. *Note:* Use Fig. 10.29.

10.22. Refer to Fig. 10.29. It is desired that the dominant poles of $C(s)/R(s)$ have $\zeta = 0.5$ and $T_s = 1$ s. For the plant of Prob 7.8(2), using method 1 with $A = 5$, determine the values of b and K_t to meet these specifications. Be careful in selecting b. Check the time response to determine the effect of the real root on the settling time.

$$H(s) = \frac{K_t s}{s + b}$$

10.23. Repeat Prob. 10.22 using method 2. Compare the results with those of Probs 10.5 and 10.22.

10.24.

$$G_x(s) = \frac{K}{s(s^2 + 4.2s + 14.4)}$$

The specifications for the closed-loop performance for a unity-feedback control system, with a step input, are $1 < M_p \leq 1.123$, $t_p \leq 1.6$ s, $t_s \leq 3.0$ s, $K_1 \geq 1.5$. Show that these specifications can be achieved without the use of a compensator. Hint: the complex poles of $G(s)H(s)$ have a $\zeta \approx 0.44$; therefore, the procedure of method 2 requires $\zeta_i < 0.44$. This value is less than the desired closed-loop value of $\zeta = 0.5$. By trial and error, select a value of $\zeta_i < 0.44$ until a satisfactory solution is achieved.

10.25. Given the system of Fig. 10.29 with

$$G_x(s) = \frac{K_x(s + 6)}{s(s + 2)} \qquad H(s) = \frac{K_h(s + 0.2)}{s + 10}$$

(a) The specification for the inner loop is that $c(t)_{ss} = 1$ for $i(t) = u_{-1}(t)$. Determine the value of K_h in order to satisfy this condition. *Hint:* Analyze $C(s)/I(s)$. (b) It is desired that the dominant poles of $C(s)/R(s)$ be exactly at $-4 \pm j4$. By use of method 1, with K_h having the value determined in part (a), determine the values of A, K_m, and K_x that yield these dominant poles.

10.26. Draw the root locus for the system containing the PID cascade compensator of Sec. 10.14. Verify the closed-loop transfer function of Eq. (10.45) and the figures of merit.

Chapter 11

For the design problems of Chapter 11, obtain the time and frequency domain figures of merit $M_p, t_p, t_s, K_m, M_m, \omega_m, \omega_\phi$, and γ for the uncompensated and compensated systems. Where appropriate, use a CAD package in solving these problems.

11.1. (*a*) For the basic control system of Prob. 8.17 adjust K to achieve an $M_m = 1.16$ and determine the corresponding values of ω_{ϕ_x} and γ, and all other figures of merit. The value of γ is to be maintained for parts (*b*) through (*d*). (*b*) Add a lag compensator, with $\alpha = 10$. (*c*) Add a lead compensator with $\alpha = 0.1$. (*d*) Add a lag-lead compensator with $\alpha = 10$. (*e*) Form a table comparing both frequency and time domain figures of merit for all parts of this problem. Specify the time constants of the compensator used in each case.

11.2. A control system has the forward transfer function

$$G_x(s) = \frac{K_x}{s^2(1 + 0.1s)}$$

The closed-loop system is to be made stable by adding a compensator G_c and an amplifier A in cascade with $G_x(s)$. An $M_m = 1.5$ is desired with $\omega_m \approx 1.4$ rad/s. (*a*) What kind of compensator is needed? (*b*) Select an appropriate α and T for the compensator. (*c*) Select the necessary value of amplifier gain A. (*d*) Plot the compensated curve. (*e*) Compare the results of this problem with those obtained in Prob. 10.2.

11.3. A basic (uncompensated) control system with unity feedback has a forward open-loop transfer function

$$G_x(s) = \frac{K_1}{s(1 + 0.2s)(1 + 0.1s)^2}$$

(*a*) For the basic system, find the gain K_1 for $\gamma = 45°$, and determine the corresponding phase-margin frequency ω_{ϕ_x}. (*b*) For the same phase-margin angle as in (*a*), it is desired to increase the phase-margin frequency to a value of $\omega_\phi = 3.0$ with the maximum possible improvement in gain. To accomplish this, a lead compensator is to be used. Determine the values of α and T that will satisfy these requirements. For these values of α and T, determine the new value of gain. (*c*) Repeat part (*b*) with the lag-lead compensator of Fig. 11.16. Select an appropriate value for T. With $G_c(s)$ inserted in cascade with $G_x(s)$, find the gain needed for 45° phase margin. (*d*) Show how the compensator has improved the system performance, i.e., determine all figures of merit (see note for Chapter 11 problems) for each part of the problem.

11.4. (*a*) Using the $G_x(s)$ of Prob. 11.3 with $K_x = 1.964$, determine ω_{ϕ_x} and γ_x for the basic system. (*b*) Design a cascade lead compensator, using $\alpha = 0.1$, in order to achieve an $\omega_\phi = 3$ rad/s, for this value of $\gamma = \gamma_x$. Determine the required value of T and A. (*c*) For the basic (uncompensated) unity-feedback control system it is desired to increase K_m by a factor of approximately 10 while maintaining $\gamma = \gamma_x$. Design a cascade compensator that yields this increase in K_m. Specify the values of α and T so that $\left|\underline{/G_c(j\omega)}\right| \leq 0.6°$ at $\omega = \omega_{\phi_x}$. Determine the values of A and K_m for the compensated system. (*d*) It is desired that a phase margin angle $\gamma = 45°$ and a phase-margin frequency $\omega_\phi = 3.0$ rad/s be achieved by use of the lag-lead compensator

$$G_c'(s) = \frac{(1 + 100s)(1 + 0.5s)}{(1 + 1000s)(1 + 0.5\alpha s)}$$

Determine the values of α and A that yield these values of γ and ω_ϕ for the compensated system. Determine the ramp error coefficient of the compensated system. (*e*) It is desired, for the feedback compensated system of Fig. 11.20, that $\gamma = 45°$ at $\omega_\phi = 3$ rad/s and the phase margin angle be determined *essentially* by the characteristics of the feedback unit. It is also specified for the compensated system that $K_1 = 100s^{-1}$ and that $\gamma = 45°$ occur at *exactly* $\omega = \omega_\phi = 3$. Initially, by use of the straight-line approximation method, determine the values of A, K_H, and α for the feedback compensator

$$H(s) = \frac{K_H(s + 0.1)}{s + 10/\alpha} \qquad 0.1 \le \alpha \le 10$$

Then, by trial and error, with the use of a CAD program, select an α and adjust K_H until $C(s)/I(s)$ is stable and yields $\gamma = 45°$. (*f*) It is desired, for the compensated system of Fig. 11.20, that $\gamma = 45°$ and the phase-margin angle be determined *essentially* by the characteristics of a feedback unit. Determine if the following proposed feedback unit will result in a stable system. If a stable response can be achieved, determine the value of A that yields the desired value of γ and the corresponding value of ω_ϕ.

$$H(s) = \frac{K_H s^2}{1 + s/2} \qquad K_H = 100$$

Note: For all parts determine the figures of merit: M_p, t_p, t_s, K_m, and M_m, and the actual values of ω_ϕ and γ that are achieved.

11.5. The mechanical system shown has been suggested for use as a compensating component in a mechanical system. (*a*) Determine whether it will function as a lead or a lag compensator by finding $X_2(s)/F(s)$. (*b*) Sketch $\mathbf{X}_2(j\omega)/\mathbf{F}(j\omega)$.

11.6. A unity-feedback control system has the transfer function

$$G(s) = \frac{K}{s(s + 2)^3}G_c(s)$$

(*a*) What is the value of the gain of the basic system for an $M_m = 1.12$? (*b*) Design a cascade compensator $G_c(s)$ that will increase the step-error coefficient by a factor of 8 while maintaining the same M_m. (*c*) What effect does the compensation have on the closed-loop response of the system? (*d*) Repeat the design using a minor-loop feedback compensator.

11.7. A unity-feedback control system has the transfer function

$$G(s) = G_c(s)G_1(s)$$

where $G_1(s) = 1/s^2$. It is desired to have an $M_m = 1.5$ with damped natural frequency of about 10.0 rad/s. Design a compensator $G_c(s)$ that will help to meet these specifications.

11.8. A system has

$$G_2(s) = \frac{2(1 + 0.05s)}{1 + 0.01s} \qquad G_3(s) = \frac{1}{s}$$

For the control system shown in Fig. (a), determine the cascade compensator $G_c(s)$ required to make the system meet the desired open-loop frequency-response characteristics shown in Fig. (b). What is the value of M_m for the compensated system?

(a)

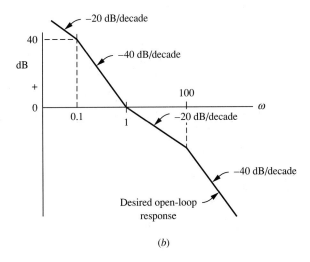

(b)

11.9. A system has the transfer function

$$G_x(s) = \frac{K_1(1 + 2s)}{s(1 + 0.2s)(1 + s + 0.8s^2)}$$

(a) For $M_m = 1.26$ determine ω_{ϕ_x}, γ, K_1, and ω_m for the original system. For parts (b) and (c) this value of γ is to be maintained. (b) Add a lag compensator with $\alpha = 10$, determine T, and adjust the gain to achieve the desired value of γ. (c) Add a lead compensator with $\alpha = 0.1$, determine T, and adjust the gain to achieve the desired value of γ. (*d*) Determine both frequency and time domain figures of merit, for all parts of this problem, and compare these values in a tabular format.

11.10. Repeat Prob. 10.1(2), using $M_m = 1.16$ as the basis of design. Compare the results.

11.11. Repeat Prob. 10.16 but solve by the use of Bode plots. *Note:* Limit G_1 and G_2 to less than 100. (*a*) Assume that the forward transfer function for the uncompensated unity-feedback system is given by $G_2G_3G_4$, with $G_1 = A_1 = 1$. Thus the control ratio is

$$\frac{\Phi(s)}{R(s)} = \frac{A_2 G_3(s)G_4(s)}{1 + A_2 G_3(s)G_4(s)}$$

Solve for the value of A_2 that yields the desired value of $M_m \approx M_p$. Determine the values of ω_{ϕ_x}, γ, M_p, t_p, t_s, and K_1. (*b*) For the compensated system, with $A_1 \neq 1$, incorporate G_4 into the minor loop. Determine $H_2(s)$ and A_1 in order to achieve the value of γ determined in (*a*) and determine all figures of merit for the compensated system, including ω_ϕ. (*c*) Compare the results of this problem with those of Prob. 10.15.

For Probs. 11.12 through 11.18, analyze the 0-dB crossing slope of the straight-line Bode plot of Lm **C/E** for system stability.

11.12. A system has

$$G_1(s) = 25 \qquad G_2(s) = 5 \qquad G_3(s) = \frac{4}{s + 2} \qquad G_4(s) = \frac{1}{s} \qquad H_1(s) = \frac{s}{s + 1}$$

System specifications are $K_1 = 200s^{-1}$, $\gamma \geq +40°$, and $\omega_\phi = 8$ rad/s. Using approximate techniques, (*a*) determine whether the feedback-compensated system satisfies all the specifications; (*b*) find a cascade compensator to be added between $G_1(s)$ and $G_2(s)$, with the feedback loop omitted, to produce the same $\mathbf{C}(j\omega)/\mathbf{E}(j\omega)$ as in part (*a*); (*c*) determine whether the system of part (*b*) satisfies all the specifications.

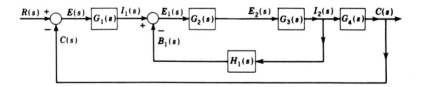

11.13. In the figure of Prob. 11.12,

$$G_1(s) = A_1 \qquad G_2(s) = A_2 \qquad G_3(s) = \frac{1}{(1 + 0.1s)(1 + 0.01s)} \qquad G_4(s) = \frac{1}{s}$$

$H_1(s)$ is a passive network. Choose A_1, A_2, and $H_1(s)$ to meet the desired open-loop response characteristic shown in the figure of Prob. 11.8 for $\omega < 80$.

11.14. For the feedback-compensated system of Fig. 11.20,

$$G_x(s) = \frac{4(1 + s/4)}{s(1 + s)(1 + s/8)} \qquad H(s) = \frac{K_t T s^2}{1 + Ts}$$

(*a*) For the original system, without feedback compensation, determine the values of A and ω_{ϕ_x} for a phase-margin angle $\gamma = 54°$. (*b*) Add the minor-loop compensator $H(s)$ and determine, for $\gamma = 54°$, $\omega_\phi = 2.3$ rad/s, and $K_1 = 140$, the new value of A and the values of K_t and T. The specifications are to be dictated by $H(s)$. *Hint:* Adjust K_t so that the following conditions are satisfied:

$$\omega_1 \le \omega_\phi/20 \qquad \omega_2 \ge 20\omega_\phi$$

where ω_1 and ω_2 are the intersection frequencies of Lm $\mathbf{G}_x$ vs. ω and Lm[$1/\mathbf{H}(j\omega)$]. (*c*) Determine the frequency and time domain figures of merit for parts (*a*) and (*b*). (*d*) Compare the results.

11.15. The system represented in the figure is an ac voltage regulator. (*a*) Calculate the response time of the uncompensated system, when the switch S is open, to an output disturbance $d_2 = u_{-1}(t)$. Calculate $v_{out}(t)$ for $t = T_s = 4/\sigma$, where σ is with the switch S open. (*b*) Repeat part (*a*) with the switch S closed. (*c*) Compare the values obtained in parts (*a*) and (*b*) and comment on any difference in performance.

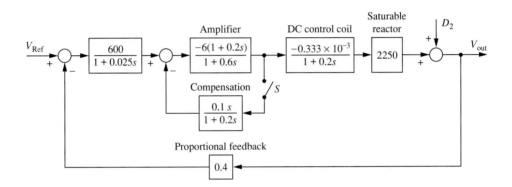

11.16. The simplified pitch-attitude-control system used in a space-launch vehicle is represented in the block diagram. (*a*) Determine the upper and lower gain margins. (*b*) Is the closed-loop system stable?

11.17. Repeat Prob. 10.21 using the log plots for $M_m = 1.16$. (*a*) For the uncompensated system adjust K to yield the value of $M_m = 1.16$ and determine the resulting values of ω_{ϕ_x}, γ, and all figures of merit. (*b*) In determining the value of K_t to use, first determine the range of values it can have in order to have an intersection between Lm $\mathbf{G}_x$ and Lm$[1/\mathbf{H}]$ and to maintain a stable open-loop $C(s)/I(s)$ transfer function. A root-locus sketch for $G_x(s)H(s) = -1$ reveals that $C(s)/I(s)$ can have right-half plane poles. Thus, for a stable $C(s)/I(s)$ the 0 dB crossing slope of Lm $\mathbf{G}_x\mathbf{H}$ must be greater than –40 dB/dec. (c) Select a value of K_t in the allowable range and determine the value of A that yields the desired value of γ. (d) Obtain the exact expressions for $C(s)/E(s)$ and $C(s)/R(s)$. (e) Determine all figures of merit for the compensated system. (f) Compare with the results in Prob. 10.3. Obtain the time response. *Hint:* Since $\angle 1/\mathbf{H}(j\omega) = -90°$, then, for $\gamma < 90°$, the new ω_ϕ is close to the intersection of $1/\mathbf{H}(j\omega)$ and $\mathbf{G}(j\omega)$. *Note:* For Prob. 10.21 (1), for $K_t = 1$, the system is unstable if the value of K is set at the value obtained for the uncompensated system to meet the desired value of M_m.

11.18. (a) For the uncompensated unity-feedback control system where

$$G_x(s) = \frac{K_x(1 + s)}{s^2(1 + s/8)}$$

determine, for $\gamma = 50°$, K_{2_x} and ω_{ϕ_x}. (b) Assume K_{2_x} has the value determined in part (a) and is not adjustable. For the feedback system of Fig. 11.20 the feedback compensator is given by $\mathbf{H}(j\omega) = K_H\mathbf{H}''(j\omega)$, where $K_H = K_tT$ and $\mathbf{H}''(j\omega) = (j\omega)^2/(1 + j\omega T)$. Draw plots of Lm$[1/\mathbf{H}''(j\omega)]$ and $\angle 1/\mathbf{H}''(j\omega)$. (c) It is desired, for the compensated system, that $\gamma = 50°$ and the phase-margin frequency $\omega_\phi = 4\omega_{\phi_x}$ should be determined *essentially* by the characteristics of the feedback unit. Using straight-line approximations, determine graphically, as a first trial, the values of T, K_t, and A that yield this value of γ and the corresponding value of ω_ϕ. Analyze the slope of Lm $\mathbf{C}(j\omega)/\mathbf{E}(j\omega)$ at the 0 dB crossing for system stability. (d) Compare the values of M_p, M_m, ω_m, t_p, t_s, and K_2 of part (c) with part (a). For the value of A determined graphically in part (c), what are the actual values of ω_ϕ and γ of part (c)?

Chapter 12

Use a CAD package in solving the design problems of Chapter 12, where appropriate.

12.1. For the control system of Prob. 10.24 the desired specifications are: $1 < M_p \le 1.16$, $t_s \le 1.35$ s, $t_p \le 0.65$ s, and $K_1 \ge 4\,\text{s}^{-1}$. If these specifications cannot be achieved by the basic system, then

(a) synthesize a desired control ratio that satisfies these specifications.

(b) Using the Guillemin-Truxal method (see Sec. 12.3), determine the required compensator $G_c(s)$.

12.2. The desired closed-loop time response for a step input is achieved by the system having the control ratio:

$$M_T(s) = \frac{23.1(s + 20)}{(s^2 + 2s + 10)(s + 2.2)(s + 21)}$$

Given:

$$G_x(s) = \frac{10}{s(s^2 + 4.2s + 14.4)} \qquad G_c(s) = A\frac{s + a}{s + b}$$

(a) Determine the required passive cascade compensator $G_c(s)$ for the unity-feedback system.

(b) Compare M_p, t_p, and t_s for the desired and actual $M(s)$.

12.3. The desired closed-loop time response for a step input is achieved by the system having the control ratio

$$M_T(s) = \frac{2.5(s + 1.6)}{(s^2 + 2s + 2)(s + 2)} \qquad G_x(s) = \frac{1}{s(s + 3)} \qquad G_c(s) = A\frac{A(s + a)}{s + b}$$

(a) Determine, by the Guillemin-Truxal method, a passive cascade compensator $G_c(s)$ for a unity-feedback system with the $G_x(s)$ shown.

(b) It is desired to simplify the compensator to the form shown. Determine the values of A, a, and b that will yield the best results.

(c) Compare M_p, t_p, t_s, and K_1 for the desired and the actual systems.

12.4. Determine a cascade compensator for a unity-feedback system that has the open-loop transfer function $G_x(s) = 10/[s(s + 4)(s + 6)]$. Determine the simplest possible approximate compensator transfer function that yields the time response, for a step input, that is achieved by the following desired control ratio:

$$M_T(s) = \frac{29.94(s + 2.1)}{s^4 + 12s^3 + 43.99s^2 + 79.42s + 62.86}$$

12.5. For the nonunity-feedback control system of Fig. 10.3,

$$G_x(s) = \left[\frac{K_x \prod_{i=1}^{w}(s + a_i)}{\prod_{j=1}^{n}(s + b_j)}\right]_{n \geq w}$$

$$M_T(s) = \frac{G_x(s)}{1 + G_x(s)H(s)} = \left[\frac{K' \prod_{m=1}^{w'}(s - z_m)}{\prod_{k=1}^{n'}(s - p_k)}\right]_{n' \geq w'}$$

Determine whether a feedback compensator $H(s)$ (other than just gain) can be evaluated by use of the Guillemin-Truxal method such that the order of its denominator is equal to or greater than the order of its numerator.

12.6. Determine a cascade compensator by the Guillemin-Truxal method where

$$G_x(s) = \frac{2}{s(s + 5)(s + 7)} \qquad M_T(s) = \frac{167.96}{(s + 1.9)(s + 20)(s^2 + 4s + 8.84)}$$

12.7. Repeat Prob. 4.17, where, for disturbance rejection, it is desired that $y(t)_{ss} = 0$ for $D(s) = D_o/s$ with $a_1 = 0$. For this problem $Y(s)$ represents the output of a nonunity-feedback control system, where $G_x(s) = K_x G'(s) = N_1/D_1$ is fixed (the degree of the numerator is w_1 and the degree of the denominator is n_1) and $H(s) = K_h H'(s) = N_2/D_2$ is to be determined (the degree of the numerator is w_2 and the degree of the denominator is n_2). *Hint:* First determine $Y(s)$ in terms of N_1, N_2, D_1, and D_2 and the degree of its numerator and denominator in terms of w_1, w_2, n_1, and n_2. For case (2) of Prob. 4.17, consider $n = w + 1$ and $n > w + 1$.

12.8. The figure below represents a disturbance-rejection control system, where

$$G_x(s) = \frac{10}{s(0.25s + 1)}$$

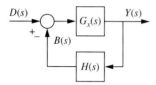

(a) The following feedback unit

$$H(s) = \frac{K_H(s + 4)(s + 8)}{s(s + a)} \qquad K_H > 0$$

is proposed to meet the specifications of $y(t)_{ss} = 0$ for $d(t) = u_{-1}(t)$, $|y(t_p)| \leq 0.01$, and $\mathrm{Lm}|Y(j\omega)/D(j\omega)| \leq -40\ \mathrm{dB} = -\mathrm{Lm}\,\alpha_p$ for $0 \leq \omega \leq 10$ rad/s. Select the value of a that satisfies these specifications.

(b) *Graphically*, by use of straight-line asymptotes, determine a value of K_H that meets these specifications with the specified $H(s)$.

(c) For your value of K_H, determine whether the system is stable by use of Routh's stability criterion. Correlate this with the slope of Lm $\mathbf{G}_x\mathbf{H}$ at the 0-dB crossing.

(d) Obtain a plot of $y(t)$ vs. t and determine α_p and t_p. *Note:* Adjust K_H so that $|y(t_p)| = \alpha_p$.

12.9. For the disturbance-rejection control system of Prob. 12.8

$$G_x(s) = \frac{200}{s(s + 2)} \qquad \text{and} \qquad H(s) = H_x(s)H_y(s)$$

where

$$H_x(s) = \frac{200}{s + 200} \qquad H_y(s) = K_y\frac{N_y(s)}{D_y(s)} = \frac{K_y(s - z_1)(s - z_2)}{s(s - p_1)}$$

Determine the feedback compensator $H_y(s)$ such that the control system can satisfy the following specifications for a unit step disturbance input $d(t)$.

$$|y(t_p)| \leq 0.005 = \alpha_p \qquad t_p \leq 0.1\mathrm{s} \qquad y(t)_{ss} = 0$$

and

$$\mathrm{Lm}\left|\frac{Y(j\omega)}{D(j\omega)}\right| \leq -46\ \mathrm{dB} = -\mathrm{Lm}\,\alpha_p$$

within the bandwidth of $0 < \omega \leq 16$ rad/s. Analyze the 0-dB crossing slope of the straight-line Bode plot of Lm $\mathbf{G}_x\mathbf{H}$ for system stability and correlate this to a root-locus sketch of $G_x(s)H(s) = -1$.

12.10. Analyze the time response characteristics of the control ratio

$$\frac{Y(s)}{R(s)} = \frac{K(s + a)}{(s + 2)(s + 5)}$$

with a unit step forcing function, for the following values of a: 0.01, 0.1, 1, 10, 50. For each value of a, adjust the value of K so that $Ka = 10$. [See Eq. (4.68).] Discuss the characteristics of overdamped versus underdamped responses for an all pole control ratio.

Chapter 13

Where appropriate use a CAD package to solve these problems.

13.1. For a single-input single-output system the open-loop matrix state and output equations are $\dot{\mathbf{x}} = \mathbf{A}\mathbf{x} + \mathbf{b}u$ and $y = \mathbf{c}^T\mathbf{x}$. Derive the overall closed-loop control ratio for a state-variable feedback system with $u = \mathbf{k}^T\mathbf{x}$:

$$\frac{Y(s)}{R(s)} = \mathbf{c}^T[s\mathbf{I} - (\mathbf{A} - \mathbf{b}\mathbf{k}^T)]^{-1}\mathbf{b}$$

13.2. For the system described by

$$D^3 y + 2D^2 y + 3Dy + 3y = 5u$$

(a) Draw the simulation diagram.

(b) Write the matrix state and output equations for the simulation diagram.

(c) Rewrite the state equation and the output equation in uncoupled form.

(d) From part (c) determine whether the system is completely controllable, observable, or both.

13.3. (a) Determine whether the following systems are completely observable.

(b) Determine whether they are completely controllable.

(c) Obtain the transfer functions.

(d) Determine how many observable states and how many controllable states are present in each system.

(e) Are the systems stable?

(1) $\dot{\mathbf{x}} = \begin{bmatrix} 1 & 0 \\ 2 & 2 \end{bmatrix}\mathbf{x} + \begin{bmatrix} 1 \\ 0 \end{bmatrix}u$ $y = [2 \quad 1]\mathbf{x}$

(2) $\dot{\mathbf{x}} = \begin{bmatrix} 0 & 1 & 0 \\ 0 & 0 & 1 \\ 0 & -2 & -3 \end{bmatrix}\mathbf{x} + \begin{bmatrix} 0 \\ 1 \\ 1 \end{bmatrix}u$ $y = [1 \quad 0 \quad 1]\mathbf{x}$

(3) $\dot{\mathbf{x}} = \begin{bmatrix} 1 & 0 & 0 \\ 0 & 1 & 1 \\ 0 & -2 & -1 \end{bmatrix}\mathbf{x} + \begin{bmatrix} 1 \\ 0 \\ 1 \end{bmatrix}u$ $y = [0 \quad 1 \quad 1]\mathbf{x}$

(4) $\dot{\mathbf{x}} = \begin{bmatrix} -1 & 0 & 0 \\ 0 & -1 & 0 \\ 0 & -2 & -2 \end{bmatrix}\mathbf{x} + \begin{bmatrix} 0 \\ 1 \\ 1 \end{bmatrix}u$ $y = [1 \quad 1 \quad 0]\mathbf{x}$

13.4. A single-input single-output system is described by the transfer function

$$G(s) = \frac{2s^2 + 20s + 48}{s^4 + 8s^3 + 19s^2 + 12s}$$

(a) Derive the phase-variable state equation. Draw the block diagram. The $\mathbf{A}_c$ matrix is singular. How is this interpreted physically?

(b) Represent the system in terms of canonical variables. Draw the corresponding block diagram.

(c) Is the system observable and controllable?

13.5. A control system is to use state-variable feedback. The plant has the given transfer function. The system is to meet the following requirements: (1) it must have zero steady-state error with a step input, (2) dominant poles of the closed-loop control ratio are to be $-2 \pm j3$, (3) the system is to be stable for all values $A > 0$, and (4) the control ratio is to have low sensitivity to increase in gain A. A cascade element $G_c(s)$ must be added to meet this requirement.

(a) Draw the block diagram showing the state-variable feedback.

(b) Determine $H_{eq}(s)$.

(c) Find $Y(s)/R(s)$ in terms of the state-variable feedback coefficients.

(d) Determine the desired control ratio $Y(s)/R(s)$.

(e) Determine the necessary values of the feedback coefficients.

(f) Sketch the root locus for $G(s)H_{eq}(s) = -1$ and show that the system is insensitive to variations in A.

(g) Determine $G_{eq}(s)$ and K_1.

(h) Determine M_p, t_p, and t_s with a step input.

13.6. A control system is to use state-variable feedback. The plant has the given transfer function. The system is to meet the following requirements: (1) it must have zero steady-state error with a step input, (2) the dominant poles of the closed-loop control ratio are supposed to be $-1 \pm j1$, (3) the system is to be stable for all values $A > 0$, and (4) the control ratio is to have low sensitivity to increase in gain A and is to contain a zero at $s = -1.6$. A cascade element may be added to meet this requirement.

$$G_x(s) = \frac{A(s + 4)}{s(s + 2)(s + 5)}$$

(a) Determine the desired control ratio $Y(s)/R(s)$.

(b) Draw the block diagram showing the state-variable feedback. Use the form illustrated in Fig. 13.17 where the term $(s + 2)/(s + 1)$ in the figure is changed to $(s + 4)/(s + 2)$ and the transfer function $1/(s + 5)$ is replaced by $(s + 1.6)/(s + 5)$ for this problem.

(c) Determine $H_{eq}(s)$.

(d) Find $Y(s)/R(s)$ in terms of the state-variable feedback coefficients.

(e) Determine the necessary values of the feedback coefficients.

(f) Sketch the root locus for $G(s)H_{eq}(s) = -1$ and show that the system is insensitive to variations in A.

(g) Determine $G_{eq}(s)$ and K_1.

(h) Determine M_p, t_p, and t_s with a step input.

13.7. Design a state-variable feedback system for the given plant $G_x(s)$. The desired complex dominant roots are to have a $\zeta = 0.45$. For a unit-step function the approximate specifications are $M_p = 1.2$, $t_p = 0.15$ s, and $t_s = 0.4$ s.

   (a) Determine a desirable $M(s)$ that will satisfy these specifications. Use steps 1 to 6 of the design procedure given in Sec. 13.6 to determine **k**. Draw the root locus for $G(s)H_{eq}(s) = -1$. From it show the "good" properties of state feedback.

   (b) Obtain $y(t)$ for the final design. Determine the values of the figures of merit and the ramp-error coefficient.

13.8. For the plants of Prob. 10.3 design a state-variable feedback system utilizing the phase-variable representation. The specifications are that dominant roots must have $\zeta = 0.5$, a zero steady-state error for a step input, and $t_s \leq 1/3$ s.

   (a) Determine a desirable $M(s)$ that will satisfy the specifications. Use steps 1 to 6 of the design procedure given in Sec. 13.6 to determine **k**. Draw the root locus for $G(s)H_{eq}(s) = -1$. From it show the "good" properties of state feedback.

   (b) Obtain $y(t)$ for the final design. Determine the values of the figures of merit and the ramp-error coefficient.

13.9. Design a state-variable feedback system that satisfies the following specifications: (1) for a unit-step input, $M_p = 1.15$ and $T_s \leq 0.8$ s; and (2) the system follows a ramp input with zero steady-state error.

   (a) Determine the expression for $y(t)$ for a step and for a ramp input.

   (b) For the step input determine M_p, t_p, and t_s.

   (c) What are the values of K_p, K_v, and K_a?

$$U(s) \quad \boxed{A} \quad \boxed{\frac{1}{s+4}} \xrightarrow{X_3(s)} \boxed{\frac{s+b}{s+1}} \xrightarrow{X_2(s)} \boxed{\frac{1}{s}} \xrightarrow{X_1(s)=Y(s)}$$

13.10. A closed-loop system with state-variable feedback is to have a control ratio of the form

$$\frac{Y(s)}{R(s)} = \frac{A(s+2)(s+0.5)}{(s^2+2s+2)(s-p_1)(s-p_2)}$$

where p_1 and p_2 are not specified.

   (a) Find the two necessary relationships between A, p_1, and p_2 so that the system has zero steady-state error with both a step and a ramp input.

   (b) With $A = 100$, use the relationships developed in (a) to find p_1 and p_2.

   (c) Use the system shown in Fig. 13.17 with the transfer function $1/(s+5)$ replaced by $(s+0.5)/(s+5)$. Determine the values of the feedback coefficients.

   (d) Draw the root locus for $G_x(s)H_{eq}(s) = -1$. Does this root locus show that the system is insensitive to changes of gain A?

(e) Determine the time response with a step and with a ramp input. Discuss the response characteristics.

13.11. For the plants shown design a state-variable feedback system that satisfies the following specifications: dominant roots must have a $\zeta = 0.6$, a zero steady-state error for a step input, and $T_s = 0.8$ s. Determine $M_p, t_s, t_p,$ and K_1.

(1) $\xrightarrow{\ U(s)\ }$ $\boxed{A}$ $\longrightarrow$ $\boxed{\dfrac{1}{s^2+20s+200}}$ $\xrightarrow{\ X_3(s)\ }$ $\boxed{\dfrac{1}{s+10}}$ $\xrightarrow{\ X_2(s)\ }$ $\boxed{\dfrac{1}{s}}$ $\xrightarrow{\ X_1(s)=Y(s)\ }$

(2) $\xrightarrow{\ U\ }$ $\boxed{\dfrac{A}{s+4}}$ $\longrightarrow$ $\boxed{\dfrac{6}{s^2+2s+13}}$ $\xrightarrow{\ X_1=Y\ }$

13.12. For the system of Fig. 13.15 it is desired to improve t_s from that obtained with the control ratio of Eq. (13.119) while maintaining an underdamped response with a smaller overshoot and faster settling time. A desired control ratio that yields this improvement is

$$\frac{Y(s)}{R(s)} = \frac{1.467(s + 1.5)}{(s + 1.1)(s^2 + 2s + 2)}$$

The necessary cascade compensator can be inserted only at the output of the amplifier A.

(a) Design a state-variable feedback system that yields the desired control ratio.

(b) Draw the root locus for $G(s)H_{eq}(s)$ and comment on its acceptability for gain variation.

(c) For a step input determine $M_p, t_p, t_s,$ and K_1. Compare with the results of Examples 1 to 3 of Sec. 13.12.

13.13. For Example 3, Sec. 13.12, determine $G_{eq}(s)$ and the system type. Redesign the state-variable feedback system to make it Type 2.

13.14. The desired control ratio $M_T(s)$ is to satisfy the following requirements: the desired dominant poles are given by $s^2 + 2s + 2$; there are no zeros; $e(t)_{ss} = 0$ for a step input; and the system is insensitive to variations in gain A. Determine an $M_T(s)$ that achieves these specifications for the plant

$$G_x(s) = \frac{A(s + 2)}{s(s + 1.1)(s + 2.4)(s + 4)}$$

13.15. Repeat Prob. 13.14 with the added stipulation that $e(t)_{ss} = 0$ for a ramp input. A cascade compensator may be added to achieve the desired specifications. A zero may be required in $M_T(s)$ to achieve $e(t)_{ss}$ for a ramp input. Compare the figures of merit with those of Prob. 13.14.

13.16. For the following plant

$$G_x(s) = \frac{A(s + 2)}{s(s^2 + 4s + 3.84)}$$

(a) Determine a desired control ratio $M_T(s)$ such that the following specifications are met: (1) the dominant roots are at $-1 \pm j1$, (2) the zero at -2 does not affect the output response, (3) $e(t)_{ss} = 0$ for a step input, and (4) the system sensitivity to gain variation is minimized.

(b) To achieve the desired $M_T(s)$, will a cascade unit in conjunction with $G_x(s)$ be needed? If your answer is yes, specify the transfer function(s) of the unit(s) that must be added to the system to achieve the desired system performance.

13.17. Given the following closed-loop control ratio

$$M(s) = \frac{Y(s)}{R(s)} = \frac{A(s + a)^2}{(s^2 + 2s + 2)(s + 2)(s + b)}$$

(a) With A, a, and b unspecified, determine $G_{eq}(s)$.

(b) For part (a), determine the condition that must be satisfied for $G_{eq}(s)$ to be a Type 2 transfer function.

(c) It is desired that $e(t)_{ss} = 0$ for step and ramp inputs. Determine the values of A, a, and b that will satisfy these specifications.

13.18. An open-loop plant transfer function is

$$G(s) = \frac{10}{s(s + 1)(s + 5)}$$

(a) Write state and output equations in control canonical (phase-variable) form.

(b) Compute the feedback matrix $\mathbf{k}_p^T$, which assigns the set of closed-loop eigenvalues $\{-0.708 \pm j0.706, -50\}$.

(c) Compute the state and output responses for the unit-step input $r(t) = u_{-1}(t)$ and $\mathbf{x}(0) = [1 \quad 0 \quad 1]^T$. Plot the responses.

13.19. For the plant represented by

$$\dot{\mathbf{x}} = \begin{bmatrix} 0 & 1 \\ -2 & -4 \end{bmatrix} \mathbf{x} + \begin{bmatrix} 0 \\ 2 \end{bmatrix} u$$

Design the feedback matrix $\mathbf{K}$ that assigns the set of closed-loop plant eigenvalues as $\{-3, -6\}$. Determine the state responses with the initial values $\mathbf{x}(0) = [1 \quad -1]^T$.

13.20. A SISO system represented by the state equation $\dot{\mathbf{x}}_p = \mathbf{A}_p\mathbf{x}_p + \mathbf{b}_p u$ has the matrices

$$\mathbf{A}_p = \begin{bmatrix} -1 & 0 & 0 \\ 0 & -2 & -2 \\ -1 & 0 & -3 \end{bmatrix} \quad \mathbf{b}_{p_1} = \begin{bmatrix} 1 \\ 1 \\ 1 \end{bmatrix} \quad \mathbf{b}_{p_2} = \begin{bmatrix} 1 \\ 1 \\ 0 \end{bmatrix}$$

For each $\mathbf{b}_p$:

(a) Determine controllability.

(b) Find the eigenvalues.

(c) Find the transformation matrix $\mathbf{T}$ that transforms the state equation to the control canonical (phase-variable) form.

(d) Determine the state-feedback matrix $\mathbf{k}_p^T$ required to assign the eigenvalue spectrum $\sigma(\mathbf{A}_{cl}) = \{-2, -4, -6\}$.

13.21. For the system of Prob. 13.20, determine the feedback matrix $\mathbf{k}$ that assigns the closed-loop eigenvalue spectrum $\sigma(\mathbf{A}_{cl}) = \{-1 + j1, -1 - j1, -2\}$.

13.22. Repeat Prob. 13.20 with

$$\mathbf{A}_p = \begin{bmatrix} 0 & 1 & 0 \\ 0 & -1 & -4 \\ 0 & 0 & -5 \end{bmatrix} \qquad \mathbf{b}_p = \begin{bmatrix} 0 \\ 1 \\ 2 \end{bmatrix} \qquad \sigma\{\mathbf{A}_{cl}\} = \{-1 + j1, -1 - j1, -4\}$$

13.23. $\dot{\mathbf{x}} = \mathbf{A}\mathbf{x} + \mathbf{B}\mathbf{u}$, where

$$\mathbf{A} = \begin{bmatrix} -2 & 1 & 0 & 0 \\ -2 & 0 & 1 & 0 \\ 0 & 1 & 1 & 0 \\ 0 & 0 & 0 & 1 \end{bmatrix} \qquad \mathbf{B} = \begin{bmatrix} 0 \\ 0 \\ 1 \\ 1 \end{bmatrix}$$

(a) Determine controllability.

(b) Transform the state equation to the control canonical form. Determine the matrix T.

(c) Design the feedback matrix **K** that assigns the closed-loop eigenvalue spectrum

$$\sigma(\mathbf{A}_{cl}) = \{-4 + j2, -4 - j2, -6, -10\}$$

(d) Obtain the time response of the closed-loop system with zero input and the initial conditions $\mathbf{x}(0) = [1 \quad 0 \quad 0 \cdot 1]^T$.

Chapter 14

14.1. A unity-feedback control system has the following forward transfer function:

$$G(s) = \frac{K(s + a)}{s^m(s + b)(s^2 + cs + d)}$$

The nominal values are $K = 15, a = 5, b = 3, c = 4$, and $d = 13$. (a) With $m = 1$, determine the sensitivity with respect to (1) K, (2) a, (3) b, (4) c, and (5) d. (b) Repeat with $m = 0$.

14.2. The range of values of a is $3 \geq a \geq 1, M_p \leq 1.12$ for a unit-step input, and $T_s \leq 0.8$s. Design a state-variable feedback system that satisfies the given specifications and is insensitive to variation of a. Assume that all states x_i are accessible and are fed back directly through the gains k_i. Obtain plots of $y(t)$ for $a = 1$, $a = 2$, and $a = 3$ for the final design. Compare M_p, t_p, t_s, and K_1. Determine the sensitivity functions for variation of A and a (at $a = 1, 2$, and 3).

14.3. Repeat Prob. 14.2 but with state X_3 inaccessible.

14.4. Determine the sensitivity function for Example 1 in Sec. 13.12 for variations in A. Insert the values of k_i in Eq. (13.118) and then differentiate with respect to A to obtain S_A^M.

14.5. For Example 2, Sec. 13.12, obtain $Y(s)/R(s)$ as a function of K_G [see Fig. 13.17 with $a = 1$]. Determine the sensitivity function $S_{K_G}^M$.

14.6. For the state-variable feedback system shown, the zero term a varies about the nominal value of 4. The feedback gains are $k_1 = 1$ and $k_2 = -0.16$. With $b = 1$, determine $S_{\delta=a}^M(s)$ when (a) the system is to be implemented as shown and (b) the system's structure is modified to minimize the sensitivity function. Evaluate $S_{\delta=5}^M(j0)$ for each case and compare.

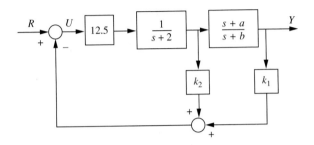

14.7. Repeat Prob. 14.6 where $a = 4 = $ constant and b varies about the nominal value of 3.

14.8. The desired control ratio for a tracking system is

$$M_T(s) = \frac{K_G(s + 1.5)}{(s + 1.1)(s^2 + 2s + 2)(s - p)}$$

This control ratio must satisfy the following specifications: $e(t)_{ss} = 0$ for $r(t) = R_0 u_{-1}(t)$, and $|S_{K_G}^M(j\omega)|$ must be a minimum over the range $0 \leq \omega \leq \omega_b$. ($a$) Determine the values of K_G and p that satisfy these specifications. The transfer function of the basic plant is

$$G_x(s) = \frac{K_x(s + 3)}{s(s + 1)(s + 5)} = \frac{K_x(s + 3)}{s^3 + 6s^2 + 5s}$$

A controllable and observable state-variable feedback-control system is to be designed for this plant so that it produces the desired control ratio above. (b) Is it possible to achieve $M_T(s)$ using only the plant in the final design? If not, what simple unit(s) must be added to the system to achieve the desired system performance? Specify the parameters of these unit(s). Using phase-variable representation, determine $\mathbf{k}^T$ for the state-variable feedback-control system. Assume appropriate values to perform the calculations.

14.9. For the state-feedback system shown, all the states are accessible, the parameters a and b vary, and sensors are available that can sense all state variables. The values of

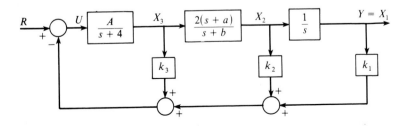

k_1, k_2, k_3, and A are known for a desired $M_T(s)$ based on nominal values for a and b. It is desired that the system response $y(t)$ be insensitive to parameter variations. (*a*) By means of a block diagram indicate an implementation of this design to minimize the effect of parameter variations on $y(t)$ by use of physically realizable networks. (*b*) Specify the required feedback transfer functions, ignoring sensor dynamics.

14.10. For the state-feedback control system of Prob. 14.9 with $a = 3$, $b = 2$, and a and b do not vary, the specifications are as follows: $e(t)_{ss} = 0$ for $r(t) = u_{-1}(t)$, $|S_A^M(j\omega)|$ must be a minimum over the range $0 \le \omega < \omega_b$, and the dominant poles of $M_T(s)$ are given by $s^2 + 4s + 13$. Determine (*a*) an $M_T(s)$ that achieves these specifications and (*b*) the values for K_G, k_1, k_2, and k_3 that yield this $M_T(s)$.

14.11. Obtain the isocline equation for each of the following differential equations. Draw sufficient isoclines to draw trajectories starting from the point indicated. Determine the kind of singularity and the stability in each case.

(*a*) $\ddot{x} + 5\dot{x} + 6x = 0$ $x_1 = 0, x_2 = 4$
(*b*) $\ddot{x} + 8\dot{x} + 12x = 0$ $x_1 = 0, x_2 = 5$
(*c*) $\ddot{x} + 6\dot{x} + 8x = 6$ $x_1 = 0, x_2 = 3$
(*d*) $\ddot{x} - 2\dot{x} + 5x = 0$ $x_1 = 2, x_2 = 2$
(*e*) $\ddot{x} + 4\dot{x} + 64x = 0$ $x_1 = 4, x_2 = 0$
(*f*) $\ddot{x} - 8\dot{x} + 17x = 34$ $x_1 = 0, x_2 = 6$
(*g*) $\dot{x} - 9x = 0$ $x_1 = 1, x_2 = 2$

14.12. Construct a phase portrait using the isocline method for the following second-order differential equations. Determine the kind of singularity and the stability for each case.

(*a*) $\ddot{x} + \dot{x}/|\dot{x}| + 4x = 0$ (*b*) $\ddot{x} + 8\dot{x} + x^2 = 0$
(*c*) $\ddot{x} + 2\dot{x}^2 + x = 0$ (*d*) $\ddot{x} + 2\dot{x}|\dot{x}| + 2x = 0$
(*e*) $\ddot{x} + \dot{x} + 4x|x| = 0$ (*f*) $\ddot{x} + \dot{x} - 8x|x| = 0$

14.13. Write the state equations for Prob. 14.11 and convert to normal form. Draw the trajectory in the normal plane. For Eqs. (*c*) and (*f*) it is simpler to change variables to move the equilibrium point to the origin.

14.14. Use the Jacobian matrix to determine the stability in the vicinity of the equilibrium points for the system described by:

(*a*) $\dot{x}_1 = 2x_1 - x_1^2 - x_1 x_2 + u_1$
     $\dot{x}_2 = -x_1 + x_2 + u_1 u_2$ where $\mathbf{u}_0 = [0 \quad 1]^T$
(*b*) $\dot{x}_1 = 4x_1 - 2x_1 x_2$
     $\dot{x}_2 = -x_1^2 + 2x_2$
(*c*) $\dot{x}_1 = -x_1^2 + x_2$
     $\dot{x}_2 = 4x_1 - 2x_2$

For parts (*b*) and (*c*), determine the equation for the slope of the phase-plane trajectories.

14.15. For the systems described by the state equations

(1) $\dot{x}_1 = x_1 - x_1^2 + x_1 x_2 - 2u_1$ $\dot{x}_2 = -x_1 + x_2 - u_1 u_2$ $\mathbf{u}_0 = [1 \quad 1]^T$
(2) $\dot{x}_1 = 2x_2$ $\dot{x}_2 = x_1 + 3x_2 + 0.25x_3^2$ $\dot{x}_3 = x_1 x_3 + x_3$
(3) $\dot{x}_1 = x_2$ $\dot{x}_2 = -3x_2 - 2\sin x_1 + u_2$

(*a*) Determine the equilibrium points for each system. (*b*) Obtain the linearized equations about the equilibrium points. (*c*) Determine whether the system is asymptotically stable or unstable in the neighborhood of its equilibrium points. (*d*) Identify the type of equilibrium points.

Chapter 15

15.1. By use of Eqs. (15.13) and (15.14), verify the following z transforms in Table 15.1: (*a*) entry 5, (*b*) entry 8, (*c*) entry 11, (*d*) entry 14.

15.2. Determine the inverse z transform, in closed form, of

$$\text{(a) } F(z) = \frac{z(z + 0.5)}{(z - 1)(z^2 - 0.5z + 0.3125)} \qquad \text{(b) } F(z) = \frac{-10(z^2 - 0.25z)}{(z - 4)(z - 1)^2}$$

15.3. Repeat Prob. 15.2 but use the power-series method to obtain the open form for $f(kT)$.

15.4. Determine the initial and final values of the z transforms given in Prob 15.2, i.e., solve for $c(0)$ and $c(\infty)$, where $R(z) = z/(z - 1)$.

15.5. Let $F(z) = C(z)/E(z)$ in Prob 15.2. Determine the difference equation for $c(kT)$ by use of the translation Theorem 1 of Sec. 15.4.

15.6. (CAD Problem) For the sampled-data control system of Fig. 15.11, $G(s)$ is given by Eq. (15.15), where $a = 3$ and T is unspecified. (*a*) Determine $G(z)$ and $C(z)/R(z)$. (*b*) Assume a sufficient number of values of T between 0.01 and 10 s, and for each value of T determine the maximum value that K can have for a stable response. (*c*) Plot $K_{\max}$ vs. T and indicate the stable region. (*d*) Obtain the root locus for this system with $T = 0.1$ s. (*e*) Use Eq. (15.19) to locate the roots on the root locus that have a $\zeta = 0.6$ and determine the corresponding value of K. (*f*) For a unit step input determine $c(kT)$, M_p, t_p, and t_s.

15.7. (CAD Problem) The sampled-data control system of Prob. 15.6 is to be compensated in the manner shown in Fig. 15.21. Determine a digital compensator that reduces t_s, of Prob. 15.6, by one-half. Determine M_p, t_p, t_s, and K_m.

15.8. For $T = 0.1$ s obtain the PCT system of Prob. 15.6. Obtain a root locus for this pseudo control system and compare it to the root locus for a unity-feedback continuous-time control system whose forward transfer function is $K/s(s + 3)$. This problem illustrates the degradation in system stability that results in converting a continuous-time system into a sampled-data system.

15.9. Repeat Prob. 15.8 for $T = 0.04$ s.

15.10. Given

$$G(s) = \frac{K(s + 3)(s + 20)}{s(s + 1.5)(s + 2 \pm j3)(s + 10 \pm j25)}$$

and $T = 0.04$ s. Map each pole and zero of $G(s)$ by use of Eq. (15.47) and by use of $z = e^{sT}$. Compare the results with respect to Fig 15.16 and analyze the warping effect.

15.11. Repeat Prob. 15.10, with $T = 0.01$ s, for:

(*a*) $s + 500$

(*b*) $s^2 + 2000s + 1,250,000 = s + 1000 \pm j500$

15.12. The poles and zeros of the following desired control ratio of the PCT system have been specified:

$$\left[\frac{C(s)}{R(s)}\right]_M = \frac{K(s+8)}{(s+5 \pm j7)(s+10)}$$

(a) What must be the value of K in order for $y(t)_{ss} = r(t)$ for a step input? (b) A value of $T = 0.1$ s has been proposed for the sampled-data system whose design is to be based on the above control ratio. Do all the poles and zeros lie in the good Tustin approximation region of Fig. 15.16? If the answer is no, what should be the value of T so that all the poles and zeros do lie in this region?

15.13.[†] (CAD Problem) For the basic system of Fig. 15.17, where $G_x(s) = K_x/s(s+2)$: (a) Determine $M_p, t_p, t_s,$ and K_1 for $\zeta = 0.65$ and $T = 0.01$ s by the DIR method. (b) Obtain the PCT control system of part (a) and the corresponding figures of merit. (c) Generate a digital compensator $D_c(z)$ (with no ZOH) that reduces the value of t_s of part (a) by approximately one-half for $\zeta = 0.65$ and $T = 0.01$ s. Do first by the DIG method (s plane) and then by the DIR method. Obtain $M_p, t_p,$ and t_s for $c(t)$ of the PCT and $c^*(t)$ (for both the DIG and the DIR designs) and K_m. Draw the plots of $c^*(t)$ vs. kT for both designs. (d) Find a digital compensator $D_c(z)$ (with no ZOH) that increases the ramp error coefficient of part (a), with $\zeta = 0.65$ and $T = 0.01$ s with a minimal degradation of the transient-response characteristics of part (a). Do first by the DIG method (s plane) and then by the DIR method. Obtain $M_p, t_p,$ and t_s for $c(t)$ of the PCT and $c^*(t)$, for both the DIG and the DIR designs, and K_m. Draw a plot of $c^*(t)$ vs. kT for both designs. (e) Summarize the results of this problem in tabular form and analyze. (f) Obtain the frequency-response plots of parts (a) to (d) and evaluate.

15.14. Repeat Prob. 15.13 for

$$G_x = K_x/[s(s+1)(s+3)]$$

15.15. Derive Eq. (15.51) from Eq. (15.50).

Chapter 16

16.1. $\dot{x} = Ax + Bu$

$$(1) \ A = \begin{bmatrix} 1 & 0 & 0 \\ 0 & 1 & 0 \\ 0 & -2 & 1 \end{bmatrix} \qquad B = \begin{bmatrix} 0 & 1 \\ 1 & 0 \\ 0 & 1 \end{bmatrix}$$

$$(2) \ A = \begin{bmatrix} -1 & 0 & 0 \\ -2 & -2 & -2 \\ -1 & 0 & -4 \end{bmatrix} \qquad B = \begin{bmatrix} 1 & 0 \\ 0 & 2 \\ 0 & 2 \end{bmatrix}$$

(a) Identify the permissible subspaces in $\ker S(\lambda)$ in which the eigenvectors must be located. Do not use specific values of λ. (See Sec. 16.4, Ex. 2.)

(b) Select linearly independent eigenvectors from the subspaces determined in (a) for:

$$\sigma(A_{cl}) = \{-4, -5, -6\}$$

[†] For part (b) use 4-decimal-digit accuracy: for parts (c) and (d) use 4-decimal-digit accuracy, then repeat using 8-digit accuracy. Compare the results.

(c) Determine the required state-feedback matrix **K** which assigns the closed-loop eigenvalue spectrum.

(d) Compute $\mathbf{A_{cl}} = \mathbf{A} + \mathbf{BK}$.

(e) Obtain the time response with $\mathbf{x}(0) = [1 \quad 0 \quad 1]^T$.

16.2. Repeat Prob. 16.1 for

(1) $\quad \mathbf{A} = \begin{bmatrix} 1 & 0 & 0 \\ 0 & 1 & 1 \\ 0 & -1 & 1 \end{bmatrix} \quad \mathbf{B} = \begin{bmatrix} 1 & 0 \\ 0 & 1 \\ 1 & 0 \end{bmatrix}$

(2) $\quad \mathbf{A} = \begin{bmatrix} 1 & 1 & 0 \\ 0 & 1 & 0 \\ 0 & -1 & 1 \end{bmatrix} \quad \mathbf{B} = \begin{bmatrix} 0 & 1 \\ 0 & 1 \\ 1 & 0 \end{bmatrix}$

Find $\mathbf{A_{cl}} = \mathbf{A} + \mathbf{BK}$.

16.3. Given:

$$\dot{\mathbf{x}} = \mathbf{Ax} + \mathbf{Bu} \qquad \mathbf{A} = \begin{bmatrix} -1 & 4 & 0 \\ 0 & -3 & 0 \\ 0 & 0 & -2 \end{bmatrix} \qquad \mathbf{B} = \begin{bmatrix} 1 & 0 \\ 0 & 0 \\ 0 & 1 \end{bmatrix}$$

(a) Determine the uncontrollable mode. (b) Identify the permissible subspace in $\ker \mathbf{S}(\lambda)$ in which the eigenvector must be located. (c) Assign the eigenvalue spectrum $\sigma(\mathbf{A_{cl}}) = \{\lambda_1, -4, -5\}$. (d) Compute K and $\mathbf{A_{cl}}$.

16.4. Repeat Prob. 16.1, using the method of Ex. 1, Sec. 16.4 or MATLAB, for the system shown with $\sigma(\mathbf{A_{cl}}) = \{-1, -2, -3, -4\}$ and $\mathbf{x}(0) = [1 \quad 0 \quad 0 \quad 1]^T$.

$$\mathbf{A} = \begin{bmatrix} 1 & 1 & 0 & 0 \\ 0 & 1 & 0 & 0 \\ 1 & 0 & 0 & 1 \\ 0 & 1 & 0 & 1 \end{bmatrix} \qquad \mathbf{B} = \begin{bmatrix} 1 & 0 \\ 1 & 0 \\ 0 & 0 \\ 0 & 1 \end{bmatrix}$$

16.5. Repeat Prob. 16.1 with

$$\mathbf{A} = \begin{bmatrix} 1 & 1 & 1 \\ 0 & 0 & 1 \\ 0 & -2 & 2 \end{bmatrix} \qquad \mathbf{B} = \begin{bmatrix} 1 & 1 \\ 0 & 0 \\ 2 & 0 \end{bmatrix} \qquad \sigma(\mathbf{A_{cl}}) = \{-1 + j1, -1 - j1, -6\}$$

In order to avoid the use of complex quantities, use Eq. (16.46) to compute **K**.

16.6. The system represented by

$$\dot{\mathbf{x}} = \begin{bmatrix} 1 & 0 & -3 \\ 0 & -1 & 1 \\ -2 & 2 & 0 \end{bmatrix} \mathbf{x} + \begin{bmatrix} 1 \\ 1 \\ 1 \end{bmatrix} u \qquad y = [1 \quad 0 \quad 1] \mathbf{x}$$

is required to follow the step input $r(t) = u_{-1}(t)$. (a) Determine the composite system matrices. (b) Is this system controllable? (c) Determine the feedback matrix $\mathbf{\bar{K}}$ required to assign the eigenvalue spectrum $\sigma(\mathbf{\bar{A}} + \mathbf{\bar{B}\bar{K}}) = \{-2, -3, -4, -5\}$. (d) For the initially quiescent system compute the time response and show that $y(t)_{ss}$ does follow $r(t)$.

16.7. Repeat Prob. 16.6 for the system of Prob. 16.1 with

$$\mathbf{r}(t) = [1 \quad -1]^T$$

$$y = \begin{bmatrix} 1 & 0 & 1 \\ 0 & 1 & 1 \end{bmatrix} \mathbf{x}$$

$$\sigma(\bar{\mathbf{A}} + \bar{\mathbf{B}}\bar{\mathbf{K}}) = \{-4, -5, -6, -7, -8\}.$$

16.8. Repeat Prob. 16.6 for the system of Prob. 16.2 with $\mathbf{r}(t) = [1 \quad -1]^T$, $\sigma(\bar{\mathbf{A}} + \bar{\mathbf{B}}\bar{\mathbf{K}}) = \{-2, -3, -4, -5, -6\}$, and

$$y = \begin{bmatrix} 0 & 1 & 1 \\ 1 & 1 & 0 \end{bmatrix} \mathbf{x}$$

16.9. Repeat Prob. 16.6 for the system of Prob. 16.3 with $r(t) = [1 \quad -1]^T$, $\sigma(\bar{\mathbf{A}} + \bar{\mathbf{B}}\bar{\mathbf{K}}) = \{-2, -3, -4, -5, -6\}$, and

$$y = \begin{bmatrix} 1 & 1 & 0 \\ 1 & 0 & 1 \end{bmatrix} \mathbf{x}$$

Even though the plant is uncontrollable, the tracking design can be achieved since the uncontrollable mode with $\lambda = -3$ is retained in the closed-loop system.

16.10. Repeat Prob. 16.6 for the system of Prob. 13.22 with $r(t) = u_{-1}(t)$ and $y = [1 \quad 0 \quad 10]\mathbf{x}$.

16.11. Draw a simulation diagram that represents the tracking system of Eq. (16.88) and the feedback-control law of Eq. (16.71).

16.12. For the system given below the specified closed-loop eigenvalue spectrum is $\sigma(\mathbf{A} + \mathbf{BK}) = \{-2, -3\}$. Find $\ker \mathbf{S}(\lambda_i)$.

$$\mathbf{A} = \begin{bmatrix} 1 & 1 \\ 2 & 0 \end{bmatrix} \qquad \mathbf{B} = \begin{bmatrix} 1 & 0 \\ 0 & 2 \end{bmatrix}$$

16.13. For Example 1 of Sec. 16.4 it is desired that the mode e^{-3t} does not appear in $y_2(t)$, where

$$\mathbf{C} = \begin{bmatrix} 0 & 1 & 0 \\ 1 & 0 & 1 \end{bmatrix}$$

(a) Select the eigenvectors that may produce this result.

(b) Does a permissible matrix $\mathbf{V}$ exist?

16.14. Repeat Prob. 16.13 for Example 2 in Sec. 16.4, where it is desired that the modes $(\cdot)e^{\lambda_2 t}$ and $(\cdot)e^{\lambda_3 t}$ do not appear in $y_1(t)$ and the mode $(\cdot)e^{-2t}$ does not appear in $y_2(t)$.

$$\mathbf{C} = \begin{bmatrix} 1 & 0 & 0 \\ 0 & 1 & 1 \end{bmatrix}$$

Chapter 17

17.1. Verify the values of the set of zeros S_2 as given by Eq. (17.60) for the system of Sec. 17.6.

17.2. For the system of Prob. 16.1, the output matrix is

$$\mathbf{C} = \begin{bmatrix} 1 & 1 & 0 \\ 0 & 0 & 1 \end{bmatrix}$$

(*a*) Design a PI controller. (*b*) Determine the asymptotic transfer function. (*c*) Determine the time responses with $\mathbf{r}_1(t) = [1 \quad 0]^T$ and $\mathbf{r}_2(t) = [0 \quad 1]^T$. Use a CAD program for the solution of this problem.

17.3. $\dot{\mathbf{x}} = \mathbf{Ax} + \mathbf{Bu}$, where

$$\mathbf{A} = \begin{bmatrix} -2 & 1 & 0 & 0 \\ -2 & 0 & 1 & 0 \\ 0 & 1 & 1 & 0 \\ 0 & 0 & 0 & 1 \end{bmatrix} \qquad \mathbf{B} = \begin{bmatrix} 0 & 0 \\ 0 & 0 \\ 1 & 1 \\ 0 & 1 \end{bmatrix}$$

$$\mathbf{C} = \begin{bmatrix} 1 & 0 & 1 & 0 \\ 0 & 0 & 0 & 1 \end{bmatrix}$$

(*a*) Design a PI controller. (*b*) Determine the transmission zeros. (*c*) Determine the asymptotic transfer function. (*d*) Determine the closed-loop time responses with $\mathbf{r}_1(t) = [1 \quad 0]^T$ and $\mathbf{r}_2(t) = [0 \quad 1]^T$.

17.4. The state and output equations of a system are

$$\dot{\mathbf{x}} = \begin{bmatrix} 0 & 0 & 1 \\ 1 & -1 & 1 \\ -2 & 0 & 2 \end{bmatrix} \mathbf{x} + \begin{bmatrix} 0 & 0 \\ 1 & 1 \\ 3 & 0 \end{bmatrix} \mathbf{u}$$

$$\mathbf{y} = \begin{bmatrix} 0 & -1 & 1 \\ 1 & 1 & 0 \end{bmatrix} \mathbf{x}$$

(*a*) Design a PI controller. (*b*) Determine the asymptotic transfer function. (*c*) Determine the time responses with $\mathbf{r}_1(t) = [1 \quad 0]^T$ and $\mathbf{r}_2(t) = [0 \quad 1]^T$. Use a CAD program for the solution of this problem.

17.5. Repeat Prob. 17.4 with the output matrix

$$\mathbf{y} = \begin{bmatrix} 0 & 1 & 0 \\ 1 & 0 & 0 \end{bmatrix} \mathbf{x}$$

17.6. Repeat Prob. 17.3 with the output matrix

$$\mathbf{y} = \begin{bmatrix} 1 & 1 & 0 & 0 \\ 0 & 0 & 0 & 1 \end{bmatrix} \mathbf{x}$$

17.7. Derive Eq. (17.25) by inserting Eq. (17.24) into Eq. (17.20) and taking the limit as $g \to \infty$.

17.8. Show that Eq. (17.47) can be put into the form of Eq. (17.48). *Hint:* Premultiply the matrix $[\lambda \mathbf{I}_m + g\mathbf{B}_2\mathbf{K}_1\mathbf{C}_2]$ by $\mathbf{C}_2^{-1}\mathbf{C}_2$ and factor to the form $[\mathbf{C}_2^{-1} (\ ) \mathbf{C}_2]$. Then apply the property on the inverse of a product of matrices given in Sec. 4.14. Note that the eigenvalues are not changed by the matrix operations, and therefore Eq. (17.44) yields the same eigenvalues as Eq. (17.43).

Chapter 18

Problems 18.1 through 18.3 are intended to give the reader experience in using a trial-and-error procedure, with the aid of straight-line Bode plots, in synthesizing a compensator $G_c(s)$ that will satisfy the performance specifications for the closed-loop system.

18.1. The plant for a unity-feedback control system is

$$P(s) = \frac{K_x}{s(s + 6)} \qquad K_x = 10$$

(a) For $K_x = 10$ determine ω_{ϕ_x} for $\gamma = 45°$. The desired closed-loop performance specifications are $K_1 = 24 \text{ s}^{-1}$ and $\gamma = 45°$. Design a lag compensator $G_c(s)$ [see Eq. (11.3)] that will achieve these closed-loop system specifications at $\omega_\phi = \omega_{\phi_x}$. Determine the values of α, T, and A of $G_c(s)$, where $1 < \alpha \le 100$. (b) For the basic system, adjust K_x to yield $K_1 = 24 \text{ s}^{-1}$ and determine the corresponding values of the phase margin angle and frequency γ and ω_ϕ, respectively. (c) Based upon the values of K_x and ω_ϕ of part (b), design a lead compensator such that $K_1 = 24 \text{ s}^{-1}$ and $\gamma = 45°$ for this value of ω_ϕ.

18.2. For the plant of Prob. 18.1 it is desired that $K_1 = 24 \text{ s}^{-1}$, $\gamma = 45°$, and $\omega_\phi = 15$ rad/s. Design a lag-lead compensator $G_c(s)$ that meets these closed-loop system performance specifications. *Hint:* Adjust K_x of the basic system to yield the desired value of K_1. Based upon the values of $\text{Lm}P(j15)$ and $\underline{/P(j15)}$, for this value of K_x, determine the location of the poles and zeros of $G_c(s)$ with the aid of a straight-line Bode plot that satisfies the desired performance specifications.

18.3. For the plant of Prob. 18.1 it is desired that $K_1 = 24 \text{ s}^{-1}$, $\gamma = 45°$, $\omega_\phi = 15$ rad/s, the gain margin $a = 3.17$ (10 dB), and $\underline{/L(j\omega)} \ge -140°$ for $\omega \le \omega_\phi$. Design a lag-lead compensator $G_c(s)$ that will achieve these closed-loop system specifications. Locate the poles and zeros of $G_c(s)$ with the aid of a straight-line Bode plot.

18.4. A design has been made for the following figure:

where the transfer functions $P(s)$, $F(s)$, and $G(s)$ are known. It turns out to be more convenient to implement the system by using the following figure:

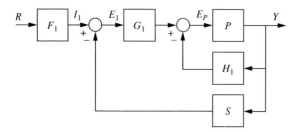

$H_1(s)$ is a tachometer and $S(s)$ is a sensor, and these transfer functions are known. Find $F_1(s)$ and $G_1(s)$ as functions of the known transfer functions $F(s)$, $G(s)$, $H_1(s)$, and $S(s)$ so that the two figures are equivalent. *Note:* the loop transmissions for both control systems must be identical.

18.5. In the following Fig. (a) $P(s) = 1/s$ and $G_1(s) = 1/(s+2)$. In Fig. (b) below $P(s) = 1/s$ and $H(s) = s/(s+1)$. (a) Find $G_2(s)$ so that $Y(s)/R(s)$ is the same for both configurations. (b) Find $S_P^T(s)$, the system sensitivity to variations in P, for each configuration. If these are different, which one yields smaller values of $|S_P^T(j\omega)|$ over a range of finite values of ω? (c) Which configuration is the best for minimizing the parameter variation effects on $Y(s)$?

(a)   (b)

18.6. For Prob. 18.4 the plant transfer function is $P(s) = K/[s(s+a)]$ where $1 \le K \le 5$ and $1 \le a \le 10$. [*Note:* see Fig. (b) of Prob. 18.9.] (a) Obtain templates for $\mathbf{P}(j\omega_i)$ and for

$$P_e(j\omega_i) = \frac{\mathbf{P}(j\omega_i)}{1 + \mathbf{P}(j\omega_i)\mathbf{H}_1(j\omega_i)}$$

which is defined as the effective plant and where $H_1(s) = s$, for values of $\omega_i = 0.5$, 1, and 2. Obtain template data for points A, B, C, and D of Fig. (b). (b) Explain why the sizes of the templates for $\mathcal{T}\mathbf{P}(j\omega_i)$ and $\mathcal{T}\mathbf{P}_e(j\omega_i)$ are different. See the discussion on the sensitivity function in Chap. 14.

18.7. For the control system shown in Fig. (a), $d(t) = u_{-1}(t)$. It is desired that $|\lim_{t\to\infty}[y(t)]| \le 10^{-2}$. Also, the maximum value of $|y(t)| < 0.3$ must be reached within 0.8 s, and $y(t)$ must not exceed this value. Design $G(s)$ so that these specifications are satisfied, where $\lim_{\omega\to\infty}\mathbf{G}(j\omega) = 0$ and $|\mathbf{G}(j\omega)|$ must be as small as possible for all ω.

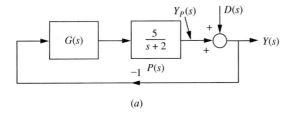

(a)

Hint: It is best to design the system by analyzing $y(t) = d(t) + y_p(t)$, where $Y_p(s) = -\{L(s)/[1 + L(s)]\}D(s)$. Figure (b) shows the desired straight-line approximations of the upper and lower bounds on $y_P(t)$.

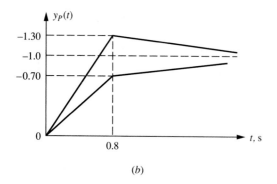

(*b*)

18.8. Verify the design of the prefilters of the two examples in Secs. 18.16 and 18.17 as follows: (*a*) Obtain the templates. (*b*) For each example, plot $\mathbf{L}_o(j\omega)$ vs. $\underline{/\mathbf{L}_0}$ on the NC. (*c*) Obtain the data from these plots that is necessary to verify the design of the prefilters.

In Probs. 18.9 and 18.10, for the control system shown, $P(s) = Ka/s(s+a)$ and the parameter variations shown in Fig. (*b*) are $K_x \leq K \leq K_y$ and $a_w \leq a \leq a_z$. (*Note:* Additional parameter variation values, other than the ones for these problems, may be supplied by the instructor.) For each of the two problems: (*a*) Design $G(s)$ and $F(s)$ so that the given performance specifications are met. (*b*) For the design achieved, for each point A, B, C, and D in Fig. (*b*), obtain $y(t)$ due to $r(t) = u_{-1}(t)$ and $d(t) = 0$. Repeat for $d(t) = u_{-1}(t)$ and $r(t) = 0$.

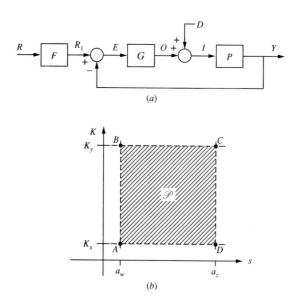

(*a*)

(*b*)

18.9. The plant parameter variations are $1 \leq a \leq 10$ and $1 \leq K \leq 12.6$. The performance specifications are, respectively,

For T_R : $T_U \to M_p = 1.2(\mathrm{Lm}M_m = 1.58$ dB) and $t_s = 1.6$ s
        : $T_U \to$ overdamped response and $t_s = 1.6$ s
For T_D : $\alpha_p = |y(t_p)_{\max}| = 0.1$ and $y(\infty) = 0$

where $r(t) = d(t) = u_{-1}(t)$.

18.10. The plant parameter variations are $1 \le a \le -15$ and $2 \le K \le 7$. The performance specifications are, respectively,

For T_R : $T_U \to M_p = 1.12(\mathrm{Lm}M_m = 1.984$ dB) and $t_s = 1.62$ s
        : $T_U \to$ overdamped response and $t_s = 1.62$ s
For T_D : $\alpha_p = |y(t_p)_{\max}| = 0.1$ and $y(\infty) = 0$

where $r(t) = d(t) = u_{-1}(t)$.

18.11. In Prob. 18.4 assume that $P = P_A P_p$, where P_A represents an actuator or a power amplifier with no uncertainty and P_p represents a plant with uncertainty. One designer made a QFT design with $P = P_A P_p$ that satisfied the system performance specifications. Another designer forgot to include the actuator in his QFT design, i.e., he assumed $P = P_p$. Is it necessary for him to redo his design?

18.12. Repeat Prob. 18.9 where now the control system is a regulator problem, i.e., $r(t) = 0$, $d(t) = u_{-1}(t)$, $\mathbf{B}_o(j\omega) = \mathbf{B}_D(j\omega)$. It is desired that $\alpha_p = |y(t_p)_{\max}| = 0.05$ and $y(\infty) = 0$.

18.13. Repeat Prob. 18.12 for Prob. 18.10.

18.14. Repeat Prob. 18.9 for $3 \le a \le 9$, $2 \le K \le 8$, $M_p = 1.3$, and $M_p = 1.0$ for T_U and T_L, respectively, and $t_s = 1.7$ s for both T_U and T_L.

ANSWERS
TO SELECTED
PROBLEMS

Chapter 2

2.1. (a) *Loop equations:* Let $R = R_2 + R_3$.

$$e(t) = \left(R_1 + LD + \frac{1}{CD}\right)i_1 - \left(LD + \frac{1}{CD}\right)i_2 \tag{1}$$

$$0 = -\left(LD + \frac{1}{CD}\right)i_1 + \left(R + LD + \frac{1}{CD}\right)i_2 \tag{2}$$

(b) *Node equations:*
   Node v_a:

$$\left(\frac{1}{R_1} + \frac{1}{R_2} + \frac{1}{LD}\right)v_a - (CD)v_b - \left(\frac{1}{R_2}\right)v_c = \left(\frac{1}{R_1}\right)e \tag{3}$$

   Node v_b:

$$-(CD)v_a + \left(\frac{1}{LD} + CD\right)v_b = 0 \tag{4}$$

   Node v_c:

$$-\left(\frac{1}{R_2}\right)v_a + \left(\frac{1}{R_2 + R_3}\right)v_c = 0 \tag{5}$$

(c) *State equations:* Let $x_1 = y_1 = i/CD$, $x_2 = i_L$, $x_2 = Di_L$, $i_c = CDy_1 = Cx_1$,
   $i_1 - i_2 = i_c = i_L$

$$\mathbf{A} = \begin{bmatrix} 0 & \dfrac{1}{C} \\ -\dfrac{1}{L} & -\dfrac{R'}{(R'+1)L} \end{bmatrix} \tag{6}$$

$$\mathbf{b}^T = \begin{bmatrix} 0 \\ 1 \\ \dfrac{1}{(R'+1)L} \end{bmatrix} \qquad \mathbf{c}^T = \begin{bmatrix} 1 & 0 \end{bmatrix} \tag{7}$$

(*d*) *Note:* Utilizing Eqs. (1), (2), (8), and (9), obtain

$$x_2 = C\dot{x}_1 \tag{8}$$

$$\dot{x}_2 = C\ddot{x}_2 \tag{9}$$

$$G(D) = \frac{y_1(t)}{u(t)} = \frac{1}{[(R'+1)LC]D^2 + [R'C]D + [R'+1]} \tag{10}$$

2.2. (*a*) *Loop equations:*

$$E = \left(R_1 + \frac{1}{C_1 D}\right)i_1 - \left(\frac{1}{C_1 D}\right)i_2 \tag{1}$$

$$0 = -\left(\frac{1}{C_1 D}\right)i_1 + \left(R_2 + LD + \frac{1}{C_1 D} + \frac{1}{C_2 D}\right)i_2 \tag{2}$$

(*b*) Node 1 between R_1 and R_2, Node 2 between R_2 and L, and Node 3 between L and C_2

$$\text{Node 1:} \quad \left(\frac{1}{R_1} + C_1 D + \frac{1}{R_2}\right)v_1 - \left(\frac{1}{R_2}\right)v_2 = \left(\frac{1}{R_1}\right)E$$

$$\text{Node 2:} \quad -\left(\frac{1}{R_2}\right)v_1 + \left(\frac{1}{R_2} + \frac{1}{LD}\right)v_2 - \left(\frac{1}{LD}\right)v_3 = 0$$

$$\text{Node 3:} \quad -\left(\frac{1}{LD}\right)v_2 + \left(\frac{1}{LD} + C_2 D\right)v_3 = 0$$

(*c*) Let $x_1 = v_1 = v_{c1}$, $x_2 = v_3 = v_{c2}$, $x_3 = i_L$

$$\mathbf{A} = \begin{bmatrix} -\dfrac{1}{R_1 C_1} & 0 & -\dfrac{1}{C_1} \\ 0 & 0 & \dfrac{1}{C_2} \\ \dfrac{1}{L} & -\dfrac{1}{L} & -\dfrac{R_2}{L} \end{bmatrix} \qquad \mathbf{b} = \begin{bmatrix} \dfrac{1}{R_1 C_1} \\ 0 \\ 0 \end{bmatrix}$$

2.3. Circuit 2: (*b*)

$$\dot{\mathbf{x}} = \begin{bmatrix} -\dfrac{R_1 + R_2}{L} & -\dfrac{1}{L} \\ \dfrac{1}{C} & 0 \end{bmatrix} \mathbf{x} + \begin{bmatrix} \dfrac{1}{L} \\ 0 \end{bmatrix} \mathbf{u}$$

2.6. (c)

$$G(D) = \frac{K_2}{D^4 + BD^3 + KD^2 + BK_aD + (K_aK_b - K^2)}$$

where $K_a = K_1 + K_2$, $K_b = K_2 + K_3$, $K = K_a + K_b$

(d) Let

$$x_1 = x_a \qquad x_2 = \dot{x}_1 \qquad x_3 = x_b$$

$$x_4 = \dot{x}_3 \qquad u = f(t) \qquad y(t) = x_b$$

$$\mathbf{A} = \begin{bmatrix} 0 & 1 & 0 & 0 \\ -K & K_2 & 0 & 0 \\ 0 & 0 & 0 & 1 \\ K_2 & 0 & K_b & -1 \end{bmatrix} \qquad \mathbf{b} = \begin{bmatrix} 0 \\ 1 \\ 0 \\ 0 \end{bmatrix} \qquad \mathbf{c} = \begin{bmatrix} 0 \\ 0 \\ 1 \\ 0 \end{bmatrix}$$

2.10. (a) $(LD + R)i = e$, $\quad K_i i = f$, $\quad \dfrac{x_a}{l_1} = \dfrac{x_b}{l_2}$

$(M_2 D^2 + B_2 D + K_2)x_b = f_2$

$(M_1 D^2 + B_1 D + K_1)x_a + \dfrac{l_2}{l_1} f_2 = f = k_i i$

Let $\dfrac{l_2}{l_1} = a$, $\dfrac{M_1}{a} + aM_2 = b$, $\dfrac{B_1}{a} + aB_2 = c$, and $\dfrac{K_1}{a} + aK_2 = d$

$(bD^2 + cD + d)x_b = K_i i$

(b) There are only three independent states:

$$x_1 = x_b, \quad x_2 = \dot{x}_b, \quad x_3 = i, \quad \text{and } u = e$$

$$\dot{x}_1 = x_2$$

$$\dot{x}_2 = -\frac{d}{b}x_1 - \frac{c}{b}x_2 + \frac{K_1}{b}x_3$$

$$\dot{x}_3 = \frac{R}{L}x_3 + \frac{l}{L}u$$

2.11. Traction force $= MD^2 x + BDx$

2.12. (b) $T(t) = (B_1 D)\theta_a - (B_1 D)\theta_b$

$$0 = -(B_1 D)\theta_a + [JD^2 + (B_2 - B_1)D + K_1]\theta_b$$

(c) Let $J = J_1 + J_2$, $y_1 = x_1 = \theta_a$, $y_2 = x_2 = \theta_b$, $x_3 = x_2$, and $u = T$

$$\dot{\mathbf{x}} = \begin{bmatrix} 0 & 0 & 1 \\ 0 & 0 & 1 \\ 0 & -\dfrac{K}{J} & \dfrac{B_1 - B_2}{J} \end{bmatrix} \mathbf{x} + \begin{bmatrix} \dfrac{1}{B_1} \\ 0 \\ \dfrac{1}{J} \end{bmatrix} \mathbf{u} \qquad \mathbf{y} = \begin{bmatrix} 1 & 0 & 0 \\ 0 & 1 & 0 \end{bmatrix} \mathbf{x}$$

(d)
$$\frac{y_1}{u} = \frac{JD^2 + B_2D + K}{B_1D[JD^2 + (B_2 - B_1)D + K]}$$

$$\frac{y_2}{u} = \frac{1}{JD^2 + (B_2 - B_1)D + K}$$

2.18. (a)
$$\left[\frac{C_1 ad}{(a + b)(c + d)} + D\right] y_1 = \frac{C_1 b}{a + b} x_1$$

2.21. $K_T v_1(t) = JL_a D^3 \theta_m + (JR_a + BL_a)D^2\theta_m + (BR_a + K_T K_b)\theta_m$

Chapter 3

3.2.
$$\theta_b(t) = -2 + 2t + A_1 e^{m_1 t} + A_2 e^{m_2 t}$$

where $A_1 = 2.01015$ $A_2 = -0.01015$

Therefore
$$\theta_b(t) = -2 + 2t + 2.010105 e^{-1.071t} - 0.01015 e^{-15.06t}$$

Check at $t = 0$: $\theta_b(0) = -2 + 2.01015 - 0.01015 = 0$

3.5. (b) $c(t) = -0.36 + 0.2t + A_1 e^{m_1 t} + A_2 e^{m_2 t} + A_3 e^{m_3 t}$

Utilizing the initial conditions yields
$$A_1 = 0.5, \qquad A_2 = (\sqrt{2}/10)\underline{/-171.9°} = A_3^*$$

3.7. (a) Physical variables: $x_1 = v_c$ and $x_2 = i_l$.
$$\mathbf{A} = \begin{bmatrix} -10^4 & -10^6 \\ 0.1 & -465 \end{bmatrix} \qquad \mathbf{b} = \begin{bmatrix} 10^4 \\ 0 \end{bmatrix} \qquad \mathbf{c}^T = [1 \quad 0]$$

(b) $v_c(t) = 48.95 + A_1 e^{m_1 t} + A_2 e^{m_2 t}$ where $Dv_c = m_1 A_1 e^{m_1 t} + m_2 A_2 e^{m_2 t}$

Inserting the initial conditions yields
$$0 = 48.95 + A_1 + A_2, \qquad 0.5 \times 10^6 = -475.5 A_1 - 9989.5 A_2,$$
$$A_1 = 1.15, \qquad A_2 = -50.1$$

3.8. (b) $x(t) = 3 + 1.5 e^{-3t} - 4.5 e^{-t}$

(c) $x(t) = 0.18 + 0.2353t + e^{-0.5t}[-0.18\cos(2t) - 0.16\sin(2t)]$
$$= 0.18 + 0.2353t + 0.242 e^{-0.5t}\sin(2t - 132°)$$

3.13.

$$\mathbf{\Phi}(t) = \begin{bmatrix} -e^{-2t} + 2e^{-4t} & 2e^{-2t} - 2e^{-4t} \\ -e^{-2t} + e^{-4t} & 2e^{-2t} - e^{-4t} \end{bmatrix} \qquad \mathbf{x}(t) = \begin{bmatrix} \frac{1}{2} - 3e^{-2t} + \frac{9}{2}e^{-4t} \\ \frac{3}{4} - 3e^{-2t} + \frac{9}{4}e^{-4t} \end{bmatrix}$$

$$y(t) = x_1 = \tfrac{1}{2} - 3e^{-2t} + \tfrac{9}{2}e^{-4t}$$

3.16. (a)

$$\mathbf{x}(t)_h = \mathbf{\Phi}(t)\mathbf{x}(0) = \begin{bmatrix} e^{-2t} \\ -2e^{-2t} \end{bmatrix}$$

(b)

$$\mathbf{x}(t) = \begin{bmatrix} 1 - 2e^{-t} + 2e^{-2t} \\ -2e^{-t} \end{bmatrix}$$

3.17. (1) (a) $\lambda_1 = -1, \lambda_{2,3} = -1 \pm j2$

$$\mathbf{\Phi}(t) = \mathbf{Z}_1 e^{-t} + \mathbf{Z}_2 e^{-t} e^{jt} + \mathbf{Z}_3 e^{-t} e^{-jt}$$

Chapter 4

4.1. (a) $f(t) = e^{-t}[t + 0.333 \sin(3t + 180°)]$

(b) $f(t) = 1.3846 - e^{-5t} + 1.9612e^{-2t} \sin(3t + 191.3°)$

4.2. (c) $x(t) = 1 - 0.9898e^{-t} + 0.01701e^{-2t} \sin(3t + 216.9°)$

4.4. (b) $x_b(t) = 4.167 \times 10^{-3} + 0.6445 \times 10^{-4}e^{-10t} - 4.81 \times 10^{-3}e^{-1.2874t}$
$- 1.128 \times 10^{-6}e^{-223.7t}$

4.5. (d) $F(s) = \dfrac{1}{s^2} - \dfrac{7}{6s} + \dfrac{5}{4(s+1)} - \dfrac{0.5}{6(s+3)}$

(h) $F(s) = \dfrac{1}{s} + \dfrac{9}{s+3} + \dfrac{28}{s+3-j1} + \dfrac{30}{s+3+j1}$

4.6. (b) $i_L(t) = 0.4833e^{-3.684t} - 0.5534 \sin(\omega_d t + 61°)$

4.7. (b) $(s^2 + 2.8s + 4)X(s) = \dfrac{2(s^2 + 4.3s + 6.13)}{s}$

$x(t) = 3.0669 + 1.5e^{-1.4t} \sin(1.416t - 45.32°)$

(d) $(s^5 + 4s^4 + 34s^3 + 110s^2 + 225s + 250)X(s) + 4s^2 - s - 56 = \dfrac{5}{s^2 + 25}$

$x(t) = 10.837e^{-t} \sin(2t - 85.233°) + 0.99974 \sin(5t + 180°) + 14.799e^{-2t}$

4.8. (a) Prob. 4.1(a): 0,   Prob. 4.1(b): 18/13

(b) Prob. 4.2(a): 0,   Prob. 4.2(c): 1

4.11. (1) (b) $x(t) = 3 + 3.8348e^{-t} \sin(t + 237.53°) + 0.2353e^{-5t}$

(c) $x(t) = 0.30t - 0.260 + 0.2712e^{-t} \sin(t + 102.53°) - 0.0047e^{-5t}$

4.12. (1) (a)

$$X(s) = \begin{bmatrix} \dfrac{1}{6s} + \dfrac{1}{2(s+2)} - \dfrac{2}{3(s+3)} \\[2mm] \dfrac{1}{3s} + \dfrac{2}{3(s+3)} \end{bmatrix}$$

(b) $G(s) = \dfrac{1}{(s+2)(s+3)}$

4.15. (1) $\sigma(\mathbf{A}) = \{-1 + j2, -1 - j2, -1\}$

4.19. (1) (a) $Q(\lambda) = \lambda^3 + 6\lambda^2 + 11\lambda + 6 = (s+1)(s+2)(s+3)$

(b)

$$\mathbf{A}^{-1} = \begin{bmatrix} -1 & 0 & 0 \\ 1/3 & -1/2 & 1/3 \\ 1/3 & 0 & -1/3 \end{bmatrix}$$

Chapter 5

5.5. $G(s) = \dfrac{-(60s + 3.8)}{(s+10)(s^2 + 0.15s + 8.0054)}$

5.11. (2) $\lambda = -2, \quad \lambda = -3, \quad \lambda = -4, \quad \lambda = -5$

(a) $y = [8 \quad 0 \quad 4 \quad 0]\mathbf{x}$

(b) $y = [4 \quad -22 \quad 36 \quad -18]\mathbf{x}$

5.13. For system (2):

(b) $\lambda_2 = -1, \quad \lambda_2 = -3, \quad \lambda_3 = -4$

(c)

$$\mathbf{T} = \begin{bmatrix} 11 & 1 & 1 \\ -1 & -3 & -4 \\ 1 & 9 & 16 \end{bmatrix}$$

(d)

$$\dot{\mathbf{z}} = \begin{bmatrix} -1 & 0 & 0 \\ 0 & -3 & 0 \\ 0 & 0 & -4 \end{bmatrix} \mathbf{z} + \begin{bmatrix} 1/3 \\ -1 \\ -2/3 \end{bmatrix} \mathbf{u} \qquad y = [0 \quad -6 \quad 3]\mathbf{x}$$

(f) The mode with $\lambda = -1$ is unobservable.

5.17. For system (1):

$$\mathbf{\Lambda} = \begin{bmatrix} -1 & 2 & 0 \\ -2 & -1 & 0 \\ 0 & 0 & -3 \end{bmatrix}$$

5.18. (2) (a) $G(s) = \dfrac{-10}{(s+2)(s+3)}$ \qquad (c) $x(t) = \begin{bmatrix} -\frac{5}{3}(1 - 3e^{-2t} + 2e^{-3t}) \\ \frac{10}{3}(1 - e^{-3t}) \end{bmatrix}$

5.19.

(1) (a) $\mathbf{T} = \begin{bmatrix} 15 & 8 & 1 \\ 5 & 1 & 0 \\ 1 & 4 & 1 \end{bmatrix}$ \qquad (b) $\mathbf{T} = \begin{bmatrix} 3 & 2 & 0 \\ 3 & 1 & 0 \\ -1 & -1 & -1 \end{bmatrix}$

5.22. (1) (a) $\lambda_1 = -1, \lambda_2 = -2, \lambda_3 = -4$

$$\mathbf{v}_1 = \begin{bmatrix} 2 \\ -3 \\ 3 \end{bmatrix} \qquad \mathbf{v}_2 = \begin{bmatrix} 1 \\ -1 \\ 2 \end{bmatrix} \qquad \mathbf{v}_3 = \begin{bmatrix} 1 \\ 0 \\ 0 \end{bmatrix}$$

Chapter 6

6.2. (a) $0 < K < 1280/9$; $s_{1,2} = \pm j2.5827$; for $K = 142.22$
   (c) $0 < K < 30$; $s_{1,2} = \pm j2.2356$; for $K = 30$
6.3. (a) $(s + 2)(s + 2.2 \pm j2.2)$ (c) $(s + 1 \pm j1.414)(s - 1 \pm j1.414)$
   (f) $(s - 1)(s + 2)(s - 3)(s + 5)$
6.4. (a) None   (c) None   (f) 2
6.10. (a) For (1): $K_p = \infty$, $K_v = 3.0$, $K_a = 0$ (b) For $r_1(t)$: (1) $e(t)_{ss} = 2/3$
   (2) $e(t)_{ss} = 6.25$
6.12. (a) 20, 0, 0   (d) ∞, 36, 0
6.13. (a): (a) 5/2,   (b) ∞,   (c) ∞
   (d): (a) 0,   (b) 1/3,   (c) ∞
6.19. $K_G c_i = a_i + K_h K_G c_i > 0$ for a stable system and $K_G c_0 = a_0 + K_h K_G c_0$ for $e(t)_{ss} = 0$
6.25. (a) $0 < K < 14$ (b) $e(t)_{ss} = 21/70$

Chapter 7

7.2. (c) For $K > 0$, real-axis branches: none; four branches; $\gamma = \pm 45°, \pm 135°$; angles of departure: $\pm 45°, \pm 135°$; imaginary-axis crossing: $\pm j2$ with $K = 0$. Branches are straight lines and coincide with the asymptotes.
   (e) Real-axis branch: 0 to $-\infty$; five branches; $\gamma = \pm 36°, \pm 108°, 180°$; $\sigma_0 = -1.6$; no breakaway or break-in points; angles of departure: $\pm 71.7°, \pm 33.8°$; and the imaginary-axis crossing: $\pm j1$ with $K = 30$. Locus crosses the $\pm 36°$ asymptotes.
7.4. (a) For $K > 0$, real-axis branches: 0 to -10; three branches; $\gamma = \pm 90°$; $\sigma_0 = 0$; break-in point: -3.9113 ($K = 3.33095$); breakaway point: -2.78698 ($K = 3.4388$); and imaginary-axis crossing: none.
   (b) $K = 0.05427$
   (c) $e(t) = 1.5951 e^{-2.2164t} \sin(2.1992t + 207.274°)$
   (d)

ω	M	α, deg	ω	M	α, deg
0	1	0	1.4	0.9586	-44.7
0.2	0.9995	-6.1	1.8	0.9153	-58.5
0.4	0.998	-12.3	2.0	0.8854	-65.5
0.6	0.9951	-18.5	2.4	0.8103	-79.3
0.8	0.9903	-24.9	2.8	0.7202	-92.3
1.0	0.9832	-31.3	3.2	0.6252	-104.1

7.7. Poles: $p_1 = -1$, $p_2 = -7$, $p_{3,4} = 4 \pm j3$

(a) For $K > 0$, real-axis branch: -1 to -7; four branches; $\gamma = \pm45°, \pm135°$; angles of departure: $0°, -180°, \pm90°$; imaginary-axis crossing: $\pm j4$ with $K = 1105$; break-in and breakaway points coincide at $s = -4$. For $K < 0$, real-axis branches: -1 to $+\infty$ and -7 to $-\infty$; $\gamma = \pm90°, 180°$; angles of departure: $0°, \pm90°, 180°$; imaginary-axis crossing: $s = 0$ with $K = -175$.

(b) $-175 < K < 1105$ (c) $K = 81$

7.8. (1) (a) For $K = -50K_1$: three branches; $\gamma = \pm90°$ for $K_1 < 0$; $\gamma = 180°$ for $K_1 > 0$; $\sigma_0 = -9.5$; breakaway points: -7.72 for $K > 0$, -3.139 for $K < 0$; break-in point: 5.362 for $K < 0$; imaginary-axis crossings: at $s = 0$ for $K = 50$ and $\pm j3.6993$ for $K = -66.32$.

(b) Stable for $-66.32 < K < 50$

7.11. (a) *Real-axis branches:* -3.2 to $-\infty$ (Note: the canceled pole and zero at $s = -5$ is a root of the characteristic equation and is not utilized in determining the root locus for locating the remaining roots.) *6-branches:* $\gamma = \pm36°, \pm108°, 180°$; $\sigma_0 = -22.5$; *Break-in point:* $s = -16.5$; *Angle of departures* (given in order of closeness to the imaginary axis in the 2nd quadrant): $144.4°, -94.4°, 4.2°$; *Imaginary-axis crossing:* $\pm j17.20$ for $K = 1.869 \times 10^7$.

(b) $K_G K_H = 1.0735$; $p_{1,2} = \pm j17.21$, $p_{3,4} = 33.28 \pm j28.05$, $p_5 = -4.634$, $p_6 = -44.46$, $p_7 = -5$

(c) $$\frac{C(s)}{R(s)} = \frac{1.869 \times 10^7(s + 3.2)}{(s + 4.634)(s + 5)(s + 44.46)(s \pm j17.21)(s + 33.28 \pm j28.05)}$$

7.12. (1) (a) $A = 1.44$

Chapter 8

8.1. (b) (1) $\mathbf{G}(j0+) = \infty\angle{-90°}$, $\mathbf{G}(j\infty) = 0\angle{-270°}$, $\omega_x = \omega_c = 2.236$, $\mathbf{G}(j\omega_x) = -0.2223$, $\omega_\phi = 0.867$

(2) $\mathbf{G}(j0+) = \infty\angle{-90°}$, $\mathbf{G}(j\infty) = 0\angle{-360°}$, $\omega_x = \omega_c = 15.82$, $\mathbf{G}(j\omega_x) = -0.0952$, $\omega_\phi = 1.972$

8.4. (1) (b) $\mathbf{G}(j0+) = \infty\angle{-90°}$, $\mathbf{G}(j\infty) = 0\angle{-270°}$, $\omega_x = 23$, $\mathbf{G}(j\omega_x) = -0.049226$

8.6. (a) It must have an initial slope of -40 dB/decade

(b) The phase-angle curve approaches an angle of $180°$ as $\omega \to 0$

8.9. Case II:

$$G(s) = \frac{20(1 + 0.5s)}{s(1 + 5s)(1 + 0.125s)}$$

8.10. (a) $K_0 = 31.7$ (b) $K_2 = 2.56 = (\omega_y)^2$

8.11. (a)

$$G(s) = \frac{0.4}{s(s + 1 \pm j4)}$$

8.13. (a) $N = -2$, $Z_R = 2$, unstable   (c) $N = -1$, $Z_R = 2$, unstable

8.14. (a) For stability $K > 1.26759$

8.15. (a) Stable for $K > -2$ (e) stable for $8 < K < 90$

8.16. (1) (a) $K_1 = 5.5654, \omega_\phi = 4.845$ (b) $K_1 = 3.345, \omega_\phi = 3.15$
     (c) maximum $K_1 = 1.9429 \times 10^5$
     (2) (a) $K_0 = 20.13, \omega_\phi = 8.3$ (b) $K_0 = 8.58, \omega_\phi = 4.61$
     (c) Stable for $0 < K_0 < \infty$

8.19. (a) $K_1 \approx 0.372 \ (\omega_x \approx 1.075)$

Chapter 9

9.1. (2) $K_1 = 7.94, M_m = 1.326, \omega_m = 7.81, \omega_\phi = 6.94, \omega_c = 15.81$
     (4) $K_1 = 0.7075, M_m = 1.318, \omega_m = 0.658, \omega_\phi = 0.6054, \omega_c = 1.508$

9.3. $M_m = 1.30, K_1 = 2.975, \omega_m = 2.75$

9.5. (1) (a) $M_m = 1, \omega_m = 0$ (b) $M_m = 1.326, \omega_m = 5.4$ (c) $M_m = 1.016$,
     $\omega_m = 2.24$

9.8. (2) (a) $M_m = 9.574, \omega_m = 0.689$ (c) $K_0 = 0.18$ for $M_m = 1.12$
     (d) $\omega_m = 0.39, \gamma = 56.16°$
     (3) Completely unstable for all positive values of gain.

9.10. (c) $K_1 = 3.046$

9.12. Using a CAD program: (a) $M_m = 1.11293, \omega_m = 0.435$ (b) $K = 1.338$
     (c) $a = 6$

Chapter 10

10.3, 10.20.

System (1)	Dominant Roots	Other roots	$K_{1,s^{-1}}$	T_p, s	t_s, s	M_o
Basic	$-1.25 \pm j2.166$	-6.501	1.72	1.67	3.36	0.14
Lag-compensated: $\alpha = 10, T = 2$	$-1.03 \pm j1.787$	-5.267 -0.726	16.16	1.8	5.37	0.41
Lead-compensated: $\alpha = 0.1, T = 0.25$	$-2.23 \pm j3.86$	-3.123 -40.42	3.13	0.916	1.78	0.10
Lag-lead-compensated: $\alpha = 10, T_1 = 2,$ $T_2 = 0.25$	$-2.08 \pm j3.6$	-0.598 -2.897 -40.40	30.18	0.967	4.15	0.244
Tachometer-compensated: $A = 1.61, K_t = 1,$ $\zeta = 0.5$	$-3.40 \pm j5.89$	-1.2	1.02	—	3.39	0.0
$A = 4.79, K_t = 1,$ $\zeta = 0.3$	$-1.86 \pm j5.92$	-4.28	2.97	0.756	1.89	0.13

10.7. (a) $a = 4, b = 2.6653$ (b) $\alpha = 1.126$ (c) lag compensator

10.8. (a) For the uncompensated system: from the performance specifications, for $\zeta = 0.5$, are $M_p = 1.16$, $t_p = 2$ s, $t_s = 4.39$ s, $K = 2.933$, $K_1 = 1.76$ s^{-1}, and the control ratio poles are $-0.9235 \pm j1.601$ and -5.153.

(b) $|\sigma| \geq 4/T_s = 2$; assuming $a = 2$, $\zeta = 0.5$, and $|\sigma| = 2$ yields $b = 4.726$

(c) $A = 5.21$ and $\alpha = 0.4232$

(d) A lead compensator; $M_p = 1.59$, $t_p = 0.917$ s, $t_s = 2.02$ s, and $K_1 = 3.88$ s^{-1}

10.20. (a) For a $\zeta = 0.5$, the uncompensated system yields the following data: $K = 25.01$, $K_1 = 5$ s^{-1}, $M_p = 1.163$, $t_p = 0.725$ s, $t_s = 1.615$ s.

(b) Based upon the desired values of ζ and $\sigma = -7$, the desired dominant poles are $P_{1,2D} = -7 \pm j12.124$. One possible solution is: let $a = 40$, $b = 0.01$, $c = 0.1$; by applying the angle condition and the use of a CAD package the following results are obtained: $A = 6.0464$, $K_h = 11.18$, $K_1 = 126.1$ s^{-1}, $M_p = 1.161$, $t_p = 0.27$ s, $t_s = 0.426$ s.

10.22. See Prob. 10.3 for basic system performance data. Based on the desired performance specifications, the desired dominant poles are: $P_{1,2D} = -4 \pm j6.928$. Let $K_x = 1250K_1$ be the gain term of $G(s)$. By applying method 1, the following results are obtained: $K_x = 760.71$, $K_t = 0.5619$, $b = 11.895$, $M_p = 1.092$, $t_p = 0.725$ s, $t_s = 0.97$ s, $K_1 = 2.369$ s^{-1}. The other poles are: $-6.296 \pm j4.486$ and -11. The plot of $c(t)$ has the desired simple second-order time response characteristics due to the effect of the real root.

Chapter 11

11.1. (a) $K_x = 1286$, $\omega_{\phi x} = 2.12$, $\gamma = 51.13°$, $M_p = 1.164$, $t_p = 1.26$ s, $t_s = 2.69$ s, $K_1 = 2.364$ s^{-1}.

(b) $1/T = 0.02$, $K_x = 1042$, $M_m = 1.356$, $M_p = 1.20$, $t_p = 1.53$ s, $t_s = 8.14$ s, $K_1 = 19.2$ s^{-1}; for $M_m = 1.16$: $K_x = 1014$, $t_p = 1.56$ s, $t_s = 8.28$ s, $K_1 = 18.64$ s^{-1}, $\omega_\phi = 2.13$, $\gamma = 40.12°$.

(c) $1/T = 4$, $K_x = 27{,}190$, $M_m = 1.46$, $M_p = 1.25$, $t_p = 0.49$ s, $t_s = 1.62$ s, $K_1 = 5$ s^{-1}; for $M_m = 1.16$: $M_p = 1.16$, $t_p = 0.52$ s, $t_s = 1.32$ s, $K_1 = 4.35$ s^{-1}, $\omega_\phi = 4.62$, $\gamma = 57.6°$.

(d) $1/T_1 = 0.2$, $1/T_2 = 4$, $K_x = 26{,}140$, $M_m = 1.397$, $M_p = 1.27$, $t_p = 0.51$ s, $t_s = 3.52$ s, $K_1 = 48$ s^{-1}; for $M_m = 1.16$: $K_x = 23{,}270$, $M_p = 1.2$, $t_p = 0.53$ s, $t_s = 4.11$ s, $K_1 = 42.8$ s^{-1}, $\omega_\phi = 4.53$, $\gamma = 56°$.

11.5. (a) $G(s) = \dfrac{1}{8K}\left[\dfrac{1 + Ts}{1 + \alpha Ts}\right]$, where $T = \dfrac{B}{K}$ and $\alpha = 0.5$

11.12. (a) At $\omega = 8$: $G(j\omega) = C(j\omega)/E(j\omega) \approx 6.3\underline{/-119°}$, thus $K_1 = 250$, $\omega_\phi = 8$, and $\gamma \geq 40°$.

(b)

$$G_c = \frac{(s + 1)(0.5s + 1)}{(10s + 1)(0.05s + 1)}$$

11.18. (a) $K_{2x} = 2.02$, $\omega_{\phi x} = 2.15$, $M_m = 1.398$, $M_p = 1.29$, $t_p = 1.39$ s, $t_s = 3.03$ s

(c) For $\gamma = 50°: u = \omega_\phi T = 8.6T = 1.19$ yields $T = 0.1384$. From the graphical construction, with $\omega_1 \le \omega_\phi/20$ and $\omega_2 = 20\omega_\phi = 172$, obtain $K_t = 24.28$ and $A = 180$, thus

$$\frac{C(s)}{E(s)} = \frac{2907(s + 1)(s + 7.225)}{s^2(s + 1.108)(s + 406.3)}$$

The 0-dB crossing slope is -12 dB/oct at $\omega = 6$, but the zero at -7.225 will have a moderating effect on the degree of stability.

(d) $M_m = 1.47$, $M_p = 1.30$, $t_p = 0.34$ s, $t_s = 1.03$ s, $K_2 = 46.65$; since the values of M_p are essentially the same, a good improvement in system performance has been achieved. $\gamma = 51°$ and $\omega_\phi = 9.12$.

Chapter 12

12.2.

$$G_c(s) = \frac{2.31(s + 20)}{s + 21}$$

With this cascade compensator $M_p = 1.12$, $t_p = 1.493$ s, $t_s = 2.941$ s, $K_1 = 1.53$. The specified $M_T(s)$ yields $M_p = 1.123$, $t_p = 1.493$, $t_s = 2.948$, $K_1 = 1.424$.

12.4. As a first trial, use

$$G_c \approx \frac{2.994(s + 2.1)}{s + 2.223}$$

With this cascade compensator $M_p = 1.03$, $t_p = 2.37$ s, $t_s = 2.94$ s, $K_1 = 0.75$. The specified $M_T(s)$ yields $M_p = 1.06$, $t_p = 2.2$ s, $t_s = 3.04$ s, $K_1 = 0.79$.

12.8. (a) The pole at the origin of $H(s)$ ensures $c(t)_{ss} = 0$; make pole $-b$ nondominant: $b = 100$

(b) $K_H = 2500$

(c) Characteristic equation: $s^4 + 104s^3 + 10,400s^2 + 120,000s + 320,000 = 0$. Routh stability criterion reveals that all roots lie in the left-half s plane. A plot of Lm $\mathbf{G}_x\mathbf{H}$ reveals that its 0-dB axis crossing slope is -6 dB/oct, indicating a stable system. The system poles are $p_1 = -4$, $p_2 = -8.06$, $p_{3,4} = -45.97 \pm j311.7$. $|y(t_p)| = 0.0025$, $t_p = 0.163$ s. For $|y(t_p)| = 0.01$, $K_H = 630$, which produces very little change in the value of t_p.

Chapter 13

13.3. (1) (a) Not observable, (b) controllable,

(c) $G(s) = \dfrac{2}{s - 2}$

(d) one observable and two controllable states, (e) unstable

(4) (a) Observable, (b) controllable

(c) $G(s) = \dfrac{s - 1}{(s + 1)(s + 2)}$

(d) Two observable and two controllable states, (e) stable

13.5. (a) Add a cascade compensator $G_c(s) = 1/(s + 1)$

(b) $H_{eq}(s) = k_3 s^2 + (2k_3 + k_2)s + k_1$

(d) A desired control ratio is

$$M_T(s) = \frac{Y(s)}{R(s)} = \frac{A}{(s^2 + 4s + 13)(s + 25)}$$

(e) With $A = 162.5$, $\mathbf{k}^T = [1 \quad 0.16923 \quad 0.08]$

(g) $G_{eq}(s) = 325/[s(s^2 + 29s + 113)]$, $K_1 = 2.876$

(h) $M_p = 1.12$, $t_p = 1.09$ s, $t_s = 1.66$ s

13.8. For system (1) (a)

$$M(s) = \frac{11,700}{(s + 12 \pm j21)(s + 20)}$$

$$k_1 = 1, \qquad k_2 = .091026, \qquad k_3 = .003077$$

(b) $y(t) = 1 + 1.025e^{-12t}\sin(21t + 171.1°) - 1.1584e^{-20t}$

$M_p = 1.057$, $t_p = 0.215$ s, $t_s = 0.273$ s, $K_1 = 11$ s^{-1}

13.12. (a)

$$M_T(s) = \frac{Y(s)}{R(s)} = \frac{1.467(s + 1.5)(s + 2)}{(s^2 + 2s + 2)(s + 1.1)(s + 2)}$$

where $\quad G_c(s)\dfrac{s + 1.5}{s + 1}, \qquad A = 1.467$

$$Y(s)/R(s) = \frac{A(s + 1.5)(s + 2)}{s^4 + [7 + A(0.5k_4 + k_3 + k_2)]s^3 + [11 + A(3k_4 + 2.5k_3 + 3.5k_2 + k_1)]s^2}$$
$$+ [5 + A(2.5k_4 + 1.5k_3 + 3k_2 + 3.5k_1)]s + 3Ak_1$$

where $k_1 = 1$, $k_2 = 0.8635$, $k_3 = -2.43541$, $k_4 = 0.5521$

(c) $M_p = 1.015$, $t_p = 3.75$ s, $t_s = 2.79$ s, $K_1 = 0.524$ s^{-1}

13.16. (a)

$$M_T(s) = \frac{195(s + 2)}{(s^2 + 2s + 2)(s + 1.95)(s + 100)}$$

(b) By applying the Guillemin-Truxal method, a reduced-order compensator is obtained by dividing the denominator of $G_c(s)$ by its numerator to obtain $G_c(s) \approx 1/(s + 99.9)$. The figures of merit for $M_T(s)$ are $M_p = 1.042$, $t_p = 3.18$ s, and $t_s = 4.23$ s, where for the system implemented with the reduced-order compensator $M_p = 1.056$, $t_p = 3.12$ s, and $t_s = 4.44$ s.

13.19. $\mathbf{k}^T = [16 \quad 5]$

$$x_1(t) = \frac{5}{3}e^{-3t} - \frac{1}{3}e^{-6t} \qquad x_2(t) = -5e^{-3t} + 4e^{-6t} \qquad \mathbf{x}(0) = \begin{bmatrix} 1 \\ -1 \end{bmatrix}$$

13.20. (1) (a) Completely controllable, (b) $\sigma(\mathbf{A}_p) = \{-1, -2, -3\}$,

(c) $\mathbf{T} = \begin{bmatrix} 6 & 5 & 1 \\ 3 & 2 & 1 \\ 0 & 2 & 1 \end{bmatrix}$ \qquad (d) $\mathbf{k}^T = [-42 \quad -33 \quad -6]$

Chapter 14

14.1. For $m = 1$: (a)

$$S_k^T(s)|_{K=15} = \frac{s^4 + 7s^3 + 25s^2 + 39s}{s^4 + 7s^3 + 25s^2 + 54s + 75}$$

14.6. (a)

$$S_\delta^M = \frac{4}{s+4}\left(\frac{s(s+1)}{s^2 + 13.5s + 50}\right) \qquad \text{where } \delta = \delta_n = a = 4$$

14.8. (a) $p = -21$, $K_G = 30.8$ (b) No; need a compensator of the form

$$G_c = \frac{A(s+a)}{(s+b)(s+c)}$$

14.11. Let $x_1 = \dot{x}_1$; equilibrium points:

(a) $x_1 = x_2 = 0$ (b) $x_1 = x_2 = 0$ (c) $x_1 = \frac{3}{4}$, $x_2 = 0$

The isocline equations are:

(a) $x_2 = \dfrac{-6x_1}{5+N}$ stable node   (c) $x_2 = \dfrac{6-8x_1}{5+N}$ stable node

(f) $x_2 = \dfrac{-17x_1 + 34}{N-8}$ unstable node   (g) $\dfrac{9x_1}{N}$ saddle point

14.12. Let $x_1 = x$, $x_2 = \dot{x}_1$:

(c) $N = \dfrac{-2x_2^2 - x_1}{x_2}$ vortex

(d) $N = \dfrac{2(-x_2|x_2| - x_1)}{x_2}$ vortex

14.15. For system (2), (a) Three equilibrium points are

$$\mathbf{x}_a = \begin{bmatrix} 0 \\ 0 \\ 0 \end{bmatrix} \qquad \mathbf{x}_b = \begin{bmatrix} -1 \\ 0 \\ 2 \end{bmatrix} \qquad \mathbf{x}_c = \begin{bmatrix} -1 \\ 0 \\ -2 \end{bmatrix}$$

(b) For $\mathbf{x}_b$,

$$\mathbf{J}_{\mathbf{x}_b} = \begin{bmatrix} 0 & 2 & 0 \\ 1 & 3 & 1 \\ 2 & 0 & -1 \end{bmatrix}$$

Chapter 15

15.3. (a) $z^{-1} + 2z^{-2} + 2.1875z^{-3} + 1.34375z^{-4} + \cdots$

15.4. (a) $c(0) = 0$, $c(\infty) = 1.846$

15.5. (a) $c(kT) = e[(k-1)T] + 0.5e[(k-2)T] + 1.5c[(k-1)T] - 0.8125c[(k-2)T]$
$\qquad + 0.3125c[(k-3)T]$

15.7. From Prob. 15.6:

$$|z| = e^{\sigma_x T} = 0.87305 \rightarrow \sigma_x = -1.36 \rightarrow \text{thus } \sigma = -(2)(1.36) = -2.72$$

For $\zeta = 0.6 \rightarrow$ in s-plane: $p_{1,2} = -2.72 \pm j3.627$;
in z-plane: $p_{1,2} = 0.7122 \pm j0.2706$; thus

$$D(z) = K_c \frac{(z - 0.7408)}{z - 0.28787} \quad \text{where } KK_c = 0.2235$$

$M_p = 1.096, t_p = 0.7+ \text{ s}, t_s = 1.1+ \text{ s}, K_1 = 3.1384 \text{ s}^{-1}$

Chapter 16

16.1. (1) (*a*)

$$\ker S(\lambda) = \left(\begin{bmatrix} 0 \\ \lambda - 1 \\ -2 \\ \overline{\quad} \\ (\lambda - 1)^2 \\ 0 \end{bmatrix} \begin{bmatrix} 2 \\ \lambda - 1 \\ 0 \\ \overline{\quad} \\ (\lambda - 1)^2 \\ 2(\lambda - 1) \end{bmatrix} \right)$$

16.3. (*a*) Rank $\mathbf{M}_c = \text{Rank}[\mathbf{B} \quad \mathbf{AB}] = 2 < n$. There is one uncontrollable mode, e^{-3t}.

16.6. (*a*)

$$\overline{\mathbf{A}} = \begin{bmatrix} \mathbf{A} & \mathbf{0} \\ -\mathbf{C} & \mathbf{0} \end{bmatrix} = \begin{bmatrix} 1 & 0 & -3 & \vdots & 0 \\ 0 & -1 & 1 & \vdots & 0 \\ -2 & 2 & 0 & \vdots & 0 \\ \hdashline -1 & 0 & -1 & \vdots & 0 \end{bmatrix}, \quad \overline{\mathbf{B}} = \begin{bmatrix} \mathbf{B} \\ \mathbf{0} \end{bmatrix} = \begin{bmatrix} 1 \\ 1 \\ 1 \\ 0 \end{bmatrix}$$

$$\overline{\mathbf{C}} = [\mathbf{C} \quad \mathbf{0}] = [1 \quad 0 \quad 1 \quad 0]$$

(*b*)

$$\text{Rank } \overline{\mathbf{M}}_c = \text{Rank} \begin{bmatrix} \mathbf{A} & \mathbf{0} \\ \mathbf{0} & -\mathbf{C} \end{bmatrix} = 4$$

(*c*)

$$\overline{\mathbf{K}} = \overline{\mathbf{Q}} \overline{\mathbf{V}}^{-1} = [-3 \quad -3 \quad -16 \quad -19.1] = \begin{bmatrix} 3 & 3 & 6.5 & 5.4510 \\ 5 & 0 & 3.5 & 3.6347 \\ 1 & 3 & 5.5 & 4.5429 \\ 2 & 2 & 3.0 & 0 \end{bmatrix}^{-1}$$

$$= [37.5767 \quad -14.4422 \quad -36.1056 \quad -3.7067]$$

$$\mathbf{A}_{cl} = \overline{\mathbf{A}} + \overline{\mathbf{B}} \overline{\mathbf{K}} = \begin{bmatrix} 38.5767 & -14.4422 & -39.1056 & -3.7067 \\ 37.5767 & -15.4422 & -35.1056 & -3.7067 \\ 35.5767 & -12.4422 & -36.1056 & -3.7067 \\ -1.0000 & 0 & -1.0000 & 0 \end{bmatrix}$$

(*d*) $y(t) = 1 - 2.5e^{-2t} - 10e^{-3t} + 22.5e^{-4t} - 11e^{5t}$

Chapter 17

17.4 $\sigma(\mathbf{A}) = \{-1, 1 + j1, 1 - j1\}$. Plant is unstable.

1. Rank $\mathbf{M}_c = 3$, Rank $\mathbf{M}_O = 3$; therefore, the plant is completely controllable and observable.

2. The augmented system is controllable, since

$$\text{Rank} \begin{bmatrix} \mathbf{B} & \mathbf{A} \\ \mathbf{0} & \mathbf{C} \end{bmatrix} = 5$$

3. Since $\mathbf{CB}$ has full rank, there is one transmission zero, $\lambda = -1$.
4. Use $\mathbf{\Sigma} = \mathbf{I}$:

$$\mathbf{K}_1 = (\mathbf{C}_2 \mathbf{B}_2)^{-1} \mathbf{\Sigma} \begin{bmatrix} 1/3 & 1/3 \\ -1/3 & 2/3 \end{bmatrix}$$

5. Let $\lambda_O = -2$
6.

$$\mathbf{\Gamma}(\lambda) = \text{diag}\left(\frac{g}{\lambda + g} \quad \frac{g}{\lambda + g} \right)$$

17.5.   Plant is unstable.

1. Plant is controllable and observable.
2. Augmented system is controllable.
3.

$$\text{Rank } \mathbf{C}_2 \mathbf{B}_2 = \text{Rank} \begin{bmatrix} 1 & 1 \\ 0 & 0 \end{bmatrix} = 1$$

(*a*) Using $d_1 = d_2 = 0$: $\mathbf{B}^* = \begin{bmatrix} 0 & 0 \\ 0 & 1 \end{bmatrix}$, $\mathbf{M} = \begin{bmatrix} 0 \\ m \end{bmatrix}$

(*b*) $\mathbf{F}_2 = \mathbf{C}_2 + \mathbf{MA}_{12} = \begin{bmatrix} 1 & 0 \\ 0 & m \end{bmatrix}$, $\mathbf{F}_2^{-1} = \begin{bmatrix} 1 & 0 \\ 0 & 1/m \end{bmatrix}$

(*c*) $\mathbf{C}_2 \mathbf{F}_2^{-1} = \begin{bmatrix} 1 & 0 \\ 0 & 0 \end{bmatrix}$

(*d*) $\mathbf{F}_1 = [0 \quad 1]^T$

4. Using $\mathbf{\Sigma} = \mathbf{I}$,

$$\mathbf{K}_1 = (\mathbf{F}_2 \mathbf{B}_2)^{-1} \mathbf{\Sigma} = \begin{bmatrix} 0 & (1/3)m \\ 1 & (-1/3)m \end{bmatrix}$$

5. Using $\lambda_O = -2$.
6.

$$\mathbf{\Gamma}(\lambda) = \begin{bmatrix} \dfrac{g}{\lambda + g} & 0 \\ 0 & \dfrac{1/m}{\lambda + 1/m} \end{bmatrix}$$

Chapter 18

18.1. (a) $\alpha = 5.477, T = 2.98, A = 15.69$

(b) For the basic system: $K_x = 96, \omega_{\phi x} = 9.4$ rad/s

(c) For $\gamma = 45°; \alpha = 0.24565, T = 0.06142, A = 4.0567$

18.5. (a) From Figures (a) and (b), respectively,

$$T_a = \frac{G_1 P}{1 + G_1 P} = \frac{1}{(s+1)^2} \tag{1}$$

$$T_b = \frac{G_2 P}{1 + P(H + G_2)} = \frac{G_2(s+1)}{s(s+1) + s + G_2(s+1)} \tag{2}$$

Setting (1) = (2) yields $G_2 = 1/(s+1)$

18.9. (a)

$$T_{R_U}(s) = \frac{s + 25.73}{s + 2.34 \pm j4.5} \qquad T_{R_L}(s) = \frac{891}{(s+2.7)(s+10)(s+33)}$$

$$G(s) = \frac{8750(s+0.5)(s+2)(s+3.5)(s+50)(s+300)(s+450)}{s(s+3)(s+20)(s+200)(s+750)^2}$$

$$P_o(s) = \frac{1}{s+1} \qquad F(s) = \frac{0.793(s+30)}{(s+2.8)(s+8.5)}$$

INDEX

Ac (alternating-current) servomotor,
  57–59
Acceleration, 35, 41, 65, 182–184
  error coefficient (parabolic), 192
  step function of, 65, 69
Accelerometer, 693
Actuating signal, 3–4, 132, 136–140,
  184–190, 192–195
  definition, 7, 139
A/D (analog-to-digital) converter, 493–494
Adder (summer), 144
Addition of matrices, 26
Adjoint matrix, 122, 164
Aircraft digital control, 12–13
Aircraft flight control system, 12–13,
  133–135, 207, 571–577, 665,
  681–682, 694–695
  heading control, 682
  landing configuration, 666
  longitudinal, 207, 571–577
  photographic, 65
  reconfigurable, 582
  reconfiguration, 12–13
  shock input, 117–119
  variable parameter, 580–634
  yaw rate, 681
Airplane control surfaces, 134, 571–572
Algebra:
  matrix, 14, 25–28, 120–126
  signal flow graphs, 149–158
α contours, 296–309
  log magnitude–angle diagram,
    307–309
  values for constructing, 300, 302
Ammeters, 41
Amplifier:
  ideal, 144
  isolation, 320, 333
  power, 7

Amplifier: power (Cont.)
  hydraulic, 48–53
  rotating, 54–55, 141
Analog computer simulation diagram, 144
Analog of electric circuits, 36, 41
  gear train, 44–45
  hydraulic system, 48–53
  liquid level system, 53–54
  mechanical rotational systems, 41–46
  mechanical translational systems, 35–40
  thermal systems, 46–48
Analog-to-digital (A/D) converter, 493–494
Analogous circuits, 41
Angle condition:
  near complex poles, 220–221
  for root locus, 210–214
Antiaircraft radar tracking control system, 9–11
Applied force, 35, 60
Applied torque, 41–46
Applied voltage, 22
Armature control, dc motor, 56–57
Attenuation compensators, 375
Augmented matrix, 124, 532, 535
Augmented state variable, 30
Automatic elevator control system, 4
Automatic plotting:
  frequency response, 245
  root locus, 215, 231–233
  time response, 109, 231, 546–551
Automatic toaster, 1–3
Automobile suspension, 653
Autopilot, 207
Auxiliary equation, 180–182, 222, 227

$B(s)$, 271–273
  change of reference for, 274
  effect of poles at origin on rotation of, 275–277
  plot of, 274

739

Back emf, 56

Backlash (dead zone), 45, 321

Bandwidth, 246

Basic sampled-data control system, 512, 514–519

Basic servo characteristics, 13–14, 176–199
error coefficients, 192–198

Basic (uncompensated) system, 15, 319

Block diagonal matrix, 172

Block diagonalization, 557–559, 566–568

Block-diagram representation, 34
of control systems, 20
overall, 38

Block diagrams, 131–140
cascade, 38, 44, 142
complex, 149
construction of, 34
control ratio from, 136–137
examples of, 131–144, 153, 190–191
for feedback compensation, 355, 390
of feedback control system, 136, 137
functional, 3, 131
G-equivalent unity-feedback, 444
matrix, 148
sampled system, 493, 512, 522
simulation diagrams, 144–160
standard terminology for, 138–140
summation points, 141
symbols and definitions, 6–8, 138–140
transfer-function matrix, 128

Bode plots, 247–256
of disturbance-rejection models, 589–590
drawing, 249–256
example, 256–259
(*See also* Log magnitude and phase diagram)
MATLAB, 309–311

Boundary, frequency response:
composite, 607–608
construction of, 624–625
procedure for determining, 605–607
universal high-frequency (UHFB), 595–597

Branches, 149
dominant, root locus, 224
equations, for state equations, 30
number of, of root locus, 215
signal flow graphs, 149

Break-in point, 218–220, 230

Break point (corner frequency), 115, 251

Breakaway point, 218–220, 241–242
example, location of, 226

Bulk modulus of elasticity, 51

CAD (computer-aided-design) packages, 6, 15, 16, 200, 215, 241, 290, 321, 331, 368, 424, 641–650

CADAC (*see* Computer-aided-design accuracy checks)

Calibration of root locus, 202, 211–217

Camera platform, 65, 190–191

Canonical form, 159, 173
matrix, 159
(*See also* Control canonical form)

Canonical state variables, 159

Capacitance:
electrical, 21
thermal, 46

Capacitive coupling, electromechanical system with, 60–62

Capacitors, 21, 30
fluid, 30
thermal, 30

Carrier systems (ac):
servomotor, 56–59

CAS (command augmentation system), 12

Cascade:
blocks in, 38
elements in, 44

Cascade compensation, 320, 372–390, 405–406, 454
comparison of, 349–352
comparison, with feedback compensation root locus, 320–352
digital, 519–520
direct polar plots, 378, 384
dominant plant complex poles, 364–365
frequency-response method for, 372–390, 399
Guillemin-Truxal method, 406–408
ideal derivative, 339–340
ideal integral, 339–340
(*See also* Lag compensators, cascade; Lead compensators, cascade; Lag-lead compensators, cascade)
logarithmic plots, 372–390
objectives, 320–321, 331, 368
performance of, 339
selection of, 369–372
vs. state-variable feedback, 438

Cascade lag compensators, 333–339, 372–378, 399
cascade lag-lead compensators (*see* Lag-lead cascade compensators)
frequency-response method for applying, 372–378
root-locus method for applying, 332–339, 378

Cascade lead compensators, 340–346, 379–385, 399
Cascade nodes, 150
Cascade transfer functions, 38, 44, 142–144
Cayley-Hamilton method, 426
Cayley-Hamilton theorem, 85
Centers, phase plane, 486
Characteristic equations, 70, 84
  from control ratio, 137, 201, 209, 211
  roots of, 234–237
    complex, standard form, 73
    imaginary, 221–222, 227
    plotting, 201–204
    from root locus, 224–233
   Routh's stability criterion, 176–182
   stability, 116, 177, 270
    by Nyquist's criterion, 270–283
Characteristic function, 97
Characteristic polynomials, 85
  evaluation of, 125
Characteristic roots, plot of, 234–237
Characteristic values, 84–85
Closed-loop control systems, 3–8
  characteristic equations, 70, 137, 201, 209, 211
  environmental conditions, 321
Closed-loop feedback system, simple, with unity feedback, 183–185
Closed-loop frequency response, 246–247, 397
  from direct polar plot, 291–302
  gain adjustment, 304–307, 312
  from log magnitude–angle diagram, 312
  M and α contours, 296–302, 307–311
    examples of, 302
  maximum value of, 255, 292
  resonant frequency, 254–256, 292
Closed-loop pole-zero assignment, 423–461
Closed-loop systems:
  performance, analysis of, 559–561
  time response of, 344, 519
Closed-loop tracking performance, 290–318
Closed-loop zero cancellation, 455, 456
Cofactor matrix, 121–122
Coil, current in, 62
Column matrix, 25
Column vector, 26, 31
Command, definition, 139
Command augmentation system (CAS), 12
Command guidance interceptor system, 133
Command input, 3
Common input, definition, 7
Companion matrix, 146–147
  transforming matrix into, 172–175
Comparison units, 3, 139, 541
  gyroscopes, 134, 333, 674–701

Comparison units *(Cont.)*
  potentiometers, 140, 190–191
  synchros, 134, 489
Compensating networks, 263–264, 332–335, 339–341
Compensation, 208, 319
  cascade *(see* Cascade compensation)
  digital-computer, 519–525
  feedback *(see* Feedback compensation)
  feedforward, 451–453
  MIMO, 590–592
  root-locus, 319–366
  for stability, 319, 368
Compensators, 319–321
  attenuation, 375
  cascade *(see* Cascade compensation)
  derivative *(see* Lead compensators)
  digital, 519
  feedback *(see* Feedback compensators)
  ideal, 332, 339, 356
  lag *(see* Lag compensators)
  lag-lead *(see* Lag-lead compensators)
  lead *(see* Lead compensators)
  location of, 320
  mechanical, 698
  parallel, 320–321
  passive *(see* Passive compensators)
  polar plots, 369, 378, 384
  sensitivity functions and, 476
Complementary function, 64
Complete controllability, 424, 539
Complex-conjugate poles, 103–106
  multiple-order, 106
Complex differentiation theorem, 94
Complex integration theorem, 95
Complex number, logarithm of, 247
Complex poles, dominant, compensation, 364–365
Complex roots, 71, 84
Composite boundary, 607–608
Compressibility of oil, 50
Compression of spring, 37
Computer-aided design, 15
Computer-aided-design accuracy checks (CADAC), 82, 97, 196, 225, 262
Computer-aided-design (CAD) packages, 6, 15, 16, 243, 494, 641–650
Computer compensation, digital, 519–525
Computer computation, 6
Computer literacy, 16
Computer solution:
  digital *(see* Digital computer solution)
  of the state equation, 82, 89
Computers:
  analog, simulation, 144

Computers *(Cont.)*
   digital: in control systems, 493
      (See also Digital computer solution)
   Conditionally stable system, 280, 286
Configurations, root locus, 205–206
Conformable matrix, 27
Conservation of energy, principle of, 75
Constant-coefficient differential equations, 19
Constituent matrix, 87
Continuous control systems, 7
Continuous signal reconstruction, 503–504
Continuous state transition equation, 88
Control:
   continuous, 7
   discontinuous, 7
Control canonical form, 173
   state-feedback design for SISO systems
      using, 433–435
Control elements, definition, 140
Control matrix, 32, 559
Control problem:
   engineering, general nature of, 15–16
Control ratio, 136, 201–203
   acceptable, 588
   disturbance, 589–590, 600, 605
   disturbance-rejection model, 418–422
   modeling of, 401–422
   nonunity feedback, 198, 208, 225–233, 322
   plots, 312–317
   poles of, 209–211
   tracking, 402–406, 592
   Tustin model of, 517
   (See also Closed-loop frequency response)
Control surfaces, airplane, 134, 571–572
Control systems:
   aircraft flight, 12–13, 571–577
   antiaircraft radar tracking, 9–10
   automatic, 1
   automatic elevator, 4
   basic sampled-data, 512, 514–519
   block-diagram representation of, 20
   camera platform, 190–191
   characteristics of, 176–199
   closed-loop, 3–8
   command guidance interceptor, 133
   computer-aided-design (CAD) packages, 6,
      15, 16, 201, 243, 309, 641–650
   continuous, 7
   digital, 12–13, 493–526
   digital computers in, 493
   digital flight (DFCS), 12–13
   discontinuous, 7
   disturbance-rejection, 408, 584–633
   feedback *(see* Feedback control system)
   fly-by-wire (FBW), 12–13

Control systems *(Cont.)*
   frequency response, 290
   home heating, 3
   input to, 65–66
   introduction to, 1–18
   linear *(see* Linear systems)
   linear time-invariant *(see* Linear
      time-invariant control systems)
   liquid-level, 53–54
   mathematical background of, 13–15
   missile, 133
   multiple-input multiple-output *(see*
      Multiple-input multiple-output control
      systems)
   multiple-input single-output *(see*
      Multiple-input single-output control
      systems)
   multivariable *(see* Multivariable control
      systems)
   nonlinear, 16, 468–491
   nuclear-reactor power-level, 10
   open-loop, 2–12
   optimal *(see* Optimal control system)
   overdamped, 80
   performance specifications, 80–81, 192–199,
      319–321, 462–466
   photographic, 664
   pitch-attitude, 574–577, 701
   position, 140–144
   power-by-wire (PBW), 13
   production-line robot, 4–6
   pseudo-continuous-time (PCT), 493, 512–514
   radar tracking, 9–10, 133, 674
   reconfigurable, 582
   robot, 660
   robust control, 554, 580–633
   sampled-data (SD), 493
   satellite, 664
   single-input single-output *(see* Single-input
      single-output closed-loop control
      system)
   solar-energy, 6
   stable, 71
   state feedback, 423–476, 433–435
   sun-following, 6
   temperature, 132
   tracking *(see* Tracking systems)
   uncompensated sampled-data, 512,
      514–519
   underdamped, 80, 81
   unity-feedback, 184
   unstable, 71
   water tank, 694
Control theory:
   analysis of, 13–15

Control theory *(Cont.)*
  conventional, 11
  modern, 11–16
Control variables, 30
Control vector, 31
Controllability, 424–430
  from canonical matrix, 426
  MATLAB diary file, 430
  optimal control system, 424
Controllability matrix, 173, 426, 429
Controllable system, 424–430
Controlled system:
  definition, 140
  indirectly, definition, 140
Controlled variable, 182
  definition, 6, 139
  indirectly, definition, 139
  ultimately, definition, 139
Controller:
  digital *(see* Digital controllers)
  flight, 134
Conventional control theory, 11
Conventional steady-state error analysis,
    176–199, 444–448
Convolution integral, 119
Corner frequency, 115, 251
Correlation:
  for higher-order system, 295–296,
    312–317
  of pole-zero diagram with frequency and
    time responses, 312–317
  s plane to z plane, 498
  for second-order system, 292–295
  of sinusoidal and time responses, 245–246,
    295–296
Coulomb friction, 35, 321
Critical damping, 72, 75, 327
Crossover point, imaginary-axis, 221–222
CTRL-C program, 650
Current(s), 21
  armature, 56
  in coil, 62
  field, 54–55

D operator notation, 20
D/A (digital-to-analog) converter, 493
Damped natural frequency (ω_d), 71
  increase in, 204
Damped sinusoid, exponentially, 71
Damping action:
  of gear train, 44–46
  network representation of, 36
Damping coefficient, 71
  critical, 72, 75

Damping constant, 72
Damping force, 36
Damping ratio (ζ), 72
  constant, s plane, 211, 236
  decrease in, 203
  dominant, 237
  effective, 295
  minimum, 595
  relation to resonant frequency, 294
  variation of roots with, 234, 238
Damping torque, 42
Dashpots, 36
Dc (direct-current) generators, 54–55
Dc (direct-current) servomotor, 56–57
  armature control, 56–57
  transfer function, 57
Dc (direct-current) shunt motor, 1
Dead time, 238–240
  Nyquist's stability criterion applied to
    systems having, 281–283
  polar plot, 281
  root locus, 239–240, 321
  (See also Transport lag)
Dead zone, 45
Decades, definition of, 248
Decaying exponential, Laplace transform
    of, 93
Decibel (unit), 248
  per decade, 248
  per octave, 248
Definitions:
  block diagram, 136–139
  controllability, 424
  flow graphs, 149–151
  gain and phase margin, 283
  IEEE, 6–7
    regulator, 8
    servomechanisms, 8, 9
  Laplace transform, 92
  logarithmic terms, 247–248
  minimum-phase system, 262
  nonminimum-phase system, 262
  observability, 424
  phase and gain margin, 283–288
  poles, 99, 110
  root locus, 201–204, 210–211
  signal flow graphs, 149–151
  steady-state error coefficients *(see* Error
    coefficients)
  system components, 140
  transfer function, 34
  variables in system, 139
  zeros, 99, 110
Degeneracy of matrix, 123
Degrees of freedom (DOF), 585

δ (delta) (*see* Impulse *entries*)
Departure angle, 220–221
  determination of, 227
Dependent nodes, 150
Derivative compensators (*see* Lead
    compensators)
Design policy, 18
Desired performance, 16
Determinants of square matrix, 121
  product of, 122
Deterministic inputs, 9
DFCS (digital flight control systems),
    12–13
Diagonal, principal, 120
Diagonal matrix, 120
Diagonal transfer function, 570
Diagonalization, block, 557–559
Diagonalizing the **A** matrix, 160–168, 169–172
Difference equations, 522
Differential equations, 13–15, 185
  constant-coefficient, 19
  electric circuits, 21–25
  general, 7
  hydraulic linear, 48–53
  Laplace transform of, 97–98
  linear constant-coefficient, 19
  liquid-level control system, 53–54, 656
  mechanical rotational systems, 41–46
  mechanical translation systems, 35–40
  simple second-order, 78
  solution of, 64–90
    classical, 13, 64–82
    Laplace transform, 13, 97–98
    state variable, 126–128
  thermal systems, 46–48
Differentiation of a matrix, 28
DIG (digitization) digital control analysis
    technique, 493, 512–514, 523–525
Digital cascade compensators, 519–525
Digital-computer compensation, 519–525
Digital-computer solution:
  factoring of polynomials, 99
  frequency response, 246
  partial fraction coefficients, 109–113
  root locus, 215, 225
  state-variable feedback, 461
Digital computers, 89, 493, 519
  in control systems, 493
Digital control analysis techniques, 493
Digital control systems, 12–13, 493–526
Digital controllers, 89, 493, 519
Digital flight control systems (DFCS), 12
Digital-to-analog (D/A) converter, 493
Digitization (DIG) digital control analysis
    technique, 493, 512–514, 523–525

Direct (DIR) digital control analysis technique,
    493, 520–523
Direct polar plots, 263–270, 291–292
  correlation to time response, 290–296
  examples of Nyquist's stability criterion
    using, 278–283
  gain adjustment, 304–307
  summary of, 269
Directed line segments, 110–112, 211
Discontinuity, jump, 117–119
Discontinuous control systems, 7
Discrete-data feedback control system, 7,
    493–526
Discrete digital control analysis technique, 493
Dissipation function, 59
Disturbance, definition, 139
Disturbance bounds, 600–607
Disturbance control ratio, 408–410, 589–590,
    600, 605
Disturbance element, definition, 140
Disturbance input, 580, 585, 592
  definition, 6
Disturbance-rejection control systems,
    408–409, 508
Disturbance-rejection model, 418–422, 589–590
  Bode plots of, 590
  control ratio, 408, 409, 420
  design example, 415–418
  design principles for SISO systems, 411–415
  second-order, 120, 409–410
Disturbance-rejection problem, 408–409, 580
Disturbance-rejection time response, 419–421
Disturbance response characteristic, 601, 605
Disturbance uncertainty, 585
DOF (degrees of freedom), 585
Dominant branch, 224
Dominant complex poles, 321–326
  rate feedback compensation and, 364–365
Dominant-pole pair, 240–241, 321–326
  (*See also* Dominant roots)
Dominant roots, 224, 589
Driving function, 97, 245
Driving transform, 97
Duplicating time, 81
Dynamic unit, 1–4, 7
Dynamical modes of systems, 530

Effective damping ratio, 295
Effective ω_n, 238
Eigenstructure, time response, 528–530
Eigenstructure assignment:
  effect on time response, 528–530, 534–535,
    537, 541, 545–546
  entire (*see* Entire eigenstructure assignment)

Eigenvalue spectrum, 530
Eigenvalues, 70, 84
    associated eigenvectors, 530–537
    complex, transforming matrix with, 169–172
    multiple, 168
    product of, 85
    self-conjugate, 84, 531
    sum of, 85
Eigenvectors, 17
    MATLAB diary file, 537–538
    in null space, 164, 166, 531
    orthogonal, 529
    reciprocal or left, 529
    right, 164 fn
    spaces, 166–168
        evaluation, 531–539
Elastance, 35, 42
Elasticity, bulk modulus of, 51
Electric circuits and components, 21–25
    loop method, 23, 76–77
    nodal method, 23
    symbols and units, 22
Electrical analogs, 41, 54
Electrical resistance, 21
Electrical second-order systems, 76–77
Electrical symbols and units, 22
Electromechanical elements, energy functions
        for, 62
Electromechanical system with capacitive
        coupling, 60–63
Elementary matrix, 124
Elements:
    energy storage (*see* Energy-storage
        elements)
    feedback, 7, 136
    network: electrical, 21–25
        mechanical rotational, 41–46
        mechanical translation, 35–41
        thermal, 46
    emf, back, 56
Energy:
    hydraulic, 52
    kinetic, 60
    potential, 60
    principle of conservation of, 75
    stored, 52
Energy functions for electromechanical
        elements, 62
Energy level, 320, 354
Energy-storage elements, 30
    hydraulic system, 52
    liquid-level system, 53–54
Engineering:
    human, 11
    systems, 7

Engineering control problem, general nature of,
        15–16
Engineers, system, 17
Entire eigenstructure assignment, 528–530
    associated eigenvectors, 528–530
    effect on time response, 533–535, 537, 541,
        546
    method of, 551–552
    for regulators, examples of, 531–541
Envelope, 71
Environmental conditions, 321
    closed-loop system, 462
    open-loop system, 463
Equilibrium position, 35, 480–489
Equivalent method, MISO, justification of,
        591–592
Error:
    squared: integral of (ISE), 245
    steady-state, 176, 186–190, 192–199
    system, 137, 139, 192–199, 292, 442
Error analysis, steady-state, 442–444
Error coefficients, 192–198, 518–519
    conventional systems, 196–199
    from logarithmic plots, 259–262
    generalized, 447
    parabolic, 195, 198
    ramp, 194, 197, 198
    state-variable feedback systems, 444–448
    step, 192–193, 196, 198
    table of, 198
    use of, 196–198, 444–448
Error equations, 79
Error-measuring devices (*see* Comparison
        units)
Euler's identity, 66, 71
Experimental determination:
    damped natural frequency, 262
    frequency response, 244, 262
    gain constant, 262
    system constants, by frequency response,
        244, 262
    time constants, 262
    transfer functions, 244, 262
Exponentially damped sinusoid, 71
    envelopes, 71

Factoring of polynomials:
    computer solution, 99
    partition method, 355
Faraday's law, 21
Fast modes, 554, 561
FBW (fly-by-wire) control system, 12–13
Feedback, 7
    dominant plant complex poles, 364

Feedback *(Cont.)*
  minor-loop, 137
  negative, summary of root-locus construction
    rules for, 241–242
  output *(see* Output feedback)
  primary, definition, 139
  quantitative theory *(see* Quantitative feedback
    theory)
  rate, 134, 355–364, 473
  of second derivative of output, 362–364
  state *(see* State feedback)
  state-variable *(see* State-variable feedback)
  unity, 183–185
Feedback action, 3, 7
Feedback coefficients, 436, 442–444
  state, 436
  steady-state analysis, 442–448
Feedback compensation, 319–321, 355–365,
    408–422
  block diagram for, 355, 390
  vs. cascade, comparison, 320, 365, 399
  design procedures, 355
  frequency-response method for, 390–398
  location of compensator, 353
  logarithmic plots, 390–398
    use of, 390–397
  maintaining system type, 353
  procedure, 355–364, 392–398
  rate *(see* Rate feedback compensation)
  results of, 364
  root-locus, 355–365
    comparisons, 363
    design procedure, 355
    objectives, 331
    rate feedback examples, 355–364
Feedback compensators, 320–321, 399
  application guidelines, 397–398
  basic, 397
  comparison of cascade and, using root
    locus, 363
Feedback control system, 7
  aircraft, 133–135
  analysis of types of, 185–190
  block diagrams of, 135, 138
  closed-loop, simple, with unity feedback,
    183–185
  definition, 7
  discrete-data, 7
  flyball governor, 8
  historical background, 8–12
  management, 11
  nonlinear, 489
  nonunity, 198–199
  MATLAB m-file, 231–233
  on-off, 8

Feedback control system *(Cont.)*
  performance based on frequency response,
    290–318
  relay, 8
  representation of, 131–175
  sampled-data, 7
  terminology, 138–140
  Type 0, 184, 259–260
  Type 1, 184, 260–261
  Type 2, 184, 190–191, 261–262
  Type 3, 184
  Type −1, 184
  types of, 183–185
  unity, 183–185
  various forms of, 353
  (See also Control systems)
Feedback element, definition, 7, 140
Feedback loop, 151
Feedback systems:
  performance, 290
  terminology, 7, 138–140
  transforming-system type, 353
  type, 183–185
Feedback theory, quantitative *(see* Quantitative
    feedback theory)
Feedback vector, 436
Feedforward compensation, 453
Feedforward matrix, 32, 148
Field current, 54–55
Figures of merit, 192, 292, 317, 322, 423
Filters *(see* Compensators)
Final value theorem:
  applications, 185, 191–195
  s plane, 96,
  z plane, 500
First-order real poles, 99–103
  multiple, 101–103
Flight controller, 134, 571
Flow equation, 50
Flow graphs, signal *(see* Signal flow graphs)
Fluid capacitor, 30
Fluid compressibility, 30
Flux, 55
Fly-by-wire (FBW) control system, 12–13
Focus, phase plane, 483
  stable, 484
  unstable, 484
Force:
  applied, 35, 59
  damping, 36
  reaction, 35–36
  reaction damping, 36
Force equations, 35–36
Forward element, definition, 7
Forward paths, 150

Forward transfer function, 137
  general form, 183–184, 269
Fourier coefficients, 495
Fourier series, 495
Fourier spectra, 495
Fourier transform 245, 495–496
Frequency:
  bounds on prefilter, 617
  control ratio versus, 312
  corner, 115, 251
  damped natural (*see* Damped natural
      frequency, *s* plane)
  effective, undamped natural, 236, 295
  gain-margin, 283
  Lm values for, 252
  of oscillation ω_d, 71, 79, 81, 237
  phase-margin, 283
  resonant (*see* Resonant frequency)
  undamped natural (*see* Undamped natural
      frequency)
Frequency-domain tracking techniques, 367
Frequency response, 244–289
  automatic plotting, 245
    (*see* MATLAB, diary file)
  cascade compensation, 367–390
  closed-loop, 245–246, 397
  closed-loop tracking performance based on,
      290–318
  computer solution, 245
  control system, 290
  and correlation of pole-zero diagram with
      time response, 312–317
  correlation of sinusoidal and time
      responses, 245–246
  design procedure, 317–318
  direct polar plots, 263–270, 378, 384
  effect of additional poles and zeros on,
      312–317
  figures of merit, 292
  higher-order system, 295
  log magnitude–angle diagram, 250,
      286–288
  logarithmic plots, 247–262, 367–399
  maximum (*see* Maximum frequency
      response)
  with parameter variations, 580–633
  from pole-zero diagram, 113–116
  reshaping, 368
  resonant frequency, 292, 296
  vs. root locus, 399
  second-order disturbance-rejection model,
      409–411
  second-order tracking system, 292
  (*See also* Correlation, of sinusoidal and
      time responses)
Frequency-response method, 244–289
  cascade and feedback compensation
      using, 398
  cascade lag compensators applied by,
      375–378
  lag-lead compensators applied by, 385–390
  lead compensators applied by, 379–385
  procedure used in, 317–318
Frequency-response plots, 246–247, 367–399
  reasons for reshaping, 368
Frequency spectra:
  complementary, 495, 504
  fundamental, 495
Frequency transfer function, 67
Frequency-transfer-function relationships,
      general, 249–250
Friction:
  damping, viscous, 36
  nonlinear, 35
  static (coulomb), 35, 321
Friction coefficient, viscous, 42
Functional block diagram, 3, 131
Fundamental matrix, 84

G-equivalent unity-feedback block
      diagram, 198–199, 403, 444
Gain:
  log magnitude curves and, 259–262
Gain adjustment:
  root locus, 201–204, 331
  of unity-feedback system: direct polar plot,
      304–307
  by use of log magnitude–angle diagram,
      312
Gain constant, 250
Gain crossover, 283
  slope at, 285
Gain margin, 283–286
  log magnitude and phase diagram
      showing, 284
Gain-margin frequency, 283–285
Gain sensitivity, time response and, 473
Gain variation, sensitivity due to, 466–468
Gear train:
  damping of, 44–46
  mechanical network, analog, 44–46
General nodes, 149
Generalized error coefficients, 447
Generator-motor control, 140–144, 191
Generators, 21
  dc, 54–55
Geometrical properties, root-locus, 215–225,
      241–242
$G_{eq}(s)$, 198–199, 403, 445

Governors, 8
Grant's rule, 225, 242
Graph transmittances, overall, 152–153
Graphical interpretation of partial-fraction
    coefficients, 109–113
Graphs:
  linear, 34
  signal flow (*see* Signal flow graphs)
Guidance system, 133–135, 183
Guillemin-Truxal (GT) method, 402, 406–408
Gyro, integrating, 333
  rate, 134
Gyroscope, 134, 333, 674, 701

H-equivalent reduction, 438
Hazony and Riley rule, 109
Heat flow, 46–48
Heaviside partial-fraction expansion,
    99–107, 504
  examples, 104, 106–107, 187
$H_{eq}(s)$, 438
Hermite normal form (HNF), 123–124
  basic operations, 104, 106, 107, 123–124
Hero's device, 8
HNF (*see* Hermite normal form)
Hold circuits, 504–506
Home heating control system, 3
Homogeneous equations, 70
  autonomous system, 488
Homogeneous state equations, 83
Hooke's law, 35
Human engineering, 11
Hurwitz's criterion, 182
Hydraulic amplifier, 48–53
Hydraulic analogs, 54
Hydraulic boost system, 12
Hydraulic energy, 52
Hydraulic linear actuator, 48–53
Hydraulic symbols and units, 48–51

ICECAP-PC, 15, 176, 201, 245, 321, 369, 424,
    494, 528, 646
Ideal derivative cascade compensation,
    339–340, 356
Ideal integral cascade compensation, 332–333
Ideal sampler, 496–498
Ideal systems, 247
Ideal value, definition, 139
Idealized system, definition, 139
Idempotency property, 87
Identity matrix, 28
IEEE (Institute of Electrical and Electronics
    Engineers), definitions, 7–8

Imaginary poles, 105–106
Impulse excitation, second-order system with,
    119–120
Impulse function, 117–119
  initial conditions, 118
  Laplace transform of, 117–119
  maximum overshoot, 119
  response to, 118–120
Impulse function input, 65
Impulse response transfer function, 119
Impulse train, 496–498
Inaccessible states, 473–476
Independent nodes, 149
Independent state variables, 29–30
Inductance, 21–25
Induction motors, two-phase, 57–58
Inductors, 21–23, 30
Industrial-process temperature-control system,
    132–133
Inertia, 41
  moment of: effective, of gear train,
    44–46
  (*See also* Moment of inertia)
Initial condition component, 97
Initial conditions, 66, 71, 76–77, 97, 477
  impulse functions, 118
  in Laplace transforms, 95–98
  in state equations, 83
Initial value theorem, 96
  s plane, 96
  z plane, 500
Initial values, 21, 27
Input:
  command, 3
  common, definition, 7
  to control systems, 65–66
  -decoupling zeros, 427, 429, 539
  disturbance (*see* Disturbance input)
  impulse function, 65, 118–120, 496–499
  irregular, 182
  multiple, 11, 83
  parabolic, 188–190, 443–444, 518
  parabolic-function, 65, 69
  phasor representation of, 66
  polynomial, 67–70, 198
  power-series function, 65
  ramp, 187–188, 443, 518
  ramp-function, 65, 66, 69
  random, 65, 182–183
  reference, definition, 7, 139
  shock, 117–119
  sinusoidal, 65–67
  standardized, 65–66, 192
  step, 186–190, 442, 518
  step-function, 65, 66, 68–69

Input *(Cont.)*
TOTAL-PC input modes, 643, 644, 649
unit-step, time response for, 517
Input-decoupling zeros, 427
Integral compensation (*see* Lag compensators)
Integral of squared error (ISE), 245
Integration of a matrix, 28
Integrator, 145, 541
Interactive computer-aided-design
    programs, 109, 641–650
Internal nodes, 151
Inverse Laplace transformation, 98–113
Inverse matrix, 112
Inverse polar plane, 303–304
Inverse *z* transform, 503–504
Inversion, matrix, by row operation,
    124–125
Irregular plants, 566–568
ISE (integral of squared error), 245
Isoclines, 479–480
trajectory using, 479

Jacobian matrix, 488–491
Jacobians, matrix of, 489
Jump discontinuity, 117–119

Kernel of matrix, 531–536
Kinetic energy, 59
Kirchhoff's laws, 21, 22, 59
analogy: of mechanical systems to electric
    circuits, 41
of thermal systems to electric circuits,
    46–48
loop method, 21–24
nodal method, 21–25

Lag compensators, 349
applied to Type 1 system, 335–339
cascade, 332–339, 372–378
comparison, 399
log plots, 372–378
rules for application, 372–373
network, 333
polar plots, 369–372, 378
root locus, 332–339
comparison with other types, 349
example of, 335–339
ideal integral, 332
passive, 333
rules for application, 334–335
transfer function, 333
Lag-lead compensators, 346–348, 387–390, 399

Lag-lead compensators *(Cont.)*
applied to Type 1 system, 347–348
cascade, 346–348, 385–390
log plots: comparison, 389
example of, 387–390
qualitative effects, 369–372,
    385–387
rules for application, 385–387
network, 263
polar plots, 264
root locus, 346–348
comparison with other types, 349
example of, 347
rules for application, 347
transfer function, 263, 346
Lagrange's equation, 59–63
application, 59–63
table of energy functions, 62
Laplace integral, 92
Laplace transform, 13, 91–129
additional matrix operations and properties
    using, 120–126
advantages of, 91
application of, to differential equations,
    97–99
of decaying exponential, 93
definition, 92
derivation of, of simple function, 92–94
direct, 92
disadvantages of, 91
of impulse function, 117–119
impulse sequence, 496–499
inverse, 98–99
of ramp function, 93
second-order system with impulse excitation,
    119–120
of sinusoid, 93
of step function, 92
theorems in applying, 94–96
Laplace transform pairs, 17
table of, 637–640
Lead compensators (derivative compensators),
    349, 379–381
applied to Type 1 system, 341–345
cascade, 381–385, 399
feedback path, 362, 393
logarithmic plots, 379–385
rules for application, 379–381
network, 341
polar plots, 369–372, 384
rules for application, 379–381
root locus, 339–343
comparison with other types, 349
complex types, 349–350
general design, 345–346

Lead compensators *(Cont.)*
  ideal derivative, 339, 359
  rules for application, 344
  specified root and gain, 345–346
  transfer function, 340
Leakage coefficient, oil, 50
Left eigenvectors, 529
L'Hospital's theorem, 117
Line segments, directed, 110, 111
Linear actuator, hydraulic, 48–53
Linear algebra, 17, 25–28
Linear constant-coefficient differential
        equations, 19
Linear control systems *(see* Linear systems)
Linear graphs, 34
Linear independence of vectors, 126
  regular basis, 527–553
Linear multivariable control systems, 527–552
Linear superposition, 94, 152
Linear systems, 16
  multivariable, 527–552
Linear time-invariant (LTI) control systems, 19
  optimum in, 608–615
  uncertain plants, 580–633
Linearity theorem, 94
Linearization, 488–491
Liquid-level control system, 53–54
Liquid-level symbols and units, 53
Lm (log magnitude), 247–248
  values for frequencies, 252
Load, inertia, 51–52
Load-induced pressure, 49
Log magnitude (Lm), 247–248
  values for frequencies, 252
Log magnitude and phase diagram, 250
  closed-loop response, 390–392
  drawing of, 250–259
    example, 256–259
  feedback approximations, 391
  gain crossover, 283, 286
  gain margin, 283
    frequency, 283
  phase crossover, 283
  phase margin:
    angle, 283
    frequency, 283, 286
  stability characteristics of, 285–286
  system types, 259–262
  (See also Bode plots)
Log magnitude–angle diagram, 250, 287–288
  characteristics, 257
  closed-loop frequency response, 312
  for conditionally stable system, 288
  constant M and α contours, 307–309
  gain adjustment by use of, 312

Log magnitude–angle diagram *(Cont.)*
  original and lag-compensated systems, 377
  original and lead-compensated systems, 381
  stability from, 286–288
Log magnitude curves, gain and, 259–262
Logarithm of complex number, 247
Logarithmic plots, 247–249
  asymptotes, 251, 254
  compensation: cascade, 372–390
    feedback, 390–398
  corner frequency, 251
  corrections: first-order factor, 251, 252
    quadratic factor, 254
  definitions, 247–248
  drawing, 250–259
  error coefficients, 259–262
  experimental determination of transfer
        function, 262
  feedback compensation using, 390–397
  lead compensator, 379–385
  reshaping, 368
  system types, 259–262
Longitudinal aircraft control, 207, 571–574, 665
Loop:
  major, 353
  minor, 139, 353
Loop equations, 31
Loop gain, 151
Loop method, electric circuits, 23–24, 76–77
Loop sensitivity, 202–204, 208
  evaluation, 212, 216, 334
  negative values, 203
  static, 202
Loop shaping, 608–615
  data for, 626
Loop transmission, 582–584
Loops, MISO, 591–592
  effective, of MIMO system, 591–592
LTI *(see* Linear time-invariant control systems)
Luenberger, D. G., 473
Luenberger observer theory, 473

M_m, 255, 296
  adjustment, 304, 312
  derivation, 292–295
$M(\omega)$ contours, 296–302
  construction values, 302
  direct polar plots, 296
  gain setting, 304, 312
  inverse polar plane, 303
  log magnitude–angle diagram, 307
  tangents, 301
MACSYMA program, 650
Magnetization curve, 55

Magnitude condition for root locus, 210–214
Major loop, 353
Manipulated variable, definition, 139
Mapping, 279
Markov parameter, first, 563
Mason gain rule, 152–153
Mass, 30, 35
   network representation of, 35
Mathematical and physical forms, 182–183
Mathematical models, 17, 19
MATLAB, 15, 82, 176, 201, 245, 290, 321,
     369, 424, 494, 528, 555
   diary file:
     Bode and Nyquist plots, 309–311
     controllability and observability, 430
     eigenvectors, 537–538
     tracker, 546–551
   m-file,
     partial-fraction, 106–107
     zetaex1 (root locus), 231–233
Matrix (matrices), 25–28
   addition, 26
   adjoint, 122
   augmented, 124, 532, 535
   block diagonal, 173
   cofactor, 121–122
   column, 26
   companion, (*see* Companion matrix)
   conformable, 27
   constituent, 87
   control, 32, 559
   controllability, 173, 426, 429
   degeneracy of, 123
   determinant, 120
   diagonal, 120
   diagonalizing, 160–168, 169–172
   differentiation, 28
   elementary, 123–124
   evaluation of the transfer function, 128
   feedforward, 32, 148
   fundamental, 84
   identity, 28
   integration, 28
   inverse, 122
     by row operation, 124
   Jacobian, 488, 490
   kernel of, 531
   measurement, 569–571
   minor, 121
   modal, 161, 528
   mode-controllability, 427
   mode-observability, 428
   modified, 171
   multiplication, 26–28
   multiplication of, by scalar, 27

Matrix *(Cont.)*
   nonconformable, 27
   nonsingular, 121
   normal form, 159
   null space of, 531, 532
   nullity of, 123
   observability, 427, 429
   operations and properties using Laplace
     transform, 120–126
   output, 32
   plant coefficient, 31
   polynomial, 175
   postmultiplication, 27
   prefilter, 591
   premultiplication, 27
   principal diagonal, 120
   proportional, 571, 573
   rank of, 122–123
   rectangular, 25
   resolvent, 127
     from signal flow graph, 156
   row, 26
   scalar, 121
   similarity transformation, 160–168, 173–175
   singular, 121
   square, (*see* Square matrix)
   state-feedback, 440
   state transition (*see* State transition matrix)
   striped, 173fn
   subscript notation, 26
   subtraction, 26
   superdiagonal, 146
   symmetric, 123
   trace, 121
   transfer-function, 128, 425
   transformation, 160, 557
   transforming into companion matrix,
     172–175
   transforming with complex eigenvalues,
     169–172
   transmission, 591
   transpose, 26
   transpose of product of, 123
   uncertain plant, 591
   unit, 28, 87
   Vandermonde, 162
Matrix block diagram, general, 148
Matrix equations, 17, 85
Matrix inversion by row operations, 124–125
Matrix polynomial, 85
MATRIX$_X$ program, 650
Maximum frequency response, 255, 292
   determination of, 115, 292, 296
   in relation to damping ratio, 255, 294
   in relation to peak overshoot, 294, 299

Maximum overshoot (*see* Peak overshoot)
Maximum value (M_m), 292
  for simple second-order system, 292–295
Measurement matrix, 569–571
Mechanical analogs, electrical and, 41
Mechanical network representation, 37
  of gear train, 45
  of rotational system, 43–44
  of simple rotational system, 42
Mechanical rotational symbols and units, 41
Mechanical rotational systems:
  multiple-element, 39–40
  simple, 37–39
Mechanical second-order systems, 74–76
Mechanical translation symbols and units, 35
Mechanical translation systems:
  multiple-element, 39–40
  simple, 37–39
  solution of, 74–76, 97–98
Mercury thermometer, simple, 47–48
Minimum phase, 256, 262, 278, 284, 287
  transfer function, 262
Microprocessors, 11
MIMO (*see* Multiple-input multiple-output
    control systems)
Minimum-phase factors, 262
Minimum-phase QFT system performance
    specifications, 581, 586–590
Minimum-phase system, 262
Minor, 129
  principal, 121
Minor loop, 137, 353
Minor of square matrix, 121
MISO (*see* Multiple-input single-output control
    systems)
Missile control system, 133
Mixed nodes, 150
mks (meter-kilogram-second) units, 22, 37,
    43, 47
Modal matrix, 161, 528
  modified, 172
Mode(s):
  dynamical, 530
  fast, 554, 561
  shape of, 530
  slow, 560
  of system, 70
  uncontrollable, 531, 539–541
Mode-controllability matrix, 427
Mode-observability matrix, 427
Modeling:
  desired control ratio, 402–406, 441
    z domain, 499
Moment of inertia, 30, 44–46
  effective, 44–46

Motor:
  ac (alternating-current) servo, 57–59
  dc (direct-current) servo, 56–57, 68, 141, 191
  position control, 140–144, 190–191
  shunt, 1–2
  torque, 56–59
Motor armature control, 56–57
Motor-generator control, 140–144, 190–191
  transfer function, 140–144
MULTI program, 650
Multiloop electric circuits, 23–24, 76–77
Multiple eigenvalues, 70, 162, 164, 168
Multiple inputs, 11, 83
Multiple-input multiple-output (MIMO) control
    systems, 15, 32, 129, 175, 424, 527,
    554, 555
  compensation, 590–592
  effective MISO loops of, 590-592
  uncertain plants, 580, 591–592
Multiple-input single-output (MISO) control
    systems, 401–402, 418–422
  basic QFT design procedure for, 617–618
  design example (QFT), 618–630
  equivalent method, justification of QFT,
    591–592
  loop (*see* Loops, MISO)
  uncertain plants, 580–634
Multiplication of transfer functions, 38
Multiplication of matrices, 26–28
Multivariable control systems, 527–634
  eigenstructure assignment, 528–552
  quantitative feedback theory (QFT),
    580–634
  regulators, 531–541
  synthesis of, 527
  trackers, 541–551
  uncontrollable, 539–541
  variable parameters, 580–634

N, number of oscillations, 322
Natural frequency:
  damped (*see* Damped natural frequency)
  undamped (*see* Undamped natural frequency)
Network:
  electric, 21–25
  mechanical, 35–42, 44–48
  thermal, 46–48
Network elements (*see* Elements, network)
Network representation:
  of damping action, 36
  of mass, 35, 42
  of spring, 35–36
Network synthesis, 352, 369, 406
Newton's laws, 35, 59

Nichols chart, 250
  applications, 312, 396, 582, 597–633
  data points for, 602, 603
  regular, 602
  rotated, 603
  tracking bounds on, 597–600
  (*See also* Log magnitude–angle diagram)
Nichols plot, 286
Nodal method:
  electrical circuits, 24–25
  mechanical rotational system, 41–46
  mechanical translation system, 35–40
  thermal system, 46–48
Node absorption, 151
Node equations, 30, 39
Node method, 20–25
Node voltage, 24
Nodes, 14, 22, 480
  phase plane:
  types of, 149–150
  unstable, 482
Noise, 246–247
Nominal plants, 582–584
Nonlinear control system, 16, 489
Nonlinear elements, 9, 712–713
Nonlinear feedback control system, 489
Nonlinear friction, 35
Nonlinear plants, 580
Nonlinearity:
  backlash, 321
  coulomb friction, 35, 321
  dead time (transport lag) (*see* Dead time)
  dead zone, 321
  limiting, 340
  linearization, 488–491
  relay, 3, 8
  saturation, 340
Nonminimum-phase factors, 262
Nonminimum-phase plants, 586
Nonminimum-phase systems, 262
Nonsingular matrix, 121
Nonunity-feedback control system,
    198–199, 322
  environmental conditions, 321
  root locus, 225–233, 322
Normal form, 159
Nuclear-reactor power-level control system, 10
Null space, 164, 166
  eigenvectors in, 166, 531
  evaluation, 531–539
  of matrix, 531, 532
Null-space vectors, 531–539
Nullity of matrix, 123
Nyquist plots, 201, 245
  MATLAB, 309–311

Nyquist's stability criterion, 270–283
  applied to systems having dead time,
    281–283
  examples of, using direct polar plot,
    278–283
  generalizing, 273–274
  limitations, 271, 289
  log magnitude–angle diagram, 286
  mathematical basis for, 271–273
  radial line, 278

Observability, 424–425, 427
  from canonical matrix, 427
  matrix, 427
  mode, 428
  optimal control system, 424
Octaves, 248
Ohm's law, 21
Oil:
  bulk modulus of, 51
  compressibility of, 50
  leakage, 50–51
ω_d (*see* Damped natural frequency)
ω_m (*see* Resonant frequency)
ω_n (*see* Undamped natural frequency)
On-off feedback control systems, 8
Open-loop control systems, 2–3, 7
  unstable, 280
Open-loop SISO system, 401–423
Open-loop systems, 1–2, 7, 433
  dc shunt motor, 1–2
  Hero's device, 8
  poles and zeros, 208
  transfer function, 137, 208, 296
Open-loop transfer function, 137, 207, 323
Operator notation, D, 20
Optimal control system, 424
  controllability, 424
  observability, 424
Optimum response criteria:
  ISE, 245
Orifice coefficient, 49
Orthogonal eigenvectors, 529
Orthogonality property, 87
Oscillations, sustained (*see* Sustained
    oscillation)
Outline of text, 16–18
Output, 34
  definition, 6–7
  multiple, 11, 83
  phasor representation of, 66
  second derivative feedback of, 362–364
  vector, 32
Output-decoupling zeros, 428

Output feedback, 554
   tracking system design using, 554–579
   tracking systems, 555–557
Output matrix, 32
Output vector, 32
Output voltage, 24
Overall block-diagram representation, 38
Overall graph transmittances, 152–153
Overdamped response, 72
Overdamped systems, 80
Overshoot:
   maximum, 81
   peak, 79, 81, 324
   per unit, 81

Padé approximation, 494, 512–514
Parabolic error coefficients, 195, 198
   from log plot, 261
Parabolic-function input, 65, 69
Parabolic input, 189–190, 443–444, 518
Parallel compensators, 319
   (*See also* Feedback compensation)
Parallel paths, 151
Parameter sensitivity, 462–476
   examples, 472
Parameter space, 585
Parameter variations, frequency response
      with, 582–634
Partial-fraction expansion:
   computer solution of coefficients, 109
   graphical interpretation of coefficients,
      109–113
   Heaviside theorems, 99–107
   MATLAB m-file, 106–107
   shortcuts in, 107–109
Particular integral, 64
Partition method:
   factoring of polynomials, 355
   examples, 356, 358, 365
Passband, 244, 246
Passive compensators, 263, 333, 340, 346,
      351, 406
Path gain, 151
Path Q, analysis of, 275
Paths, signal flow graph, 150–151
PBW (power-by-wire) control systems, 13
PCT (pseudo-continuous-time) control systems,
      493, 512–514
PD (proportional plus derivative) controller,
      339–340
Peak overshoot, 79, 81, 324
   effect of additional poles on, 312–317,
      328, 331
   effect of zeros on, 312–317, 328

Peak overshoot *(Cont.)*
   impulse response, 119–120
   second-order system, 78–80, 294
Peak time, 81, 324
Pendulum, 487
Performance:
   of cascade compensators, 339
   characteristics, 234–238
   closed-loop system, analysis of, 559–561
   comparison of systems, 358
   criteria for evaluation, 321–326
   irregular system, 568–569
   specifications, 15, 192, 199, 201–204, 245,
      331, 370, 401–461
      minimum-phase QFT system, 586–590
   systems, 200–201
   tracking, closed-loop, 290–318, 541–551,
      580–634
Phase, nonminimum (*see* Nonminimum phase)
Phase crossover, 283
Phase margin, 283–285
   on log magnitude and phase diagram, 284
Phase-margin angle, 283
Phase-margin frequency, 283
Phase plane, 29, 477, 478
Phase space, 477
Phase trajectory, 29, 477
Phase variables, 146, 439–441, 477
   general properties of state feedback,
      439–441
Phasor, 66, 246
Phasor representation:
   of input, 66
   of output, 67
Physical and mathematical forms, 182–193
Physical state variables, 29
PI (proportional plus integral) control law, 555
PI (proportional plus integral) controller,
      332–333
PID (proportional plus integral plus derivative)
      controller, 352–353
Pitch-attitude control system, 701
Pitch pointing maneuver, 574–576
Placement technique, pole-zero, 406
Plant coefficient matrix, 31
Plant complex poles, 451
Plant templates, 592–595
   construction of, 620–622
   unstable, generation of, 630–632
Plants:
   all-pole, 448–451
   irregular, 555
      characteristics, 566–568
   nominal, 582–584
   nonminimum-phase, 262, 586

Plants *(Cont.)*
  pole-zero *(see* Pole-zero plants)
  regular, 555
    design procedure for, 562–563
  uncertain, 584
    matrix, 591
    MIMO, 590–591
  with variable parameters, 584–586
Polar plots:
  compensators, 370
  direct *(see* Direct polar plots)
  gain adjustment, 304
  lead compensator *(see* Lead compensators,
      polar plots)
  from logarithmic plots, 247, 263
  Nyquist's criterion, 270–283
  reshaping, 368
  time-response correlation, 290–295
  of Type 0 system, 264
  of Type 1 system, 266, 378, 384
  of Type 2 system, 267
Pole variation, sensitivity due to, 468
Pole-zero assignment, closed-loop, 423–460
Pole-zero diagrams:
  correlation of, with frequency and time
      responses, 312–317
  frequency response from, 113–116
  stability, 116
Pole-zero locations:
  closed loop, 236–237
  effect on performance, 312–317
  improvement of performance, 331
Pole-zero placement method, 402–406
Pole-zero plants: state-variable feedback,
    453–460
Poles:
  additional, root locus with, 207
  additional significant, 326–329
  complex:
    angle condition near, 220–221
    plant, 451
  of control ratio, 209–211
  definitions, 99, 110
  dominant, 224, 240, 312–317, 321–326
  dominant complex *(see* Dominant complex
      poles)
  location of desired dominant, 403
  nondominant, effect of, 322–326
  at origin, effect on rotation of $B(s)$,
      275–277
  placement by state feedback, 423–461
  plant complex, 451
  of response transform, 99
    complex-conjugate *(see* Complex-conjugate
        poles)

Poles *(Cont.)*
  first-order real *(see* First-order real poles)
  imaginary, 105–106
  location of, 116
  stability of, 116
  time response and, 327
Polynomial input, 67–70, 198
Polynomial matrix, 175
Polynomials:
  characteristic *(see* Characteristic polynomials)
  computer solution, 99
  general, 85
  matrix, 85
  remainder, 86
Position control of space-vehicle camera,
    190–191
Position-control system, 140–144, 489
Position error coefficients, 192
Postmultiplication matrix, 27
Potential energy, 54, 62, 487
Potentiometer, 132, 134, 140, 190
Power amplifiers, rotating, 54–55
Power-by-wire (PBW) control systems, 13
Power gain, 55
Power-series function input, 65
Power-series method, 403
Prefilter:
  design of, 615–617
  frequency bounds on, 617
Prefilter matrix, 591
Premultiplication matrix, 27
Pressure, load-induced, 49
Pressure drop, 55
Primary feedback, definition, 139
Principal diagonal, 120
Principal minor:
  of square matrix, 121
Product:
  of determinants, 122
Production-line robot, 4–6
Program CC, 650
Proportional matrix, 571, 573
Proportional plus derivative (PD) controller,
    339–340
  (See also Lead compensators)
Proportional plus integral (PI) control law, 555
Proportional plus integral (PI) controller,
    333–334
Proportional plus integral plus derivative (PID)
    controller, 352–353
  (See also Lag compensators)
Pseudo-continuous-time (PCT) control systems,
    493, 512–514
Pulse train, 494
Pyrheliometer, 5, 6

QFT (*see* Quantitative feedback theory)
Quadratic factor, standard forms of, 73
Quantitative feedback theory (QFT), 402
    minimum-phase system performance
        specifications, 586–590
    problem classes, 581
    single-loop system, 584–586
    technique, 580–634
    (*See also* MIMO and MISO)

Radar tracking control system, antiaircraft, 9–11
Ramp error coefficients, 193–194, 197, 198
    effect of lag compensation on, 335
    from log plot, 260
    related to control ratio, 447
Ramp function, Laplace transform of, 93
Ramp-function input, 65, 66, 169
    (*See also* Input, ramp)
Ramp input, 187–190, 443, 518
Random input, 65, 182
Rank:
    of matrix, 122–123
Rank deficiency, 164
Rao, S. N., 224
Rate feedback compensation, 358–362, 393–398
    plants with dominant complex poles,
        364–365
    simplified, 355–358
Ratio:
    control (*see* Control ratio)
    damping (*see* Damping ratio)
Reaction damping force, 36
Reaction force, 35–36
Reaction torque, 41–42
Reactor, nuclear, 10
Real-axis root locus, 215–216
Real differentiation theorem, 95
Real integration theorem, 95–96
Real part, 66
Reciprocal eigenvectors, 529
Reconfigurable control systems, 582
Reconstruction, continuous signal, 503
Rectangular matrix, 25
Reference input, definition, 7, 139
Reference input elements, definition, 140
Reference positions, 37
Reference selector, 1, 3
    definition, 6, 7
Regular plants, 555
Regulators, 8
    entire eigenstructure assignment for,
        530–539
Relay feedback control systems, 8
Remainder polynomial, 86

Reshaping:
    of frequency response, 368
    of polar plots, 368
    of root locus, 204–207, 331, 355
Residues of response transform, 99
Resistance:
    electrical, 21
    thermal, 46
Resistor-inductor-capacitor circuits, series, 23,
        30–32
Resistor-inductor circuits, series, 22–23, 30
Resistors, 21
Resolvent matrix, 127
    from signal flow graph, 156
Resonant frequency (ω_m), 292
    determination of, 292
    related to damping ratio, 255, 293–295
    for simple second-order system,
        292–295
Response:
    overdamped, 72, 327
    steady-state (*see* Steady-state response)
    underdamped, 326–327, 402–406
    zero-input, 89
    zero-state, 89
Response characteristics, 80–82
    effect of real pole on, 326
    second-order, 80–82, 292–295
Response function, 97
Response transform, 97, 176
    poles of (*see* Poles, of response transform)
    residues of, 99
Riley and Hazony rule, 109
Rise time, 81
Robot:
    arm control, 666
    production-line, 4–6
Robust control systems, 555, 580
Root locus, 201, 203
    with additional pole, 207
    with additional zero, 204
    angle condition for, 210–215
    angle of departure, 220–221
    asymptotes of, as *s* approaches infinity,
        217–218
    automatic plotting, 215, 225
        (*See also* MATLAB zetaex1 m-file)
    of basic transfer function, 205
    branches: intersection of, 222–223
        number of, 215
    break-in point, 218–220
    breakaway point, 218–220
    calibration of, 202, 212–215
    comparison of cascade and feedback
        compensation using, 363

Root locus *(Cont.)*
  compensation *(see* Cascade compensation;
      Feedback compensation)
  computer solution, 207, 215
  constant damped natural frequency
      contour ω_d, 236
  constant damping ratio, 236
  constant undamped natural frequency
      contour ω_n, 236
  construction of, 213
  construction rules, 215–225
    tabulation, 241–242
  configurations, 204–207
  design considerations, 331
  determination of roots on, 224–225
  dominant roots, 224, 236, 321
  end points, 216–217
  examples, 225–234
  factoring of polynomials, 355
  general shape of, 223–224
  geometrical properties, 215–225
  Grant's rule, 225, 228
  imaginary axis crossing, 221–222, 227
  lead compensator *(see* Lead compensators,
      root locus)
  loop sensitivity, 202, 208
    for imaginary roots, 221–222
  magnitude condition for, 210–215
  MATLAB:
    zetaex1 m-file, 231–233
  nonunity feedback, 225–233, 245, 322
  number of branches of, 215
  procedure outline, 207–208
  qualitative analysis of, 205–207
  real-axis locus, 215–216
  reshaping, 331
  selection of best roots, 202, 321
  sum of roots, 223
  symmetry about real axis, 215
  synthesis, 240–241
  time response from analysis of, 312–317,
      321–329
  transport lag, 238–240
  vs. frequency response, 399
  z domain, 499, 500–503, 505, 521–523
Root-locus branches, nonintersection or
      intersection of, 222
Root-locus compensation, 319–366
Root-locus method, 200–243
  cascade lag compensators applied by, 378
  construction rules, 215–225
  geometrical properties, 215–225
  lag-lead compensators applied by, 390
  lead compensators applied by, 385
  procedure outline, 207–208

Root-locus method *(Cont.)*
  summary of construction rules for negative
      feedback, 241–242
Root-locus theory, 13
Root solving, 200–243
  partition method, 355–356, 358, 362
Roots:
  characteristic *(see* Characteristic roots)
  of characteristic equation:
    distinct (simple), 99–106, 502
    multiple (repeated), 84, 99–106
  complex, 71–73
  determination of, on root locus, 224–225
  dominant, 224
  multiple, 70, 75
  plotting, 201–204, 245
  real, 71, 73
  from root locus, 225–233, 355
  of second-order system, 234–238
  variation of, with damping ratio, 236–238
Rotating power amplifiers, 54–55
Rotational systems, mechanical *(see* Mechanical
      rotational systems)
Routh's stability criterion, 175–182
  auxiliary equation, 180–181, 222, 227
Row matrix, 26
Row vector, 26

s plane, 214
  asymptotes of root locus as *s* approaches
      infinity, 217–218
  prohibited region in, 271
  to *z* plane transformation methods,
      509–511
Saddle point, phase plane, 486
  unstable, 490
Sampled function, 494
Sampled-data (SD) feedback control system,
      7, 493
  basic or uncompensated, 512, 514–519
Sampler:
  ideal, 496–498
  pulse, 494–496
Sampling, 494–499
  frequency spectra, 495–497
SAS (stability augmentation system), 12
Satellite-tracking system, 664
Saturation (magnetization) curve, 55
Saturation signal, 340, 664
Scalar, multiplication of matrix by, 27–28
Scalar matrix, 121
Schauder fixed point theorem, 592
SD *(see* Sampled-data control system)
Sears Tower building elevator, 4

Second-order systems:
  correlation of sinusoidal and time responses, 292–295
  determination of M_m and ω_m, 115, 292–294
  effect of gain change, 201–204
  electrical, 76–77
  equivalent of, 237
  frequency of oscillation, 79
  frequency response, 292
  with impulse excitation, 119–120
  mechanical, 74–76
  peak overshoot, 78–79, 292
  performance characteristic, 234–238
  response characteristics, 80–82
  transients, 77–80
Selector, reference, 1–3, 7
Self-conjugate sets, 531
  of roots, 84
Self-loop, 151
Sensing element (see Comparison units)
Sensitivity analysis, 466–473
  due to pole and gain variation, 468
  gain, time response and, 473
  parameter (see Parameter sensitivity)
  system, 469–470
  time-response, 471
Sensitivity functions, system, 462–466, 476
Sensitivity of state-variable-feedback systems, 450, 455, 462–476
Separatrices, phase plane, 487
Series compensator (see Cascade compensation)
Series paths, 150
Series resistor-inductor-capacitor circuits, 23, 30–32
Series resistor-inductor circuits, 22–23, 30
Servo characteristics, basic, 13–14, 176–199
  error coefficients, 192–199
Servomechanisms, 8, 9
Servomotors:
  ac, 57–59
  dc, 56–57
  two-phase, 57–59
Settling time, 80–81
  effect of poles on, 312–317, 326–329
  effect of zeros on, 328
  improvement, 343–345
Sewage system, 656
SFG (see Signal flow graphs)
Shape of modes, 530
Shock inputs, 117–119
Shunt motor, dc, 1–2
Signal flow graphs (SFG), 131, 149–154
  algebra, 150–151
  definitions, 149–150
  general analysis of, 151–153

Signal flow graphs (Cont.)
  resolvent matrix from, 156
  state transition, 154–158
Similarity transformation, 160
Simplified rate feedback compensation, 355–358
Simulation:
  diagrams, 144–148
  phase variable, 146–147
  tracking system, 574–577
Single-input single-output (SISO) closed-loop control system, 3, 6, 17, 32, 129, 175, 401, 423, 527
  disturbance-rejection design principles for, 411–415
Single-loop QFT system, 584-586
Singular matrix, 121
Singularity functions, 65
Singularity of function, 110fn
Singularity points, phase plane, 477, 480
Sink nodes, 150
Sinusoid:
  exponentially damped, 71
  Laplace transform of, 93
Sinusoidal function, 65–67, 245, 262
Sinusoidal input, 65–67
Sinusoidal responses, correlation of time responses and, 244–246, 295
SISO (see Single-input single-output closed-loop control system)
Slow modes, 560
Solar-energy control systems, 6
Solenoid, 132, 653, 667
Source nodes, 149
Space:
  null (see Null space)
  state, 29, 477
Space shuttle, 1
Space-vehicle camera, position control of, 190–191
Space vehicles, 1, 191, 321
Spaces, eigenvector, 17, 166–168, 531
  evaluation, 531–539
Specific heat, 47
Specifications, performance (see Performance, specifications)
Speed control, dc shunt motor, 1–2
Spring, 35, 42, 487
  compression of, 37
  network representation of, 35
Square matrix, 25
  determinant of, 121
  minor of, 121
  principal minor of, 121
  trace of, 121

Stability, 176–182, 477
  addition of pole or zero, 204–207
  characteristics of log magnitude and phase
     diagram, 285
  conditional, 280, 286
  from impulse response, 119
  location of poles and, 116–117
  from log magnitude–angle diagram, 286–288
  Nyquist's criterion (*see* Nyquist's stability
     criterion)
  polar plots, 278–283
  of pole-zero diagram, 116–117
  Routh's criterion (*see* Routh's stability
     criterion)
  z domain, 499, 502
Stability augmentation system (SAS), 12
Stabilization (*see* Compensation)
Stabilized platform, 674–675
Stable equilibrium, 480–492
Standard forms:
  of quadratic factor, 73
Standard symbols, 136–140
Standardized input, 65–66, 183
State, 28–29
  inaccessible (*see* Inaccessible states)
State concepts, 28–34
State deviation, 489
State diagrams, 154–158
  parallel, from transfer functions, 158–160
  resolvent matrix from, 156
State equations, 19, 31–34
  computer solution, 91
  definition, 30
  homogeneous, 83–84
  initial conditions, 83
  multiple-input multiple-output, 32, 83
  single-input single-output, 32, 433
  solution of, 126–128
  uncoupled, 160
  use of state transformation for solution of,
     168–169
State feedback:
  control system, 423–476, 433–435,
     528–552
  pole placement by, 423–460
  for SISO systems, 431–433
State-feedback matrix, 440
State reconstruction, 504–505
State plane, 477
State space, 29, 477
State trajectory, 29, 477
State transformation used for state equation
     solution, 168–169
State transition equations:
  continuous, 88

State transition matrix (STM), 66, 84, 478
  evaluating, 84–86
  properties, 84
State transition signal flow graph, 154–158
State-variable control theory, 11
State-variable equations, 82–84
  complete solution of, 88–89
  computer solution, 82, 89, 461
State-variable feedback, 436–439
  accessible states, 423
  all-pole plant, 448–451
  cascade compensator, 453, 456
  comparison with cascade compensation,
     438
  inaccessible states, 423, 473–476
  phase variables: design procedure, 441
    general properties, 439–441
    root locus, 440
  plant with complex poles, 451
  pole-zero plant, 453–460
  position control system, 436–439
  sensitivity analysis, 450, 466–472
  steady-state error analysis, 442–444
  steady-state error coefficients, 444–448
State-variable method, 82
State variables, 6, 28–29
  augmented, 30
  canonical, 29, 158–160
  independent, 29–30
  phase, 29, 146
  physical, 29
State vector, 29, 32
States:
  accessible, 423, 462
  inaccessible, 423, 462
Static friction (Coulomb), 35, 321
Static loop sensitivity, 202
Steady-state component, 64–66
Steady-state error analysis, 442–444
  conventional, 176–199, 446–448
  feedback coefficient, 442–448
  state-variable feedback, 442–448
Steady-state error coefficients (*see* Error
     coefficients)
Steady-state response:
  correlation with polar plot, 295
  error, 186, 192–199, 442–448
  improvement, 332
  parabolic input, 190
  polynomial input, 67–70
  sinusoidal, 244–289
    determination of system constants, 262
  sinusoidal input, 66–67
  for stable unity-feedback systems, 190
Steady-state solution, 64

Step error coefficients, 192–193, 196, 198
  effect of lag compensation, 334
  from log plots, 259
Step function, 65, 68–79, 92, 182, 186–187, 189–190, 541
  of acceleration, 65, 69
  Laplace transform of, 92
  of velocity, 65, 69
Step-function input, 64, 65, 66, 68–69
Step input, 186–190, 442, 518
Step-zero-error system, 442
Stiffness, 35, 41
Stiffness coefficient, 41
STM (*see* State transition matrix)
Stored energy, 52
Straight climb maneuver, 576–577
Striped matrix, 173fn
Subscript notation, matrix, 26
Subtraction matrix, 26
Summation points, 141
Summer (adder), 144
Sun-following control systems, 6
Superdiagonal, 146
Superposition theorem, 94
Suspension, automobile, 653
Sustained oscillation, 72, 234
  Nyquist criterion, 277
Sylvester expansion, 87, 478
Symbols and units:
  electrical, 22
  hydraulic, 48–53
  liquid level, 53
  mechanical rotational, 41
  mechanical translation, 35
  thermal, 46
Symmetric matrix, 123
Synchros, 134
Synthesis:
  complex multiloop system, 353
  network, 352, 369, 406
  root locus, 240–241
  in *z* domain, 500–503
System:
  antiaircraft, 9
  basic, 15–16, 319
  conditionally stable, 280, 288
  constants, determination of, 259, 683–685
  controllable, 424–430
  discrete-data feedback control, 7, 493–526
  experimental and design procedures, 262
  feedback (*see* Feedback systems)
  linear, 16, 19
  observable, 424–430
  representation, 131–175
  response model, 402–406, 441

System (*Cont.*)
  solar-energy, 6
  stable, 71, 278–281, 285–288
  sun-following, 6
  thermal, 46–48, 132–133
  unstable, 71, 222, 285–288
  variable parameters, 580–634
System components, definitions, 140
System dynamics, definition, 6–7
System engineers, 17
System equations, writing, 19–63
System error, 137
  definition, 139
System output equation, 32
System response, 119
System transfer function, 97
Systems:
  closed-loop (*see* Closed-loop systems)
  control (*see* Control systems)
  with dead time, Nyquist's stability criterion applied to, 281–283
  definition, 6–7
  dynamical modes of, 530
  ideal, 247
  irregular: design example, 571–574
    performance, 568–569
  MIMO (*see* Multiple-input multiple-output control systems)
  minimum-phase, 262
  MISO (*see* Multiple-input single-output control systems)
  nonminimum-phase, 262
  optimal (*see* Optimal control systems)
  performance, 200–201
  regular, design example, 563–566
  sensitivity analysis, 466–472
  SISO (*see* Single-input single-output closed-loop control system)
  state of (*see* State)
  tracking (*see* Tracking systems)
  type (*see* Type *entries*)
  uncontrollable, 539–541
  unity-feedback (*see* Unity-feedback systems)
  variables in, definitions of, 139–140
Systems engineering, 7

Tachometer, 362–364, 393
  feedback compensation, 353–365, 390–398, 431–476
Taylor series expansion, 58, 491
Temperature, 46
Temperature-control system, 132
Terminology, feedback systems, 7, 138–140
Thermal analog, 47

Thermal capacitance, 47
Thermal capacitor, 47
Thermal resistance, 47
Thermal symbols and units, 47
Thermal systems, 46–48
Thermocouple, 132
Thermostat, 3
"Through" variables, 41
Thumbprint specification, 586
Time:
  dead (*see* Dead time)
  duplicating, 81
  peak, 324
  rise, 81
  settling (*see* Settling time)
Time constant, definition, 74
  experimental determination of, 262
Time invariance, 19
Time response, 67
  best, 383
  of closed-loop system, 519
  and correlation of pole-zero diagram with
    frequency response, 312–317
  and correlation of sinusoidal responses,
    245–246, 295
  disturbance-rejection, 419–421, 580, 585,
    589–590, 600–607
  gain sensitivity and, 473
  second-order disturbance-rejection model,
    409–411
  sensitivity, 471
  terms, 209
  for unit-step input, 517
Time responses:
  automatic plotting, 109
  computer solution, 109, 231–233, 546–551
  effect of eigenstructure (*see* Eigenstructure
    assignment, effect on time response)
  peak time (t_ρ), 78, 81, 321–329
    determination from pole-zeros, 324
    effect of poles on, 326
    effect of zeros on, 328
  from pole-zero diagram, 109–113, 312–317,
    321–329
  settling time (*see* Settling time)
  specifications (*see* Correlation, of sinusoidal
    and time responses)
  time to first zero (t_0), 81
Time-response specifications, 81–82
Time, rise, 81
Toaster, automatic, 1–3
Torque, 41–42
  developed, 56
  motor, 56–59
  reaction, 41–42

Torque equations, 41
Torque-speed curves, 58
TOTAL-PC, 15, 91, 176, 321, 369, 424,
    642–646
  input modes, 643
  options, 643–645
  QFT, MISO, 645
Trace of square matrix, 121
Tracker system, 183
Trackers, multivariable control system,
    541–551
Tracking bounds, 597–600
Tracking control ratio, 402–406, 586
Tracking control system, radar, 9, 133, 674
Tracking models, 402–406, 586–590
Tracking performance, closed-loop, based
    on frequency response, 290–318,
    586–588
Tracking systems, 76, 183, 541–551
  design example, 543–551, 618–630
  design of, using output feedback, 554–579
    using QFT, 580–634
  output feedback, 555–557
  simulation, 574–577
  (*See also* Mechanical translation system)
Tracking techniques, frequency-domain, 367,
    580–634
Trajectory:
  phase, 29
  state, 29
Transfer function(s), 34, 38, 44, 57
  ac servomotor, 57–59
  basic, root locus of, 205
  cascade, blocks in, 38, 44, 142–144
  dc servomotor, 56
  definition of, 34, 97
  diagonal, 570–571
  environmental conditions, 321
  experimental determination of, 262
  forward, 137, 269
  frequency, 67
  impulse, 496
  impulse response, 119
  lag compensator, 333
  lag-lead compensator, 263
  lead compensator, 340
  matrix, 128
  minimum phase, 262
  motor-generator, 140–144
  multiplication of, 38
  nonminimum phase, 262
  open-loop, 137, 208
  overall, 136
    determination of, 136–137
  parallel state diagrams from, 158–160

Transfer functions *(Cont.)*
standard form, 182–183, 208, 249
Type 0 system, 259–260
Type 1 system, 260–261
Type 2 system, 261–262
z-domain, 507
Transfer-function matrix, 128, 425
Transformation:
s plane to *z* plane, 507–511
Transformation matrix, 162, 557
Transforms:
Fourier, 245, 495–496
Laplace *(see* Laplace transform)
response, 97, 117, 119, 176
z *(see z* transform)
Transient component, 64–66
Transient response:
classical method, 70–73
correlation to frequency response, 245–246,
290–318
dominant complex poles, 321–326
sinusoidal and time: higher-order system, 295
second-order system, 234–237, 292
Transient-response curves, 80–82
Transient solution, 64, 70–73, 201, 234–238
critically damped, 72
frequency of oscillation, 72–76
overdamped, 72
from pole and zero locations, 290–318,
321–329
second-order, 77–80
underdamped, 71–82
Translation in time theorem, 94
Translation systems, mechanical *(see*
Mechanical translation systems)
Transmission:
loop, 582–584
Transmission matrix, 591
Transmission zeros, 560, 562, 568
Transmittances, overall graph, 152–154
Transport lag, 238–239, 281
frequency response, 281
root-locus, 239–240
(See also Dead time)
Transpose:
of matrix, 26
of product of matrices, 122
Trolley, 654
Tustin model of control ratio, 517
Tustin transformation, 507–511
Two-phase servomotor, 57–59
Type 0 feedback control system, 264–266
Type 0 system, 186–187, 190
conversion to, 353–354
log magnitude curves, 259

Type 0 system *(Cont.)*
polar plots, 264–266
steady-state response, 186–187
Type 1 feedback control system, 266–267
polar plots of, 378, 384
Type 1 system, 187–188, 196–197
lag compensation applied to, 335–339
lag-lead cascade compensators applied to,
347–348
lead compensators applied to, 341–345
log magnitude curves, 260–261
polar plots, 266–267, 378, 384
steady-state response, 186–187
transfer function, 260
Type 2 feedback control system, 267–269
Type 2 system, 189–190
example, 190–191
log magnitude curves, 261–262
polar plots, 267–269
steady-state response, 189–190
transfer function, 261
Type 3 feedback control system, 184
Types:
conversion to type *m*, 446
system: analysis of, 185–190
characteristics of, 190, 275–281
conversion to Type 0, 353–354
feedback, 183–199
increase of, 332
related to log magnitude curves,
259–262
related to polar plots, 269
state-variable feedback, 442–448

U-contour, 595–597
UHF (universal high-frequency boundary),
595–597
Uncertain plant matrix, 591
Uncertain plants, MIMO, 590–591
Uncertainty, plant, 585
Uncompensated sampled-data control system,
512, 514–519
Uncontrollable mode, 528, 539–541
Uncontrollable multivariable control systems,
539–541
Uncontrollable systems, 539–541
Uncoupled state equation, 160
Undamped natural frequency (ω_n), 72
definition, 72
effective, 238, 295
experimental determination of, 262
increase in, 203
Underdamped response, 72
effect of real poles on, 326–329, 402–408

Underdamped systems, 80–82
Unit circle, z domain, 499, 502
Unit matrix, 28, 87
Unit-step input, time response for, 517
Units (*see* Symbols and units)
Units, comparison (*see* Comparison units)
U.S. customary units, 37, 43, 49
Unity-feedback block diagram, G-equivalent, 444–448
Unity-feedback control system, 183–185
Unity-feedback systems:
   gain adjustment of: direct polar plot, 304–307
   stable, steady-state response for, 190
Universal high-frequency boundary (UHF), 595–597

Valve-controlled hydraulic actuator, 48–53
Valve displacement, 49
Vandermonde matrix, 162
Variable parameters, plant with, 580–634
Variables, 20–21
   canonical, 159
   control, 30
   controlled (*see* Controlled variable)
   manipulated, definition, 139
   phase, 146, 439–441
   state (*see* State variables)
   in system, definitions of, 139–140
   "through," 41
Vector(s):
   column, 26, 31
   feedback, 436
   linear independence of, 126
   null-space, 531, 533
   output, 32
   row, 26
   state, 29, 32
Velocity, 35, 60, 182
Velocity (ramp) error coefficient (*see* Ramp error coefficients)
   step function of, 65, 69
Vertical translation maneuver, 575–577
Viscous friction, 36
Viscous friction coefficient, 42
Voltage:
   applied, 22
   capacitor, 21

Voltage (*Cont.*)
   node, 24
   output, 24
Voltage drops, 21–22
Voltage sources, 21, 22
Voltmeters, 41

$W(s)$ plane, 215
Water tank control system, 694
Weighting function, 119

z domain:
   inverse transform, 503
   root-locus, 499, 500–503, 520–523
   stability, 499, 502
   synthesis in, 500–503
   transfer function, 508
z plane, 494
   s plane to z plane transformation methods, 509–511
z-plane to s-plane correlation, 498
z transform, 500
   inverse, 503
   table, 501
   theorems, 500
   unit circle, 499, 502
Zero cancellation, closed-loop, 455–456
Zero-error system:
   parabolic, 443–444
   ramp, 443
   step, 407, 438–442
Zero-input response, 89
Zero-order hold (ZOH), 504–506
Zero-state response, 89
Zeros:
   definition, 99
   of denominator, 99
   effect of: on $c(t)$, 328
     on root locus, 204–207
   input-decoupling, 427, 539
   by Nyquist's criterion, 270–274
   output-decoupling, 428
   transmission, 560, 562, 568
ζ (zeta), 72, 103–104, 236, 245
   (*See also* Damping ratio)
ZOH (zero-order hold), 504–506